Handbook of Survival Analysis

Chapman & Hall/CRC Handbooks of Modern Statistical Methods

Series Editor

Garrett Fitzmaurice
Department of Biostatistics
Harvard School of Public Health
Boston, MA, U.S.A.

Aims and Scope

The objective of the series is to provide high-quality volumes covering the state-of-the-art in the theory and applications of statistical methodology. The books in the series are thoroughly edited and present comprehensive, coherent, and unified summaries of specific methodological topics from statistics. The chapters are written by the leading researchers in the field, and present a good balance of theory and application through a synthesis of the key methodological developments and examples and case studies using real data.

The scope of the series is wide, covering topics of statistical methodology that are well developed and find application in a range of scientific disciplines. The volumes are primarily of interest to researchers and graduate students from statistics and biostatistics, but also appeal to scientists from fields where the methodology is applied to real problems, including medical research, epidemiology and public health, engineering, biological science, environmental science, and the social sciences.

Published Titles

Longitudinal Data Analysis
Edited by Garrett Fitzmaurice, Marie Davidian, Geert Verbeke, and Geert Molenberghs

Handbook of Spatial Statistics
Edited by Alan E. Gelfand, Peter J. Diggle, Montserrat Fuentes, and Peter Guttorp

Handbook of Markov Chain Monte Carlo
Edited by Steve Brooks, Andrew Gelman, Galin L. Jones, and Xiao-Li Meng

Handbook of Survival Analysis
Edited by John P. Klein, Hans C. van Houwelingen, Joseph G. Ibrahim, and Thomas H. Scheike

Chapman & Hall/CRC
Handbooks of Modern
Statistical Methods

Handbook of Survival Analysis

Edited by

John P. Klein
Hans C. van Houwelingen
Joseph G. Ibrahim
Thomas H. Scheike

CRC Press is an imprint of the
Taylor & Francis Group, an informa business
A CHAPMAN & HALL BOOK

CRC Press
Taylor & Francis Group
6000 Broken Sound Parkway NW, Suite 300
Boca Raton, FL 33487-2742

First issued in paperback 2019

ISBN-13: 978-1-4665-5566-2 (hbk)
ISBN-13: 978-0-367-33096-5 (pbk)

Library of Congress Cataloging-in-Publication Data

Handbook of survival analysis / [edited by] John P. Klein, Hans C. van Houwelingen, Joseph G. Ibrahim, Thomas Scheike.
pages cm -- (Chapman & Hall/CRC handbooks of modern statistical methods)
Summary: "This handbook focuses on the analysis of lifetime data arising from the biological and medical sciences. It deals with semiparametric and nonparametric methods. For investigators new to this field, the book provides an overview of the topic along with examples of the methods discussed. It presents both classical methods and modern Bayesian approaches to the analysis of data"-- Provided by publisher.
Summary: "Preface This volume examines modern techniques and research problems in the analysis of life time data analysis.
Includes bibliographical references and index.
ISBN 978-1-4665-5566-2 (hardback)
1. Medical sciences--Statistical methods--Computer programs. 2. Survival analysis (Biometry) 3. Regression analysis--Data processing. 4. Prognosis--Statistical methods. I. Klein, John P., 1950-

R853.S7.H36 2013
610.72'7--dc23 2013019090

Visit the Taylor & Francis Web site at
http://www.taylorandfrancis.com

and the CRC Press Web site at
http://www.crcpress.com

Contents

Preface ix

About the Editors xi

List of Contributors xiii

I Regression Models for Right Censoring 1

1 Cox Regression Model 5
Hans C. van Houwelingen and Theo Stijnen

2 Bayesian Analysis of the Cox Model 27
Joseph G. Ibrahim, Ming-Hui Chen and Danjie Zhang, and Debajyoti Sinha

3 Alternatives to the Cox Model 49
Torben Martinussen and Limin Peng

4 Transformation Models 77
Danyu Lin

5 High-Dimensional Regression Models 93
Jennifer A. Sinnott and Tianxi Cai

6 Cure Models 113
Yingwei Peng and Jeremy M. G. Taylor

7 Causal Models 135
Theis Lange and Naja H. Rod

II Competing Risks 153

8 Classical Regression Models for Competing Risks 157
Jan Beyersmann and Thomas H. Scheike

9 Bayesian Regression Models for Competing Risks 179
Ming-Hui Chen, Mário de Castro, Miaomiao Ge, and Yuanye Zhang

10 Pseudo-Value Regression Models 199
Brent R. Logan and Tao Wang

11 Binomial Regression Models 221
Randi Grøn and Thomas A. Gerds

12 Regression Models in Bone Marrow Transplantation – A Case Study 243
Mei-Jie Zhang, Marcelo C. Pasquini, and Kwang Woo Ahn

III Model Selection and Validation 263

13 Classical Model Selection 265
Florence H. Yong, Tianxi Cai, L.J. Wei, and Lu Tian

14 Bayesian Model Selection 285
Purushottam W. Laud

15 Model Selection for High-Dimensional Models 301
Rosa J. Meijer and Jelle J. Goeman

16 Robustness of Proportional Hazards Regression 323
John O'Quigley and Ronghui Xu

IV Other Censoring Schemes 341

17 Nested Case-Control and Case-Cohort Studies 343
Ørnulf Borgan and Sven Ove Samuelsen

18 Interval Censoring 369
Jianguo Sun and Junlong Li

19 Current Status Data: An Illustration with Data on Avalanche Victims 391
Nicholas P. Jewell and Ruth Emerson

V Multivariate/Multistate Models 413

20 Multistate Models 417
Per Kragh Andersen and Maja Pohar Perme

21 Landmarking 441
Hein Putter

22 Frailty Models 457
Philip Hougaard

23 Bayesian Analysis of Frailty Models 475
Paul Gustafson

24 Copula Models 489
Joanna H. Shih

25 Clustered Competing Risks 511
Guoqing Diao and Donglin Zeng

26 Joint Models of Longitudinal and Survival Data 523
Wen Ye and Menggang Yu

27 Familial Studies 549
Karen Bandeen-Roche

VI Clinical Trials 569

28 Sample Size Calculations for Clinical Trials 571
Kristin Ohneberg and Martin Schumacher

29 Group Sequential Designs for Survival Data 595
Chris Jennison and Bruce Turnbull

30 Inference for Paired Survival Data 615
Jennifer Le-Rademacher and Ruta Brazauskas

Index 633

Preface

This volume examines modern techniques and research problems in the analysis of lifetime data analysis. This area of statistics deals with time-to-event data which is complicated not only by the dynamic nature of events occurring in time but also by censoring where some events are not observed directly, but rather they are known to fall in some interval or range.

Historically survival analysis is one of the oldest areas of statistics dating its origin to classic life table construction begun in the 1600s. Much of the early work in this area involved constructing better life tables and long tedious extensions of non-censored nonparametric estimators. Modern survival analysis began in the late 1980s with pioneering work by Odd Aalen on adapting classical Martingale theory to these more applied problems. Theory based on these counting process martingales made the development of techniques for censored and truncated data in most cases easier and opened the door to both Bayesian and classical statistics for a wide range of problems and applications.

In this volume we present a series of chapters which provide an introduction to the advances in survival analysis techniques in the past thirty years. These chapters can serve four complementary purposes. First, they provide an introduction to various areas in survival analysis for graduates students and other new researchers to this field. Second, they provide a reference to more-established investigators in this area of modern investigations into survival analysis. Third, with a bit of supplementation on counting process theory this volume is useful as a text for a second or advanced course in survival analysis. We have found that the instructor of such a course can pick and choose chapters in areas he/she deems most useful to the students or areas of interest to the instructor. Lastly, these chapters can help practicing statisticians pick the best statistical method to analyze their survival data experiment.

To help with reading the volume we have grouped chapters into six parts, each with a brief introduction by the editor. These parts are:

I Regression Models for Right Censoring

II Competing Risks

III Model Selection and Validation

IV Other Censoring Schemes

V Multivariate/Multistate Models

VI Clinical Trials

We believe that the chapters and topics presented here provide a good overview of the current status of survival analysis and will further inspire research into this area.

We would like to express our thanks to all the authors who contributed their time and effort to this volume. Without these contributions the volume would not be possible. A special thank you goes to Professor van Houwelingen who took upon himself much of the work of putting the authors' LaTeX and pdf files into a book format.

About the Editors

John P. Klein received his Ph.D. in 1980 from the University of Missouri. His thesis work was on dependent competing risks and safe dose estimation with dependent competing risks. While a student, he was a summer intern for two summers at the Oak Ridge National Labs working on modeling the effects of low-level radiation on mice.

Upon graduation, Dr. Klein joined the faculty in the Department of Statistics at the Ohio State University (OSU) where he rose to the rank of professor. While at OSU he was the statistician for the Clinical Research Center and Director of Statistics for the OSU Comprehensive Cancer Center. In 1987 he was a Gæsteprofessor at the University of Copenhagen which led to a long and fruitful collaboration with faculty there.

In 1993 Dr. Klein moved to Medical College of Wisconsin to be professor and director of the Division of Biostatistics. Since that time, he has been the statistical director of the Center for Blood and Marrow Transplantation. Recently, he has been appointed statistical director of the Clinical and Translational Research Institute of Southeast Wisconsin and the statistical director of the MCW Cancer Center. He is recognized as a leading expert in the analysis of bone marrow transplant data.

Dr. Klein is the author of 230 research papers, about half of which are in the statistical literature. He is a co-author of *Survival Analysis: Techniques for Censored and Truncated Data*, a standard graduate text. He is an associate editor of *Biometrics, Life Time Data Analysis, Dysphagia*, and *The Iranian Journal of Statistics.* He is an elected member of the International Statistical Institute and a fellow of the American Statistical Institute.

Hans C. van Houwelingen received his Ph.D. in Mathematical Statistics from the University of Utrecht, The Netherlands, in 1973. He stayed at the Mathematics Department in Utrecht until 1986. During that time his theoretical research interest was Empirical Bayes methodology as developed by Herbert Robbins. His main contribution was the finding that Empirical Bayes rules could be improved by monitorization.

On the practical side he was involved in collaborations with researchers in Psychology, Chemistry and Medicine. The latter brought him to Leiden in 1986 where he was appointed chair and department head in Medical Statistics at the Leiden Medical School, which was transformed into the Leiden University Medical Center (LUMC) in 1996.

Together with his Ph.D. students he developed several research lines in logistic regression, survival analysis, meta-analysis, statistical genetics and statistical bioinformatics. In the meantime the department grew into the Department of Medical Statistics and Bioinformatics, which also includes the chair and staff in Molecular Epidemiology.

He was editor-in-chief of *Statistica Neerlandica* and served on the editorial boards of *Statistical Methods In Medical Research*, *Lifetime Data Analysis*, *Biometrics*, *Biostatistics*, *Biometrical Journal*, and *Statistics & Probability Letters.* He is an elected member of ISI, fellow of ASA, honorary member of the International Society for Clinical Biostatistics (ISCB), Dutch Statistical Society (VVS) and ANed, the Dutch Region of the International Biometric Society (IBS).

Dr. van Houwelingen retired on January 1, 2009. On that occasion he was appointed Knight in the Order of the Dutch Lion. After his retirement he wrote, together with Hein

Putter, the Chapman & Hall monograph *Dynamic Prediction in Clinical Survival Analysis*, published in 2012.

Joseph G. Ibrahim is alumni distinguished professor of biostatistics at the University of North Carolina, Chapel Hill (UNC). Dr. Ibrahim's areas of research focus on Bayesian inference, missing data problems, and genomics. He received his Ph.D. in statistics from the University of Minnesota in 1988. With over 19 years of experience working in cancer clinical trials, Dr. Ibrahim directs the UNC Center for Innovative Clinical Trials. He is also the director of graduate studies in UNC's Department of Biostatistics, as well as the program director of the cancer genomics training grant in the department. He has directed or co-directed 27 doctoral students and 7 post-doctoral fellows. He is currently the editor of the Applications and Case Studies Section of the *Journal of the American Statistical Association* and currently serving as the associate editor for several other statistical journals. Dr. Ibrahim has published over 230 research papers, mostly in top statistical journals. He also has published two advanced graduate-level books on Bayesian survival analysis and Monte Carlo methods in Bayesian computation. He is an elected fellow of the American Statistical Association and the Institute of Mathematical Statistics, and an elected member of the International Statistical Institute.

Thomas H. Scheike received his Ph.D. in Mathematical Statistics from the University of California at Berkley in 1993, and received a doctoral degree in Statistics (Dr. Scient) from the University of Copenhagen in 2002. He has been in the Department of Biostatistics at the University of Copenhagen since 1994.

His research interest is in biostatistics with particular emphasis on survival analyses and longitudinal data. He has written the Springer monograph, *Dynamic Regression Models for Survival Data* (2006) with Torben Martinussen. Dr. Scheike has been involved in several R-packages (timereg, HaploSurvival, mets) that have been developing recent methods in survival analyses for the biostatistical community.

Dr. Scheike was editor-in-chief of *Scandinavian Journal of Statistics*, and served on the editorial boards of *Journal of the Royal Statistical Society*, *Lifetime Data Analysis*, and *Statistica Sinica*.

List of Contributors

Editors

John P. Klein
Division of Biostatistics
Medical College of Wisconsin
Milwaukee, WI, USA

Hans C. van Houwelingen
Leiden University Medical Center
Leiden, The Netherlands

Joseph G. Ibrahim
Department of Biostatistics
University of North Carolina
Chapel Hill, NC, USA

Thomas H. Scheike
University of Copenhagen
Copenhagen, Denmark

Authors

Kwang Woo Ahn
Division of Biostatistics
Medical College of Wisconsin
Milwaukee, WI, USA

Per Kragh Andersen
Department of Biostatistics
University of Copenhagen
Copenhagen, Denmark

Karen Bandeen-Roche
Department of Biostatistics
Johns Hopkins Bloomberg School of Public Health
Baltimore, MD, USA

Jan Beyersmann
Freiburg University Medical Center and Ulm University
Freiburg and Ulm, Germany

Ørnulf Borgan
University of Oslo
Oslo, Norway

Ruta Brazauskas
Division of Biostatistics
Medical College of Wisconsin
Milwaukee, WI, USA

Tianxi Cai
Department of Biostatistics
Harvard School of Public Health
Boston, MA, USA

Ming-Hui Chen
Department of Statistics
University of Connecticut
Storrs, CT, USA

Mário de Castro
Instituto de Ciências Matemáticas e de Computação
Universidade de São Paulo
São Carlos-SP, Brazil

Guoqing Diao
George Mason University
Fairfax, VA, USA

Ruth Emerson
University of California
Berkeley, CA, USA

Miaomiao Ge
Clinical Bio Statistics
Boehringer Ingelheim Pharmaceuticals, Inc.
Ridgefield, CT, USA

Thomas A. Gerds
Department of Biostatistics
University of Copenhagen
Copenhagen, Denmark

Jelle J. Goeman
Leiden University Medical Center
Leiden, The Netherlands

Randi Grøn
Department of Biostatistics
University of Copenhagen
Copenhagen, Denmark

Paul Gustafson
Department of Statistics
University of British Columbia
Vancouver, BC, Canada

Philip Hougaard
Biometric Division, Lundbeck
Valby, Denmark

Joseph G. Ibrahim
Department of Biostatistics
University of North Carolina
Chapel Hill, NC, USA

Chris Jennison
University of Bath
Bath, UK

Nicholas P. Jewell
University of California
Berkeley, CA, USA

Theis Lange
Section of Social Medicine
Department of Public Health, University of Copenhagen
Copenhagen, Denmark

Purushottam W. Laud
Division of Biostatistics
Medical College of Wisconsin
Milwaukee, WI, USA

Jennifer Le-Rademacher
Division of Biostatistics
Medical College of Wisconsin
Milwaukee, WI, USA

Junlong Li
Department of Biostatistics
Harvard University
Boston, MA, USA

Danyu Lin
Department of Biostatistics
University of North Carolina
Chapel Hill, NC, USA

Brent R. Logan
Division of Biostatistics
Medical College of Wisconsin
Milwaukee, WI, USA

Torben Martinussen
Department of Biostatistics
University of Copenhagen
Copenhagen, Denmark

Rosa J. Meijer
Leiden University Medical Center
Leiden, The Netherlands

Kristin Ohneberg
Institute of Medical Biometry and Medical Informatics
University Medical Center Freiburg
Freiburg, Germany

John O'Quigley
Université Pierre et Marie Curie - Paris VI
Paris, France

Marcelo C. Pasquini
Division of Hematology and Oncology
Medical College of Wisconsin
Milwaukee, WI, USA

Limin Peng
Department of Biostatistics and Bioinformatics
Emory University
Atlanta, GA, USA

Yingwei Peng
Queen's University
Kingston, Ontario, Canada

Maja Pohar Perme
Department of Biostatistics and Medical Informatics
University of Ljubljana
Ljubljana, Slovenia

Hein Putter
Department of Medical Statistics and Bioinformatics
Leiden University Medical Center
Leiden, The Netherlands

Naja H. Rod
Section of Social Medicine
Department of Public Health, University of Copenhagen
Copenhagen, Denmark

Sven Ove Samuelsen
University of Oslo
Oslo, Norway

Thomas H. Scheike
University of Copenhagen
Copenhagen, Denmark

Martin Schumacher
Institute of Medical Biometry and Medical Informatics
University Medical Center Freiburg
Freiburg, Germany

Joanna H. Shih
National Cancer Institute
Bethesda, MD, USA

Debajyoti Sinha
Department of Statistics
Florida State University
Tallahassee, FL, USA

Jennifer A. Sinnott
Department of Biostatistics
Harvard School of Public Health
Boston, MA, USA

Theo Stijnen
Leiden University Medical Center
Leiden, The Netherlands

Jianguo Sun
Department of Statistics
University of Missouri
Columbia, MO, USA

Jeremy M. G. Taylor
University of Michigan
Ann Arbor, MI, USA

Lu Tian
Stanford University School of Medicine
Stanford, CA, USA

Bruce Turnbull
Cornell University
Ithaca, NY, USA

Hans C. van Houwelingen
Leiden University Medical Center
Leiden, The Netherlands

Tao Wang
Division of Biostatistics
Medical College of Wisconsin
Milwaukee, WI, USA

L.J. Wei
Harvard School of Public Health
Boston, MA, USA

Ronghui Xu
University of California
San Diego, CA, USA

Wen Ye
Department of Biostatistics
University of Michigan
Ann Arbor, MI, USA

Florence H. Yong
Harvard School of Public Health
Boston, MA, USA

Menggang Yu
Department of Biostatistics and Medical Informatics
University of Wisconsin
Madison, WI, USA

Donglin Zeng
University of North Carolina
Chapel Hill, NC, USA

Danjie Zhang
Department of Statistics
University of Connecticut
Storrs, CT, USA

Mei-Jie Zhang
Division of Biostatistics
Medical College of Wisconsin
Milwaukee, WI, USA

Yuanye Zhang
Novartis Institutes for BioMedical Research Inc.
Cambridge, MA, USA

Part I

Regression Models for Right Censoring

In Part 1 we present statistical methods for right-censored and left-truncated survival data. For this type of data it is assumed that there is a single event which causes death or failure of an individual sampling unit. These sample units could be humans or animals subjected to some type of treatment or they could be mechanical or electronic units. Here if the time-to-failure is T then we are interested in making an inference about the survival function $S(t) = Pr[T > t]$. In engineering applications this function is called the "reliability function." In the sequel we focus on the biological applications of methods.

For many applications the time-to-failure is not observable for all individuals in the study but partial information that the event time is longer than some censoring time is all that is available. Such observations are called right-censored observations and the information they give us is simply that the failure time for an individual is beyond their censoring time.

There are many types of right censoring. There is type I censoring where each subject has assigned to it a fixed censoring time after which observation on the subject stops. This censoring scheme is used in reliability applications to shorten the on-study time or in biological studies when the observational window is fixed. For type II censoring the number of failures is fixed and the censoring time is random. This scheme is used most often in engineering applications to reduce the on-study time. Finally there is progressive censoring where what is observed is the smaller of the event time T and a random censoring time C. The censoring time C, commonly the lost-to-follow-up time, reflects when the individual drops out of the study or stops being followed. Most analysis assumes that this censoring time is non-informative or independent of the survival time. Censoring is non-informative if there is no information on the survival time, T, available from the magnitude of the censoring time.

In some cases in addition to being right-censored the data is left-truncated. Data is said to be left-truncated if only those potential subjects that have had some truncating event occur at a time τ are at risk at any point in time beyond τ. The classic example is the Channing house study which examined the survival probabilities for senior citizens in a nursing home. In that study at a given time, t_0, only subjects who enter the home at an age prior to t_0, are considered at risk for death at this age, and any patient with an entry age greater than t_0 is not included in the risk set. Since left truncation simply modifies the risk set, the methods developed for right-censored data are usually the same as those for left-truncated right-censored data.

Inference techniques for survival data can trace their roots to three key publications. The first is the paper by Kaplan and Meier (1958) which developed a non-parametric estimator of the survival function (see Section 1.1.2). This "Kaplan-Meier" estimator was a modern version of the classical life tables used in actuary science and developed by Edmond Halley in 1693 (yes the famous comet discoverer). The second work is the paper by David Cox (1972) which is the centerpiece of this first set of chapters and discussed briefly below.

The third and perhaps most important work is the development of the theory of counting processes and their use in survival analysis by Odd Aalen. His pioneering work on counting processes and martingales, starting with his 1975 Ph.D. thesis, has had profound influence on survival analysis techniques. Inferences for fundamental quantities associated with cumulative hazard rates in survival analysis and models for analysis of event histories are typically based on Aalen's work.

The Cox model or the proportional hazards model is perhaps the most common method in survival analysis. This model is based on modeling the hazard rate $\lambda(t|\mathbf{Z})$. Here the hazard rate is the rate at which individuals are experiencing the event, namely

$$\lambda(t|\mathbf{Z}) = -\frac{d\ln(S(t|\mathbf{Z}))}{dt} = \frac{f(t|\mathbf{Z})}{S(t|\mathbf{Z}))}.$$

Here $S(T|\mathbf{Z})$ ($f(T|\mathbf{Z})$) is the survival probability (density) given a vector $\mathbf{Z}$ of covariates.

The Cox model in its most widely used formulation assumes that we can write $\lambda(t|\mathbf{Z})$ as

$$\lambda(t|\mathbf{Z}) = \lambda_0(t)\exp\{\boldsymbol{\beta}^t\mathbf{Z}\}$$

where $\boldsymbol{\beta}$ is a vector of parameters and $\lambda_0(t)$ is a baseline hazard. Note that for this model if we have two individuals with two sets of covariates that

$$\frac{\lambda(t|\mathbf{Z}_1)}{\lambda(t|\mathbf{Z}_2)} = \frac{\lambda_0(t)\exp\{\boldsymbol{\beta}^t\mathbf{Z}_1\}}{\lambda_0(t)\exp\{\boldsymbol{\beta}^t\mathbf{Z}_2\}} = \exp\{\boldsymbol{\beta}^t(\mathbf{Z}_1 - \mathbf{Z}_2)\}$$

which is independent of t. Hence this is called the "proportional hazards model."

The Cox model is the most popular regression model for survival data. Its properties can be derived using the counting process techniques of Aalen. There are inference packages for it in almost every statistical package. It can be extended quite easily to time-dependent covariates and models for multistate models or models with random effects.

In Chapter 1 we present a paper by van Houwelingen and Stijnen on the classical estimation of the Cox model and in Chapter 2 a paper by Ibrahim et al. on the Bayesian approach to estimation for this model. In Chapter 5 we present a paper by Sinnott and Cai on how to make inference about model parameters in the Cox model when the number of parameters is very large.

Chapters 3 and 4 examine alternatives to the Cox model for right-censored data. In Chapter 3 we present a paper by Martinussen and Peng which surveys non-proportional hazards models including Aalen's additive hazards model, the accelerated failure time model and Quartile regression models. In Chapter 4 Lin surveys a class of transformation models for right-censored data. For such models we assume that the model is given by $H(T) = \boldsymbol{\beta}^t\mathbf{Z} + \epsilon$ where $H()$ is an unspecified increasing function, $\boldsymbol{\beta}$ a set of regression coefficients and ϵ a random error with a parametric distribution such as the extreme-value distribution or the standard logistic error distribution. Special cases include the proportional hazards and proportional odds model. Lin examines this class of models in a variety of sampling and modeling situations.

Chapters 6 and 7 show how the basic ideas of the Cox model can be extended to more complicated models. In Chapter 6 Peng and Taylor look at cure models. In such models an unknown fraction of the population cannot fail and as such is cured of the disease under study. Finally, Chapter 7 by Lange and Rod looks at the use of the Cox model as a causal model.

Bibliography

Aalen, O. O. (1975), Statistical Inference for a Family of Counting Processes, PhD thesis, University of California, Berkeley.

Cox, D. R. (1972), 'Regression models and life-tables', *Journal of the Royal Statistical Society - Series B* **34**, 187–220.

Halley, E. (1693), 'An estimate of the degrees of the mortality of mankind, drawn from curious tables of the births and funerals at the city of Breslaw with an attempt to ascertain the price of annuities upon lives', *Philosophical Transactions* **17**, 596–610.

Kaplan, E. and Meier, P. (1958), 'Nonparametric estimation from incomplete observations', *Journal of the American Statistical Association* **43**, 457–481.

1

Cox Regression Model

Hans C. van Houwelingen and Theo Stijnen

Leiden University Medical Center

CONTENTS

1.1 Basic statistical concepts 5
  1.1.1 Survival time and censoring time 5
  1.1.2 The Kaplan-Meier estimator 6
  1.1.3 The hazard function 7
1.2 The proportional hazards (Cox) model 8
1.3 Fitting the Cox model 9
1.4 Example: NKI breast cancer data 11
1.5 Martingale residuals, model fit 14
  Continued example 15
  Influential data points 15
1.6 Extensions of the data structure 16
  1.6.1 Delayed entry, left truncation 16
  1.6.2 Time-dependent covariates 17
  1.6.3 Continued example 17
1.7 Beyond proportional hazards assumption 19
  1.7.1 Stratified models 19
  1.7.2 Time-varying coefficients, Schoenfeld residuals 20
  1.7.3 Continued example 21
  1.7.4 Final remarks 21
  Bibliography 23

1.1 Basic statistical concepts

1.1.1 Survival time and censoring time

To set up the framework for survival data with right censoring, two random variables need to be defined

$$T_{\text{surv}} : \text{the survival time}, \quad T_{\text{cens}} : \text{the censoring time}.$$

The censoring time T_{cens} is often denoted by C. Different censoring mechanisms can be distinguished. A common one for clinical data is administrative censoring, where the censoring time is determined by the termination of the study. For most purposes it suffices to consider the censoring to be random. The crucial condition for statistical analysis is that survival time T_{surv} and censoring time T_{cens} are independent. In the presence of explanatory variables this condition can be weakened to independence of T_{surv} and T_{cens} conditional on the explanatory variables.

For both random variables the cumulative distribution functions $F_{\text{surv}}(t) = P(T_{\text{surv}} \leq t)$ and $F_{\text{cens}}(t) = P(T_{\text{cens}} \leq t)$ can be defined. The distribution function of the survival time is called the "failure function." In survival analysis it is often more convenient to use the complimentary functions, the survival (or survivor) function $S(t)$, and the censoring function $G(t)$ defined by

$$S(t) = 1 - F_{\text{surv}}(t) = P(T_{\text{surv}} > t),$$
$$G(t) = 1 - F_{\text{cens}}(t) = P(T_{\text{cens}} > t).$$

It is assumed that T_{surv} has a continuous distribution, implying that the survival function $S(t)$ is continuous and differentiable.

In summarizing a survival dataset the most important information is given by (an estimate of) the survival function, but it is also relevant to show (an estimate of) the censoring function. The censoring function describes the distribution of the follow-up times if no individual would have died.

In practice it is mostly impossible to observe both T_{surv} and T_{cens}. The observed "survival time" T is the smallest of the two,

$$T = \min(T_{\text{surv}}, T_{\text{cens}}) .$$

Moreover, it is known whether T_{surv} or T_{cens} has been observed. This is indicated by the event indicator D. The usual definition is

$$D = \begin{cases} 0, & \text{if } T = T_{\text{cens}} ; \\ 1, & \text{if } T = T_{\text{surv}} . \end{cases}$$

So, the information on the survival status is summarized in the pair (T, D).

1.1.2 The Kaplan-Meier estimator

The starting point for the statistical analysis is a sample of n independent observations

$$(t_1, d_1), (t_2, d_2), ..., (t_n, d_n)$$

from (T, D). Of the observed survival times $t_1, ..., t_n$, those with $d_i = 1$ are called the event times. The observed survival times with $d_i = 0$ are called the censoring times.

It is convenient to use the notation from the field of counting processes

$$Y_i(t) = \mathbf{1}\{t_i \geq t\}, \ \ \overline{Y}(t) = \textstyle\sum_{i=1}^{n} Y_i(t),$$

$$N_i(t) = \mathbf{1}\{t_i \leq t, d_i = 1\}, \ \ \overline{N}(t) = \textstyle\sum_{i=1}^{n} N_i(t).$$

The risk set $R(t)$ is defined as $R(t) = \{i; t_i \geq t\}$. Its size is given by $\overline{Y}(t)$. Assuming that T_{surv} and T_{cens} are independent, both $S(t)$ and $G(t)$ can be estimated by versions of the Kaplan-Meier estimator (Kaplan and Meier, 1958). The estimator of the survival function is given

$$\hat{S}_{KM}(t)) = \prod_{s \leq t} (1 - \frac{\Delta \overline{N}(s)}{\overline{Y}(s)}),$$

where $\Delta \overline{N}(t)$ is the number of events at time t. The estimator $\hat{G}_{KM}(t)$ is defined similarly.

The formula also covers the case of tied event times. The standard error of the estimate is given by Greenwood's formula (Greenwood, 1926)

$$se^2(\hat{S}_{KM}(t)) = \hat{S}^2_{KM}(t) \cdot \sum_{s \le t} \frac{\Delta \overline{N}(s)}{\overline{Y}(s)(\overline{Y}(s) - \Delta \overline{N}(s))}$$

The estimate $\hat{S}_{KM}(t)$ together with its standard error $se(\hat{S}_{KM}(t))$ can be used to construct $(1-\alpha) \cdot 100\%$ pointwise confidence intervals for the survival function $S(t)$. The simplest confidence interval is $\hat{S}_{KM}(t) \pm z_{1-\alpha/2}\, se(\hat{S}_{KM}(t))$, with $z_{1-\alpha/2}$ the $1-\alpha/2$ percentile of the standard normal distribution. It might however fall outside the interval $[0,1]$. This can be remedied by using transformations such as $\ln(S(t))$ or $\ln(-\ln(St))$. Such transformed confidence intervals are needed for smaller sample sizes. For more detail see Borgan and Liestol (1990).

1.1.3 The hazard function

Under the assumption of a continuous distribution with differentiable survival function the hazard function, also known as the "force of mortality" is defined by

$$\lambda(t)dt = P(T < t + dt | T \ge t) \; .$$

From $P(T > t + dt | T \ge t) = S(t + dt)/S(t)$ the following alternative definition can be obtained

$$\lambda(t) = -S'(t)/S(t) = -\frac{d \ln(S(t))}{dt} \; .$$

A related concept is the cumulative hazard function denoted by $\Lambda(t)$ and defined by

$$\Lambda(t) = \int_0^t \lambda(s) ds \; ,$$

Obviously, $\Lambda(t)$ and $S(t)$ are closely related: $\Lambda(t) = -\ln(S(t))$, $S(t) = \exp(-\Lambda(t))$

Since the hazard function $\lambda(t)$ is only well defined if the survival function $S(t)$ is differentiable, it is hard to estimate the hazard function properly because the Kaplan-Meier estimate of the survival function is a non-differentiable step function and some smoothing is needed before a proper estimate of the hazard function can be obtained.

It is much easier to estimate the cumulative hazard function $\Lambda(t)$. One way to do that is to use the link between $\Lambda(t)$ and $S(t)$ and to define

$$\hat{\Lambda}_{KM}(t) = -\ln(\hat{S}_{KM}(t)) = \sum_{s \le t} \ln(1 - \frac{\Delta \overline{N}(s)}{\overline{Y}(s)})$$

An alternative is the Nelson-Aalen estimator (Nelson, 1969; Aalen, 1975). The estimator and its standard error are given by

$$\hat{\Lambda}_{NA}(t) = \int_0^t \frac{d\overline{N}(s)}{\overline{Y}(s)}, \; se^2(\hat{\Lambda}_{NA}(t)) = \int_0^t \frac{d\overline{N}(s)}{\overline{Y}(s)^2}$$

If the sample size is large, there is very little difference between $\hat{\Lambda}_{KM}(t)$ and $\hat{\Lambda}_{NA}(t)$ or, similarly, between $\hat{S}_{KM}(t)$ and $\hat{S}_{NA}(t) = \exp(-\hat{\Lambda}_{NA}(t))$. The jumps in $\hat{\Lambda}_{NA}(t)$ define a discrete estimate of the hazard concentrated in the event times: $\hat{\lambda}_{NA}(t) = \Delta\overline{N}(t)/\overline{Y}(t)$. This definition also applies in the presence of ties.

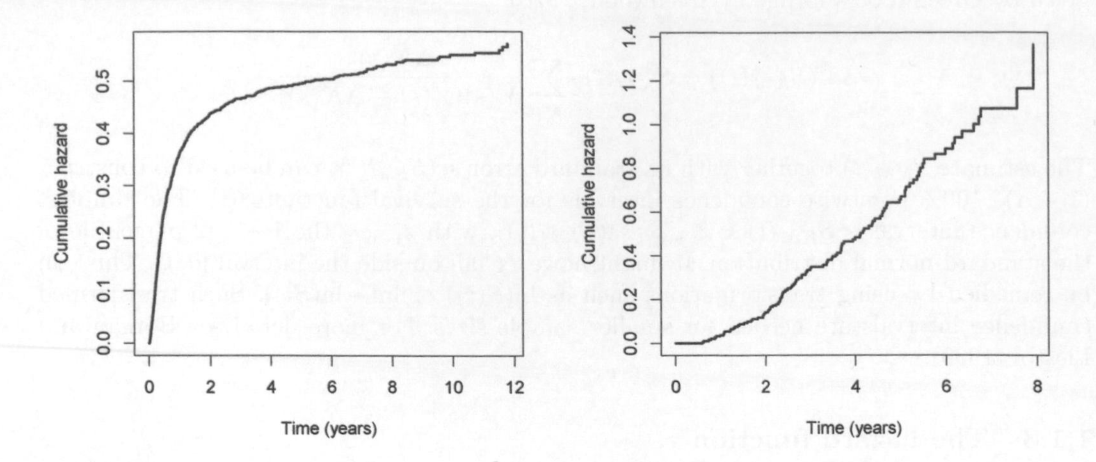

FIGURE 1.1
Nelson-Aalen estimates of the cumulative hazard for ALL patients (left) and CML patients (right).

Although the hazard function is hard to estimate, it plays an essential conceptual role in thinking about the process of survival. In the clinical setting, the shape of the hazard function determines the long-term prospect for a patient. A decreasing hazard function implies that prognosis gets better as you live longer ("old better than new"). An increasing hazard function implies that prognosis gets worse as you live longer. Plotting the (estimated) cumulative hazard function for a dataset can be a convenient way of detecting an increasing or decreasing hazard function. A convex cumulative hazard function points towards an increasing hazard, while a concave cumulative hazard function goes with a decreasing hazard. An example borrowed from van Houwelingen and Putter (2012) is given in Figure 1.1 showing plots of the Nelson-Aalen estimates for death or relapse in ALL patients (left) and for death in CML patients (right). In case of a cure, the cumulative hazard will reach a ceiling. For a further discussion of the interpretation of hazard curves see Klein and Moeschberger (2003).

1.2 The proportional hazards (Cox) model

To develop models for survival data in a population of individuals, one needs a simple way of describing the variation in survival among individuals. A popular model is to consider the individual specific hazard function $\lambda_i(t)$ and to make the proportional hazards assumption that

$$\lambda_i(t) = c_i \lambda_0(t)$$

where c_i is a constant and $\lambda_0(t)$ is a hazard function which is left unspecified.

The effect of covariates on the hazard can conveniently be modeled by taking $c_i = \exp(X_i^\top \beta)$ leading to the proportional hazards regression model, better known as the Cox regression model, introduced in Cox (1972),

$$\lambda(t|X) = \lambda_0(t)\exp(X^\top \beta) .$$

Here, $\lambda_0(t)$ is the baseline hazard that determines the shape of the survival function, X is the column vector of the covariates of an individual and β is a column vector of regression coefficients. It is common practice not to define a parametric model for the baseline hazard. This is in line with the practice to show the Kaplan-Meier estimate of the survival function as a summary of the data. As in regression models for other types of data, the covariate vector X can contain transformations and interactions of the risk factors. It should be noted that there is no constant term in the regression vector. The constant is absorbed in the baseline hazard : $\ln(\lambda_0(t))$ can be seen as a time-dependent intercept in the linear model for $\ln(\lambda(t|X))$. The implication is that centering the covariates, replacing X by $X - \mathrm{E}[X]$, would change the baseline, but not the regression coefficients. In some software such centering is applied and it is not always easy to figure out what a reported baseline hazard stands for. The survival function implied by the model is given by

$$S(t|X) = \exp(-\exp(X^\top \beta)\Lambda_0(t)) = S_0(t)^{\exp(X^\top \beta)} .$$

Here, $\Lambda_0(t) = \int_0^t \lambda_0(s)ds$ is the cumulative baseline hazard, and $S_0(t) = \exp(-\Lambda_0(t))$ the baseline survival function. The linear predictor $X^\top \beta$ is known as the prognostic index and denoted by PI.

The marginal survival function is obtained by taking the expected survival function $\mathrm{E}[S(t|X)]$ in the population. Since X appears in the exponent, $\mathrm{E}[S(t|X)]$ is not the same as the estimated survival for the average person: $S(t) = \mathrm{E}[S(t|X)] \neq S(t|\mathrm{E}[X])$. The difference between $\mathrm{E}[S(t|X)]$ and $S(t|\mathrm{E}[X])$ can be expected to be small if the variance of the prognostic index $PI = X^\top \beta$ is small and $S(t|X)$ is not too far from 1.

1.3 Fitting the Cox model

It is most interesting to read the original paper by Cox (1972) and the written discussion following it. The focus of Cox is on the estimation of the regression coefficients using what is called the "partial likelihood." The estimation of the baseline hazard has long been neglected. However, the effect of the hazard ratio can only be fully understood if the baseline hazard is known as well. (The best way of understanding the model is by visualizing the estimated survival curves for representative values of the covariate vector X).

To emphasize the importance of both components (baseline hazard and regression coefficients), the full likelihood of the data will be taken as the starting point for fitting the model.

The available data is a sample of n independent observations from the triple (T, D, X), that is

$$(t_1, d_1, x_1), (t_2, d_2, x_2)..., (t_n, d_n, x_n) .$$

The log-likelihood of the data given the covariates is given by

$$l(\lambda_0, \beta) = \sum_{i=1}^{n}(-\Lambda_0(t_i)\exp(x_i^\top \beta) + d_i(\ln(\lambda_0(t_i)) + x_i^\top \beta) .$$

This expression will be maximized by concentrating all the risk in the event times. This leads to a discrete version of the hazard as discussed in Section 1.1.3 for which the cumulative hazard is defined as

$$\Lambda_0(t) = \sum_{s \le t} \lambda_0(s) .$$

Plugging this into the expression for the log-likelihood and rearranging some terms gives

$$l(\lambda_0, \beta) = \sum_t \left(-\lambda_0(t) \sum_i Y_i(t) \exp(x_i^\top \beta) + \ln(\lambda_0(t)) \Delta \overline{N}(t) + \sum_i \Delta N_i(t) x_i^\top \beta \right) .$$

This formula allows for ties. For fixed value of β this expression is maximal for what is called the "Breslow estimator" (Breslow, 1974) given by

$$\hat{\lambda}_0(t|\beta) = \frac{\Delta \overline{N}(t)}{\sum_i Y_i(t) \exp(x_j^\top \beta)}$$

the resulting maximized (or profile) log-likelihood is

$$l(\hat{\lambda}_0(\beta), \beta) = pl(\beta) + \sum_t (-\Delta \overline{N}(t) + \ln(\Delta \overline{N}(t)) .$$

Here, $pl(\beta)$ is Cox's partial log-likelihood defined as

$$pl(\beta) = \sum_{i=1}^n \int_0^\infty \ln \left(\frac{\exp(x_i^\top \beta)}{\sum_j Y_j(t) \exp(x_j^\top \beta)} \right) dN_i(t) . \tag{1.1}$$

Cox did not obtain this expression as a profile likelihood, but used a conditioning argument. The term $\exp(x_i^\top \beta) / \sum_j Y_j(t) \exp(x_j^\top \beta)$ for $t = t_i$ can be interpreted as the probability that individual i is the one that died at event time t_i given the risk set $R(t_i)$ of people still alive and in follow-up just prior to t_i. The conditional argument gets complicated in the presence of ties. The definition above allows for the presence of ties and was suggested by Breslow in the discussion of Cox's paper (Breslow, 1972). It is perfectly valid if ties are incidental and only due to rounding of the observed times.

So, the computational procedure is to estimate β by maximizing the partial log-likelihood and estimating the baseline-hazard by the Breslow estimator with $\beta = \hat{\beta}$.

The survival function given the covariate x can be estimated either by the analogue of the Nelson-Aalen estimator

$$\hat{S}_{NA}(t|x, \hat{\beta}) = \exp \left(-\hat{\Lambda}_0(t) \exp(x^\top \hat{\beta}) \right) ,$$

or the analogue of the Kaplan-Meier estimator, the product-limit estimator

$$\hat{S}_{PL}(t|x, \hat{\beta}) = \prod_{s \le t} \left(1 - \exp(x^\top \hat{\beta}) \hat{\lambda}_0(s) \right) .$$

Most software packages provide $\hat{S}_{NA}$. In R, both $\hat{S}_{NA}$ and $\hat{S}_{PL}$ can be calculated, through the type argument of the function survfit in the **survival** package. Note that $\hat{S}_{NA}$ always yields a proper survival function, while $\hat{S}_{PL}$ will yield weird results if $\exp(x^\top \hat{\beta}) \hat{\lambda}_0(t_i) > 1$ for some t_i. In practice, however, there is very little difference between the two methods if the sample size is rather large. The small sample behavior of these estimators and additional variants is discussed in Andersen et al. (1996).

It has been shown in Tsiatis (1981) and Andersen and Gill (1982) that the partial likelihood can be treated as a regular likelihood, in the sense that the estimate $\hat{\beta}$ has an asymptotic normal distribution with mean β and covariance matrix given by the inverse observed Fisher information matrix. The first derivative, known as the score function or the estimation equation for β, is given by

$$\frac{\partial pl(\beta)}{\partial \beta} = \sum_i \int_0^\infty Y_i(t)(x_i - \bar{x}(\beta,t) dN_i(t) ,$$

with $\bar{x}(\beta,t)$ the weighted average of the x_j's in the risk set $R(t)$, that is

$$\bar{x}(\beta,t) = \frac{\sum_j Y_j(t) x_j \exp(x_j^\top \beta)}{\sum_j Y_j(t) \exp(x_j^\top \beta)} .$$

The Fisher information of the partial likelihood is given by

$$I_{pl}(\beta) = -\frac{\partial^2 pl(\beta)}{\partial \beta^2} = \int_0^\infty var(x|\beta,t) d\bar{N}(t),$$

with

$$var(x|\beta,t) = \frac{\sum_j Y_j(t)(x_j - \bar{x}(\beta,s)(x_j - \bar{x}(\beta,t))^\top \exp(x_j^\top \beta)}{\sum_j Y_j(t) \exp(x_j^\top \beta)}$$

the weighted covariance matrix in $R(t)$.

Similarly, it is shown in the same papers (Andersen and Gill, 1982; Tsiatis, 1981) that estimates of individual survival probabilities $\hat{S}(t|x)$ are asymptotically normal with mean $S(t|x)$ and a covariance matrix that can be obtained from the observed Fisher information of the full likelihood $l(\lambda_0, \beta)$. It is immaterial which method (NA or PL) is used to estimate the probabilities, because they are asymptotically equivalent. The asymptotic variance of $\hat{S}(t|x) = \hat{S}(t|x,\hat{\beta})$ is complicated by the fact that it depends on $\hat{\beta}$ both directly and indirectly through the dependence of $\hat{\Lambda}_0(t)$ on $\hat{\beta}$. The asymptotic variance of $-\ln(\hat{S}(t|x)) = \hat{\Lambda}_0(t)\exp(x^\top \hat{\beta})$ may be estimated consistently by

$$\int_0^t \left(\frac{\exp(x^\top \hat{\beta})}{\sum_j Y_j(s) \exp(x_j^\top \hat{\beta})} \right)^2 d\bar{N}(s) + \hat{q}(t|x)^\top I_{pl}^{-1}(\hat{\beta})\, \hat{q}(t|x) ,$$

with

$$\hat{q}(t|x) = \int_0^t (x - \bar{x}(\hat{\beta},s)) \frac{\exp(x^\top \hat{\beta})}{\sum_j Y_j(s) \exp(x_j^\top \hat{\beta})} d\bar{N}(s) ,$$

The formula is based on the finding that $\hat{\beta}$ and $\hat{\Lambda}_0(t)$ are asymptotically independent if X is dynamically centered at $\bar{x}(\hat{\beta},t)$ and $\hat{\Lambda}_0(t)$ is replaced by $\hat{\Lambda}_0(t)\exp(\bar{x}(\hat{\beta},t)^\top \hat{\beta})$. The formula can be used to construct confidence intervals for $\hat{S}(t|x)$ on the ln-scale, or on the probability scale. However, not all software packages have the option for computing such confidence interval.

1.4 Example: NKI breast cancer data

The example used throughout this chapter is the Dutch breast cancer dataset that contains data on the overall survival of breast cancer patients as collected in the Dutch Cancer Institute (NKI) in Amsterdam. This dataset became very well-known because it was used in

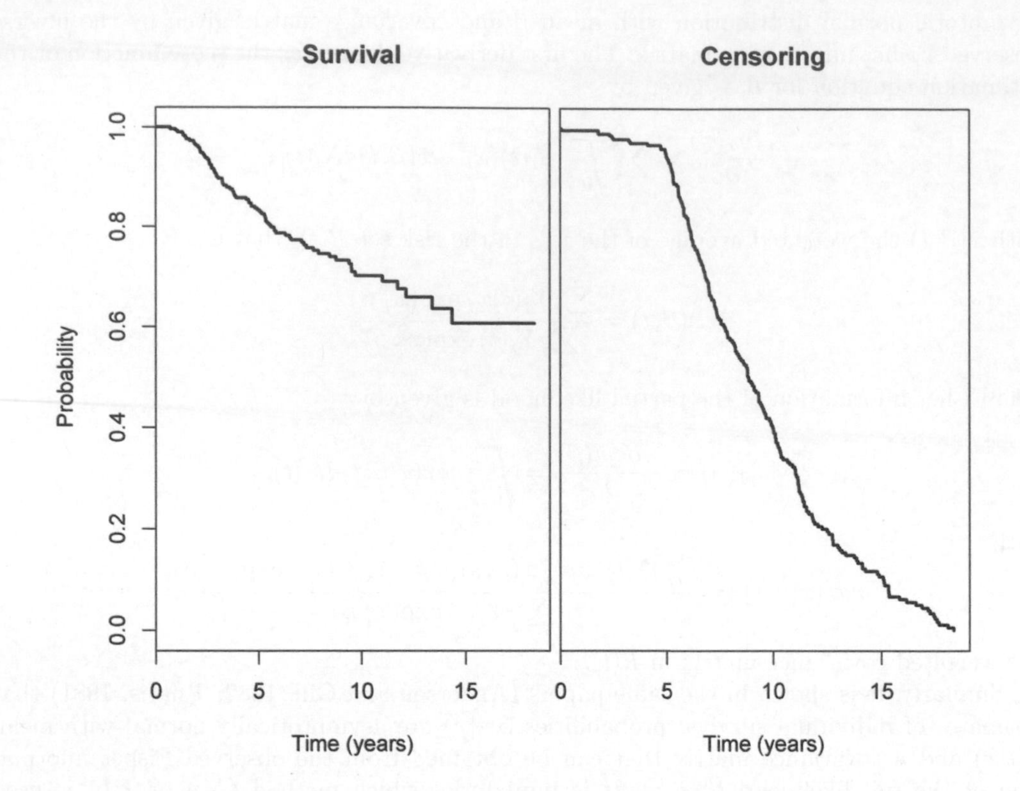

FIGURE 1.2
Survival and censoring functions for the breast cancer dataset.

one of the first successful studies that related the survival of breast cancer to gene expression. The findings of this study were reported in two highly cited and highly influential papers: van't Veer et al. (2002) van de Vijver et al. (2002).

This dataset contains the clinical and genomic data on 295 patients with 79 events as reported in van de Vijver et al. (2002). The data was reanalyzed by van Houwelingen in cooperation with the statisticians of the NKI and published in van Houwelingen et al. (2006). The survival and censoring functions of this dataset are shown in Figure 1.2. The survival curve appears to stabilize at a long-term survival rate of about 60%. The censoring curve shows that the median follow-up in the dataset is about 9 years. The data are available through www.msbi.nl/DynamicPrediction, the website of van Houwelingen and Putter (2012).

The dataset contains clinical and genomic information on the patients. In this chapter only the clinical information is used. Using high-dimensional genomic information is discussed in Chapter 5 and Chapter 15. The information about the clinical risk factors available after surgery is given in Table 1.1.

Of the categorical covariates, Histological Grade and Vascular Invasion appear to have a significant univariate effect. For the continuous covariates, the univariate Cox regression coefficients are given in Table 1.1 as well. No attempt is made at this stage to optimize the functional form of these covariates. A simple model is applied with a linear effect of each continuous covariate. Apparently Tumor Diameter, Age of the Patient and Estrogen Level have a significant univariate effect. Table 1.1 also gives the regression coefficients for the

TABLE 1.1
The clinical risk factors and their effects on survival in the breast cancer dataset. Shown are the estimated regression coefficients (B) and their standard errors (SE) in separate Cox models for each risk factor ("univariate") and in a Cox model including all covariates ("multivariate").

				univariate		multivariate	
Covariate	Category		Freq.	B	SE	B	SE
Chemotherapy	No		185	0	0	0	0
	Yes		110	-0.235	0.240	-0.423	0.298
Hormonal surgery	No		255	0	0	0	0
	Yes		40	-0.502	0.426	-0.172	0.442
Type of surgery	Excision		161	0	0	0	0
	Mastectomy		134	0.185	0.225	0.154	0.249
Histological grade	Intermediate		101	0	0	0	0
	Poorly differentiated		119	0.789	0.248	0.266	0.281
	Well differentiated		75	-1.536	0.540	-1.308	0.547
Vascular invasion	-		185	0	0	0	0
	+		80	0.682	0.234	0.603	0.253
	+/-		30	-0.398	0.474	-0.146	0.491
Covariate	Min	Max	Mean	SD	B	SE	B	SE
Diameter (mm)	2	50	22.54	8.86	0.037	0.011	0.020	0.013
# positive nodes	0	13	1.38	2.19	0.064	0.046	0.074	0.052
Age at diagnosis	26	53	43.98	5.48	-0.058	0.020	-0.039	0.020
Estrogen level	-1.591	0.596	-0.260	0.567	-1.000	0.183	-0.750	0.211

Note: SD stands for standard deviation.

Cox model including all covariates. Estrogen level and histological grade seem to be the most important predictors.

For categorical covariates, the effect is often expressed as the hazard ratio with respect to the baseline category $HR = \exp(B)$ and the corresponding 95%-confidence interval $(\exp(B - 1.96 \cdot SE), \exp(B + 1.96 \cdot SE))$. For example, the univariate hazard ratio Mastectomy:Excision equals 1.203 with 95%-confidence interval (0.774,1.870). For continuous covariates, the $\exp(B)$ will give the hazard ratio per unit increase. It depends on the scaling of the covariate. For example Age at Diagnosis in Table 1.1 is measured in years. The hazard ratio per year is very close to one (0.944). It makes more sense to look at the hazard ratio per 10 years. The univariate effect hazard ratio per 10 years is given by $\exp(10 \cdot (-0.058)) = 0.560$ with 95%-confidence interval (0.378,0.829).

The variation in survival is directly related to the standard deviation of the prognostic index $PI = X^\top \hat{\beta}$. In this data it equals 1.125. The variation in survival is shown in the left panel of Figure 1.3 using percentiles of the prognostic index and the uncertainty of the estimated survival curves is shown in the right-hand panel of Figure 1.3 contrasting two patients. Patient 1 has negative vascular invasion and well-differentiated histology and Patient 2 has positive vascular invasion and poorly differentiated histology. Both have the continuous covariates at the mean value and the other categorical covariates at the baseline value.

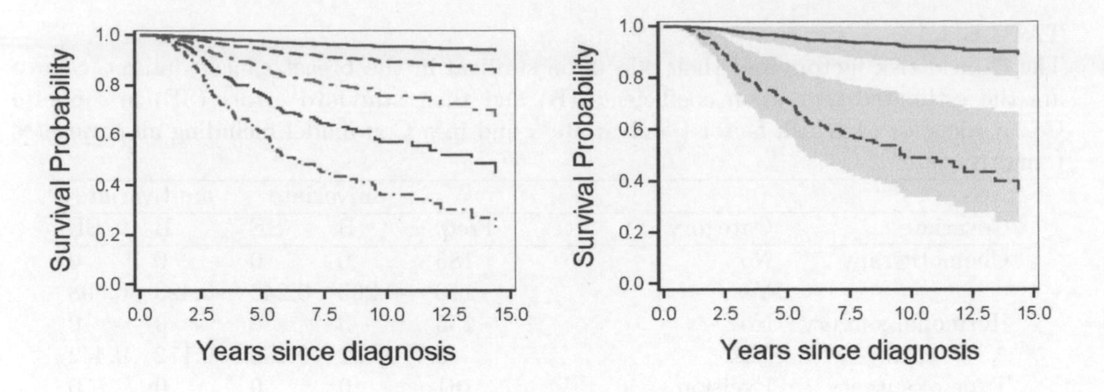

FIGURE 1.3
Predicted survival curves for percentiles (10 (top), 25, 50, 75, 90 (bottom)) of the prognostic index (left) and predicted survival curves with 95% pointwise confidence intervals for two patients (right).

1.5 Martingale residuals, model fit

The process $N_i(t)$ defined in Section 1.1.2 can be seen as a Poisson process with intensity $\lambda_i(t)$ that is stopped when an event or censoring occurs. The process $Y_i(t)$ indicates whether an individual is still under observation. When the individual is still under observation at t ($Y_i(t) = 1$), the probability of an event between t and $t + dt$ is given by $\lambda_i(t)dt$, or more general $d\Lambda_i(t)$. This is summarized in the definition of the martingale process

$$M_i(t) = \int_0^t Y_i(s)d(N_i(s) - \Lambda_i(s))$$

Roughly speaking a martingale is a stochastic process with increments that have mean zero given the past of the process. For a formal definition of a martingale and its relevance for the Cox model, see the following books: Andersen et al. (1993), Therneau and Grambsch (2000), Fleming and Harrington (1991) and O'Quigley (2008). The two important properties of the martingale process are

$$E[M_i(t)] = 0 \ , \ var(M_i(t)) = E[\int_0^t Y_i(s)d\Lambda_i(s)] \ .$$

A special case is $M_i = M_i(\infty) = D_i - \Lambda_i(T_i)$ with $var(M_i) = E[\Lambda_i(T_i)]$. The empirical counterpart of the martingale process

$$\widehat{M_i}(t) = \int_0^t Y_i(s)d(N_i(s) - \hat{\Lambda}_i(s))$$

is known as the "martingale residual process" and $\widehat{M_i} = \widehat{M_i}(\infty) = d_i - \hat{\Lambda}_i(t_i)$ as the martingale residual. The estimating equations of the Cox model can be described in terms of the martingale residual process and the martingale residual:

$$\sum_i \widehat{M_i}(t) = 0 \ \forall t \ , \ \sum_i x_i \widehat{M_i} = 0 \ .$$

The proof is left as an exercise. The right-hand side gives the estimating equations of the partial likelihood described in Section 2.1, while the left-hand side yields the Breslow estimator of the baseline hazard $\hat{\lambda}_0(t)$.

The martingale residuals can be used to identify outlying observations or groups of observations and to check the correctness of the functional form of continuous covariate and interactions between covariates. Detection of individual outliers can be based on $\widehat{M}_i/\sqrt{\hat{\Lambda}_i(t_i)}$. However, this residual is very skewed and there is no reference distribution for it. Outlying groups could be identified by considering $\sum_{i\in G}\widehat{M}_i/\sqrt{\sum_{i\in G}\hat{\Lambda}_i(t_i)}$, where G is a subgroup of observations defined by their covariate values. (See also Verweij et al. (1998)). Correctness of the functional form of a covariate in the model may be checked by plotting $\widehat{M}$ against X. The estimating equation forces the linear trend to be zero. Local smoothers like LOESS may help to detect nonlinear relations.

Continued example

An example for the data of Section 5.5 is given in Figure 1.4 and Figure 1.5. (The martingale residuals are obtained by Proc PHREG from SAS). Figure 1.5 shows how interactions can be detected by martingale residuals. Figure 1.4 shows the martingale residuals for the prognostic index and estrogen level. The LOESS smoother shows that there is no indication of a nonlinear effect. Figure 1.5 shows how interactions can be detected by martingale residuals. The figure shows scatter plots of martingale residual versus estrogen level in the three histological grade categories together with a simple linear fit. The different slopes in the three groups might be an indication of interaction between estrogen level and histological grade.

Influential data points

Residuals can also be used to detect influential data points. The influence function is well-defined in ordinary linear regression, but less straightforward in Cox regression. Some software produce `dfbeta residual`'s that show the change in regression coefficients if an observation is left out. For more detail see Chapter 7 of Therneau and Grambsch (2000).

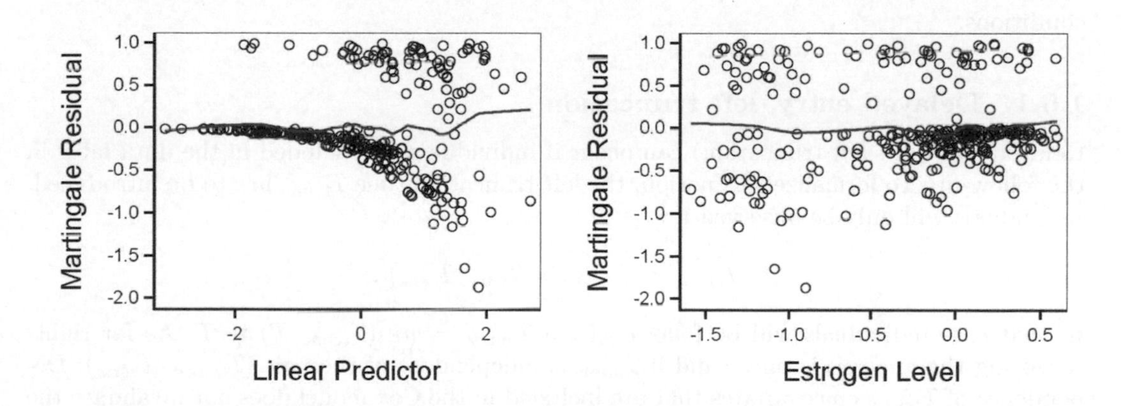

FIGURE 1.4
Martingale residuals for the prognostic index (left) and estrogen level (right) with LOESS smoothing.

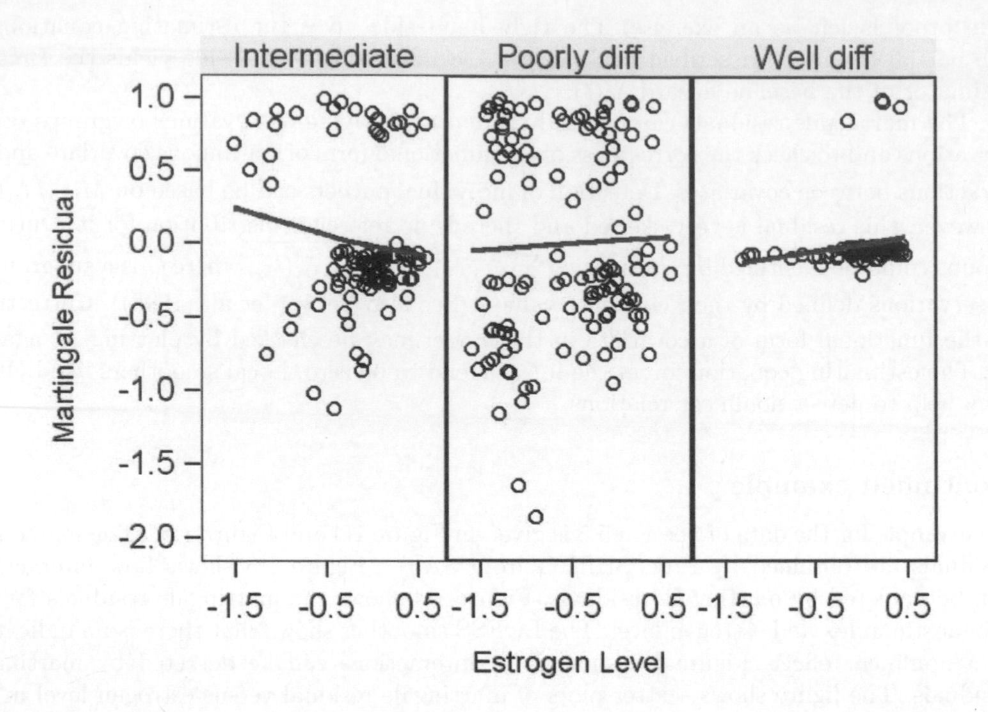

FIGURE 1.5
Martingale residuals for estrogen level by histological grade categories with linear fits.

1.6 Extensions of the data structure

The data structure of the Cox model considered so far is rather rigid. All individuals have to be followed from $t = 0$ onwards and the covariates are not allowed to change over time. However, the estimation procedure described in Section 2.1 allows relaxation of both conditions.

1.6.1 Delayed entry, left truncation

Delayed entry (or left truncation) can occur if individuals are included in the data later in the follow-up. To formalize this notion, the left truncation time T_{trunc} has to be introduced. Individuals will only be observed if

$$T_{\text{trunc}} < T = min(T_{\text{surv}}, T_{\text{cens}}) \ .$$

In that case individuals will be followed from $T_{\text{entry}} = min(T_{\text{trunc}}, T)$ to T. As for right-censoring the analysis is only valid if T_{surv} is independent of the pair (T_{trunc}, T_{cens}). Dependence of T_{trunc} on covariates that are included in the Cox model does not invalidate the analysis. Delayed entry implies that t_{entry} has to be added as an extra piece of information. The data format is (t_{entry}, t, d, x).

The main consequence of delayed entry is a change in $Y_i(t)$ which is now defined as

$$Y_i(t) = \mathbf{1}\{t_{entry,i} < t \leq t_i\}$$

The risk set $R(t)$ consists of all individuals with $t_{entry,i} < t \leq t_i$. Consequently, the number at risk $\overline{Y}(t)$ is no longer a monotonically decreasing function of t. Formally, the estimating procedures are exactly the same as in Sections 1.1 and 2.1. This is one of the advantages of modeling the hazard. Due to the left-truncation, there is a danger that the early risk sets are very empty and the early hazards cannot be estimated with enough precision, which has consequences for the precision of the whole survival curve. A minimal number is required of individuals with follow-up from $t = 0$ onwards. The formal requirement is $P(T_{\text{trunc}} = 0) > 0$. Under that condition the distribution of T_{trunc} can be estimated from the data as well, but that is hardly ever done in practice.

The programming is getting slightly more complicated because one needs to take into account when the individuals enter the follow-up. To analyze such data, software is needed that allows delayed entry. Examples of such software are SAS, Stata, and R.

The interpretation of the analysis is getting complicated if the truncation time is not independent of the survival time. See the continued example of Section 1.6.3.

1.6.2 Time-dependent covariates

The second extension is to allow for time-dependent covariates, the value of which may change over time. These covariates are denoted by $X(t)$. In principal, the formal definitions and estimation procedures remain valid if the covariates are allowed to be time-dependent. The model modifies into

$$h(t|X(t)) = \lambda_0(t)\exp(X(t)^\top \beta) \ .$$

Such data are easy to analyze if $X(t)$ is piecewise constant (and does not change too often), by creating separate records for each period of constant $X(t)$ and using software that allows delayed entry. Suppose that an individual enters at t_0 with covariate vector $X(t_0) = x_0$. That value changes consecutively into x_i at time t_i for $i = 1, 2, ...$ At the observed survival time t_{obs} the current value is x_{last} that started at t_{last} . To cover that individual, the following records for $(t_{\text{entry}}, t, d, x)$ have to be entered:

Begin interval	End interval	Status	Covariate
t_0	t_1	0	x_0
t_1	t_2	0	x_1
$\vdots$	$\vdots$	$\vdots$	$\vdots$
t_{last}	t_{obs}	d	x_{last}

It is not hard to check that this will give the correct contribution to the log-likelihood for this individual. If $X(t)$ changes very often, or indeed if it is a continuous function of t, one should realize that in an actual dataset only the values of $X(t)$ at the event times are needed. This means that for an individual with survival time t one record is needed for each event time $\leq t$. The **survival** package in R contains the function survSplit to create such expanded datasets. In Stata this is achieved by using stsplit.

Although models with time-dependent covariates can be fitted rather easily, it should be stressed that a Cox model with time-dependent covariates is of no prognostic use, unless the distribution of future values of $X(t)$ is known. Ways to obtain prognostic models involving time-dependent covariates will be discussed in Chapters 20, 21 and 26 of this handbook.

1.6.3 Continued example

In the example of Section 5.5 the time scale as in most clinical studies is time since diagnosis (or start of treatment) and age at diagnosis is an important risk factor. For older patient

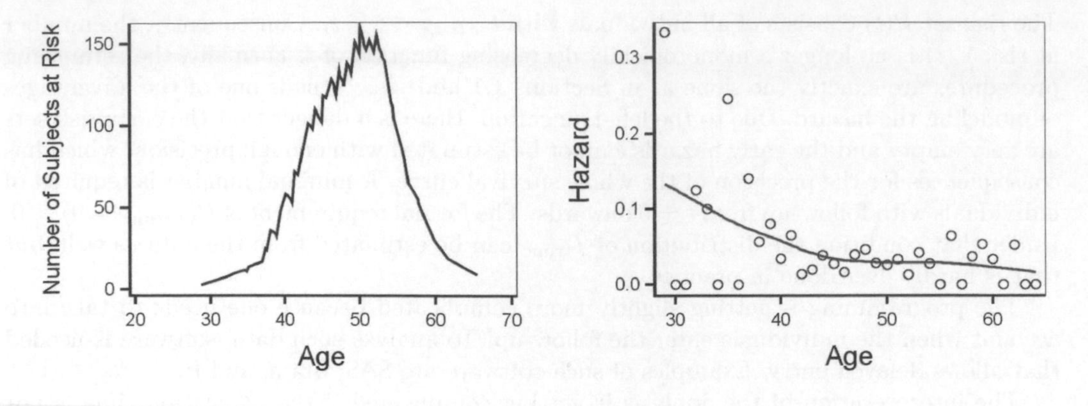

FIGURE 1.6
Number at risk (left) and smoothed estimate of the hazard (right) versus age.

TABLE 1.2
Cox models using different time scales.

	time since diagnosis		age	
Covariate	B	SE	B	SE
Chemotherapy: Yes	-0.423	0.298	-0.370	0.300
Hormonal therapy: Yes	-0.172	0.442	-0.106	0.443
Type of surgery: Mastectomy	0.154	0.249	0.143	0.248
Histological grade: Poorly differentiated	0.266	0.281	0.262	0.279
Histological grade: Well differentiated	-1.308	0.547	-1.281	0.550
Vascular invasion: +	0.603	0.253	0.624	0.259
Vascular invasion: +/-	-0.146	0.491	-0.187	0.495
Diameter (mm)	0.020	0.013	0.016	0.013
Number of positive nodes	0.074	0.052	0.074	0.057
Age at diagnosis (years)	-0.039	0.020	-0.051	0.042
Estrogen level (mlratio)	-0.750	0.211	-0.771	0.215

populations it can be relevant to take age as time scale. Age at diagnosis, or more generally age at entering the study, will then act as left-truncation time. This is demonstrated for the breast cancer data of Section 5.5 although it might be less relevant there. Figure 1.6 shows the number at risk over the age range (left) and the Nelson-Aalen estimate of the hazard. Table 1.2 compares the Cox model of Section 5.5 with the Cox model on the age scale. The coefficients are very similar. It should be noted, however, that in this model age at diagnosis is completely confounded with the time-dependent covariate time since diagnosis. The similarity of the coefficients for age at diagnosis in the two approaches is coincidental. Figure 1.7 shows predicted survival curves for different values of age at diagnosis (35, 45, 55) with all categorical covariates at the reference value and all continuous covariates at their mean value. The left panel is based on the analysis of this section, while the right panel uses the analysis from Section 5.5. The striking difference is that the analysis of this section enables predictions for all patients up to the age of 65, while the analysis from Section 5.5 allows predictions up to 15 years after diagnosis. Predicting up to the age of 65 for all patients does not make. If the graphs show only the predictions for the first 15 years since diagnosis, there is little difference between the two graphs. A minor difference is the presence of larger jumps for the age=35 patient in the left panel. This shows that,

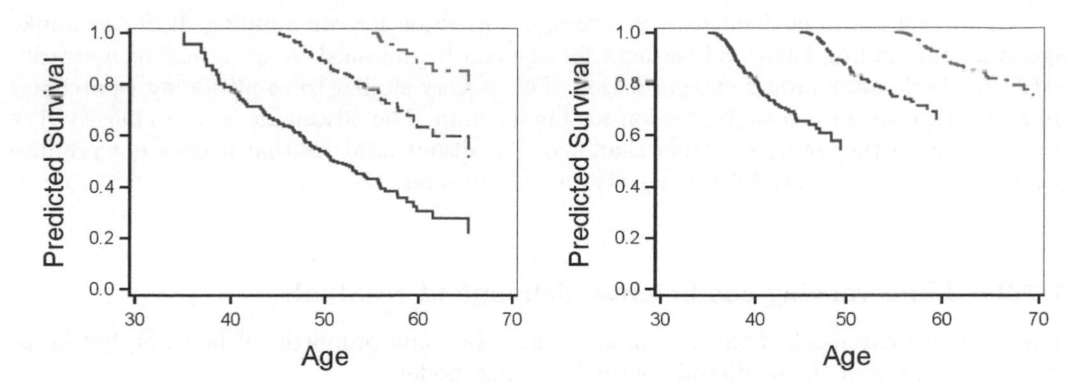

FIGURE 1.7
Predictions for different values of age at diagnosis (35, 45 ,55), with age as time scale (left) and time since diagnosis as time scale (right).

compared with analysis of Section 5.5, the analysis of this section has more uncertainty for the younger patients at the start of follow-up and for the elder patients at the end of follow-up.

1.7 Beyond proportional hazards assumption

The proportional hazards (PH) assumption is crucial in the Cox model. However, as for all models, there is no a priori reason why the model should hold true. It might be a plausible model on data with short follow-up, but it becomes questionable in the case of long-term follow-up. There are two ways to extend the model beyond PH: stratification and allowing time-varying effects.

1.7.1 Stratified models

The Cox model can be extended into a stratified Cox model by considering a categorical stratification variable, G say, with values $1, ..., K$. The stratified model allows the baseline hazard to depend on the stratum, that is

$$h(t|X, G) = h_{G0}(t) \exp(X^\top \beta) .$$

This model assumes that the effect of the covariates in X is the same in each stratum, but allows the baseline hazard to depend on the stratum. The model can be fitted in the same way as the Cox model. The technical difference is that now the risk sets are stratum specific. The main application of the stratified model is to include categorical covariates for which it can be expected a priori that the PH assumption will not hold. In that situation the stratum-specific baseline hazard is an essential part of the model and the model can only be reliable if each stratum has enough events to obtain a stable estimate of the baseline hazard. A simple check is to draw stratum-specific Kaplan-Meiers. There is little hope to obtain good models for the strata in which the Kaplan-Meiers are too "jumpy." It is possible to allow the effect of a covariate to be different across strata by adding the interaction of the covariate with the stratification variable.

An application in epidemiological setting is correction for confounding. If, for example, age is a confounding factor, adjustment for age can be obtained by using age as a stratification variable after proper categorization. This is very similar to conditioning (matching) on age categories in logistic regression for binary data. The advantage is an unbiased view on the effect of the covariates other than age. The disadvantage is that it does not produce a reliable prediction model if the age strata are too small.

1.7.2 Time-varying coefficients, Schoenfeld residuals

The standard extension of the Cox model that allows non-proportional hazards, but keeps the linear effects of the predictors, is the following model

$$\lambda(t|x) = \lambda_0(t)\exp(x^\top \beta(t)) \ .$$

As pointed out in Schoenfeld (1982), the contributions of each risk set to the score function of Section 2.1 can be used to get an impression of the variability of $\beta(t)$ and the validity of the model with $\beta_j(t) \equiv \beta_j$ for each covariate X_j. For risk set $R(t)$, the Schoenfeld residual is defined as

$$score(t) = x_{ev}(t) - \bar{x}(\hat{\beta}, t)$$

Here $x_{ev}(t)$ is the (mean) x-value of the individual(s) with an event at time t, that is $x_{ev}(t) = \sum_i x_i \cdot \Delta N_i(t)/\Delta\overline{N}(t)$. If the proportional hazards model holds true, the expected value of $score(t)$ equals zero. The validity of the model for the jth covariate can be checked by plotting the jth component of $score(t)$ versus t. The mean value of the residual equals zero by construction. Time trends in the residuals are an indication of violation of the proportional hazards assumption. Visual inspection of the plots can be followed by a formal test of the proportional hazards assumption.

This approach has been refined in Grambsch and Therneau (1994, 1995) and is described in Section 6.2 of their book (Therneau and Grambsch, 2000). Their proposal is to plot the components of the so-called "scaled Schoenfeld residual"

$$score^*(t) = \hat{\beta} + V(t)^{-1} score(t) \ ,$$

where

$$V(t) = var(x|\hat{\beta}, t))$$

is the weighted covariance in $R(t)$ as defined in Section 2.1. They point out that $score^*(t)$ is an estimate of the local value of the regression coefficient at t in the time-varying coefficients model.

For the sake of robustness, it is wiser to use a parametric model for the time-varying regression coefficients. This can be done by considering a set of m basis functions $f_1(t), ..., f_m(t)$ and taking

$$\beta(t) = \sum_{j=1}^{m} \gamma_j f_j(t) \ .$$

Here, each γ_j is a vector of the same length as β, namely the number of covariates. It is helpful for the interpretation of the parameters if the basis functions are defined in such a way that

$$f_1(t) \equiv 1 \ , \quad f_j(0) = 0, \text{ for } j = 2, \ldots, m \ .$$

The interpretation is then that the effect of the covariates at $t = 0$ is given by γ_1 ($\beta(0) = \gamma_1$), while the other γ's describe the violation of the PH-model for each of the covariates.

A popular choice for the second basis function is $f_2(t) = \ln(1 + t)$, which starts with $f_2(0) = 0$, has derivative $f_2'(0) = 1$ and slows down later in the follow-up. If the interest is only in testing the PH-assumption for each covariate, it suffices to take $m = 2$ and $f_2(t) = \ln(1 + t)$. This is very close to the suggestion in Cox's original paper (Cox, 1972), who takes $f_2(t) = \ln(t)$ itself. If very early events can occur, Cox's proposal could put too much emphasis on early events, because $\ln(t)$ "explodes" for t close to zero.

Models with time-varying coefficients can be handled in the same way as time-dependent covariates. They are in fact equivalent because the set of time-dependent covariates $Xf_1(t), \ldots, Xf_m(t)$ yields exactly the same model. Some packages allow one to do this internally, SPSS and SAS, while others, such as R and Stata, require restructuring the database as discussed in Section 26.3. For most packages, it is not easy to get predicted survival curves with confidence bounds.

The problem with the time-varying effects model is the danger of overfitting. The number of parameters has doubled while the partial likelihood has not improved much. The model could be pruned by significance testing. The danger of starting from a time-constant model and checking the significance of the extension to a time-varying model, is that covariates showing a time-varying effect that changes from positive to negative (or the other way around) might be missed. Inspired by a clinical example, a strategy for model building that would detect such switching effects is presented in Putter et al. (2005). The strategy proposed there consists of a forward selection procedure in which each of the covariates, together with their interaction with time, is considered. The covariate is included in the model together with the covariate-by-time interaction if the likelihood-ratio test for the model with both covariate and covariate-by-time indicates a significantly better fit compared to the model without. In a subsequent pruning step, each of the covariate by time interactions is considered and removed from the model in case the interaction was not significant. Similar strategies are developed in Sauerbrei et al. (2007).

1.7.3 Continued example

Stratification can be a useful tool in exploratory data analysis. To get more insight in possible time-varying effects of age at diagnosis, it was categorized into $< 42, 42 - 48, > 48$. An analysis with stratification on those age at diagnosis groups and inclusion of all covariates in the same way as before yields the baseline survival curves of Figure 1.8.

There is some indication of non-proportionality. This confirmed by the scaled Schoenfeld residual plot for age shown in Figure 1.9.

A systematic check for non-proportionality was carried out by starting from the model of Section 5.5 and allowing an interaction with $\ln(1+t)$ in a stepwise extension of the model. That retained only the interaction of estrogen level with $\chi^2_{[1]} = 4.96$, $P - value = 0.026$. The effect of estrogen level in a model with all other covariates included is given by $\hat{\beta}_{Estrogen}(t) = -2.28 + 0.95 \ln(1 + t)$. The scaled Schoenfeld residual plot of Figure 1.10 suggests a quadratic interaction with time. Fitting this gives a marginal improvement of the model ($\chi^2_{[2]} = 6.25$, $P - value = 0.044$) and $\hat{\beta}_{Estrogen}(t) = -2.36 + 0.58 * t - 0.036 * t^2$ which is very close to the model based on $\ln(1 + t)$.

1.7.4 Final remarks

The number of parameters in the time-varying coefficients model can be greatly reduced by exploiting the fact that most covariates show very similar patterns over time.

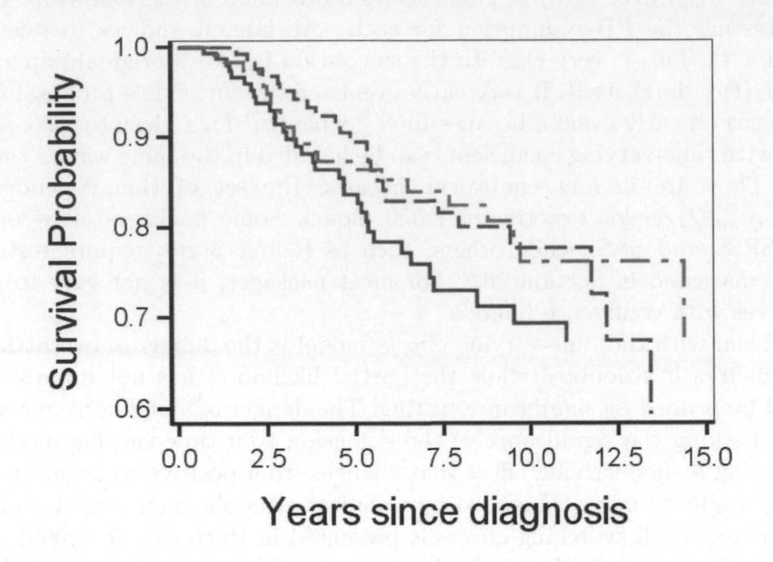

FIGURE 1.8
Baseline survival curves for different strata of age at diagnosis: < 42 (solid), 42–48 (dashed), > 48 (dots and dashes).

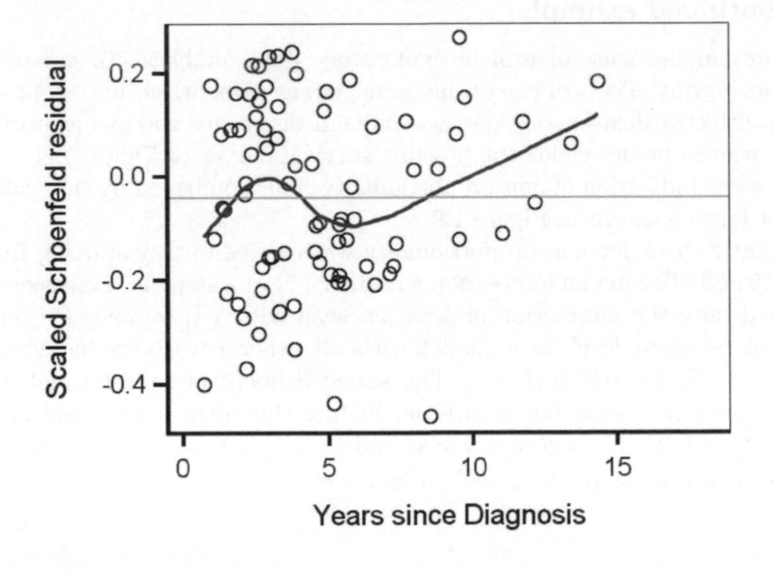

FIGURE 1.9
Scaled Schoenfeld residuals for age. The horizontal reference line corresponds with the estimated regression for age.

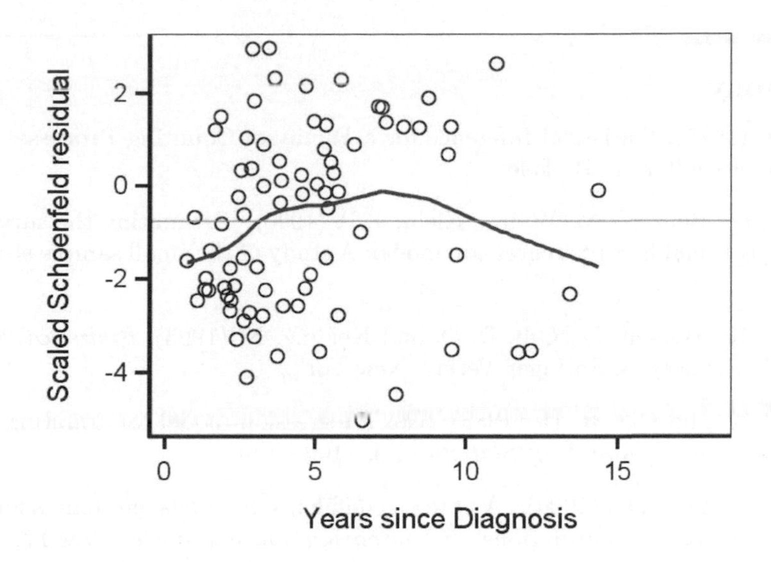

FIGURE 1.10
Scaled Schoenfeld residuals for estrogen level. The horizontal reference line corresponds with the estimated regression coefficient of estrogen level.

A starting point that gives an impression how such a model is obtained by the following very simple two-stage procedure:

1. Fit a simple Cox yielding $PI_{fixed} = x^\top \hat{\beta}_{fixed}$;
2. Fit a time-varying model with PI_{fixed} as single covariate.

Application to the breast cancer data with $\ln(1+t)$ as interaction term gives an improvement of the model by $\chi^2_{[1]} = 7.21$, $P - value = 0.007$ with estimated effect $\hat{\beta}_{PI} = 2.27 - 0.72 * \ln(1+t)$, which declines from 2.27 at $t = 0$ to 0.44 at $t = 10$.

The slightly heuristic two-stage approach can be formalized by the model

$$\lambda(t|x) = \lambda_0(t) \exp\left(x^\top \beta \cdot \left(\sum_{i=1}^{m} \gamma_i f_i(t)\right)\right) .$$

Here, the same restriction applies to the time-functions as above, namely $f_1(t) \equiv 1$ and $f_j(0) = 0$, for $j = 2, \ldots, m$. This model is known as a reduced rank model of $rank = 1$. Reduced rank is an established methodology for parsimonious interaction models in analysis of variance and linear regression. It has been introduced in the present context in Perperoglou et al. (2006*a*,*b*), who also developed the software to fit such models.

The time-varying coefficients models make it possible to explore the effect of violation of the PH assumption when fitting a simple Cox model. This is amply discussed in Xu and O'Quigley (2000). For further details see Chapter 16 of this handbook.

Bibliography

Aalen, O. O. (1975), Statistical Inference for a Family of Counting Processes, PhD thesis, University of California, Berkeley.

Andersen, P. K., Bentzon, M. W. and Klein, J. P. (1996), 'Estimating the survival function in the proportional hazards regression model: A study of the small sample size properties', **23**, 1–12.

Andersen, P. K., Borgan, Ø., Gill, R. D. and Keiding, N. (1993), *Statistical Models Based on Counting Processes*, Springer-Verlag, New York.

Andersen, P. K. and Gill, R. D. (1982), 'Cox's regression model for counting processes: A large sample study', *Annals of Statistics* **10**, 1100–1120.

Borgan, Ø. and Liestol, K. (1990), 'A note on confidence-intervals and bands for the survival function based on transformations', *Scandinavian Journal of Statistics* **17**, 35–41.

Breslow, N. E. (1972), 'Discussion of Professor Cox's paper', *Journal of the Royal Statistical Society - Series B* **34**, 216–217.

Breslow, N. E. (1974), 'Covariance analysis of censored survival data', *Biometrics* **30**, 89–99.

Cox, D. R. (1972), 'Regression models and life-tables', *Journal of the Royal Statistical Society - Series B* **34**, 187–220.

Fleming, T. R. and Harrington, D. P. (1991), *Counting Processes & Survival Analysis*, Applied probability and statistics, Wiley, New York.

Grambsch, P. M. and Therneau, T. M. (1994), 'Proportional hazards tests and diagnostics based on weighted residuals', *Biometrika* **81**, 515–526.

Grambsch, P. M. and Therneau, T. M. (1995), 'Diagnostic plots to reveal functional form for covariates in multiplicative intensity models', *Biometrics* **51**, 1469–1482.

Greenwood, M. (1926), The natural duration of cancer, *in* 'Reports on Public Health and Medical Subjects 33', London: His Majesty's Stationery Office, pp. 1–26.

Kaplan, E. and Meier, P. (1958), 'Nonparametric estimation from incomplete observations', *Journal of the American Statistical Association* **43**, 457–481.

Klein, J. P. and Moeschberger, M. L. (2003), *Survival Analysis: Techniques for Censored and Truncated Data*, Statistics for Biology and Health, 2nd ed., Springer, New York.

Nelson, W. (1969), 'Hazard plotting for incomplete failure data', *Journal of Quality Technology* **1**, 27–52.

O'Quigley, J. (2008), *Proportional Hazards Regression*, Springer, New York.

Perperoglou, A., le Cessie, S. and van Houwelingen, H. C. (2006*a*), 'A fast routine for fitting Cox models with time varying effects of the covariates', *Computer Methods and Programs in Biomedicine* **81**, 154–161.

Perperoglou, A., le Cessie, S. and van Houwelingen, H. C. (2006*b*), 'Reduced-rank hazard regression for modeling non-proportional hazards', *Statistics in Medicine* **25**, 2831–2845.

Putter, H., Sasako, M., Hartgrink, H. H., van de Velde, C. J. H. and van Houwelingen, J. C. (2005), 'Long-term survival with non-proportional hazards: results from the Dutch Gastric Cancer Trial', *Statistics in Medicine* **24**, 2807–2821.

Sauerbrei, W., Royston, P. and Look, M. (2007), 'A new proposal for multivariable modeling of time-varying effects in survival data based on fractional polynomial time-transformation', *Biometrical Journal* **49**, 453–473.

Schoenfeld, D. (1982), 'Partial residuals for the proportional hazards regression model', *Biometrika* **69**, 239–241.

Therneau, T. M. and Grambsch, P. M. (2000), *Modeling Survival Data: Extending the Cox Model*, Springer, New York.

Tsiatis, A. A. (1981), 'A large sample study of Cox's regression model', *Annals of Statistics* **9**, 93–108.

van de Vijver, M. J., He, Y. D., van 't Veer, L. J., Dai, H., Hart, A. A. M., Voskuil, D. W., Schreiber, G. J., Peterse, J. L., Roberts, C., Marton, M. J., Parrish, M., Atsma, D., Witteveen, A., Glas, A., Delahaye, L., van der Velde, T., Bartelink, H., Rodenhuis, S., Rutgers, E. T., Friend, S. H. and Bernards, R. (2002), 'A gene-expression signature as a predictor of survival in breast cancer', *New England Journal of Medicine* **347**, 1999–2009.

van Houwelingen, H. C., Bruinsma, T., Hart, A. A. M., van't Veer, L. J. and Wessels, L. F. A. (2006), 'Cross-validated Cox regression on microarray gene expression data', *Statistics in Medicine* **25**, 3201–3216.

van Houwelingen, H. C. and Putter, H. (2012), *Dynamic Prediction in Clinical Survival Analysis*, Monographs on statistics and applied probability, CRC Press, Boca Raton.

van't Veer, L. J., Dai, H. Y., van de Vijver, M. J., He, Y. D. D., Hart, A. A. M., Mao, M., Peterse, H. L., van der Kooy, K., Marton, M. J., Witteveen, A. T., Schreiber, G. J., Kerkhoven, R. M., Roberts, C., Linsley, P. S., Bernards, R. and Friend, S. H. (2002), 'Gene expression profiling predicts clinical outcome of breast cancer', *Nature* **415**, 530–536.

Verweij, P. J. M., van Houwelingen, H. C. and Stijnen, T. (1998), 'A goodness-of-fit test for Cox's proportional hazards model based on martingale residuals', *Biometrics* **54**, 1517–1526.

Xu, R. and O'Quigley, J. (2000), 'Estimating average regression effect under non-proportional hazards', *Biostatistics* **1**, 423–439.

2

Bayesian Analysis of the Cox Model

Joseph G. Ibrahim

Department of Biostatistics, University of North Carolina

Ming-Hui Chen and Danjie Zhang

Department of Statistics, University of Connecticut

Debajyoti Sinha

Department of Statistics, Florida State University

CONTENTS

2.1	Introduction	27
2.2	Fully parametric models	29
2.3	Semiparametric models	32
2.3.1	Piecewise constant hazard model	32
2.3.2	Models using a gamma process	33
2.3.3	Gamma process prior with continuous-data likelihood	33
2.3.4	Relationship to partial likelihood	33
2.3.5	Gamma process on baseline hazard	34
2.3.6	Beta process models	35
2.3.7	Correlated prior processes	37
2.3.8	Dirichlet process models	38
2.4	Prior elicitation	39
2.5	Other topics	40
2.6	A case study: an analysis of melanoma data	40
2.7	Discussion	42
	Acknowledgments	43
	Bibliography	44

Great strides in the analysis of survival data using Bayesian methods have been made in the past ten years due to advances in Bayesian computation and the feasibility of such methods. In this chapter, we review Bayesian advances in survival analysis and discuss the various semiparametric modeling techniques that are now commonly used. We review parametric and semiparametric approaches to Bayesian survival analysis, with a focus on proportional hazards models. References to other types of models are also given.

2.1 Introduction

Nonparametric and semiparametric Bayesian methods for survival analysis have witnessed extensive development since 1980s. Many of these methods are now easily implementable using standard-free software (e.g., OpenBUGS) as well as popular commercial statistical software (e.g., MCMC and PHREG Procedures in SAS). The literature on semiparametric Bayesian methods is too enormous to list all the important references here. In this chapter, we discuss several types of Bayesian survival models, including parametric models as well as models involving nonparametric prior processes for the baseline functions, typically either the cumulative baseline hazard or the baseline survival function. Instead of providing the comprehensive review of all the available methods, we will focus on various versions of Cox's (Cox, 1972, 1975) semiparametric relative risk model

$$h(t|\boldsymbol{x}(t)) = h_0(t)\exp\{\boldsymbol{x}(t)'\boldsymbol{\beta}\}\ , \tag{2.1}$$

where $h_0(t)$ is the baseline hazard and $h(t|\boldsymbol{x}(t))$ is the hazard at time t for a subject with possibly time-dependent covariates $\boldsymbol{x}(t)$ and $\boldsymbol{\beta}$ is the corresponding vector of regression coefficients. Unless mentioned otherwise, we will present our methods for the special case of time-constant (fixed) covariates $\boldsymbol{x}(t) = \boldsymbol{x}$. However, these methods can be easily extended to accommodate time-dependent covariates $\boldsymbol{x}(t)$ and in some places we will explicitly specify the extensions.

We consider the piecewise constant hazards, the gamma process, the beta process, the correlated prior processes, and the Dirichlet process for the baseline functions. In each case, we give a development of the prior process, construct the likelihood function, derive the posterior distributions, and discuss MCMC sampling techniques for inference. We also give references to other types of Bayesian models, including frailty models, joint models for longitudinal and survival data, flexible classes of hierarchical models, accelerated failure time models, multivariate survival models, spatial survival models, and Bayesian model diagnostics.

The semiparametric Bayesian survival analysis of right-censored survival data based on any model of survival function $S(t|\boldsymbol{x};\boldsymbol{\theta})$ can use two fundamental approaches for the likelihood function, where $\boldsymbol{\theta}$ represents the set of all model parameters and $\boldsymbol{x}$ is a vector of covariates. Let $f(t|\boldsymbol{x};\boldsymbol{\theta})$ denote the density under the survival model $S(t|\boldsymbol{x};\boldsymbol{\theta})$. The continuous data likelihood based on the observed right-censored data $D = (n, \boldsymbol{y}, \boldsymbol{\nu}, X)$ is

$$L(\boldsymbol{\theta}|D) = \prod_{i=1}^{n} f(y_i|\boldsymbol{x}_i;\boldsymbol{\theta})^{\nu_i} S(y_i|\boldsymbol{x}_i;\boldsymbol{\theta})^{1-\nu_i}, \tag{2.2}$$

where $\boldsymbol{y} = (y_1, y_2, \ldots, y_n)'$, $\boldsymbol{\nu} = (\nu_1, \nu_2, \ldots, \nu_n)'$, ν_i is the indicator of censoring, y_i is the observed survival time subject to right-censoring, and X is the $n \times p$ matrix of covariates with the ith row $\boldsymbol{x}_i'$.

When the survival model follows the Cox model of (2.1) with fixed covariates $\boldsymbol{x}$, after some algebra, the likelihood of (2.2) can be expressed as

$$L(\boldsymbol{\theta}|D) = \prod_{i=1}^{n} \exp\Big\{ - H_{0(i)} \sum_{j\in R_{(i)}} \exp(\boldsymbol{x}_j'\boldsymbol{\beta})\Big\}\{dH_0(y_{(i)})\exp(\boldsymbol{x}_{(i)}'\boldsymbol{\beta})\}^{\nu_{(i)}}, \tag{2.3}$$

where $\boldsymbol{\theta} = (\boldsymbol{\beta}, H_0(\cdot))$, $H_0(t) = \int_0^t h_0(u)du$ is the cumulative baseline hazard, $0 = y_{(0)} < y_{(1)} < \cdots < y_{(n)}$ are the ordered observed survival times, $x_{(i)}$ is the covariates associated with $y_{(i)}$, $H_{0(i)} = H_0(y_{(i)}) - H_0(y_{(i-1)})$ is the increment in cumulative hazard between consecutive observed survival times $y_{(i-1)}$ and $y_{(i)}$, and $R_{(i)} = \{j : \ y_{(j)} \geq y_{(i)}\}$. Some key references for semiparametric Bayesian survival analysis using the continuous-time likelihood include Susarla and Van Ryzin (1976), Kalbfleisch (1978), Dykstra and Laud (1981), and Hjort (1990). References discussing computational implementation of Bayesian inference using continuous time likelihood include Damien et al. (1996), Laud et al. (1996), Laud et al. (1998), Walker et al. (1999) and the references therein.

In practice, survival data is often either observed or recorded via a design with monitoring grid, that is, intervals of time recording the approximate time of an event or censoring. Hence, the observed survival time y_i of any subject is only known up to an interval (however small the interval may be) where the length of the interval Δ_i depends either on the coarseness of the data-recording grid or on the design of monitoring the survival times. Let D_G denote the observed data, where we only know that the subject i has failed within an interval $(y_i, y_i + \Delta_i)$ when $\nu_i = 1$ and (y_i, ∞) when $\nu_i = 0$. To construct the likelihood corresponding to this observed discrete-time (or grouped) data D_G, we first use a partition of the time axis, $0 < a_1 < a_2 < \ldots < a_J < \infty$, where $a_J \geq$ maximum monitoring time, each $y_i = a_k$ and $y_i + \Delta_i = a_{k+1}$ for some k. The assumption about monitoring of subjects is that the grid interval is equal to the monitoring interval of each subject. For this situation, the likelihood is

$$L(\boldsymbol{\theta}|D_G) = \prod_{i=1}^{n}\{S(y_i|\boldsymbol{\theta}) - S(y_i + \Delta_i|\boldsymbol{\theta})\}^{\nu_i} S(y_i|\boldsymbol{\theta})^{(1-\nu_i)} = \prod_{j=1}^{J} G_j \ , \tag{2.4}$$

where $G_j = \exp[-h_j \sum_{k\in\mathcal{R}_j - \mathcal{D}_j} \exp(\boldsymbol{x}_k'\boldsymbol{\beta})]\ \{1 - \exp[-h_j \sum_{k\in\mathcal{D}_j} \exp(\boldsymbol{x}_k'\boldsymbol{\beta})]\}$, the risk-set $\mathcal{R}_k$ at time a_k is the set of people "at risk" in the interval $I_k = (a_{k-1}, a_k]$, $\mathcal{D}_k$ is the set of people failing at that interval I_k, and $h_j = H_0(a_j) - H_0(a_{j-1})$. This likelihood of (2.4) has more fidelity to the actual observed data. In spite of the ability of semiparametric Bayesian method to handle the likelihood of (2.4), the continuous data likelihood of (2.2) is often used as an approximation to (2.4) provided that the monitoring grid is not too coarse compared to the scale of the survival time. The continuous time likelihood can be viewed as a limiting case of the discrete time likelihood of (2.4).

There is the general perception that not much is gained in using a strictly nonparametric $h_0(t)$ along with the discrete-time likelihood of (2.4). In practice, the implementation of the Bayesian inference with nonparametric $h_0(t)$ is much more complicated than that of using a discretized parametric $h_0(t)$. We will later discuss how a discrete approximation of nonparametric $h_0(t)$ can be made arbitrarily accurate to approximate the nonparametric $h_0(t)$. Thus, in practice, a discrete approximation of $h_0(t)$ used with the discrete-time likelihood can give a very justifiable Bayesian inference.

The rest of this chapter is organized as follows. In Section 2.2, we review Bayesian parametric survival models. In Section 2.3, we discuss semiparametric Bayesian methods for survival analysis and focus on the proportional hazards model of Cox (1972). We examine the piecewise constant, gamma, beta, and Dirichlet process models. In Section 2.4, we discuss the prior elicitation process by incorporating historical data from a previous study. In Section 2.5, we give several references on other types of models and applications in Bayesian survival analysis. Section 2.6 presents a detailed analysis of the melanoma data. We conclude this chapter with a brief discussion in Section 2.7.

2.2 Fully parametric models

Bayesian approaches to fully parametric survival analysis have been considered by many in the literature. The statistical literature in Bayesian parametric survival analysis and life-testing is too enormous to list here, but some references dealing with applications to medicine or public health include Grieve (1987), Achcar et al. (1987), Achcar et al. (1985), Chen et al. (1985), Dellaportas and Smith (1993), and Kim and Ibrahim (2001). The most common types of parametric models used are the exponential, Weibull, and log-normal models.

The exponential model with survival function $S(t|\lambda) = P[T > t|\lambda] = \exp(-\lambda t)$ and density $f(t|\lambda) = \lambda \exp(-\lambda t)$ is the most basic parametric model in survival analysis. For observed survival data $D = (n, \boldsymbol{y}, \boldsymbol{\nu})$, subject to right-censoring, we can write the likelihood function of λ as

$$L(\lambda|D) = \prod_{i=1}^{n} f(y_i|\lambda)^{\nu_i} S(y_i|\lambda)^{(1-\nu_i)} = \lambda^d \exp\left(-\lambda \sum_{i=1}^{n} y_i\right), \tag{2.5}$$

where $d = \sum_{i=1}^{n} \nu_i$. The conjugate prior for λ with hyperparameters (α_0, λ_0) is the gamma prior $\mathcal{G}(\alpha_0, \lambda_0)$, with density

$$\pi(\lambda|\alpha_0, \lambda_0) \propto \lambda^{\alpha_0 - 1} \exp(-\lambda_0 \lambda).$$

The resulting posterior distribution of λ is given by

$$\pi(\lambda|D) \propto \lambda^{\alpha_0 + d - 1} \exp\left\{-\lambda(\lambda_0 + \sum_{i=1}^{n} y_i)\right\}. \tag{2.6}$$

Recognizing the kernel of the posterior distribution in (2.6) as a $\mathcal{G}(\alpha_0 + d, \lambda_0 + \sum_{i=1}^{n} y_i)$ distribution, the posterior mean and variance of λ are obtained as

$$E(\lambda|D) = \frac{\alpha_0 + d}{\lambda_0 + \sum_{i=1}^{n} y_i} \quad \text{and} \quad \text{Var}(\lambda|D) = \frac{\alpha_0 + d}{(\lambda_0 + \sum_{i=1}^{n} y_i)^2}.$$

The posterior predictive distribution of a future failure time y_{new} is given by

$$\begin{aligned}
\pi(y_{new}|D) &= \int_0^\infty f(y_{new}|\lambda)\pi(\lambda|D)\, d\lambda \\
&\propto \int_0^\infty \lambda^{\alpha_0 + d + 1 - 1} \exp\left\{-\lambda(y_{new} + \lambda_0 + \sum_{i=1}^{n} y_i)\right\} d\lambda \\
&\propto \left(\lambda_0 + \sum_{i=1}^{n} y_i + y_{new}\right)^{-(d+\alpha_0+1)}.
\end{aligned} \tag{2.7}$$

The normalized posterior predictive distribution is thus given by

$$\pi(y_{new}|D) = \begin{cases} \frac{(d+\alpha_0)(\lambda_0 + \sum_{i=1}^{n} y_i)^{(\alpha_0+d)}}{(\lambda_0 + \sum_{i=1}^{n} y_i + y_{new})^{(\alpha_0+d+1)}} & \text{if } y_{new} > 0, \\ 0 & \text{otherwise.} \end{cases} \tag{2.8}$$

The derivation of (2.7) above needs the use of a gamma integral. The predictive distribution in (2.8) is known as an *inverse beta* distribution (Aitchison and Dunsmore, 1975).

To build a regression model, we introduce covariates through λ, and write $\lambda_i = \varphi(\boldsymbol{z}_i'\boldsymbol{\beta})$, where $\boldsymbol{z}_i$ is a $(p+1) \times 1$ vector of covariates including an intercept, $\boldsymbol{\beta}$ is a $(p+1) \times 1$ vector of regression coefficients, and $\varphi(.)$ is a known function. A common form of φ is to take $\varphi(\boldsymbol{z}_i'\boldsymbol{\beta}) = \exp(\boldsymbol{z}_i'\boldsymbol{\beta})$. Another form of φ is $\varphi(\boldsymbol{z}_i'\boldsymbol{\beta}) = (\boldsymbol{z}_i'\boldsymbol{\beta})^{-1}$ (Feigl and Zelen, 1965). Using $\varphi(\boldsymbol{z}_i'\boldsymbol{\beta}) = \exp(\boldsymbol{z}_i'\boldsymbol{\beta})$, we are led to the likelihood function

$$L(\boldsymbol{\beta}|D) = \exp\left\{\sum_{i=1}^{n} \nu_i \boldsymbol{z}_i'\boldsymbol{\beta}\right\} \exp\left\{-\sum_{i=1}^{n} y_i \exp(\boldsymbol{z}_i'\boldsymbol{\beta})\right\}, \tag{2.9}$$

where $D = ((n, \boldsymbol{y}, \boldsymbol{\nu}, Z)$ and Z is a $n \times (p+1)$ matrix with the ith row $\boldsymbol{z}_i$. Common prior distributions for $\boldsymbol{\beta}$ include an improper uniform prior, i.e., $\pi(\boldsymbol{\beta}) \propto 1$, and a normal prior. Suppose we specify a p-dimensional normal prior for $\boldsymbol{\beta}$, denoted by $N_p(\boldsymbol{\mu}_0, \Sigma_0)$, where $\boldsymbol{\mu}_0$ denotes the prior mean and Σ_0 denotes the prior covariance matrix. Then the posterior distribution of $\boldsymbol{\beta}$ is given by

$$\pi(\boldsymbol{\beta}|D) \propto L(\boldsymbol{\beta}|D)\pi(\boldsymbol{\beta}|\boldsymbol{\mu}_0, \Sigma_0), \tag{2.10}$$

where $\pi(\boldsymbol{\beta}|\boldsymbol{\mu}_0, \Sigma_0)$ is the multivariate normal density with mean $\boldsymbol{\mu}_0$ and covariance matrix Σ_0. The closed forms for the posterior distribution of $\boldsymbol{\beta}$ are generally not available. However, due to the availability of statistical packages such as `OpenBUGS`, or LIFEREG and MCMC Procedures in SAS, the regression model in (2.9) can easily be fitted using Markov chain Monte Carlo (MCMC) sampling methods to sample from the posterior in (2.10). Before the advent of MCMC, numerical integration techniques were employed by Grieve (1987).

The Weibull model with density

$$f(y|\alpha, \mu) = \alpha y^{\alpha-1} \exp(\mu - \exp(\mu) y^{\alpha}). \tag{2.11}$$

and survival function $S(y|\alpha, \mu) = \exp\{-\exp(\mu) y^{\alpha}\}$ is perhaps the most widely used parametric survival model. We can write the likelihood function of (α, μ) based on right-censored data D as

$$L(\alpha, \mu|D) = \alpha^d \exp\left\{ d\mu + \sum_{i=1}^{n} (\nu_i(\alpha - 1)\log(y_i) - \exp(\mu) y_i^{\alpha}) \right\}.$$

When α is assumed known, the conjugate prior for $\exp(\mu)$ is the gamma prior. No joint conjugate prior is available when (α, μ) are both assumed unknown. In this case, a typical joint prior specification is to take α and μ to be independent, with a gamma $\mathcal{G}(\alpha_0, \kappa_0)$ prior for α and a normal $N(\mu_0, \sigma_0^2)$ prior for μ. The joint posterior distribution of (α, λ) is given by

$$\begin{aligned}\pi(\alpha, \mu|D) &\propto L(\alpha, \mu|D)\pi(\alpha|\alpha_0, \kappa_0)\pi(\mu|\mu_0, \sigma_0^2) \\ &\propto \alpha^{\alpha_0+d-1} \exp\left\{ d\mu + \sum_{i=1}^{n} (\nu_i(\alpha-1)\log(y_i) - \exp(\mu) y_i^{\alpha}) - \kappa_0 \alpha - \frac{(\mu - \mu_0)^2}{2\sigma_0^2} \right\}.\end{aligned}$$

The joint posterior distribution of (α, μ) does not have a closed form, but it can be shown that the conditional posterior distributions $[\alpha|\mu, D]$ and $[\mu|\alpha, D]$ are log-concave, and thus Gibbs sampling is straightforward for this model.

To build the Weibull regression model, we introduce covariates through μ, and write $\mu_i = \boldsymbol{z}_i'\boldsymbol{\beta}$. Common prior distributions for $\boldsymbol{\beta}$ include the uniform improper prior, i.e., $\pi(\boldsymbol{\beta}) \propto 1$, and a normal prior. Assuming a $N_p(\boldsymbol{\mu}_0, \Sigma_0)$ prior for $\boldsymbol{\beta}$ and a gamma prior for α, we are led to the joint posterior

$$\begin{aligned}\pi(\boldsymbol{\beta}, \alpha|D) \propto \alpha^{\alpha_0+d-1} \exp\Bigg\{ &\sum_{i=1}^{n} (\nu_i \boldsymbol{z}_i'\boldsymbol{\beta} + \nu_i(\alpha - 1)\log(y_i) - y_i^{\alpha} \exp(\boldsymbol{z}_i'\boldsymbol{\beta})) \\ &- \kappa_0 \alpha - \frac{1}{2}(\boldsymbol{\beta} - \boldsymbol{\mu}_0)\Sigma_0^{-1}(\boldsymbol{\beta} - \boldsymbol{\mu}_0) \Bigg\}.\end{aligned}$$

Closed forms for the posterior distribution of $\boldsymbol{\beta}$ are generally not available, and therefore one needs to use numerical integration or MCMC methods. Due to the availability of statistical packages such as `OpenBUGS` and LIFEREG and MCMC Procedures in SAS, the Weibull regression model can easily be fitted using MCMC sampling techniques. The development for the log-normal model, gamma models, extreme value model, and other parametric models is similar to that of the Weibull model. A multivariate extension of the Weibull model includes the Poly-Weibull model of Berger and Sun (1993).

2.3 Semiparametric models

2.3.1 Piecewise constant hazard model

One of the most convenient and popular discrete time models for semiparametric survival analysis is the piecewise constant hazard model. To construct this model, we first construct a finite partition of the time axis, $0 < s_1 < s_2 < \ldots < s_J$, with $s_J > y_i$ for all $i = 1, 2, \ldots, n$. Thus, we have the J intervals $(0, s_1]$, $(s_1, s_2]$, ..., $(s_{J-1}, s_J]$. In the jth interval, we assume a constant baseline hazard $h_0(y) = \lambda_j$ for $y \in I_j = (s_{j-1}, s_j]$. Letting $\boldsymbol{\lambda} = (\lambda_1, \lambda_2, \ldots, \lambda_J)'$, we can write the likelihood function of $(\boldsymbol{\beta}, \boldsymbol{\lambda})$ for the n subjects as

$$L(\boldsymbol{\beta}, \boldsymbol{\lambda}|D) = \prod_{i=1}^{n}\prod_{j=1}^{J}\Big(\lambda_j \exp(\boldsymbol{x}_i'\boldsymbol{\beta})\Big)^{\delta_{ij}\nu_i} \exp\Big\{-\delta_{ij}\Big[\lambda_j(y_i - s_{j-1}) + \sum_{g=1}^{j-1}\lambda_g(s_g - s_{g-1})\Big]\exp(\boldsymbol{x}_i'\boldsymbol{\beta})\Big\}, \tag{2.12}$$

where $\delta_{ij} = 1$ if the ith subject failed or was censored in the jth interval, and 0 otherwise. The indicator δ_{ij} is needed to properly define the likelihood over the J intervals. The semiparametric model in (2.12), sometimes referred to as a "piecewise exponential model," is quite general and can accommodate various shapes of the baseline hazard over the intervals. Moreover, we note that if $J = 1$, the model reduces to a parametric exponential model with failure rate parameter $\lambda \equiv \lambda_1$. The piecewise constant hazard model is a useful and simple model for modeling survival data. It serves as the benchmark for comparisons with other semiparametric or fully parametric models for survival data.

A common prior of the baseline hazard $\boldsymbol{\lambda}$ is the independent gamma prior $\lambda_j \sim \mathcal{G}(\alpha_{0j}, \lambda_{0j})$ for $j = 1, 2, \ldots, J$. Here α_{0j} and λ_{0j} are prior parameters which can be elicited through the prior mean and variance of λ_j. Another approach is to build a prior correlation among the λ_j's (Leonard, 1978; Sinha, 1993) using a correlated prior $\boldsymbol{\psi} \sim N(\boldsymbol{\psi}_0, \Sigma_\psi)$, where $\psi_j = \log(\lambda_j)$ for $j = 1, 2, \ldots, J$.

The likelihood in (2.12) is based on continuous survival data. The likelihood function based on grouped/discretized survival data D_G of (2.4) for the piecewise constant hazard model is given by $L(\boldsymbol{\beta}, \boldsymbol{\lambda}|D_G) \propto \prod_{j=1}^{J} G_j$, where

$$G_j = \exp\Big\{-\lambda_j\Delta_j \sum_{k\in\mathcal{R}_j-\mathcal{D}_j} \exp(\boldsymbol{x}_k'\boldsymbol{\beta})\Big\} \times \prod_{l\in\mathcal{D}_j}\left[1 - \exp\{-\lambda_j\Delta_j \exp(\boldsymbol{x}_l'\boldsymbol{\beta})\}\right], \tag{2.13}$$

$\Delta_j = a_j - a_{j-1}$, $\mathcal{R}_j$ is the set of patients at risk, and $\mathcal{D}_j$ is the set of patients having failures in the interval I_j. Using a $N_p(\boldsymbol{\mu}_0, \Sigma_0)$ prior for $\boldsymbol{\beta}$, the marginal posterior of $\boldsymbol{\beta}$ given the data D_G and the hyperparameters $\tau_0 = (\boldsymbol{\mu}_0, \Sigma_0, \alpha_0, c_0)$ is given as

$$\pi(\boldsymbol{\beta}|D_G; \tau_0) \propto \prod_{j=1}^{J}\left[G_j^*\right] \exp\left\{-\frac{1}{2}(\boldsymbol{\beta} - \boldsymbol{\mu}_0)\Sigma_0^{-1}(\boldsymbol{\beta} - \boldsymbol{\mu}_0)\right\}, \tag{2.14}$$

where $G_J^* = E[G_j|c_{0j}, \alpha_{0j}]$ is the expectation taken with respect to the independent Gamma prior density of each h_j.

2.3.2 Models using a gamma process

The gamma process is perhaps the most commonly used nonparametric prior process for the Cox model. The seminal paper by Kalbfleisch (1978) describes the gamma process prior for the baseline cumulative hazard function $H_0(\cdot)$ (see also Burridge (1981)). The gamma process can be described as follows. Let $\mathcal{G}(\alpha, \lambda)$ denote the gamma distribution with shape parameter $\alpha > 0$ and scale parameter $\lambda > 0$. Let $\alpha(t), t \geq 0$, be an increasing left continuous function such that $\alpha(0) = 0$, and let $Z(t), t \geq 0$, be a stochastic process with the properties: (i) $Z(0) = 0$; (ii) $Z(t)$ has independent increments in disjoint intervals; and (iii) for $t > s$, $Z(t) - Z(s) \sim \mathcal{G}(c(\alpha(t) - \alpha(s)), c)$. Then the process $\{Z(t) : t \geq 0\}$ is called a gamma process and is denoted by $Z(t) \sim \mathcal{GP}(c\alpha(t), c)$. We note here that $\alpha(t)$ is the mean of the process and c is a weight or confidence parameter about the mean. The sample paths of the gamma process are almost surely increasing functions. It is a special case of a Levy process whose characteristic function is given by

$$E[\exp\{iy(Z(t) - Z(s)\}] = (\phi(y))^{c(\alpha(t)-\alpha(s))},$$

where ϕ is the characteristic function of an infinitely divisible distribution function with unit mean. The gamma process is the special case $\phi(u) = \{c/(c - iu)\}^c$.

2.3.3 Gamma process prior with continuous-data likelihood

For the semiparametric Bayesian analysis of the Cox model of (2.1), a gamma process prior

$$H_0 \sim \mathcal{GP}(c_0 H^*, c_0), \tag{2.15}$$

is often used as a prior for the cumulative baseline hazard function $H_0(t)$. The prior mean "prior guess" $H^*(t)$ of the unknown function $H_0(t)$ is an increasing function, and H^* is often assumed to be a known parametric function with hyperparameter vector γ_0. For example, if $H^*(t) = \gamma_0 t$ corresponds to the constant γ_0 as a prior guess for baseline hazard. Similarly, $H^*(t) = \eta_0 t^{\kappa_0}$ corresponds to a Weibull as a prior guess, where $\gamma_0 = (\eta_0, \kappa_0)'$ is a vector of hyperparameters. The joint survival function of $\mathbf{T} = (T_1, \cdots, T_n)$ after marginalizing the prior process is given by

$$P(\mathbf{T} > \boldsymbol{y}|\boldsymbol{\beta}, X, \gamma_0, c_0) = \prod_{j=1}^{n} [\phi(iV_j)]^{c_0(H^*(y_{(j)}) - H^*(y_{(j-1)}))}, \tag{2.16}$$

where $V_j = \sum_{l \in \mathcal{R}_j} \exp(x_l'\boldsymbol{\beta})$, $\mathcal{R}_j$ is the risk set at time $y_{(j)}$ and $y_{(1)} < y_{(2)} < \ldots, < y_{(n)}$ are distinct ordered times. For continuous data, when the ordered survival times are all distinct, the marginal likelihood of $(\boldsymbol{\beta}, \gamma_0, c_0)$ can be obtained by differentiating (2.16). Note that this likelihood, used by Kalbfleisch (1978), Clayton (1991), and among others, is defined only when the observed survival times $\boldsymbol{y} = (y_1, \cdots, y_n)$ are distinct. In the next subsection, we present the likelihood and prior associated with grouped survival data using a gamma process prior for the baseline hazard.

2.3.4 Relationship to partial likelihood

Kalbfleisch (1978) and more recently, Sinha et al. (2003) show that the partial likelihood of Cox (1975) can be obtained as a limiting case of the marginal posterior of $\boldsymbol{\beta}$ in the

Cox model under a gamma process prior for the cumulative baseline hazard H_0. As it is mentioned earlier, the marginal joint survival function of $\mathbf{T}$ given $(\boldsymbol{\beta}, H^*, c_0)$ for Gamma process prior of (2.15) is given in (2.16) with $\phi(u) = \{c_0/(c_0 - iu)\}$.

Now let $\boldsymbol{\theta} = (\boldsymbol{\beta}', h_0^*, c_0)'$, where $h_0^*(y) = \frac{d}{dy}H^*(y)$. Via differentiating the joint survival function of (2.16) at y_i when $\nu_i = 1$, we can obtain the continuous-time data likelihood function of $\boldsymbol{\theta}$ as

$$L(\boldsymbol{\theta}|D) = \prod_{j=1}^{n} \Big[\exp\Big\{ H^*(y_{(j)}) \log\Big(1 - \frac{\exp(\boldsymbol{x}_j'\boldsymbol{\beta})}{c_0 + A_j}\Big)^{c_0} \Big\} \times \Big\{ - c_0 h_0^*(y_{(j)}) \log\Big(1 - \frac{\exp(\boldsymbol{x}_j'\boldsymbol{\beta})}{c_0 + A_j}\Big) \Big\}^{\nu_i} \Big],$$

where $A_j = \sum_{l\in\mathcal{R}_j} \exp(\boldsymbol{x}_l'\boldsymbol{\beta})$, $d = \sum_{i=1}^{n} \nu_i$ and $h^* = \prod_{j=1}^{n} [h_0(y_{(j)})]^{\nu_i}$. Now we have

$$\lim_{c_0\to 0} \exp\left\{ H_0(y_{(j)}) \log\left(1 - \frac{\exp(\boldsymbol{x}_j'\boldsymbol{\beta})}{c_0 + A_j}\right)^{c_0} \right\} = 0$$

for $j = 1, 2, \ldots, n$, and

$$\lim_{c_0\to 0} \log\left(1 - \frac{\exp(\boldsymbol{x}_j'\boldsymbol{\beta})}{c_0 + A_j}\right) = \log\left(1 - \frac{\exp(\boldsymbol{x}_j'\boldsymbol{\beta})}{A_j}\right) \approx -\frac{\exp(\boldsymbol{x}_j'\boldsymbol{\beta})}{A_j}$$

for $j = 1, 2, \ldots, n-1$. Thus, we have

$$\lim_{c_0\to 0} \frac{L(\boldsymbol{\theta}|D)}{c_0^d \log(c_0) h^*} \approx \prod_{j=1}^{n} \left[\frac{\exp(\boldsymbol{x}_j'\boldsymbol{\beta})}{A_j} \right]^{\nu_i}. \tag{2.17}$$

We see that the right-hand side of (2.17) is precisely Cox's partial likelihood.

Now if we let $c_0 \to \infty$, we get the likelihood function based on $(\boldsymbol{\beta}, h_0)$. We can show that

$$\lim_{c_0\to\infty} L(\boldsymbol{\beta}, c_0, h_0|D) = \prod_{j=1}^{n} \left(\exp\left\{-H^*(y_{(j)}) \exp(\boldsymbol{x}_j'\boldsymbol{\beta})\right\}\right) \left\{h_0(y_{(j)}) \exp(\boldsymbol{x}_j'\boldsymbol{\beta})\right\}^{\nu_j}. \tag{2.18}$$

Thus, we see that (2.18) is the likelihood function of $(\boldsymbol{\beta}, h_0)$ based on the proportional hazards model.

2.3.5 Gamma process on baseline hazard

An alternative specification of the semiparametric Cox model is to specify a gamma process prior on the hazard rate itself. Such a formulation is considered by Dykstra and Laud (1981) in their development of the extended gamma process. Here, we consider a discrete approximation of the extended gamma process with the piecewise constant baseline hazard $h_0(t)$ within the partition $0 = a_0 < a_1 < \ldots < a_J$ of the time axis. Let

$$\delta_j = h_0(a_j) - h_0(a_{j-1})$$

denote the increment in the baseline hazard in the interval $I_j = (a_{j-1}, a_j]$, $j = 1, 2, \ldots, J$, and $\boldsymbol{\delta} = (\delta_1, \delta_2, \ldots, \delta_J)'$. We follow Ibrahim et al. (1999) for constructing the approximate likelihood function of $(\boldsymbol{\beta}, \boldsymbol{\delta})$. For an arbitrary individual in the population, the survival function for the Cox model at time y is given by

$$S(y|\boldsymbol{x}) = \exp\Big\{-\eta\int_0^y h_0(u)\,du\Big\} \approx \exp\Big\{-\eta\Big(\sum_{i=1}^{J}\delta_i(y-a_{i-1})^+\Big)\Big\}, \tag{2.19}$$

where $h_0(0) = 0$, $(u)^+ = u$ if $u > 0$, 0 otherwise, and $\eta = \exp(\boldsymbol{x}'\boldsymbol{\beta})$. This first approximation arises since the specification of $\boldsymbol{\delta}$ does not specify the entire hazard rate, but only the δ_j. For purposes of approximation, we take the increment in the hazard rate, δ_j, to occur immediately after a_{j-1}. Let p_j denote the probability of a failure in the interval I_j, $j = 1, 2, \ldots, J$. Using (2.19), we have

$$\begin{aligned} p_j &= S(a_{j-1}) - S(a_j) \\ &\approx \exp\Big\{-\eta\sum_{l=1}^{j-1}\delta_l(a_{j-1}-a_{l-1})\Big\}\Big[1-\exp\Big\{-\eta(a_j-a_{j-1})\sum_{l=1}^{j}\delta_l\Big\}\Big]. \end{aligned}$$

Thus, in the jth interval I_j, the contribution to the likelihood function for a failure is p_j, and $S(a_j)$ for a right-censored observation. For $j = 1, 2, \ldots, J$, let d_j be the number of failures, $\mathcal{D}_j$ be the set of subjects failing, c_j be the number of right-censored observations, and $\mathcal{C}_j$ is the set of subjects that are censored. The grouped data likelihood function is thus given by

$$L(\boldsymbol{\beta}, \boldsymbol{\delta}|D) = \prod_{j=1}^{J}\Big\{\exp\{-\delta_j(A_j+B_j)\}\prod_{k\in\mathcal{D}_j}[1-\exp\{-\eta_k T_j\}]\Big\}, \tag{2.20}$$

where $\eta_k = \exp(\boldsymbol{x}_k'\boldsymbol{\beta})$, $T_j = (s_j - s_{j-1})\sum_{l=1}^{j}\delta_l$, and

$$A_j = \sum_{l=j+1}^{J}\sum_{k\in\mathcal{D}_l}\eta_k(a_{l-1}-a_{j-1}),\quad B_j = \sum_{l=j}^{J}\sum_{k\in\mathcal{C}_l}\eta_k(a_l-a_{j-1}),$$

We note that this likelihood involves a second approximation. Instead of conditioning on exact event times, we condition on the set of failures and set of right-censored events in each interval, and thus we approximate continuous right-censored data by grouped data. Prior elicitation and Gibbs sampling for this model has been discussed in Ibrahim et al. (2001) in detail.

2.3.6 Beta process models

We first discuss time-continuous right-censored survival data without covariates. Kalbfleisch (1978) and Ferguson and Phadia (1979) used the definition of the cumulative hazard $H(t)$ as

$$H(t) = -\log(S(t)), \tag{2.21}$$

where $S(t)$ is the survival function. The gamma process can be defined on $H(t)$ when this definition of the cumulative hazard is appropriate. A more general way of defining the hazard function, which is valid even when the survival time distribution is not continuous, is to use the definition of Hjort (1990). General formulae for the cumulative hazard function $H(t)$ are

$$H(t) = \int_{[0,t]}\frac{dF(u)}{S(u)}, \tag{2.22}$$

where

$$F(t) = 1 - S(t) = 1 - \prod_{[0,t]} \{1 - dH(t)\}. \tag{2.23}$$

The cumulative hazard function $H(t)$ defined here is equal to (2.21) when the survival distribution is absolutely continuous. Hjort (1990) presented what he called a beta process with independent increments as a prior for $H(.)$. A beta process generates a proper cdf $F(t)$, as defined in (2.22), and has independent increments of the form

$$dH(s) \sim \mathcal{B}(c(s)dH^*(s), c(s)(1 - dH^*(s))), \tag{2.24}$$

where $\mathcal{B}(a, b)$ denotes the beta distribution with parameters (a, b). Due to the complicated convolution property of independent beta distributions, the exact distribution of the increment $H(s)$ is only approximately beta over any finite interval, regardless of how small the length of the interval might be. See Hjort (1990) for formal definitions of the beta process prior and for properties of the posterior with right-censored time-continuous data. It is possible to deal with the beta process for the baseline cumulative hazard appropriately defined under a Cox model with time continuous data, but survival data in practice is commonly grouped within some grid intervals, where the grid size is determined by the data and trial design. So for practical purposes, it is more convenient and often sufficient to use a discretized version of the beta process (Hjort, 1990; Sinha, 1997) along with grouped survival data. The beta process prior for the cumulative baseline hazard in (2.24) has been discussed by many authors, including Hjort (1990), Damien et al. (1996), Laud et al. (1996), Sinha (1997), and Florens et al. (1999). Here we focus only on the discretized beta process prior with a grouped data likelihood.

Within the spirit of the definition of the cumulative hazard function $H(t)$ defined in (2.22), a discretized version of the Cox model can be defined as

$$S(s_j|\boldsymbol{x}) = P(T > s_j|\boldsymbol{x}) = \prod_{k=1}^{j} (1 - h_k)^{\exp(\boldsymbol{x}'\boldsymbol{\beta})},$$

where h_k is the discretized baseline hazard rate in the interval $I_k = (s_{k-1}, s_k]$. The likelihood can thus be written as

$$L(\boldsymbol{\beta}, \boldsymbol{h}) = \prod_{j=1}^{J} \left((1 - h_j)^{\sum_{i \in \mathcal{R}_j - \mathcal{D}_j} \exp(\boldsymbol{x}_i'\boldsymbol{\beta})} \right) \prod_{l \in D_j} \left(1 - (1 - h_j)^{\exp(x_l'\boldsymbol{\beta})} \right),$$

where $\boldsymbol{h} = (h_1, h_2, \ldots, h_J)'$. To complete the discretized beta process model, we specify independent beta priors for the h_k's. Specifically, we take $h_k \sim \mathcal{B}(c_{0k}\alpha_{0k}, c_{0k}(1 - \alpha_{0k}))$, and independent for $k = 1, 2, \ldots, J$. Though it is reasonable to assume that the h_k's are independent from each other a priori, the assumption of an exact beta distribution of the h_k's is only due to an approximation to the true time-continuous beta process. Thus, according to the time-continuous beta process, the distribution of the h_k's is not exactly beta, but it can be well approximated by a beta distribution only when the width of I_k is small.

Under the discretized beta process defined here, the joint prior density of $\boldsymbol{h}$ is thus given by

$$\pi(\boldsymbol{h}) \propto \prod_{j=1}^{J} h_j^{c_{0j}\alpha_{0j}-1} (1 - h_j)^{c_{0j}(1-\alpha_{0j})-1}.$$

A typical prior for $\boldsymbol{\beta}$ is a $N_p(\boldsymbol{\mu}_0, \Sigma_0)$ prior, which is independent of $\boldsymbol{h}$. Assuming an arbitrary prior for $\boldsymbol{\beta}$, the joint posterior of $(\boldsymbol{\beta}, h)$ can be written as

$$\pi(\boldsymbol{\beta}, \boldsymbol{h}|D) \propto \prod_{j=1}^{J} \left((1-h_j)^{\sum_{i \in \mathcal{R}_j - \mathcal{D}_j} \exp(\boldsymbol{x}_i'\boldsymbol{\beta})} \right) \prod_{l \in \mathcal{D}_j} \left(1 - (1-h_j)^{\exp(\boldsymbol{x}_l'\boldsymbol{\beta})} \right)$$
$$\times \prod_{j=1}^{J} h_j^{c_{0j}\alpha_{0j}-1} (1-h_j)^{c_{0j}(1-\alpha_{0j})-1} \pi(\boldsymbol{\beta}).$$

Now define $h_j = P[s_{j-1} < Y \leq s_j | Y > s_{j-1})]$. Therefore, the survival curve is given by $S(s_j) = \prod_{k=1}^{j}(1-h_k)$. The prior distribution of h_j is given by

$$h_j \sim \mathcal{B}\Big(c_{0j}\alpha_{0j},\ c_{0j}(1-\alpha_{0j}) \Big) \quad \text{for} \quad j = 1, 2, \ldots, J, \tag{2.25}$$

where the h_j's are independent, and each with mean α_{0j} and variance $\alpha_{0j}(1-\alpha_{0j})/(c_{0j}+1)$. Therefore, c_{0j} is the measure of confidence around the prior mean α_{0j} of the hazard rate h_j in I_j. Given the prior structure of (2.25), the posterior distribution of the h_j's given grouped survival data is also independent beta with

$$h_j | D \sim \mathcal{B}\Big(c_{0j}\alpha_{0j} + d_j,\ c_{0j}(1-\alpha_{0j}) + r_j - d_j \Big), \tag{2.26}$$

where $D = \{(d_j, r_j), j = 1, 2, \ldots, J\}$ denotes the complete grouped data. The joint posterior of the h_j's given interval-censored data is not as straightforward as (2.26), and is discussed in Ibrahim et al. (2001).

2.3.7 Correlated prior processes

The gamma process prior of Kalbfleisch (1978) assumes independent cumulative hazard increments. This is unrealistic in most applied settings, and does not allow for borrowing of strength between adjacent intervals. A correlated gamma process for the cumulative hazard yields a natural smoothing of the survival curve. Although the idea of smoothing is not new (Arjas and Gasbarra, 1994; Aslanidou et al., 1998; Sinha, 1998; Gamerman, 1991; Berzuini and Clayton, 1994), its potential has not been totally explored in the presence of covariates. Modeling dependence between hazard increments has been discussed by Gamerman (1991) and Arjas and Gasbarra (1994). Gamerman (1991) proposed a Markov prior process for the $\{\log(\lambda_k)\}$, by modeling

$$\log(\lambda_k) = \log(\lambda_{k-1}) + \epsilon_k, \qquad E(\epsilon_k) = 0, \quad \text{and} \quad \text{Var}(\epsilon_k) = \sigma_k^2.$$

Arjas and Gasbarra (1994) introduced a first-order autoregressive structure on the increment of the hazards by taking

$$\lambda_k | \lambda_{k-1} \sim \mathcal{G}(\alpha_k, \alpha/\lambda_{k-1})$$

for $k > 1$. Nieto-Barajas and Walker (2002) proposed dependent hazard rates with a Markovian relation, given by

$$\lambda_1 \sim \mathcal{G}(\alpha_1, \gamma_1), \;\; u_k | \lambda_k, v_k \sim \mathcal{P}(v_k \lambda_k), \;\; v_k | \xi_k \sim \mathcal{E}\left(1/\xi_k\right), \tag{2.27}$$

$$\lambda_{k+1} | u_k, v_k \sim \mathcal{G}(\alpha_{k+1} + u_k, \gamma_{k+1} + v_k), \tag{2.28}$$

and

$$\boldsymbol{\beta} \sim \pi(\boldsymbol{\beta}),$$

for $k \geq 1$, where $\pi(\boldsymbol{\beta})$ denotes the prior for $\boldsymbol{\beta}$, which can be taken to be a normal distribution, for example.

2.3.8 Dirichlet process models

The Dirichlet process (Ferguson, 1974) is an important nonparametric prior. There is a rich literature on the Dirichlet process in various applications; see Ibrahim et al. (2001) for a list of references and a nice section in Klein and Moeschberger (2003). The Dirichlet process specifies a nonparametric prior over a class of possible distribution functions $F(y)$ for a random variable Y, where $F(y) = P(Y \leq y)$. To define the Dirichlet process, let Ω be the sample space, and suppose $\Omega = B_1 \cup B_2 \cup \ldots \cup B_k$, where the B_j's are disjoint. Then a stochastic process P indexed by elements of a particular partition $B = \{B_1, B_2, \ldots, B_k\}$ is said to be a Dirichlet process on (Ω, B) with parameter vector α, if for any partition of Ω, the random vector $(P(B_1), P(B_2), \ldots, P(B_k))$ has a Dirichlet distribution with parameter $(\alpha(B_1), \alpha(B_2), \ldots, \alpha(B_k))$.

The parameter vector α is a probability measure so that $\alpha = F_0(\cdot)$, where $F_0(\cdot)$ is the prior hyperparameter for $F(\cdot)$, and thus $\alpha(B_j) = F_0(b_{2j}) - F_0(b_{1j})$. The hyperparameter $F_0(.)$ is called the base measure of the Dirichlet process prior. We define a weight parameter c_0 ($c_0 > 0$) that gives prior weight to $F_0(\cdot)$, so that $(F(B_1), F(B_2), \ldots, F(B_k))$ has a Dirichlet distribution with parameters $(c_0F_0(B_1), c_0F_0(B_2), \ldots, c_0F_0(B_k))$. We say that F has a Dirichlet process prior with parameter c_0F_0 if $(F(B_1), F(B_2), \ldots, F(B_k))$ has a Dirichlet distribution with parameters $(c_0F_0(B_1), c_0F_0(B_2), \ldots, c_0F_0(B_k))$ for every possible partition of the sample space $\Omega = B_1 \cup B_2 \cup \ldots \cup B_k$. The earliest work on Dirichlet processes in the context of survival analysis is based on Ferguson and Phadia (1979) and Susarla and Van Ryzin (1976). Susarla and Van Ryzin derived the Bayes estimator of the survival function under the Dirichlet process prior and also derived the posterior distribution of the cumulative distribution function with right-censored data. In this section, we briefly summarize the fundamental results of Ferguson and Phadia (1979) and Susarla and Van Ryzin (1976). Letting $S(t)$ denote the survival function, Susarla and Van Ryzin (1976) derived the Bayes estimator of $S(t)$ under the squared error loss

$$L(\hat{S}, S) = \int_0^\infty (\hat{S}(t) - S(t))^2 dw(t),$$

where w is a nonnegative decreasing weight function on $(0, \infty)$ and $\hat{S}(t)$ is an estimator of $S(t)$. Then, the Bayes estimator $\hat{S}(u)$ under squared error loss is given by

$$\hat{S}(u) = \frac{c_0(1 - F_0(u)) + N^+(u)}{c_0 + n} \prod_{j=k+1}^{l} \left(\frac{(c_0(1 - F_0(y_{(j)})) + N(y_{(j)})}{c_0(1 - F_0(y_{(j)})) + N(y_{(j)}) - \lambda_j} \right)$$

in the interval $y_{(j)} \leq u \leq y_{(l+1)}$, $l = k, k+1, \ldots, m$, with $y_{(k)} = 0$, $y_{(m+1)} = \infty$. The Kaplan-Meier estimator of $S(u)$ (Kaplan and Meier, 1958) is a limiting case of $\hat{S}(u)$ when $F_0 \to 1$. Other work on the Dirichlet process in survival data includes Kuo and Smith (1992), where they used the Dirichlet process in problems in doubly censored survival data. Generalization of the Dirichlet process have also been used in survival analysis. Mixture of Dirichlet Process (MDP) models have been considered by the MDP model (Escobar, 1994; MacEachern, 1994) removes the assumption of a parametric prior at the second stage, and replaces it with a general distribution G. The distribution G has a Dirichlet process prior, leading to **Stage 1:** $[\boldsymbol{y}_i|\theta_i] \sim \Pi_{n_i}(h_1(\boldsymbol{\theta}_i))$, **Stage 2:** $\boldsymbol{\theta}_i|G \overset{i.i.d.}{\sim} G$; and **Stage 3:** $[G|c_0, \boldsymbol{\psi}_0] \sim \mathcal{DP}(c_0 \cdot G_0(h_2(\boldsymbol{\psi}_0)))$, where G_0 is a w-dimensional parametric distribution, often called the *base measure*, and c_0 is a positive scalar.

There are two special cases in which the MDP model leads to the fully parametric case. As $c_0 \to \infty$, $G \to G_0(\cdot)$, so that the base measure is the prior distribution for $\boldsymbol{\theta}_i$. Also, if $\boldsymbol{\theta}_i \equiv \boldsymbol{\theta}$ for all i, the same is true. For a more hierarchical modeling approach, it is possible to place prior distributions on (c_0, ψ_0). The specification in Stage 2 results in a semiparametric specification in that a fully parametric distribution is given in Stage 1 and a nonparametric distribution is also given in Stages 2 and 3. Doss (1994), Doss and Huffer (1998), and Doss and Narasimhan (1998) discussed the implementation of MDP priors for $F(t) = 1 - S(t)$ in the presence of right-censored data using the Gibbs sampling algorithm. A Bayesian nonparametric approach based on mixtures of Dirichlet priors offers a reasonable compromise between purely parametric and purely nonparametric models.

Other generalizations of the Cox model have been examined by Sinha et al. (1999), and problems investigating interval censored data have been investigated by Sinha (1997). A nice review paper in Bayesian survival analysis is given in Sinha and Dey (1997). Further details on Bayesian semiparametric methods can be found in Ibrahim et al. (2001).

2.4 Prior elicitation

To ease the presentation, we discuss how to use the power prior of Ibrahim and Chen (2000) to incorporate historical data from a previous study only for the piecewise constant hazard model since the development of the power prior for other models presented in Sections 2.2 and 2.3 is similar. Let $D_0 = (n_0, \boldsymbol{y}_0, X_0, \boldsymbol{\nu}_0)$ denote the data from the previous study (i.e., historical data), where n_0 denotes the sample size of the previous study, $\boldsymbol{y}_0$ denotes a right-censored vector of survival times with censoring indicators $\boldsymbol{\nu}_0$, and X_0 denotes the $n \times p$ matrix of covariates. For most problems, there are no firm guidelines on the method of prior elicitation. The use of D_0 for the current study in survival analysis largely depends on the similarity of the two studies. In most clinical trials, for example, no two studies are ever identical. In clinical trials, the patient populations typically differ from study to study and other factors may also make the two studies heterogeneous. Due to these differences, an analysis which simply pools the data from both studies may not be desirable. In this case, it may be more appropriate to "weight" the data from the previous study so as to control its impact on the current study. Thus, it is desirable for the investigators to have a prior distribution that allows them to tune or weight D_0 in order to control its impact on the current study.

Let $\pi_0(\boldsymbol{\beta}, \boldsymbol{\lambda})$ denote the initial prior for $(\boldsymbol{\beta}, \boldsymbol{\lambda})$. The power prior of Ibrahim and Chen (2000) for $(\boldsymbol{\beta}, \boldsymbol{\lambda})$ is given by

$$\pi(\boldsymbol{\beta}, \boldsymbol{\lambda} | D_0, a_0) \propto \{L(\boldsymbol{\beta}, \boldsymbol{\lambda} | D_0)\}^{a_0} \pi_0(\boldsymbol{\beta}, \boldsymbol{\lambda}), \tag{2.29}$$

where $L(\boldsymbol{\beta}, \boldsymbol{\lambda} | D_0)$ is the likelihood function of $(\boldsymbol{\beta}, \boldsymbol{\lambda})$ based on the historical data D_0 and thus, $L(\boldsymbol{\beta}, \boldsymbol{\lambda} | D_0)$ is (2.12) with D and δ_{ij}'s replaced by $D_0 = (n_0, \boldsymbol{y}_0, X_0, \boldsymbol{\nu}_0)$ and δ_{0ij}'s, where $\delta_{0ij} = 1$ if $y_{0i} \in I_j$ and 0 otherwise for $i = 1, 2, \ldots, n_0$ and $j = 1, 2 \ldots, J$. In (2.29), the parameter a_0 can be interpreted as a relative precision parameter for the historical data. It is reasonable to restrict the range of a_0 to be between 0 and 1, and thus we take $0 \leq a_0 \leq 1$. One of the main roles of a_0 is that it controls the heaviness of the tails of the prior for $(\boldsymbol{\beta}, \boldsymbol{\lambda})$. As a_0 becomes smaller, the tails of (2.29) become heavier. When $a_0 = 1$, (2.29) corresponds to the update of $\pi_0(\boldsymbol{\beta}, \boldsymbol{\lambda})$ using Bayes theorem. That is, with $a_0 = 1$, (2.29) corresponds to the "posterior distribution" of $(\boldsymbol{\beta}, \boldsymbol{\lambda})$ based on the historical data D_0. If $a_0 = 0$, then the prior does not depend on the historical data. That is, $a_0 = 0$ is equivalent to a prior specification with no incorporation of historical data. Thus, the parameter a_0 controls the influence of the historical data on the current study. Such control is important in cases where there is heterogeneity between the historical data and the data from the current study, or when the sample sizes of the historical data and the current data are quite different.

In this chapter, we consider a_0 to be fixed. Following Ibrahim et al. (2012), we use the Deviance Information Criterion (DIC) of Spiegelhalter et al. (2002) to determine the optimal choices of (J, a_0). We carry out a detailed analysis of melanoma data in Section 2.6 to demonstrate the use of the power prior (2.29).

2.5 Other topics

Fully parametric Bayesian approaches to frailty models are examined in Sahu et al. (1997), where they considered a frailty model with a Weibull baseline hazard. Semiparametric approaches have also been examined. Clayton (1991) and Sinha (1993) considered a gamma process prior on the cumulative baseline hazard in the proportional hazards frailty model. Qiou et al. (1999) examined a positive stable frailty distribution, and Sargent (1998) examined frailty models using Cox's partial likelihood. Gustafson (1997) discussed Bayesian hierarchical frailty models for multivariate survival data. For detailed summaries of these models and additional references, see the book by Ibrahim et al. (2001).

Bayesian approaches to joint models for longitudinal and survival data have been considered by Faucett and Thomas (1996), Wang and Taylor (2001), Law et al. (2002), Brown and Ibrahim (2003a), Brown and Ibrahim (2003b), Ibrahim et al. (2004), and Brown et al. (2005). Other topics in Bayesian methods in survival analysis include proportional hazards models built from monotone functions (Gelfand and Mallick, 1995), accelerated failure time models (Kuo and Mallick, 1997; Walker and Mallick, 1999; Johnson and Christensen, 1989). Survival models using Multivariate Adaptive Regression Splines (MARS) has been considered by Mallick et al. (1999). Changepoint models have been considered by Sinha et al. (2002). Bayesian methods for model diagnostics in survival analysis were considered in Shih and Louis (1995), Gelfand and Mallick (1995), and Sahu et al. (1997). Bayesian latent residual methods were given in Aslanidou et al. (1998), and the prequential methods were discussed in Arjas and Gasbarra (1997). Bayesian spatial survival models were considered by Carlin and S. (2003). Bayesian methods for missing covariate data in survival analysis include Chen et al. (2002, 2006, 2009). Other work on Bayesian survival analysis with specific applications in epidemiology and related areas include Chen, Dey and Sinha (2000), Dunson (2001), Dunson et al. (2003), and the references therein. Books discussing Bayesian survival analysis include Ibrahim et al. (2001), Chen, Shao and Ibrahim (2000), Carlin and Louis (2000), and Congdon (2003, 2006). Bayesian methods for frailty models and Bayesian methods for model selection are considered by the authors in Chapter 23 and Chapter 14, respectively.

2.6 A case study: an analysis of melanoma data

In this section, we consider two Eastern Cooperative Oncology Group (ECOG) phase III melanoma clinical trials, E1684 and E1690. The first trial, E1684, was a two-arm clinical trial comparing high-dose interferon (IFN) to Observation (OBS). E1690 was a subsequent phase III clinical trial involving identical treatments and patient populations as E1684. E1690 was intended as a confirmatory trial to E1684. There were a total of $n_0 = 285$ and $n = 427$

patients in the E1684 and E1690 trials, respectively. Results of the E1684 and E1690 trials have been published in Kirkwood et al. (1996, 2000). E1690 serves as the current study while E1684 serves as the historical study for our example here. The response variable is the relapse-free survival (RFS), which is defined as the time from randomization until progression of tumor or death, whichever comes first. The covariates we consider include treatment ($x_1 = 0$ if OBS and 1 if IFN), age (x_2 in years), and gender ($x_3 = 0$ if male and 1 if female). We standardized x_2 by subtracting the sample mean and then dividing the sample standard deviation from the pooled E1690 and E1684 data for numerical stability in the posterior computation.

We carry out a Bayesian analysis of E1690 under the piecewise constant hazard model in (2.12). We use the E1684 historical data to construct our prior as in (2.29). The model fit is assessed via the DIC measure. Let $\boldsymbol{\theta} = (\boldsymbol{\beta}, \boldsymbol{\lambda})$. For our example, $\boldsymbol{\beta} = (\beta_1, \beta_2, \beta_3)'$. We first define the deviance function as $\text{Dev}(\boldsymbol{\theta}) = -2 \log L(\boldsymbol{\beta}, \boldsymbol{\lambda}|D)$, where $L(\boldsymbol{\beta}, \boldsymbol{\lambda}|D)$ is given by (2.12). Then, according to Spiegelhalter et al. (2002), the DIC measure can be calculated as follows:

$$\text{DIC} = \text{Dev}(\bar{\boldsymbol{\theta}}) + 2p_D,$$

where $p_D = \overline{\text{Dev}} - \text{Dev}(\bar{\boldsymbol{\theta}})$ is the effective number of model parameters, and $\bar{\boldsymbol{\theta}} = E[\boldsymbol{\theta}|D, D_0, a_0]$ and $\overline{\text{Dev}} = E[\text{Dev}(\boldsymbol{\theta})|D, D_0, a_0]$ denote the posterior means of $\boldsymbol{\theta}$ and $\text{Dev}(\boldsymbol{\theta})$ with respect to the posterior distribution corresponding to the likelihood given in (2.12) and the power prior in (2.29). The initial prior in (2.29) is specified as follows: $\pi_0(\boldsymbol{\beta}, \boldsymbol{\lambda}) = \pi_0(\boldsymbol{\beta})\pi_0(\boldsymbol{\lambda})$, where $\pi_0(\boldsymbol{\beta}) \propto \exp\left\{-\frac{1}{2\sigma_0^2}\boldsymbol{\beta}'\boldsymbol{\beta}\right\}$ with $\sigma_0^2 = 10{,}000$, and $\pi_0(\boldsymbol{\lambda}) \propto \prod_{j=1}^{J} \lambda_j^{b_{01}-1} \exp(-b_{02}\lambda_j)$ with $b_{01} = 0.001$ and $b_{02} = 0.001$.

One of the critical issues is the choice of J and a_0 in the Bayesian analysis under the piecewise constant hazard model in (2.12) using the power prior, that is, what value of (J, a_0) should the investigator use in the analysis? To address this issue, we use the value (J, a_0) that yields the best model fit according to the DIC. Figure 2.1 shows that the DIC is optimized when $J = 5$ and $a_0 = 0.5$. From Figure 2.1, it is interesting to see that (i) the

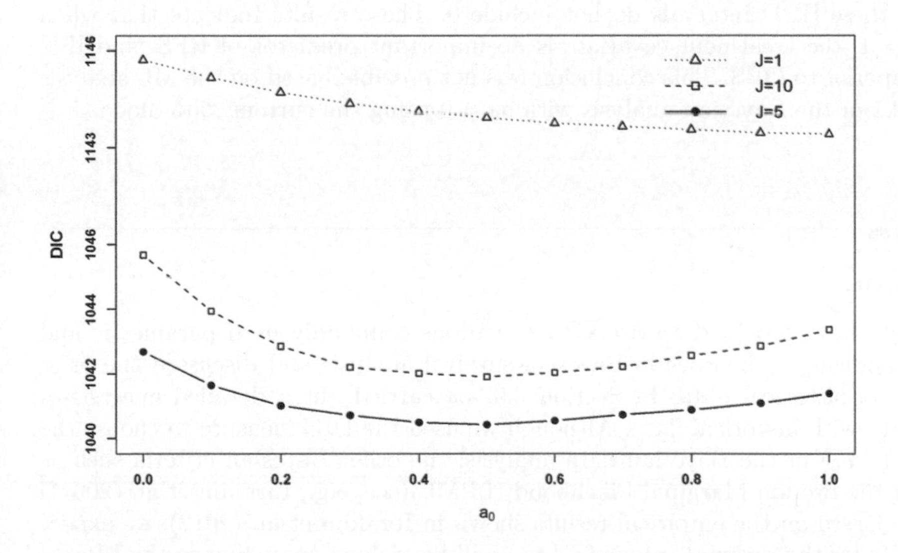

FIGURE 2.1
DIC plots for $J = 1, 5, 10$ for E1690 using E1684 as the historical data.

TABLE 2.1
The maximum likelihood estimates for E1690.

Variable	Parameter	EST	SE	95% CI
Treatment	β_1	-0.213	0.130	(-0.468, 0.042)
Age	β_2	0.116	0.065	(-0.013, 0.244)
Gender	β_3	-0.145	0.136	(-0.411, 0.122)

model with $J = 5$ fits the data much better than the ones with $J = 1$ or $J = 10$ for all a_0 values according to the DIC; (ii) the DIC decreases in a_0 when $J = 1$; and (iii) both DIC curves are convex in a_0 with the minimum values attained at $a_0 = 0.5$ when $J = 5$ and $J = 10$. These results indicate that (i) when a simple baseline hazard function is used, which does not fit the data well, the more borrowing from the historical data, the better fit the model, and (ii) when a more complex baseline hazard function is used, which fits the data much better, the best model fit is achieved with a moderate amount of borrowing from the historical data according to the DIC. Note that when $J = 5$, the values of DIC were 1042.68, 1040.42, and 1041.39 for $a_0 = 0, 0.5, 1$, respectively. Based on Figure 2.1, $a_0 = 0.5$ appears to be a reasonable choice for the Bayesian analysis, along with other a_0 values used as a sensitivity analysis such as $a_0 = 0$ and $a_0 = 1$.

Table 2.1 shows the maximum likelihood (ML) estimates (ESTs), the standard errors (SEs), and 95% confidence intervals (CIs) of $\boldsymbol{\beta}$ for the E1690 data alone based on the partial likelihood approach. From Table 2.1, all three covariates were not significant at the significance level of 0.05 since all 95% CIs contain 0. This result indicates that the treatment IFN does not have a significant impact on relapse-free survival (RFS) based on the E1690 data alone. Table 2.2 shows the posterior means (EST), the posterior standard deviations (SD), and 95% highest posterior density (HPD) intervals of $\boldsymbol{\beta}$ and $\boldsymbol{\lambda}$ for $a_0 = 0, 0.5, 1$ under $J = 5$. When $a_0 = 0$ (no incorporation of historical data), the posterior estimates of $\boldsymbol{\beta}$ were very similar to those ML estimates. From Table 2.2, we see that the posterior standard deviations decrease when a_0 increases. We further see from this table that the HPD intervals of β_1 (treatment) were $(-0.479, -0.063)$ when $a_0 = 0.5$ and $(-0.478, -0.108)$ when $a_0 = 1$. Both of these HPD intervals do not include 0. These results indicate that when $a_0 = 0.5$ or $a_0 = 1$, the treatment covariate is an important predictor of RFS, and IFN may in fact be superior to OBS. This conclusion was not possible based on the ML analysis shown in Table 2.1 or the Bayesian analysis with $a_0 = 0$ using the current data alone.

2.7 Discussion

In this chapter, we have provided an overview of various commonly used parametric and semiparametric modeling techniques in Bayesian survival analysis and discussed the prior elicitation based on historical data. In Section 2.6, we carried out a detailed analysis of the melanoma data with historical data. Although we used the DIC measure to choose the optimal value of (J, a_0) in the Bayesian data analysis, the other Bayesian criteria such as the Logarithm of the Pseudo Marginal Likelihood (LPML)(see, e.g., Ibrahim et al. (2001)) can also be used. Based on the empirical results shown in Ibrahim et al. (2012), we expect that similar results for the optimal value of (J, a_0) will be yielded according to the LPML.

We wrote the FORTRAN 95 code with double precision and IMSL subroutines as well as the SAS macro using the MCMC and IML Procedures to carry out all of the posterior

TABLE 2.2
The posterior estimates for E1690 using E1684 as the historical data under $J = 5$.

a_0	Parameter	EST	SD	95% HPD Interval
0	β_1	-0.218	0.128	(-0.465, 0.029)
	β_2	0.119	0.065	(-0.008, 0.247)
	β_3	-0.150	0.134	(-0.412, 0.112)
	λ_1	0.509	0.087	(0.343, 0.680)
	λ_2	0.641	0.108	(0.441, 0.860)
	λ_3	0.562	0.084	(0.406, 0.732)
	λ_4	0.314	0.050	(0.219, 0.412)
	λ_5	0.112	0.019	(0.077, 0.149)
0.5	β_1	-0.269	0.107	(-0.479, -0.063)
	β_2	0.101	0.055	(-0.013, 0.206)
	β_3	-0.104	0.111	(-0.411, 0.112)
	λ_1	0.607	0.083	(0.447, 0.774)
	λ_2	0.729	0.102	(0.532, 0.930)
	λ_3	0.546	0.072	(0.407, 0.690)
	λ_4	0.331	0.044	(0.245, 0.419)
	λ_5	0.111	0.015	(0.082, 0.141)
1.0	β_1	-0.294	0.095	(-0.478, -0.108)
	β_2	0.093	0.048	(-0.001, 0.187)
	β_3	-0.080	0.099	(-0.272, 0.119)
	λ_1	0.667	0.079	(0.516, 0.824)
	λ_2	0.783	0.097	(0.604, 0.979)
	λ_3	0.534	0.064	(0.409, 0.658)
	λ_4	0.340	0.041	(0.262, 0.423)
	λ_5	0.110	0.013	(0.085, 0.137)

computations in Section 2.6. Both the FORTRAN 95 code and the SAS macro gave very similar results. We generated 20,000 Gibbs iterations after 2,000 "burn-in" iterations to compute the DIC and posterior estimates for each combination of (J, a_0) shown in Table 2.2. The HPD intervals were computed using the Monte Carlo method proposed by Chen and Shao (1999). For each combination of (J, a_0), the computing time was less than 30 seconds for the FORTRAN 95 code on an Intel i7 processor machine with 8 GB of RAM memory using a GNU/Linux operating system while the computing time was about 30 minutes for the SAS code using STAT 12.1 with slice option in the MCMC Procedure on an Intel Core(TM)2 Duo CPU machine with 1.96 GB of RAM memory. We further note that the PHREG Procedure in SAS does fit the Bayesian piecewise constant hazard model, but it does not have an option to incorporate historical data using the power prior. Both the FORTRAN 95 code and the SAS macro are available from the authors upon request.

Acknowledgments

Drs. Ibrahim and Chen's research was partially supported by U.S. National Institutes of Health (NIH) grants #GM 70335 and #CA 74015.

Bibliography

Achcar, J. A., Bolfarine, H. and Pericchi, L. R. (1987), 'Transformation of survival data to an extreme value distribution', *The Statistician* **36**, 229–234.

Achcar, J. A., Brookmeyer, R. and Hunter, W. G. (1985), 'An application of Bayesian analysis to medical follow-up data', *Statistics in Medicine* **4**, 509–520.

Aitchison, J. and Dunsmore, I. R. (1975), *Statistical Prediction Analysis*, New York: Cambridge University Press.

Arjas, E. and Gasbarra, D. (1994), 'Nonparametric Bayesian inference from right censored survival data, using the gibbs sampler', *Statistica Sinica* **4**, 505–524.

Arjas, E. and Gasbarra, D. (1997), 'On prequential model assessment in life history analysis', *Biometrika* **84**, 505–522.

Aslanidou, H., Dey, D. K. and Sinha, D. (1998), 'Bayesian analysis of multivariate survival data using Monte Carlo methods', *Canadian Journal of Statistics* **26**, 33–48.

Berger, J. O. and Sun, D. (1993), 'Bayesian analysis for the Poly-Weibull distribution', *Journal of the American Statistical Association* **88**, 1412–1418.

Berzuini, C. and Clayton, D. G. (1994), 'Bayesian analysis of survival on multiple time scales', *Statistics in Medicine* **13**, 823–838.

Brown, E. R. and Ibrahim, J. G. (2003a), 'A Bayesian semiparametric joint hierarchical model for longitudinal and survival data', *Biometrics* **59**, 221–228.

Brown, E. R. and Ibrahim, J. G. (2003b), 'Bayesian approaches to joint cure rate and longitudinal models with applications to cancer vaccine trials', *Biometrics* **59**, 686–693.

Brown, E. R., Ibrahim, J. G. and DeGruttola, V. (2005), 'A flexible b-spline model for multiple longitudinal biomarkers and survival', *Biometrics* **61**, 64–73.

Burridge, J. (1981), 'Empirical Bayes analysis for survival time data', *Journal of the Royal Statistical Society, Series B* **43**, 65–75.

Carlin, B. P. and Louis, T. A. (2000), *Bayes and Empirical Bayes Methods for Data Analysis*, Second Edition. Boca Raton, FL: Chapman & Hall.

Carlin, B. P. and Banerjee, S. (2003), Hierarchical multivariate CAR models for spatio-temporally correlated survival data (with discussion), *in* J. M. Bernardo, M. J. Bayarri, J. O. Berger, A. P. Dawid, D. Heckerman, A. F. M. Smith and M. West, eds., 'Bayesian Statistics, 7', Oxford: Clarendon Press, pp. 45–63.

Chen, M.-H., Dey, D. K. and Sinha, D. (2000), 'Bayesian analysis of multivariate mortality data with large families', *Applied Statistics* **49**, 129–144.

Chen, M.-H., Ibrahim, J. G. and Lipsitz, S. R. (2002), 'Bayesian methods for missing covariates in cure rate models', *Lifetime Data Analysis* **8**, 117–146.

Chen, M.-H., Ibrahim, J. G. and Shao, Q.-M. (2006), 'Posterior propriety and computation for the Cox regression model with applications to missing covariates', *Biometrika* **93**, 791–807.

Chen, M.-H., Ibrahim, J. G. and Shao, Q.-M. (2009), 'Maximum likelihood inference for the cox regression model with applications to missing covariates', *Journal of Multivariate Analysis* **100**, 2018–2030.

Chen, M.-H. and Shao, Q.-M. (1999), 'Monte Carlo estimation of Bayesian credible and HPD intervals', *Journal of Computational and Graphical Statistics* **8**, 69–92.

Chen, M.-H., Shao, Q.-M. and Ibrahim, J. G. (2000), *Monte Carlo Methods in Bayesian Computation*, New York: Springer-Verlag.

Chen, W. C., Hill, B. M., Greenhouse, J. B. and Fayos, J. V. (1985), Bayesian analysis of survival curves for cancer patients following treatment, *in* 'Bayesian Statistics, 2', Amsterdam: North-Holland, pp. 299–328.

Clayton, D. G. (1991), 'A Monte Carlo method for Bayesian inference in frailty models', *Biometrics* **47**, 467–485.

Congdon, P. (2003), *Applied Bayesian Modeling*, Second Edition. New York: John Wiley & Sons.

Congdon, P. (2006), *Bayesian Statistical Modeling*, New York: John Wiley & Sons.

Cox, D. R. (1972), 'Regression models and life tables (with discussion)', *Journal of the Royal Statistical Society, Series B* **34**, 187–220.

Cox, D. R. (1975), 'Partial likelihood', *Biometrika* **62**, 269–276.

Damien, P., Laud, P. W. and Smith, A. F. M. (1996), 'Implementation of Bayesian nonparametric inference based on beta processes', *Scandinavian Journal of Statistics* **23**, 27–36.

Dellaportas, P. and Smith, A. F. M. (1993), 'Bayesian inference for generalized linear and proportional hazards models via gibbs sampling', *Applied Statistics* **42**, 443–459.

Doss, H. (1994), 'Bayesian nonparametric estimation for incomplete data via successive substitution sampling', *Annals of Statistics* **22**, 1763–1786.

Doss, H. and Huffer, F. (1998), 'Monte Carlo methods for Bayesian analysis of survival data using mixtures of dirichlet priors', *Technical Report* **22**, 1763–1786.

Doss, H. and Narasimhan, B. (1998), Dynamic display of changing posterior in Bayesian survival analysis, *in* D. Dey, P. Müller and D. Sinha, eds., 'Practical Nonparametric and Semiparametric Bayesian Statistics', New York: Springer-Verlag, pp. 63–84.

Dunson, D. B. (2001), 'Bayesian modeling of the level and duration of fertility in the menstrual cycle', *Biometrics* **57**, 1067–1073.

Dunson, D. B., Chulada, P. and Arbes, S. J. (2003), 'Bayesian modeling of time-varying and waning exposure effects', *Biometrics* **59**, 83–91.

Dykstra, R. L. and Laud, P. W. (1981), 'A Bayesian nonparametric approach to reliability', *The Annals of Statistics* **9**, 356–367.

Escobar, M. D. (1994), 'Estimating normal means with a dirichlet process prior', *Journal of the American Statistical Association* **89**, 268–277.

Faucett, C. J. and Thomas, D. C. (1996), 'Simultaneously modeling censored survival data and repeatedly measured covariates: A gibbs sampling approach', *Statistics in Medicine* **15**, 1663–1685.

Feigl, P. and Zelen, M. (1965), 'Estimation of exponential survival probabilities with concomitant information', *Biometrics* **21**, 826–838.

Ferguson, T. S. (1974), 'Prior distributions on spaces of probability measures', *Annals of Statistics* **2**, 615–629.

Ferguson, T. S. and Phadia, E. G. (1979), 'Bayesian nonparametric estimation based on censored data', *Annals of Statistics* **7**, 163–186.

Florens, J. P., Mouchart, M. and Rolin, J. M. (1999), 'Semi- and non-parametric Bayesian analysis of duration models with dirichlet priors: A survey', *International Statistical Review* **67**, 187–210.

Gamerman, D. (1991), 'Dynamic Bayesian models for survival data', *Applied Statistics* **40**, 63–79.

Gelfand, A. E. and Mallick, B. K. (1995), 'Bayesian analysis of proportional hazards models built from monotone functions', *Biometrics* **51**, 843–852.

Grieve, A. P. (1987), 'Applications of Bayesian software: Two examples', *The Statistician* **36**, 283–288.

Gustafson, P. (1997), 'Large hierarchical Bayesian analysis of multivariate survival data', *Biometrics* **53**, 230–242.

Hjort, N. L. (1990), 'Nonparametric Bayes estimators based on beta processes in models of life history data', *Annals of Statistics* **18**, 1259–1294.

Ibrahim, J. G. and Chen, M.-H. (2000), 'Power prior distributions for regression models', *Statistical Sciences* **15**, 46–60.

Ibrahim, J. G., Chen, M.-H. and Chu, H. (2012), 'Bayesian methods in clinical trials: a bayesian analysis of ecog trials e1684 and e1690', *BMC Medical Research Methodology* **12**, DOI: 10.1186/1471–2288–12–183.

Ibrahim, J. G., Chen, M.-H. and MacEachern, S. N. (1999), 'Bayesian variable selection for proportional hazards models', *Canadian Journal of Statistics* **27**, 701–717.

Ibrahim, J. G., Chen, M.-H. and Sinha, D. (2001), *Bayesian Survival Analysis*, New York: Springer-Verlag.

Ibrahim, J. G., Chen, M.-H. and Sinha, D. (2004), 'Bayesian methods for joint modeling of longitudinal and survival data with applications to cancer vaccine studies', *Statistica Sinica* **14**, 863–883.

Johnson, W. and Christensen, R. (1989), 'Nonparametric Bayesian analysis of the accelerated failure time model', *Statistics and Probability Letters* **7**, 179–184.

Kalbfleisch, J. D. (1978), 'Nonparametric Bayesian analysis of survival time data', *Journal of the Royal Statistical Society, Series B* **40**, 214–221.

Kaplan, E. L. and Meier, P. (1958), 'Nonparametric estimation from incomplete observations', *Journal of the American Statistical Association* **53**, 457–481.

Kim, S. W. and Ibrahim, J. G. (2001), 'On Bayesian inference for parametric proportional hazards models using noninformative priors', *Lifetime Data Analysis* **6**, 331–341.

Kirkwood, J. M., Ibrahim, J. G., Sondak, V. K., Richards, J., Flaherty, L. E., Ernstoff, M. S., Smith, T. J., Rao, U., Steele, M. and Blum, R. (2000), 'The role of high- and low-dose interferon alfa-2b in high-risk melanoma: First analysis of intergroup trial e1690/s9111/c9190', *Journal of Clinical Oncology* **18**, 2444–2458.

Kirkwood, J. M., Strawderman, M. H., Ernstoff, M. S., Smith, T. J., Borden, E. C. and Blum, R. H. (1996), 'Interferon alfa-2b adjuvant therapy of high-risk resected cutaneous melanoma: The eastern cooperative oncology group trial est 1684', *Journal of Clinical Oncology* **14**, 7–17.

Klein, J. P. and Moeschberger, M. L. (2003), *Survival Analysis: Techniques for Censored and Truncated Data*, New York: Springer-Verlag.

Kuo, L. and Mallick, B. K. (1997), 'Bayesian semiparametric inference for the accelerated failure-time model', *Canadian Journal of Statistics* **25**, 457–472.

Kuo, L. and Smith, A. F. M. (1992), Bayesian computations in survival models via the Gibbs sampler, *in* J. P. Klein and G. P. K., eds., 'Survival Analysis: State of the Art', Boston: Kluwer Academic, pp. 11–24.

Laud, P. W., Damien, P. and Smith, A. F. M. (1998), Bayesian nonparametric and covariate analysis of failure time data, *in* 'Practical Nonparametric and Semiparametric Bayesian Statistics', New York: Springer, pp. 213–225.

Laud, P. W., Smith, A. F. M. and Damien, P. (1996), 'Monte Carlo methods for approximating a posterior hazard rate process', *Statistics and Computing* **6**, 77–83.

Law, N. J., Taylor, J. M. G. and Sandler, H. (2002), 'The joint modeling of a longitudinal disease progression marker and the failure time process in the presence of cure', *Biostatistics* **3**, 547–563.

Leonard, T. (1978), 'Density estimation, stochastic processes and prior information', *Journal of the Royal Statistical Society, Series B* **40**, 113–146.

MacEachern, S. N. (1994), 'Estimating normal means with a conjugate style dirichlet process prior', *Communications in Statistics – Theory and Methods* **23**, 727–741.

Mallick, B. K., Denison, D. G. T. and Smith, A. F. M. (1999), 'Bayesian survival analysis using a MARS model', *Biometrics* **55**, 1071–1077.

Nieto-Barajas, L. E. and Walker, S. G. (2002), 'Markov beta and gamma processes for modeling hazard rates', *Scandinavian Journal of Statistics* **29**, 413–424.

Qiou, Z., Ravishanker, N. and Dey, D. K. (1999), 'Multivariate survival analysis with positive frailties', *Biometrics* **55**, 637–644.

Sahu, S. K., Dey, D. K., Aslanidou, H. and Sinha, D. (1997), 'A Weibull regression model with gamma frailties for multivariate survival data', *Lifetime Data Analysis* **3**, 123–137.

Sargent, D. J. (1998), 'A general framework for random effects survival analysis in the Cox proportional hazards setting', *Biometrics* **54**, 1486–1497.

Shih, J. A. and Louis, T. A. (1995), 'Assessing gamma frailty models for clustered failure time data', *Lifetime Data Analysis* **1**, 205–220.

Sinha, D. (1993), 'Semiparametric Bayesian analysis of multiple event time data', *Journal of the American Statistical Association* **88**, 979–983.

Sinha, D. (1997), 'Time-discrete beta process model for interval-censored survival data', *Canadian Journal of Statistics* **25**, 445–456.

Sinha, D. (1998), 'Posterior likelihood methods for multivariate survival data', *Biometrics* **54**, 1463–1474.

Sinha, D., Chen, M.-H. and Ghosh, S. K. (1999), 'Bayesian analysis and model selection for interval-censored survival data', *Biometrics* **55**, 585–590.

Sinha, D. and Dey, D. K. (1997), 'Semiparametric Bayesian analysis of survival data', *Journal of the American Statistical Association* **92**, 1195–1212.

Sinha, D., Ibrahim, J. G. and Chen, M.-H. (2002), 'Bayesian models for survival data from cancer prevention studies', *Journal of the Royal Statistical Society, Series B* **63**, 467–477.

Sinha, D., Ibrahim, J. G. and Chen, M.-H. (2003), 'A Bayesian justification of Cox's partial likelihood', *Biometrika* **90**, 629–641.

Spiegelhalter, D. J., Best, N. G., Carlin, B. P. and van der Linde, A. (2002), 'Bayesian measures of model complexity and fit (with discussion)', *Journal of the Royal Statistical Society, Series B* **64**, 583–639.

Susarla, V. and Van Ryzin, J. (1976), 'Nonparametric Bayesian estimation of survival curves from incomplete observations', *Journal of the American Statistical Association* **71**, 897–902.

Walker, S. G., Damien, P., Laud, P. W. and Smith, A. F. M. (1999), 'Bayesian nonparametric inference for random distributions and related functions (with discussion)', *Journal of the Royal Statistical Society, Series B* **61**, 485–528.

Walker, S. G. and Mallick, B. K. (1999), 'A Bayesian semiparametric accelerated failure time model', *Biometrics* **55**, 477–483.

Wang, Y. and Taylor, J. M. G. (2001), 'Jointly modeling longitudinal and event time data, with applications to AIDS studies', *Journal of the American Statistical Association* **96**, 895–905.

3

Alternatives to the Cox Model

Torben Martinussen

Department of Biostatistics, University of Copenhagen

Limin Peng

Department of Biostatistics and Bioinformatics, Emory University

CONTENTS

3.1 Additive hazards regression 49
3.1.1 Model specification and inferential procedures 50
3.1.2 Goodness-of-fit procedures 53
3.1.3 Further results on additive hazard models 54
3.1.3.1 Structural properties of the additive hazard model 54
3.1.3.2 Clustered survival data and additive hazard model 55
3.1.3.3 Additive hazard change point model 55
3.1.3.4 Additive hazard and high-dimensional regressors 56
3.1.3.5 Combining the Cox model and the additive model 56
3.1.3.6 Gastrointestinal tumour data 57
3.2 The accelerated failure time model 58
3.2.1 Parametric models 58
3.2.2 Semiparametric models 59
3.2.2.1 Inference based on hazard specification 59
3.2.2.2 Inference using the additive mean specification 61
3.3 Quantile regression for survival analysis 62
3.3.1 Introduction 62
3.3.2 Estimation under random right censoring with C always known 63
3.3.3 Estimation under covariate-independent random right censoring 63
3.3.4 Estimation under standard random right censoring 64
3.3.4.1 Self-consistent approach 64
3.3.4.2 Martingale-based approach 66
3.3.5 Variance estimation and other inference 67
3.3.6 Extensions to other survival settings 68
3.3.7 An illustration of quantile regression for survival analysis 68
Bibliography 71

3.1 Additive hazards regression

The Cox regression model is clearly the most used hazards model when analyzing survival data as it provides a convenient way of summarizing covariate effects in terms of relative risks. The proportional hazards assumption may not always hold, however. A typical violation of the assumption is time-changing covariate effects which is often encountered in bio-medical applications. A typical example is a treatment effect that varies with time such as treatment efficacy fading away over time due to, for example, tolerance developed by the patient. Under such circumstances, the Aalen additive hazard model (Aalen, 1980) may provide a useful alternative to the Cox model as it easily incorporates time-varying covariate effects. For Aalen's model, the hazard function for an individual with a p-dimensional covariate X vector takes the form

$$\alpha(t|X) = X^T \beta(t). \tag{3.1}$$

The *time-varying regression function* $\beta(t)$ is a vector of locally integrable functions, and usually the X will have 1 as its first component allowing for an intercept in the model. Model (3.1) is very flexible and can be seen as a first order Taylor series expansion of a general hazard function $\alpha(t|X)$ around the zero covariate:

$$\alpha(t|X) = \alpha(t, 0) + X^T \alpha'(t, X^*)$$

with X^* on the line segment between 0 and X. As we shall see below, it is the cumulative regression function

$$B(t) = \int_0^t \beta(s)\, ds$$

that is easy to estimate and its corresponding estimator converges at the usual $n^{1/2}$-rate. Estimation and asymptotical properties were derived by Aalen (Aalen, 1980), and further results for this model were given in Aalen (1989), Aalen (1993), and Huffer and McKeague (1991) derived efficient estimators.

The full flexibility of the model (3.1) may not always be needed and it is usually also of interest to try to simplify the model investigating whether or not a specific covariate effect really varies with time. A very useful model in this respect is the semiparametric additive hazards model of McKeague and Sasieni (1994). It assumes that the hazard function has the form

$$\alpha(t|X, Z) = X^T \beta(t) + Z^T \gamma, \tag{3.2}$$

where X and Z are p-dimensional and q-dimensional covariate vectors, respectively. Apart from its use for testing time-varying effects, this model is useful in its own right because it is then possible to make a sensible bias-variance trade-off, where effects that are almost constant can be summarized as such and effects that are strongly time-varying can be described as such. Note also that model (3.2) covers model (3.1) simply by leaving out the last part of model (3.2) by leaving Z empty. On the other hand, in practice, one may start out with model (3.1) and then test for a specific covariate whether its effect varies significantly with time. If this is not the case one may then simplify to model (3.2) with Z being that specific covariate, and one can then proceed in that way with some or all of the remaining covariates; the starting point, though, will now be model (3.2). Below we show how to do such inference in model (3.2). Lin and Ying (1994) considered a special case of (3.2), where only the intercept term is allowed to depend on time; this model is often referred to as the Lin and Ying model. A nice discussion of pros and cons of the additive hazards model can be found in Aalen et al. (2008).

3.1.1 Model specification and inferential procedures

We present estimation and inferential procedures based on model (3.2) since, as mentioned above, model (3.1) is covered by model (3.2) simply by leaving Z empty and therefore the below results also cover model (3.1).

Let T be the survival time of interest with conditional hazard function $\alpha(t|X, Z)$ given the covariate vectors X and Z. In practice T may be right-censored by U so that we observe $(\tilde{T} = T \wedge U, \Delta = I(T \leq U), X, Z)$. We assume that T and U are conditionally independent given (X, Z). Let $(\tilde{T}_i, \Delta_i, X_i, Z_i)$ be n iid replicates from this generic model. Under the above independent right-censoring scheme, the ith counting process $N_i(t) = I(\tilde{T}_i \leq t, \Delta_i = 1)$ has intensity

$$\lambda_i(t) = Y_i(t) \left[X_i^T \beta(t) + Z_i^T \gamma\right],$$

where $Y_i(t) = I(t \leq \tilde{T}_i)$ is the at risk indicator. We assume that all counting processes are observed in the time-interval $[0, \tau]$, where $\tau < \infty$. Each counting process has compensator $\Lambda_i(t) = \int_0^t \lambda_i(s)ds$ such that $M_i(t) = N_i(t) - \Lambda_i(t)$ is a martingale. Define the n-dimensional counting process $N = (N_1, ..., N_n)^T$ and the n-dimensional martingale $M = (M_1, ..., M_n)^T$. Let also $X = (Y_1X_1, \ldots, Y_nX_n)^T$, $Z = (Y_1Z_1, \ldots, Y_nZ_n)^T$. The results presented below hold under some regularity conditions; see Martinussen and Scheike (2006). Writing the model in incremental form

$$dN(t) = X(t)dB(t) + Z(t)\gamma dt + dM(t),$$

and since $E\{dM(t)\} = 0$, this suggests the following (unweighted) least squares estimator of $\{B(t) = \int_0^t \beta(s)\, ds, \gamma)\}$ McKeague and Sasieni (1994):

$$\hat{\gamma} = \left(\int_0^\tau Z^T H Z\, dt\right)^{-1} \int_0^\tau Z^T H dN(t), \tag{3.3}$$

$$\hat{B}(t) = \int_0^t X^- dN(t) - \int_0^t X^- Z\, dt\hat{\gamma}, \tag{3.4}$$

where X^- denotes the generalized inverse $(X^TX)^{-1}X^T$ and $H = I - XX^-$ assuming here that the required inverses exist. Considering the model with only non-parametric terms, model (3.1), means that Z is empty and is readily seen from the latter display that the estimator of $B(t)$ in that case is given by

$$\hat{B}(t) = \int_0^t X^- dN(t),$$

which is the estimator initially proposed by Aalen (1980). Weighted estimators, and in theory efficient estimators, exist. However, they depend on the unknown parameters, and therefore, to calculate them, some kind of smoothing is needed which may compromise efficiency in practice; see McKeague and Sasieni (1994) for more on the efficient estimators. Here we will only deal with the unweighted estimators that can be calculated without any use of smoothing. It is quite obvious that these simple and direct estimators have sound properties as the following arguments show. Considering $\hat{\gamma}$, the counting process integral can be written as

$$\begin{aligned}\int_0^\tau Z^T H dN(t) &= \int_0^\tau Z^T H d\{XdB(t) + Z\gamma dt\} + \int_0^\tau Z^T H dM(t) \\ &= \int_0^\tau Z^T H Z dt\gamma + \int_0^\tau Z^T H dM(t)\end{aligned}$$

since $HX = 0$, and therefore $\hat{\gamma}$ is an essentially unbiased estimator of γ since the martingale term has zero-mean. Similar arguments concerning $\hat{B}(t)$ are readily given. The limit distributions of the estimators are

$$n^{1/2}\{\hat{\gamma} - \gamma\} = C^{-1}(\tau)n^{-1/2} \int_0^\tau Z^T H dM(t),$$

$$n^{1/2}\{\hat{B}(t) - B(t)\} = n^{1/2} \int_0^\tau X^- dM(t) - P(t)n^{1/2}\{\hat{\gamma} - \gamma\},$$

where

$$C(t) = n^{-1} \int_0^t Z^T H Z\, ds, \quad P(t) = \int_0^t X^- Z\, dt.$$

These limit distributions may be written as sums of essentially iid terms:

$$n^{1/2}\{\hat{\gamma} - \gamma\} = n^{-1/2} \sum_{i=1}^n \epsilon_i^\gamma + o_P(1), \quad n^{1/2}\{\hat{B}(t) - B(t)\} = n^{-1/2} \sum_{i=1}^n \epsilon_i^B(t) + o_P(1),$$

where

$$\epsilon_i^\gamma = c^{-1}(\tau) \int_0^\tau \{Z_i - (z^T x)(x^T x)^{-1} X_i\} dM_i(t),$$

$$\epsilon_i^B(t) = \int_0^\tau (x^T x)^{-1} X_i dM_i(t) - p(t)\epsilon_i^\gamma,$$

where $c(t)$ and $p(t)$ denote the limits in probability of $C(t)$ and $P(t)$, respectively. Also, $x^T x$ is used as notation for the limit in probability of $n^{-1}X^T X$, and similarly with $z^T x$. The limit distributions may be simulated likewise what Lin et al. (1993) suggested in a Cox-model setting:

$$n^{1/2}\{\hat{\gamma} - \gamma\} \sim n^{-1/2} \sum_{i=1}^n \hat{\epsilon}_i^\gamma G_i, \quad n^{1/2}\{\hat{B}(t) - B(t)\} \sim n^{-1/2} \sum_{i=1}^n \hat{\epsilon}_i^B(t) G_i,$$

where we let $\sim$ indicate that two quantities have the same limit distribution. In the latter display, $G_1, \ldots, G_n$ are independent standard normals and $\hat{\epsilon}_i^\gamma$ is obtained from ϵ_i^γ by replacing deterministic quantities with their empirical counterparts and by replacing $M_i(t)$ with $\hat{M}_i(t)$, $i = 1, \ldots, n$, and similarly with $\hat{\epsilon}_i^B(t)$. The result is that, conditional on the data,

$$\left(n^{-1/2} \sum_{i=1}^n \hat{\epsilon}_i^\gamma G_i, n^{-1/2} \sum_{i=1}^n \hat{\epsilon}_i^B(t) G_i \right)$$

will have the same limit distribution as

$$\left(n^{1/2}\{\hat{\gamma} - \gamma\}, n^{1/2}\{\hat{B}(t) - B(t)\} \right).$$

The hypothesis of time-invariance

$$\mathrm{H}_0 : \ \beta_p(t) = \beta_p,$$

focusing without loss of generality on the pth regression coefficient, may be reformulated in terms of the cumulative regression function $B_p(t) = \int_0^t \beta_p(s)\, ds$ as

$$\mathrm{H}_0 : \ B_p(t) = \beta_p \cdot t.$$

Martinussen and Scheike (2006) studied the test process

$$V_n(t) = n^{1/2}(\hat{B}_p(t) - \hat{B}_p(\tau)\frac{t}{\tau}) \tag{3.5}$$

that is easy to compute and considered test statistics as

$$\sup_{t \leq \tau} |V_n(t)|.$$

Under H_0, we have

$$V_n(t) = n^{1/2} \left\{ (\hat{B}_p(t) - B_p(t)) - (\hat{B}_p(\tau) - B_p(\tau)) \frac{t}{\tau} \right\}.$$

Clearly, the limit distribution of $V_n(t)$ is not a martingale due to the term $\hat{B}_p(\tau)$, but one may use the above resampling technique of Lin et al. (1993) to approximate its limit distribution. Specifically, the limit distribution of $V_n(t)$ may be approximated by

$$\hat{V}_n(t) = n^{-1/2} \sum_{i=1}^{n} \left[\{\hat{\epsilon}_i^B(t)\}_p - \{\hat{\epsilon}_i^B(\tau)\}_p \frac{t}{\tau} \right] G_i$$

where v_k is the kth element of a given vector v and where we are fixing the data. The above limit distributions and the conditional multiplier approach may also be used to test other hypothesis such as overall effect of a given covariate. Such inferential procedures and estimation are implemented in the contributed R package *timereg*; for examples of its use, see Martinussen and Scheike (2006).

3.1.2 Goodness-of-fit procedures

The additive hazard model is very versatile but it is still of importance to check whether the model provides an adequate fit to the data at hand. The starting point will then usually be model (3.1) and if this model is found to give an adequate fit then one may proceed to investigate the time-dynamics of the covariate effects. The goodness-of-fit procedures presented below are based on the martingale residuals

$$\hat{M}(t) = N(t) - \int_0^t X(s) d\hat{B}(s) = \int_0^t H(s) dM(s).$$

Under the model, the martingale residuals are thus themselves martingales. To validate the model fit, one possibility is to sum these residuals depending on the level of the covariates (Aalen, 1993; Grønnesby and Borgan, 1996). A K-cumulative residual process may then be defined as

$$M_K(t) = \int_0^t K^T(s) d\hat{M}(s) = \int_0^t K^T(s) H(s) dM(s),$$

where $K(t) = \{K_1^T(t), \ldots, K_n^T(t)\}^T$ is an $n \times m$ matrix, possibly depending on time. A typical choice of K is to let it be a factor with levels defined by the quartiles of the considered covariates. Based on a iid representation of $M_K(t)$ one may calculate p-values using for instance a supremum-test, see Martinussen and Scheike (2006). With a continuous covariate one may also target a test to investigate whether the specific covariate is included in the model on the right scale, a typical alternative is to use a log-transformed version of the covariate. This construction is similar to what Lin et al. (1993) proposed under the Cox model and is based on

$$M_c(z) = \sum_{i=1}^{n} I(X_{i1} \leq z) \hat{M}_i(\tau)$$

here assuming that the covariate under investigation is X_1. This process in z has a decomposition similarly to M_K developed above, and resampling may thus be used to validate the fit of the model (Martinussen and Scheike, 2006). These goodness-of-fit procedures are implemented in the R-package *timereg*. For other goodness-of-fit testing concerning the additive hazard model, see Gandy and Jensen (2005*a*) and Gandy and Jensen (2005*b*).

3.1.3 Further results on additive hazard models

The additive hazard is applied in some more specialized areas of survival analysis. Below we describe some of these. But first we present some structural properties of the model that underline that this model may indeed provide an appealing alternative to the proportional hazard model.

3.1.3.1 Structural properties of the additive hazard model

Consider the additive hazard model

$$\alpha(t|X, Z) = \beta_0(t) + \beta_X(t)X + \beta_Z(t)Z \tag{3.6}$$

and let $B_X(t) = \int_0^t \beta_X(s)\,ds$ and similarly with $B_Z(t)$. We will explore when the additive hazard structure is preserved when only conditioning on X which was first investigated in Aalen (1993). The model above is formulated with X and Z both being scalar valued but the following can easily be generalized to cover the vector situation also. The hazard function, when only conditioning on X, is

$$\alpha(t|X) = \beta_0(t) + \beta_X(t)X + \beta_Z(t) + \psi_{Z|X}(B_Z(t)),$$

where $\psi_{Z|X}(\cdot)$ denotes minus the derivative of the log Laplace transform of Z given X. It is therefore readily seen that additive structure is preserved if X and Z are independent as only the intercept term is changed. This also holds true if (X, Z) has a normal distribution also allowing for dependence between the two. This is in contrast to the proportional hazard model where the proportional structure is not preserved even when X and Z are independent Struthers and Kalbfleisch (1986).

When dealing with data from a non-randomized study it may be of interest to calculate the causal hazard function as it corresponds to the hazard function under a randomized study. To do so we briefly introduce Pearl's (Pearl, 2000) do operator, $do(X = x)$ (or $\hat{X} = x$), which is an intervention setting variable X to x. Therefore, the conditional distribution of T given X is set to x, $P(T|\hat{X} = x)$, is different from $P(T|X = x)$ where the latter is the conditional distribution of T given that we find X equal to x. Taking again model (3.6) as the starting point it may be seen, using the so-called "G-computation formula" (Robins, 1986; Pearl, 2000), that

$$\alpha(t|\hat{X} = x) = \tilde{\beta}_0(t) + \beta_X(t)x, \tag{3.7}$$

$$\tilde{\beta}_0(t) = \beta_0(t) + \beta_Z(t)\psi_Z(B_Z(t)),$$

where $\psi_Z(\cdot)$ denotes minus the derivative of the log Laplace transform of Z. It is seen from display (3.7) that the additive hazard structure is again preserved and therefore the causal effect of X can be estimated from model (3.6) formulated for the observed data. Such a property does not hold true for the proportional hazard model; see Martinussen and Vansteelandt (2012) for more details on these issues. The additive hazard model also forms the basis for investigation of so-called "direct effects" in Fosen et al. (2006); Martinussen (2010); Lange and Hansen (2011); and Martinussen, Vansteelandt, Gerster and Hjelmborg (2011). In a situation as depicted in Figure 3.1, the direct effect of X, formulated via the hazard function on absolute scale, is

$$\alpha(t|\hat{X} = x + 1, \hat{K}) - \alpha(t|\hat{X} = x, \hat{K})$$

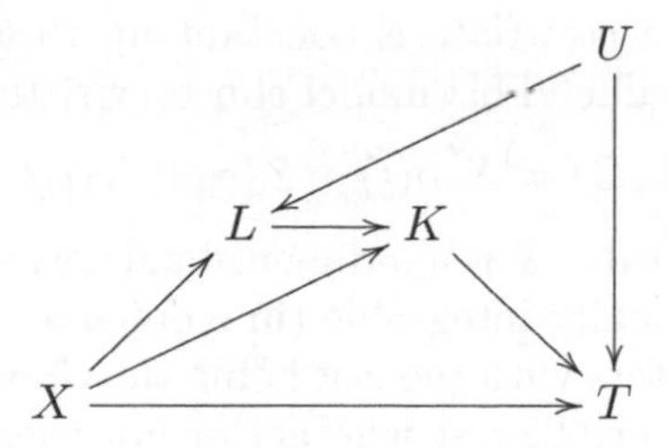

FIGURE 3.1
Situation with L and K being observed confounders and U being unobserved.

3.1.3.2 Clustered survival data and additive hazard model

Clustered failure time data arise in many research areas, for example, in studies where time to disease onset is recorded for members of a sample of families. One approach to analyze such data is to use a frailty model, and recently Martinussen, Scheike and Zucker (2011) proposed the use of the additive hazard model in such a setting. The Aalen additive gamma frailty hazard model specifies that individual i in the kth cluster has conditional hazard function

$$\alpha_{ik}(t|Z_k, X_{ik}) = Z_k X_{ik}^T \beta(t), \tag{3.8}$$

where Z_k is the unobserved frailty variable, taken to be gamma distributed with mean 1 and variance θ, X_{ik} is the covariate vector and $\beta(t)$ is the vector of time-dependent regression coefficients. The counting process associated with individual i in the kth cluster has the following intensity with respect to the marginal history:

$$\lambda_{ik}^m(t) = \{1 + \theta X_{ik}^T B(t-)\}^{-1} Y_{ik}(t) X_{ik}^T \beta(t), \tag{3.9}$$

where $Y_{ik}(t)$ is the at-risk indicator. Thus the marginal intensity still has an additive structure, but with the design depending on $B(t-)$. Based on this, an estimator was developed that has a recursive structure, and large sample properties were presented; see Martinussen, Scheike and Zucker (2011) for more details. Another approach was taken by Cai and Zeng (2011) where the frailty effect also enters at the additive scale

$$\alpha_{ik}(t|Z_k, X_{ik}) = \beta_0(t) + X_{ik}^T \beta + Z_k, \tag{3.10}$$

using the Lin and Ying version of the additive model and where Z_k follows a one-parameter distribution with zero-mean. Estimators and large sample are provided in Cai and Zeng (2011).

As alternative to the frailty approach one may instead use a marginal regression model approach if interest centers on estimating regression effects at the population level and estimation of correlation is of less interest. For use of the fully nonparametric version of the additive hazards model in this context, see Martinussen and Scheike (2006). Instead one may also apply the Lin and Ying version of the additive model; see Pipper and Martinussen (2004) and Yin and Cai (2004).

3.1.3.3 Additive hazard change point model

As described in Section 3.1 the Aalen additive hazard model is well suited to investigate the time dynamics of the effect of a given covariate. Hence one can test the constant effects model against a model where the structure of the effect is left fully unspecified. An appealing model in between the constant effects model and the full Aalen model is the change-point model in which the effect of a covariate is constant up to an unknown point in time and changes thereafter to a new value. This model can be written as

$$\alpha(t|X, Z) = X^{'}\beta(t) + Z\{\gamma_1 + \gamma_2 I(t > \theta)\}, \tag{3.11}$$

where $\alpha(t|X, Z)$ is the hazard rate, X is a p-dimensional covariate and Z is a scalar covariate, $\beta(t)$ is a vector of unknown locally integrable time-dependent regression functions, and γ_1, γ_2 and θ are unknown parameters with the last being the change-point parameter that is also taken as an unknown parameter. To test whether or not there is evidence of a change-point effect corresponds to testing the hypothesis $H_0 : \gamma_2 = 0$, which is a nontrivial problem as the change-point parameter θ is only present under the alternative. Martinussen and Scheike (2007) derived a test tailored to investigate this hypothesis, and also gave estimators for all the unknown parameters defining the model. They also proposed a test that can be used to compare the change point model to full Aalen model with no prespecified form of the covariate effects.

3.1.3.4 Additive hazard and high-dimensional regressors

It has become of increasing interest to develop statistical methods to explore a potential relationship between a high-dimensional regressor, such as gene-expression data, to the timing of an event such as death. It has been demonstrated that the additive hazard model may also be useful in this respect. The starting point for these developments has so far been the additive hazard model with all effects being constant with time, that is, the Lin and Ying model,

$$\alpha(t|Z) = \beta(t) + Z^T\gamma,$$

where Z is p-dimensional covariate vector; in this case with p large. With gene expression data, p may be very large and often much larger than the sample size n. This is referred to as the $p >> n$ situation. Based on the Lin and Ying model Martinussen and Scheike (2009*a*) derived a partial least squares estimator via the so-called "Krylov sequence," and Ma et al. (2006) derived a principal components analysis in this context. Other popular approaches in a high-dimensional regressor setting is the Lasso and Ridge regression to mention some; see Bovelstad et al. (2009) for a comparison of methods using the Cox model as the base model. One complication with the additive hazard model compared with the proportional hazard model is, however, that there is no simple likelihood to work with for the additive model. Leng and Ma (2007) suggested to use a least squares criterion, similarly to the partial likelihood for Cox's regression model, for the Lin and Ying model:

$$L(\beta) = \beta^T\{\int Z^T(t)H(t)Z(t)dt\}\beta - 2\beta^T\{\int Z^T(t)H(t)dN(t)\}. \tag{3.12}$$

using the notation introduced in Section 3.1. This criterion was also suggested independently by Martinussen and Scheike (2009*b*) that provided further motivation. This criterion opens the route to many popular estimation procedures applied in the high-dimensional regressor setting when the underlying model is the Lin and Ying model. Recently Gorst-Rasmussen and Scheike (2012) took the same model to propose a sure independence screening method that shows good performance. For estimation in the high-dimensional case one may use the contributed R package *ahaz* that can also do other types of estimation and inference within the additive hazard model context.

3.1.3.5 Combining the Cox model and the additive model

The Cox model and the Aalen model may be combined in various ways to improve model fit or to enhance interpretation of parameters. Scheike and Zhang (2002) considered the model

$$\alpha(t|X, Z) = \{X^T\beta(t)\}\exp\{Z^T\gamma\},$$

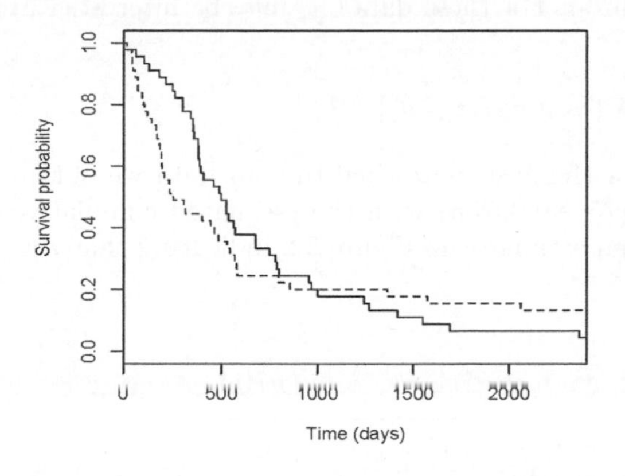

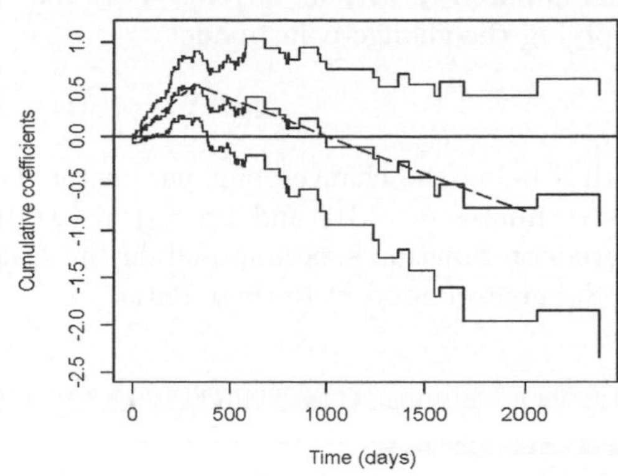

FIGURE 3.2
Gastrointestinal tumour data. Left display: Kaplan-Meier estimates of survival probabilities of the chemotherapy (solid line) and chemotherapy plus radiotherapy (dashed line) groups. Right display: effect of combined therapy - Aalens least squares estimate of $B_1(t)$ with 95% pointwise confidence bands with the dashed curve being the estimate based on a change-point model).

termed the Cox-Aalen model, where the baseline hazard function of the Cox model is replaced by an additive hazard term depending on X. Martinussen and Scheike (2002) considered a proportional excess risk model that also combines the Cox model and the additive model.

It is also of interest to decide whether the Cox model or the additive hazard is most appropriate to apply in a given application. This is a nontrivial problem as these two classes of models are non-nested. Martinussen et al. (2008) used the MizonRichard encompassing test for this particular problem, which corresponds to fitting the Aalen model to the martingale residuals obtained from the Cox regression analysis.

3.1.3.6 Gastrointestinal tumour data

A typical instance where the Cox model provides a poor fit is when crossing survival curves is observed as shown in Figure 3.2 (left display) that displays the Kaplan-Meier estimates of survival probabilities of the chemotherapy and chemotherapy plus radiotherapy groups of gastric cancer patients in a randomized clinical trial (Stablein and Koutrouvelis, 1985); these data were also considered in Zeng and Lin (2007*b*). The crossing of the survival curves can be explained by the patients receiving the more aggressive treatment (radiotherapy) who are at elevated risk of death initially but may benefit from the treatment in the long run if they are able to tolerate the treatment initially. That the Cox model provides a poor fit to the data is confirmed by the supremum test based on the score process Lin et al. (1993) with $p < 0.001$. We fitted Aalens additive hazards model to the gastrointestinal tumour data:

$$\alpha(t|X) = \beta_0(t) + \beta(t)X,$$

where X is the indicator of chemotherapy plus radiotherapy treatment. The cumulative regression coefficient corresponding to the combined therapy group is depicted in Figure 3.2 (right display), showing nicely that a time varying effect is indeed the case for these data with a negative effect of the combined treatment in the first 300 days or so, and then apparently with an adverse effect thereafter. For these data one may be interested in applying the change-point model

$$\alpha(t|X) = \beta_0(t) + X\{\gamma_1 + \gamma_2 I(t > \theta)\},$$

with θ being the change-point parameter. For the gastrointestinal tumour data we obtain the estimates $\hat{\theta} = 315$ and $(\hat{\gamma}_1, \hat{\gamma}_2) = (0.00175, -0.00255)$ with the estimated cumulative regression function superimposed on the Aalen estimator in Figure 3.2 indicating that this model gives a good fit to these data.

3.2 The accelerated failure time model

Another useful model in survival analysis is the accelerated failure time (AFT) model that can be formulated using survival functions. Let T be a survival time and Z a covariate vector that does not depend on time and let $S(t|Z)$ be the conditional survival function. The AFT model assumes that

$$S(t|Z) = S_0(te^{Z^T\beta}), \tag{3.13}$$

where S_0 is a baseline survival function. From (3.13) it is seen that covariates act multiplicatively on time so that their effect is to accelerate (or decelerate) time-to-failure relative to S_0. An equivalent and popular formulation of AFT-model is the following linear regression for the log-transformed event time, $\log(T)$, given Z such that

$$\log(T) = -Z^T\beta + \varepsilon, \tag{3.14}$$

where the distribution of e^{ε} is given by S_0 and ε is assumed to be independent of Z. The AFT model has received considerable attention in the statistical literature in the last two decades or so. Although, in its semiparametric version (see below) it is computationally intensive, it is now considered as an alternative to the Cox model. It is appealing due to its direct relationship between the failure time and covariates, which was also noted by David Cox in Reid (1994). One may extend the model to cover the case where covariates may change with time by considering

$$e^{\varepsilon} = \int_0^T e^{Z^T(t)\beta}\, dt$$

see for example Zeng and Lin (2007*a*), but we will only consider the simpler case here. The AFT model can also be formulated in terms of the hazard function for T given Z:

$$\lambda(t|Z) = \lambda_0(te^{Z^T\beta})e^{Z^T\beta}, \tag{3.15}$$

where $\lambda_0(t)$ is the hazard associated with the distribution of e^{ε}.

3.2.1 Parametric models

If we are willing to specify a specific form of S_0 then the AFT model is fully parametric. For instance, the Weibull model is given by

$$S_0(t) = e^{-kt^{\alpha}}, \quad k, \alpha > 0.$$

In such a case, estimation and statistical inference is straightforward using standard likelihood methods for survival data. For the Weibull model we have the hazard function $\lambda_0(t) = k\alpha t^{\alpha-1}$ and hence, by (3.15),

$$\lambda(t|Z) = k\alpha(te^{Z^T\beta})^{\alpha-1}e^{Z^T\beta} = \lambda_0(t)e^{Z^T\alpha\beta}$$

being also a proportional hazard function with a Weibull baseline hazard function, so in this specific example the two models coincide.

3.2.2 Semiparametric models

We will now look into the case where $S_0(t)$ is left unspecified, or, equivalently, ε in (3.14) is not specified, which is referred to as the "semiparametric AFT model." There exist two approaches to tackle this situation. One develops estimation based on the hazard specification (3.15), while the other one uses the additive mean model specification (3.14) as a starting point.

3.2.2.1 Inference based on hazard specification

Let C be the censoring time for T, and put $\tilde{T} = T \wedge C$ and $\Delta = I(T \leq C)$. Hence the intensity of the counting process $N(t) = I(\tilde{T} \leq t)\Delta$ is thus given by $Y(t)\lambda(t)$ with $Y(t) = I(t \leq \tilde{T})$ being the at-risk indicator and $\lambda(t)$ given in (3.15). Assume that n i.i.d. counting processes are being observed subject to this generic hazard model. We thus consider $N(t) = \{N_1(t), .., N_n(t)\}$ the n-dimensional counting process of all subjects. Define also the time-transformed counting process

$$N^*(t) = (N_1(te^{-Z_1^T\beta}), .., N_n(te^{-Z_n^T\beta}))$$

with associated at-risk process $Y_i^*(t,\beta) = Y_i(te^{-Z_i^T\beta})$, $i = 1, \ldots, n$. This time-transformation is made because then the intensity of $N_i(te^{-Z_i^T\beta})$ is

$$\lambda_i^*(t) = Y_i^*(t)\lambda_0(t),$$

which immediately suggests that $\Lambda_0(t) = \int_0^t \lambda_0(s)ds$ should be estimated by the Breslow-type estimator

$$\hat{\Lambda}_0(t,\beta) = \int_0^t \frac{1}{S_0^*(s,\beta)} dN_\cdot^*(s) \tag{3.16}$$

where $dN_\cdot^*(t) = \sum_{i=1}^n dN_i^*(t)$, and

$$S_0^*(t,\beta) = \sum_{i=1}^n Y_i^*(t,\beta)$$

if β were known. The efficient score function for β is

$$\begin{aligned}
&\sum_{i=1}^n \int_0^\infty \frac{\partial}{\partial\beta}(\lambda_i(t,\beta))\lambda_i^{-1}(t,\beta)\,(dN_i(t) - Y_i(t)\lambda_i(t)dt) \qquad (3.17)\\
&= \sum_{i=1}^n \int_0^\infty \left(\frac{\lambda_0'(te^{Z_i^T\beta})te^{Z_i^T\beta}}{\lambda_0(te^{Z_i^T\beta})} + 1\right) Z_i(dN_i(t) - Y_i(t)\lambda_i(t)dt)\\
&= \sum_{i=1}^n \int_0^\infty \left(\frac{\lambda_0'(u)u}{\lambda_0(u)} + 1\right) Z_i(dN_i^*(u) - Y_i^*(u,\beta)d\Lambda_0(u)),
\end{aligned}$$

and inserting $d\hat{\Lambda}_0(u,\beta)$ for $d\Lambda_0(u)$ gives

$$U_W(\beta) = \sum_{i=1}^n \int_0^\infty W(u) Z_i \left(dN_i^*(u) - \frac{Y_i^*(u,\beta)}{S_0^*(u,\beta)} dN_\cdot^*(u) \right)$$

$$= \sum_{i=1}^n \int_0^\infty W(u) \left(Z_i - E^*(u,\beta)\right) dN_i^*(u), \tag{3.18}$$

where

$$S_1^*(u,\beta) = \sum_{i=1}^n Y_i^*(u,\beta) Z_i, \quad E^*(u,\beta) = \frac{S_1^*(u,\beta)}{S_0^*(u,\beta)},$$

and

$$W(u) = \left(\frac{\lambda_0'(u) u}{\lambda_0(u)} + 1 \right) \tag{3.19}$$

is the efficient weight function. We cannot use (3.18) directly for estimation purposes since the weight function $W(u)$ involves the unknown baseline hazard function $\lambda_0(u)$ and its derivative $\lambda_0'(u)$. These can be estimated and inserted into (3.18) but it is not recommendable since it is hard to get reliable estimates of especially $\lambda_0'(t)$. A way around this is to take (3.18) and replace the weight function $W(u)$ with one that can be computed as for example $W(u) = 1$ or $W(u) = n^{-1} S_0^*(u,\beta)$ referred to as the "log-rank and Gehan weight functions," respectively. A practical complication is that the score function $U_W(\beta)$ is a step function of β so $U_W(\beta) = 0$ may not have a solution. The score function may furthermore not be component-wise monotone in β. It is actually monotone in each component of β if the Gehan weight is chosen (Fygenson and Ritov, 1994). The estimator $\hat{\beta}$ is usually chosen as the one which minimizes $||U_W(\beta)||$. The contributed R package *lss* calculates the Gehan rank estimator. It has been established, under regularity conditions, that $n^{1/2}(\hat{\beta} - \beta)$ is asymptotically zero-mean normal with covariance matrix $A_W^{-1} B_W A_W^{-1}$, where A_W and B_W are the limits in probability of

$$\frac{1}{n} \sum_{i=1}^n \int_0^\infty W(u) \left(Z_i - E^*(u,\beta)\right)^{\otimes 2} \left(\frac{\lambda_0'(u) u}{\lambda_0(u)} + 1 \right) dN_i^*(u),$$

$$\frac{1}{n} \sum_{i=1}^n \int_0^\infty W(u)^2 \left(Z_i - E^*(u,\beta)\right)^{\otimes 2} dN_i^*(u),$$

respectively; see Tsiatis (1990) and Ying (1993). It is seen that A_W and B_W coincide in the case where $W(u)$ is taken as the efficient weight function (3.19). The asymptotic covariance matrix depends on λ_0', which is difficult to estimate. One may, however, apply a resampling technique avoiding estimation of λ_0'; see Lin et al. (1998) and Jin et al. (2003).

The score function can also be written on the (log)-transformed time scale

$$U_{\tilde{W}}(\beta) = \sum_{i=1}^n \int_{-\infty}^\infty \tilde{W}(t) \left(Z_i - E^*(e^t,\beta)\right) dN_i^*(e^t)$$

$$= \sum_{i=1}^n \int_{-\infty}^\infty \tilde{W}(t) \left(Z_i - \tilde{E}(t - Z_i^T\beta, \beta)\right) d\tilde{N}_i(t - Z_i^T\beta), \tag{3.20}$$

where $\tilde{N}_i(t) = I(\log(\tilde{T}_i) \le t)\Delta_i$, $\tilde{Y}_i(t) = I(t \le \log(\tilde{T}_i))$,

$$\tilde{E}(t,\beta) = \sum_{i=1}^n Z_i \tilde{Y}_i(t) / \sum_{i=1}^n \tilde{Y}_i(t), \quad \tilde{W}(t) = \frac{\lambda_{0\varepsilon}'(t)}{\lambda_{0\varepsilon}(t)},$$

with $\lambda_{0\varepsilon}(t)$ the hazard function for ε. Another way of motivating (3.20) is by classical non-parametric testing theory; see Tsiatis (1990) and Kalbfleisch and Prentice (2002) for much more details on that approach.

An interesting variant of (3.15), considered by Chen and Jewell (2001), is

$$\lambda(t) = \lambda_0(te^{Z^T\beta_1})e^{Z^T\beta_2}, \tag{3.21}$$

which contains the proportional hazard model ($\beta_1 = 0$) and the accelerated failure time model ($\beta_1 = \beta_2$), and for $\beta_2 = 0$ what is called "the accelerated hazard model" (Chen and Wang, 2000). Chen and Jewell (2001) suggested estimating equations for estimation of $\beta = (\beta_1, \beta_2)$ and showed for the resulting estimator, $\hat{\beta}$, that $n^{1/2}(\hat{\beta} - \beta)$ is asymptotically zero-mean normal with a covariance that also involves the unknown baseline hazard (and its derivative). They also suggested an alternative resampling approach for estimating the covariance matrix without having to estimate the baseline hazard function or its derivative. With these tools at hand, one may then investigate whether it is appropriate to simplify (3.21) to either the Cox model or the AFT model.

3.2.2.2 Inference using the additive mean specification

There exist other ways of estimating the regression parameters β of the semiparametric AFT-model, which build more on classical linear regression models estimation. Starting with (3.14), let $V = \log(T)$, $U = \log(C)$ and $\tilde{V} = V \wedge U$, and write model (3.14) as

$$V = -\beta_0 - Z^T\beta + \varepsilon$$

assuming that ε is independent of $Z = (Z_1, \ldots, Z_p)^T$ and has zero mean. If V was not right-censored, then it is of course an easy task to estimate the regression parameters using least squares regression. A number of authors (Miller, 1976; Buckley and James, 1979; and Koul et al., 1981) have extended the least-squares principle to accommodate censoring; we describe the approach of Buckley and James. The idea is to replace V with a quantity that has the same mean as V, and which can be computed based on the right-censored sample. With

$$V^* = V\Delta + (1 - \Delta)E(V \mid V > U, Z),$$

and $\Delta = I(V \leq U)$, then $E(V^* \mid Z) = E(V \mid Z)$. Still, V^* is not observable but it can be estimated as follows. Since

$$E(V \mid V > U, Z) = -Z^T\beta + \frac{\int_{U+Z^T\beta}^{\infty} v dF(v)}{1 - F(U + Z^T\beta)}$$

with F the distribution of $V + Z^T\beta$, one can construct the so-called "synthetic data points:"

$$\hat{V}_i^*(\beta) = V_i\Delta_i + (1 - \Delta_i)\left(-Z_i^T\beta + \frac{\int_{U_i+Z_i^T\beta}^{\infty} v d\hat{F}(v)}{1 - \hat{F}(U_i + Z_i^T\beta)}\right),$$

where $\hat{F}$ is the Kaplan-Meier estimator based on $(\tilde{V}_i + Z_i^T\beta, \Delta_i)$, $i = 1, \ldots, n$. One may then estimate the parameters from the normal equations leading to the following estimating equation for the regression parameter vector β:

$$U(\beta) = \sum_{i=1}^{n}(\hat{V}_i^*(\beta) + Z_i^T\beta)(Z_i - \overline{Z}) = 0, \tag{3.22}$$

where $\overline{Z} = n^{-1} \sum_i Z_i$. Equation (3.22) needs to be solved iteratively if it has a solution. This estimator is referred to as the "Buckley-James estimator." The large sample properties of the resulting estimator were studied by Ritov (1990). Equation (3.22) can also be written as, with $S(v) = v$,

$$\sum_{i=1}^{n} \left(\Delta_i S(U_i + Z_i^T \beta) + (1 - \Delta_i) \frac{\int_{U_i + Z_i^T \beta}^{\infty} S(v) d\hat{F}(v)}{1 - \hat{F}(U_i + Z_i^T \beta)} \right) (Z_i - \overline{Z}) = 0, \tag{3.23}$$

that may be derived from a likelihood principle; the efficient choice of $S(v)$ being $S(v) = f'(v)/f(v)$ with $f(v) = F'(v)$ the density function. Ritov (1990) also established an asymptotic equivalence between the two classes of estimators given by (3.20) and (3.23).

The estimating function $U(\beta)$ in (3.22) is neither continuous nor monotone in β, which makes it difficult to calculate the estimator in practice. Jin et al. (2006) proposed an iterative solution to (3.22) with a preliminary consistent estimator as starting value; this estimator is also available in the R package *lss*.

3.3 Quantile regression for survival analysis

3.3.1 Introduction

In survival analysis, quantile regression (Koenker and Bassett, 1978) offers a significant extension of the accelerated failure time (AFT) model. Let T denote the failure time of interest and $\boldsymbol{Z} \equiv (Z_1, \ldots, Z_p)^{\top}$ denote a $p \times 1$ vector of covariates. Define $\tilde{\boldsymbol{Z}} = (1, \boldsymbol{Z}^{\top})^{\top}$ and $V = \log(T)$. For the response V, the τ-th conditional quantiles of V given $\tilde{\boldsymbol{Z}}$ is defined as $Q_V(\tau|\tilde{\boldsymbol{Z}}) \equiv \inf\{t : \Pr(V \leq t|\tilde{\boldsymbol{Z}}) \geq \tau\}$, where $\tau \in (0,1)$. We adopt the same quantile definition for other response random variables. With survival data, a common quantile regression modeling strategy is to link the conditional quantile $Q_V(\tau|\tilde{\boldsymbol{Z}})$ to $\tilde{\boldsymbol{Z}}$ through a linear model:

$$Q_V(\tau|\tilde{\boldsymbol{Z}}) = \tilde{\boldsymbol{Z}}^{\top} \boldsymbol{\beta}_0(\tau), \quad \tau \in [\tau_L, \tau_U], \tag{3.24}$$

where $\boldsymbol{\beta}_0(\tau)$ is a $(p+1) \times 1$ vector of unknown regression coefficients possibly depending on τ, and $0 \leq \tau_L \leq \tau_U \leq 1$. Like the AFT model, the quantile regression model (3.24) offers an easy coefficient interpretation as a covariate effect on event time. Specifically, a non-intercept coefficient in $\boldsymbol{\beta}_0(\tau)$ represents the change in the τth quantile of $\log(T)$ given one unit change in the corresponding covariate.

It is important to note that the AFT model,

$$\log(T) = \boldsymbol{Z}^{\top} \boldsymbol{b} + \varepsilon, \quad \varepsilon \perp \boldsymbol{Z}, \tag{3.25}$$

is a special case of model (3.24) with $\boldsymbol{\beta}_0(\tau) = (Q_{\varepsilon}(\tau), \boldsymbol{b}^{\top})^{\top}$. Here $\perp$ stands for independence, and $Q_{\varepsilon}(\tau)$ represents the τth quantile of ε. From this view, we see that the AFT model (3.25) requires the effects of covariates on $Q_V(\tau|\tilde{\boldsymbol{Z}})$ be constant for all $\tau \in [\tau_L, \tau_U]$ (i.e., location shift effects). In contrast, the quantile regression model (3.24) flexibly formulates coefficients as functions of τ, thereby permitting covariate effects to vary across different quantile levels. As a result, it can accommodate a more general relationship between T and $\boldsymbol{Z}$ compared to the AFT model (3.25).

Note, when $\tau_L = \tau_U$, model (3.24) would only assert "local" linearity between the quantile of $\log(T)$ and $\tilde{\boldsymbol{Z}}$ at a single quantile level, and therefore imposes weaker restrictions than a "global" quantile regression model corresponding to a case with $\tau_L < \tau_U$. A practical advantage of conducting "global" quantile regression is the capability of investigating the dynamic pattern of covariate effects.

In this section, we shall primarily focus on quantile regression with randomly right-censored data. Let C denote time to censoring. Define $\tilde{T} = T \wedge C$ and $\Delta = I(T \leq C)$. The observed data under right censorship consists of n i.i.d. replicates of $(\tilde{T}, \Delta, \tilde{\boldsymbol{Z}})$, denoted by $(\tilde{T}_i, \Delta_i, \tilde{\boldsymbol{Z}}_i)$, $i = 1, \ldots, n$. In addition, we define $\tilde{V} = \log(\tilde{T})$, $\tilde{V}_i = \log(\tilde{T}_i)$, $U = \log(C)$, and $U_i = \log(C_i)$. In Sections 3.3.2, 3.3.3 and 3.3.4, we consider three different random censoring scenarios, with general inference procedures discussed in Section 3.3.5. Extensions to more complex survival settings are briefly described in Section 3.3.6.

3.3.2 Estimation under random right censoring with C always known

Early work on quantile regression with censored data by Powell (1984, 1986) was targeted at type I censoring cases where the censoring time C is fixed and prespecified. Motivated by the fact that $Q_{\tilde{V}}(\tau|\tilde{\boldsymbol{Z}}) = \{\tilde{\boldsymbol{Z}}^{\intercal}\boldsymbol{\beta}_0(\tau)\} \wedge U$, an estimator of $\boldsymbol{\beta}_0(\tau)$ is defined as the minimizer of

$$r(\boldsymbol{b}, \tau) = \sum_{i=1}^{n} \rho_\tau\{\tilde{V}_i - (\tilde{\boldsymbol{Z}}_i^{\intercal}\boldsymbol{b}) \wedge U_i\}$$

with respect to $\boldsymbol{b}$, where $\rho_\tau(x) = x\{\tau - I(x < 0)\}$. This estimation method can also be applied to a more general case where C is independent of T given $\tilde{\boldsymbol{Z}}$, and is always known but not necessarily fixed. In the absence of right censoring (i.e., $C_i = \infty$), $r(\boldsymbol{b}, \tau)$ reduces to the check function of Koenker and Bassett (1978), the objective function for defining sample regression quantiles with complete data without censoring. Note that $r(\boldsymbol{b}, \tau)$ is not convex in $\boldsymbol{b}$ and thus it may have multiple local minima. Further efforts have been made to improve the numerical performance of this approach by several authors, for example: Fitzenberger (1997); Buchinsky and Hahn (1998); and Chernozhukov and Hong (2001). An implementation of Powell's method is available in the *crq* function in the contributed R package *quantreg* by Koenker (2012).

3.3.3 Estimation under covariate-independent random right censoring

Under the assumption that T and C are independent and C is independent of $\tilde{\boldsymbol{Z}}$ (i.e., covariate-independent random censoring), a natural estimating equation for $\boldsymbol{\beta}_0(\tau)$, derived from Ying et al. (1995)'s work, is given by

$$n^{-1/2} \sum_{i=1}^{n} \tilde{\boldsymbol{Z}}_i \left[\frac{I\{\tilde{V}_i - \tilde{\boldsymbol{Z}}^{\intercal}\boldsymbol{\beta}(\tau) > 0\}}{\widehat{G}\{\tilde{\boldsymbol{Z}}^{\intercal}\boldsymbol{\beta}(\tau)\}} - (1 - \tau) \right] = 0, \tag{3.26}$$

where $\widehat{G}(\cdot)$ is the Kaplan-Meier estimate for $G(\cdot)$, the survival function of $\tilde{C}$. The estimating function, the left-hand side (LHS) of (3.26), is not continuous in $\boldsymbol{\beta}(\tau)$, and thus an exact zero-crossing may not exist. The solution to Equation (3.26) may be alternatively defined as a minimizer of the L_2 norm of the estimating function. Such an objective function is discontinuous and may have multiple minima. To solve Equation (3.26), as suggested by Ying et al. (1995), the grid search method may be used for cases with low-dimensional $\tilde{\boldsymbol{Z}}$ and simulated annealing algorithm (Lin and Geyer, 1992) may be used for high-dimensional cases.

For the same right censoring scenario, an alternative estimating equation for $\boldsymbol{\beta}_0(\tau)$ is suggested by the work of Peng and Fine (2009). Specifically, the inverse probability of censoring weighting (IPCW) technique (Robins and Rotnitzky, 1992) can be used to estimate $\beta_0(\tau)$ based on the fact that

$$E\left\{\frac{I(\tilde{V} \leq t, \Delta = 1)}{G(\tilde{V})}|\tilde{\boldsymbol{Z}}\right\} = \Pr(V \leq t|\tilde{\boldsymbol{Z}}).$$

This leads to the following estimating equation for $\beta_0(\tau)$:

$$n^{-1/2}\sum_{i=1}^{n}\tilde{\boldsymbol{Z}}_i\left[\frac{I\{\tilde{V}_i \leq \tilde{\boldsymbol{Z}}_i^{\mathsf{T}}\beta(\tau), \Delta_i = 1\}}{\widehat{G}(\tilde{V}_i)} - \tau\right] = 0. \tag{3.27}$$

Note, the estimating function in (3.27) is not continuous but monotone (Fygenson and Ritov, 1994). By the monotonicity of (3.27), the solution to Equation (3.27) can be reformulated as the minimizer of the L_1-type convex function (of $\boldsymbol{b}$),

$$\sum_{i=1}^{n}\left\{I(\Delta_i = 1)\left|\frac{\tilde{V}_i}{\widehat{G}(\tilde{V}_i)} - \boldsymbol{b}^{\mathsf{T}}\frac{\tilde{\boldsymbol{Z}}_i}{\widehat{G}(\tilde{V}_i)}\right|\right\} + \left|M - \boldsymbol{b}^{\mathsf{T}}\sum_{l=1}^{n}\left(-\frac{\tilde{\boldsymbol{Z}}_l I(\Delta_l = 1)}{\widehat{G}(\tilde{V}_l)} + 2\tilde{\boldsymbol{Z}}_l\tau\right)\right|,$$

where M is an extremely large positive number selected to bound $\boldsymbol{b}^{\mathsf{T}}\sum_{l=1}^{n}\left(-\frac{Z_l I(\Delta_l=1)}{\widehat{G}(\tilde{V}_l)} + 2Z_l\tau\right)$ from the above for all $\boldsymbol{b}$'s in the compact parameter space for $\beta_0(\tau)$. This minimization problem can be readily solved by the *l1fit* function in S-PLUS or the *rq()* function in the contributed R package *quantreg* (Koenker, 2012).

3.3.4 Estimation under standard random right censoring

Standard random right censoring refers to a censoring mechanism where C is only assumed to be independent of T given $\tilde{\boldsymbol{Z}}$. It is less restrictive than the censorship considered in Sections 3.3.2 and 3.3.3.

Two major types of approaches have been developed for quantile regression with survival data subject to standard random right censoring; one type employs the principle of self-consistency (Efron, 1967) and the other type utilizes the martingale structure associated with randomly right-censored data. Hereafter, we shall refer to them as a self-consistent approach and martingale-based approach, respectively. Here we focus on approaches that are oriented to the following global linear quantile regression model,

$$Q_V(\tau|\tilde{\boldsymbol{Z}}) = \tilde{\boldsymbol{Z}}^{\mathsf{T}}\beta_0(\tau), \quad \tau \in [0, \tau_U], \tag{3.28}$$

which is a special case of model (3.24) with $\tau_L = 0$. Of note, model (3.28) entails $\exp\{\tilde{\boldsymbol{Z}}^{\mathsf{T}}\beta_0(0)\} = 0$, which is useful boundary information to be employed in the estimation of $\beta_0(\tau)$ with $\tau > 0$. Some efforts, for example, by Wang and Wang (2009), have been made to address a local linear quantile regression model (i.e., $\tau_L = \tau_U$) by incorporating nonparametric estimation of the distribution function of C given $\tilde{\boldsymbol{Z}}$.

3.3.4.1 Self-consistent approach

Portnoy (2003) made the first attempt to tackle the estimation of model (3.28) under the standard random right censoring assumption by employing the principle of self-consistency (Efron, 1967). The critical idea behind his algorithm is about splitting a censored event time, $\tilde{V}_i$ with $\Delta_i = 0$, between U_i and ∞ with appropriately designed weights, sharing the same spirit as that adopted for Efron's self-consistent estimator of survival function (Efron, 1967). The initial iterative self-consistent algorithm (Portnoy, 2003) was simplified into a grid-based sequential estimation procedure (Neocleous et al., 2006), which is implemented by the *crq* function in the contributed R package *quantreg* (Koenker, 2012). The corresponding asymptotic studies were established by Portnoy and Lin (2010).

The self-consistent approach can be formulated through stochastic integral equations (Peng, 2012). Define $N_i(t) = I(\tilde{V}_i \leq t, \Delta_i = 1)$, $R_i(t) = I(\tilde{V}_i \leq t,\ \Delta_i = 0)$, and $F_V(t) \equiv \Pr(V \leq t)$. First, consider Efron's (Efron, 1967) self-consistent estimating equation for $F_V(t)$ in the one-sample case:

$$F_V(t) = n^{-1}\sum_{i=1}^{n}\left\{N_i(t) + R_i(t)\frac{F_V(t) - F_V(\tilde{V}_i)}{1 - F_V(\tilde{V}_i)}\right\}. \tag{3.29}$$

Expressing the right-hand side (RHS) of (3.29) by a stochastic integral and further applying stochastic integral by parts, one can rewrite Equation (3.29) as

$$F_V(t) = n^{-1}\sum_{i=1}^{n}\left[N_i(t) + R_i(t)\{1 - F_V(t)\}\int_0^t \frac{R_i(u)}{\{1 - F_V(u)\}^2}dF_V(u)\right]. \tag{3.30}$$

With t replaced by $\tilde{\boldsymbol{Z}}_i^{\intercal}\boldsymbol{\beta}(\tau)$, Equation (3.30) evolves into an estimating equation for $\boldsymbol{\beta}_0(\tau)$:

$$n^{1/2}\boldsymbol{S}_n^{(SC)}(\boldsymbol{\beta}, \tau) = 0, \tag{3.31}$$

where $\boldsymbol{S}_n^{(SC)}(\boldsymbol{\beta}, \tau)$ equals

$$n^{-1}\sum_{i=1}^{n}\tilde{\boldsymbol{Z}}_i\left[N_i\{\tilde{\boldsymbol{Z}}_i^{\intercal}\boldsymbol{\beta}(\tau)\} + R_i\{\tilde{\boldsymbol{Z}}_i^{\intercal}\boldsymbol{\beta}(\tau)\}(1-\tau)\int_0^{\tau}\frac{R_i\{\tilde{\boldsymbol{Z}}_i^{\intercal}\boldsymbol{\beta}(u)\}}{(1-u)^2}du - \tau\right].$$

Define a grid of τ-values, $\mathcal{G}$, as $0 < \tau_1 < \tau_2 < \ldots < \tau_M = \tau_U$. Let $\|\mathcal{G}\|$ denote the size of $\mathcal{G}$, $\max_{k=0,\ldots,M}(\tau_{k+1} - \tau_k)$. Without further mentioning, $\mathcal{G}$ will be adopted throughout Section 3.3.4. A self-consistent estimator, $\widehat{\boldsymbol{\beta}}_{SC}(\cdot)$, is defined as a cadlag step function that only jumps at the grid points of $\mathcal{G}$ and approximates the solution to Equation (3.31). The algorithm taken to obtain $\widehat{\boldsymbol{\beta}}_{SC}(\cdot)$ is outlined as follows.

1. Set $\exp\{\tilde{\boldsymbol{Z}}_i\widehat{\boldsymbol{\beta}}_{SC}(0)\} = 0$ for all i. Set $k = 0$.

2. Given $\exp\{\tilde{\boldsymbol{Z}}_i\widehat{\boldsymbol{\beta}}_{SC}(\tau_l)\}$ for $l \leq k$, obtain $\widehat{\boldsymbol{\beta}}_{SC}(\tau_{k+1})$ as the minimizer of the following weighted check function:

$$\left[\sum_{\Delta_i=1}\rho_\tau(\tilde{V}_i - \tilde{\boldsymbol{Z}}_i^{\intercal}\boldsymbol{b}) + \sum_{\Delta_i=0}\{\tilde{w}_{k+1,i}\rho_\tau(\tilde{V}_i - \tilde{\boldsymbol{Z}}_i^{\intercal}\boldsymbol{b}) \right.$$
$$\left. +(1 - \tilde{w}_{k+1,i})\rho_\tau(Y^* - \tilde{\boldsymbol{Z}}_i^{\intercal}\boldsymbol{b})\}\right], \tag{3.32}$$

   where $\tilde{w}_{k+1,i} = \sum_{l=0}^{k} R_i\{\tilde{\boldsymbol{Z}}_i^{\intercal}\widehat{\boldsymbol{\beta}}_{SC}(\tau_l)\}\left(\frac{1-\tau_{k+1}}{1-\tau_{l+1}} - \frac{1-\tau_{k+1}}{1-\tau_l}\right)$, and Y^* is an extremely large value.

3. Replace k by $k+1$ and repeat step 2 until $k = M$ or only censored observations remain above $\exp\{\tilde{\boldsymbol{Z}}_i^{\intercal}\widehat{\boldsymbol{\beta}}_{SC}(\tau_{k-1})\}$.

Large sample studies for $\widehat{\boldsymbol{\beta}}_{SC}(\cdot)$ are facilitated by the stochastic integral equation representation of (3.31). Specifically, under certain regularity conditions and given $\lim_{n\to\infty}\|\mathcal{G}\| = 0$, $\sup_{\tau\in[\nu,\tau_U]}\|\widehat{\boldsymbol{\beta}}_{SC}(\tau) - \boldsymbol{\beta}_0(\tau)\| \to_p 0$, where $0 < \nu < \tau_U$. If $n^{1/2}\lim_{n\to\infty}\|\mathcal{G}\| = 0$ is further satisfied, then $n^{1/2}\{\widehat{\boldsymbol{\beta}}_{SC}(\tau) - \boldsymbol{\beta}_0(\tau)\}$ converges weakly to a Gaussian process for $\tau \in [\nu, \tau_U]$. Peng (2012) also investigated several variants of $\widehat{\boldsymbol{\beta}}_{SC}(\cdot)(\cdot)$, showing their asymptotic equivalence to $\widehat{\boldsymbol{\beta}}_{SC}$ as well as their connection with the self-consistent estimator proposed by Portnoy (Portnoy, 2003; Neocleous et al., 2006).

3.3.4.2 Martingale-based approach

Model (3.28) can also be estimated based on the martingale structure of randomly right-censored data (Peng and Huang, 2008). Define $\Lambda_V(t|\tilde{\boldsymbol{Z}}) = -\log\{1 - \Pr(V \le t|\tilde{\boldsymbol{Z}})\}$, $N(t) = I(\tilde{V} \le t, \Delta = 1)$, and $M(t) = N(t) - \Lambda_V(t \wedge Y|\tilde{\boldsymbol{Z}})$. Let $N_i(t)$ and $M_i(t)$ be sample analogs of $N(t)$ and $M(t)$, respectively, $i = 1, \ldots, n$. Since $M_i(t)$ is the martingale process associated with the counting process $N_i(t)$, $E\{M_i(t)|\tilde{\boldsymbol{Z}}_i\} = 0$ for all $t > 0$. This implies

$$E\left\{\sum_{i=1}^{n} \tilde{\boldsymbol{Z}}_i \left[N_i\{\tilde{\boldsymbol{Z}}_i^\intercal \boldsymbol{\beta}_0(\tau)\} - \Lambda_V\{\tilde{\boldsymbol{Z}}_i^\intercal \boldsymbol{\beta}_0(\tau) \wedge \tilde{V}_i|\tilde{\boldsymbol{Z}}_i\}\right]\right\} = 0. \tag{3.33}$$

By the monotonicity of $\tilde{\boldsymbol{Z}}_i^\intercal \boldsymbol{\beta}_0(\tau)$ in $\tau \in [0, \tau_U]$ under model (3.28), we have

$$\Lambda_V\{\tilde{\boldsymbol{Z}}_i^\intercal \boldsymbol{\beta}_0(\tau) \wedge \tilde{V}_i|\tilde{\boldsymbol{Z}}_i\} = \int_0^\tau I\{\tilde{V}_i \ge \tilde{\boldsymbol{Z}}_i^\intercal \boldsymbol{\beta}_0(u)\} dH(u), \tag{3.34}$$

where $H(x) = -\log(1-x)$. Coupling the equalities (3.33) and (3.34) suggests the estimating equation,

$$n^{1/2} \boldsymbol{S}_n^{(PH)}(\boldsymbol{\beta}, \tau) = 0, \tag{3.35}$$

where

$$\boldsymbol{S}_n^{(PH)}(\boldsymbol{\beta}, \tau) = n^{-1} \sum_{i=1}^{n} \tilde{\boldsymbol{Z}}_i \left[N_i\{\tilde{\boldsymbol{Z}}_i^\intercal \boldsymbol{\beta}(\tau)\} - \int_0^\tau I\{\tilde{V}_i \ge \tilde{\boldsymbol{Z}}_i^\intercal \boldsymbol{\beta}(u)\} dH(u)\right].$$

An estimator of $\boldsymbol{\beta}_0(\tau)$, denoted by $\widehat{\boldsymbol{\beta}}_{PH}(\tau)$, can be obtained through approximating the stochastic solution to Equation (3.35) by a cadlag step function. The sequential algorithm for obtaining $\widehat{\boldsymbol{\beta}}_{PH}(\tau)$ is outlined as follows:

1. Set $\exp\{\tilde{\boldsymbol{Z}}_i^\intercal \widehat{\boldsymbol{\beta}}_{PH}(\tau_0)\} = 0$ for all i. Set $k = 0$.

2. Given $\exp\{\tilde{\boldsymbol{Z}}_i^\intercal \widehat{\boldsymbol{\beta}}_{PH}(\tau_l)\}$ for $l \le k$, obtain $\widehat{\boldsymbol{\beta}}_{PH}(\tau_{k+1})$ as the minimizer of the following L_1-type convex objective function:

$$l_{k+1}(\boldsymbol{h}) = \sum_{i=1}^{n} \left|\Delta_i \tilde{V}_i - \delta_i \tilde{\boldsymbol{Z}}_i^\intercal \boldsymbol{h}\right| + \left|Y^* - \sum_{l=1}^{n} (-\Delta_l \tilde{\boldsymbol{Z}}_l^\intercal \boldsymbol{h})\right|$$
$$+ \left|Y^* - \sum_{r=1}^{n} \left[(2\tilde{\boldsymbol{Z}}_r^\intercal \boldsymbol{h}) \sum_{l=0}^{k} I\{\tilde{V}_r \ge \tilde{\boldsymbol{Z}}_r^\intercal \widehat{\boldsymbol{\beta}}_{PH}(\tau_l)\}\{H(\tau_{l+1}) - H(\tau_l)\}\right]\right|,$$

   where Y^* is an extremely large value.

3. Replace k by $k + 1$ and repeat step 2 until $k = M$ or no feasible solution can be found for minimizing $l_k(\boldsymbol{h})$.

The *crq* function in the contributed R package *quantreg* (Koenker, 2012) provides an implementation of $\widehat{\boldsymbol{\beta}}_{PH}(\tau)$ based on an algorithm slightly different from the one presented above. More recently, Huang (2010) derived a grid-free estimation procedure for model (3.28) by using the concept of quantile calculus.

Peng and Huang (2008) established the uniform consistency and weak convergence of $\widehat{\boldsymbol{\beta}}_{PH}(\cdot)$. Moreover, $\widehat{\boldsymbol{\beta}}_{PH}(\cdot)$ was shown to be asymptotically equivalent to the self-consistent estimator $\widehat{\boldsymbol{\beta}}_{SC}(\cdot)$ (Peng, 2012). The numerical results reported in Koenker (2008) and Peng (2012) confirm this theoretical result and show comparable computational performance between these two approaches.

3.3.5 Variance estimation and other inference

The estimators of $\beta_0(\tau)$ discussed in Sections 3.3.2–3.3.4 generally have asymptotic variances that involve unknown density functions. Under random right censoring with known censoring time or covariate-independent censoring, we can adapt Huang (2002)'s technique to avoid density estimation. Specifically, let $\widehat{\beta}(\tau)$ be general notation for an estimator of $\beta_0(\tau)$, and $\boldsymbol{S}_n(\beta(\tau),\tau)$ denote the estimating function associated with $\widehat{\beta}(\tau)$, for example, the LHS of (3.26) and (3.27). Asymptotic theory may show that $\boldsymbol{S}_n\{\beta_0(\tau),\tau\}$ converges to a multivariate normal distribution with variance matrix $\boldsymbol{\Sigma}(\tau)$. Suppose one can obtain a consistent estimator of $\boldsymbol{\Sigma}(\tau)$, denoted by $\widehat{\boldsymbol{\Sigma}}(\tau)$. The following are the main steps to estimate the asymptotic variance of $\widehat{\beta}(\tau)$:

1. Find a symmetric and nonsingular $(p+1)\times(p+1)$ matrix $\boldsymbol{E}_n(\tau) \equiv \{\boldsymbol{e}_{n,1}(\tau),\ldots,\boldsymbol{e}_{n,p+1}(\tau)\}$ such that $\widehat{\boldsymbol{\Sigma}}(\tau) = \{\boldsymbol{E}_n(\tau)\}^2$.

2. Calculate $\boldsymbol{D}_n(\tau) = \left(\boldsymbol{S}_n^{-1}\{\boldsymbol{e}_{n,1}(\tau),\tau\} - \widehat{\beta}(\tau),\ldots,\boldsymbol{S}_n^{-1}\{\boldsymbol{e}_{n,p+1}(\tau),\tau\} - \widehat{\beta}(\tau)\right)$, where $\boldsymbol{S}_n^{-1}(\boldsymbol{e},\tau)$ is defined as the solution to $\boldsymbol{S}_n(\boldsymbol{b},\tau) - \boldsymbol{e} = 0$.

3. Estimate the asymptotic variance matrix of $n^{1/2}\{\widehat{\beta}(\tau) - \beta_0(\tau)\}$ by $n\{\boldsymbol{D}_n(\tau)\}^{\otimes 2}$.

Bootstrapping procedures are also frequently used for variance estimation under quantile regression. Resampling methods that follow the idea of Parzen and Ying (1994) were shown to yield valid variance estimates and other inference (Peng and Huang, 2008). Simple bootstrapping procedures based on resampling with replacement also seem to have satisfactory performance (Portnoy, 2003; Peng, 2012).

One practical appeal of quantile regression with survival data is its capability of accommodating and exploring varying covariate effects. Second-stage inference can be employed to serve this need. Given $\widehat{\beta}(\tau)$ for a range of τ, it is often of interest to investigate: (1) how to summarize the information provided by these estimators to help understand the underlying effect mechanism; and (2) how to determine whether some covariates have constant effects so that a simpler model may be considered.

A general formulation of problem (1) may correspond to the estimation of some functional of $\beta_0(\cdot)$, denoted by $\boldsymbol{\Psi}(\beta_0)$. A natural estimator for $\boldsymbol{\Psi}(\beta)$ is given by $\boldsymbol{\Psi}(\widehat{\beta}_0)$. Such an estimator may be justified by the functional delta method provided that $\boldsymbol{\Psi}$ is compactly differentiable at β_0 (Andersen et al., 1998).

Addressing question (2) can be formulated as a testing problem for the null hypothesis $\tilde{H}_{0,j}:\ \beta_0^{(j)}(\tau) = \rho_0,\ \tau \in [\tau_L,\tau_U]$, where the superscript ${}^{(j)}$ indicates the jth component of a vector, and ρ_0 is an unspecified constant, $j = 2,\ldots,$ or $p+1$. An example test procedure is presented in Peng and Huang (2008). Of note, accepting $\tilde{H}_{0,j}$ for all $j \in \{2,\ldots,p+1\}$ may indicate the adequacy of an AFT model. This naturally renders a procedure for testing the goodness-of-fit of an AFT model.

Model checking is often of practical importance. When the interest only lies in checking the linearity between covariates and conditional quantile at a single quantile level, one can adapt the approaches developed for uncensored data, for example, the work by Zheng (2000), Horowitz and Spokoiny (2002), and He and Zhu (2003). When the focus is to test a global linear relationship between conditional quantiles and covariates, a natural approach is to use a stochastic process which has mean zero under the assumed model. For example, a diagnostic process for model (3.28) (Peng and Huang, 2008) may take the form,

$$K_n(\tau) = n^{-1/2}\sum_{i=1}^{n} q(\tilde{\boldsymbol{Z}}_i)M_i(\tau;\widehat{\beta}),$$

where $q(\cdot)$ is a known bounded function, and

$$M_i(\tau; \boldsymbol{\beta}) = N_i\big(\exp\{\tilde{\boldsymbol{Z}}_i^{\intercal}\boldsymbol{\beta}(\tau)\}\big) - \int_0^{\tau} I\{\tilde{V}_i \geq \tilde{\boldsymbol{Z}}_i^{\intercal}\boldsymbol{\beta}(u)\} dH(u).$$

It can be shown that $K(\tau)$ converges weakly to a zero-mean Gaussian process, whose distribution can be approximated by that of

$$K^*(\tau) = n^{-1/2}\sum_{i=1}^{n} q(\tilde{\boldsymbol{Z}}_i) M_i(\tau; \hat{\boldsymbol{\beta}})(1-\zeta_i) + n^{-1/2}\sum_{i=1}^{n} q(\tilde{\boldsymbol{Z}}_i)\{M_i(\tau; \boldsymbol{\beta}^*) - M_i(\tau; \hat{\boldsymbol{\beta}})\}.$$

Here $\boldsymbol{\beta}^*(\tau)$ denote the resampling estimator obtained by perturbing the estimating equation (3.35) by $\{\zeta_i\}_{i=1}^n$, which are independent variates from a nonnegative known distribution with mean 1 and variance 1. An unusual pattern of $K(\cdot)$ compared to that of $K^*(\cdot)$ would suggest a lack-of-fit of model (3.28). A formal lack of fit test is given by the supremum statistic, $\sup_{\tau \in [l,u]} |K(\tau)|$, where $0 < l < u < \tau_U$. The p-value may be approximated by the empirical proportion of $\sup_{\tau \in [l,u]} |K^*(\tau)|$ exceeding $\sup_{\tau \in [l,u]} |K(\tau)|$.

3.3.6 Extensions to other survival settings

In practice, survival data may involve more complex censoring mechanism than random right censoring. Truncation can also present, for example, in many observational studies. There are recent method developments for quantile regression in more complex survival scenarios. For example, Ji et al. (2012) proposed a modification of Peng and Huang (2008)'s martingale-based approach for survival data subject to known random left censoring and/or random left truncation in addition to random right censoring. Quantile regression with competing risks or semicompeting risks data was addressed by the work of Peng and Fine (2009) and Li and Peng (2011).

There are continued research efforts to address quantile regression for other important survival scenarios, such as interval censored data and dependently censored data. As a promising regression tool for survival analysis, quantile regression may be recommended in a greater extent in real applications to provide complementary and yet useful secondary analysis.

3.3.7 An illustration of quantile regression for survival analysis

To illustrate quantile regression for survival analysis, we use a dataset from a dialysis study that investigated predictors of mortality risk in a cohort of 191 incident dialysis patients (Kutner et al., 2002). Analysis covariates included patient's age (AGE), the indicator of fish consumption over the first year of dialysis (FISHH), the indicator of baseline HD dialysis modality (BHDPD), the indicator of moderate to severe symptoms of restless leg symptoms (BLEGS), the indicator of education equal or higher than college (HIEDU), and the indicator of being black (BLACK). We first fit the data with AFT model (3.25), where T stands for time-to-death. In Table 3.1, we present the estimation results based on the log-rank estimator, Gehan's estimator, and the least-squares estimator. All covariates except for BLEGS are consistently shown to have significant effects on survival by all three different estimators. For the coefficient of BLEGS, Gehan's estimator and the least-squares estimator yield significant p values while the log-rank estimator does not.

We next conduct quantile regression based on model (3.28) for the same dataset. Figure 3.3 displays Peng and Huang (2008)'s estimator of $\beta_0(\tau)$ along with 95% pointwise confidence intervals. In Figure 3.3, we observe that the coefficient for BLEGS diminishes

TABLE 3.1
Results from fitting AFT model to the dialysis dataset. Coef: coefficient estimate; SE: standard error.

	Gehan's estimator			Log-rank estimator			Least-squares estimator		
	Coef	SE	p value	Coef	SE	p value	Coef	SE	p value
AGE	−.031	.005	< .001	−.035	.004	<.001	.033	.006	< .001
FISHH	.402	.139	.004	.485	.128	<.001	.507	.163	.002
BHDPD	−.505	.156	.001	−.473	.136	< .001	−.509	.164	.002
BLEGS	.340	.166	.040	.173	.145	.232	−.412	.176	.019
HIEDU	−.352	.133	.008	−.364	.161	.024	.335	.139	.016
BLACK	.640	.144	<.001	.591	.138	< .001	.643	.153	< .001

gradually with τ whereas estimates for the other coefficients seem to be fairly constant. We apply the second-stage inference to formally investigate the constancy of each coefficient. The results confirm our observation from Figure 3.3, suggesting a varying effect of BLEGS and constant effects of the other covariates. This may lead to an interesting scientific implication that BLEGS may affect the survival experience of dialysis patients with short survival times, but may have little impact on that of long-term survivors. In addition, the evidence for the nonconstancy of BLEGS coefficient may indicate some degree of lack-of-fit of the AFT model for the dialysis data.

We further estimate the average quantile effects defined as $\int_l^u \beta_0^{(i)}(u)du$ $(i = 2, \ldots, 7)$. The results are given in Table 3.2. We observe that the estimated average effects are similar to the AFT coefficients obtained by Gehan's estimator and the least-squares estimator, but have relatively larger discrepancies with the log-rank estimates. This may indirectly reflect the presence of some nonconstant covariate effect. When quantile regression model (3.28) holds without any varying covariate effects, we would expect to see more consistent results between Gehan's estimator, which emphasizes early survival, and the log-rank estimator, which treats early and late survival information equally.

TABLE 3.2
Estimation of average covariate effects based on quantile regression. AveEff: Estimated average effect; SE: standard error.

	AveEff	SE	p value
AGE	−.030	.003	< .001
FISHH	.327	.116	.005
BHDPD	−.489	.162	.003
BLEGS	−.369	.161	.022
HIEDU	−.350	.137	.011
BLACK	.654	.144	< .001

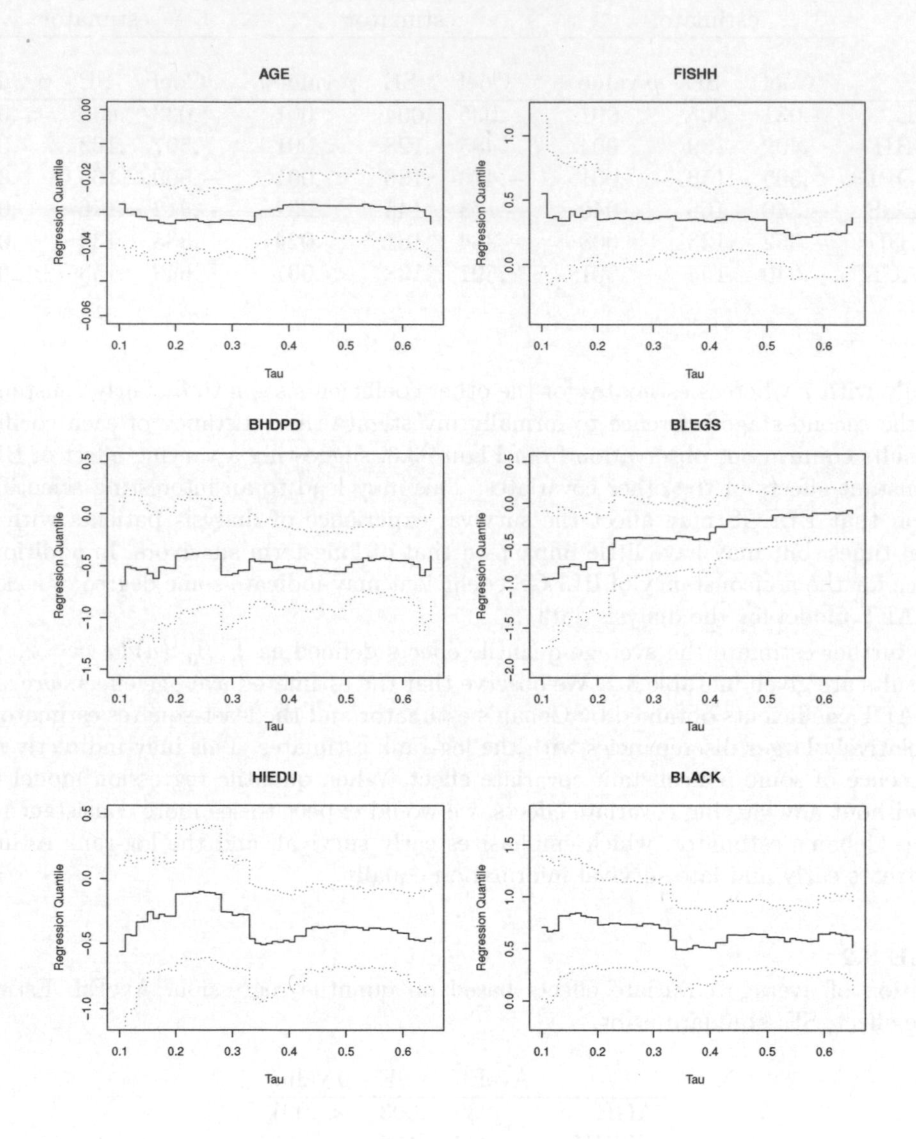

FIGURE 3.3
Peng and Huang's estimator (solid lines) and 95% pointwise confidence intervals (dotted lines) of regression quantiles in the dialysis example.

Bibliography

Aalen, O. O. (1980), A model for non-parametric regression analysis of counting processes, *in* W. Klonecki, A. Kozek and J. Rosinski, eds., 'Lecture Notes in Statistics-2: Mathematical Statistics and Probability Theory', New York: Springer-Verlag, pp. 1–25.

Aalen, O. O. (1989), 'A linear regression model for the analysis of life times', *Statist. Med.* **8**, 907–925.

Aalen, O. O. (1993), 'Further results on the non-parametric linear regression model in survival analysis', *Statist. Med.* **12**, 1569–1588.

Aalen, O. O., Borgan, Ø. and Gjessing, H. K. (2008), *Survival and event history analysis*, Statistics for Biology and Health, Springer, New York. A process point of view.
URL: http://dx.doi.org/10.1007/978-0-387-68560-1

Andersen, P. K., Borgan, Ø., Gill, R. D. and Keiding, N. (1998), *Statistical Models Based on Counting Processes*, 2nd edn, New York: Springer-Verlag.

Bovelstad, H., Nygard, S. and Borgan, Ø. (2009), 'Survival prediction from clinico-genomic models - a comparative study', *BMC Bioinformatics* **10**(1), 413.

Buchinsky, M. and Hahn, J. (1998), 'An alternative estimator for censored quantile regression', *Econometrica* **66**, 653–671.

Buckley, J. and James, I. R. (1979), 'Linear regression with censored data', *Biometrika* **66**, 429–436.

Cai, J. and Zeng, D. (2011), 'Additive mixed effect model for clustered failure time data', *Biometrics* **67**(4), 1340–1351.

Chen, Y. and Jewell, N. (2001), 'On a general class of semiparametric hazards regression models', *Biometrika* **88**, 687–702.

Chen, Y. Q. and Wang, M.-C. (2000), 'Analysis of accelerated hazards models', *J. Amer. Statist. Assoc.* **95**(450), 608–618.

Chernozhukov, V. and Hong, H. (2001), 'Three-step censored quantile regression and extramarital affairs', *J. Amer. Statist. Assoc.* pp. 872–882.

Efron, B. (1967), 'The two-sample problem with censored data', *Proc. Fifth Berkley Symposium in Mathematical Statistics, IV*, 831–553.

Fitzenberger, B. (1997), 'A guide to censored quantile regressions', *Handbooks of Statistics: Robust Inference* **15**, 405–437.

Fosen, J., Ferkingstad, E., Borgan, Ø. and Aalen, O. (2006), 'Dynamic path analysis: a new approach to analyzing time-dependent covariates', *Lifetime Data Anal.* **12**(2), 143–167.

Fygenson, M. and Ritov, Y. (1994), 'Monotone estimating equations for censored data', *Ann. Statist.* **22**, 732–746.

Gandy, A. and Jensen, U. (2005*a*), 'Checking a semiparametric additive risk model', *Lifetime Data Anal.* **11**, 451–472.

Gandy, A. and Jensen, U. (2005*b*), 'On goodness-of-fit tests for Aalen's additive risk model', *Scand. J. Statist.* **32**, 425–445.

Gorst-Rasmussen, A. and Scheike, T. (2012), 'Independent screening for single-index hazard rate models with ultra-high dimensional features', *J. Roy. Statist. Soc. Ser. B* **75**, 217–245.

Grønnesby, J. K. and Borgan, Ø. (1996), 'A method for checking regression models in survival analysis based on the risk score', *Lifetime Data Anal.* **2**, 315–328.

He, X. and Zhu, L.-X. (2003), 'A lack-of-fit test for quantile regression', *J. Amer. Statist. Assoc.* **98**, 1013–1022.

Horowitz, J. and Spokoiny, V. (2002), 'An adaptive, rate-optimal test of linearity for median regression models', *J. Amer. Statist. Assoc.* **97**, 822–835.

Huang, Y. (2002), 'Calibration regression of censored lifetime medical cost', *J. Amer. Statist. Assoc.* **98**, 318–327.

Huang, Y. (2010), 'Quantile Calculus and Censored Regression', *Ann. Statist.* **38**, 1607–1637.

Huffer, F. W. and McKeague, I. W. (1991), 'Weighted least squares estimation for Aalen's additive risk model', *J. Amer. Statist. Assoc.* **86**, 114–129.

Ji, S., Peng, L., Cheng, Y. and Lai, H. (2012), 'Quantile regression for doubly censored data', *Biometrics* **68**, 101–112.

Jin, Z., Lin, D. Y., Wei, L. J. and Ying, Z. (2003), 'Rank-based inference for the accelerated failure time model', *Biometrika* **90**, 341–353.

Jin, Z., Lin, D. Y. and Ying, Z. (2006), 'On least-squares regression with censored data', *Biometrika* **93**, 147–161.

Kalbfleisch, J. D. and Prentice, R. L. (2002), *The Statistical Analysis of Failure Time Data*, Wiley, New York.

Koenker, R. (2008), 'Censored quantile regression redux', *Journal of Statistical Software* **27**, http://www.jstatsoft.com.

Koenker, R. (2012), *Quantile Regression*, R package version 4.81, http://cran.r-project.org/web/packages/quantreg/quantreg.pdf.

Koenker, R. and Bassett, G. (1978), 'Regression quantiles', *Econometrica* **46**, 33–50.

Koul, H., Susarla, V. and Van Ryzin, J. (1981), 'Regression analysis with randomly right censored data', *Ann. Statist.* **9**, 1276–1288.

Kutner, N. G., Clow, P. W., Zhang, R. and Aviles, X. (2002), 'Association of fish intake and survival in a cohort of incident dialysis patients', *American Journal of Kidney Diseases* **39**, 1018–1024.

Lange, T. and Hansen, J. (2011), 'Direct and indirect effects in a survival context', *Epidemiology (Cambridge, Mass.)* **22**, 575–581.

Leng, C. and Ma, S. (2007), 'Path consistent model selection in additive risk model via lasso', *Statist. Med.* **26**, 3753–3770. **URL:** http://dx.doi.org/10.1002/sim.2834

Li, R. and Peng, L. (2011), 'Quantile regression for left-truncated semi-competing risks data', *Biometrics* **67**, 701–710.

Lin, D. and Geyer, C. (1992), 'Computational methods for semi-parametric linear regression with censored data', *Journal of Computational and Graphical Statistics* **1**, 77–90.

Lin, D. Y., Wei, L. J. and Ying, Z. (1993), 'Checking the Cox model with cumulative sums of martingale-based residuals', *Biometrika* **80**, 557–572.

Lin, D. Y., Wei, L. J. and Ying, Z. (1998), 'Accelerated failure time models for counting processes', *Biometrika* **85**, 605–618.

Lin, D. Y. and Ying, Z. (1994), 'Semiparametric analysis of the additive risk model', *Biometrika* **81**, 61–71.

Ma, S., Kosorok, M. R. and Fine, J. P. (2006), 'Additive risk models for survival data with high-dimensional covariates', *Biometrics* **62**, 202–210.

Martinussen, T. (2010), 'Dynamic path analysis for event time data: large sample properties and inference', *Lifetime Data Anal.* **16**, 85–101.

Martinussen, T., Aalen, O. O. and Scheike, T. H. (2008), 'The Mizon-Richard encompassing test for the Cox and Aalen additive hazards models', *Biometrics* **64**, 164–171. **URL:** http://www.jstor.org/stable/25502033

Martinussen, T. and Scheike, T. H. (2002), 'A flexible additive multiplicative hazard model', *Biometrika* **89**, 283–298. **URL:** http://www.jstor.org/stable/4140577

Martinussen, T. and Scheike, T. H. (2006), *Dynamic Regression Models for Survival Data*, Springer-Verlag, New York.

Martinussen, T. and Scheike, T. H. (2007), 'Aalen additive hazards change-point model', *Biometrika* **94**, 861–772.

Martinussen, T. and Scheike, T. H. (2009*a*), 'The additive hazards model with high-dimensional regressors', *Lifetime Data Anal.* **15**, 330–342.

Martinussen, T. and Scheike, T. H. (2009*b*), 'Covariate selection for the semiparametric additive risk model', *Scand. J. Statist.* **36**, 602–619.

Martinussen, T., Scheike, T. H. and Zucker, D. M. (2011), 'The Aalen additive gamma frailty hazards model', *Biometrika* **98**, 831–843. **URL:** http://biomet.oxfordjournals.org/content/early/2011/10/21/biomet.asr049.abstract

Martinussen, T. and Vansteelandt, S. (2012), A note on collapsibility and confounding bias in Cox and Aalen regression models, Technical report, Department of Biostatistics, University of Copenhagen.

Martinussen, T., Vansteelandt, S., Gerster, M. and Hjelmborg, J. V. (2011), 'Estimation of direct effects for survival data by using the Aalen additive hazards model', *J. Roy. Statist. Soc. Ser. B* **73**, 773–788. **URL:** http://dx.doi.org/10.1111/j.1467-9868.2011.00782.x

McKeague, I. W. and Sasieni, P. D. (1994), 'A partly parametric additive risk model', *Biometrika* **81**, 501–514.

Miller, R. G. (1976), 'Least squares regression with censored data', *Biometrika* **63**, 449–464.

Neocleous, T., Vanden Branden, K. and Portnoy, S. (2006), 'Correction to censored regression quantiles by S. Portnoy, 98 (2003), 1001–1012', *J. Amer. Statist. Assoc.* **101**, 860–861.

Parzen, M. I., Wei, L. J. and Ying, Z. (1994), 'A resampling method based on pivotal estimating functions', *Biometrika* **81**, 341–350.

Pearl, J. (2000), *Causality: Models, Reasoning, and Inference*, New York: Cambridge University Press.

Peng, L. (2012), 'A note on self-consistent estimation of censored regression quantiles', *J. Multivariate Analysis* **105**, 368–379.

Peng, L. and Fine, J. (2009), 'Competing risks quantile regression', *J. Amer. Statist. Assoc.* **104**, 1440–1453.

Peng, L. and Huang, Y. (2008), 'Survival analysis with quantile regression models', *J. Amer. Statist. Assoc.* **103**, 637–649.

Pipper, C. B. and Martinussen, T. (2004), 'An estimating equation for parametric shared frailty models with marginal additive hazards', *J. Roy. Statist. Soc. Ser. B* **66**, 207–220.

Portnoy, S. (2003), 'Censored regression quantiles', *J. Amer. Statist. Assoc.* **98**, 1001–1012.

Portnoy, S. and Lin, G. (2010), 'Asymptotics for censored regression quantiles.', *Journal of Nonparametric Statistics* **22**, 115–130.

Powell, J. (1984), 'Least absolute deviations estimation for the censored regression model', *Journal of Econometrics* **25**, 303–325.

Powell, J. (1986), 'Censored regression quantiles', *Journal of Econometrics* **32**, 143–155.

Reid, N. (1994), 'A conversation with Sir David Cox', *Statistical Science* **9**, 439–455.

Ritov, Y. (1990), 'Estimation in a linear regression model with censored data', *Ann. Statist.* **18**, 303–328.

Robins, J. (1986), 'A new approach to causal inference in mortality studies with sustained exposure periods - application to control of the healthy worker survivor effect', *Mathematical Modeling* **7**, 1393–1512.

Robins, J. and Rotnitzky, A. (1992), Recovery of information and adjustment for dependent censoring using surrogate markers, *in* N. Jewell, K. Dietz and V. Farewell, eds., 'AIDS Epidemiology-Methodological Issues', Boston: Birkhauser, 24–33.

Scheike, T. H. and Zhang, M.-J. (2002), 'An additive-multiplicative Cox-Aalen regression model', *Scand. J. Statist.* **29**, 75–88. **URL:** http://www.jstor.org/stable/4616700

Stablein, D. M. and Koutrouvelis, I. A. (1985), 'A two-sample test sensitive to crossing hazards in uncensored and singly censored data', *Biometrics* **41**, 643–652. **URL:** http://www.jstor.org/stable/2531284

Struthers, C. A. and Kalbfleisch, J. D. (1986), 'Misspecified proportional hazard models', *Biometrika* **73**, 363–369.

Tsiatis, A. A. (1990), 'Estimating regression parameters using linear rank tests for censored data', *Ann. Statist.* **18**, 354–372.

Wang, H. and Wang, L. (2009), 'Locally weighted censored quantile regression', *J. Amer. Statist. Assoc.* **104**, 1117–1128.

Yin, G. and Cai, J. (2004), 'Additive hazards model with multivariate failure time data', *Biometrika* **91**, 801–818.

Ying, Z. (1993), 'A large sample study of rank estimation for censored regression data', *Ann. Statist.* **21**.

Ying, Z., Jung, S. H. and Wei, L. J. (1995), 'Survival analysis with median regression models', *J. Amer. Statist. Assoc.* **90**, 178–184.

Zeng, D. and Lin, D. (2007*a*), 'Efficient estimation for the accelerated failure time model', *J. Amer. Statist. Assoc.* **102**, 1387–1396.

Zeng, D. and Lin, D. (2007*b*), 'Maximum likelihood estimation in semiparametric regression models with censored data', *J. Roy. Statist. Soc. Ser. B* **69**, 507–564.

Zheng, J. (2000), 'A consistent test of conditional parametric distributions', *Econometric Theory* **16**, 667–691.

4

Transformation Models

Danyu Lin

Department of Biostatistics, University of North Carolina

CONTENTS

4.1	Introduction	77
4.2	Data, models and likelihoods	78
	4.2.1 Transformation models for counting processes	78
	4.2.2 Transformation models with random effects for dependent failure times	80
	4.2.3 Joint models for repeated measures and failure times	80
4.3	Estimation	81
4.4	Asymptotic properties	82
4.5	Examples	83
	4.5.1 Lung cancer study	83
	4.5.2 Colon cancer study	84
	4.5.3 HIV study	86
4.6	Discussion	87
	Acknowledgments	88
	Bibliography	89

4.1 Introduction

The proportional hazards model (Cox, 1972) specifies that the hazard function for the failure time T conditional on a set of covariates Z takes the form

$$\lambda(t|Z) = e^{\beta^{\mathrm{T}} Z}\lambda_0(t),$$

where β is a set of unknown regression parameters, and $\lambda_0(\cdot)$ is an arbitrary baseline hazard function. The partial likelihood (Cox, 1975) is used to estimate β. Breslow (1972) provided an estimator for the cumulative baseline hazard function $\Lambda_0(t) \equiv \int_0^t \lambda_0(s)ds$. The asymptotic properties of the maximum partial likelihood estimator and the Breslow estimator were established by Tsiatis (1981) and Andersen and Gill (1982) among others.

Because the proportional hazards assumption may be violated in practice, it is useful to consider nonproportional hazards models. Under the proportional odds model (Bennett, 1983), the hazard ratio between two sets of covariate values converges to unity, rather than staying constant, as time increases. Both the proportional hazards and proportional odds models belong to the class of linear transformation models, which relates an unknown transformation of T linearly to Z:

$$H(T) = -\beta^{\mathrm{T}} Z + \epsilon, \tag{4.1}$$

where $H(\cdot)$ is an unspecified increasing function, β is a set of unknown regression parameters, and ϵ is a random error with a parametric distribution (Kalbfleisch and Prentice, 2002, p. 241). The choices of the extreme-value and standard logistic error distributions yield the proportional hazards and proportional odds models, respectively. Dabrowska and Doksum (1988), Cheng et al. (1995) and Chen et al. (2002) proposed inefficient estimators for this class of models. Murphy et al. (1997) proposed efficient nonparametric maximum likelihood estimators (NPMLEs) for the proportional odds model, while Zeng and Lin (2006) developed NPMLEs for the whole class of linear transformation models.

Although the class of linear transformation models generalizes the proportional hazards model, it is less flexible than the latter in that it is confined to traditional survival time (i.e., single-event) data and time-independent covariates. By contrast, the proportional hazards model can easily incorporate time-dependent covariates and has been extended to recurrent event data through the counting process framework (Andersen and Gill, 1982). To accommodate time-dependent covariates and recurrent events, Zeng and Lin (2006) formulated transformation models through the intensity function for the counting process. In addition, Zeng and Lin (2007) studied transformation models with random effects for multivariate failure time data and joint models for repeated measures and failure time.

In this chapter, we provide an overview of transformation models, adopting the general framework of Zeng and Lin (2006; 2007). In the next section, we describe transformation models for various data structures and construct the corresponding likelihood functions. In Section 4.3, we show how to calculate the NPMLEs. In Section 4.4, we present the asymptotic results. In Section 4.5, we provide illustrations with three real examples. In Section 4.6, we discuss some related problems.

4.2 Data, models and likelihoods

4.2.1 Transformation models for counting processes

Let counting process $N^*(t)$ denote the number of events that have occurred by time t, and let $Z(\cdot)$ denote a vector of possibly time-dependent covariates. We specify that the cumulative intensity function for $N^*(t)$ conditional on $\{Z(s); s \leq t\}$ takes the form

$$\Lambda(t|Z) = G\left\{\int_0^t R^*(s) e^{\beta^{\mathrm{T}} Z(s)} d\Lambda(s)\right\}, \tag{4.2}$$

where G is a specific increasing function, $R^*(\cdot)$ is an indicator process, β is a vector of unknown regression parameters, and $\Lambda(\cdot)$ is an unspecified increasing function. For survival data, $R^*(t) = I(T \geq t)$, where $I(\cdot)$ is the indicator function. For recurrent events, $R^*(\cdot) = 1$. The choice of $G(x) = x$ yields the proportional hazards/intensity model (Cox, 1972; Andersen and Gill, 1982). For survival data with time-independent covariates, Equation (4.2) implies that $G\{\Lambda(t)e^{\beta^{\mathrm{T}} Z}\}$ is a cumulative hazard function so that $\Lambda(T)e^{\beta^{\mathrm{T}} Z} = G^{-1}(-\log \epsilon_0)$, where ϵ_0 has a uniform distribution. Thus, $\log \Lambda(T) = -\beta^{\mathrm{T}} Z + \log G^{-1}(-\log \epsilon_0)$, which is equivalent to (4.1).

We consider the class of Box-Cox transformations $G(x) = \{(1+x)^\rho - 1\}/\rho$ $(\rho \geq 0)$ with $\rho = 0$ corresponding to $G(x) = \log(1+x)$ and the class of logarithmic transformations $G(x) = \log(1 + rx)/r$ $(r \geq 0)$ with $r = 0$ corresponding to $G(x) = x$.

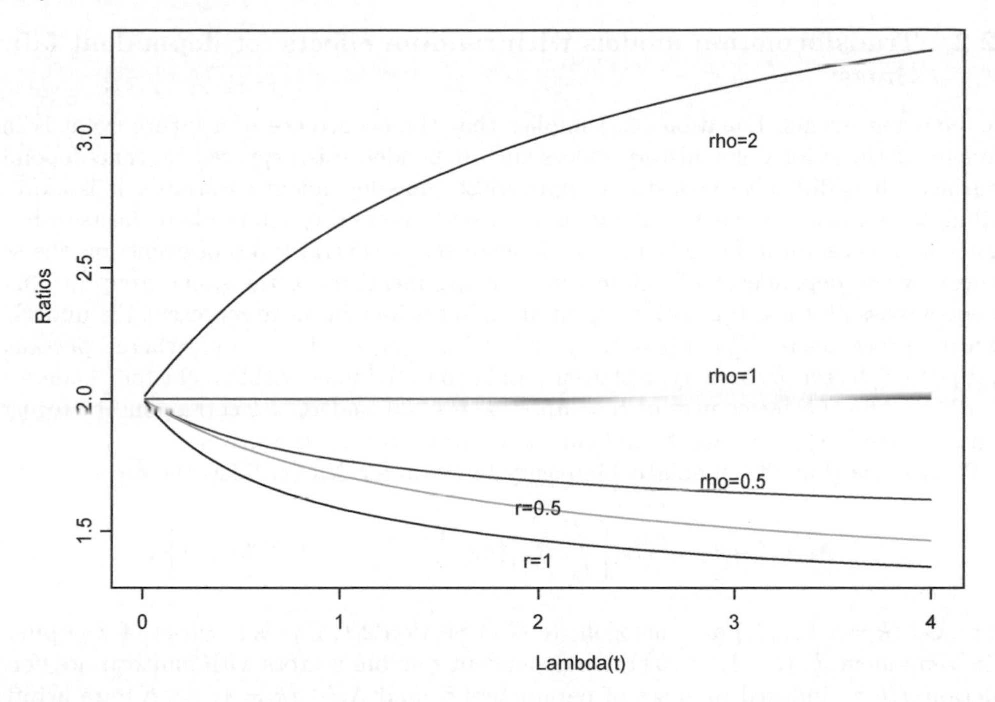

FIGURE 4.1
Plots of the ratios $\Lambda(t|Z = z)/\Lambda(t|Z = 0)$ against $\Lambda(t)$ at $e^{\beta^{\mathrm{T}} z} = 2$ under the transformation models $\Lambda(t|Z) = G\{\Lambda(t)e^{\beta^{\mathrm{T}} Z}\}$ with $G(x) = \{(1 + x)^{\rho} - 1\}/\rho$ and $G(x) = \log(1 + rx)/r$.

Figure 4.1 shows the patterns of covariate effects over time for these two classes of transformations. For the first class, covariate effects increase over time if $\rho > 1$ and decrease over time if $\rho < 1$. For the second class, covariate effects always decrease over time, the rate of decrease being higher for larger r.

Let C denote the censoring time, which is assumed to be independent of $N^*(\cdot)$ conditional on $Z(\cdot)$. For a random sample of n subjects, the data consist of $\{N_i(t), R_i(t), Z_i(t); t \in [0, \tau]\}$ $(i = 1, \ldots, n)$, where $R_i(t) = I(C_i \geq t)R_i^*(t)$, $N_i(t) = N_i^*(t \wedge C_i)$, $a \wedge b = \min(a, b)$, and τ is the duration of the study. To allow general censoring/truncation patterns, we redefine $N_i(t)$ as the number of events observed by time t on the ith subject, and $R_i(t)$ as the indicator of whether the ith subject is at risk at t.

Write $\lambda(t|Z) = \Lambda'(t|Z)$. Assume that censoring is noninformative about β and $\Lambda(\cdot)$. Then the likelihood for β and $\Lambda(\cdot)$ is proportional to

$$\prod_{i=1}^{n} \prod_{t \leq \tau} \{R_i(t)\lambda(t|Z_i)\}^{dN_i(t)} \exp\left\{-\int_0^{\tau} R_i(t)\lambda(t|Z_i)dt\right\}, \tag{4.3}$$

where $dN_i(t)$ is the increment of N_i over $[t, t + dt)$. For survival data, (3) reduces to

$$\prod_{i=1}^{n} \lambda(t|Z_i)^{\delta_i} \exp\left\{-\int_0^{\widetilde{T}_i} \lambda(t|Z_i)dt\right\},$$

where $\widetilde{T}_i = T_i \wedge C_i$, and $\delta_i = I(T_i \leq C_i)$.

4.2.2 Transformation models with random effects for dependent failure times

For recurrent events, Equation (4.2) implies that the occurrence of a future event is independent of the prior event history unless such dependence is captured by time-dependent covariates. It is difficult to construct appropriate time-dependent covariates. It is more appealing to characterize the dependence of recurrent events through random effects or frailty. Frailty is also useful in formulating the dependence of several types of events on the same subject or the dependence of failure times among members of the same group or cluster. To encompass all these types of multivariate failure time data, we represent the underlying counting processes by $N^*_{ikl}(\cdot)$ $(i = 1, \ldots, n; k = 1, \ldots, K; l = 1, \ldots, n_{ik})$, where i pertains to a subject or cluster, k to the type of event, and l to individuals within a cluster (Andersen et al., 1992). The specific choices of $K = n_{ik} = 1$, $n_{ik} = 1$ and $K = 1$ correspond to recurrent events, multiple types of events and clustered failure times, respectively.

We assume that the cumulative intensity function for $N^*_{ikl}(t)$ takes the form

$$\Lambda_k(t|Z_{ikl}; b_i) = G_k\left\{\int_0^t R^*_{ikl}(s)e^{\beta^{\mathrm{T}} Z_{ikl}(s)+b_i^{\mathrm{T}}\widetilde{Z}_{ikl}(s)}d\Lambda_k(s)\right\}, \tag{4.4}$$

where G_k $(k = 1, \ldots, K)$ are analogous to G of Section 2.1, $\widetilde{Z}_{ikl}$ is a subset of Z_{ikl} plus the unit component, b_i $(i = 1, \ldots, n)$ are independent random vectors with multivariate density function $f(b; \gamma)$ indexed by a set of parameters γ, and $\Lambda_k(\cdot)$ $(k = 1, \ldots, K)$ are arbitrary increasing functions. Equation (4.4) accommodates nonproportional hazards/intensity models and multiple random effects. We recommend to use normal random effects, which have unrestricted covariance matrices. In light of the linear transformation model representation, normal random effects are more natural than gamma frailty, even for the proportional hazards model. Computationally, normal distributions are more tractable than others, especially for multiple random effects.

Write $\theta = (\beta^{\mathrm{T}}, \gamma^{\mathrm{T}})^{\mathrm{T}}$. Let C_{ikl}, $N_{ikl}(\cdot)$ and $R_{ikl}(\cdot)$ be defined analogously to C_i, $N_i(\cdot)$ and $R_i(\cdot)$ of Section 2.1. Assume that C_{ikl} is independent of $N^*_{ikl}(\cdot)$ and b_i conditional on $Z_{ikl}(\cdot)$ and noninformative about θ and Λ_k $(k = 1, \ldots, K)$. The likelihood for θ and Λ_k $(k = 1, \ldots, K)$ is

$$\begin{aligned}\prod_{i=1}^{n}\int_b \prod_{k=1}^{K}\prod_{l=1}^{n_{ik}}\prod_{t\le\tau}&\Bigg[R_{ikl}(t)\lambda_k(t)e^{\beta^{\mathrm{T}} Z_{ikl}(t)+b^{\mathrm{T}}\widetilde{Z}_{ikl}(t)}G_k\\ &\times\left\{\int_0^t R_{ikl}(s)e^{\beta^{\mathrm{T}} Z_{ikl}(s)+b^{\mathrm{T}}\widetilde{Z}_{ikl}(s)}d\Lambda_k(s)\right\}\Bigg]^{dN_{ikl}(t)}\\ &\times\exp\left[-G_k\left\{\int_0^\tau R_{ikl}(t)e^{\beta^{\mathrm{T}} Z_{ikl}(t)+b^{\mathrm{T}}\widetilde{Z}_{ikl}(t)}d\Lambda_k(t)\right\}\right]f(b;\gamma)db,\end{aligned} \tag{4.5}$$

where $\lambda_k(t) = \Lambda'_k(t)$ $(k = 1, \ldots, K)$.

4.2.3 Joint models for repeated measures and failure times

Let Y_{ij} denote a response variable and X_{ij} a vector of covariates observed at time t_{ij}, for observation $j = 1, \ldots, n_i$ on subject $i = 1, \ldots, n$. We formulate their relationships through generalized linear mixed models (Diggle et al., 2002, §7.2). The random effects b_i $(i = 1, \ldots, n)$ are independent zero-mean random vectors with multivariate density function $f(b; \gamma)$ indexed by a set of parameters γ. Given b_i, the responses $Y_{i1}, \ldots, Y_{in_i}$ are independent and follow a generalized linear model with density $f_y(y|X_{ij}; b_i)$. The conditional means satisfy

$$g\{E(Y_{ij}|X_{ij};b_i)\} = \alpha^{\mathrm{T}}X_{ij} + b_i^{\mathrm{T}}\widetilde{X}_{ij}, \tag{4.6}$$

where g is a known link function, α is a set of regression parameters, and $\widetilde{X}$ is a subset of X.

As in Section 2.1, let $N_i^*(t)$ denote the number of events the ith subject has experienced by time t and $Z_i(\cdot)$ a vector of covariates. We allow $N_i^*(\cdot)$ to take multiple jumps so as to accommodate recurrent events. If one is interested in adjusting for informative drop-out in the repeated measures analysis, then $N_i^*(\cdot)$ will take a single jump at the drop-out time. To account for the correlation between $N_i^*(\cdot)$ and the Y_{ij}, we incorporate the random effects b_i into Equation (4.2)

$$\Lambda(t|Z_i;b_i) = G\left\{\int_0^t R_i^*(s)e^{\beta^{\mathrm{T}}Z_i(s)+(\psi\circ b_i)^{\mathrm{T}}\widetilde{Z}_i(s)}d\Lambda(s)\right\},$$

where $\widetilde{Z}_i$ is a subset of Z_i plus the unit component, ψ is a vector of unknown constants, and $v_1 \circ v_2$ is the component-wise product of two vectors v_1 and v_2. Normally, we set $X_{ij} = Z_i(t_{ij})$. It is assumed that $N_i^*(\cdot)$ and the Y_{ij} are independent given b_i, Z_i and X_{ij}.

Write $\theta = (\alpha^{\mathrm{T}}, \beta^{\mathrm{T}}, \gamma^{\mathrm{T}}, \psi^{\mathrm{T}})^{\mathrm{T}}$. Assume that censoring and measurement times are non-informative (Tsiatis and Davidian, 2004). Then the likelihood for θ and $\Lambda(\cdot)$ can be written as

$$\prod_{i=1}^{n}\int_b \prod_{t\le\tau}\{R_i(t)\lambda(t|Z_i;b)\}^{dN_i(t)}\exp\left\{-\int_0^\tau R_i(t)\lambda(t|Z_i;b)dt\right\}\prod_{j=1}^{n_i} f_y(Y_{ij}|X_{ij};b)f(b;\gamma)db, \tag{4.7}$$

where $\lambda(t|Z;b) = \Lambda'(t|Z;b)$.

4.3 Estimation

The likelihood functions given in (4.3), (4.5) and (4.7) can all be written in the following form

$$L_n(\theta,\mathcal{A}) = \prod_{i=1}^{n}\prod_{k=1}^{K}\prod_{l=1}^{n_{ik}}\prod_{t\le\tau}\lambda_k(t)^{dN_{ikl}(t)}\Psi(\mathcal{O}_i;\theta,\mathcal{A}), \tag{4.8}$$

where $\mathcal{A} = (\Lambda_1, \ldots, \Lambda_K)$, $\mathcal{O}_i$ is the observation on the ith subject or cluster, and Ψ is a functional of random process $\mathcal{O}_i$, infinite-dimensional parameter $\mathcal{A}$ and d-dimensional parameter θ. To obtain the NPMLEs of θ and $\mathcal{A}$, we treat $\mathcal{A}$ as right-continuous and replace $\lambda_k(t)$ by the jump size of Λ_k at t. Under model (4.2) with $G(x) = x$, the NPMLEs are identical to the maximum partial likelihood estimator of β and the Breslow estimator of Λ.

To calculate the NPMLEs, we maximize $L_n(\theta, \mathcal{A})$ with respect to θ and the jump sizes of $\mathcal{A}$ at the observed event times. This maximization can be carried out in many scientific computing programs. For example, the Optimization Toolbox of MATLAB contains an algorithm *fminunc* for unconstrained nonlinear optimization. One may choose between large-scale and medium-scale optimization. The large-scale optimization algorithm is a subspace trust region method based on the interior-reflective Newton algorithm of Coleman and Li (1994; 1996). Each iteration involves approximate solution of a large linear system using the technique of preconditioned conjugate gradients. The gradient of the function is required.

The Hessian matrix is not required, and is estimated numerically when it is not supplied. We recommend to provide the Hessian matrix, so that the algorithm is faster and more reliable. The medium-scale optimization is based on the BFGS Quasi-Newton algorithm with a mixed quadratic and cubic line search procedure. This algorithm is also available in *Numerical Recipes in C* (Press et al., 1992). MATLAB also contains an algorithm *fmincon* for constrained nonlinear optimization, which is similar to *fminunc*.

The optimization algorithms do not guarantee a global maximum and may be slow for large sample sizes. Our experience, however, shows that these algorithms perform very well for small and moderate sample sizes provided that the initial values are appropriately chosen. One may use the estimates from the Cox proportional hazards model or a parametric model as the initial values. One may also use some other sensible initial values, such as zero for the regression parameters and Y for $H(Y)$. To gain more confidence in the estimates, one may try different initial values.

It is natural to fit random-effects models through the expectation-maximization (EM) algorithm (Dempster et al., 1977), in which random effects pertain to missing data. The EM algorithm is particularly convenient for the proportional hazards model with random effects because, in the M-step, the estimator of the regression parameter is the root of an estimating function that takes the same form as the partial likelihood score function and the estimator for $\mathcal{A}$ takes the form of the Breslow estimator; see Nielsen et al. (1992), Klein (1992) and Andersen et al. (1997) for the formulas in the special case of gamma frailty.

For transformation models without random effects, we may use the Laplace transformation to convert the problem into the proportional hazards model with a random effect. Let ξ be a random variable whose density $f(\xi)$ is the inverse Laplace transformation of $e^{-G(t)}$, i.e., $e^{-G(t)} = \int_0^\infty e^{-t\xi} f(\xi) d\xi$. If $P(T > t|\xi) = \exp\{-\xi \int_0^t e^{\beta^{\mathrm{T}} Z(s)} d\Lambda(s)\}$, then $P(T > t) = \exp\left[-G\{\int_0^t e^{\beta^{\mathrm{T}} Z(s)} d\Lambda(s)\}\right]$. Thus, we can turn the estimation of the general transformation model into that of the proportional hazards frailty model. This strategy also works for general transformation models with random effects, although there will be two sets of random effects in the likelihood; see Appendix A.1 of Zeng and Lin (2007) for details.

There is another simple and efficient approach. Using either the forward or the backward recursion described in Appendix A.2 of Zeng and Lin (2007), we can reduce the task of solving equations for θ and all the jump sizes of Λ to that of solving equations for θ and only one of the jump sizes. This method is more efficient and more stable than direct optimization.

4.4 Asymptotic properties

We consider the general likelihood given in (4.8). Denote the true values of θ and $\mathcal{A}$ by θ_0 and $\mathcal{A}_0$ and their NPMLEs by $\widehat{\theta}$ and $\widehat{\mathcal{A}}$. Under mild regularity conditions (Zeng and Lin, 2007; 2010), $\widehat{\theta}$ is strongly consistent for θ_0 and $\widehat{\mathcal{A}}(\cdot)$ uniformly converges to $\mathcal{A}_0(\cdot)$ with probability one. In addition, the random element $n^{1/2}\{\widehat{\theta} - \theta_0, \widehat{\mathcal{A}}(\cdot) - \mathcal{A}_0(\cdot)\}$ converges weakly to a zero-mean Gaussian process, and the limiting covariance matrix of $\widehat{\theta}$ achieves the semiparametric efficiency bound (Bickel et al., 1993).

To estimate the variances and covariances of $\widehat{\theta}$ and $\widehat{\mathcal{A}}(\cdot)$, we treat (4.8) as a parametric likelihood with θ and the jump sizes of $\mathcal{A}$ as the parameters and then invert the observed information matrix for all these parameters. This approach not only allows one to estimate the covariance matrix of $\widehat{\theta}$, but also the covariance function for any functional of $\widehat{\theta}$ and $\widehat{\mathcal{A}}(\cdot)$. The latter is obtained by the delta-method (Andersen et al., 1992, §II.8) and is useful in predicting event occurrences. A limitation of this approach is that it requires inverting a potentially large-dimensional matrix and thus may not work well when there are a large number of observed failure times.

When the interest lies primarily in θ, one can use the profile likelihood approach (Murphy and van der Vaart, 2000). Let $pl_n(\theta)$ be the profile log-likelihood function for θ, i.e., $pl_n(\theta) = \max_{\mathcal{A}} \log L_n(\theta, \mathcal{A})$. Then the (s,t)th element of the inverse covariance matrix of $\widehat{\theta}$ can be estimated by $-\epsilon_n^{-2}\{pl_n(\widehat{\theta}+\epsilon_n e_s+\epsilon_n e_t)-pl_n(\widehat{\theta}+\epsilon_n e_s-\epsilon_n e_t)-pl_n(\widehat{\theta}-\epsilon_n e_s+\epsilon_n e_t)+pl_n(\widehat{\theta})\}$, where ϵ_n is a constant of order $n^{-1/2}$, and e_s and e_t are the sth and tth canonical vectors, respectively. The profile likelihood function can be easily calculated through the algorithms described in the previous section. Specifically, $pl_n(\theta)$ can be calculated via the EM algorithm by holding θ fixed in both the E-step and the M-step. In this way, the calculation is very fast due to the explicit expression of the estimator of $\mathcal{A}$ in the M-step. In the recursive formulas, the profile likelihood function is a natural product of the algorithm.

4.5 Examples

4.5.1 Lung cancer study

We first consider survival data from the Veterans' Administration lung cancer trial (Prentice, 1973). The subset of data for the 97 patients without prior therapy has been analyzed by many authors, including Bennett (1983), Pettitt (1984), Cheng et al. (1995), Murphy et al. (1997) and Chen et al. (2002). Chen et al. related the survival time to the performance status and tumour type through linear transformation models with $G(x) = \log(1 + rx)/r$, where r = 0, 1, 1.5 and 2. For comparisons, we fitted the same models and display the results in Table 4.1.

For $r = 0$, our numbers agree with the standard software output. For $r = 1$, our results are similar to those of Murphy et al. (1997). Small tumour is significantly different from large tumour under r = 1, 1.5 and 2, but not under $r = 0$.

To determine which model best fits the data, we plot in Figure 4.2 the observed values of the loglikelihood functions for the Box-Cox and logarithmic transformations. The likelihood is maximised at $r = 0.83$. Since the likelihood at $r = 1$ is only slightly smaller, one would choose $r = 1$ to obtain the familiar proportional odds model. The prediction of the subject-specific survival experience under the proportional odds model is illustrated in Figure 4.3.

TABLE 4.1
Estimates of regression parameters for the Veteran's Administration lung cancer data.

	$r = 0$	$r = 1$	$r = 1.5$	$r = 2$
Performance status	-0.024 (0.006)	-0.053 (0.010)	-0.063 (0.012)	-0.072 (0.014)
Adeno vs. large tumour	0.851 (0.348)	1.314 (0.554)	1.497 (0.636)	1.679 (0.712)
Small vs. large tumour	0.547 (0.321)	1.383 (0.524)	1.605 (0.596)	1.814 (0.661)
Squam vs. large tumour	-0.215 (0.347)	-0.181 (0.588)	-0.075 (0.675)	0.045 (0.749)

Note: Standard error estimates shown in parentheses.

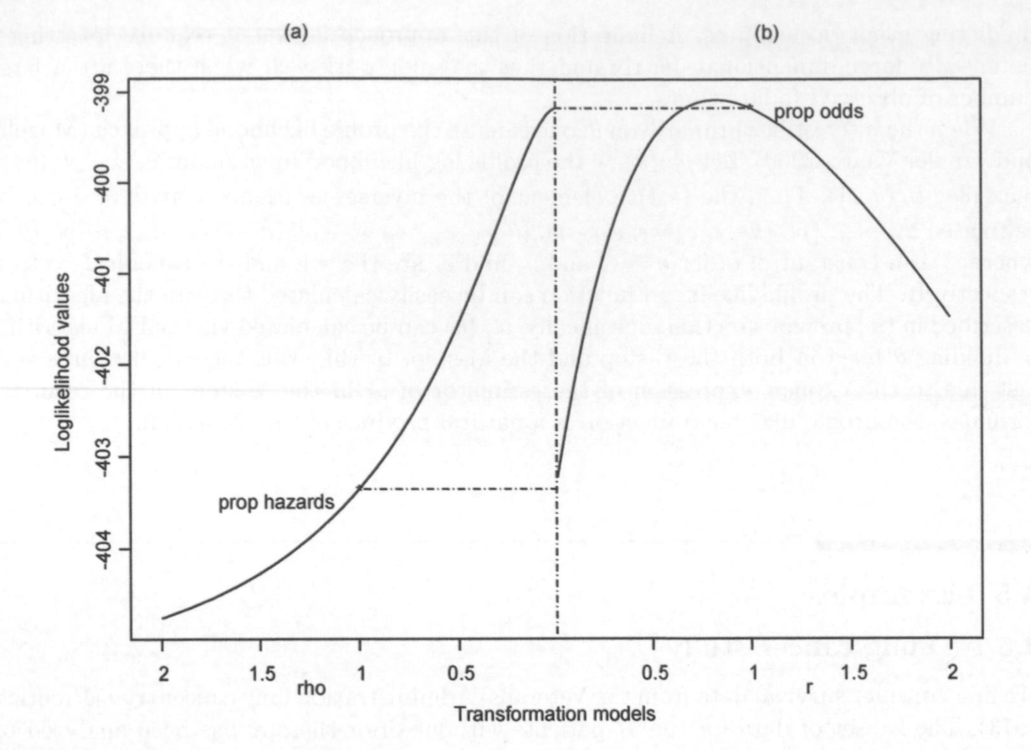

FIGURE 4.2
The observed values of the loglikelihood functions for the lung cancer data: (a) pertains to the Box-Cox transformations $G(x) = \{(1+x)^\rho - 1\}/\rho$; (b) pertains to the logarithmic transformations $G(x) = \log(1+rx)/r$.

4.5.2 Colon cancer study

In a clinical trial on colon cancer, 315, 310 and 304 patients with Stage C disease received observation, levamisole alone, and levamisole combined with fluorouracil (Lev+5-FU), respectively (Lin, 1994). By the end of the study, 155 patients in the observation group, 144 in the levamisole alone group and 103 in the Lev+5-FU group had cancer recurrences, and there were 114, 109 and 78 deaths in the observation period, levamisole alone and Lev+5-FU groups, respectively. Lin (1994) fitted separate proportional hazards models to cancer recurrence and death. That analysis ignored the informative censoring on cancer recurrence and did not explore the joint distribution of the two endpoints.

Following Lin (1994), we focus on the comparison between the observation and Lev+5-FU groups. We treat cancer recurrence as the first type of failure and death as the second, and consider four covariates:

$$Z_{1i} = \begin{cases} 0 & \text{if the } i\text{th patient was on observation,} \\ 1 & \text{if the } i\text{th patient was on Lev+5-FU;} \end{cases}$$

$$Z_{2i} = \begin{cases} 0 & \text{if the surgery for the } i\text{th patient took place } \leq 20 \text{ days prior to randomization,} \\ 1 & \text{if the surgery for the } i\text{th patient took place } > 20 \text{ days prior to randomization;} \end{cases}$$

$$Z_{3i} = \begin{cases} 0 & \text{if the depth of invasion for the } i\text{th patient was submucosa or muscular layer,} \\ 1 & \text{if the depth of invasion for the } i\text{th patient was serosa;} \end{cases}$$

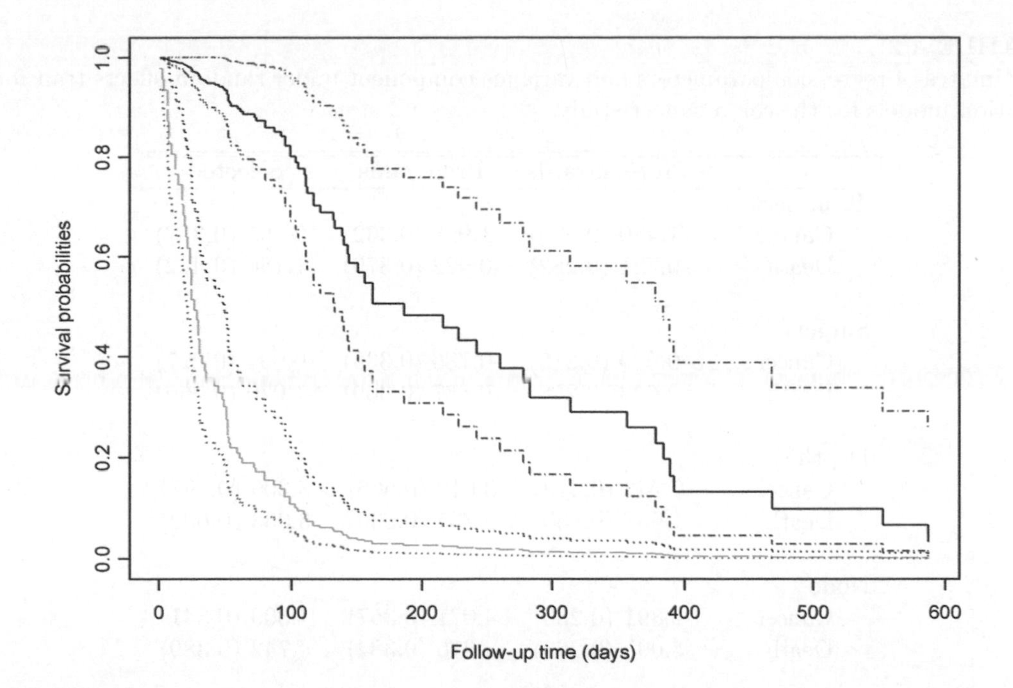

FIGURE 4.3
Estimated survival curves for the lung cancer patients: the upper three curves pertain to the point estimate and 95% confidence limits for a patient with a large tumour and performance status of 80, and the lower three curves to those of a patient with a small tumour and performance status of 40.

$$Z_{4i} = \begin{cases} 0 & \text{if the number of nodes involved in the } i\text{th patient was 1–4,} \\ 1 & \text{if the number of nodes involved in the } i\text{th patient} > 4. \end{cases}$$

We fit the class of models in (4.4) with a normal random effect (to capture the dependence between cancer recurrence and death) and the Box-Cox transformations $\{(1+x)^\rho - 1\}/\rho$ and logarithmic transformations $r^{-1}\log(1+rx)$ through the EM algorithm. The combination of $G_1(x) = 2\{(1+x)^{1/2} - 1\}$ and $G_2(x) = \log(1+1.45x)/1.45$ maximizes the likelihood function. Thus, we select this bivariate model.

Table 4.2 presents the results under the selected model and the proportional hazards and proportional odds models. All three models show that Lev+5-FU is effective in preventing cancer recurrence and death. The interpretation of treatment effects and the prediction of events depend on which model is used.

We can predict an individual's future events based on his/her event history. The survival probability at time t for a patient with covariate values z and with cancer recurrence at t_0 is

$$\frac{\int_b \exp\{-G_2(\Lambda_2(t)e^{\beta_2^{\mathrm{T}} z+b})\}G_1'(\Lambda_1(t_0)e^{\beta_1^{\mathrm{T}} z+b})\exp\{-G_1(\Lambda_1(t_0)e^{\beta_1^{\mathrm{T}} z+b})\}d\Phi(b/\sigma_b)}{\int_b \exp\{-G_2(\Lambda_2(t_0)e^{\beta_2^{\mathrm{T}} z+b})\}G_1'(\Lambda_1(t_0)e^{\beta_1^{\mathrm{T}} z+b})\exp\{-G_1(\Lambda_1(t_0)e^{\beta_1^{\mathrm{T}} z+b})\}d\Phi(b/\sigma_b)}, \quad t \geq t_0,$$

where Φ is the standard normal distribution function. We estimate this probability by replacing all the unknown parameters with their sample estimators, and estimate the standard error by the delta method. An example of this kind of prediction is given in Figure 4.4.

To test the global null hypothesis of no treatment effect on cancer recurrence and death,

TABLE 4.2
Estimates of regression parameters and variance component under random-effects transformation models for the colon cancer study.

	Prop. hazards	Prop. odds	Selected
Treatment			
Cancer	-1.480 (0.236)	-1.998 (0.352)	-2.265 (0.357)
Death	-0.721 (0.282)	-0.922 (0.379)	-1.186 (0.422)
Surgery			
Cancer	-0.689 (0.219)	-0.786 (0.335)	-0.994 (0.297)
Death	-0.643 (0.258)	-0.837 (0.369)	-1.070 (0.366)
Depth			
Cancer	2.243 (0.412)	3.012 (0.566)	3.306 (0.497)
Death	1.937 (0.430)	2.735 (0.630)	3.033 (0.602)
Node			
Cancer	2.891 (0.236)	4.071 (0.357)	4.309 (0.341)
Death	3.095 (0.269)	4.376 (0.384)	4.742 (0.389)
σ_b^2	11.62 (1.22)	24.35 (2.46)	28.61 (3.06)
log-likelihood	-2895.1	-2895.0	-2885.7

Note: Standard error estimates are shown in parentheses.

one may impose the condition of a common treatment effect while allowing separate effects for the other covariates. The estimates of the common treatment effects are -1.295, -1.523 and -1.843, with standard error estimates of 0.256, 0.333 and 0.318 under the proportional hazards, proportional odds and selected models. Thus, one would conclude that Lev+5-FU is highly efficacious.

4.5.3 HIV study

A clinical trial was conducted to evaluate the benefit of switching from zidovudine (AZT) to didanosine (ddI) for HIV patients who have tolerated AZT for at least 16 weeks (Lin and Ying, 2003). A total of 304 patients were randomly chosen to continue the AZT therapy while 298 patients were assigned to ddI. The investigators were interested in comparing the CD4 cell counts between the two groups at weeks 8, 16 and 24. A total of 174 AZT patients and 147 ddI patients dropped out of the study due to patient's request, physician's decision, toxicities, death and other reasons.

To adjust for informative drop-out in the analysis of CD4 counts, we consider the joint modeling described in Section 2.3. For CD counts, we use a special case of (4.6)

$$E\{\log(Y_{ij})|X_i, t_{ij}; b_i\} = \alpha_0 + \alpha_1 X_i + \alpha_2 t_{ij} + b_i, \tag{4.9}$$

where X_i is the indicator for ddI, $t_{ij} = 8$, 16 and 24 weeks, and b_i is zero-mean normal with variance σ_b^2. For drop-out time, we use the proportional hazards model with covariate X_i and random effect b_i. Table 4.3 summarizes the results of this analysis.

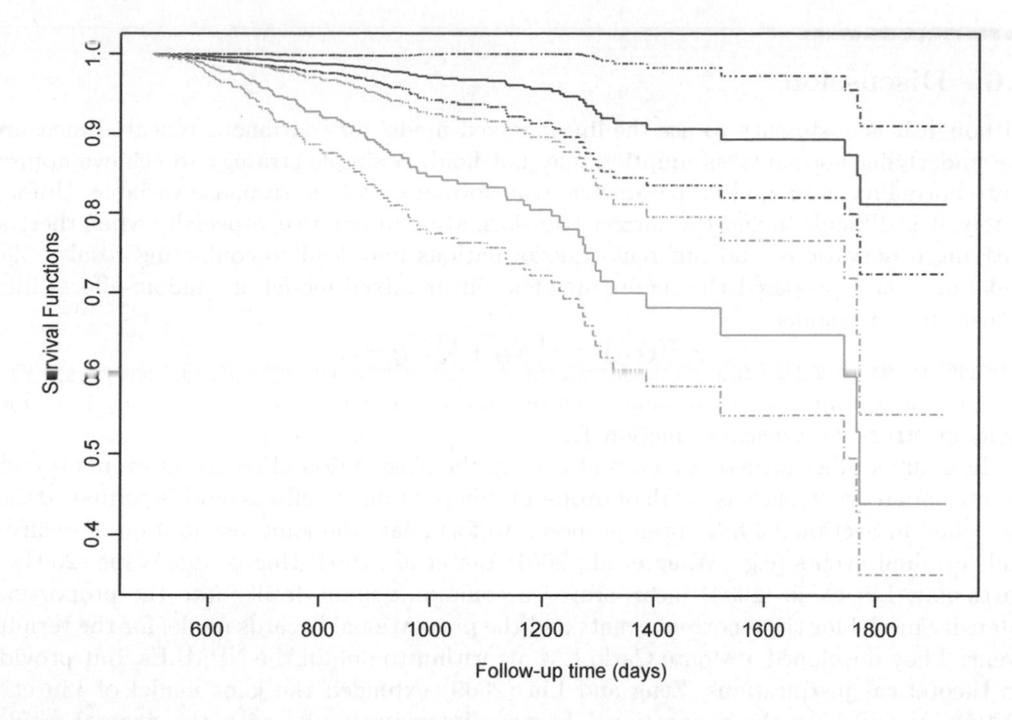

FIGURE 4.4
Estimated survival probabilities of the colon cancer patients with cancer recurrences at days 500 under the selected model: the point estimates and pointwise 95% confidence limits are shown by the solid and dot-dashed curves, respectively. The upper and lower sets of curves pertain to $z = (1, 1, 0, 0)$ and $z = (0, 0, 1, 1)$, respectively.

TABLE 4.3
Joint analysis of CD4 counts and drop-out time for the HIV study.

	Est	SE
CD4 Counts		
ddI	0.506	0.215
Time	-0.041	0.005
Dropout Time		
ddI	-0.316	0.116
σ_b^2	7.421	0.575
ψ	-0.132	0.021

Note: Est and SE denote the parameter estimate and standard error estimate, respectively.

These results account for informative drop-out and show that ddI slowed down the decline of CD4 counts over time. The strong significance of the variance component indicates that drop-out was highly informative.

4.6 Discussion

Although it is customary to use the linear mixed model for continuous repeated measures, the underlying normality assumption may not hold. A simple strategy to achieve approximate normality is to apply a parametric transformation to the response variable. Unfortunately, it is difficult to find the correct transformation in practice, especially when there are outlying observations, and different transformations may lead to conflicting results. Zeng and Lin (2007) proposed the semiparametric linear mixed model or random-effects linear transformation model

$$\widetilde{H}(Y_{ij}) = \alpha^{\mathrm{T}} X_{ij} + b_i^{\mathrm{T}} \widetilde{X}_{ij} + \epsilon_{ij},$$

where $\widetilde{H}$ is an unknown increasing function, and ϵ_{ij} $(i = 1, \ldots, n; j = 1, \ldots, n_{ij})$ are independent errors with density function f_ϵ.

In many studies involving recurrent events, the observation of recurrent events is ended by a terminal event, such as death or drop-out. Shared random effects models similar to those described in Section 2.3 have been proposed to formulate the joint distribution of recurrent and terminal events (e.g., Wang et al., 2001; Liu et al., 2004; Huang and Wang, 2004). In particular, Liu et al. (2004) incorporated a common gamma frailty into the proportional intensity model for the recurrent events and the proportional hazards model for the terminal event. They developed a Monte Carlo EM algorithm to obtain the NPMLEs, but provided no theoretical justifications. Zeng and Lin (2009) extended the joint model of Liu et al. (2004) by replacing the proportional hazards/intensity model with the general random effects transformation models and established the asymptotic properties of the NPMLEs.

The methods described in this chapter have been implemented in MATLAB by my colleague Donglin Zeng; see http://www.bios.unc.edu/~dzeng/Transform.html. The class of linear transformation models was recently implemented by Stata. There is an ongoing effort to implement the transformation models described in Section 2 in Stata.

Cohort sampling, such as case-cohort (Prentice, 1986) and nested case-control sampling (Thomas, 1977), provides a cost-effective strategy to conduct large epidemiological cohort studies with expensive exposure variables. Kong et al. (2004) and Lu and Tsiatis (2006) proposed weighted estimators for the class of linear transformation models under case-cohort sampling, while Zeng et al. (2006) studied NPMLEs for general transformation models under both case-cohort and nested case-control sampling.

We have focused on right-censored data. Interval censoring arises when the failure time is only known to fall in some interval. It is much more challenging to apply the NPMLE to interval censored data than right-censored data. So far the asymptotic theory is only available for the proportional hazards model with current status data (Huang, 1996), which arise when the failure time is only known to be less than or greater than a single monitoring time. We are currently working in this area.

Acknowledgments

This work was supported by the National Institutes of Health. Much of this work is a summary of the joint research of Danyu Lin with Donglin Zeng.

Bibliography

Andersen, P. K., Borgan, Ø., Gill, R. D. and Keiding, N. (1992). *Statistical Models Based on Counting Processes.* New York, Springer.

Andersen, P. K. and Gill, R. D. (1982). Cox's regression model for counting processes: A large sample study. *Ann. Statist.*, **10**, 1100–1120.

Andersen, P. K., Klein, J. P., Knudsen, K. M. and Tabanera y Placios, R. (1997). Estimation of Cox's regression model with shared gamma frailties. *Biometrics*, **53**, 1475–1484.

Bennett, S. (1983). Analysis of survival data by the proportional odds model. *Statist. Med.*, **2**, 273–277.

Bickel, P. J., Klaassen, C. A. J., Ritov, Y. and Wellner, J. A. (1993). *Efficient and Adaptive Estimation for Semiparametric Models.* Baltimore, Johns Hopkins Univ.

Breslow, N. E. (1972). Discussion of the paper by D. R. Cox. *J. R. Statist. Soc. B*, **34**, 216–217.

Chen, K., Jin, Z. and Ying, Z. (2002). Semiparametric analysis of transformation models with censored data. *Biometrika*, **89**, 659–668.

Cheng, S. C., Wei, L. J. and Ying, Z. (1995). Analysis of transformation models with censored data. *Biometrika*, **82**, 835–845.

Coleman, T. F. and Li, Y. (1994). On the convergence of reflective Newton methods for large-scale nonlinear minimization subject to bounds. *Math. Programming*, **67**, 189–224.

Coleman, T. F and Li, Y. (1996). An interior, trust region approach for nonlinear minimization subject to bounds. *SIAM J. Optim.*, **6**, 418–445.

Cox, D. R. (1972). Regression models and life-tables (with discussion). *J. R. Statist. Soc. B*, **34**, 187–220.

Cox, D. R. (1975). Partial likelihood. *Biometrika*, **62**, 269–276.

Dabrowska, D. M. and Doksum, K. A. (1988). Partial likelihood in transformation models with censored data. *Scand. J. Statist.*, **18**, 1–23.

Dempster, A. P., Laird, N. M., and Rubin, D. B. (1977). Maximum likelihood estimation from incomplete data via the EM algorithm (with discussion). *J. R. Statist. Soc. B*, **39**, 1–38.

Diggle, P. J., Heagerty, P., Liang, K.-Y. and Zeger, S. L. (2002). *Analysis of Longitudinal Data*, 2nd Ed. Oxford University Press.

Huang, J. (1996). Efficient estimation for the proportional hazards model with interval censoring. *Ann. Statist.*, **24**, 540–568.

Huang, C. and Wang, M. (2004). Joint modeling and estimation for recurrent event processes and failure time data. *J. Am. Statist. Ass.*, **99**, 1153–1165.

Kalbfleisch, J. D. and Prentice, R. L. (2002). *The Statistical Analysis of Failure Time Data*, 2nd Ed. Hoboken, Wiley.

Klein, J. (1992). Semiparametric estimation of random effects using the Cox model based on the EM algorithm. *Biometrics*, **48**, 795–806.

Kong, L., Cai, J. and Sen, P. K. (2004). Weighted estimating equations for semiparametric transformation models with censored data from a case-cohort design. *Biometrika*, **91**, 305–319.

Lin, D. Y. (1994). Cox regression analysis of multivariate failure time data: The marginal approach. *Statist. Med.*, **13**, 2233–2247.

Lin, D. Y. and Ying, Z. (2003). Semiparametric regression analysis of longitudinal data with informative drop-outs. *Biostatistics*, **4**, 385–398.

Liu, L., Wolfe, R. A., and Huang, X. (2004). Shared frailty models for recurrent events and a terminal event. *Biometrics*, **60**, 747–756.

Lu, W. and Tsiatis, A. A. (2006). Semiparametric transformation models for the case-cohort study. *Biometrika*, **93**, 207–214.

Murphy, S. A., Rossini, A. J., and van der Vaart, A. W. (1997). Maximal likelihood estimation in the proportional odds model. *J. Am. Statist. Ass.*, **92**, 968–976.

Murphy, S. A. and van der Vaart, A. W. (2000). On profile likelihood. *J. Am. Statist. Ass.*, **95**, 449–485.

Nielsen, G. G., Gill, R. D., Andersen, P. K. and Sorensen, T. I. A. (1992). A counting process approach to maximum likelihood estimation in frailty models. *Scand. J. Statist.*, **19**, 25–43.

Pettitt, A. N. (1984). Proportional odds models for survival data and estimates using ranks. *Appl. Statist.* **33**, 169–175.

Prentice, R. L. (1973). Exponential survivals with censoring and explanatory variables. *Biometrika*, **60**, 278–288.

Prentice, R. L. (1986). A case-cohort design for epidemiologic cohort studies and disease prevention trials. *Biometrika*, **73**, 1–11.

Press, W. H., Teukolsky, S. A., Vetterling, W. T. and Flannery, B. P. (1992). *Numerical Recipes in C.* New York, Cambridge University Press.

Thomas, D. (1977). Addendum to: Methods of cohort analysis: appraisal by application to asbestos mining, by F. D. K. Liddell, J. C. McDonald and D. C. Thomas. *J. Roy. Statist. Soc., Series A*, **140**, 469–491.

Tsiatis, A. A. (1981). A large sample study of Cox's regression model. *Ann. Statist.*, **9**, 93–108.

Tsiatis, A. A. and Davidian, M. (2004). Joint modeling of longitudinal and time-to-event data: an overview. *Statistica Sinica*, **14**, 793–818.

Wang, M. C., Qin, J. and Chiang, C. (2001). Analyzing recurrent event data with informative censoring. *J. Am. Statist. Ass.*, **96**, 1057–1065.

Zeng, D. and Lin, D. Y. (2006). Efficient estimation of semiparametric transformation models for counting processes. *Biometrika*, **93**, 627–640.

Zeng, D. and Lin, D. Y. (2007). Maximum likelihood estimation in semiparametric regression models with censored data (with discussion). *J. Roy. Statist. Soc., Series B*, **69**, 507–564.

Zeng, D. and Lin, D. Y. (2009). Semiparametric transformation models with random effects for joint analysis of recurrent and terminal events. *Biometrics*, **65**, 746–752.

Zeng, D. and Lin, D. Y. (2010). A general asymptotic theory for maximum likelihood estimation in semiparametric regression models with censored data. *Statistica Sinica*, **20**, 871–910.

Zeng, D., Lin, D. Y., Avery, C. L., North, K. E. and Bray, M. S. (2006). Efficient semiparametric estimation of haplotype-disease associations in case-cohort and nested case-control studies. *Biostatistics*, **7**, 486–502.

5

High-Dimensional Regression Models

Jennifer A. Sinnott and Tianxi Cai
Department of Biostatistics, Harvard School of Public Health

CONTENTS

5.1 Introduction 93
5.2 Methods based on feature selection 95
5.2.1 Discrete feature selection 95
5.2.2 Shrinkage methods 96
5.2.3 Methods based on group structure 98
5.2.4 Selection of tuning parameters 99
5.3 Methods based on derived variables 100
5.3.1 Principal components regression 101
5.3.2 Approaches based on partial least squares 102
5.4 Other models 102
5.4.1 Nonparametric hazard model 103
5.4.2 Additive risk model 103
5.4.3 Accelerated failure time model 103
5.4.4 Semiparametric linear transformation models 104
5.5 Data analysis example 104
5.6 Remarks 106
Acknowledgments 107
Bibliography 107

5.1 Introduction

An increasingly important goal in medical research is to extract information from a large number of variables measured on patients in order to make predictions of disease-related outcomes. When the outcome of interest is a possibly censored time to event, such as the time to disease recurrence or death, statistical methods that account for censoring must be used. Most classical statistical methods that relate covariates to outcome assume that the number of covariates, p, is less than the number of observations, n; to work well, most methods require p to be significantly less than n. In current research, however, it is common for p to be large relative to n, a data structure usually described as *high-dimensional data.* Sometimes, as in the case of most genomic studies, the covariates vastly outnumber the sample size. This setting is often denoted by $p \gg n$, and is occasionally referred to as *ultra-high dimensional data.* In these settings, most classical statistical methods are not applicable without modification. Here, we review methods that have been developed for relating high-dimensional data to survival outcomes, focusing on methods which can be used to make risk predictions for new observations; for methods related to testing in the high-dimensional setting, see Chapter 15.

The methods we discuss here are applicable in any setting in which the goal of the study is to relate a possibly censored survival time to a large number of predictors, but for illustration purposes, we will focus on the particular setting of relating gene expression data to recurrence-free survival of breast cancer patients. Let T_i be the time to disease recurrence or death for the ith patient; because this time may not be observed if the patient withdraws from the study or dies of another cause, we instead observe $Y_i = \min\{T_i, C_i\}$, where C_i is the time of censoring, and $\Delta_i = I[T_i \leq C_i]$, the indicator of whether the observed time corresponds to disease recurrence or censoring. Let $\mathbf{X}_i$ be a low-dimensional vector of variables we want to insist are included as linear effects in any model under consideration; these could include routine clinical variables like age and tumor grade, or established biomarkers. Let $\mathbf{Z}_i$ be a p-dimensional vector of variables subject to possible dimension reduction, such as genomic predictors or other covariates under investigation. For simplicity, we will refer to the components of $\mathbf{X}_i$ as "clinical variables" and the components of $\mathbf{Z}_i$ as "genomic variables," though the actual variables in these two groups will vary depending on the application. Letting $\mathbf{W}_i = (\mathbf{X}_i^\top, \mathbf{Z}_i^\top)^\top$, the observed data thus consist of n independent and identically distributed vectors, $\mathcal{O} = \{(Y_i, \Delta_i, \mathbf{W}_i^\top), i = 1, ..., n\}$. The censoring C_i is assumed to be independent of T_i conditional on $\mathbf{W}_i$, which is the conventional assumption required to obtain consistent estimators of model parameters for censored survival data. However, we note that under potential model misspecification, a stronger assumption such as C_i being independent of both T_i and $\mathbf{W}_i$ is often imposed to ensure proper inference. More details and related discussions can be found in Robins et al. (1997) and Uno et al. (2007).

The most commonly used model for analyzing survival data is the Cox proportional hazards (PH) model (Cox, 1972) introduced in Chapter 1, which assumes that the hazard of an event at time t for an individual with covariates $\mathbf{W}$ can be modeled by the product of a baseline hazard function $\lambda_0(t)$ and a regression function of the covariates:

$$\lambda_{\mathbf{W}}(t) = \lambda_0(t) \exp(\mathbf{W}^\top \boldsymbol{\theta}). \tag{5.1}$$

The standard classical approach to fit this model is to maximize the log-partial likelihood (LPL)

$$\ell(\theta) = \sum_{i=1}^{n} \Delta_i \left(\mathbf{W}_i^\top \boldsymbol{\theta} - \log \left\{ \sum_{j \in R_i} \exp\{\mathbf{W}_j^\top \boldsymbol{\theta}\} \right\} \right) \tag{5.2}$$

with respect to θ, where R_i is the *risk set* of individuals still at risk at time Y_i. However, when p is large relative to n, this approach may be infeasible, and a number of different strategies have been proposed for adapting this approach to the high-dimensional setting. Some methods proceed by *feature selection*, in which only a subset of the covariates are selected for inclusion in the model. Feature selection can be done discretely, by developing a strategy to assess whether individual features should enter the model, or by shrinkage, in which penalization on the magnitude of the coefficients in the model leads to some coefficients being set identically to zero (e.g., Tibshirani et al., 1997; Fan and Li, 2002; Zhang and Lu, 2007; Antoniadis et al., 2010). Hybrids of marginal screening and shrinkage, such as sure independence screening (SIS), have been proposed to handle ultra-high dimensional survival data (e.g., Fan et al., 2010; Zhao and Li, 2011). In recent years, there has been growing interest in leveraging external information about groupings of predictors, such as pathway information about genomic markers. Structured regularization based on predetermined groups has been proposed to incorporate such information (Kim et al., 2012; Wang et al., 2009). These methods will be discussed in Section 5.2.

Another set of approaches, discussed in Section 5.3, focus on summarizing the feature space with a smaller number of derived variables, without necessarily reducing the number of features involved in the model. These methods develop constructed covariates that summarize information in the original feature space, and aim to capture the original feature space using a small number of these constructed covariates.

For settings where the PH assumption fails to hold, a wide range of survival models such as the accelerated failure time model (introduced in Chapter 3) and the semiparametric transformation model have been proposed as useful alternatives. Methods for fitting these models have also been extended to incorporate high-dimensional predictors, and we discuss these extensions in Section 5.4.

In Section 5.5, we apply a selection of these methods to breast cancer datasets in Wang et al. (2005) and Sotiriou et al. (2006) and compare the resulting models with respect to prediction performance, which we quantify using a C-statistic (5.7). A common secondary goal beyond predictiveness is *parsimony*, or *sparsity*. This may be desirable for cost effectiveness if the goal of the study is to develop a tool for physicians to use to predict patient risk; it may also be desirable for interpretability, since developing and validating biological hypotheses based on genomic associations can be easier if there are fewer of them. Additional methods and remarks about the strengths and weaknesses of various methods will be discussed in Section 5.6. Other helpful recent reviews of methods for survival analysis when p is large relative to n may be found in Witten and Tibshirani (2010) and Binder et al. (2011).

5.2 Methods based on feature selection

One strategy for dealing with high dimensionality in the covariate space is to assume that only a subset of the features, say $k < p$ features, are associated with survival, and attempt to select those features for inclusion in the model. This assumption of *sparsity* would be reasonable if only a small number of features are expected to be predictive of survival. In practice, a sparsity assumption may also be imposed if a parsimonious solution is desired due to the cost associated with measuring a large number of features. The existing methods which assume the true model is sparse may work well with respect to prediction accuracy regardless of whether the sparsity assumption holds, but the established theoretical results on the proposed estimators typically require sparsity. Some major approaches for feature selection include *discrete feature selection* (Section 5.2.1) and *shrinkage-based feature selection* (Section 5.2.2). Of increasing interest are methods that select groups of features or otherwise take into account group structure; such methods could be used in genomic data to make use of existing biological annotation of pathways and networks of genes. Some methods for integrating prior group information into risk prediction are described in Section 5.2.3. For most regularization procedures, tuning parameters play an important role in the performance of the final model. In Section 5.2.4, we discuss methods for selecting the tuning parameters.

5.2.1 Discrete feature selection

A simple approach for selecting features is univariate selection, in which we screen variables one at a time for association with survival. For example, for each feature we could perform a univariate Cox score test, and include the top k features based on the ranking of the corresponding p-value. The number k is a tuning parameter and can be selected to achieve a certain error rate such as the family-wise error rate or the false discovery rate (Benjamini and Hochberg, 1995). This approach is easy to implement; however, in settings where covariates are correlated, as in gene expression data, it may select highly correlated features which do not lead to a multivariate model that improves over the univariate models.

To improve over univariate selection, we could account for correlation between genes by including genes in a multivariate model sequentially, analogous to forward stepwise linear regression. That is, we could start with the null model (or the model with clinical covariates alone), calculate the score statistics for all features, and include the feature with the largest score statistic. Then, with that first feature already in the model, we can use a score test to select which of the remaining features can be added to most improve the model. We continue this process until our model includes k genes, where k is once again a tuning parameter. This approach is also easy to implement, and better accounts for correlation between genes; however, it leads to a locally optimal model rather than the best model with k genes.

Bøvelstad et al. (2007) compared the performance of these discrete selection methods with the performance of methods based on shrinkage (ridge and Lasso, discussed in Section 5.2.2) and summary variables (supervised and unsupervised principal components regression and partial least squares, discussed Section 5.3). They demonstrated that methods based on shrinkage and derived variables tended to outperform discrete variable selection.

5.2.2 Shrinkage methods

Feature selection methods based on discrete screening may not capture well the joint effects of multiple genes, and hence may result in prediction models with low prediction accuracy. On the other hand, fitting a joint model with p features may not be feasible or stable when p is not small relative to n. To overcome such difficulties, various regularization procedures, aiming to maximize a penalized LPL with a penalty accounting for the model complexity, have been proposed. In this section, unless noted otherwise, we assume the correct specification of the Cox model (5.1):

$$\lambda_{\mathbf{W}}(t) = \lambda_0(t)\exp(\mathbf{W}^\top \boldsymbol{\theta}) = \lambda_0(t)\exp(\mathbf{X}^\top \boldsymbol{\alpha} + \mathbf{Z}^\top \boldsymbol{\beta}), \tag{5.3}$$

where $\lambda_{\mathbf{W}}(t)$ is the conditional hazard function given $\mathbf{W}$, $\lambda_0(t)$ is an unknown baseline hazard function, and $\boldsymbol{\theta} = (\boldsymbol{\alpha}^\top, \boldsymbol{\beta}^\top)^\top$ is the unknown vector of regression parameters to be estimated. An L_2-penalty on the magnitude of the $\boldsymbol{\beta}$ coefficients yields the ridge-regularized estimator $\widehat{\boldsymbol{\theta}}_{\text{ridge}}$ (Verweij and Van Houwelingen, 1994), which is the minimizer of:

$$-\ell(\boldsymbol{\theta}) + \lambda \sum_{j=1}^{p} \beta_j^2 \tag{5.4}$$

where $\ell(\cdot)$ is the LPL function and, throughout, $\lambda \geq 0$ denotes a tuning parameter that needs to be selected. In the high-dimensional setting, with a properly chosen λ, $\widehat{\boldsymbol{\theta}}_{\text{ridge}}$ is likely to outperform the standard LPL estimator even when $p < n$. The asymptotic properties of the ridge estimator can be found in Huang and Harrington (2002) for the case with fixed p. Importantly, this approach does not do feature selection because all components of $\widehat{\boldsymbol{\theta}}_{\text{ridge}}$ will in general be nonzero.

When a sparse solution is desired, a natural approach is to use an L_1-penalty to regularize the LPL, which yields the Lasso solution $\widehat{\boldsymbol{\theta}}_{\text{Lasso}} = (\widehat{\boldsymbol{\alpha}}_{\text{Lasso}}^\top, \widehat{\boldsymbol{\beta}}_{\text{Lasso}}^\top)^\top$ (Tibshirani et al., 1997; Gui and Li, 2005; Park and Hastie, 2007) which minimizes:

$$-\ell(\boldsymbol{\theta}) + \lambda \sum_{j=1}^{p} |\beta_j|.$$

The non-smoothness of the L_1 penalty leads to sparse solutions with large λ's. Using similar arguments as given in Knight and Fu (2000) for the Lasso regularized least squares estimator, one may show that with a properly chosen $\lambda = O(n^{\frac{1}{2}})$, $\widehat{\boldsymbol{\theta}}_{\text{Lasso}}$ is $\sqrt{n}$-consistent for $\boldsymbol{\theta}$ under correct specification of (5.3) for fixed p. However, such a $\sqrt{n}$-consistent estimator is not consistent in variable selection in that $P(\widehat{\boldsymbol{\beta}}_{\text{Lasso},\mathcal{A}^c} = 0)$ converges to a constant $\mathcal{P}_0 < 1$, where $\mathcal{A} = \{j : \beta_j \neq 0\}$, $\mathcal{A}^c$ is the complement set of $\mathcal{A}$ and throughout, we use notation $\boldsymbol{\beta}_{\mathcal{A}}$ to denote the subvector of $\boldsymbol{\beta}$ corresponding to the index set $\mathcal{A}$.

The inconsistency in variable selection and the bias towards zero of the nonzero coefficients estimated in finite sample are undesirable features of the standard Lasso which motivated the development of methods in which coefficients receive different amounts of penalization depending on the magnitude of their values. One such approach is the adaptive Lasso which uses weighted L_1-penalties to apply less penalization to larger coefficients and more penalization to variables that are potentially non-informative (Zou, 2006). The adaptive Lasso penalized LPL estimator (Zhang and Lu, 2007) is defined as $\widehat{\boldsymbol{\theta}}_{\text{Lasso}}$ the minimizer of:

$$-\ell(\boldsymbol{\theta}) + \lambda \sum_{j=1}^{p} |\beta_j| \tau_j$$

where the τ_j are positive data-driven weights. In the low-dimensional setting, the τ_j may be chosen as $\tau_j = 1/|\widetilde{\beta}_j|$ where the $\widetilde{\beta}_j$ are from the unpenalized LPL estimator. In the high-dimensional setting, the $\widetilde{\beta}_j$ may be taken from the ridge estimator $\widehat{\boldsymbol{\theta}}_{\text{ridge}}$ (5.4). Another approach for reducing the bias in estimating the nonzero β_j is to employ alternative bounded penalty functions. Fan and Li (2002) noted that a "good" penalty function should aim to produce an estimator that is (i) unbiased, in that the estimator is nearly unbiased when the true parameter is large avoiding model bias; (ii) sparse, in that small coefficients are automatically set to 0 to reduce model complexity; and (iii) continuous, in that the estimator is continuous in data to avoid instability in model prediction. One penalty function that satisfies all these properties is the smoothly clipped absolute deviation penalty (SCAD) (Fan and Li, 2002), which leads to $\widehat{\boldsymbol{\theta}}_{\text{SCAD}}$, the minimizer of:

$$-\ell(\boldsymbol{\theta}) + \sum_{j=1}^{p} \left\{ I\{|\beta_j| \leq \lambda\} + \frac{(a\lambda - |\beta_j|)}{(a-1)\lambda} I\{|\beta_j| > \lambda\} \right\}$$

for tuning parameters λ and a. For fixed p, both $\widehat{\boldsymbol{\theta}}_{\text{aLasso}}$ and $\widehat{\boldsymbol{\theta}}_{\text{SCAD}}$ are $\sqrt{n}$-consistent with properly chosen λ (Fan and Li, 2002; Zhang and Lu, 2007). Furthermore, these estimators possess the *oracle property* in that asymptotically they perform as well as if the active set $\mathcal{A}$ is known. Specifically, these procedures attain model selection consistency with $P(\widehat{\boldsymbol{\beta}}_{\mathcal{A}^c} = 0) \to 1$, and $n^{\frac{1}{2}}(\widehat{\boldsymbol{\beta}}_{\mathcal{A}} - \boldsymbol{\beta}_{\mathcal{A}})$ is asymptotically normal with mean 0 and variance $\mathcal{I}_{\mathcal{A},\mathcal{A}}(\boldsymbol{\beta})^{-1}$, where $\mathcal{I}_{\mathcal{A},\mathcal{A}}(\boldsymbol{\beta})$ is the sub-matrix of the information matrix corresponding to the index set $(\mathcal{A}, \mathcal{A})$.

Another potential problem with the Lasso is that when two highly correlated features are associated with the outcome of interest, the Lasso will tend to identify only one of the features, which can be undesirable for interpretability and replicability. To counteract this problem, Zou and Hastie (2005) proposed the elastic net (EN) for linear regression. The EN adds a ridge-type penalty to the Lasso, minimizing

$$-\ell(\boldsymbol{\theta}) + \lambda_2 \sum_{j=1}^{p} \beta_j^2 + \lambda_1 \sum_{j=1}^{p} |\beta_j|,$$

which improves Lasso's ability to identify sets of correlated genes associated with outcome.

The EN penalty is applied to the Cox model with an algorithm adapted to the high-dimensional setting in Engler and Li (2007). An adaptive version of the EN, similar to the adaptive Lasso, was proposed in Zou and Zhang (2009) to overcome the inconsistency in model selection and the large bias in estimating non-zero coefficients of the EN estimator. One may employ such a penalty for the LPL to improve the estimation of $\boldsymbol{\theta}$. The asymptotic properties of the adaptive EN estimator for linear regression were derived in Zou and Zhang (2009) for the case when p grows with n but at a slower rate. One may expect that similar properties hold for the adaptive EN penalized LPL estimator.

For the ultra-high dimensional setting, when p is much larger than n, an approach that combines univariates selection (5.2.1) with shrinkage estimation (5.2.2) is sure independence screening (SIS) developed in Fan et al. (2010) and Zhao and Li (2011). In Zhao and Li (2011), a marginal Cox model is fit for every feature to get estimates of the univariate parameter and its variance. Then covariates are retained whose magnitude (standardized by its variance) passes a threshold which is defined in terms of the desired false positive rate. Finally, the retained variables are combined into a full model using a shrinkage approach such as the Lasso, Adaptive Lasso, or SCAD.

In addition to the aforementioned procedures, a variety of other methods that can produce sparse solutions have been developed for the Cox model, such as the Dantzig selector (Antoniadis et al., 2010); Cox univariate shrinkage (Tibshirani et al., 2009); and covariance regularized regression (Witten and Tibshirani, 2009). In general, the relative performance of these methods depends on the sparsity of the signal, the dimension p and the signal-to-noise ratio. Benner et al. (2010) compare the performance of the Cox model with the ridge, Lasso, adaptive Lasso, elastic net, and SCAD penalties on high-dimensional data, and find that the traditional Lasso and the elastic net appear to perform best in real data applications. On the other hand, Waldron et al. (2011) compared the performance of ridge, Lasso, and the elastic net on genomic data, and preferred ridge and the elastic net.

5.2.3 Methods based on group structure

Extensive work has been done to annotate the human genome, and there are numerous databases that list pathways and networks of genes thought to work together (e.g., the Molecular Signature Database (Subramanian et al., 2005)). In addition, accumulated knowledge about genetic architecture can be useful for grouping markers into sets of markers that have similar magnitudes of effects. Gene-set based analysis could also potentially increase power because the number of pathways is generally much smaller than the number of genes and, if properly grouped, the joint effects of multiple markers in a gene-set can be easier to detect than their individual effects. A number of procedures have been proposed to improve estimation by leveraging information about the group structure.

The group Lasso, developed for linear regression (Yuan and Lin, 2006), can be adapted to the Cox model (e.g., Kim et al., 2012). Specifically, writing the model with K groups of features and p_k features in the kth group as:

$$\lambda_{\mathbf{W}}(t) = \lambda_0(t) \exp\left\{ \boldsymbol{\alpha}^\top \mathbf{X} + \sum_{k=1}^{K} \sum_{j=1}^{p_k} \beta_{kj} Z_{kj} \right\},$$

the group Lasso estimator would minimize:

$$-\ell(\boldsymbol{\theta}) + \lambda \sum_{k=1}^{K} \sqrt{\boldsymbol{\beta}_k^\top \mathbb{K}_k \boldsymbol{\beta}_k},$$

where $\boldsymbol{\beta}_k = (\beta_{k1}, ..., \beta_{kp_k})^\top$ and the $\mathbb{K}_k$ are pre-specified positive definite weight matrices.

Examples of $\mathbb{K}_k$ include the identity matrix $\mathbb{I}_{p_k \times p_k}$, and the identity matrix scaled to account for group size, $p_k \mathbb{I}_{p_k \times p_k}$. For certain values of λ, some of the pathway effects will be set identically to zero; for pathways which are not set to zero, the estimated effects of all markers in the pathway will in general be nonzero. Thus, this penalty selects groups, but does not do feature selection within a group; one advantage of this approach is that it is independent of the basis used within a group. The asymptotic properties of the group Lasso estimator as well as an adaptive version of the group LASSO with data dependent choices of $\mathbb{K}_k$ were studied in Bach (2008) and Nardi and Rinaldo (2008) for the linear regression setting. One may expect similar properties of the estimators corresponding to the PH model. In general, many annotated pathways may overlap extensively; fitting the group Lasso penalty for linear regression when groups overlap is discussed in Jacob et al. (2009), and the computational method described there can be easily extended to the Cox model.

The group Lasso penalty treats a group as a unit and does not allow feature selection within group. However, in settings where only a fraction of the genes within a group are associated with the survival outcome, it would be desirable to only include these features in risk prediction. To enable such within-group feature selection, Wang et al. (2009) propose a hierarchical approach that both selects important pathways and selects important genes within pathway. Specifically, they propose imposing an L_1-type penalty within pathway by minimizing

$$-\ell(\boldsymbol{\beta}) + \lambda \sum_{k=1}^{K} \left(\sum_{j=1}^{p_k} |\beta_{kj}| \right)^{\frac{1}{2}}$$

for an appropriate tuning parameter λ; they also propose an adaptive version that allows different penalties for different coefficients, and provide asymptotic results for a general set of estimators defined by minimizing penalized objective functions of the form $-\ell_n(\boldsymbol{\theta}) + \sum_{k=1}^{K} p_{\lambda_n}^{(k)}(|\boldsymbol{\beta}_k|)$. They establish variable selection consistency for both groups and individual features, as well as oracle properties with respect to estimation efficiency when $p_{\lambda_n}^{(k)}(|\boldsymbol{\beta}_k|)$ satisfy a set of criteria. Finally, they describe how to apply their method when pathways overlap.

5.2.4 Selection of tuning parameters

Tuning plays an important role in almost all regularization procedures. When p is large, it can be easy to find a model that accurately predicts the response in the data used to fit the model – the *training data* – but which has poor predictive performance when applied to a new dataset – the *validation data.* This is because with so many predictors available, the model can take advantage of random correlations between predictors and response in the training data to improve its apparent fit in that data; this problem is often referred to as *overfitting.* The goal of tuning, then, is to select a model with sufficient complexity to capture the signal in the training data and that provides useful predictions in a new dataset.

To choose a good value for a tuning parameter, we need to identify a range of possible values for the parameter, and select a criterion for evaluating the model corresponding to each value. In shrinkage procedures such as the ridge in which we minimize (5.4), small values of λ yield a fit close to an unconstrained fit, while large values of λ yield a fit with all components of β near 0; thus, we search for values of λ which provide a range of candidate models from a nearly full model to a nearly null model. To evaluate the model fit associated with each value of the tuning parameter, standard approaches include *cross-validation, AIC,* and *BIC,* but these criteria may require some additional care in the censored data setting.

When performing cross-validation, the data are repeatedly partitioned into training data and validation data, and for each partition, the model is built on the training data and its fit is evaluated on the validation data. This procedure is performed for each value of λ under consideration, and the value of λ which optimizes the model fit (averaged across partitions) is selected. In linear and logistic regression, the likelihood is often used as the model fit criterion in cross-validation, and it is natural to try to use the LPL as the model fit criterion for the Cox model. However, as described in Verweij and Van Houwelingen (2006), this needs to be done with care because unlike the likelihood which is the sum of independent terms, the LPL is the sum of dependent terms. Other criteria that could be used within cross-validation are measurements such as the C-statistic discussed in Section 5.5.

The AIC and BIC criteria were developed for linear regression and in that setting, each takes the form of the log-likelihood penalized by a term accounting for model complexity; to select a model, we would choose the tuning parameter which minimizes our selected criterion (either AIC or BIC). While the initial development of these criteria and their interpretation depended on the log-likelihood being a sum of independent terms, they have been successfully used for model selection in the Cox model by replacing the log-likelihood by the LPL, and defining

$$\text{AIC} = -2\ell\{\widehat{\boldsymbol{\theta}}(\lambda)\} + 2\text{DF}(\lambda)$$

and

$$\text{BIC} = -2\ell\{\widehat{\boldsymbol{\theta}}(\lambda)\} + \log(n)\text{DF}(\lambda),$$

where $\text{DF}(\lambda)$ are the effective degrees of freedom (DF) for the model corresponding to $\widehat{\boldsymbol{\theta}}(\lambda)$. One issue that has been raised when applying the BIC is whether the penalty term should be a function of the overall sample size n or the number of events $D = \sum_{i=1}^{n} \Delta_i$, since there are only D nonzero terms in the LPL (5.2). Using D in place of n is discussed and advocated in Volinsky and Raftery (2000). The effective DF for a given model with a regularized estimate of $\boldsymbol{\theta}$ can be defined in various ways including counting the number of nonzero coefficients, estimating the trace of a projection matrix, and estimating the covariance between the predicted response and observed response (Hastie et al., 2009). For the Lasso penalty, Zou et al. (2007) showed that the effective DF can be approximated well by the number of nonzero coefficients.

The choice of the criteria may depend on the goal of regularization. BIC-type criteria tend to select overly sparse models, which can hinder risk prediction; while AIC-type criteria and cross-validation tend to include more features. This can improve prediction performance, but can also lead to problems due to overfitting to the training data.

5.3 Methods based on derived variables

Feature selection is an appealing way to reduce model complexity, and is particularly effective when only a small number of candidate features relate to the outcome of interest. An alternative way to reduce the complexity of the feature space is to project the original space to a lower dimensional subspace and derive prediction models within the subspace. We can conceptualize a standard linear regression of $\mathbf{Y}$ on $\mathbf{Z}$ as a projection of the n-vector $\mathbf{Y}$ onto the subspace $V \in \mathbb{R}^n$ spanned by the p columns of the matrix $\mathbf{Z}$. When the columns of $\mathbf{Z}$ are collinear, the dimension of V will be less than p; when the columns of $\mathbf{Z}$ are correlated, we may be able to well-approximate V by a subspace V' with even smaller dimension c, in the sense that the projection of $\mathbf{Y}$ onto V' is "close" to the projection of $\mathbf{Y}$ onto V in some way. This intuition motivates a number of dimension reduction techniques, in which we seek to summarize information about the p variables $Z_1, \ldots, Z_p$ using a smaller number, c, of constructed variables $\mathcal{U}_1, \ldots, \mathcal{U}_c$, which are linear combinations of the original components of $\mathbf{Z}$. Letting $\mathbb{U}$ be the $n \times c$ matrix whose rows are the values of the constructed variables $\vec{\mathcal{U}} = (\mathcal{U}_1, \ldots, \mathcal{U}_c)^\top$ for each of the n individuals, we have:

$$\mathbb{U} = \mathbf{Z}\mathbb{V} \tag{5.5}$$

where $\mathbb{V}$ is a $p \times c$ matrix of weights. When $c \ll n$, standard low-dimensional regression methods can be implemented directly with the constructed variables $\vec{\mathcal{U}}$. For example, in linear regression, we could simply regress $\mathbf{Y}$ on the columns of $\mathbb{U}$ using ordinary least squares. Similarly, for survival data, we could use the constructed variables $\vec{\mathcal{U}}$ in the Cox model:

$$\lambda_{\mathbf{W}}(t) = \lambda_0(t) \exp(\boldsymbol{\alpha}' \mathbf{X} + \boldsymbol{\beta}' \vec{\mathcal{U}}). \tag{5.6}$$

What is left, then, is to specify a method for choosing the weight matrix $\mathbb{V}$ (and with it, the number of constructed variables c). We discuss principal components regression in Section 5.3.1, as well as methods based on partial least squares, a method developed for linear regression, in Section 5.3.2.

5.3.1 Principal components regression

A principal components (PC) decomposition of $\mathbf{Z}$ constructs covariates $\vec{\mathcal{U}}$ which sequentially capture directions of greatest variability in the data while being themselves uncorrelated. Specifically, we let the c columns of $\mathbb{V}$, $\{\mathbf{v}_1, .., \mathbf{v}_c\}$ be:

$$\mathbf{v}_l = \text{argmax}_{\mathbf{v}} \{\mathbf{v}^\top \mathbf{Z}^\top \mathbf{Z} \mathbf{v}\} \text{ such that } \mathbf{v}_l^\top \mathbf{v}_l = 1 \text{ and } \mathbf{v}_l^\top \mathbf{v}_{l'} = 0 \text{ for } l' < l.$$

The number of PCs to use in the model, c, may be any number less than $\min\{p, n\}$, and is a tuning parameter. Note that if tuning c by cross-validation, the principal components should be recalculated in the training set each time the data is partitioned. The PCs can be found by taking $\mathbb{V} = \mathbf{V}$ in the singular value decomposition of the data matrix

$$\mathbf{Z} = \mathbf{U}\mathbf{D}\mathbf{V}^\top.$$

PC regression (Massy, 1965) proceeds, as suggested above, by including these derived variables $\vec{\mathcal{U}}$ as covariates in the model of interest – for us, the Cox model. Note that PC regression based on standard singular value decomposition is never sparse in the original data, regardless of the choice of c, since the columns of $\mathbf{V}$ generally have nonzero components. However, sparse PC analysis has been proposed that selects only a subset of variables for each PC and can lead to a sparse model (Zou et al., 2006; Witten et al., 2009).

One main feature of PC regression is that the dimension reduction is completely unsupervised, in that the derived variables are only constructed using information on the predictors $\mathbf{Z}$ without considering how the Z_j relate to the outcome. To incorporate the information on the association between $\mathbf{Z}$ and the outcome into constructing the derived variables, supervised PC regression has been proposed (Bair and Tibshirani, 2004; Bair et al., 2006). In this approach, variables are screened univariately first, and those that pass a screening threshold are collected into a smaller data matrix, from which principal components are calculated as above. These PCs are sparse in the original data, and may encode more information relevant to the outcome of interest. Methods based on partial least squares (Section 5.3.2) also try to construct summary variables while incorporating information about outcome.

5.3.2 Approaches based on partial least squares

As described, standard PC regression is unsupervised and hence it is possible that while the top PCs capture variability in the feature space well, they are not associated with outcome. An alternative to PC regression methods is the partial least squares (PLS) method for constructing derived variables $\vec{\mathcal{U}}$, which has been previously proposed for linear regression (Wold et al., 1993). In linear regression when the outcome of interest $\mathbf{Y}$ is a continuous variable, PLS approaches seek directions that capture the largest amount of covariance between $\mathbf{Y}$ and $\mathbf{Z}$ instead of finding directions that capture the largest amount of variance in $\mathbf{Z}$. If we let the c columns $\{\mathbf{v}_1, ..., \mathbf{v}_c\}$ of $\mathbb{V}$ in (5.5) be

$$\mathbf{v}_l = \text{argmax}_{\mathbf{v}}\{\mathbf{v}^\top \mathbf{Z}^\top \mathbf{Y}\mathbf{Y}^\top \mathbf{Z}\mathbf{v}\} \text{ such that } \mathbf{v}_l^\top \mathbf{v}_l = 1 \text{ and } \mathbf{v}_l^\top \mathbf{Z}^\top \mathbf{Z}\mathbf{v}_{l'} = 0 \text{ for } l' < l,$$

then, as before, we can use the constructed variables $\mathbb{U} = \mathbf{Z}\mathbb{V}$ in a regression model.

The above formulation cannot be directly used in the survival setting since the weights $\mathbf{v}_l$ involve the covariance between $\mathbf{Y}$ and $\mathbf{Z}$ which is not directly available with censored data. A number of approaches have been proposed. Nguyen and Rocke (2002) suggest implementing PLS using the observed time to event T in place of Y regardless of an individual's censoring status, but this approach could produce misleading covariates not associated with survival if censoring is extensive or related to covariates. Park et al. (2002) reformulate the survival problem using Poisson regression in a generalized linear model framework. A comparison of these methods as well as a modification of the method of Park et al. (2002) are presented in Nygård et al. (2008), who also discuss explicitly how to include non-genomic covariates in their models in such a way that the non-genomic covariates are not subject to any dimension reduction; doing so is not always obvious in other PLS-based methods.

Another PLS-based approach, partial Cox regression, in which constructed variables are built up sequentially, is proposed by Li and Gui (2004). A first summary variable, say $\mathcal{U}_1$, is defined as the linear combination of genes with coefficients given by the coeffcients from univariate Cox models. To construct the next summary variable, all genes are regressed on $\mathcal{U}_1$, and the residuals from those regressions are used as new covariates in Cox models which include $\mathcal{U}_1$ as a covariate as well. The coefficients on the residual covariates from those Cox models are used as weights in $\mathcal{U}_2$. A final PLS-inspired approach is the sliced inverse regression method of Li and Li (2004). They begin by extracting the first q PCs of the data, where q is chosen so that the associated PCs capture a reasonable proportion of the variability structure of the predictors. They then use sliced inverse regression adapted to censored data (Li et al., 1999) to try to identify a small number of linear combinations of the principal components that capture all the information in the covariate space related to survival time.

5.4 Other models

While the Cox model has enjoyed a predominance of use in the medical literature for analyzing censored data, a number of survival models have been proposed as useful alternatives for settings when the Cox's proportional hazards assumption may fail to hold. Ma et al. (2010) compare the performance of gene signatures selected by Lasso in high-dimensional genomic settings in three commonly used survival models: the Cox model, the accelerated failure time (AFT) model, and the additive risk model. They found that the predictiveness and reproducibility of the gene signatures varied by model, and stressed the importance of considering multiple models when developing prognosis models. Here, we briefly describe methods developed for the nonparametric hazard model (Section 5.4.1), the additive risk model (Section 5.4.2), the AFT model (Section 5.4.3), and the semiparametric linear transformation model (Section 5.4.4).

5.4.1 Nonparametric hazard model

The nonparametric hazard model relaxes the assumption that the relationship between the covariates and the log-hazard is linear – i.e., it states that:

$$\lambda_{\mathbf{W}}(t) = \lambda_0(t)\exp\{g(\mathbf{W})\}$$

where the form of the function $g(\mathbf{W})$ is not specified. Li and Luan (2002) propose fitting this model by assuming $g(\mathbf{W})$ belongs to a reproducing kernel Hilbert space, and different choices of kernel lead to different amounts of flexibility in the form of $g(\mathbf{W})$. Leng and Zhang (2006) propose another method for fitting this model using kernel methods, by proposing a particular kernel with known basis functions and penalizing the sum of the norms of those basis functions to induce sparsity.

5.4.2 Additive risk model

The additive risk model, as introduced in Chapter 3, assumes that the hazard function is the sum of a baseline hazard function and covariate effects:

$$\lambda_{\mathbf{W}_i}(t) = \lambda_0(t) + \boldsymbol{\theta}^\top \mathbf{W}.$$

Ma and Huang (2007) propose methods for fitting this model with a Lasso-type penalty on the coefficients. More flexible formulations of the model have also been considered. Ma et al. (2006) examine several such formulations in the high-dimensional setting, and propose fitting them using PC regression for dimension reduction, with an additional step of test-based PC selection. Martinussen and Scheike (2009) develop ridge, Lasso, adaptive Lasso, and Dantzig selector methods for this model.

5.4.3 Accelerated failure time model

The accelerated failure time (AFT) model, as introduced in Chapter 3, relates the covariates directly to survival time via

$$\log T = g(\mathbf{W}) + \epsilon,$$

where ϵ is error with completely unspecified distribution. For linear effects with $g(\mathbf{W}) = \boldsymbol{\theta}^\top \mathbf{W}$, Huang et al. (2006) and Cai et al. (2009) develop methods for fitting the AFT model with a Lasso-type penalty on the coefficients. PC regression methods are described in Ma (2007). Methods that account for group structure are discussed in Wei and Li (2007) and Luan and Li (2008). Kernel machine methods that can allow $g(\mathbf{W})$ to take a nonlinear form are discussed in Liu et al. (2010).

A number of methods implement an imputation-based approach for fitting this model. They first impute the missing survival times for those whose times are censored, and then apply a method developed for linear regression on the imputed data. Using this idea, Huang and Harrington (2005) propose an adaptation of PLS, as do Datta et al. (2007), who compare Lasso and PLS procedures after imputation.

5.4.4 Semiparametric linear transformation models

The semiparametric linear transformation model states that:

$$h(T) = \boldsymbol{\theta}^\top \mathbf{W} + \epsilon$$

where ϵ now has a prespecified distribution but h is an unknown monotone increasing function. This model includes both the PH and proportional odds models as special cases. While this model lacks a natural loss function, Zhang et al. (2010) developed a variable selection procedure with Lasso-type penalty by constructing a "profiled score" loss function using estimating equations.

5.5 Data analysis example

Genomic information has already improved our understanding of breast cancer. A number of gene expression signatures have been introduced into clinical practice to better identify cancers with high and low risk of recurrence (Desmedt et al., 2011). Despite these advances, approximately 60% of patients with early-stage breast cancer are given adjuvant therapy in addition to local treatment, while only a small proportion are thought to benefit (Reis-Filho and Pusztai, 2011). Better markers of aggressive disease would help physicians predict which patients could safely avoid adjuvant therapy and its negative side effects, and which patients should be treated with more aggressive therapy.

To compare the performance of some of the methods described here, we applied the methods to a gene expression study of 286 lymph node negative breast cancer patients who received no systemic adjuvant therapy (Wang et al., 2005). Our goal was to derive gene expression signatures for predicting time to breast cancer progression or death using various methods, and to compare their predictive performance. First, to show the performance of the methods within a single dataset, we randomly partitioned the 286 patients into a training set of 150 patients and an internal validation set of 136 patients; each model was built in the training set and tuned using AIC and BIC within that dataset, and then applied to the internal validation set. To demonstrate the portability of each approach to independent data, we also applied each model to an external validation set made up of 119 lymph node negative patients with no adjuvant therapy, with gene expression assessed on the same chip (Sotiriou et al., 2006). Among the entire group of 286 patients in Wang et al. (2005), 107 deaths or recurrences were observed, with follow-up time ranging between 2 months and 14.3 years (median 7.2 years); 63% of observations were censored. Among the 119 patients in Sotiriou et al. (2006), 27 deaths or recurrences were observed, with follow-up time ranging between 2 months and 14.5 years (median 7.7 years); 77% of observations were censored. Both datasets were standardized so that the genes had mean 0 and standard deviation 1.

To assess the prediction performance of each model, we estimated a C-statistic using the approach proposed in Uno et al. (2011). A C-statistic is a measure of the concordance between an estimated risk score and the survival times (Harrell et al., 1996). If $g(\mathbf{W})$ is the risk score calculated from a model (e.g., $g(\mathbf{W}) = \theta^\top \mathbf{W}$ in model (5.1)) then a possible C-statistic is $\Pr(g(\mathbf{W}_1) > g(\mathbf{W}_2) | T_2 > T_1)$, which captures how well the ordering of the survival times matches the ordering of the estimated risk scores; however, when the survival times are subject to censoring and the censoring time may have support shorter than that of T, that C-statistic may not be estimable. We may instead define a modified version which is estimable, denoted by C_τ, which captures information only over a prespecified follow-up

TABLE 5.1
Results of breast cancer data analysis. Six methods, tuned with either the AIC or BIC criteria, are shown with an indication of their tuning parameter values (number of genes, estimated degrees of freedom, or number of principal components). Presented are C-statistics (C_τ) and 95% confidence interval lower and upper bounds (LB and UB) of the resulting models *within* the data used to build the model (Training); on an excluded subset of the same dataset (Internal); and on an independent dataset (External).

		Training			Internal			External		
	Tuning	C_τ	LB	UB	C_τ	LB	UB	C_τ	LB	UB
Univariate AIC	30 genes	89.8	85.6	94	57.8	47.5	68.1	63.5	47.9	79.2
Univariate BIC	11 genes	83	77.1	88.8	60.7	50.8	70.5	63.2	48.5	78
Ridge AIC	29.5 est DF	87.9	83.3	92.5	69.8	60.7	78.8	65.9	50.9	80.8
Ridge BIC	5.71 est DF	82.5	76.5	88.5	69.4	59.7	79.2	66.1	51.7	80.6
Lasso AIC	6 genes	80.1	73.7	86.6	58	47.9	68.1	60.2	45.7	74.6
Lasso BIC	0 genes	50	30.3	69.7	50	28.3	71.7	50	21.8	78.2
Elastic Net AIC	8.63 est DF	83.3	77.5	89.1	69.6	59.9	79.2	66.7	52	81.3
Elastic Net BIC	4.79 est DF	82.3	76.3	88.4	69.4	59.5	79.2	66.1	51.7	80.5
Adaptive Lasso AIC	5 genes	67	58.4	75.6	58.2	48.2	68.3	57.1	41.4	72.8
Adaptive Lasso BIC	0 genes	50	30.3	69.7	50	28.3	71.7	50	21.8	78.2
PCR AIC	104 PCs	100	100	100	63.7	54.4	73	61.9	47.1	76.7
PCR BIC	8 PCs	86.1	81.1	91.1	74.5	66.7	82.3	62.1	45.6	78.6

period $(0, \tau)$:

$$C_\tau = \Pr(g(\mathbf{W}_1) > g(\mathbf{W}_2) | T_2 > T_1, T_1 < \tau). \tag{5.7}$$

For example, in a study with approximately 5 years of follow-up after subject accrual, we might take $\tau = 5$. In our example, we take $\tau = 3$ years, and calculate the estimate of C_τ provided in Uno et al. (2011) which is a consistent, nonparametric estimate of C_τ under the assumption that censoring is independent of survival and covariates; we also provide 95% confidence intervals calculated using perturbation resampling.

The datasets initially had 12,774 genes, but the methods can run into computational difficulties when implemented using the whole data. Therefore, we did a preliminary screening using only the 150 patients in the training set, retaining genes with univariate Cox model p-value of 0.005 or less. This reduced the data to 394 genes. The results from fitting a selection of methods (univariate selection; penalization with ridge, Lasso, elastic net, and adaptive Lasso penalties; and principal components regression) and two tuning methods (AIC and BIC) are shown in Table 5.1. As presented, we used n in the BIC criterion; however, changing n to be D, the number of deaths in the data, did not substantively change the results. The analysis was performed in `R` using the packages `survival`, `genefilter`, `survcomp`, `penalized` and `survC1` (R Development Core Team, 2011; Therneau and Lumley, 2011; Gentleman et al., 2012; Haibe-Kains et al., 2010; Goeman, 2011; Uno, 2011).

The effects on model size of tuning using AIC versus BIC are evident: minimizing the BIC produces a much sparser model, even selecting 0 genes in the two Lasso fits. The AIC selects many more factors into the model and can lead to models that are potentially overfitted, but have higher predictive values. The methods which perform best with respect to predictive performance in this data are ridge regression and the elastic net, and the performance of these procedures is not overly dependent on whether AIC or BIC is used to tune the model, despite different resulting model complexity as quantified by the estimated degrees of freedom. Contrasting the C-statistics from the training dataset and the validation datasets, we can see alarmingly high overfitting biases, suggesting the importance of correcting for such biases using independent datasets when analyzing high-dimensional data. Results from the internal and external validation are fairly comparable, indicating that the gene signatures derived from these methods have reasonably good portability.

5.6 Remarks

A wide range of methods have become available to construct regression models for survival outcomes in the presence of high-dimensional predictors. Dimension reduction, a key component of these methods, can be achieved through feature selection and/or using lower dimensional derived predictors.

For the feature selection-based methods, it is crucial to select an appropriate tuning parameter to achieve an optimal balance between model complexity and estimated prediction accuracy with available data. The relative performance of different approaches is largely dependent on the particular setting, including aspects such as the signal-to-noise ratio, the sparsity of the underlying model as well as the goal of the model building. When the goal is to identify informative predictors for discovery purposes, procedures aiming to achieve variable selection consistency may be preferable. For such cases, the tuning parameters of the regularization procedures could be selected based on the BIC criterion. On the other hand, less stringent rules for variable selection, while resulting in larger models, often provides better prediction performance. Hence for prediction purposes, one may choose AIC or cross-validation for tuning parameter selection.

In general, for both prediction and interpretability, it is important to develop models that include low-dimensional routine clinical covariates in addition to functions of the high-dimensional genomic data (Bøvelstad et al., 2009; van Houwelingen and Putter, 2012). For a model to be useful, its predictions should order individuals correctly according to their true risk, a quality often referred to as *discrimination* and captured by the C-statistic. Additionally, the model should provide accurate predictions of the actual survival time for each patient. This latter quality is often referred to as *calibration*, and a different measure of model performance should be used to assess it, such as the Brier score (Brier, 1950).

In addition to the methods described above, clustering techniques are commonly used to assist in analyzing genomic data, and some have been proposed for implementation with survival outcomes (e.g., Hastie et al., 2001). In clustering methods, genes whose expression patterns are similar across individuals can be gathered together into a "meta-gene" in some manner, and then this meta-gene can be used as a predictor. Clustering methods can be somewhat unstable because the structure of the clusters can heavily depend on technique and parameter choices. One potentially fruitful use for cluster information is in building gene sets for use in methods that can take advantage of group structure, such as those mentioned in Section 5.2.3. Such an approach is proposed in Ma et al. (2007). They first divide genes into clusters using a K-means clustering approach. Important variables are selected within cluster using Lasso, and then important gene clusters are selected using the group Lasso.

When the primary goal is prediction, ensemble methods may be useful to achieve a good bias-variance tradeoff. Ensemble methods are methods which take a fitting procedure (such as a procedure for estimating $\widehat{\beta}$ in a Cox model) and perform that procedure repeatedly on perturbations of the original data, producing ultimately a final fit which is averaged over those perturbations (Bühlmann, 2004). For example, one such ensemble method is called "bagging" and aims to improve stability of a possibly unstable estimation approach such as variable selection by repeating the procedure on bootstrapped samples of the original data and averaging over the resulting $\widehat{\beta}$ estimates to create a final model. These methods can improve prediction by reducing variance or reducing bias, depending on the base procedure, but can result in models with some loss of interpretability. Several ensemble methods have been proposed for survival analysis (Hothorn et al., 2006); a comparison of some ensemble methods with standard approaches may be found in Van Wieringen et al. (2009).

Acknowledgments

JAS was supported by the National Institutes of Health (NIH) grants T32 GM074897 and T32 CA09001. TC was supported by the NIH grant R01 GM079330 and the National Science Foundation (NSF) grant DMS-0854970.

Bibliography

Antoniadis, A., Fryzlewicz, P. and Letué, F. (2010), 'The dantzig selector in Cox's proportional hazards model', *Scandinavian Journal of Statistics* **37**(4), 531–552.

Bach, F. (2008), 'Consistency of the group lasso and multiple kernel learning', *The Journal of Machine Learning Research* **9**, 1179–1225.

Bair, E., Hastie, T., Paul, D. and Tibshirani, R. (2006), 'Prediction by supervised principal components', *Journal of the American Statistical Association* **101**(473), 119–137.

Bair, E. and Tibshirani, R. (2004), 'Semi-supervised methods to predict patient survival from gene expression data', *PLoS Biology* **2**(4), e108.

Benjamini, Y. and Hochberg, Y. (1995), 'Controlling the false discovery rate: a practical and powerful approach to multiple testing', *Journal of the Royal Statistical Society. Series B (Methodological)* pp. 289–300.

Benner, A., Zucknick, M., Hielscher, T., Ittrich, C. and Mansmann, U. (2010), 'High-dimensional cox models: The choice of penalty as part of the model building process', *Biometrical Journal* **52**(1), 50–69.

Binder, H., Porzelius, C. and Schumacher, M. (2011), 'An overview of techniques for linking high-dimensional molecular data to time-to-event endpoints by risk prediction models', *Biometrical Journal* **53**(2), 170–189.

Bøvelstad, H., Nygård, S. and Borgan, Ø. (2009), 'Survival prediction from clinico-genomic models-a comparative study', *BMC bioinformatics* **10**(1), 413.

Bøvelstad, H., Nygård, S., Størvold, H., Aldrin, M., Borgan, Ø., Frigessi, A. and Lingjærde, O. (2007), 'Predicting survival from microarray data: a comparative study', *Bioinformatics* **23**(16), 2080–2087.

Brier, G. (1950), 'Verification of forecasts expressed in terms of probability', *Monthly Weather Review* **78**(1), 1–3.

Bühlmann, P. (2004), 'Bagging, boosting and ensemble methods', *Handbook of Computational Statistics*, 877–907.

Cai, T., Huang, J. and Tian, L. (2009), 'Regularized estimation for the accelerated failure time model', *Biometrics* **65**(2), 394–404.

Cox, D. (1972), 'Regression models and life-tables', *Journal of the Royal Statistical Society. Series B (Methodological)*, 187–220.

Datta, S., Le-Rademacher, J. and Datta, S. (2007), 'Predicting patient survival from microarray data by accelerated failure time modeling using partial least squares and lasso', *Biometrics* **63**(1), 259–271.

Desmedt, C., Michiels, S., Haibe-Kains, B., Loi, S. and Sotiriou, C. (2011), 'Time to move forward from first-generation prognostic gene signatures in early breast cancer', *Breast Cancer Research and Treatment*, 1–3.

Engler, D. and Li, Y. (2007), 'Survival analysis with large dimensional covariates: an application in microarray studies', *Harvard University Biostatistics Working Paper Series*, 68.

Fan, J., Feng, Y. and Wu, Y. (2010), 'High-dimensional variable selection for Cox's proportional hazards model', *Borrowing Strength: Theory Powering Applications-A Festschrift for Lawrence D. Brown, Institute of Mathematical Statistics, Beachwood, OH*, 70–86.

Fan, J. and Li, R. (2002), 'Variable selection for cox's proportional hazards model and frailty model', *The Annals of Statistics* **30**(1), 74–99.

Gentleman, R., Carey, V., Huber, W. and Hahne, F. (2012), *genefilter: methods for filtering genes from microarray experiments.* R package version 1.32.0.

Goeman, J. J. (2011), *Penalized R package.* R package version 0.9-37.

Gui, J. and Li, H. (2005), 'Penalized cox regression analysis in the high-dimensional and low-sample size settings, with applications to microarray gene expression data', *Bioinformatics* **21**(13), 3001–3008.

Haibe-Kains, B., Sotiriou, C. and Bontempi, G. (2010), *survcomp: Performance Assessment and Comparison for Survival Analysis.* R package version 1.1.6. **URL:** http://CRAN.R-project.org/package=survcomp

Harrell, F., Lee, K. and Mark, D. (1996), 'Tutorial in biostatistics multivariable prognostic models: issues in developing models, evaluating assumptions and adequacy, and measuring and reducing errors', *Statistics in Medicine* **15**, 361–387.

Hastie, T. J., Tibshirani, R. and Friedman, J. (2009), *The Elements of Statistical Learning: Data Mining, Inference, and Prediction*, Springer, New York.

Hastie, T., Tibshirani, R., Botstein, D. and Brown, P. (2001), 'Supervised harvesting of expression trees', *Genome Biology* **2**(1), research 003.12.

Hothorn, T., Bühlmann, P., Dudoit, S., Molinaro, A. and Van Der Laan, M. (2006), 'Survival ensembles', *Biostatistics* **7**(3), 355–373.

Huang, J. and Harrington, D. (2002), 'Penalized partial likelihood regression for right-censored data with bootstrap selection of the penalty parameter', *Biometrics* **58**(4), 781–791.

Huang, J. and Harrington, D. (2005), 'Iterative partial least squares with right-censored data analysis: A comparison to other dimension reduction techniques', *Biometrics* **61**(1), 17–24.

Huang, J., Ma, S. and Xie, H. (2006), 'Regularized estimation in the accelerated failure time model with high-dimensional covariates', *Biometrics* **62**(3), 813–820.

Jacob, L., Obozinski, G. and Vert, J. (2009), Group lasso with overlap and graph lasso, *in* 'Proceedings of the 26th Annual International Conference on Machine Learning', ACM, pp. 433–440.

Kim, J., Sohn, I., Jung, S., Kim, S. and Park, C. (2012), 'Analysis of survival data with group lasso', *Communications in Statistics-Simulation and Computation* **41**(9), 1593 1605.

Knight, K. and Fu, W. (2000), 'Asymptotics for lasso-type estimators', *Annals of Statistics* **28**, 1356–1378.

Leng, C. and Zhang, H. (2006), 'Model selection in nonparametric hazard regression', *Nonparametric Statistics* **18**(7-8), 417–429.

Li, H. and Gui, J. (2004), 'Partial Cox regression analysis for high-dimensional microarray gene expression data', *Bioinformatics* **20**(suppl 1), i208–i215.

Li, H. and Luan, Y. (2002), 'Kernel cox regression models for linking gene expression profiles to censored survival data,' *Pacific Symposium on Biocomputing*, vol. **8**, 65.

Li, K., Wang, J. and Chen, C. (1999), 'Dimension reduction for censored regression data', *The Annals of Statistics* **27**(1), 1–23.

Li, L. and Li, H. (2004), 'Dimension reduction methods for microarrays with application to censored survival data', *Bioinformatics* **20**(18), 3406–3412.

Liu, Z., Chen, D., Tan, M., Jiang, F. and Gartenhaus, R. B. (2010), 'Kernel based methods for accelerated failure time model with ultra-high dimensional data', *BMC Bioinformatics* **11**, 606.

Luan, Y. and Li, H. (2008), 'Group additive regression models for genomic data analysis', *Biostatistics* **9**(1), 100–113.

Ma, S. (2007), 'Principal component analysis in linear regression survival model with microarray data', *J. Data Sci.* **5**, 183–98.

Ma, S. and Huang, J. (2007), 'Additive risk survival model with microarray data', *BMC Bioinformatics* **8**(1), 192.

Ma, S., Huang, J., Shi, M., Li, Y. and Shia, B. (2010), 'Semiparametric prognosis models in genomic studies', *Briefings in Bioinformatics* **11**(4), 385–393.

Ma, S., Kosorok, M. and Fine, J. (2006), 'Additive risk models for survival data with high-dimensional covariates', *Biometrics* **62**(1), 202–210.

Ma, S., Song, X. and Huang, J. (2007), 'Supervised group lasso with applications to microarray data analysis', *BMC Bioinformatics* **8**(1), 60.

Martinussen, T. and Scheike, T. (2009), 'Covariate selection for the semiparametric additive risk model', *Scandinavian Journal of Statistics* **36**(4), 602–619.

Massy, W. (1965), 'Principal components regression in exploratory statistical research', *Journal of the American Statistical Association* **60**, 234–256.

Nardi, Y. and Rinaldo, A. (2008), 'On the asymptotic properties of the group lasso estimator for linear models', *Electronic Journal of Statistics* **2**, 605–633.

Nguyen, D. and Rocke, D. (2002), 'Partial least squares proportional hazard regression for application to DNA microarray survival data', *Bioinformatics* **18**(12), 1625–1632.

Nygård, S., Borgan, Ø., Lingjærde, O. and Størvold, H. (2008), 'Partial least squares Cox regression for genome-wide data', *Lifetime Data Analysis* **14**(2), 179–195.

Park, M. and Hastie, T. (2007), 'L1-regularization path algorithm for generalized linear models', *Journal of the Royal Statistical Society: Series B (Statistical Methodology)* **69**(4), 659–677.

Park, P., Tian, L. and Kohane, I. (2002), 'Linking gene expression data with patient survival times using partial least squares', *Bioinformatics* **18**(suppl. 1), S120–S127.

R Development Core Team (2011), *R: A Language and Environment for Statistical Computing*, R Foundation for Statistical Computing, Vienna, Austria. ISBN 3-900051-07-0. **URL:** http://www.R-project.org/

Reis-Filho, J. and Pusztai, L. (2011), 'Gene expression profiling in breast cancer: classification, prognostication, and prediction', *The Lancet* **378**(9805), 1812–1823.

Robins, J., Ritov, Y. et al. (1997), 'Toward a curse of dimensionality appropriate(coda) asymptotic theory for semi-parametric models', *Statistics in Medicine* **16**(3), 285–319.

Sotiriou, C., Wirapati, P., Loi, S., Harris, A., Fox, S., Smeds, J., Nordgren, H., Farmer, P., Praz, V., Haibe-Kains, B. et al. (2006), 'Gene expression profiling in breast cancer: understanding the molecular basis of histologic grade to improve prognosis', *Journal of the National Cancer Institute* **98**(4), 262.

Subramanian, A., Tamayo, P., Mootha, V. K., Mukherjee, S., Ebert, B. L., Gillette, M. A., Paulovich, A., Pomeroy, S. L., Golub, T. R., Lander, E. S. and Mesirov, J. P. (2005), 'Gene set enrichment analysis: a knowledge-based approach for interpreting genome-wide expression profiles', *Proc Natl Acad Sci U S A* **102**(43), 15545–15550.

Therneau, T. and Lumley, T. (2011), *survival: Survival analysis, including penalised likelihood.* R package version 2.36-5. **URL:** http://CRAN.R-project.org/package=survival

Tibshirani, R. et al. (1997), 'The lasso method for variable selection in the Cox model', *Statistics in Medicine* **16**(4), 385–395.

Tibshirani, R. et al. (2009), 'Univariate shrinkage in the Cox model for high dimensional data', *Statistical Applications in Genetics and Molecular Biology* **8**(1), 21.

Uno, H. (2011), *survC1 R package.* R package version 1.0-1.

Uno, H., Cai, T., Pencina, M. J., D'Agostino, R. B. and Wei, L. J. (2011), 'On the c-statistics for evaluating overall adequacy of risk prediction procedures with censored survival data', *Stat Med* **30**(10), 1105–1117.

Uno, H., Cai, T., Tian, L. and Wei, L. (2007), 'Evaluating prediction rules for t-year survivors with censored regression models', *Journal of the American Statistical Association* **102**(478), 527–537.

van Houwelingen, J. and Putter, H. (2012), *Dynamic Prediction in Clinical Survival Analysis*, CRC Press, Boca Raton, FL.

Van Wieringen, W., Kun, D., Hampel, R. and Boulesteix, A. (2009), 'Survival prediction using gene expression data: a review and comparison', *Computational statistics & data analysis* **53**(5), 1590–1603.

Verweij, P. and Van Houwelingen, H. (1994), 'Penalized likelihood in Cox regression', *Statistics in Medicine* **13**(23-24), 2427–2436.

Verweij, P. and Van Houwelingen, H. (2006), 'Cross-validation in survival analysis', *Statistics in Medicine* **12**(24), 2305–2314.

Volinsky, C. and Raftery, A. (2000), 'Bayesian information criterion for censored survival models', *Biometrics* **56**(1), 256–262.

Waldron, L., Pintilie, M., Tsao, M., Shepherd, F., Huttenhower, C. and Jurisica, I. (2011), 'Optimized application of penalized regression methods to diverse genomic data', *Bioinformatics* **27**(24), 3399–3406.

Wang, S., Nan, B., Zhu, N. and Zhu, J. (2009), 'Hierarchically penalized cox regression with grouped variables', *Biometrika* **96**(2), 307–322.

Wang, Y., Klijn, J., Zhang, Y., Sieuwerts, A., Look, M., Yang, F., Talantov, D., Timmermans, M., Meijer-van Gelder, M., Yu, J. et al. (2005), 'Gene-expression profiles to predict distant metastasis of lymph-node-negative primary breast cancer', *The Lancet* **365**(9460), 671–679.

Wei, Z. and Li, H. (2007), 'Nonparametric pathway-based regression models for analysis of genomic data', *Biostatistics* **8**(2), 265–284.

Witten, D. and Tibshirani, R. (2009), 'Covariance-regularized regression and classification for high dimensional problems', *Journal of the Royal Statistical Society: Series B (Statistical Methodology)* **71**(3), 615–636.

Witten, D. and Tibshirani, R. (2010), 'Survival analysis with high-dimensional covariates', *Statistical Methods in Medical Research* **19**(1), 29–51.

Witten, D., Tibshirani, R. and Hastie, T. (2009), 'A penalized matrix decomposition, with applications to sparse principal components and canonical correlation analysis', *Biostatistics* **10**(3), 515–534.

Wold, S., Johansson, E. and Cocchi, M. (1993), 'PLS: Partial Least Squares projections to latent structures', *3D QSAR in Drug Design* **1**, 523–550.

Yuan, M. and Lin, Y. (2006), 'Model selection and estimation in regression with grouped variables', *Journal of the Royal Statistical Society: Series B (Statistical Methodology)* **68**(1), 49–67.

Zhang, H. and Lu, W. (2007), 'Adaptive lasso for Cox's proportional hazards model', *Biometrika* **94**(3), 691–703.

Zhang, H., Lu, W. and Wang, H. (2010), 'On sparse estimation for semiparametric linear transformation models', *Journal of Multivariate Analysis* **101**(7), 1594–1606.

Zhao, S. and Li, Y. (2011), 'Principled sure independence screening for Cox models with ultra-high-dimensional covariates', *Journal of Multivariate Analysis* **105**(1), 397–411.

Zou, H. (2006), 'The adaptive lasso and its oracle properties', *Journal of the American Statistical Association* **101**(476), 1418–1429.

Zou, H. and Hastie, T. (2005), 'Regularization and variable selection via the elastic net', *Journal of the Royal Statistical Society: Series B (Statistical Methodology)* **67**(2), 301–320.

Zou, H., Hastie, T. and Tibshirani, R. (2006), 'Sparse principal component analysis', *Journal of Computational and Graphical Statistics* **15**(2), 265–286.

Zou, H., Hastie, T. and Tibshirani, R. (2007), 'On the degrees of freedom of the lasso', *The Annals of Statistics* **35**(5), 2173–2192.

Zou, H. and Zhang, H. (2009), 'On the adaptive elastic-net with a diverging number of parameters', *Annals of Statistics* **37**(4), 1733.

6

Cure Models

Yingwei Peng
Queen's University

Jeremy M. G. Taylor
University of Michigan

CONTENTS

6.1 Introduction 113
6.2 Mixture cure models 114
6.2.1 Model formulation 114
6.2.2 Estimation methods 116
6.2.3 Tonsil cancer example 117
6.2.4 Identifiability 118
6.2.5 Mixture cure model for clustered survival data 120
6.3 Proportional hazards cure model 122
6.3.1 Model formulation 122
6.3.2 Estimation methods 123
6.3.3 Proportional hazards cure model for clustered survival data 124
6.4 Unifying cure models based on transformations 125
6.5 Joint modeling of longitudinal and survival data with a cure fraction 127
6.6 Cure models and relative survival in population studies 128
6.7 Software for cure models 128
Bibliography 129

6.1 Introduction

The cure model, or sometimes called "the cure rate model," refers to a class of models for censored survival data from subjects when some of them will not develop the event of interest, however long they are followed. Those who are not going to develop the event of interest are often referred to as "cured subjects," or "long-term survivors." Clinical studies in cancer is a situation where there is a strong rationale for the existence of cured subjects because if the treatment is successful, the original cancer is removed and the subject will not experience recurrence of the disease. This is particularly true for patients in early cancer stages. Cured subjects can also be found in other disciplines, such as economics and social studies.

Let T be the time of the event of interest, then a characteristic of a cure model is that the limit of $P(T > t)$ is non-zero as t tends to infinity. If a simple estimate of $P(T > t)$, obtained for example from a Kaplan-Meier plot, does suggest a non-zero asymptote, then this is the type of situation where a cure model may be appropriate and useful.

The challenge in analyzing survival data from subjects with a possibility of being cured is how to handle censoring. A subject is censored either because he or she is cured or because he or she is not cured but has not been followed up for long enough for the event to occur. In many situations these two possibilities cannot be distinguished. In some cases it may be possible to define a threshold time, either from the scientific context or for pragmatic reasons, beyond which the event cannot happen, then a censored subject is considered cured if the censored time is greater than the threshold.

In cure models the distribution of T itself is usually not of interest, but rather it is the relationship between T and other covariates that is of primary interest. In this chapter we will discuss the formulation of cure models that involve covariates, such covariates are typically measured at time zero, but could also be time-dependent. Maller and Zhou (1996) provided a nice review of research in cure models prior to 1996. Other brief reviews can be found in Yakovlev et al. (1996); Ibrahim et al. (2001*b*); Lawless (2003). The purpose of this chapter is to review recent advances in cure models for analyzing survival data with a cured fraction. Particularly, the review will focus on general descriptions of cure models discussed in the last fifteen years in the literature. We will omit the more theoretical aspects which can be found in the references at the end of the chapter. The relationship among the models, their properties, estimation methods and software will be presented.

6.2 Mixture cure models

6.2.1 Model formulation

Because it is considered that some subjects would never experience the event of interest, it is natural to consider a mixture model where the population is a mixture of two groups, the cured and the not cured. Let T be the time to the event of interest and let Y be the cure indicator with $Y = 1$ if the subject is not cured and $Y = 0$ otherwise. Let $P(Y = 1) = \pi$. Define $S_u(t)$ and $S_c(t)$ as the survival functions of the uncured and cured populations respectively, i.e., $S_u(t) = P(T > t|Y = 1)$ and $S_c(t) = P(T > t|Y = 0)$. Following the discussion in Section 6.1, $S_c(t) \equiv 1$, i.e., it is a degenerate survival function. Thus the mixture cure model is defined by the following unconditional survival function of T:

$$P(T > t) = S(t) = \pi S_u(t) + 1 - \pi \tag{6.1}$$

Even though the mixture cure model was first proposed to analyze survival data with a cured fraction more than 60 years ago (Boag, 1949), it is still attracting a great deal of attention, because of its easy-to-use mixture model structure, its appealing interpretation and the ease of generalization to more complex situations. Since the mixture cure model is really a combination of two models, one sometimes called "the incidence model" for the probability of cure and one sometimes called "the latency model" for the event time, this facilitates the separate consideration of the effect of the covariates on the cure probability and the effect of covariates on the distribution of the time to the event for those who are not cured.

The effect of z, a set of covariates, on π is often modeled using a logistic link

$$\text{logit}[\pi(z)] = z'\gamma \tag{6.2}$$

where $\text{logit}(\pi) = \log(\pi/(1 - \pi))$ and γ is a vector of the coefficients of the covariates in z, which includes an intercept term.

The covariate effects on $S_u(t)$ can be modeled in a number of ways. Let x be the set of covariates that may have effects on $S_u(t)$ and $S_{u0}(t)$ be the baseline survival function when $x = 0$, which could be described either parametrically or nonparametrically. Typically there would be considerable overlap between x and z, and they may be identical except for the extra intercept term in z. The most widely used model for the effects of x on $S_u(t)$ is based on the proportional hazards (PH) assumption (Kuk and Chen, 1992; Peng and Dear, 2000; Sy and Taylor, 2000; Peng, 2003*b*; Fang et al., 2005):

$$S_u(t) = S_u(t|x) = S_{u0}(t)^{\exp(x'\beta)} \tag{6.3}$$

where β is a vector of the coefficients of the covariates in x. Whether β includes an intercept term depends on whether the baseline distribution is specified or not. The corresponding mixture cure model is referred to as the PHMC model . It is easy to use and interpret due to its similarity with the popular Cox PH model (Cox, 1972).

The accelerated failure time (AFT) model (Cox and Oakes, 1984) can also be used to model the effects of x on $S_u(t)$ (Li and Taylor, 2002*a*,*b*; Zhang and Peng, 2007*a*,*b*; Lu, 2010), which leads to the following equation

$$S_u(t|x) = S_{u0}(te^{-x'\beta}) \tag{6.4}$$

It is equivalent to assuming $\log T = x'\beta + \epsilon$ and $P(e^{\epsilon} > t) = S_{u0}(t)$ for uncured subjects. The corresponding mixture cure model is referred to as the "AFTMC model." Unlike the PH assumption, the AFT assumption allows crossing hazards and direct interpretation of the effects of x on the $\log T$ scale.

Another useful alternative model assumption for $S_u(t|x)$ is the accelerated hazards (AH) model (Chen and Wang, 2000; Zhang and Peng, 2009):

$$S_u(t|x) = S_{u0}(te^{-x'\beta})^{\exp(x'\beta)} \tag{6.5}$$

The corresponding cure model is referred to as the "AHMC model." Unlike the PH and AFT assumptions above, the AH assumption allows a gradual effect of x on the distribution of T for uncured subjects. That is, if the baseline hazard function is monotone but not a constant, the hazard functions of two groups differ only when $t > 0$, and the larger the time t, the greater the differences in hazard. This can be seen easily if the model is rewritten based on the corresponding hazard functions: $h_u(t|x) = h_{u0}(te^{x'\beta})$.

The proportional odds (PO) assumption can also be used to model the effects of x on $S_u(t|x)$ as follows:

$$S_u(t|x) = \frac{1}{1 + [S_{u0}(t)^{-1} - 1]e^{-\beta' x}}$$

The corresponding cure model is referred to as the "POMC model." Contrary to the AH model, the PO assumption implies that the hazard ratio of the uncured subjects approaches one as $t \to \infty$. That is, the differences in hazard will fade away under the PO assumption.

For the interpretation for all of these models it is important to recognize that $S_u(t|x)$ is the conditional distribution of T given not cured, and the interpretation in terms of hazards or odds does not apply to the unconditional distribution $S(t)$.

The major appeal of the mixture cure model is the flexibility it provides in how the covariates can affect the event time distribution. The association is given by the two sets of parameters γ and β. The γ's describe whether the subject is cured and the β's describe when the event will happen amongst those who are not cured. Because there are more parameters needed to describe the relationship between the covariates and the event time than in a typical survival model, the data may need to be richer or its size larger to permit reliable estimation. If the data are not very informative about the effect of a covariate, then there is a danger that the corresponding γ and β are competing with each other to describe its effect. However, if the data are sufficiently informative then the potential for a clearer interpretation makes the model very attractive.

6.2.2 Estimation methods

The EM algorithm (Dempster et al., 1977) can be conveniently used to obtain the maximum likelihood estimates of the parameters in the mixture cure models. Let C be the non-informative censoring time variable. Suppose that there are n subjects in a study, and the observed values of $\min(T, C)$, $I(T \leq C)$, x and z for the subject i are denoted as $(t_i, \delta_i, x_i, z_i)$, $i = 1, \ldots, n$, where $I(A)$ is an indicator function with $I(A) = 1$ if A is true and 0 otherwise. Let y_i be the value of the partially latent variable Y for subject i. Given y_i, $i = 1, \ldots, n$, the log-likelihood function is $\ell(\gamma, \beta, S_{u0}|y_1, \cdots, y_n) = \ell_1(\gamma|y_1, \cdots, y_n) + \ell_2(\beta, S_{u0}|y_1, \cdots, y_n)$, where

$$\ell_1(\gamma|y_1, \cdots, y_n) = \sum_{i=1}^{n} \{y_i \log[\pi(z_i)] + (1 - y_i) \log[1 - \pi(z_i)]\}$$

$$\ell_2(\beta, S_{u0}|y_1, \cdots, y_n) = \sum_{i=1}^{n} \{y_i \delta_i \log[h_u(t_i|x_i)] + y_i \log[S_u(t_i|x_i)]\}$$

Given the current estimates of γ, β, S_{u0}, the E-step of the EM algorithm calculates the posterior expectation of y_i as follows:

$$w_i = E(y_i|\gamma, \beta, S_{u0}) = \delta_i + (1 - \delta_i) \frac{\pi(z_i) S_u(t_i|x_i)}{1 - \pi(z_i) + \pi(z_i) S_u(t_i|x_i)}$$

and replaces y_i with w_i in ℓ_1 and ℓ_2. The M-step updates the estimates of γ, β, S_{u0} by maximizing $\ell_1(\gamma|w_1, \ldots, w_n)$ and $\ell_2(\beta, S_{u0}|w_1, \ldots, w_n)$. The E-step and M-step iterate until a convergence is achieved. After the algorithm has converged the estimates of w_i have a nice interpretation as the probability the subject is not cured given the observed data for that subject.

Maximizing $\ell_1(\gamma|w_1, \ldots, w_n)$ can be carried out easily by the Newton-Raphson method. Maximizing $\ell_2(\beta, S_{u0}|w_1, \ldots, w_n)$, however, relies on how $S_{u0}(t)$ is parametrized. If $S_{u0}(t)$ is specified up to a few unknown parameters, $\ell_2(\beta, S_{u0}|w_1, \ldots, w_n)$ can be easily maximized in a similar way as $\ell_1(\gamma|w_1, \ldots, w_n)$. We will primarily consider the semiparametric mixture cure model, in which $S_{u0}(t)$ is nonparametrically specified. Given the fact that $w_i \delta_i = \delta_i$ and that ℓ_2 is similar to the log-likelihood functions of survival data without a cure fraction, many estimation methods for survival data without a cure fraction can be adapted to maximize $\ell_2(\beta, S_{u0}|w_1, \ldots, w_n)$ when S_{u0} is nonparametrically specified. For example, for the PHMC model, $\ell_2(\beta, S_{u0}|w_1, \ldots, w_n)$ can be treated as the regular log-likelihood function for the PH model with $w_1, \ldots, w_n$ as offset values, and it can be maximized using existing methods for Cox's PH model (Peng and Dear, 2000; Sy and Taylor, 2000, 2001; Peng, 2003*a*,*b*). The estimates are proved to be consistent and asymptotically normal (Fang et al., 2005). The EM algorithm can be easily adapted to accommodate a monotone baseline hazard function $h_{u0}(t)$ (Peng and Dear, 2004) and further heterogeneity in the failure time distribution of uncured subjects that is not captured by the existing x (Peng and Zhang, 2008*a*).

The methods above only produce a nonparametric, nonsmooth estimate of the baseline survival function $S_{u0}(t)$. If a smooth baseline survival function is preferred, the PHMC model can be estimated by allowing flexible modeling of the hazard function $h_{u0}(t)$ using M-splines

(Corbiere et al., 2009). The smoothing parameter in the M-splines can be determined using cross-validation.

Semiparametric methods for the AFT model can also be adapted to update $\ell_2(\beta, S_{u0}|w_1, \ldots, w_n)$ even though they tend to be more involved than those for the PH model. For example, the two semiparametric methods for the AFT model discussed in Ritov (1990) were both adapted to update $\ell_2(\beta, S_{u0}|w_1, \ldots, w_n)$ successfully in Li and Taylor (2002*a*); Zhang and Peng (2007*b*). However, both methods suffer from slow or lack of convergence due to nonsmooth estimating equations. Recently, a more efficient estimation method based on a smoothed likelihood function was proposed for the AFT model (Zeng and Lin, 2007). It was adapted to update $\ell_2(\beta, S_{u0}|w_1, \ldots, w_n)$ for the AFTMC model (Lu, 2010), and is more computationally efficient than the previous methods. It also produces a smooth baseline hazard function estimate.

For the AHMC model, a semiparametric estimation method for the AH model (Chen and Wang, 2000) was adapted to update $\ell_2(\beta, S_{u0}|w_1, \ldots, w_n)$ for the AHMC model (Zhang and Peng, 2009). But this method also suffers the computational issues due to nonsmooth estimating equations, and the method of Zeng and Lin (2007) was recently adapted by Zhang et al. (2013) to estimate the parameters in the semiparametric AHMC model more computationally efficiently.

Bayesian methods are also considered for the mixture cure models. For example, Zhuang et al. (2000); Cho et al. (2001) considered the parametric PHMC model with $S_{u0}(t)$ from a Weibull or a piecewise constant hazard distribution and possible missing values in x and z. The missing values are assumed to be missing at random (MAR). They employed proper priors on the parameters in the mixture cure model and in the distributions for the missing covariates and a Metropolis-Hastings algorithm within the Gibbs sampler to obtain draws from the posterior distribution of the parameters. An improper prior should be avoided for γ because otherwise the posterior distribution of β and γ will be improper (Chen et al., 1999), which may cause convergence issues in the simulation algorithm.

6.2.3 Tonsil cancer example

A head and neck cancer study (Withers et al., 1995; Sy and Taylor, 2000; Peng et al., 2007) provides a nice example where the merits of the mixture cure model can be illustrated. In this study patients with localized disease of the tonsil were treated with radiation therapy. In this situation the goal of the radiation treatment is to kill the cancerous cells within the tumor, and the endpoint of interest for this treatment is recurrence of the cancer within the tonsil region, called "local recurrence." For tonsil cancer it is well known that the majority of local recurrences occur within three years, and very rarely are they after five years. Thus patients who are followed for more than five years can effectively be considered as locally cured. In this study there were many patients with follow-up longer than five years and a Kaplan-Meier plot of time-to-recurrence did have a horizontal asymptote (see Figure 6.1). This is a situation where it is natural to consider a mixture model because the patients can be thought of as either being cured or not cured at time zero by the treatment. However, the cured status of the patient is not observed at time zero, and may only reveal itself later. In this study the time independent covariates of interest were the dose of radiation and the number of days over which it was delivered, the size of the tumor (measured by T stage), the nodal status and the patient's age. Fitting a PHMC model revealed that the dose, number of days and size of the tumor were important factors in whether the patient was cured, but not in when the recurrence occurred given not cured. In contrast the age of the patient was important for when the recurrence happens. Younger patients tended to recur earlier, which is very consistent with the concept that such patients have more aggressive or faster growing (in this case faster regrowing) tumors. The nodal status had a possible impact on

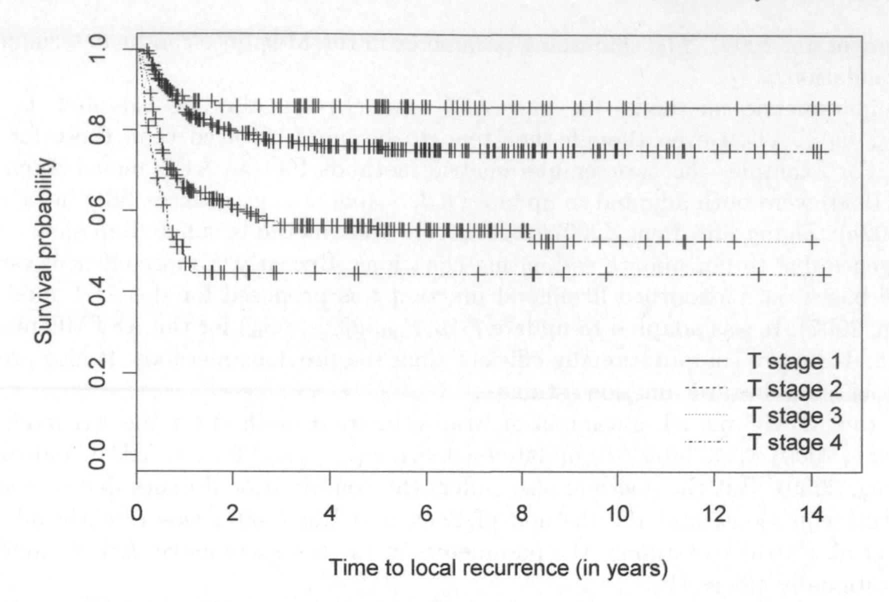

FIGURE 6.1
Kaplan-Meier survival curves of the tonsil cancer patients by T stages.

both the incidence and the latency. This is consistent with clinical knowledge that node positivity reflects both larger extent of disease, hence harder to cure, and the ability of the tumor to spread, hence earlier recurrence. The specific parameter estimates are given by Model 1 in Table 25.1 when a logistic model is assumed for the incidence and a Weibull model for the latency in the PHMC model. More details of using PHMC models for this dataset can be found in Sy and Taylor (2000) and Peng et al. (2007).

6.2.4 Identifiability

A mixture cure model can be considered a special case of a frailty model in survival analysis. Thus some of the well-known identifiability issues with frailty models for single event times may also arise for cure models. However, the cure model is not as general as the frailty model, because the frailty variable is binary and for one of the mixture groups the survival distribution is known. Because of this, the issues of identifiability are usually less of a concern.

The fundamental potential problem with identifiability arises because of uncertainty associated with the tail of the distribution $S_{u0}(t)$. The parameters of the model for $\pi(z)$ describe what happens at $t = \infty$. Since follow-up is never infinite, this may be problematic if many events could plausibly occur after the longest follow-up. In this case it may be difficult to distinguish a high cure rate with a short tail for $S_{u0}(t)$ from a low cure rate with a long tail. Thus for example, different choices for the form of $S_{u0}(t)$ can lead to different cure rates (Yu et al., 2004), also the variances of the parameter estimates tend to be large (Farewell, 1986). For these reasons the absolute value of the cure rate should be interpreted cautiously. In practice we have found that these problems do not manifest themselves if care is taken in choosing situations where a cure model is appropriate and in choosing which aspects of the fit to interpret. While the cure rates themselves need to be interpreted cautiously, the regression coefficients β and γ, except for the intercept in γ, are more stable provided there is a sufficient number of events.

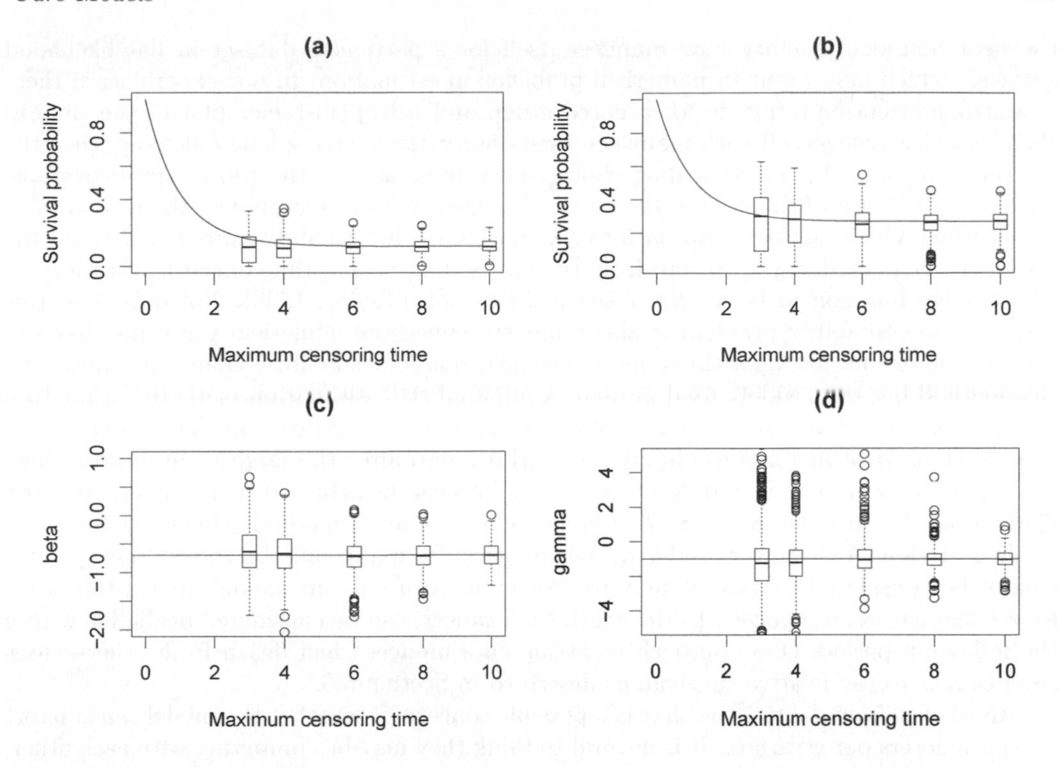

FIGURE 6.2
Boxplots of the estimated cure rates from group $z = 0$ (a), $z = 1$ (b), and boxplots of the estimates of β (c) and γ (d) under various maximum censoring times (follow-up times).

The identifiability issue can be illustrated in the following numerical example. Suppose that we generate data from model (6.1), (6.2), and (6.4) with $S_{u0}(t) = \exp(-\lambda t)$. We only consider one binary covariate that is used for both x and z in the model with $\beta = \log(0.5)$, and $\gamma' = (\gamma_0, \gamma_1) = (2, -1)$. The setting implies that the cure rate for the group with $z = 0$ is 12% and 27% for the group with $z = 1$, and that the hazard of uncured subjects in the group with $x = 1$ is half of the hazard of uncured subjects in the group with $x = 0$. We examine how the parameter estimates in the mixture cure model are affected by the identifiability issue caused by insufficient follow-up. The censoring times are generated from the uniform distribution between 0 and M, a value that determines the length of follow-up. We set $M = 3, 4, 6, 8, 10$, and for each value of M, we generated 500 datasets with each dataset containing 200 censored survival times and covariate values from the mixture cure model. The mixture cure model with $S_{u0}(t)$ from the Weibull distribution is fit to the data. The boxplots of the estimated cure rates from the two groups ($z = 0$ and $z = 1$) and of the estimates of β and γ are given in Figure 6.2 (a), (b), (c), and (d), respectively. The smooth curve in (a) and (b) are the true survival function $S(t)$ for $z = 0$ and $z = 1$, and the horizontal lines in (c) and (d) correspond to the true value of $\beta = \log(0, 5)$ and $\gamma_1 = -1$. It is easy to see that when the follow-up decreases, the variability of cure rates increases. However, the estimates of β and γ_1 are relatively stable, and the increase in variability is mild compared to that in cure rate estimates when the follow-up time decreases.

If fully nonparametric models are used for both $\pi(z)$ and $S_u(t|x)$, then identifiability is likely to be a problem. However, once some structure is assumed for these terms, such as a logistic model for $\pi(z)$ and proportional hazards for $S_u(t|x)$, then the models are identifiable (Li et al., 2001; Peng and Zhang, 2008*b*). Although technically identifiable, the near non-identifiability may manifest itself for a particular dataset in flat likelihood surfaces, which may result in numerical problems in estimation. In our experience if there is a strong scientific rationale for a cured group and a Kaplan-Meier plot of the time to the event clearly levels off with many censored observations having long follow-up, then the mixture cure model is an appealing choice and safe to use. In the tonsil cancer example described in Section 6.2.3, where the event of interest is local recurrence, these conditions are satisfied. One pragmatic approach to eliminating the identifiability problem is to set the tail of the estimated $S_{u0}(t)$ to zero from the largest uncensored time onwards, thus forcing the survival function to be a proper survival function (Taylor, 1995). Not only does this solve the identifiability problem it also tends to reduce the numerical problems that can arise in the estimation algorithms due to the near non-identifiability. Due to its simplicity, this method has been widely used in many semiparametric estimation methods for mixture cure models. A less-known but more sophisticated method to address this issue is to have the estimated survival function gradually approaching zero after the largest uncensored time, instead of being zero immediately at that time. This can be achieved by using a parametric distribution for just the tail of $S_u(t|x)$ in the semiparametric methods (Peng, 2003*a*).

In situations, in a cancer study say, where death from any cause is the endpoint, which cannot be considered curable, it may be dangerous to use a cure model, unless the death from other causes, as opposed to the death from cancer, can be considered negligible within the follow-up period. One approach to fitting cure models when death from other causes can occur is to use relative survival, as described in Section 6.6.

Another identifiability issue that is a possible concern, is whether the model can support two parameters per covariate. It is natural to think they may be competing with each other, diminishing confidence in their interpretation. In our experience, this is not a concern if there is sufficient follow-up in the data. We illustrate this with the tonsil cancer data described in Section 6.2.3. Model 1 in Table 6.1 shows the results when all covariates are included in both parts of the model, with the clear interpretation that some covariates (T-stage, Total dose and Treatment duration) are important for incidence, that age is important for latency, and that node may be important for both. Models 2 and 3 are when the covariates are included in only one part of the model. For most of the covariates the coefficients do not change that much from Model 1, thus there is a real loss of interpretation. For example, the important effects of total dose and treatment duration on incidence in Model 1 do not get compensated for in a model with only latency coefficients (Model 2), thus these effects would be missed. Similarly the effect of age on latency is missed in a model (Model 3) with only coefficients for incidence.

The results for Model 4 are for when all the observations are artificially censored at 1 year, a time that would be considered as insufficient follow-up. Some of the covariate coefficients are substantially different from those in Model 1; for example, the effect of age on latency is now lost. The difference is due to identifiability caused by insufficient follow-up as demonstrated in Figure 6.2.

6.2.5 Mixture cure model for clustered survival data

The mixture cure model has been generalized to the situation where subjects are clustered. Clustered subjects can be twins, patients from the same family, hospital or health centers, animals from the same litter, etc. Repeated failure times from one subject also forms a cluster. Due to shared environment and latent factors specific to each cluster, the survival times of uncured subjects and the cure status of all subjects from the same cluster tend to be correlated. It may be important to take this correlation into account when analyzing such data.

Let T_{ij}, Y_{ij}, δ_{ij}, x_{ij} and z_{ij} be the observed failure time, cure status, censoring indicator,

TABLE 6.1
Parameter estimates and their standard errors and p-values from four Weibull mixture cure models for the tonsil cancer data: Model 1 (covariates appeared in both latency and incidence parts), Model 2 (covariates appeared in latency part only), Model 3 (covariates appeared in incidence part only), Model 4 (covariates appeared in both latency and incidence parts, but survival times are censored at 1 year).

	Model 1			Model 2			Model 3			Model 4		
	β	s.e.	p-value	β	s.e.	p-value	β	s.e.	p-value	β	s.e.	p-value
Latency part												
Log(scale)	-0.16	0.05	.00	-0.15	0.05	.00	-0.07	0.05	.20	-0.62	0.08	.00
Intercept	-2.24	0.73	.00	-2.26	0.76	.00	-0.06	0.08	.42	-0.97	0.65	.13
T stage 2 (vs. stage 1)	0.59	0.28	.04	0.48	0.33	.14				0.20	0.32	.54
T stage 3 (vs. stage 1)	0.26	0.28	.34	0.04	0.32	.89				-0.26	0.26	.31
T stage 4 (vs. stage 1)	-0.30	0.30	.32	-0.53	0.34	.12				-0.25	0.27	.35
Node	-0.26	0.14	.07	-0.35	0.15	.02				-0.33	0.17	.05
Total dose	0.04	0.11	.69	0.14	0.12	.22				0.12	0.09	.21
Treatment duration	0.02	0.07	.79	-0.03	0.07	.61				-0.03	0.05	.46
Sex: male (vs. female)	0.09	0.15	.55	0.06	0.15	.71				-0.32	0.17	.05
Age	0.28	0.08	.00	0.25	0.08	.00				0.07	0.06	.25
Incidence part												
Intercept	-0.07	1.00	.94	-0.66	0.09	.00	0.30	0.98	.76	0.43	1.08	.69
T stage 2 (vs. stage 1)	0.82	0.35	.02				0.73	0.35	.04	0.40	0.45	.37
T stage 3 (vs. stage 1)	1.69	0.34	.00				1.69	0.34	.00	1.22	0.38	.00
T stage 4 (vs. stage 1)	2.22	0.43	.00				2.40	0.44	.00	2.07	0.46	.00
Node	0.40	0.20	.04				0.45	0.20	.02	0.29	0.26	.26
Total dose	-0.79	0.18	.00				-0.78	0.18	.00	-0.56	0.18	.00
Treatment duration	0.47	0.13	.00				0.45	0.12	.00	0.34	0.12	.00
Sex: male (vs. female)	0.16	0.21	.45				0.14	0.21	.51	-0.20	0.29	.49
Age	0.12	0.09	.21				0.06	0.09	.50	-0.05	0.10	.62

and two sets of covariates from the jth subject in the ith cluster, $i = 1, \ldots, n$, $j = 1, \ldots, n_i$. We assume that given covariates, the censoring is independent of the failure time T_{ij} and the cure status Y_{ij}. We further assume that $T_{ij}|Y_{ij} = 1$'s and Y_{ij}'s from subjects in the same cluster may be correlated, but those from different clusters are independent.

A wide range of statistical methods have been developed to handle clustered survival data with a cure fraction under the mixture cure model framework. One popular approach is to estimate subject-specific measures of effect based on random effect or frailties. Denote by u_i the random effect that induces the correlation among $T_{ij}|Y_{ij} = 1$, and denote by v_i the random effect that induces the correlation among Y_{ij}, then (6.2) and (6.4) in the mixture cure model (6.1) can be extended to include the random effects in the following conditional models (Yau and Ng, 2001; Lai and Yau, 2008; Peng and Taylor, 2011):

$$S_u(t|x_{ij}, u_i) = S_{u0}(t)^{\exp(x'_{ij}\beta + u_i)},$$
$$\text{logit}[\pi(z_{ij}, v_i)] = z'_{ij}\gamma + v_i,$$

where u_i and v_i are often assumed to follow a normal distribution with mean 0 and variance σ_u^2 and σ_v^2. The two random effects can be independent or correlated. The models above provide simple shared frailty structures for the data. A more flexible model can be assumed for both parts of the mixture cure model by replacing u_i by $\tilde{x}'_{ij}u$ and v_i by $\tilde{z}'_{ij}v$, where u and v are vectors of random effects, and $\tilde{x}_{ij}$ and $\tilde{z}_{ij}$ are the design vectors associated with u and v. The parameters can be estimated using the BLUP (best linear unbiased predictor) method and the REML (residual maximum likelihood estimator) when the baseline survival function $S_{u0}(t)$ is parametrically or nonparametrically specified (Yau and Ng, 2001; Lai and Yau, 2008). Obtaining the maximum likelihood estimates of the parameters, on the other hand, can be computationally intensive (Seppa et al., 2010; Peng and Taylor, 2011). But this approach allows non-normal random effects and can be used to estimate parameters in the AFTMC or the AHMC model with random effects.

The other approach to handle clustered survival data with a cured fraction is to estimate marginal, population-averaged measures of effects. This approach usually specifies the marginal distributions of $T_{ij}|Y_{ij}$ and Y_{ij} similar to (6.1) and (6.2) with one of (6.3), (6.4), (6.5), and leaves the correlation structures among them unspecified, thus is robust to model misspecification. To estimate the parameters in the marginal distributions, one can temporarily ignore the correlation and estimate the parameters using a method discussed in Section 6.2.2, and then adjust the variances of the estimates due to the correlation using the sandwich or Jackknife methods (Peng et al., 2007; Yu and Peng, 2008). The models of Chatterjee and Shih (2001) and Wienke et al. (2003) are marginal mixture cure models. However, the correlations among $T_{ij}|Y_{ij}$ and among Y_{ij} are explicitly estimated in the model in addition to the parameters in the marginal distributions. However, these approaches differ in that they do not allow covariate effects, and the marginal parameters are estimated by a quasi-likelihood method, instead of using a full-likelihood method.

6.3 Proportional hazards cure model

6.3.1 Model formulation

Another approach to develop a cure model is to consider the kinetics underlying the growth of a tumor in a cancer patient. One theory is to assume (Yakovlev et al., 1996)

$$T = \min\{\tilde{T}_1, \ldots, \tilde{T}_N\} = \tilde{T}_{(1)}, \tag{6.6}$$

where $\tilde{T}_i$'s are i.i.d. latent event times or activation times that can be viewed as promotion time for cancer cells to develop a detectable cancer mass, N is a discrete random variate to denote the number of $\tilde{T}_i$ and is independent of $\tilde{T}_i$'s, and $\tilde{T}_{(1)}$ is the first-order statistic of $\tilde{T}_1, \ldots, \tilde{T}_N$. This is also called the "first-activation scheme" and is one of a number of schemes developed in Cooner et al. (2007) that may be suitable for tumor kinetics. The schemes involve different distributions for N and $\tilde{T}_i$ and different order statistics $\tilde{T}_{(r)}$, $1 \leq r \leq N$, to define the failure time T, and they lead to different cure models. The most popular model is based on the first activation scheme with N following a Poisson distribution with mean $\exp(\beta_0 + z'\beta)$, and $\tilde{T}_i \sim F^H(t)$. Under this assumption, the unconditional survival function is

$$P(T > t) = S(t|z) = \exp[-e^{\beta_0+z'\beta} F^H(t)] \tag{6.7}$$

The model can be rewritten as $S(t|z) = [e^{-\exp(\beta_0)F^H(t)}]^{\exp(z'\beta)}$ or $h(t|z) = f^H(t)\exp(\beta_0 + z'\beta)$, where $f^H(t) = dF^H(t)/dt$. It is obviously similar to the classic PH model (Cox, 1972) except that the baseline survival function counterpart of the PH model in model (6.7) is an improper survival function satisfying $\lim_{t\to\infty} \exp[-e^{\beta_0}F^H(t)] = \exp(-e^{\beta_0}) \in (0, 1)$. Thus we refer to this model as the proportional hazards cure (PHC) model. Since the baseline cumulative hazard counterpart of the PH model in model (6.7), $\exp(\beta_0)F^H(t)$, is bounded, model (6.7) is also called the "bounded cumulative hazard cure model."

As an alternative to the mixture cure model, the PHC model received a great deal of interest in the last decade due to the similarity of the model to the classical PH model. However, one disadvantage of the PHC model in comparison with the mixture cure model is that the effect of z on the distribution of $T|Y = 1$ does not have a simple interpretation. This can be seen from the fact that $P(Y = 1) = \pi = \exp(-e^{\beta_0})$ and

$$P(T > t|Y = 1) = S_u(t|z) = \frac{\exp[e^{\beta_0+z'\beta}F^H(t)] - 1}{\exp[e^{\beta_0+z'\beta}] - 1}$$

in the PHC model. The proportionality restriction in z in the model also limits the wide applications of the model in practice. The restriction can be relaxed by including covariate effects in $F^H(t)$, such as

$$1 - F^H(t|x) = S^H(t|x) = S_0^H(t)^{\exp(x'\gamma)}, \tag{6.8}$$

where $S_0^H(t)$ is a baseline survival function (Tsodikov, 2002). The inclusion of x in $F^H(t)$ widens the applicability of the PHC model, because there are now two sets of parameters to describe the effects of the covariates. But the covariate effects in $S_u(t)$ are difficult to interpret compared to the PHMC model because the effects of z and x are not well separated.

6.3.2 Estimation methods

Let $D = \{t_1, \ldots, t_n, \delta_1, \ldots, \delta_n\}$ be the set of observed data. The likelihood function for the data is

$$L(\beta, \alpha|D) = \prod_{i=1}^{n} [e^{\beta_0+z_i'\beta} f^H(t_i)]^{\delta_i} \exp[-e^{\beta_0+z_i'\beta} F^H(t_i)]$$

where α is the parameter in $F^H(t)$. Maximum likelihood estimation methods can be used to estimate β and $F^H(t)$ when $F^H(t)$ is nonparametrically specified (Tsodikov, 1998, 2003). By treating N in (6.6) as a latent variable, Chen and Ibrahim (2001) proposed to use the EM algorithm to obtain the maximum likelihood estimates of the parameters in the PHC model when $F^H(t)$ is assumed to have a piecewise constant hazard with a smoothing parameter that controls the degree of parametricity in the right tail of $F^H(t)$. The method can also handle missing values in covariates when missing is MAR (Chen and Ibrahim, 2001; Herring and Ibrahim, 2002).

Bayesian methods have been described to estimate the parameters in the PHC model. The posterior distribution of β and α can be written as

$$p(\beta, \alpha|D) \propto L(\beta, \alpha|D)p_0(\beta, \alpha)$$

where $p_0(\beta, \alpha)$ is a prior of β and α. If a historical dataset D_0 is available, the prior can also depend on the historical data using the power prior formulation:

$$p(\beta, \alpha|D) \propto L(\beta, \alpha|D)[L(\beta, \alpha|D_0)]^{\alpha_0}p_0(\beta, \alpha)$$

where $0 \leq \alpha_0 \leq 1$ is prespecified and it determines the contribution of historical data to the prior of the parameters.

If a nonparametric instead of parametric specification for the baseline distribution $F^H(t)$ is preferred, Ibrahim et al. (2001*a*) proposed a method based on a piecewise constant hazard assumption for $F^H(t)$ and a smoothing parameter that controls the degree of parametricity in the right tail of $F^H(t)$.

Due to the similarity of (6.7) to the PH model, some nice Bayesian properties of the PH model are inherited by the model (6.7). For example, if a noninformative prior is used in β, i.e., $p_0(\beta, \alpha) \propto p_0(\alpha)$, the posterior distribution $p(\beta, \alpha|D)$ can still be a proper distribution (Chen et al., 1999).

The Bayesian method is also convenient to deal with missing covariates. When missing covariates are assumed to be MAR, Chen, Ibrahim and Lipsitz (2002) proposed to specify a parametric distribution for the covariates that is written as a sequence of one-dimensional conditional distributions. Then the Bayesian method can be used to estimate the parameters in the PHC model with proper priors on the parameters in the distribution for the missing covariates.

Due to the complexity of the posterior distributions in the Bayesian methods, Markov chain Monte Carlo methods are often used to approximate the posterior distributions. Details can be found in Ibrahim et al. (2001*b*) and Tsodikov et al. (2003).

6.3.3 Proportional hazards cure model for clustered survival data

The PHC model can be readily extended to analyze clustered survival data with a cure fraction in a similar way to how a frailty model extends the standard PH model. One may assume that given a frailty u_i that is shared within cluster i, the conditional survival function of T_{ij} is

$$S(t|x_{ij}, u_i) = \exp[-u_i e^{\beta_0 + z_{ij}'\beta} F^H(t)],$$

and u_i follows a prespecified frailty distribution with a fixed scale. One example of the frailty distribution is the stable distribution (Chen, Ibrahim and Sinha, 2002) because the resulting marginal model preserves the proportionality of the conditional model in z. Other frailty distributions, such as the gamma distribution, can also be considered.

Another approach to extend the PHC model for clustered data is to define the frailty term u_i to be shared by $\tilde{T}_1, \ldots, \tilde{T}_N$'s in (6.6) from all subjects in the same clusters (Yin, 2005). The resulting frailty model is

$$S(t|x_{ij}, u_i) = \exp\{-e^{\beta_0 + z_{ij}'\beta}[1 - S^H(t)^{u_i}]\},$$

Most of the existing methods to estimate the parameters in the models above are Bayesian methods. The promotion time distribution $F^H(t)$ can be specified parametrically or nonparametrically using a piecewise hazard distribution with gamma priors on the hazard segments.

6.4 Unifying cure models based on transformations

The cure models discussed in Section 6.2 and Section 6.3 are the two most common cure models in the literature. There are other cure models that are less known in the statistics community, but are useful alternatives to the two common models. One example of such a model is the proportional odds cure (POC) model (Gu et al., 2011):

$$P(T > t) = S(t|z) = [1 + e^{\beta_0 + z'\beta} F^O(t)]^{-1} \tag{6.9}$$

where $F^O(t)$ is a cumulative distribution function. It is a proportional odds model if $F^O(t)$ is treated as a baseline odds function. Of course, as an odds function, $F^O(t)$ is improper. Thus this model is an analogy of the extension of the PH model to the PHC model by using an improper baseline odds function in the proportional odds model of Pettitt (1984).

Another less-known cure model is the additive cure model (Yin and Ibrahim, 2005*b*):

$$h(t|z) = f^H(t) + \beta_0 + z'\beta \tag{6.10}$$

where $f^H(t)$ is a proper probability density function. This is an additive hazard model (Lin and Ying, 1994) except that the baseline hazard function is replaced with $f^H(t)$, thus is improper.

Given the different cure models, it becomes important to choose an appropriate cure model among them in practice for a given dataset. The biological interpretation found for the PHC models may help if it is plausible. However, the elusive nature of the latent variables $\tilde{T}_1, \ldots, \tilde{T}_N$ used in the biological interpretation often makes this task challenging. One approach to alleviate this issue is to unify different types of cure models under a new model formed based on the Box-Cox transformation (Box and Cox, 1964). For example, the unconditional survival function of T may be defined as follows (Yin and Ibrahim, 2005*a*; Taylor and Liu, 2007)

$$S(t|z,x) = \begin{cases} \left\{1 - \dfrac{\lambda \exp(\beta' z)}{1 + \lambda \exp(\beta' z)} F^H(t|x)\right\}^{1/\lambda} & 0 < \lambda \leq 1 \\ \exp(-e^{\beta' x} F^H(t|x)) & \lambda = 0 \end{cases} \tag{6.11}$$

where λ is the transformation parameter and $F^H(t|x)$ is a cumulative distribution function that may depend on a vector of covariates x. It is easy to see that model (6.11) becomes the mixture cure model (6.1) when $\lambda = 1$ and the PHC model (6.7) when $\lambda \to 0$. Therefore model (6.11) unifies the mixture cure model and the PHC model. The model can also be motivated using one of the tumor activation schemes in Section 6.3 (Peng and Xu, 2012).

Another family of cure models (Taylor and Liu, 2007) is given by

$$S(t|z,x) = \begin{cases} \{\pi(z)^\lambda + (1 - \pi(z)^\lambda) S^H(t|x)\}^{1/\lambda} & \lambda \neq 0 \\ \exp[\log(\pi(z))(1 - S^H(t|x))] & \lambda = 0 \end{cases} \tag{6.12}$$

where $S^H(t|x)$ is a proper survival distribution. In this model $\pi(z)$ could have a logistic form $\pi(z) = \exp(\beta' z)/(1+\exp(\beta' z)))$ or more conveniently be derived from a log-log link, $\pi(z) = \exp(-\exp(\beta' z))$. It is easy to see that model (6.12) is a mixture cure model when $\lambda = 1$ and a PHC model when $\lambda = 0$. Note also that this model has the appealing feature that the limit as $t \to \infty$ of $S(t|z,x)$ is $\pi(z)$ and it does not depend on the value of λ.

Another unified cure model based on the Box-Cox transformation is given as follows (Zeng et al., 2006):

$$S(t|z) = \begin{cases} [1 + ae^{z'\beta} F^H(t|x)]^{-1/\lambda} & \lambda > 0 \\ \exp[-e^{z'\beta} F^H(t|x)] & \lambda = 0 \end{cases} \tag{6.13}$$

It is easy to show that model (6.13) unifies the PHC model (6.7) (when $\lambda \to 0$) and the POC model (6.9) (when $\lambda = 1$).

The Box-Cox transformation can also be used on the two hazard functions in the additive cure model (6.10) to obtain the following cure model (Yin and Ibrahim, 2005*b*)

$$h(t|z) = \begin{cases} [f^H(t)^\lambda + \lambda(\beta_0 + z'\beta)]^{1/\lambda} & 0 < \lambda \leq 1 \\ f^H(t)\exp(\beta_0 + z'\beta) & \lambda = 0 \end{cases} \tag{6.14}$$

It is clear that this model unifies the PHC model (when $\lambda = 0$) and the additive cure model (when $\lambda = 1$).

Some transformation models are based on unspecified monotone increasing transformations on the failure time directly. That is, for $T|Y = 1$, a semiparametric transformation model (Lu and Ying, 2004) is

$$g(T) = -z'\beta + \epsilon \tag{6.15}$$

where $g(\cdot)$ is an unknown monotone increasing function and ϵ is the error term with a known continuous distribution that is independent of z and survival time. If ϵ is chosen to follow the extreme value distribution, the corresponding cure model is the PHMC model. On the other hand, if ϵ follows the logistic distribution, the corresponding cure model is the POMC model. It is easy to define a new distribution based on the Box-Cox transformation that includes the extreme value distribution and the logistic distribution as special cases. Thus, this semiparametric transformation cure model unifies the PHMC model and the POMC model.

There are a variety of estimation methods proposed in the literature to estimate the parameters in the unified cure models above, including parametric, semiparametric, and Bayesian methods. Unfortunately, due to the complexity of the unified models, the estimation methods usually fix the value of λ in the Box-Cox transformation at a prespecified value and then apply the estimation methods to estimate other parameters in the models. This approach greatly limits the ability to use the unified cure models to determine which cure model is adequate for a given dataset. This issue was studied in Peng and Xu (2012) for model (6.11), and it is found that the model has a reasonable power to select between the mixture cure model and the PHC model when $F^H(t)$ is correctly specified. However, when $F^H(t)$ is nonparametrically specified, the power can be low. It is expected that other unified cure models have similar properties.

For the mixture cure model specified by (6.2) and (6.15), Lu and Ying (2004) proposed unbiased estimating equations to estimate β, γ and $g(\cdot)$. This model was further considered by Othus et al. (2009) for a case with time-dependent covariates and dependent censoring. They proposed using an inverse censoring probability reweighting scheme to deal with the dependent censoring and to construct unbiased estimating equations to estimate the parameters.

The unified cure models can be adapted to model clustered survival data with a cured fraction. For example, a frailty term can be added to model (6.13) to produce a shared frailty model (Yin, 2008)

$$S(t|z_{ij}) = \begin{cases} [1 + \lambda u_i e^{z'_{ij}\beta} F^H(t|x)]^{-1/\lambda} & \lambda > 0 \\ \exp[-u_i e^{z'_{ij}\beta} F^H(t|x)] & \lambda = 0 \end{cases}$$

where the frailty term u_i may follow a gamma distribution with a fixed-scale parameter. Other distributions can also be considered for u_i. If a marginal model is preferred, the idea of Peng et al. (2007) and Yu and Peng (2008) may be applied directly to model (6.15) for the clustered data (Chen and Lu, 2012), and a semiparametric estimation method is proposed to estimate the parameters in the model. The model (6.15) can also be used in the marginal of a copula model to include a correlation parameter in the model for clustered data (Chen and Yu, 2012). An extension of the two-stage semiparametric estimation method of Chatterjee and Shih (2001) is available for this model.

6.5 Joint modeling of longitudinal and survival data with a cure fraction

In clinical trials and other medical studies, it is common to collect important information on some longitudinal markers, such as disease characteristics or quality of life scales, in addition to the survival status of a patient. Due to the association between the longitudinal data and survival times, a joint analysis of the longitudinal and survival data would provide more efficient estimates than separate analyses of the data (Ibrahim et al., 2010). In the earlier literature, joint models were mainly based on a classical linear mixed effect model for the longitudinal data and a Cox PH model for the survival data. When a fraction of cured subjects is evident in the survival data, a joint model incorporating a cured fraction is useful. For example, in a prostate cancer study, the data were analyzed using a joint model of the longitudinal prostate-specific antigen (PSA) values and survival times including a cured fraction (Law et al., 2002).

Let $l_i(t)$ be a longitudinal measurement from the ith subject at time t. In a joint model, the longitudinal measurement is often modeled by

$$l_i(t) = g(t, x_i, \alpha, \xi) + \epsilon_i(t) \tag{6.16}$$

where $g(\cdot)$ is a trajectory function of the longitudinal measurements, α is a fixed effect, ξ is a random effect, and $\epsilon_i(t)$ is a random error. For survival data, both the PHMC model and the PHC model have been considered in joint models. To connect the longitudinal model to the model for the survival data in a joint model, one can share the random effect ξ of (6.16) in the PHC model (Chen et al., 2004; Song et al., 2012) as follows

$$S(t|z) = \exp[-e^{\beta_0 + z'\beta + \xi} F^H(t)]$$

or use the longitudinal trajectory as the time-dependent variable in the PHC model (Brown and Ibrahim, 2003) as follows

$$1 - F^H(t) = S_0^H(t)^{\exp(g(t, x_i, \alpha, \xi))}$$

Yu et al. (2008) considered a PHMC model for the survival data and allow $g(\cdot)$ to appear in $S_u(t|x)$ as a time-dependent covariate and $g(\cdot)$ to depend on the latent cure status Y. Due to complexity of the joint models, Bayesian methods are often employed to estimate the parameters in the model and computation can be tedious. In comparison, the method of Song et al. (2012) is computationally less intensive due to the choice of the random effect and error term distributions.

6.6 Cure models and relative survival in population studies

In the cure models we have described the event of interest is observable; for example, in a cancer study we can observe recurrence of the disease after treatment. However, in some population studies only the time of death is available, and while the length of the follow-up and the nature of the disease may suggest that the majority of the deaths are due to the disease, deaths due to other causes may occur and cannot be separated. Since one cannot be cured of death, this situation typically gives Kaplan-Meier plots which do not have an obvious asymptote, and using a standard cure model may not be appropriate.

For such studies, it is convenient to define occurrence of cure when the mortality rate of those diagnosed with the disease returns to the same level as that expected in the general population. Or equivalently, the excess mortality rate due to the disease approaches zero. Let $h_p(t)$ be the observed hazard for the target group, $h^*(t)$ be the expected hazard for the general population, and $h(t)$ be the excess hazard (mortality) rate due to the disease. Then one can assume

$$h_p(t) = h^*(t) + h(t)$$

and the fact that the excess mortality rate approaches zero implies $\lim_{t\to\infty} h(t) = 0$. If the corresponding survival functions are defined as $S_p(t)$, $S^*(t)$ and $S(t)$. Then $S_p(t) = S^*(t)S(t)$. It is obvious that $S(t)$, often referred to as the "relative survival function" in population studies, is an improper survival function with $\lim_{t\to\infty} S(t) = \lim_{t\to\infty} \exp\left(-\int_0^t h(u)du\right) > 0$. Therefore, the methods discussed in Sections 6.2 and 6.3 can be employed to model the relative survival function. The estimation generally involves $h^*(t)$, the expected hazard rate for general population, which is often obtained from life tables. The relative survival cure model has been studied by a number of researchers (Lambert et al., 2007, 2010; Yu et al., 2004).

6.7 Software for cure models

For some simple cure models, no special software packages are required and standard statistical packages often can be used to fit them with minimal coding work. For example, the nonparametric estimation method proposed by Maller and Zhou (1992) for cure rate is directly based on the Kaplan-Meier survival estimator and can be obtained by most statistical software packages, such as SAS and R. A simple mixture cure model such as (6.1) with $S_u(t)$ from the exponential or Weibull distribution can be easily coded in SAS or R to obtain the maximum likelihood estimates of parameters in the model. When a complicated distribution is considered for $S_u(t)$ or covariates are considered in π or in $S_u(t)$, the necessary coding becomes complicated in standard statistical software packages and specialized software packages are required to fit such models.

GFCURE is an R package to fit the parametric AFTMC models based on the work of Peng et al. (1998). When choosing the Weibull distribution in the model, it is equivalent to fitting a parametric PHMC model. For semiparametric mixture cure models, SEMICURE is an R package to fit the semiparametric PHMC models based on the work of Peng (2003*b*). A SAS macro PSPMCM (Corbiere and Joly, 2007) implements both the parametric AFTMC models and the semiparametric PHMC models. Recently a new R package SMCURE is available at the CRAN website, and it extends SEMICURE to include the semiparametric AFTMC models.

For non-mixture cure models, an R package NLTM, that is based on the work in Tsodikov (2003); Tsodikov and Garibotti (2007), can be used to fit the semiparametric PHC model for independent data.

CUREREGR is a program that fits both the mixture cure model and the PHC model parametrically. It is available as a standalone Windows program and as a STATA module, and is based on the work of Sposto (2002).

For relative survival cure models in population studies, STATA module STRSMIX and STRSNMIX are available to fit the relative survival by the mixture cure model or by the PHC model (Lambert, 2007). CANSURV is a Windows program to estimate the relative survival using a mixture cure model (Yu et al., 2005).

Bibliography

Boag, J. W. (1949), 'Maximum likelihood estimates of the proportion of patients cured by cancer therapy', *Journal of the Royal Statistical Society* **11**, 15–53.

Box, G. E. P. and Cox, D. R. (1964), 'An analysis of transformations', *Journal of the Royal Statistical Society, Series B* **26**, 211–252.

Brown, E. R. and Ibrahim, J. G. (2003), 'Bayesian approaches to joint cure-rate and longitudinal models with applications to cancer vaccine trials', *Biometrics* **59**, 686–693.

Chatterjee, N. and Shih, J. (2001), 'A bivariate cure-mixture approach for modeling familial association in disease', *Biometrics* **57**, 779–786.

Chen, C.-M. and Lu, T.-F. C. (2012), 'Marginal analysis of multivariate failure time data with a surviving fraction based on semiparametric transformation cure models', *Computational Statistics & Data Analysis* **56**, 645–655.

Chen, C.-M. and Yu, C.-Y. (2012), 'A two-stage estimation in the Clayton-Oakes model with marginal linear transformation models for multivariate failure time data', *Lifetime Data Analysis* **18**, 94–115.

Chen, M.-H. and Ibrahim, J. G. (2001), 'Maximum likelihood methods for cure rate models with missing covariates', *Biometrics* **57**, 43–52.

Chen, M.-H., Ibrahim, J. G. and Lipsitz, S. R. (2002), 'Bayesian methods for missing covariates in cure rate models', *Lifetime Data Analysis* **8**, 117–146.

Chen, M.-H., Ibrahim, J. G. and Sinha, D. (1999), 'A new Bayesian model for survival data with a surviving fraction', *Journal of the American Statistical Association* **94**, 909–919.

Chen, M.-H., Ibrahim, J. G. and Sinha, D. (2002), 'Bayesian inference for multivariate survival data with a cure fraction', *Journal of Multivariate Analysis* **80**, 101–126.

Chen, M.-H., Ibrahim, J. G. and Sinha, D. (2004), 'A new joint model for longitudinal and survival data with a cure fraction', *Journal of Multivariate Analysis* **91**, 18–34.

Chen, Y. Q. and Wang, M.-C. (2000), 'Analysis of accelerated hazards models', *Journal of the American Statistical Association* **95**, 608–618.

Cho, M., Schenker, N., Taylor, J. M. G. and Zhuang, D. (2001), 'Survival analysis with long-term survivors and partially observed covariates', *The Canadian Journal of Statistics* **29**, 421–436.

Cooner, F., Banerjee, S., Carlin, B. P. and Sinha, D. (2007), 'Flexible cure rate modeling under latent activation schemes', *Journal of the American Statistical Association* **102**, 560–572.

Corbiere, F., Commenges, D., Taylor, J. M. G. and Joly, P. (2009), 'A penalized likelihood approach for mixture cure models', *Statistics in Medicine* **28**, 510–524.

Corbiere, F. and Joly, P. (2007), 'A SAS macro for parametric and semiparametric mixture cure models', *Computer Methods and Programs in Biomedicine* **85**, 173–180.

Cox, D. R. (1972), 'Regression models and life-tables', *Journal of the Royal Statistical Society, Series B* **34**, 187–220.

Cox, D. R. and Oakes, D. (1984), *Analysis of Survival Data*, Chapman & Hall, New York.

Dempster, A. P., Laird, N. M. and Rubin, D. B. (1977), 'Maximum likelihood from incomplete data via the EM algorithm', *Journal of the Royal Statistical Society, Series B* **39**, 1–38.

Fang, H.-B., Li, G. and Sun, J. (2005), 'Maximum likelihood estimation in a semiparametric logistic/proportional-hazards mixture model', *Scandinavian Journal of Statistics* **32**, 59–75.

Farewell, V. T. (1986), 'Mixture models in survival analysis: Are they worth the risk?', *The Canadian Journal of Statistics* **14**(3), 257–262.

Gu, Y., Sinha, D. and Banerjee, S. (2011), 'Analysis of cure rate survival data under proportional odds model', *Lifetime Data Analysis* **17**, 123–134.

Herring, A. H. and Ibrahim, J. G. (2002), 'Maximum likelihood estimation in random effects cure rate models with nonignorable missing covariates', *Biostatistics* **3**, 387–405.

Ibrahim, J. G., Chen, M.-H. and Sinha, D. (2001*a*), 'Bayesian semiparametric models for survival data with a cure fraction', *Biometrics* **57**, 383–388.

Ibrahim, J. G., Chen, M.-H. and Sinha, D. (2001*b*), *Bayesian Survival Analysis*, Springer-Verlag, New York.

Ibrahim, J. G., Chu, H. and Chen, L. M. (2010), 'Basic concepts and methods for joint models of longitudinal and survival data', *Journal of Clinical Oncology* **28**, 2796–2801.

Kuk, A. Y. C. and Chen, C. (1992), 'A mixture model combining logistic regression with proportional hazards regression', *Biometrika* **79**, 531–41.

Lai, X. and Yau, K. K. W. (2008), 'Long-term survivor model with bivariate random effects: Applications to bone marrow transplant and carcinoma study data', *Statistics in Medicine* **27**, 5692–5708.

Lambert, P. C. (2007), 'Modeling of the cure fraction in survival studies', *The Stata Journal* **7**, 351–375.

Lambert, P. C., Dickman, P. W., Weston, C. L. and Thompson, J. R. (2010), 'Estimating the cure fraction in population-based cancer studies by using finite mixture models', *Journal of the Royal Statistical Society, Series C* **59**, 35–55.

Lambert, P. C., Thompson, J. R., Weston, C. L., and Dickman, P. W. (2007), 'Estimating and modeling the cure fraction in population-based cancer survival analysis', *Biostatistics* **8**, 576–594.

Law, N. J., Taylor, J. M. G. and Sandler, H. (2002), 'The joint modeling of a longitudinal disease progression marker and the failure time process in the presence of cure', *Biostatistics* **3**, 547–563.

Lawless, J. F. (2003), *Statistical Models and Methods for Lifetime Data*, 2nd edn, John Wiley & Sons, Hoboken, New Jersey.

Li, C.-S. and Taylor, J. M. G. (2002*a*), 'A semi-parametric accelerated failure time cure model', *Statistics in Medicine* **21**, 3235–3247.

Li, C.-S. and Taylor, J. M. G. (2002*b*), 'Smoothing covariate effects in cure models', *Communications in Statistics* **31**(3), 477–493.

Li, C.-S., Taylor, J. M. G. and Sy, J. P. (2001), 'Identifiability of cure models', *Statistics & Probability Letters* **54**, 389–395.

Lin, D. Y. and Ying, Z. (1994), 'Semiparametric analysis of the additive risk model', *Biometrika* **81**, 61–71.

Lu, W. (2010), 'Efficient estimation for an accelerated failure time model with a cure fraction', *Statistica Sinica* **20**, 661–674.

Lu, W. and Ying, Z. (2004), 'On semiparametric transformation cure model', *Biometrika* **91**, 331–343.

Maller, R. A. and Zhou, S. (1992), 'Estimating the proportion of immunes in a censored sample', *Biometrika* **79**(4), 731–739.

Maller, R. A. and Zhou, X. (1996), *Survival Analysis with Long-Term Survivors*, John Wiley & Sons Ltd.

Othus, M., Li, Y. and Tiwari, R. C. (2009), 'A class of semiparametric mixture cure survival models with dependent censoring', *Journal of the American Statistical Association* **104**, 1241–1250.

Peng, Y. (2003*a*), 'Estimating baseline distribution in proportional hazards cure models', *Computational Statistics & Data Analysis* **42**, 187–201.

Peng, Y. (2003*b*), 'Fitting semiparametric cure models', *Computational Statistics & Data Analysis* **41**(3-4), 481–490.

Peng, Y. and Dear, K. B. G. (2000), 'A nonparametric mixture model for cure rate estimation', *Biometrics* **56**, 237–243.

Peng, Y. and Dear, K. B. G. (2004), An increasing hazard cure model, *in* N. Balakrishnan and C. R. Rao, eds., 'Advances in Survival Analysis', Vol. 23 of *Handbook of Statistics*, Elsevier, North-Holland, chapter 31, pp. 545–557.

Peng, Y., Dear, K. B. G. and Denham, J. W. (1998), 'A generalized F mixture model for cure rate estimation', *Statistics in Medicine* **17**, 813–830.

Peng, Y. and Taylor, J. M. G. (2011), 'Mixture cure model with random effects for the analysis of a multi-centre tonsil cancer study', *Statistics in Medicine* **30**, 211–223.

Peng, Y., Taylor, J. M. G. and Yu, B. (2007), 'A marginal regression model for multivariate failure time data with a surviving fraction', *Lifetime Data Analysis* **13**, 351–369.

Peng, Y. and Xu, J. (2012), 'An extended cure model and model selection', *Lifetime Data Analysis* **18**, 215–233.

Peng, Y. and Zhang, J. (2008*a*), 'Estimation method of the semiparametric mixture cure gamma frailty model', *Statistics in Medicine* **27**, 5177–5194.

Peng, Y. and Zhang, J. (2008*b*), 'Identifiability of mixture cure frailty model', *Statistics & Probability Letters* **78**, 2604–2608.

Pettitt, A. N. (1984), 'Proportional odds models for survival data and estimation using ranks', *Applied Statistics* **33**, 169–175.

Ritov, Y. (1990), 'Estimation in a linear regression model with censored data', *Annals of Statistics* **18**, 303–328.

Seppa, K., Hakulinen, T., Kim, H.-J. and Laara, E. (2010), 'Cure fraction model with random effects for regional variation in cancer survival', *Statistics in Medicine* **29**, 2781–2793.

Song, H., Peng, Y. and Tu, D. (2012), 'A new approach for joint modeling of longitudinal measurements and survival times with a cure fraction', *Canadian Journal of Statistics* **40**, 207–224.

Sposto, R. (2002), 'Cure model analysis in cancer: an application to data from the children's cancer group', *Statistics in Medicine* **21**, 293–312.

Sy, J. P. and Taylor, J. M. G. (2000), 'Estimation in a Cox proportional hazards cure model', *Biometrics* **56**, 227–236.

Sy, J. P. and Taylor, J. M. G. (2001), 'Standard errors for the Cox proportional hazards cure model', *Mathematical and Computer Modeling* **33**, 1237–1251.

Taylor, J. M. G. (1995), 'Semi-parametric estimation in failure time mixture models', *Biometrics* **51**, 899–907.

Taylor, J. M. G. and Liu, N. (2007), Statistical issues involved with extending standard models, *in* V. Nair, ed., 'Advances in Statistical Modeling and Inference: Essays in Honor of Kjell A Doksum', Series in Biostatistics, World Scientific Publishing Company, chapter 15, pp. 299–311.

Tsodikov, A. (1998), 'A proportional hazards model taking account of long-term survivors', *Biometrics* **54**, 1508–1516.

Tsodikov, A. (2002), 'Semi-parametric models of long-and short-term survival: an application to the analysis of breast cancer survival in Utah by age and stage', *Statistics in Medicine* **20**, 895–920.

Tsodikov, A. (2003), 'Semiparametric models: a generalized self-consistency approach', *Journal of the Royal Statistical Society, Series B* **65**, 759–774.

Tsodikov, A. D., Ibrahim, J. G. and Yakovlev, A. Y. (2003), 'Estimating cure rates from survival data: an alternative to two-component mixture models', *Journal of the American Statistical Association* **98**(464), 1063–1078.

Tsodikov, A. and Garibotti, G. (2007), 'Profile information matrix for nonlinear transformation models', *Lifetime Data Analysis* **13**, 139–159.

Wienke, A., Lichtenstein, P. and Yashin, A. I. (2003), 'A bivariate frailty model with a cure fraction for modeling familial correlation in disease', *Biometrics* **59**, 1178–1183.

Withers, H. R., Peters, L. J., Taylor, J. M. G., Owen, J. B., Morrison, W. H., Schultheiss, T. E., Keane, T., O'Sullivan, B., van Dyk, J., Gupta, N., Wang, C. C., Jones, C. U., Doppke, K. P., Myint, S., Thompson, M., Parsons, J. T., Mendenhall, W. M., Dische, S., Aird, E. G. A., Henk, J. M., Bidmean, M. A. M., Svoboda, V., Chon, Y., Hanlon, A. L., Peters, T. L. and Hanks, G. E. (1995), 'Local control of carcinoma of the tonsil by radiation therapy: an analysis of patterns of fractionation in nine institutions', *International Journal of Radiation Oncology, Biology, Physics* **33**, 549–562.

Yakovlev, A. Y., Tsodikov, A. D. and Asselain, B. (1996), *Stochastic Models of Tumor Latency and Their Biostatistical Applications*, Vol. 1 of *Mathematical Biology and Medicine*, World Scientific, Singapore.

Yau, K. K. W. and Ng, A. S. K. (2001), 'Long-term survivor mixture model with random effects: application to a multi-centre clinical trial of carcinoma', *Statistics in Medicine* **20**, 1591–1607.

Yin, G. (2005), 'Bayesian cure rate frailty models with application to a root canal therapy study', *Biometrics* **61**, 552–558.

Yin, G. (2008), 'Bayesian transformation cure frailty models with multivariate failure time data', *Statistics in Medicine* **27**, 5929–5940.

Yin, G. and Ibrahim, J. G. (2005*a*), 'Cure rate models: a unified approach', *The Canadian Journal of Statistics* **33**, 559–570.

Yin, G. and Ibrahim, J. G. (2005*b*), 'A general class of Bayesian survival models with zero and nonzero cure fractions', *Biometrics* **61**, 403–412.

Yu, B. and Peng, Y. (2008), 'Mixture cure models for multivariate survival data', *Computational Statistics & Data Analysis* **52**, 1524–1532.

Yu, B., Tiwari, R. C., Cronin, K. A. and Feuer, E. J. (2004), 'Cure fraction estimation from the mixture cure models for grouped survival data', *Statistics in Medicine* **23**, 1733–1747.

Yu, B., Tiwari, R. C., Cronin, K. A., McDonald, C. and Feuer, E. J. (2005), 'CANSURV: A windows program for population-based cancer survival analysis', *Computer Methods and Programs in Biomedicine* pp. 195–203.

Yu, M., Taylor, J. M. G. and Sandler, H. M. (2008), 'Individual prediction in prostate cancer studies using a joint longitudinal survivalcure model', *Journal of the American Statistical Association* **103**, 178–187.

Zeng, D. and Lin, D. Y. (2007), 'Efficient estimation for the accelerated failure time model', *Journal of the American Statistical Association* **102**, 1387–1396.

Zeng, D., Yin, G. and Ibrahim, J. G. (2006), 'Semiparametric transformation models for survival data with a cure fraction', *Journal of the American Statistical Association* **101**, 670–684.

Zhang, J. and Peng, Y. (2007*a*), 'An alternative estimation method for the accelerated failure time frailty model', *Computational Statistics & Data Analysis* **51**, 4413–4423.

Zhang, J. and Peng, Y. (2007*b*), 'A new estimation method for the semiparametric accelerated failure time mixture cure model', *Statistics in Medicine* **26**, 3157–3171.

Zhang, J. and Peng, Y. (2009), 'Accelerated hazards mixture cure model', *Lifetime Data Analysis* **15**, 455–467.

Zhang, J., Peng, Y. and and Li, H. (2013), 'A new semiparametric estimation method for accelerated hazards mixture cure model', *Computational Statistics & Data Analysis* **59**, 95–102.

Zhuang, D., Schenker, N., Taylor, J. M. G., Mosseri, V. and Dubray, B. (2000), 'Analysing the effects of anaemia on local recurrence of head and neck cancer when covariate values are missing', *Statistics in Medicine* **19**, 1237–1249.

7

Causal Models

Theis Lange and Naja H. Rod

Section of Social Medicine, Department of Public Health, University of Copenhagen

CONTENTS

7.1 Introduction 135
7.2 Tools for formalizing cause and effect 136
7.3 Analysis in the absence of feedback 139
 7.3.1 Causal interpretation of the classic Cox modeling approach 140
 7.3.2 Mimicking an actual randomized trial 141
7.4 Analysis with exposure-confounder feedback 142
 7.4.1 The medical background of the HAART example 143
 7.4.2 Defining causal effects in the presence of feedback 143
 7.4.3 Estimating causal effects from observational data in the presence of feedback 144
 7.4.4 Assumptions for drawing causal conclusions in the presence of feedback 145
 7.4.5 Implementing the mini trials approach 146
7.5 Appendix: R code 149
 Bibliography 150

7.1 Introduction

Anybody working professionally with data and statistics knows (or should know) that one should never deduce the existence of a cause and effect relationship between two variables just because they are statistically associated. Indeed, most professionals will almost without provocation provide the catch-phrase “Association is not causation.” Whilst this phrase is undoubtedly true it is a bit like writing “May contain traces of nuts” on all food packages; it protects the statisticians (or food producer), but also lessens the value of the conducted analysis (or food label). In this chapter we will explore when and how association can indeed be interpreted as causation in particular in the context of survival analysis.

The chapter is structured as follows. Section 7.2 describes the basic concepts and ideas in causal inference. The content of this section applies not only to survival outcomes, but to any type of outcome. Section 7.3 explains how to draw causal conclusions from survival outcomes in the absence of time-dependent covariates. In Section 7.4 the methods are extended to also allow for time-dependent covariates and in particular feedback between exposure and other variables over time. Finally, Section 7.5 presents advice on how to implement some of the discussed methods in the software package R.

7.2 Tools for formalizing cause and effect

The gold standard for drawing conclusions about cause and effect is large, well-designed randomized controlled trials (Concato et al., 2000; Hernan, 2004) in which participants are randomly allocated to treatment or no treatment. In a sufficiently large sample, such random treatment allocation will ensure comparability in risk of the outcome between the treatment groups, formally known as exchangeability. Assuming that the randomized controlled trial is ideal in all other respects, i.e., full compliance with the assigned treatment regime and no loss to follow-up, any difference between treatment groups can be interpreted as an effect of treatment. However, for obvious ethical, financial, and practical reasons we cannot address all scientific questions by conducting a randomized trial; often we must resort to observational data. The overarching goal of *causal inference* is to analyze when and how statistical associations can be given a causal interpretation. Often this corresponds to analyzing observational data in a way that mimics specific randomized trials.

To fix these concepts consider the following problem from public health: *how does physical activity affect the risk of developing type 2 diabetes?*

Type 2 diabetes is a major and increasing public health problem and there is strong evidence that physical activity is effective in type 2 diabetes prevention (Malkawi, 2012; Jeon et al., 2007). Several large intervention trials have addressed how physical intervention affects the risk of diabetes especially in people with pre-diabetic conditions including impaired fasting glucose and impaired glucose tolerance. Taking the Diabetes Prevention Program Outcomes Study as an example (Knowler et al., 2002), they randomly assigned 3234 nondiabetic persons with elevated fasting and post-load plasma glucose concentrations to placebo or a lifestyle-modification program with the goals of at least a 7 percent weight loss and at least 150 minutes of physical activity per week. For simplicity we will only focus on the physical activity intervention in the following.

Randomized trials to address this question are conducted by initially recruiting a suitably large number of non-diabetic people, or in this example pre-diabetic people, satisfying some additional inclusion criteria and then randomly assigning them to either a physical active group (e.g., >2.5 hours/week of moderate or >1 hour/week of vigorous physical activity) or an inactive group (e.g., <1 hour/week of moderate and <1 hour/week of vigorous physical activity). Clearly it would be unethical to randomize persons to inactivity as this is known to have adverse health effects, but as a thought experiment it would be possible. Next, it must be ensured that the participants follow their assigned physical activity regime and are regularly monitored for development of type 2 diabetes for the decided follow-up time, e.g., 10 years. For each person in the trial, the outcome of interest would be time to development of type 2 diabetes; the potentially important issue of competing risks, e.g., death will for simplicity not be considered in this chapter; see instead Chapter 6. While such a study is practically possible to conduct, indeed as mentioned above several related randomized trials have been conducted (Stringhini et al., 2012), it does require much logistic work to ensure compliance and it takes very long before results are obtained due to the long follow-up. Furthermore, it would as mentioned not be ethically acceptable to randomize people to inactivity. Despite these logistic and ethical challenges when assessing the casual effect of physical activity on type 2 diabetes risk, this is what is meant.

The mathematical framework of counterfactual variables — see Pearl (2009) and Hernn and Robins (2012) and the many references therein — was developed to formalize cause and effect. In essence, the counterfactual framework builds on the idea that to establish causality one should ideally compare the risk of outcome in the same individual or the same group of individuals when both treated and not treated. As a person can only be either treated or not treated at a specific time, one of these risks will inevitably be counter-to-the-fact and therefore not observable. The overarching aim of causal inference is to design studies and model data in a way that mimics this ideal of comparing the same group of individuals under two different treatment regimes. In the type 2 diabetes example, we would for each person let $T^{(a)}$ denote the time until onset of type 2 diabetes if, perhaps contrary to fact, physical activity regime was set to a; in the following we will refer this time to type 2 diabetes onset as a survival time. Thus, for each person we imagine that there are not one, but in fact many possible survival times corresponding to the different physical activity regimes. Naturally, we will only get to observe one of these survival times namely the one corresponding to actual physical activity regime (and indeed we might only get to see a censored version of the survival time). If the randomized study was conducted and analyzed by the means of a Cox model, the resulting hazard ratio would precisely be the rate between the hazards for the counterfactual variables under the assumptions of exchangeability, full treatment compliance and no selective loss to follow-up.

In the type 2 diabetes example a randomized trial with fully adequate follow-up time, say 10 years, is not likely to be conducted, instead observational data can be used to mimic the randomized trial. However, in observational studies there might be other variables affecting both physical activity and disease onset (e.g., socioeconomic position, and gender). Collectively such variables are referred to as confounders. A useful tool to understand the whole network of causal influences is the Directed Acyclic Graphs (DAGs), which is a way to visualize and clarify one's prior beliefs about the underlying causal structure and thereby identify a minimal sufficient set of confounders; see Greenland et al. (1999). Figure 7.1 shows how a DAG could look for the type 2 diabetes example (for ease of presentation not all relevant variables are included; for instance if age is an important predictor for both exposure and outcome). The arrows in a DAG indicate that there is reason to believe that a causal link exists between two variables. While DAGs are useful for understanding the whole causal network of interconnections it is important to stress that they are only as valid as the assumptions on which they are based. In particular it is important that the researcher includes all relevant variables including those that are unobserved.

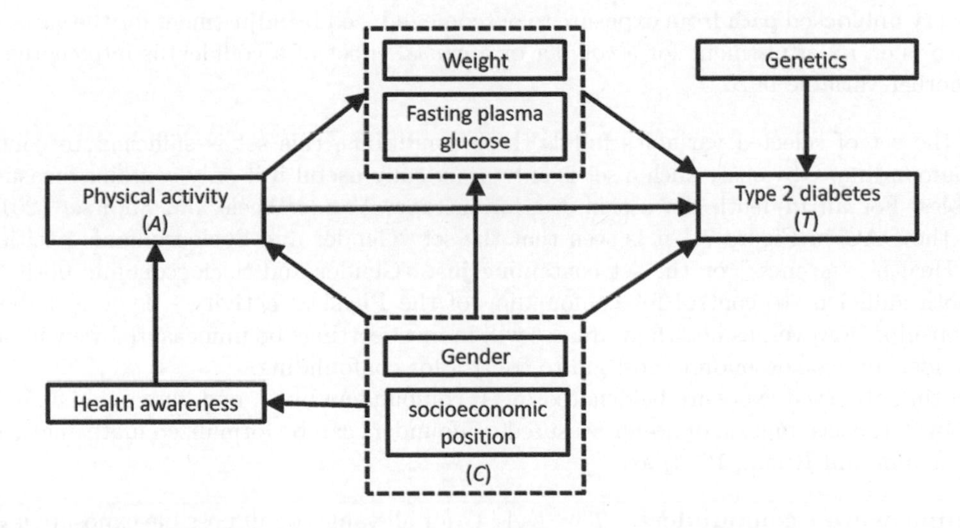

FIGURE 7.1
DAG depicting beliefs regarding the causal network surrounding physical activity (exposure) and time to onset of type 2 diabetes (outcome).

The following terminology is in line with the definitions and arguments presented in Greenland et al. (1999). Each variable in a DAG is called a node, and single-headed arrows represent direct causal links (i.e., links not involving other variables in the DAG) between the nodes. A causal path is one that can be traced through a sequence of single-headed arrows entering an arrow through the tail and leaving through the head. Thus, causal paths represent for instance how the effect of exposure cascades to the outcome, e.g., Physical activity $\rightarrow$ Weight $\rightarrow$ Type 2 diabetes. A backdoor path between exposure and outcome is defined as a path that has an arrowhead pointing towards exposure, e.g., Physical activity $\leftarrow$ Gender $\rightarrow$ Type 2 diabetes. Any unblocked backdoor paths from exposure to outcome will allow for these to be statistically associated even if they are not causally related. An unblocked backdoor path therefore represents potential confounding.

If one were to conduct a naive statistical analysis including only exposure and outcome, all paths between exposure and outcome (i.e., physical activity and type 2 diabetes) would be unblocked no matter the direction of the arrows. The only exception being paths containing colliding arrows (e.g., Physical activity $\rightarrow$ Weight $\leftarrow$ Socioeconomic position *to* Type 2 diabetes, where Weight is a so-called collider). Unblocked backdoor paths (i.e., those not containing a collider) can be closed by adjusting for one of the variables on the path, such as for example adjustment for gender and socioeconomic position will close the open backdoor path Physical activity $\leftarrow$ Gender & socioeconomic position $\rightarrow$ Type 2 diabetes. If such adjustment requires adjustment for a variable that is a collider (or a causal effect of a collider) on a different pathway, it is important to be aware that adjustment for a collider will open the path by creating a non-causal association between its two causes. It should also be emphasized that lack of a directed path from one variable to another is based on the assumption of no causal effect linking one with the other. That a necessary subset of variables S has been selected to prevent confounding can be verified by the following algorithm:

1. Every unblocked backdoor path from exposure to outcome is intercepted by a variable in S.

2. Every unblocked path from exposure to outcome induced by adjustment for the variables in S (i.e., by adjustment for a collider or a causal effect of a collider) is intercepted by another variable in S.

If the set of selected variables fulfills these conditions, this set is sufficient to control for confounding. Obviously such a set is only practically useful if it only contains measured variables. For an in-depth discussion, see for instance VanderWeele and Shpitser (2011). From the DAG in Figure 7.1 it is seen that the set "Gender and Socioeconomic position" and "Health awareness" or the set containing just "Gender and Socioeconomic position" are both sufficient to control for confounding of the Physical activity - Type 2 diabetes relationship. However, as health awareness will in most settings be unmeasured we will only use gender and socioeconomic position to control for confounding.

Letting observed exposure be denoted by A, confounders by C, and outcome (i.e., event time) by T the assumption of no-unmeasured confounders can be formalized mathematically (Rosenbaum and Rubin, 1983) as

No unmeasured confounders: $T^{(a)} \perp A \mid C$ for all values of all possible exposure levels a, where $T^{(a)} \perp A \mid C$ denotes that the counterfactual variable $T^{(a)}$ is independent of A conditional on C.

While the assumption of no-unmeasured confounders is arguably the most well-known assumption for deducing causal effects, it is not the only assumption that must be met in order to deduce causal effects from observational studies. Furthermore, the so-called *positivity* and *consistency* assumptions must be satisfied. Mathematically these assumptions can be stated as

Positivity: $P(A = a|C = c) > 0$ for all values of a and all values of c, that is all possible values of the baseline variables. If exposure is continuous the probability is replaced by a density.

Consistency: If $A = a$ for a given person then $Y^{(a)} = Y$ for that person.

In the context of the type 2 diabetes example, the assumption of positivity implies that for any combination of socioeconomic position and gender there is strictly positive probability for each of the physical activity regimes. Note that the assumption does not imply that the dataset used in the planed analysis must contain persons with all possible combinations of baseline variables and exposures (which would become infeasible with many baseline variables due to the curse of dimensionality); the assumption only implies that there could have been a person with any combination of baseline variables and exposure. For the type 2 diabetes example, this assumption would seem unproblematic to adopt. In the context of the type 2 diabetes example, the assumption of consistency implies that a person who was randomized to the same physical activity regimes as the person would have chosen such regimes on his own if the trial had not existed will not change his survival as a consequence of the trial. Intuitively, the assumption of consistency requires a sufficient amount of overlap between the intervention (i.e., randomized trial) one is attempting to deduce the causal effect of and the mechanisms giving rise to the dataset at hand. It is often difficult to assess if the assumption of consistency is reasonable, but as a rule of thumb it is less of a problem the more biomedical the underlying mechanisms for the causal effect in question are, e.g., the effect of an antibiotics pill is hardly affected by who suggested to take the pill; in contrast, the effect of exercise is vastly different depending on whether the exercise is being forced upon you or is conducted voluntarily. Thus, in the context of the type 2 diabetes example the assumption of consistency might be questionable.

Common for all three assumptions is that they are not testable by statistical means. Therefore, the assumptions must be justified by content matter arguments. The development of practically applicable ways of conducting sensitivity analysis assessing the consequences of failure to meet one or more of the three assumptions is an active research area within causal inference (Vanderweele, 2008; Vanderweele and Arah, 2011).

In the following section we will show how observational data on physical activity, gender, socioeconomic position, and time to onset of type 2 diabetes can be used to mimic the mentioned randomized trial and indeed estimate the causal effect of physical activity on type 2 diabetes.

7.3 Analysis in the absence of feedback

For ease of presentation we will in this section assume that gender can be dropped from the DAG in Figure 7.1 without violating the assumption of no-unmeasured confounders. Then, as argued above, socioeconomic position (measured in three levels) will be sufficient to control for confounding. We will furthermore assume that the consistency assumption holds; that is assume that type 2 diabetes development does not depend on how or why a specific level of physical activity was chosen. As any person, irrespective of the values for the confounders, has a positive risk of developing type 2 diabetes the final assumption of

TABLE 7.1
Estimates of the effect of physical activity and socioeconomic position on onset of type 2 diabetes using a Cox model. The estimates correspond to the ones in Stringhini et al. (2012), but the estimates should only be taken as illustrative.

	Hazard ratio	95% Confidence interval
Physical activity		
Active	Ref.	
Inactive	1.33	(1.13; 1.56)
Socioeconomic position		
Low	Ref.	
Middle	1.38	(1.17; 1.63)
High	1.55	(1.21; 1.97)

positivity is satisfied. Finally, it is assumed that censoring is independent; that is that the unobserved event times for the censored part of the sample are not systematically different from the observed event times in the sample. As we will only focus on the comparison of rates, the issue of competing risks (see Chapter 6) can be ignored.

7.3.1 Causal interpretation of the classic Cox modeling approach

Given data on baseline confounders (socioeconomic position), physical activity, and time to either onset of type 2 diabetes or censoring a classic Cox model can be fitted; see Chapter 1. Before interpreting the results, the fit of the Cox model should be carefully assessed. In particular, it must be ensured that the assumptions of proportional hazards and absence of interactions are satisfied. If not, the modeled should be modified accordingly. Table 7.1 below shows values for the effect of physical activity and socioeconomic position from a standard Cox regression using time since inclusion in cohort as the underlying timescale. The estimates are taken from Stringhini et al. (2012), which are based on the observational Whitehall II cohort study; however, in the original analysis more confounders were included. For the sake of clarity of presentation these additional confounders have been omitted from Table 7.1. Consequently the numbers in Tables 7.1 and 7.2 should merely be taken as illustrative.

Under the assumptions discussed in the previous paragraphs, the hazard ratio of 1.33, which corresponds to the inactive vs. physically active group, can be given the following causal interpretation: Imagine a randomized study as the one described in Section 7.2, but only including persons with a specific socioeconomic position, say "Low," then the hazard for the group randomized to the inactive group would be 1.33 times as high as the group randomized to the physically active group. Note that this interpretation holds irrespective of which level of socioeconomic position was chosen as an inclusion criterion, which is an effect of the absence of interactions in the Cox model, and irrespective of whether socioeconomic position was included as an ordinary co-variate or as a strata variable. When the fraction of onset of type 2 diabetes is small, say under 20%, the effect can also be phrased more bluntly as: every time the physically active group experiences one case of type 2 diabetes the inactive group will experience 1.33 cases of type 2 diabetes.

In conclusion, hazard ratios obtained using classic the Cox modeling approach can, under additional assumptions of no-unmeasured confounders, positivity, and consistency, be given the causal interpretation as the effects one would see in an ideal randomized trial (i.e., a

TABLE 7.2
Cross-tabulation of socioeconomic position and physical activity in both the observed population (upper half) and the pseudo-population (lower half) constructed by weighting each observation by the probability of being in the activity group that observation is in.

		Socioeconomic position		
		Low	Middle	High
Observed population, Col percentage (number)				
Physical activity	Active	57% (637)	81% (2,680)	88% (2490)
	Inactive	43% (477)	19% (612)	12% (341)
Pseudo-population, Weight (implied number)				
Physical activity	Active	1.8 (1,114)	1.2 (3,292)	1.1 (2,831)
	Inactive	2.3 (1,114)	5.3 (3,292)	8.3 (2,831)

randomized trial without selection bias, misclassification, loss to follow-up, etc.) conducted within a specific subpopulation defined by the confounders. It should be noted that the validity of the assumption of no unmeasured confounders is untestable and often violated.

7.3.2 Mimicking an actual randomized trial

If one is interested in mimicking a more traditional randomized trial, that is a randomized trial which enrolls people with all socioeconomic positions and produces an overall (i.e., unconditional) hazard ratio for the effect, the classic (Cox) modeling approach does not suffice as it only delivers effect measures conditional on having a specific socioeconomic position. Instead so-called inverse probability of treatment estimation (IPTW) can be employed (Robins et al., 2000). Here the original observations are re-weighted to create a pseudo-population, in which the randomized trial has been mimicked. Note that if the outcome was not a survival time, but a simple continuous variable, which could be modeled by a linear model, regression techniques and inverse probability of treatment estimation would yield the same result.

Consider the type 2 diabetes example again. Table 7.2 below shows that among persons with the lowest socioeconomic position, physical inactivity was much more common than among persons with the highest socioeconomic position. In a large randomized study such imbalances would not occur because of randomization. Due to the assumptions of no-unmeasured confounding and consistency, the few persons in the high socioeconomic position and inactive group are assumed to be representative of what would have happened if we had randomized all high socioeconomic position persons to this activity regime. The randomized study can therefore be mimicked by letting these few persons in the high-socioeconomic position and inactive group count for all the other persons who would also have been in this group in a randomized trial. This can be obtained by letting each observation count as one divided by the probability of being in the exposure group that the observation is actually in conditional on baseline variables. The lower half of Table 7.2 illustrates this; for instance for active/low socioeconomic position, the weight is computed as $1/0.43 = 2.33$ and the implied group size is $2.33 * 477 = 1,111$ or in fact 1,114 if no rounding had been done in the intermediate calculations. The numbers in Table 7.2 are obtained from Stringhini et al. (2012), but since important confounding has been omitted the table should only be treated as an illustration.

Once the pseudo-population has been created, the causal effect of physical activity can be estimated by for instance a Cox model including only physical activity as explanatory variable, but fitted to the pseudo-population (in practice this is done by weighting the analysis with the weights given in Table 7.2). Since the same observations are being re-used in the pseudo-population, one must employ robust standard errors, which will provide conservative confidence intervals (Robins et al., 2000). The resulting hazard ratios correspond to the ones that would be found in a randomized trial with recruitment from the general population, where persons were randomly assigned to the two activity groups assuming that the distribution of baseline variables for the persons constituting the used dataset is the same as in the general population.

If there are several baseline confounders the weights cannot be computed by simple counting methods as is done in Table 7.2 since the groups defined by combining several baseline confounders will likely be too small. Instead a model for the exposure must be fitted and probabilities extracted from the model (typically by using predict functionality in the employed software). In the type 2 diabetes example, a logistic regression could be used to model how physical activity depends on socioeconomic position, gender, and perhaps other relevant confounders. The appendix of this chapter provides `R` code for both computing weights and conducting the weighted analysis.

Fitting a model including only the exposure, but to a pseudo-population constructed by weighting instead of the original dataset is known as a Marginal Structural Model (MSM) estimated by inverse probability of treatment weighting (Robins et al., 2000). Formally, MSMs are models in which the counterfactual variables are the ones being modeled instead of the observed variables. Thus in the type 2 diabetes example, it would be a model for the counterfactual time to onset of type 2 diabetes $T^{(k)}$:

$$\lim_{dt \to 0} P(T^{(k)} \in]t; t+dt] \mid T^{(k)} > t)/dt = \lambda_0(t) \exp\{b_k\},$$

where $\lambda_0(t)$ is an unspecified baseline hazard, $b_1 = 0$ by assumption, and b_2 denotes the causal log-hazard ratio for being physically active. Thus, when employing MSMs, confounding is addressed not by including potential confounders in the model directly, but instead by fitting the model to a pseudo-population obtained by re-weighting the original dataset such that there is no association between confounders and exposure in the pseudo-population.

7.4 Analysis with exposure-confounder feedback

In the preceding section it was explored how to draw causal conclusions from survival data when both exposure and confounders were fixed from the baseline time point onwards. However, survival data by definition includes a time period from the baseline time point to the event time of interest. Thus, there is ample room for the exposure and other variables to change and even affect one another over this time interval; in the following this will be referred to as "feedback."

In this section we address (*a*) how to even define causal effects when the exposure is allowed to change over time and (*b*) how to employ these concepts in practice. Assessing causal effects in the presence of feedback is currently at the forefront of research in causal inference (Gran et al., 2010; Hernan et al., 2008; Lu, 2005). It will therefore not be possible to cover all the techniques suggested in the literature. Instead this section will consider one specific scientific problem, namely the use of highly active antiretroviral therapy (HAART) to prevent HIV progressing to AIDS. An integral part of HAART is to monitor HIV patients in order to initiate and adjust treatment in response to patient development; there is in other words feedback between treatment and other variables over time. Thus, causal conclusions must be drawn in the presence of feedback. Indeed, the causal relationship between HAART and HIV progression has been a key example of feedback mechanisms in the causal inference literature (Hernan et al., 2000; Gran et al., 2010).

It should be noted that Chapter 1 considers models with time-dependent covariates, which at first glance would seem appropriate for dealing with feedback. However, the models from Chapter 1 do not provide any definition of causal effects in the presence of feedback (that is the effect estimates obtained are not directly interpretable as expected effects in any randomized study). Why this is the case will be further explained in Section 7.4.3. Accordingly these models are not well suited to draw causal conclusions in the presence of feedback (Hernan et al., 2000).

7.4.1 The medical background of the HAART example

In this section we give a very brief introduction to how and why HAART works to prevent HIV from progressing to AIDS. For a more in-depth discussion of the medical aspects of HAART, see, e.g., Ledergerber et al. (1999).

HAART is a combination of at least three drugs, which has been shown in several studies to substantially reduce disease progression in HIV-infected patients. If left untreated, the HIV infection will gradually reduce the effectiveness of the patient's immune system, which will be reflected in decreasing CD4 counts. HAART halts disease progression and thus increases CD4 counts. For ethical reasons no placebo-controlled randomized trial of HAART has been conducted and the trials involving HAART have all had follow-up of a few years or less (Sterne et al., 2005) despite the fact the therapy must be taken for life. Thus, full treatment effect and long-term effects must be obtained form observational studies.

Since HAART must be taken for life, one does not wish to initiate treatment too soon since this will lengthen time on HAART and therefore increase risk of developing resistance to the therapy and lengthen the time the patient must tolerate the side effects of the treatment. Simultaneously, one must not initiate HAART too late since this increases the risk of the disease progressing to AIDS. To balance these opposite aims, the CD4 count of the patient is being regularly monitored to assess when to initiate HAART. Consequently, CD4 count is a time-dependent confounder for the effect of HAART, because patients with lower counts are more likely to be treated. CD4 count is also affected by HAART, and is thus intermediate on the causal pathway from such treatment to AIDS or death. In other words, there is feedback between CD4 count and treatment. Figure 7.2 below illustrates the causal structure.

7.4.2 Defining causal effects in the presence of feedback

In order to give a meaningful definition of causal effects in the presence of feedback one must try to imagine which randomized trial one would ideally have liked to perform. In the HIV example the goal of such an imaginary randomized trial would be to find the best *treatment strategy*. A treatment strategy is a set of rules which for each time point (i.e., doctor visit), and using only past and present values of treatment and CD4 count, decides whether or not treatment should be taken from this time point to the next. Examples of such a treatment strategy are the following: "First time CD4 count drops below 350 initiate HAART treatment and continue irrespectively of future CD4 counts" and "Give HAART whenever CD4 count is below 350." An ideal randomized trial would start by defining a number of such treatment strategies and subsequently recruit still healthy HIV patients (i.e., patients with a CD4 count above some predefined level, e.g., 500). These patients

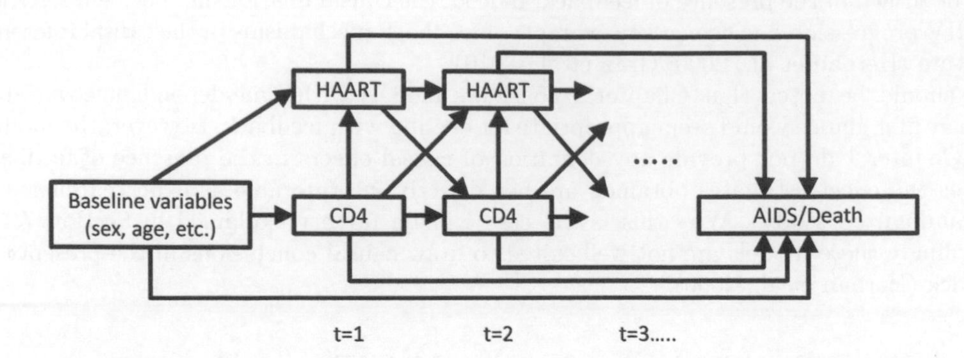

FIGURE 7.2
DAG depicting the causal relationship between the first two time periods after a patient has been diagnosed with HIV.

would then randomly be assigned to one of the treatment strategies and the causal effect assessed by comparing AIDS-free survival time between the groups.

Thus, feedback has been incorporated in the definition of causal effects by replacing the simple treatment group with a dynamic strategy, which adapts in the light of patient response. Note that this definition of causal effects closely mimics the choices faced by doctors when treating actual patients.

The above definition of causal effect in the presence of feedback can be formalized using counterfactual variables by first defining $\bar{a}$ as a treatment strategy. The treatment strategy defines whether treatment should be given at each time point, thus $\bar{a} = (a_1, a_2, ...)$ where, e.g., a_2 denotes the treatment decision taken at time 2. Thus, a_2 can depend on all past treatments (i.e., a_1) as well as past and present CD4 counts and baseline variables (e.g., sex and age). For each treatment strategy, a corresponding counterfactual AIDS-free survival time can be defined, which will be denoted $T^{(\bar{a})}$. The comparisons of the different counterfactual AIDS-free survival times can for instance be done using a Cox model leading to causal hazard ratios for the different treatment strategies as explained below.

7.4.3 Estimating causal effects from observational data in the presence of feedback

Looking at Figure 7.2 it is observed that the feedback between treatment and CD4 counts implies that time period 2 CD4 count is on the causal pathway from period 1 treatment to outcome; it is a so-called mediator of the effect from treatment at time period 1 to outcome. However, the same variable, that is CD4 count at time period 2, is also a confounder for the relationship between treatment at time period 2 and the outcome. Using standard regression techniques one could either include CD4 count as a time-dependent confounder, but this would block the effect of treatment at time period 1 and therefore lead to a bias, or not include the variable CD4 count, but this would make the relationship between treatment at time period 2 and the outcome confounded and therefore also lead to bias. It is this dual role for the same variable as both mediator and confounder that prevents standard regression-based techniques, including survival models with time-dependent covariates, from being able to estimate causal effects in the presence of feedback.

To deal with this type of feedback (which is sometimes referred to as "exposure-dependent confounding") Robins et al. in their seminal paper (Robins et al., 2000) introduced Marginal Structural Models. As discussed in the preceding sections, MSMs are fitted using IPTW techniques. Intuitively, the IPTW techniques work by constructing a pseudo-population through weighting in which the confounder, e.g., CD4 count, does not predict treatment and is therefore no longer a confounder. However, the drawback is that weights must be estimated and computed for each time period, which can lead to unstable weights in particular for individuals with unusual covariate histories.

An alternative, see e.g., Gran et al. (2010), approach is to look at a given treatment strategy, say $\bar{a}$, directly and find the individuals who, for at least some time period, by chance followed this particular treatment strategy. All these little stretches of person time can then be combined using standard survival analysis tools to a full picture of the distribution of the counterfactual event time $T^{(\bar{a})}$. Finally, different treatment strategies can be compared using for instance Cox models and causal hazard ratios obtained. The drawback of this approach is that it is only suited for fairly simple treatment strategies; in particular strategies such as "Once CD4 count drop below 350 initiate treatment and continue indefinitely." More complex treatment strategies for instance involving multiple treatment initiations and discontinuations cannot be assessed as it is not likely that a suitable number of people would follow such strategies purely by chance not even for short time intervals.

In the rest of this section the second approach will be explored in more detail. As will become apparent, this approach can be thought of as literally constructing a number of mini trials from the observational data, which hopefully should provide a more intuitive understanding of the approach as a complement to the more technical mathematics. The exploration will closely follow the study of HAART treatment to prevent HIV progression by Gran et al. (2010). By initially focusing on the more intuitively understandable mini trials approach, it should hopefully also be easier for the interested reader to understand the more technical MSM approach at a later stage. Whilst this section will not go into further detail with the MSM-based approach, this should not in any way be taken as an indication of this being a less useful model strategy. It is merely a consequence of the fact that the field of causal inference in the presence of feedback is currently so fast-moving that a complete coverage is outside the scope of this chapter. For excellent descriptions of MSM-based techniques, see Robins et al. (2000); Robins and Hernn (2009); Rotnizky (2009); Bekaert et al. (2010); Vansteelandt et al. (2009); Sterne et al. (2005) and the many references therein.

Neither the mini trials approach nor the MSM approach can deduce causal effects in the presence of feedback using just the simple assumptions discussed in Section 7.2. The necessary generalizations are the focus of the next section.

7.4.4 Assumptions for drawing causal conclusions in the presence of feedback

Faced with feedback between exposure and confounders, the assumptions from Section 7.2 no longer suffice to allow for a causal interpretation of a statistical analysis. Instead, the assumptions must be generalized as follows (Robins and Hernn, 2009).

No unmeasured confounders (extended): At each time point there are no unmeasured confounders between treatment at this time point and the final outcome. Note that current values of time varying confounders and past treatment are all included as measured confounders when formulating this assumption. Mathematically the assumption can be written as

$T^{(\bar{a})} \perp A_t \mid (A_{t-1}, ..., A_1) = (a_{t-1}, ..., a_1), (L_t, ..., L_1) = (l_t, ..., l_1)$

for all regimes $\bar{a}$ and all trajectories $l_1, l_2, ...$

Positivity (extended): Conditional on any achievable trajectory for treatment and confounders up to time point t, i.e., any values of $(a_{t-1}, ..., a_1)$ and $(l_t, ..., l_1)$ which have positive probability, then all values of treatment at time point t, i.e., A_t, have positive probability. Mathematically this can be written as

If $P((A_{t-1}, ..., A_1) = (a_{t-1}, ..., a_1), (L_t, ..., L_1) = (l_t, ..., l_1)) \neq 0$ then it must hold that $P(A_t = a_t \mid (A_{t-1}, ..., A_1) = (a_{t-1}, ..., a_1), (L_t, ..., L_1) = (l_t, ..., l_1)) > 0$ for all values of a_t.

Consistency (extended): If $\bar{A} = \bar{a}$ for a particular individual, then it must also hold that $T^{(\bar{a})} = T$ for that person.

While these assumptions mathematically are much more involved than the corresponding assumptions in the simple setup without feedback, see Section7.2, they can in fact be thought of as imposing the simple assumptions to each time point, where past treatment and past and current confounders are included in the set of measured confounders.

In the HAART example the extended assumption of no unmeasured confounders corresponds to assuming that doctors at time point t only look at baseline variables, past HAART treatment and current and past CD4 counts when deciding whether or not to prescribe HAART at time point t. Critically (and perhaps somewhat questionable), it is assumed that doctors do not take other time-varying measurements, e.g., general patient well-being, into account. Or to be precise; that none of these other time-varying measurements are predictive for AIDS-free survival. In the context of the HAART example, the implication of the extended assumption of positivity is that no matter which history of CD4 counts and HAART a given patient has, all treatment options are being actively considered and could be chosen. If all doctors had followed very strict rules, e.g., "when CD4 is below 400 give HAART," this assumption could not have been met. However, from the data it does not appear as if such strict rules have been followed (Sterne et al., 2005). The final assumption of extended consistency implies that it does not matter for AIDS-free survival how a given treatment came about; that is, it does not matter for AIDS-free survival if a treatment was given a part of predefined plan or made up along the way if in effect the same treatment was given. Since progression to AIDS or death is very much a biomedical process, the extended assumption of consistency would seem justifiable.

7.4.5 Implementing the mini trials approach

In Gran et al. (2010) the goal is to estimate the overall causal effect of HAART; i.e., to compare HAART with no treatment at all. Thus, the two considered treatment strategies are very simple either $\bar{a} = (0, 0, ...)$ or $\bar{a} = (1, 1, ...)$. The analysis is based on the SWISS Cohort Study, which is an ongoing multi-center research project following up HIV-infected adults aged 16 or older (Ledergerber et al., 1999). Patients have scheduled doctor appointments every 6 months, with additional laboratory measurements taken every 3 months, but treatment initiation can take place at any time. The analysis in Gran et al. considers each month as a time period and uses last observation carried forward for months without additional measurements. In total, 2,161 individuals contributed to the data used in the analysis; see the paper for further details.

The key idea in the mini trials approach is to imagine that a new trial was initiated each month. To illustrate this approach, consider month t. All persons who were not taking HAART in month $t-1$ are included in the mini trial starting at month t. Persons initiating HAART at month k will constitute the treatment group for this mini trial, while persons not initiating HAART at month k will constitute the control group. The treatment group will be followed to either AIDS/death or end of follow-up. The control group will be followed to AIDS/death, end of follow-up, or treatment initiation. Thus, persons in the control group

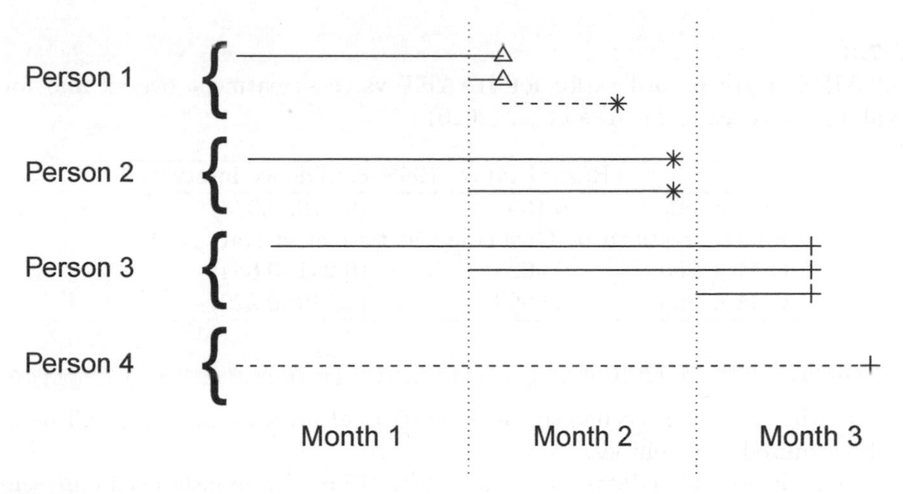

FIGURE 7.3
Illustration of the mini trials approach. Full lines correspond to time spent in the control group, while dashed lines correspond to treatment group. Stars denote onset of type 2 diabetes, vertical lines denote censoring, and triangles denote artificial censoring.

of the mini trial started at month t will be artificially censored if they initiate HAART at a later stage. Figure 7.3 illustrates the different types of individuals in the mini trials starting at month t.

Obviously, a person in the control group of the mini trial starting at month t is also eligible to participate in the mini trial starting at month $t+1$. Thus, the same person can contribute person-months to many different control groups from different mini trials at the same time. However, if that person initiates HAART he will be artificially censored in all control groups he is a part of and subsequently only contribute person-months to a single treatment group. In practical terms each person in each mini trial will produce a row in the extended pseudo-dataset for each month that person is under risk in this trial. Each row contains information about treatment/control group status, the originating mini trial, current and past CD4 count, AIDS or death in that month, as well as baseline information (age, sex, etc.).

The artificial censoring occurring if a person from a control group initiates HAART is by no means an independent censoring. On the contrary initiation of HAART is likely a consequence of disease progression. Accordingly, persons still in the control groups must be weighted to compensate for the selection happening through the dependent censoring; this is know as "inverse probability of censoring techniques" (Robins et al., 2000). In practice this is done by fitting a survival model. In Gran et al. the Aalen additive hazard model is employed, to the censoring process (both types of censoring are considered as one, but one could equally well have considered the artificial censoring separately). That is use censoring as the event of interest in a survival model conditional on current and last month's CD4 count, treatment status, and baseline variables. In Gran et al. models for probability of censoring are fitted to each mini trial separately, but these models could have been combined. Next, the model for censoring can be used to obtain estimates of the probability of not being censored up until specific month of observation for each person in the extended dataset. Each month of observation is now weighted by the inverse of the probability of not being censored before that month. The R package "timereg" contains an implementation of Aalen's additive hazard model, which can be used to obtain the required probabilities. If a Cox model is

TABLE 7.3
Estimated AIDS/death hazard ratios for HAART vs. no treatment overall and for subgroups. Values are taken from Gran et al. (2010).

	Hazard ratio	(95% confidence interval)
Overall effect	0.165	(0.079; 0.343)
Effects stratified by CD4 count at treatment start		
CD4 $\geq$ 200	0.402	(0.241; 0.684)
CD4 $<$ 200	0.066	(0.019; 0.229)

used to model the censoring mechanism, most statistical software packages will be able to compute the required probabilities.

At this point the extended dataset contains 1,201,315 person-months each represented as a single row in the extended dataset. Each row contains information about treatment/control group status, the originating mini trial, current and past CD4 count, AIDS or death in that month, baseline information (age, sex, etc.), and a weight to account for dependent censoring. The causal effect of HAART vs. no treatment can now be estimated by comparing control groups with treatment groups using a weighted Cox model stratified on mini trials and conditioning on baseline variable, CD4 count in the month of the mini trial, and 3 months before the mini trial. The actual analysis in Gran et al. included a few additional variables, but these have been omitted in this discussion for clarity of presentation. The theoretical underpinning for using the stratified Cox model to combine effect estimates from the different mini trials is that it in effect corresponds to using composite likelihood inference; see Gran et al. for further details.

The stratified and weighted Cox model can be estimated in any software package, which can handle the relatively large datasets re-sulting from the re-use of the same person in many mini trials. Confidence intervals can be obtained using jackknife estimated standard errors. Table 7.3 below presents selected results from the analysis.

The estimated overall effect hazard ratio of 0.165 (95% confidence interval 0.079 to 0.343) can be interpreted as the effect one would find in an ideal randomized trial where HIV patients were randomized to either HAART or no-treatment as soon as they entered the trial. Thus HAART provides a significant (both clinical and statistical) improvement in AIDS-free survival. If the ideal randomized trial instead had as an inclusion the criterion that CD4 count should be below 200, the effect of HAART increases even further to a hazard ratio of 0.066. However, delaying the initiation of HAART until CD4 count is below 200 might still be sub-optimal compared to an earlier initiation.

This causal interpretation hinges on the three assumptions of no unmeasured confounding, consistency and positivity discussed Section 7.4.4 but also on the correctness of the employed models. The three basic assumptions from Section 7.4.4 cannot be formally tested, but should be critically discussed using subject matter knowledge. In contrast the correctness of the employed models (e.g., the Aalen model for the censoring and the Cox model for the AIDS-free survival) can, and should be assessed using traditional model diagnostics; see e.g., Chapters 1, 3, and 13.

Whilst the analysis in Gran et al. clearly demonstrates the usefulness of the mini trials approach as well as the substantial therapeutic value of HAART, it does not provide any guidance to when it is most optimal to initiate HAART; the study merely establishes that it is always better to initiate HAART than to do nothing. Instead the mini trials approach could have been used to compare more complex treatment strategies, e.g., "Initiate HAART when the CD4 count first drop below 350." This could be achieved by redefining the treatment group in each mini trial accordingly. Indeed determining optimal treatment strategies in the presence of feedback is currently being very actively researched in the causal inference community; see for instance Hernan et al. (2006); Robins et al. (2008); van der Laan and Petersen (2007); Cain et al. (2010).

7.5 Appendix: R code

This appendix explains how to compute the weights used in IPTW for a binary exposure and how to use these weights to estimate a causal hazard ratio, i.e., fitting a MSM. The exploration follows the type 2 diabetes example and is an extension of the weights presented in Table 7.2. Assume that data is in the R data frame `myData`, which contains the following variables.

phys Binary exposure variable (0: inactive, 1: active).

SEP Socioeconomic position (1: low, 2: middle, 3: high).

Gender (0: female, 1: male).

onset Time to onset of type 2 diabetes or censoring (years).

event Indicator for type 2 diabetes (0: censoring, 1: otherwise).

Initially a logistic regression is fitted with exposure as dependent variable and including all confounders:

```
fitExp <- glm(phys $\sim$ factor(SEP) + factor(Gender),
              data=myData, family="binomial")
```

At this stage the model fit should be critically evaluated by the usual techniques for logistic regressions. In the following it is assumed that the model fit is deemed acceptable. The required weights can now be obtained by first computing probabilities of having the exposure each person actually has.

```
temp <- predict(fitExp, type="response")
prob <- (myData$phys==1)*temp + (myData$phys==0)*(1-temp)
myData$W <- 1/prob
```

Note that the predict-function only supplies probabilities of being in the active group, which is why the second line is required. Finally, the causal hazard ratio can be obtained by fitting a Cox model including only exposure as a covariate, but weighted by the just computed weights.

```
library(survival)
fitCox <- coxph(Surv(onset, event) $\sim$ factor(phys),
                data=myData, weights=myData$W, robust=T)
summary(fitCox)
```

Bibliography

Bekaert, M., Vansteelandt, S. and Mertens, K. (2010), 'Adjusting for time-varying confounding in the subdistribution analysis of a competing risk', *Lifetime Data Analysis* **16**, 45–70.

Cain, L. E., Robins, J. M., Lanoy, E., Logan, R., Costagliola, D. and Hernan, M. A. (2010), 'When to start treatment? A systematic approach to the comparison of dynamic regimes using observational data', *International Journal of Biostatistics* **6**(2), Article 18.

Concato, J., Shah, N. and Horwitz, R. I. (2000), 'Randomized, controlled trials, observational studies, and the hierarchy of research designs', *New England Journal of Medicine* **342**(25), 1887–1892.

Gran, J. M., Roysland, K., Wolbers, M., Didelez, V., Sterne, J. A., Ledergerber, B., Furrer, H., von Wyl, V. and Aalen, O. O. (2010), 'A sequential Cox approach for estimating the causal effect of treatment in the presence of time-dependent confounding applied to data from the Swiss HIV Cohort Study', *Statistics in Medicine* **29**(26), 2757–2768.

Greenland, S., Pearl, J. and Robins, J. (1999), 'Causal diagrams for epidemiologic research', *Epidemiology* **10**, 37–48.

Hernan, M. A. (2004), 'A definition of causal effect for epidemiological research', *J Epidemiol Community Health* **58**(4), 265–271.

Hernan, M. A., Alonso, A., Logan, R., Grodstein, F., Michels, K. B., Willett, W. C., Manson, J. E. and Robins, J. M. (2008), 'Observational studies analyzed like randomized experiments: an application to postmenopausal hormone therapy and coronary heart disease', *Epidemiology* **19**(6), 766–779.

Hernan, M. A., Brumback, B. and Robins, J. M. (2000), 'Marginal structural models to estimate the causal effect of zidovudine on the survival of HIV-positive men', *Epidemiology* **11**(5), 561–570.

Hernan, M. A., Lanoy, E., Costagliola, D. and Robins, J. M. (2006), 'Comparison of dynamic treatment regimes via inverse probability weighting', *Basic and Clinical Pharmacology and Toxicology* **98**(3), 237–242.

Hernn, M. and Robins, J. (2012), *Causal Inference*, Chapman & Hall. In preparation available online at http://www.hsph.harvard.edu/faculty/miguel-hernan/causal-inference-book/.

Jeon, C. Y., Lokken, R. P., Hu, F. B. and van Dam, R. M. (2007), 'Physical activity of moderate intensity and risk of type 2 diabetes: a systematic review', *Diabetes Care* **30**(3), 744–752.

Knowler, W. C., Barrett-Connor, E., Fowler, S. E., Hamman, R. F., Lachin, J. M., Walker, E. A. and Nathan, D. M. (2002), 'Reduction in the incidence of type 2 diabetes with lifestyle intervention or metformin', *New England Journal of Medicine* **346**(6), 393–403.

Ledergerber, B., Egger, M., Opravil, M., Telenti, A., Hirschel, B., Battegay, M., Vernazza, P., Sudre, P., Flepp, M., Furrer, H., Francioli, P. and Weber, R. (1999), 'Clinical progression and virological failure on highly active antiretroviral therapy in HIV-1 patients: a prospective cohort study. Swiss HIV Cohort Study', *Lancet* **353**(9156), 863–868.

Lu, B. (2005), 'Propensity score matching with time-dependent covariates', *Biometrics* **61**(3), 721–728.

Malkawi, A. (2012), 'The effectiveness of physical activity in preventing type 2 diabetes in high risk individuals using well-structured interventions: a systematic review', *Journal of Diabetology* **2**, 1.

Pearl, J. (2009), *Causality: models, reasoning, and inference*, Cambridge University Press, New York.

Robins, J. and Hernn, M. (2009), *Estimation of the causal effects of time-varying exposures*, Chapman & Hall, Boca Raton, FL, chapter 23, pp. 553–598.

Robins, J., Hernn, M. and Brumback, B. (2000), 'Marginal structural models and causal inference in epidemiology', *Epidemiology* **11(5)**, 550–560.

Robins, J., Orellana, L. and Rotnitzky, A. (2008), 'Estimation and extrapolation of optimal treatment and testing strategies', *Statistics in Medicine* **27**(23), 4678–4721.

Rosenbaum, P. and Rubin, D. (1983), 'The central role of the propensity score in observational studies for causal effects', *Biometrika* **70**, 41–55.

Rotnizky, A. (2009), *Inverse probability weighted methods* in *Longitudinal Data Analysis*, G. Fitzmaurice, M. Davidian, G. Verbeke, G. Molenberghs, Eds., chpt. 20, Chapman & Hall/CRC Press, Boca Raton, FL.

Sterne, J. A., Hernan, M. A., Ledergerber, B., Tilling, K., Weber, R., Sendi, P., Rickenbach, M., Robins, J. M. and Egger, M. (2005), 'Long-term effectiveness of potent antiretroviral therapy in preventing AIDS and death: a prospective cohort study', *Lancet* **366**(9483), 378–384.

Stringhini, S., Tabak, A. G., Akbaraly, T. N., Sabia, S., Shipley, M. J., Marmot, M. G., Brunner, E. J., Batty, G. D., Bovet, P. and Kivimaki, M. (2012), 'Contribution of modifiable risk factors to social inequalities in type 2 diabetes: prospective Whitehall II cohort study', *British Medical Journal* **345**, e5452.

van der Laan, M. J. and Petersen, M. L. (2007), 'Causal effect models for realistic individualized treatment and intention to treat rules', *International Journal of Biostatistics* **3**(1), Article 3.

Vanderweele, T. J. (2008), 'The sign of the bias of unmeasured confounding', *Biometrics* **64**(3), 702–706.

Vanderweele, T. J. and Arah, O. A. (2011), 'Bias formulas for sensitivity analysis of unmeasured confounding for general outcomes, treatments, and confounders', *Epidemiology* **22**(1), 42–52.

VanderWeele, T. and Shpitser, I. (2011), 'A new criterion for confounder selection', *Biometrics* **67**, 1406–1413.

Vansteelandt, S., Mertens, K., Suetens, C. and Goetghebeur, E. (2009), 'Marginal structural models for partial exposure regimes', *Biostatistics* **10**(1), 46–59.

Part II

Competing Risks

Competing risks occur in many medical studies. In this type of data an individual may fail from one of several causes. The investigator is interested in the distribution of the time to failure from one of these causes in the presence of all other causes. A common example of competing risks is one where a patient may die from one of several causes such as cancer, heart disease, complications of diabetes, and so forth. Another example of competing risks is found in defining failure as being either the result of the disease progressing or the toxicity of the treatment. A very special case of competing risks is the right-censored data described in Part I.

In competing risks we assume that there are two potential failure times: X_1 and X_2. Here we assume two competing risks, one of interest and the second of all other failure causes. We cannot observe (X_1, X_2) but rather we observe $T = \min(X_1, X_2)$, the failure time and $\delta = 1$ if $X_1 < X_2$ (Failure from cause 1) or 2 if $X_2 < X_1$ (Failure from cause 2). Patients may still be right censored, which occurs when observation on T stops at some censoring time. For competing risks data we cannot tell if (X_1, X_2) are independent or not. In fact Langberg et al. (1978) show that for any pair (X_1, X_2) there is an independent pair of random variables (X_1^*, X_2^*) so that (X_1, X_2) and (X_1^*, X_2^*) have the same observable distributions for (T, δ).

Competing risks can also be modeled using a multistate model. For this model, discussed in Chapter 20, there are k absorbing states, one for each failure model. No assumptions about potential failure times is needed and the Markov models discussed in Chapter 20 can be applied.

The key parameters in competing risks theory depend on the inference desired. Of interest is the crude hazard rate for a given cause. This quantity is defined as

$$h_1(t) = \lim_{\Delta t \to 0} \frac{Pr[t < T \leq t + \Delta t, \delta = 1 | T \geq t]}{\Delta t}.$$

Note that for the crude rate we are looking at how fast individuals are failing from cause 1 among those individuals who could fail from any cause at time t. This is an event rate set in the real world where an individual could fail from any of the causes. An alternative to the crude hazard rate is the crude probability of the event or the cumulative incidence function. This probability is given by

$$C_1(t) = Pr[T \leq t, \delta = 1].$$

Note that there is no function of $h_1(t)$ which returns the cumulative incidence function. Here the cumulative incidence for cause 1 is the probability that an individual fails from cause 1 at time t in the presence of all other competing risks. For two competing risks, one can show that $C_1(t) + C_2(t) + S(t) = 1$ where $S(t)$ is the survival function of the time T and is the probability that neither competing risk has occurred by time t. Note that opposed to the usual survival models the crude hazard rates and the cumulative incidence functions have no easily recognized mathematical relationship. Also note that $C_1(t)$ is not a proper distribution function since $\lim_{t \to \infty} C_1(t) \neq 1$. $C_1(t)$ is called for this reason a sub-distribution function.

The cumulative incidence function for cause 1 can be written as

$$C_1(t) = \int_0^t h_1(x) S(x) dx$$

where $h_1(t)$ is the hazard rate of cause 1. In many applications it is useful to define the so-called sub-distributional hazard by

$$\lambda_1(t) = -\frac{d \ln[1 - C_1(t)]}{dt}$$

and use this quantity in a Cox-like analysis instead of the crude hazard rates. This approach gives Fine and Gray's (1999) model.

In this part we examine models for the crude probabilities. These probabilities are for individuals in the real world who are at risk for failure from any of the competing risks. There is a body of literature (see Moeschberger and David (1978), for example) that deals with "partial crude" probabilities and rates. For partial crude probabilities we are trying to make some inference about the failure rate in a competing risks experiment conducted in a counter-factual world where some of the competing risks cannot occur. An example of historical interest of this type of inference is the pioneering work of Bernoulli (1766) who was interested in looking at death rates in London and if small pox, then the leading cause of death, could be completely prevented. Details of these calculations can be found in Moeschberger and David's (1978) monograph.

In this part we examine a number of techniques for regression modeling for competing risks data. Chapter 8, by Beyersmann and Scheike, examines classical approaches to the regression problem. They examine and compare regression modeling based on a Cox model for the crude hazard rate $h_1(t)$ and models based on the sub-distributional hazard rate $\lambda_1(t)$. They show that these two approaches may give different regression coefficients which leads to different conclusions about the covariates importance. The next chapter by Chen et al. repeats the analysis methods in Chapter 8 using Bayesian methods.

In Chapters 10 and 11 we present two relatively recent alternatives to the Cox regression-like models of Chapters 8 and 9. In Chapter 10 Logan and Wang introduce pseudo-value regression models. These models have applications to competing risks, overall survival, mean survival and multistate models. They are based on the pseudo-values from an unbiased estimator of some parameter of interest. The pseudo-value for the jth individual to estimate θ based on an unbiased estimator $\hat{\theta}$ are defined as $n\hat{\theta} - (n-1)\hat{\theta}^{(-j)}$ where $\hat{\theta}^{(-j)}$ is the estimator $\hat{\theta}$ with the jth observation removed. These pseudo-observations, usually computed on a grid of time points, are then used in a Generalized Estimating Equation regression model. For competing risks the statistic used for $\hat{\theta}$ is the univariate cumulative incidence function. This approach gives values in most cases close to the Fine and Gray models.

The next chapter by Grøn and Gerds examines the so-called binomial model. This model is based on the fact that at a series of event times we have a binomial response, namely: the competing risk has either occurred at this time or some other competing risks have occurred. We could use almost any model for these binomial probabilities such as a logistic model. In the binomial, set-up models for this function are fit to the data at these time points. An inverse of the probability of censoring weighted (IPCW) estimating equations is used to estimate parameters.

Chapter 12 by Zhang et al. presents a series of illustrations on datasets drawn from the large dataset of the Center for International Blood and Marrow Transplantation (CIBMTR) is presented. The CIBMTR is one of the leading areas of research into hematopoietic stem cell transplantation. In this chapter regression models for both overall survival and competing risks probabilities are presented. The authors present some new results which show how to construct estimates of an adjusted survival or cumulative incidence function. These are summary curves that present an estimate of survival for an average patient with an average set of covariates. The chapter also deals in detail with computation methods for this regression model and estimated survival curves.

Bibliography

Bernoulli, D. (1766), 'Essai d'une nouvelle analyse de la mortalité causée par la petite vérole, et des avantages de l'inoculation pour la prévenir', *Mém. de l'Académie Royale de Science*, pp. 1–45.

Fine, J. P. and Gray, R. J. (1999), 'A proportional hazards model for the subdistribution of a competing risk', *Journal of the American Statistical Association* **94**, 496–509.

Langberg, N., Proschan, F. and Quinzi, A. J. (1978), 'Converting dependent models into independent ones, preserving essential features', *Annals of Probability* **6**, 174–181.

Moeschberger, M. L. and David, H. A. (1978), *The theory of competing risks*, Griffin & Company LTD, London.

8

Classical Regression Models for Competing Risks

Jan Beyersmann
Freiburg University Medical Center and Ulm University

Thomas H. Scheike
University of Copenhagen

CONTENTS

8.1 Introduction 157
8.2 The competing risks multistate model 158
 8.2.1 The multistate model 158
 8.2.2 Advantages over the latent failure time model 159
8.3 Nonparametric estimation 160
8.4 Data example (I) 162
8.5 Regression models for the cause-specific hazards 163
 8.5.1 Cox's proportional hazards model 165
 8.5.2 Aalen's additive hazards model 166
8.6 Data example (II) 167
8.7 Regression models for the cumulative incidence functions 168
 8.7.1 Subdistribution hazard 169
8.8 Data example (III) 170
8.9 Other regression approaches 171
8.10 Further remarks 173
 Bibliography 173

8.1 Introduction

Analysing time until death is the archetypical example of a survival analysis, hence the name. In practice, however, time-to-event endpoints are often composites. Common examples in oncology include progression-free survival, disease-free survival and relapse-free survival (Mathoulin-Pelissier et al., 2008). Relapse-free survival is the time until relapse or death, whatever occurs first. Death can also be considered to be a composite, if we distinguish between disease-specific death, say, and death from other causes (Cuzick, 2008). In other words, the "survival time" is typically the time until the first of a number of possible events. It is often tacitly understood that sooner or later every individual experiences at least one of these events. Then the distribution of the survival time will approach one as time progresses. For instance, the cumulative event probabilities of disease-specific death and of death from other causes, respectively, add up to the cumulative all-cause mortality distribution. The latter will ultimately approach one.

Competing risks are the single components of such a composite time-to-event endpoint.

A competing risks model as discussed in this chapter considers time-until-first-event *and* type-of-first-event (Putter et al., 2007). A competing risks analysis therefore provides for more specific results, e.g., in that it allows to study a treatment effect on relapse (but not death) and the treatment effect on death without prior relapse.

Because survival data are often incompletely observed as a consequence of left truncation and right censoring, survival analysis is based on hazards. The concept of hazards is amenable to competing risks. There will now be as many hazards — often called "cause-specific or event-specific hazards" — as there are competing risks. The sum of all cause-specific hazards equals the usual hazard corresponding to the time until any first event.

It will be a key theme of the present chapter that virtually any regression model for a "usual" survival hazard can straightforwardly be used for the cause-specific hazards, too. However, interpretation of the results in terms of probabilities will be complicated by their dependence on all cause-specific hazards. This is so, because the sum of all cause-specific hazards equals the all-cause survival hazard, which in turn determines the survival distribution. As a consequence, regression models of the cumulative event probabilities of a competing risk have emerged since the late 1990s.

A classical textbook reference for competing risks in general and for Cox regression of the cause-specific hazards is the first edition of Kalbfleisch and Prentice (2002) from 1980. A rigorous mathematical treatment of semiparametric multiplicative models and nonparametric additive models for the cause-specific hazards using counting process theory is contained in Andersen et al. (1993). Applied texts include Andersen et al. (2002), Putter et al. (2007) and Beyersmann et al. (2012), the latter two putting an emphasis on using R. The overview paper Andersen and Perme (2008) includes a discussion on inference for cumulative event probabilities of a competing risk.

8.2 The competing risks multistate model

8.2.1 The multistate model

We disregard covariates for the time being. Consider competing risks data arising from a multistate model as depicted in Figure 8.1 for two competing risks. Boxes in the figure indicate states which an individual may occupy. At time 0, all individuals are in the initial

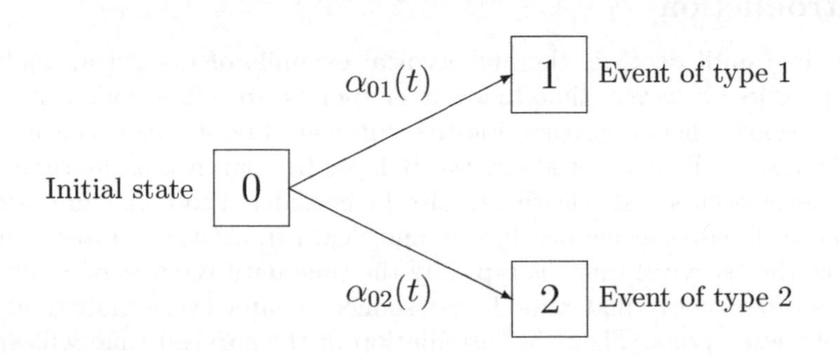

FIGURE 8.1
Competing risks multistate model for two competing risks with cause-specific hazards $\alpha_{0j}(t)$, $j = 1, 2$.

state 0. Events are modelled as transitions between the states. A competing risks model only models transitions out of the initial state, indicated by the arrows in the figure. An individual's event time can be envisaged as the waiting time in state 0. Occurrence of a competing risk of type j is modeled by making a transition from 0 to j at this time.

More formally, consider the competing risks process $(X_t)_{t\geq 0}$, with state space $\{0, 1, 2, \dots J\}$. The competing risks process moves out of the initial state 0, $P(X_0 = 0) = 1$, at time T,

$$T = \inf\{t > 0 \,|\, X_t \neq 0\}.$$

As usual, we assume $(X_t)_{t\geq 0}$ to be right-continuous. The competing risks process is then in one of the competing event states $1, 2, \dots J$ at time T. The type of the first event, often called "cause of failure," therefore is

$$X_T \in \{1, 2, \dots J\},$$

the state the process enters at time T.

The stochastic behaviour of the competing risks process is completely determined by the cause-specific hazards

$$\alpha_{0j}(t)\mathrm{d}t = P(T \in \mathrm{d}t, X_T = j \,|\, T \geq t),\ j = 1, 2, \dots J,$$

which we assume to exist. Here we have written dt both for the length of the infinitesimal interval $[t, t + \mathrm{d}t)$ and the interval itself. We also write $A_{0j}(t)$ for the cumulative cause-specific hazards, $A_{0j}(t) = \int_0^t \alpha_{0j}(u)\mathrm{d}u$, $j = 1, 2$.

The cause-specific hazards can be thought of as momentary forces of transition, moving along the arrows in Figure 8.1. They generate competing risks data as follows (Gill and Johansen, 1990; Allignol et al., 2011b). The sum of all cause-specific hazards yields the usual all-cause survival hazard $\alpha_{0\cdot}(t)$ corresponding to the distribution of T,

$$\alpha_{0\cdot}(t) = \sum_{j=1}^{J} \alpha_{0j}(t),\ A_{0\cdot}(t) = \sum_{j=1}^{J} A_{0j}(t).$$

Say, $T = t_0$. Then the competing event type is j with probability $P(X_{t_0} = j | T = t_0) = \alpha_{0j}(t_0)/\alpha_{0\cdot}(t_0)$.

The survival function of T is $P(T > t) = \exp(-A_{0\cdot}(t))$ and the cumulative event probabilities of the competing risks — often called "cumulative incidence functions"— are

$$F_j(t) = P(T \leq t, X_T = j) = \int_0^t P(T > u-)\, \alpha_{0j}(u)\, \mathrm{d}u,\ j = 1, 2, \dots J.$$

We note that $P(T > t) + \sum_{j=1}^{J} F_j(t) = 1$. Both the survival function and, via $P(T > u-)$, the cumulative incidence functions depend on all cause-specific hazards.

8.2.2 Advantages over the latent failure time model

Competing risks data are sometimes considered to arise from risk-specific latent times. Restricting ourselves to two competing risks for ease of presentation, the latent failure time model postulates the existence of random variables $T^{(1)}, T^{(2)} \in [0, \infty)$. The connection to the multistate data is

$$T = \min\left(T^{(1)}, T^{(2)}\right) \quad \text{and} \quad X_T = 1 \iff T^{(1)} < T^{(2)}.$$

The times $T^{(1)}, T^{(2)}$ are latent, because, say, $T^{(2)}$ in general remains unobserved, if $X_T = 1$. This is unlike the data (T, X_T) arising from the multistate model, which are observable save for left truncation and right censoring.

One difficulty of the additional latent failure time structure is that one has to specify the dependence of $T^{(1)}$ and $T^{(2)}$, which, however, is empirically non-identifiable, see, e.g., Chapter 17 of Crowder (2012) and the references therein.

We do not use the latent failure time model because nothing is gained from superimposing this additional structure, as, in particular, Aalen (1987) forcefully argues. As we will see below, the (cumulative) cause-specific hazards are empirically identifiable. Knowledge of these does suffice because they generate the competing risks data (T, X_T) as described in Section 8.2.1.

Two remarks are in place before dropping the subject: Firstly, an analysis of $\alpha_{01}(t)$, say, coincides with an analysis of the hazard corresponding to the distribution of $T^{(1)}$, *if* one assumes $T^{(1)}$ and $T^{(2)}$ to exist and to be independent. This is sometimes misinterpreted in the sense that the aim of analysing $\alpha_{01}(t)$ is to learn about the distribution of $T^{(1)}$. This is not the case.

Secondly, there persists an attraction towards the latent failure time approach. The reason probably lies in the fact that it suggests a way to answer "what if" questions. The distribution of $T^{(1)}$ is often interpreted as the survival distribution in a hypothetical world where the competing risk of type 2 no longer occurs. The value of such hypothetical considerations has been questioned (Andersen and Keiding, 2012), but they are feasible without the additional latent failure time structure simply by modifying the cause-specific hazards, potentially equating them with zero.

Both such hypothetical consideration and the theory of competing risks are typically traced back to Daniel Bernoulli's 18th century argument in favour of vaccination against smallpox; see, e.g., Section 3.3 of Beyersmann et al. (2012) or Appendix A of David and Moeschberger (1978). Bernoulli did not use latent times but (time-constant) hazards and he hypothesized that vaccination would equate the smallpox hazard with zero.

8.3 Nonparametric estimation

It is instructive to recapitulate nonparametric estimation in the presence of competing risks before considering regression models. The appealingly simple Nelson-Aalen estimator of the cumulative cause-specific hazards highlights in which sense competing risks act as censoring, an issue which has led to quite some confusion. The nonparametric estimators also provide a template for prediction based on regression models for the cause-specific hazards; the approach will be to replace the Nelson-Aalen estimator by its model-based counterparts.

We consider n individuals under study. Their individual competing risks data are assumed to be i.i.d. replicates of $(X_t)_{t\geq 0}$, where observation of $(X_t)_{t\geq 0}$ is subject to independent right censoring/left truncation as in Andersen et al. (1993) and Aalen et al. (2008). We aggregate the data over all individuals. In counting process notation, let

$$\begin{aligned} Y(t) &= \#\text{ Individuals observed to be in state 0 just before } t, \\ N_{0j}(t) &= \#\text{ Individuals with observed } 0 \to j\text{-transition in } [0,t],\ j = 1, 2, \ldots J, \\ N_{0\cdot}(t) &= \sum_{j=1}^{J} N_{0j}(t) = \#\text{ Individuals with an observed event in } [0,t]. \end{aligned}$$

We also write $\Delta N_{0j}(t)$ for the increment $N_{0j}(t) - N_{0j}(t-)$, i.e., the number of type j events observed exactly at time t, and $\Delta N_{0\cdot}(t) = \sum_{j=1}^{J} \Delta N_{0j}(t)$.

The Nelson-Aalen estimator $(\hat{A}_{01}(t), \hat{A}_{02}(t), \ldots \hat{A}_{0J}(t))$ of the cumulative cause-specific hazards has jth entry

$$\hat{A}_{0j}(t) = \sum_{s \leq t} \frac{\Delta N_{0j}(s)}{Y(s)},$$

where the sum is taken over all observed event times s, $s \leq t$. The Nelson-Aalen estimator of the cumulative all-cause hazard is $\hat{A}_{0\cdot}(t) = \sum_{j=1}^{J} \hat{A}_{0j}(t)$.

Note that for computation of $\hat{A}_{01}(t)$, say, the numerator $\Delta N_{01}(s)$ only counts observed type 1 events, while the denominator $Y(s)$ handles right-censored event times and observed competing events of type j alike, $j \neq 1$. This implies that for computing $\hat{A}_{01}(t)$ in some statistical software package, we may *code* both the usual censoring events and observed competing events other than type 1 as a censoring event.

However, these roles change when computing $\hat{A}_{02}(t), \ldots \hat{A}_{0J}(t)$. For $\hat{A}_{02}(t)$, only observed type 2 events are counted in the numerator, while the denominator $Y(s)$ handles right-censored event times and observed competing events other than type 2 alike. Only the usual censoring events would *always* be coded as a censoring event.

We will encounter this principle again when considering regression models for the cause-specific hazards. A formal justification can be found in Chapter III of Andersen et al. (1993).

The usual Kaplan-Meier estimator of $P(T > t)$ is a deterministic function of $\hat{A}_{0\cdot}(t)$ and, hence, of all cause-specific Nelson-Aalen estimators,

$$\hat{P}(T > t) = \prod_{s \leq t} \left(1 - \Delta\hat{A}_{0\cdot}(s)\right),$$

where we have written $\Delta\hat{A}_{0\cdot}(s)$ for the increment $\hat{A}_{0\cdot}(s) - \hat{A}_{0\cdot}(s-)$.

The Aalen-Johansen estimator of the cumulative incidence functions can be derived from the Kaplan-Meier estimator recalling that the cumulative incidence functions add up to the all-cause distribution function. Considering the increments $\widehat{P}(T \leq t) - \widehat{P}(T < t)$, one sees that

$$1 - \hat{P}(T > t) = \sum_{s \leq t} \hat{P}(T > s-) \cdot \Delta\hat{A}_{0\cdot}(s).$$

The interpretation of $\hat{P}(T > s-) \cdot \Delta\hat{A}_{0\cdot}(s)$ is that it estimates the probability to have an event at time s. Using $\hat{A}_{0\cdot}(t) = \sum_{j=1}^{J} \hat{A}_{0j}(t)$ yields the Aalen-Johansen estimator of the cumulative incidence functions,

$$\hat{P}(T \leq t, X_T = j) = \sum_{s \leq t} \hat{P}(T > s-) \cdot \Delta\hat{A}_{0j}(s),\ j = 1, 2, \ldots J.$$

The interpretation of the summands now is that they estimate the probability to have an event *of type* j at time s. The Aalen-Johansen estimator can also be obtained by plugging the Kaplan-Meier estimator and the Nelson-Aalen estimator into the representation of $P(T \leq t, X_T = j)$ given at the end of Section 8.2.1.

A detailed discussion of the Nelson-Aalen, Kaplan-Meier and Aalen-Johansen estimators is in Chapter IV of Andersen et al. (1993) and in Chapter 3 of Aalen et al. (2008).

8.4 Data example (I)

We consider a random subsample of 1,000 patients from ONKO-KISS, a surveillance program of the German National Reference Centre for Surveillance of Hospital-Acquired Infections. The dataset is part of the R package `compeir`, which is available at `http://cran.r-project.org`. The patients in the dataset have been treated by peripheral blood stem-cell transplantation, which has become a successful therapy for severe hematologic diseases. After transplantation, patients are neutropenic; that is, they have a low count of white blood cells, which are the cells that primarily avert infections. Occurrence of bloodstream infection during neutropenia is a severe complication.

The dataset contains information on the event time, i.e., a patient's time of neutropenia until occurrence of bloodstream infection, end of neutropenia or death, whatever occurs first, and on the event type. Transplants are either autologous (cells are taken from the patient's own blood) or allogeneic.

The dataset contains 564 patients with an allogeneic transplant. Of these, 120 acquired bloodstream infection. End of neutropenia, alive and without prior infection, was observed for 428 patients. These numbers are 83 and 345, respectively, for the remaining 436 patients with an autologous transplant. There were few cases of death without prior infection and few censoring events.

Figure 8.2 displays the Nelson-Aalen estimates of the cumulative cause-specific hazards within transplantation group. For ease of presentation, we have used a combined competing endpoint "end of neutropenia, alive or dead, without prior bloodstream infection," because there were few such death cases. We have also omitted pointwise confidence intervals in order not to further complicate the figure; Beyersmann et al. (2012) explain how to add such confidence intervals in practice.

The figure illustrates that the cause-specific hazard for end of neutropenia is the major hazard in both transplant groups. We also find that both cause-specific hazards are reduced by allogeneic transplants, the major effect being on the hazard for end of neutropenia. Thinking of the cause-specific hazards as momentary forces of transition, this means that the all-cause "force" is reduced for patients undergoing allogeneic transplant, and that the relative magnitude of the cause-specific forces changes in favour of infection. The interpretation is that events of *any* type are delayed for allogeneic transplants. During this prolonged time of event-free neutropenia, patients are exposed to an only slightly reduced infection hazard. As a consequence, there will *eventually* be more infections in the allogeneic group.

The figure also illustrates the importance to analyse *all* competing risks. For instance in epidemiology, researchers sometimes only compute the infection incidence density or incidence rate, i.e., an estimate of the cause-specific hazard of infection under the assumption that the hazard is constant over time. Because of Figure 8.2 (right), such an analysis would be incomplete and miss a key point if not complemented by an analysis of the other competing risk.

Figure 8.3 displays the corresponding Aalen-Johansen estimates. The figure confirms our previous conclusion that events of any type are delayed within the group of allogeneic transplants, but that there will eventually be more infections for this group. We have again omitted pointwise confidence intervals for ease of presentation and refer to Beyersmann et al. (2012) for a practical textbook account.

One may ask whether we are over-interpreting the difference of the curves in the left panel of Figure 8.3. Hieke et al. (2013) investigated this question in the full dataset and found the early difference between the cumulative incidence functions to be significant based

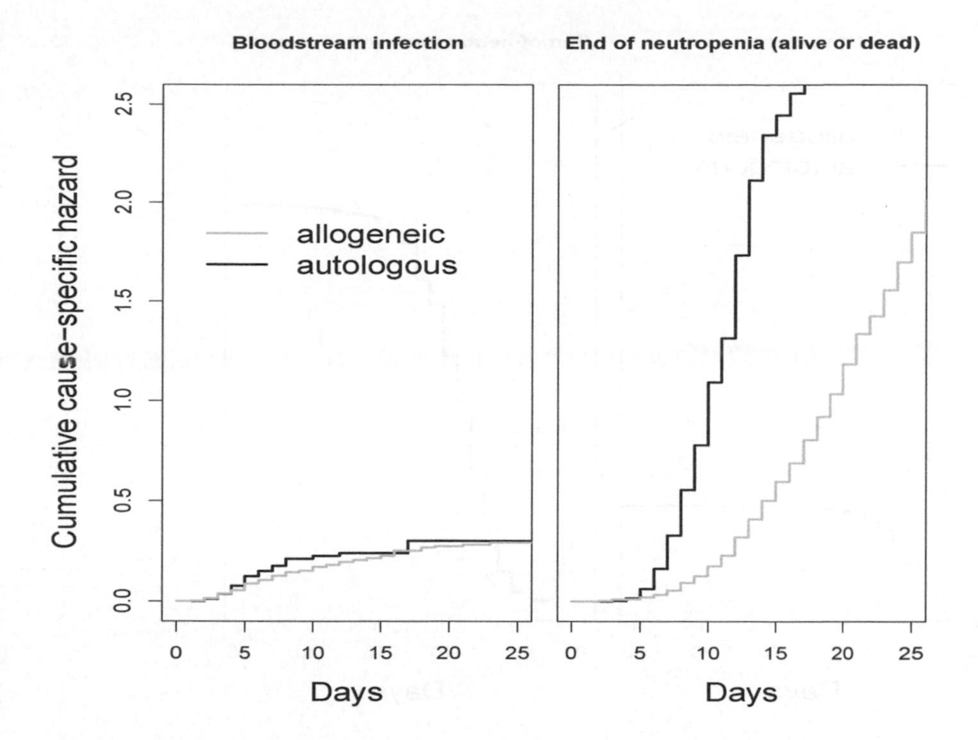

FIGURE 8.2
Nelson-Aalen estimates of the cumulative cause-specific hazards within a transplantation group.

on simultaneous confidence bands. They also discussed medical literature on why allogeneic transplants are expected to increase the proportion of infected patients.

8.5 Regression models for the cause-specific hazards

The aim is to relate the cause-specific hazards to a vector of covariates Z_i for individual i, $i = 1, \ldots, n$, known at time origin. A hazard regression model can also be formulated for time-dependent covariates, but the interpretation often becomes more difficult, see, e.g., Chapters 4, 8, and 9 of Aalen et al. (2008). Results from regression models for all cause-specific hazards and with only baseline covariates can be interpreted in terms of probabilities and can, e.g., be used to predict cumulative incidence functions. In general, this is not possible anymore with time-dependent covariates. A simple binary time-dependent covariate can be modelled by introducing an additional transient state $\tilde{0}$ into the multistate model of Figure 8.1. Transitions between states 0 and $\tilde{0}$ would reflect changes of the time-dependent covariate. Occurrence of a competing risk would still be modelled by transitions into one of the two absorbing states 1 and 2. A regression model including this time-dependent covariate and for the cause-specific hazard of event j, $j \in \{1, 2\}$, would compare the hazard of the $\tilde{0} \to j$ transition with the hazard of the $0 \to j$ transition. However, probabilities

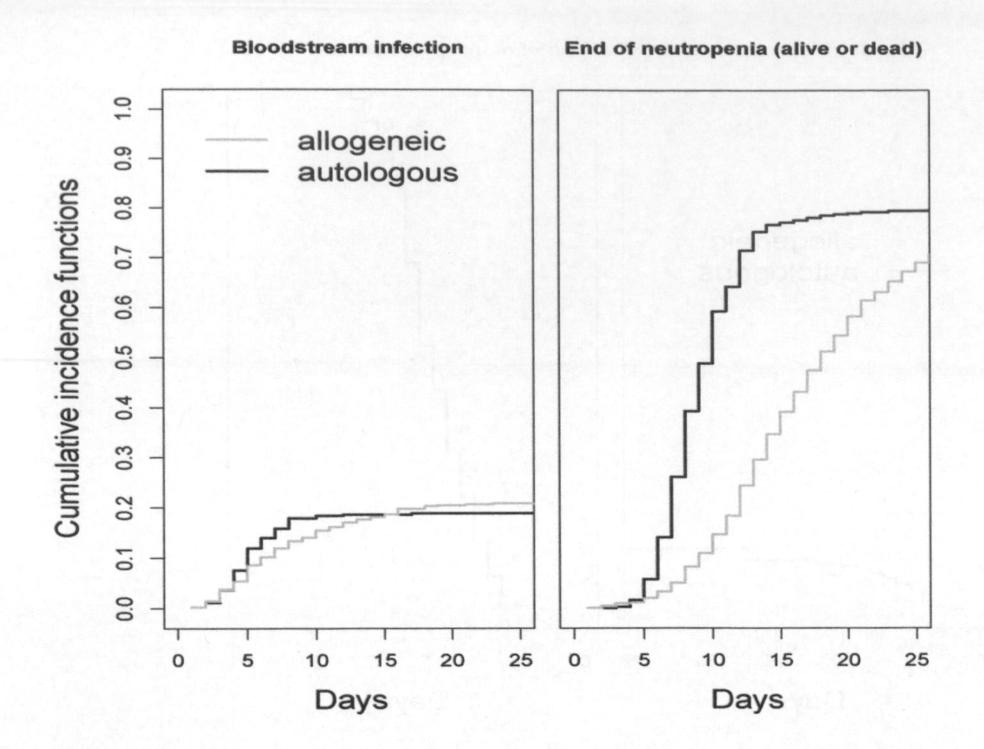

FIGURE 8.3
Aalen-Johansen estimates of the cumulative incidence functions within a transplantation group.

would also depend on the $\tilde{0} \leftrightarrow 0$ transitions, which are not modelled in this approach. The problem persists in the absence of competing risks.

We restrict ourselves to baseline covariates for ease of presentation. Particular attention to time-dependent covariates in the presence of competing risks has been given by Cortese and Andersen (2010) and Beyersmann et al. (2012), Section 11.2.

We will also assume that covariates have cause-specific, i.e., different effects on the cause-specific hazards. Models are also feasible where one covariate has a common effect on all cause-specific hazards, while another covariate has different effects on these hazards. However, these models are rarely used in practice for competing risks. They do lead to more parsimonious models, which is one reason why they are attractive for more complex multistate models. Readers are referred to Chapter VII of Andersen et al. (1993) for an in-depth treatment and to Lunn and McNeil (1995) and Andersen and Keiding (2002) for practical accounts.

Assuming cause-specific effects, virtually any hazard regression model can easily be fitted to a cause-specific hazard by coding the other competing events as censoring, as stated earlier. The data example in Section 8.4 illustrated that this approach should typically be applied to each cause-specific hazard in turn, because one might otherwise miss important aspects of the data.

Cox's proportional hazards model is one of the most common choices with competing risks and will be discussed in Section 8.5.1. A drawback of the model in the competing risks setting is that assuming all cause-specific hazards to follow Cox models usually precludes the all-cause hazard to comply with the proportional hazards assumption. If, as is common in oncology, one uses the Cox model to analyse both the composite time-to-event endpoint and the competing risks, this may lead to inconsistencies. Klein (2006) therefore argued in favour of Aalen's additive hazard model. The reason is that the cause-specific hazards add up to the all-cause hazard of the composite. We discuss the Aalen model in Section 8.5.2.

8.5.1 Cox's proportional hazards model

Proportional cause-specific hazards models assume that

$$\alpha_{0j;i}(t; Z_i) = \alpha_{0j;0}(t) \cdot \exp\left(\beta_{0j} \cdot Z_i\right), \, j = 1, 2, \ldots J, \, i = 1, \ldots, n,$$

where β_{0j} is a $1 \times p$ vector of regression coefficients, Z_i is a $p \times 1$ vector of covariates for individual i, and $\alpha_{0j;0}(t)$ is an unspecified, non-negative baseline hazard function. We also write

$$A_{0j;0}(t) = \int_0^t \alpha_{0j;0}(u) \mathrm{d}u \quad \text{and} \quad A_{0j;i}(t; Z_i) = \int_0^t \alpha_{0j;i}(u; Z_i) \mathrm{d}u$$

for the respective cumulative cause-specific hazards.

Andersen and Borgan (1985) used counting processes and the results of Andersen and Gill (1982) to study multivariate Cox models, including the present competing risks case; see Andersen and Borgan (1985) for earlier references and Chapter VII.2 of Andersen et al. (1993) for a textbook account. They derived a partial likelihood which is a product over all observed event times, all individuals and all competing risk types. Assuming cause-specific effects β_{0j}, $j = 1, 2, \ldots J$, the partial likelihood factors into J parts. The jth part is algebraically identical to the partial likelihood that one obtains by treating observed competing events of type $\tilde{\jmath}$, $\tilde{\jmath} \in \{1, 2, \ldots J\} \setminus \{j\}$, as censoring.

As a consequence, we can use any Cox routine of a statistical software package to fit a Cox model to the jth cause-specific hazard by coding both the usual censoring events and the other observed competing events as a censoring event. One analogously obtains the Breslow estimator $\hat{A}_{0j;0}(t)$ of the cumulative cause-specific baseline hazard. Writing $\hat{\beta}_{0j}$ for the estimator of β_{0j} obtained by maximizing the partial likelihood, the predicted cumulative cause-specific for a covariate vector equal to z is

$$\hat{A}_{0j}(t; z) = \hat{A}_{0j;0}(t) \cdot \exp\left(\hat{\beta}_{0j} \cdot z\right).$$

We reiterate that a Cox analysis in the presence of competing risks remains incomplete as long as this approach has not been applied to all competing risks in turn. Readers are also warned that a Cox routine of a statistical software package may additionally return Kaplan-Meier-type survival probabilities. This information, however, is typically without use because probabilities will depend on all cause-specific hazards.

Survival probabilities and cumulative incidence functions may be predicted by replacing the increments of the cause-specific Nelson-Aalen estimators in Section 8.3 with their predicted counterparts. The predicted survival probability is

$$\hat{P}(T > t \,|\, z) = \prod_{s \le t} \left(1 - \left(\sum_{j=1}^{J} \Delta \hat{A}_{0j}(s; z) \right) \right),$$

where as before Δ indicates an increment and the index s runs over all observed event times s, $s \le t$. The predicted cumulative incidence functions are

$$\hat{P}(T \le t, X_T = j \,|\, z) = \sum_{s \le t} \hat{P}(T > s - \,|\, z) \cdot \Delta \hat{A}_{0j}(t; z).$$

These predictions have been implemented in SAS Macros (Rosthøj et al., 2004) and in the R packages `mstate` (de Wreede et al., 2010, 2011) and `riskRegression` (Gerds et al., 2012). One nice property of these predictions is that they ensure non-decreasing cumulative hazards and that $\hat{P}(T > t \mid z) + \sum_j \hat{P}(T \leq t, X_T = j \mid z) = 1$.

Readers are referred to Chapter VII of Andersen et al. (1993) for a careful discussion of the properties of multivariate Cox regression and subsequent prediction.

8.5.2 Aalen's additive hazards model

Additive hazards models assume that

$$\alpha_{0j;i}(t; Z_i) = \alpha_{0j;0}(t) + \beta_{0j}(t) \cdot Z_i,\ j = 1, 2, \ldots J,\ i = 1, \ldots, n,$$

where $\beta_{0j}(t)$ is a $1 \times p$ vector of regression coefficient functions, Z_i is a $p \times 1$ vector of covariates for individual i, and $\alpha_{0j;0}(t)$ is an unspecified, non-negative baseline hazard function. The cumulative cause-specific hazards can then be computed as $A_{0j;0}(t) + Z_i B_{0j}(t)$ where

$$A_{0j;0}(t) = \int_0^t \alpha_{0j;0}(u) \mathrm{d}u \quad \text{and} \quad B_{0j}(t) = \int_0^t \beta_{0j}(u) \mathrm{d}u.$$

This model has been introduced by Aalen (1980) and has been studied in further details in a large number of papers. The model can be fitted using the R-packages `addreg`[1], `survival`, or `timereg`.

In the context of the competing risks setting one important property of the model as pointed out by for example Klein (2006) is that it is closed under addition. Therefore if the cause-specific hazards are additive, the total hazard of dying $\sum_{j=1}^J \alpha_{0j;i}(t; Z_i)$ is still additive. The standard least squares estimator of the cumulative hazards has the nice property that if the covariates are the same for all causes then the estimator of the total hazard for mortality is equivalent to the sum of the estimators of the cause specific hazards. The model is very flexible but has the problem that the standard fitting procedures does not enforce the condition that the cumulative hazard is non-decreasing.

Given the standard least-squares estimators of $A_{0j;0}(t)$ and $B_{0j}(t)$, that we denote as $\widehat{A}_{0j;0}(t)$ and $\widehat{B}_{0j}(t)$, we can predict the cumulative cause-specific hazards for a covariate vector equal to z as

$$\hat{A}_{0j}(t; z) = \hat{A}_{0j;0}(t) + z\hat{B}_{0j}(t).$$

Subsequently this model can also be used to predict survival probabilities and cumulative incidence functions by again replacing the increments of the cause-specific Nelson-Aalen estimators in Section 8.3 with their predicted counterparts. The predicted survival probability then becomes

$$\hat{P}(T > t \mid z) = \prod_{s \leq t} \left(1 - \Big(\sum_{j=1}^{J} \Delta \hat{A}_{0j}(s; z) \Big) \right).$$

The predicted cumulative incidence functions are

$$\hat{P}(T \leq t, X_T = j \mid z) = \sum_{s \leq t} \hat{P}(T > s- \mid z) \cdot \Delta \hat{A}_{0j}(t; z).$$

These predictions have been implemented in `addregmc` (Aalen et al., 2001); see the same web page as for `addreg`.

[1] `www.med.uio.no/imb/english/research/groups/causal-inference-methods/software`

8.6 Data example (II)

We revisit the ONKO-KISS data of Section 8.4 and illustrate using standard Cox regression for the cause-specific hazards. Fitting separate Cox models to the competing risks outcomes as described above, we find that allogeneic transplants decrease the cause-specific hazard of bloodstream infection by an estimated hazard ratio of 0.77 (95% confidence interval $[0.58, 1.03]$). The analysis of the competing cause-specific hazard finds that allogeneic transplants decrease it by an estimated hazard ratio of 0.27 ($[0.23, 0.31]$).

The interpretation is as before: Allogeneic transplants decrease both cause-specific hazards. Therefore, they also decrease the all-cause hazard and events of any type are delayed in this group. However, the decrease as measured by the cause-specific hazard ratio is much more pronounced for the hazard for end of neutropenia. Allogeneic transplants therefore change the relative magnitude of the cause-specific hazards in favour of infection. As a consequence, there are eventually more infections in this group.

Note that this reasoning has neglected the fact that the hazard for end of neutropenia is also the major cause-specific hazard as illustrated in Figure 8.2. Computing the Breslow estimators and subsequent predictions would capture this aspect. Briefly speaking, the very same cause-specific hazard ratio is the more important the more pronounced the corresponding cause-specific baseline hazard is. Readers are referred to Allignol et al. (2011b) who discussed this aspect via simulations from the empirical law of baseline group data.

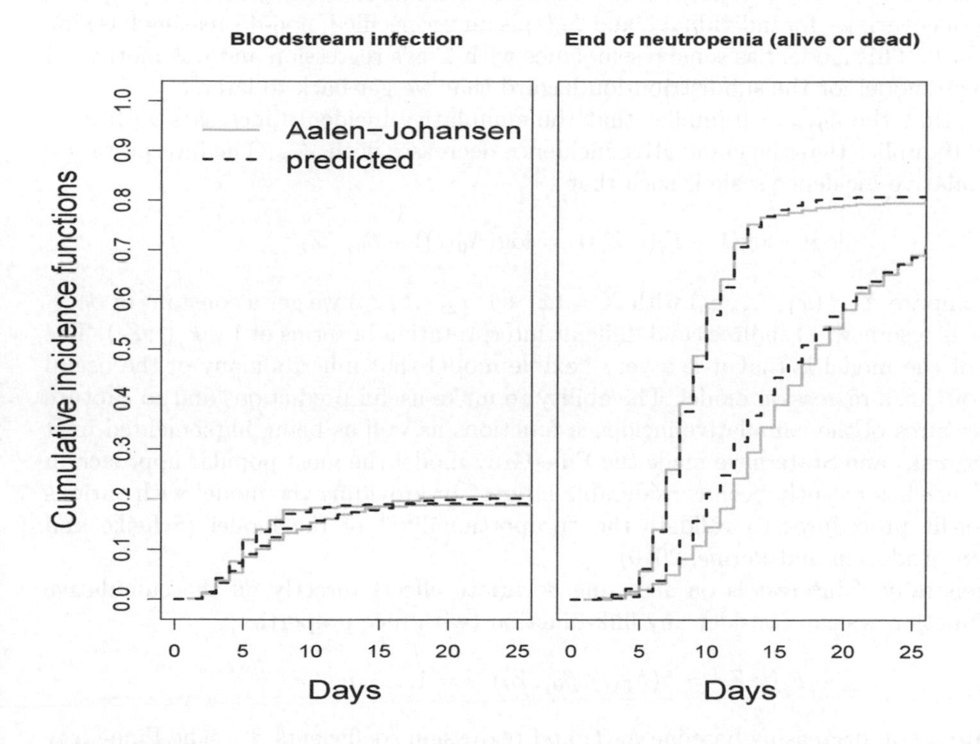

FIGURE 8.4
Aalen-Johansen estimates (dark-grey) as in Figure 8.3 and predicted cumulative incidence functions (black) based on Cox models for the cause-specific hazards.

The predicted cumulative incidence functions are compared with the Aalen-Johansen estimates in Figure 8.4. We find that the fit is reasonable, although it is not perfect for the cumulative incidence function for end of neutropenia. The reason is a time-varying effect of allogeneic transplants on the cause-specific hazard for end of neutropenia, cf. Figure 8.2 (right), which is not captured by the present Cox model with time-invariant regression coefficients.

8.7 Regression models for the cumulative incidence functions

The data example illustrated that a covariate effect on the cause-specific hazard scale does not translate without further ado into an effect on the cumulative incidence functions scale. This has motivated efforts to *directly* model the cumulative incidence functions. Gray (1988) considered a K-sample test for cumulative incidence functions, which directly makes inference about the cumulative incidence functions. This work was later extended into a regression setting where the effects were further quantified in the Fine-Gray model (Fine and Gray, 1999).

The Fine-Gray model assumes that the cumulative incidence for cause j and subject i is given by

$$F_j(t; Z_i) = 1 - \exp\{-\Lambda_0(t) \cdot \exp(\beta_{0j} \cdot Z_i)\}, \, i = 1, \ldots, n,$$

where β_{0j} is a $1 \times p$ vector of regression coefficients, related to the j'th cause, Z_i is a $p \times 1$ vector of covariates for individual i, and $\Lambda_0(t)$ is an unspecified, non-decreasing baseline with $\Lambda_0(0) = 0$. This model has some resemblance with Cox's regression and was motivated as a Cox-type model for the subdistribution hazard that we get back to later.

We note that the $\beta_{0j,k} > 0$ implies that the cumulative incidence increases with $Z_{i,k}$, and $\beta_{0j,k} < 0$ implies that the cumulative incidence decreases with $Z_{i,k}$. The interpretation on the cumulative incidence scale is such that

$$\log(-\log(1 - F_j(t; Z_i))) = \log(\Lambda_0(t)) + \beta_{0j} \cdot Z_i$$

and if we compare $X = (x_1, \ldots, x_p)$ with $\tilde{X} = (x_1 + 1, x_2, \ldots, x_p)$ we get a constant of $\beta_{0j,1}$. Clearly this is a somewhat indirect and difficult interpretation in terms of $1 - F_j(t; Z_i)$. The advantage of the model is that it is a very flexible model that inherits many of the useful properties of Cox's regression model. The ability to make useful predictions and to capture the main features of the cumulative incidence functions as well as being implemented in R (package `cmprsk`) and Stata have made the Fine-Gray model the most popular approach in practice. There has recently been considerable interest in providing the model with various Goodness-of-fit procedures to validate the "proportionality" of the model (Scheike and Zhang, 2008; Andersen and Perme, 2010).

More generally if interest is on accessing covariate effects directly on the cumulative incidence function we can consider any link-function (with nice properties)

$$F_j(t; Z_i) = h(\Lambda_0(t), \beta_{0j}, Z_i), \, i = 1, \ldots, n,$$

and estimate a non-decreasing baseline $\Lambda_0(t)$ and regression coefficients β_{0j}. The Fine-Gray model is then given by the link $h_{\text{fg}}(a, b, z) = 1 - \exp(a \exp(bz))$.

Various other link functions, known from binary data, that aim at making the interpretation of the regression coefficients easier have been suggested. Notably, one may use the logistic link-function, $h_{\text{logistic}}(a, b, z) = \exp(a + bz)/(1 + \exp(a + bz))$, absolute,

$h_{\text{absolute}}(a,b,z) = a + bz$, or relative risk measures, $h_{\text{relative}}(a,b,z) = a\exp(bz)$, as described and advocated in Fine (2001); Ambrogi et al. (2008); and Gerds et al. (2012).

We also note that the Fine and Gray model, although custom-made for one cumulative incidence function, is often used to model all cumulative incidence functions, which will typically imply that at least one of these models is misspecified, see, e.g., Section 5.3.4 in Beyersmann et al. (2012). In addition, when fitting models separately, the regression models will not satisfy the natural constraint that $\hat{P}(T > t \mid z) + \sum_j \hat{P}(T \le t, X_T = j \mid z) = 1$. Grambauer et al. (2010) found that a misspecified Fine and Gray analysis still has a quantitative interpretation in that it informs on the plateau of the cumulative incidence functions.

The direct regression models of this section are typically estimated using inverse probability of censoring weighted (IPCW) score equations or related techniques, for example the subdistribution based approach, the pseudo-value approach, see Chapter 10, or the binomial regression approach, see Chapter 11. The Fine-Gray model is typically best fitted using the subdistribution hazard that we present briefly in the next section. Ruan and Gray (2008) suggested a multiple imputation approach.

One key point that adds additional complexity to the estimation of these models is the underlying IPCW model. The typical assumption is that the IPCW is independent of covariates and then the censoring distribution can be estimated by a simple and nonparametric Kaplan-Meier estimator, but if the censoring distribution depends on covariates included in the model then one needs to correctly model this association. This point is often forgotten in practical work.

The Fine-Gray model has been extended to handle left truncation in recent work (Geskus, 2011; Zhang et al., 2011; Shen, 2011). Further methodological developments include stratified models (Zhou et al., 2011), frailties (Katsahian et al., 2006; Katsahian and Boudreau, 2011; Dixon et al., 2011; Scheike et al., 2010), marginal modeling (Scheike et al., 2010; Chen et al., 2008), time-dependent covariates (Beyersmann and Schumacher, 2008), parametric regression (Jeong and Fine, 2007), sample size calculation (Latouche and Porcher, 2007) and joint modeling (Deslandes and Chevret, 2010). Fine (2001) considered linear transformation models of the cumulative incidence function, covering both the Fine and Gray model and a proportional odds model. Sun et al. (2006) proposed a combination of Aalen's additive hazards model and the Cox model for the subdistribution hazard.

8.7.1 Subdistribution hazard

We now briefly describe the subdistribution hazard that can be used for making estimating equations for the parameters of the model. These estimating equations become an IPCW version of a Cox type score equation.

The approach of Fine and Gray (1999) was to consider a "subdistribution time" until occurrence of a certain competing risk, say, type 1,

$$\tilde{T} = \inf\{t > 0 \mid X_t = 1\}$$

which equals the real life event time T, if and only if $X_T = 1$. Otherwise, the subdistribution time equals infinity. Then, the distribution of the subdistribution time equals $P(T \le t, X_T = 1)$ for $t \in [0, \infty)$. Fine and Gray now suggested to fit a Cox model to the corresponding subdistribution hazard $\lambda(t)$,

$$\lambda(t) = -\frac{\mathrm{d}}{\mathrm{d}t}\log\left(1 - P(T \le t, X_T = 1)\right) = \frac{P(T > t)}{1 - P(T \le t, X_T = 1)}\,\alpha_{01}(t). \tag{8.1}$$

Because $P(T \le t, X_T = 1) = 1 - \exp(-\int_0^t \lambda(u)\,\mathrm{d}u)$, the result measures a direct effect on the cumulative incidence function of type 1 events.

The technical difficulty of the Fine and Gray model stems from the *observed* competing events. Their subdistribution times equal infinity, such that their *censored* subdistribution times equal the censoring times. Because observation often stops at the real life event time, these censoring times are in general unknown. In other words, the risk sets associated with the subdistribution hazard approach will be unknown for observed competing events after their real life event times. The main technical achievement of Fine and Gray (1999) was to approximate these risk sets by directly modeling the censoring distribution. They also used empirical process arguments to study the asymptotic properties of the model. In practice, one typically assumes random censoring, i.e., censoring does not depend on covariates, and uses the Kaplan-Meier estimator of the censoring survival function.

To be specific, assume i.i.d. data $(\min(T_i, C_i), X_{T_i}, Z_i)_{i=1,\dots n}$ with random censorship following the survival function $G(t) = P(C \geq t)$. Consider the "complete data" processes $\tilde{N}_i(t) = \mathbf{1}(T_i \leq t, X_{T_i} = 1)$ of type 1 events with at-risk process $\tilde{Y}_i(t) = 1 - \tilde{N}_i(t-)$. Note that $\tilde{Y}_i$ is only the usual "complete data" at-risk process in the absence of competing risks. These functions are in general unknown, but their products with an indicator function $r_i = \mathbf{1}(C_i \geq \min(T_i, t))$ denoting knowledge of vital status are computable from the observable data.

Writing $\hat{G}$ for the Kaplan-Meier estimator of G, Fine and Gray suggested to use the weights $w_i(t) = r_i(t)\hat{G}(t)/\hat{G}(\min(t, T_i, C_i))$ in the "complete data" score function,

$$U(\beta_{01}) = \sum_{i=1}^{n} \int_0^{\infty} \left(Z_i - \frac{\sum_{j=1}^{n} w_j(s)\tilde{Y}_j(s) Z_j \exp\{\beta_{01} Z_j\}}{\sum_{j=1}^{n} w_j(s)\tilde{Y}_j(s) \exp\{\beta_{01} Z_j\}} \right) w_i(s) \mathrm{d}\tilde{N}_i(s). \tag{8.2}$$

This score function reduces to a standard score function for the Cox model in the absence of competing risks, i.e., if only type 1 events are feasible. However, if there are competing risks, the reason to use (8.2) is that the "subdistribution risk set" $\mathbf{1}(\min(\tilde{T}_i, C_i) \geq t)$ is in general unknown after T_i, if $X_{T_i} \neq 1$ and $T_i \leq C_i$. The rationale is that $w_j(t)\tilde{Y}_j(t)$ approximates this subdistribution risk set. Fine and Gray also derived an estimator of the subdistribution baseline hazard along analogous lines. One considers the usual Breslow estimator for complete subdistribution data $(\tilde{N}_i, \tilde{Y}_i)$ and then introduces weights as above.

The *interpretational* difficulty of the subdistribution model also stems from the observed competing events, because, conceptually, these are kept in the subdistribution risk set *after* their real life event times. This has led to somewhat controversial views on the subdistribution hazard concept. For instance, Pintilie (2007) and Lim et al. (2010) stressed that keeping the observed competing events in the risk set accounts for or incorporates the presence of competing risks. On the other hand, Andersen and Keiding (2012) argued that regarding individuals at risk after the real life failure times compromises interpretability of the subdistribution hazard as a hazard, i.e., as an instantaneous risk of failure.

8.8 Data example (III)

We begin with a Fine and Gray analysis of the cumulative incidence function of bloodstream infection. Allogeneic transplants increase the risk of bloodstream infection by a ratio of $\log(1 - F_1(t; \text{allogeneic}))/\log(1 - F_1(t; \text{autologous}))$ at 1.09 ([0.83, 1.44]). The result is obviously different from the Cox analyses of the cause-specific hazards as illustrated in Table 8.1.

Interpreting the result, we first note that the Aalen-Johansen estimators in Figure 8.3 (left) cross, violating the proportional subdistribution hazards assumption. This can be

TABLE 8.1
Results of the data analyses of Sections 8.6 and 8.8: The left panel displays cause-specific hazard ratios. The Fine-Gray result is a subdistribution hazard ratio, logistic link yields an odds ratio. The numbers in square brackets denote 95% confidence intervals.

Cox models for the cause-specific hazards		Direct regression models for the cumulative infection probability	
infection	competing event	Fine-Gray	logistic link
0.77 ([0.58, 1.03])	0.27 ([0.23, 0.31])	1.09 ([0.83, 1.44])	1.07 ([0.78, 1.46])

formally validated by considering goodness-of-fit procedures, see for example Scheike and Zhang (2008). As a consequence, the subdistribution hazard ratio must be interpreted as a time-averaged effect on the scale of the cumulative incidence function (Claeskens and Hjort, 2008). But how the time-average is constructed depends on for example the censoring pattern, so this makes the interpretation difficult. The qualitative interpretation is that the subdistribution hazard ratio of 1.09 reflects that the plateau of the cumulative incidence function is increased for allogeneic transplants. Readers are referred to Beyersmann et al. (2007) for an extensive analysis of the full ONKO-KISS dataset. We here note that censoring was entirely administrative and did not depend on the type of transplant which is an underlying assumption.

A Fine and Gray analysis of $P(T \leq t, X_T = 1)$ and a Cox analysis of $\alpha_{01}(t)$ sometimes lead to comparable or numerically almost identical results, a fact which has led to some confusion in its own right. Grambauer et al. (2010) found that the results are comparable if a covariate has no effect on the other cause-specific hazards or if censoring is heavy. The reason for the latter fact is that the difference between $\alpha_{01}(t)$ and $\lambda(t)$ is small for early times; see (8.1). These findings also suggest an indirect and hence limited way to quantitatively interpret the subdistribution hazard ratio.

Using the `timereg` or `riskRegression` packages of R we also fitted a logistic link model to describe the effect of allogeneic transplants, and conclude that the allogeneic transplants have a increased risk of infection with an odds ratio of 1.07 ([0.78, 1.46]).

The results from all regression models are tabulated in Table 8.1. Finally, we compared the predicted cumulative incidence functions of bloodstream infection based on a Fine-Gray model with the Aalen-Johansen estimates of Figure 8.3. Because of the model misspecification discussed above, the predicted curves cannot capture the crossing of the Aalen-Johansen estimates, but they do capture their plateaus.

8.9 Other regression approaches

We have focused on the two main regression approaches in competing risks. Because hazards are the key quantities in survival analysis, it is both natural and straightforward to fit hazard regression models to the cause-specific hazards. The data example illustrated that it is via the cause-specific hazards that one understands how the cumulative event probabilities evolve, but that there is also a need for direct regression models for the cumulative incidence functions.

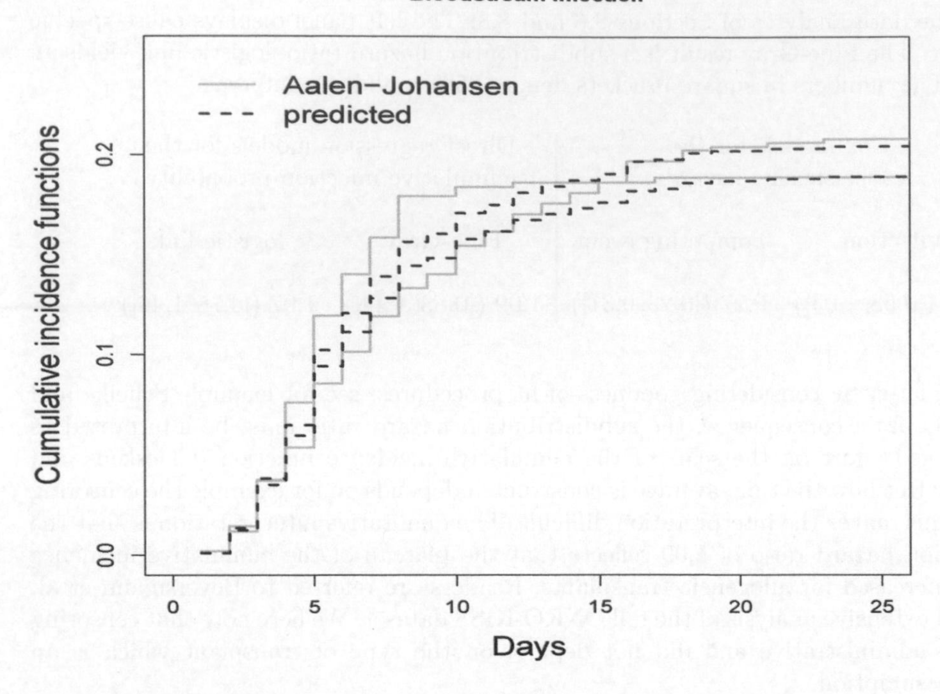

FIGURE 8.5
Aalen-Johansen estimates (dark-grey) as in Figure 8.3 (left) and predicted cumulative incidence functions (black) based on a Fine-Gray model for the outcome bloodstream infection.

There are further regression approaches. Larson and Dinse (1985) considered the decomposition

$$P(T \leq t, X_T = j) = P(T \leq t \,|\, X_T = j) P(X_T = j)$$

and suggested separate regression models for the two probabilities on the right-hand side of the above display, see also Hernandez-Quintero et al. (2011) and the references therein. One important technical difficulty of the approach is that $P(X_T = j) = \lim_{t\to\infty} P(T \leq t, X_T = j)$ will not be identifiable with many survival data. Fine (1999) therefore considered

$$P(T \leq \min(t, \tau), X_T = j) = P(T \leq t \,|\, T \leq \tau, X_T = j) P(T \leq \tau, X_T = j)$$

for a fixed time point τ inside the support of $\min(T, C)$; see also Shen (2012) and the references therein. The interpretational difficulty is that this mixture approach considers a lifetime distribution conditional on the failure cause, which is in general unknown before the event (Andersen and Keiding, 2012).

In contrast, Nicolaie et al. (2010) considered so-called "vertical modeling" via the decomposition

$$P^{T,X_T} = P^T P^{X_T \,|\, T},$$

which follows the prospective algorithm of Section 8.2.1. Similar to the decomposition used by Larson and Dinse, Nicolaie et al. used a standard survival model for P^T and a multinomial logistic regression model for the (conditional) distribution of the failure type. Because the prospective point of view considers these probabilities conditional on the event time, the technical challenge is that such a model is needed as a function of time. In their data analysis, Nicolaie et al. used splines for this purpose.

Further regression approaches include Allignol et al. (2011a) who used "temporal process regression" (Fine et al., 2004) for the so-called conditional probability function of Pepe and Mori (1993), which is defined for type 1 events as

$$P(T \leq t, X_T = 1 \,|\, T > t \text{ or } \{T \leq t, X_T = 1\}) = P(X_t = 1 \,|\, X_t \in \{0, 1\}).$$

Pepe and Mori advocated this quantity, because it is a monotone increasing function, starting at 0 and reaching 1, just like a distribution function. The interpretational difficulty is that it is not the distribution function of an obvious random variable of the competing risks setting. Andersen and Keiding (2012) argue that this function only becomes useful, if the more complex illness-death multistate model applies.

We finally mention Fiocco et al. (2005) who started from Cox models as in Section 8.5.1 in situations with many regression parameters (many competing risks and/or many covariates) but relatively few events. The idea is to decompose the matrix of all regression coefficients into two matrices. One matrix aggregates the covariates into prognostic scores and is estimated based on all events. The other matrix contains the cause-specific effects of the prognostic scores.

8.10 Further remarks

We have assumed that the data are replicates of (T, X_T), subject to independent left truncation and right censoring. This implies that the event type is known for individuals with an observed event time. Sometimes, only the time, but not the event type is known. If these data are missing completely at random, they may be removed from the analysis. Alternatively, one may introduce "event type unknown" as an additional competing risk, which would preserve the risk sets. More sophisticated methods have been discussed by, e.g., Nicolaie et al. (2011) and Lee et al. (2011); see also the references in these papers.

Earlier, we have referred to Andersen et al. (1993) for a careful mathematical treatment of hazard regression models. The asymptotic distribution of predictions may then be studied using the functional delta method, which can be used to derive pointwise confidence intervals. Simultaneous confidence bands, however, are typically based on resampling methods, see Martinussen and Scheike (2006) for a textbook account.

Bibliography

Aalen, O. (1987), 'Dynamic modeling and causality', *Scandinavian Actuarial Journal* pp. 177–190.

Aalen, O., Borgan, Ø. and Fekjær, H. (2001), 'Covariate adjustment of event histories estimated from Markov chains: The additive approach', *Biometrics* **57**, 993–1001.

Aalen, O., Borgan, Ø. and Gjessing, H. (2008), *Survival and Event History Analysis*, Springer, New York.

Aalen, O. O. (1980), 'A model for non-parametric regression analysis of counting processes', *Springer Lect. Notes in Statist.* **2**, 1–25.

Allignol, A., Latouche, A., Yan, J. and Fine, J. (2011a), 'A regression model for the conditional probability of a competing event: application to monoclonal gammopathy of unknown significance', *Journal of the Royal Statistical Society: Series C (Applied Statistics)* **60**(1), 135–142.

Allignol, A., Schumacher, M., Wanner, C., Drechsler, C. and Beyersmann, J. (2011b), 'Understanding competing risks: a simulation point of view', *BMC Medical Research Methodology* **11**, 86.

Ambrogi, A., Biganzoi, E. and Boracchi, P. (2008), 'Estimates of clinically useful measures in competing risks survival analysis', *Statistics in Medicine* **27**, 6407–6425.

Andersen, P., Abildstrøm, S. and Rosthøj, S. (2002), 'Competing risks as a multi-state model', *Statistical Methods in Medical Research* **11**(2), 203–215.

Andersen, P. and Borgan, Ø. (1985), 'Counting process models for life history data: A review', *Scandinavian Journal of Statistics* **12**, 97–140.

Andersen, P., Borgan, Ø., Gill, R. and Keiding, N. (1993), *Statistical Models Based on Counting Processes*, Springer, New York.

Andersen, P. and Gill, R. (1982), 'Cox's regression model for counting processes: A large sample study', *The Annals of Statistics* **10**, 1100–1120.

Andersen, P. and Keiding, N. (2002), 'Multi-state models for event history analysis', *Statistical Methods in Medical Research* **11**(2), 91–115.

Andersen, P. and Keiding, N. (2012), 'Interpretability and importance of functionals in competing risks and multistate models', *Statistics in Medicine* **31**(11–12), 1074–1088.

Andersen, P. and Perme, M. (2008), 'Inference for outcome probabilities in multi-state models', *Lifetime Data Analysis* **14**(4), 405–431.

Andersen, P. and Perme, M. (2010), 'Pseudo-observations in survival analysis', *Statistical Methods in Medical Research* **19**(1), 71–99.

Beyersmann, J., Allignol, A. and Schumacher, M. (2012), *Competing Risks and Multistate Models with R*, Springer, New York.

Beyersmann, J., Dettenkofer, M., Bertz, H. and Schumacher, M. (2007), 'A competing risks analysis of bloodstream infection after stem-cell transplantation using subdistribution hazards and cause-specific hazards', *Statistics in Medicine* **26**(30), 5360–5369.

Beyersmann, J. and Schumacher, M. (2008), 'Time-dependent covariates in the proportional subdistribution hazards model for competing risks', *Biostatistics* **9**, 765–776.

Chen, B., Kramer, J., Greene, M. and Rosenberg, P. (2008), 'Competing Risk Analysis of Correlated Failure Time Data', *Biometrics* **64**(1), 172–179.

Claeskens, G. and Hjort, N. (2008), *Model Selection and Model Averaging*, Cambridge University Press, Cambridge.

Cortese, G. and Andersen, P. (2010), 'Competing risks and time-dependent covariates', *Biometrical Journal* **52**(1), 138–158.

Crowder, M. J. (2012), *Multivariate Survival Analysis and Competing Risks.*, Boca Raton, FL: Chapman & Hall/ CRC.

Cuzick, J. (2008), 'Primary endpoints for randomised trials of cancer therapy', *The Lancet* **371**(9631), 2156–2158.

David, H. and Moeschberger, M. (1978), *The Theory of Competing Risks*, Griffin's Statistical Monograph No. 39, Macmillan, New York.

de Wreede, L. C., Fiocco, M. and Putter, H. (2011), 'mstate: An R package for the analysis of competing risks and multi-state models', *Journal of Statistical Software* **38**(7), 1–30.

de Wreede, L., Fiocco, M. and Putter, H. (2010), 'The mstate package for estimation and prediction in non- and semi-parametric multi-state and competing risks models', *Computer Methods and Programs in Biomedicine* **99**, 261–274.

Deslandes, E. and Chevret, S. (2010), 'Joint modeling of multivariate longitudinal data and the dropout process in a competing risk setting: application to ICU data', *BMC Medical Research Methodology* **10**(1), 69.

Dixon, S., Darlington, G. and Desmond, A. (2011), 'A competing risks model for correlated data based on the subdistribution hazard', *Lifetime Data Analysis* **17**, 473–495.

Fine, J. (1999), 'Analysing competing risks data with transformation models', *Journal of the Royal Statistical Society: Series B (Statistical Methodology)* **61**(4), 817–830.

Fine, J. (2001), 'Regression modeling of competing crude failure probabilities', *Biostatistics* **2**(1), 85–97.

Fine, J. and Gray, R. (1999), 'A proportional hazards model for the subdistribution of a competing risk', *Journal of the American Statistical Association* **94**(446), 496–509.

Fine, J., Yan, J. and Kosorok, M. (2004), 'Temporal process regression', *Biometrika* **91**(3), 683–703.

Fiocco, M., Putter, H. and Van Houwelingen, J. (2005), 'Reduced rank proportional hazards model for competing risks', *Biostatistics* **6**(3), 465–478.

Gerds, T. A., Scheike, T. H. and Andersen, P. K. (2012), 'Absolute risk regression for competing risks: interpretation, link functions, and prediction', *Statististics in Medicine* **31**(29), 3921–3930.

Geskus, R. (2011), 'Cause-specific cumulative incidence estimation and the Fine and Gray model under both left truncation and right censoring', *Biometrics* **67**, 39–49.

Gill, R. and Johansen, S. (1990), 'A survey of product-integration with a view towards application in survival analysis', *Annals of Statistics* **18**(4), 1501–1555.

Grambauer, N., Schumacher, M. and Beyersmann, J. (2010), 'Proportional subdistribution hazards modeling offers a summary analysis, even if misspecified', *Statistics in Medicine* **29**, 875–884.

Gray, R. (1988), 'A class of k-sample tests for comparing the cumulative incidence of a competing risk', *Annals of Statistics* **16**(3), 1141–1154.

Hernandez-Quintero, A., Dupuy, J. and Escarela, G. (2011), 'Analysis of a semiparametric mixture model for competing risks', *Annals of the Institute of Statistical Mathematics* **63**(2), 305–329.

Hieke, S., Dettenkofer, M., Bertz, H., M, S. and Beyersmann, J. (2013), 'Initially fewer bloodstream infections for allogeneic versus autologous stem-cell transplants in neutropenic patients', *Epidemiology and Infection* **141**, 158–164.

Jeong, J.-H. and Fine, J. P. (2007), 'Parametric regression on cumulative incidence function', *Biostatistics* **8**(2), 184–196.

Kalbfleisch, J. and Prentice, R. (2002), *The Statistical Analysis of Failure Time Data. 2nd Ed.*, Wiley, Hoboken.

Katsahian, S. and Boudreau, C. (2011), 'Estimating and testing for center effects in competing risks', *Statistics in Medicine* **30**(13), 1608–1617.

Katsahian, S., Resche-Rigon, M., Chevret, S. and Porcher, R. (2006), 'Analysing multicentre competing risks data with a mixed proportional hazards model for the subdistribution', *Statistics in Medicine* **25**(24), 4267–4278.

Klein, J. (2006), 'Modeling competing risks in cancer studies', *Statistics in Medicine* **25**, 1015–1034.

Larson, M. and Dinse, G. (1985), 'A mixture model for the regression analysis of competing risks data', *Applied statistics* **34**(3), 201–211.

Latouche, A. and Porcher, R. (2007), 'Sample size calculations in the presence of competing risks', *Statistics in Medicine* **26**(30), 5370–5380.

Lee, M., Cronin, K., Gail, M., Dignam, J. and Feuer, E. (2011), 'Multiple imputation methods for inference on cumulative incidence with missing cause of failure', *Biometrical Journal* **53**(6), 974–993.

Lim, H., Zhang, X., Dyck, R. and Osgood, N. (2010), 'Methods of competing risks analysis of end-stage renal disease and mortality among people with diabetes', *BMC Medical Research Methodology* **10**, 97.

Lunn, M. and McNeil, D. (1995), 'Applying Cox regression to competing risks', *Biometrics* **51**, 524–532.

Martinussen, T. and Scheike, T. (2006), *Dynamic Regression Models for Survival Data*, New York, NY: Springer.

Mathoulin-Pelissier, S., Gourgou-Bourgade, S., Bonnetain, F. and Kramar, A. (2008), 'Survival end point reporting in randomized cancer clinical trials: a review of major journals', *Journal of Clinical Oncology* **26**(22), 3721–3726.

Nicolaie, M., van Houwelingen, H. and Putter, H. (2010), 'Vertical modeling: A pattern mixture approach for competing risks modeling', *Statistics in Medicine* **29**(11), 1190–1205.

Nicolaie, M., van Houwelingen, H. and Putter, H. (2011), 'Vertical modeling: Analysis of competing risks data with missing causes of failure', *Statistical Methods in Medical Research* **doi: 10.1177/0962280211432067**.

Pepe, M. and Mori, M. (1993), 'Kaplan-Meier, marginal or conditional probability curves in summarizing competing risks failure time data?', *Statistics in Medicine* **12**, 737–751.

Pintilie, M. (2007), 'Analysing and interpreting competing risk data', *Statistics in Medicine* **26**(6), 1360–1367.

Putter, H., Fiocco, M. and Geskus, R. (2007), 'Tutorial in biostatistics: competing risks and multi-state models', *Statistics in Medicine* **26**(11), 2277–2432.

Rosthøj, S., Andersen, P. and Abildstrom, S. (2004), 'SAS macros for estimation of the cumulative incidence functions based on a Cox regression model for competing risks survival data', *Computer Methods and Programs in Biomedicine* **74**, 69–75.

Ruan, P. and Gray, R. (2008), 'Analyses of cumulative incidence functions via non-parametric multiple imputation', *Statistics in Medicine* **27**(27), 5709–5724.

Scheike, T. H., Sun, Y., Zhang, M. J. and Jensen, T. K. (2010), 'A semiparametric random effects model for multivariate competing risks data', *Biometrika* **97**, 133–145.

Scheike, T. and Zhang, M.-J. (2008), 'Flexible competing risks regression modeling and goodness-of-fit', *Lifetime Data Analysis* **14**(4), 464–483.

Shen, P. (2011), 'Proportional subdistribution hazards regression for left-truncated competing risks data', *Journal of Nonparametric Statistics* **23**(4), 885–895.

Shen, P. (2012), 'Regression analysis for cumulative incidence probability under competing risks and left-truncated sampling', *Lifetime Data Analysis* **18**, 1–18.

Sun, L., Liu, J., Sun, J. and Zhang, M.-J. (2006), 'Modeling the subdistribution of a competing risk', *Statistica Sinica* **16**(4), 1367–1385.

Zhang, X., Zhang, M.-J. and Fine, J. (2011), 'A proportional hazards regression model for the subdistribution with right-censored and left-truncated competing risks data', *Statistics in Medicine* . DOI: 10.1002/sim.4264.

Zhou, B., Latouche, A., Rocha, V. and Fine, J. (2011), 'Competing risks regression for stratified data', *Biometrics* **67**(2), 661–670.

9

Bayesian Regression Models for Competing Risks

Ming-Hui Chen

Department of Statistics, University of Connecticut

Mário de Castro

Instituto de Ciências Matemáticas e de Computação, Universidade de São Paulo

Miaomiao Ge

Clinical Bio Statistics, Boehringer Ingelheim Pharmaceuticals, Inc.

Yuanye Zhang

Novartis Institutes for BioMedical Research, Inc.

CONTENTS

9.1 Introduction 180
9.2 The models for competing risks survival data 181
  9.2.1 Multivariate time to failure model 181
  9.2.2 Cause-specific hazards model 183
  9.2.3 Mixture model 184
  9.2.4 Subdistribution model 185
  9.2.5 Connections between the CS, M, and S models 186
  9.2.6 Fully specified subdistribution model 187
9.3 Bayesian inference 188
  9.3.1 Priors and posteriors 189
  9.3.2 Computational development 190
  9.3.3 Bayesian model comparison 191
9.4 Application to an AIDS study 192
9.5 Discussion 195
  Acknowledgments 196
  Bibliography 196

Competing risks data are routinely encountered in various medical applications due to the fact that patients may die from different causes. By now, there is a vast literature on the development of models and methods for fitting and analyzing this type of survival data. In this chapter, we provide a comprehensive overview of Bayesian analysis of competing risks data. Specifically, several models for competing risks survival data are presented and their properties are discussed. The posterior computation and Bayesian model comparison are developed. A real dataset from an AIDS study is used to illustrate the Bayesian methodology.

9.1 Introduction

Competing risks data arise in the medical research studies when the survival data include failure time due to one of two or more terminating events or death from different causes. Let T_1 and T_2 denote two failure times and let C denote the right censoring time. Figure 9.1 illustrates three types of bivariate survival data, where a dot represents a pair of observed times and a line segment in the direction of either T_1 or T_2 indicates that either T_1 or T_2 is censored for that subject. For example, in 9.1 (a), for observation 1, T_2 is observed while T_1 is censored; for observation 2, both T_1 and T_2 are observed; and for observation 4, both T_1 and T_2 are censored.

Figure 9.1 (a) shows six observed data points of usual bivariate failure times. From this plot, we see that these data points spread out in the whole first quadrant. Figure 9.1 (b) illustrates three observed competing risks data points, which are realizations of (T_1, T_2, C). We see from this plot that all three data points fall in the 45-degree straight line in the first quadrant. For example, in Figure 9.1 (b), for observation 1, T_2 is observed, i.e., $T_2 \leq C$, while $T_1 > T_2$ and T_1 cannot be observed in this case; for observation 2, both T_1 and T_2 are censored and in this case, $T_1 > C$ and $T_2 > C$. Unlike the usual bivariate survival data shown in Figure 9.1 (a), the competing risks data are observed only along the 45-degree straight line and, therefore, the correlation between T_1 and T_2 is not identified for the competing risks data (Tsiatis, 1975). Another related type of survival data is called the "semi-competing risks data." Semi-competing risks data arise when the survival data include both the time to a nonterminating event and the time to a terminating event. An illustration of semi-competing risks data is given in Figure 9.1 (c). From 9.1 (c), we see that all observations lie in the upper wedge. A similar illustration was also given in Jiang et al. (2003) to show bivariate semi-competing risks survival data. Due to the limited space, we mainly focus on Bayesian analysis of competing risks data in this chapter. The recent development on the Bayesian analysis of semi-competing risks data and the related literature can be found in Zhang et al. (2012).

There is a rich literature on the frequentist approach for fitting and analyzing competing risks data. The multivariate model of failure times due to different causes was proposed

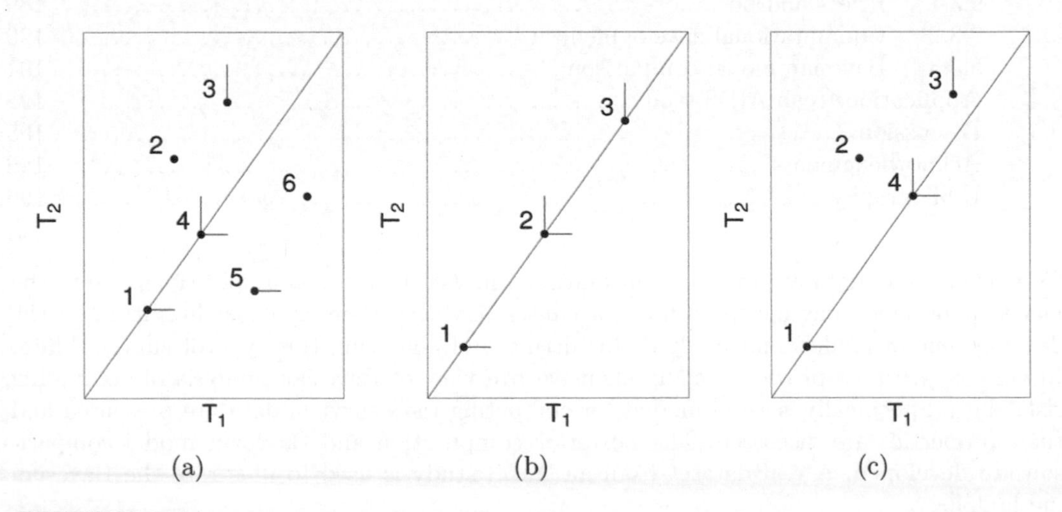

FIGURE 9.1
Illustration of bivariate survival data without special association (a), competing risks data (b), and semi-competing risks data (c).

in Gail (1975). It was shown in Tsiatis (1975) that for any joint distribution of n failure times there exists a joint distribution of n independent failure times such that the cumulative incidence functions from the two joint distributions coincide, which implies that the correlations between the failure times are not identifiable in the multivariate failure time model. The cause-specific hazards model was introduced in Prentice et al. (1978) and the mixture model with hazards function conditional on failure from a specific cause was developed in Larson and Dinse (1985). The subdistribution model with proportional hazards assumption was discussed in Fine and Gray (1999) for assessing the covariates' effect on the cumulative incidence function of the cause of interest. Recent research on competing risks data includes Fan (2008) for introducing a Bayesian nonparametric methodology based on full likelihood for the proportional subdistribution hazard model and Elashoff et al. (2007, 2008) for jointly modeling the longitudinal measurements and survival data with competing risks, where they extended respectively the cause-specific hazards model and the mixture model for survival data, and used latent random variables to link together the sub-models for longitudinal measurements and survival data. The literature on Bayesian analysis of competing risks data is still sparse. The subdistribution model of Fine and Gray (1999) for each cause specific risk was extended in Fan (2008) via Bayesian nonparametric methods. More recently, the Bayesian methods were developed in Hu et al. (2009); Huang et al. (2011) for a joint analysis of longitudinal measurements and survival data with competing risks, in which cause-specific hazards models were considered for modeling survival times. As pointed out in Fine and Gray (1999) one of the nice properties of the subdistribution model is that the effect of a covariate on the marginal probability function can be directly assessed. However, the subdistribution model of Fine and Gray (1999) cannot be compared to other models as the competing risks for other causes are not estimated in their model. Due to this reason, Ge and Chen (2012) extended the model of Fine and Gray (1999) to develop a fully specified subdistribution model, in which a subdistribution model is for the primary cause of death and conditional distributions are for other causes of death.

The rest of this chapter is organized as follows. In Section 9.2, we present several models for competing risks data. The priors and posteriors, the posterior computation, and Bayesian model comparison criteria are discussed in Section 9.3. The detailed analysis of a real dataset from an AIDS study is carried out in Section 9.4. We conclude this chapter with a brief discussion and future research in Section 9.5.

9.2 The models for competing risks survival data

In this section, we present several models and examine their properties for survival data with competing risks. The models discussed below include the multivariate time to failure model, the cause-specific hazards model, the mixture model, the subdistribution model, and the fully specified subdistribution model.

9.2.1 Multivariate time to failure model

A multivariate time to failure model was introduced in Gail (1975). Denote T_j as the random variable of time to failure from cause j, J as the total number of causes, and δ as the cause indicator with $\delta = j$ indicating the observation is failed from cause j and $\delta = 0$ indicating the observation is censored for $j = 1, 2, \ldots, J$. The joint survival function of $T_1, T_2, \ldots, T_J$

is defined by

$$S(t_1, t_2, \ldots, t_J) = \Pr(T_1 > t_1, T_2 > t_2, \ldots, T_J > t_J).$$

The sub-survival function for cause j is defined by

$$\begin{aligned} S_j(t_j) =& \Pr(T_j > t_j, \delta = j) \\ =& \int_{t_j}^{\infty} \int_{u_j}^{\infty} \cdots \int_{u_j}^{\infty} (-1)^J \frac{\partial^J S(u_1, u_2, \ldots, u_J)}{\partial u_1 \ldots \partial u_J} du_1 \ldots du_{j-1} du_{j+1} \ldots du_J du_j. \end{aligned} \tag{9.1}$$

We note that $S_j(0) = \Pr(\delta = j)$, which is not necessarily equal to 1. Let $T = \min\{T_1, T_2, \ldots, T_J\}$, then $\delta = j$ when $T = T_j$. The overall survival function of $T > t$ is

$$S_T(t) = \Pr(T_1 > t, T_2 > t, \ldots, T_J > t) = S(t, t, \ldots, t).$$

Tsiatis (1975) proved in a theorem that for any joint distribution of time to failure variables from J causes, there exists a joint distribution of J independent time to failure variables such that the sub-survival functions for any cause j from the two joint distributions coincide. This theorem implies that the correlation between time to failure variables from different causes is not identifiable in the multivariate time to failure model.

For a simple illustration, assume T_1 and T_2 are the time to failure variables due to two different causes for a patient. As shown in Figure 9.2, only the earlier failure time can be observed, and in this case T_1 is observed for failure from cause 1 while T_2 can never be observed. Thus, the correlation between T_1 and T_2 is not identifiable.

An example applying the theorems in Tsiatis (1975) is also shown here. Assume that the joint survival function of two dependent time to failure variables T_1 and T_2 has the form

$$S(t_1, t_2) = \exp(-\lambda t_1 - \mu t_2 - \vartheta t_1 t_2),$$

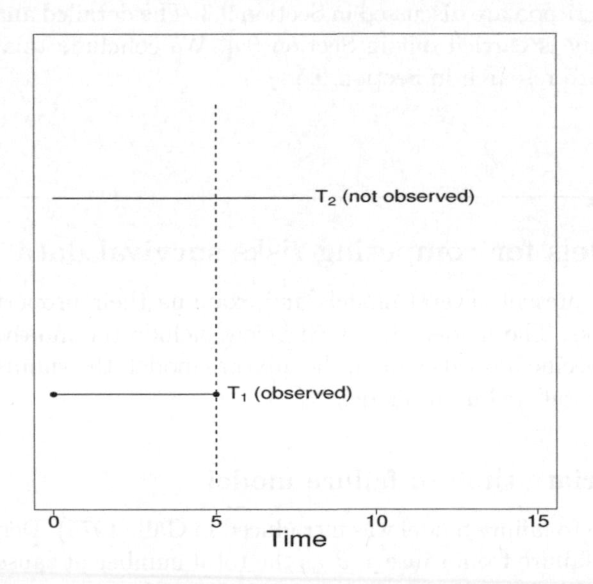

FIGURE 9.2
Simple illustration of failure times of two causes for one patient.

with $\lambda > 0$, $\mu > 0$, and $\vartheta > 0$. The marginal survival function of T_j for cause j is $MS_j(t_j) = \Pr(T_j > t_j)$ for $j = 1, 2$. Thus, for cause 1 and cause 2, the marginal survival functions are given by $MS_1(t_1) = \exp(-\lambda t_1)$ and $MS_2(t_2) = \exp(-\mu t_2)$. The joint density function of T_1 and T_2 is given by

$$f(t_1, t_2) = (\vartheta^2 t_1 t_2 + \lambda\vartheta t_1 + \mu\vartheta t_2 + \lambda\mu - \vartheta)\exp(-\lambda t_1 - \mu t_2 - \vartheta t_1 t_2).$$

From (9.1), the sub-survival functions for the two causes can be obtained as follows:

$$\begin{aligned} S_1(t_1) = \exp\Big\{\frac{(\lambda+\mu)^2}{4\vartheta}\Big\}\Big[&\frac{1}{2}\exp\Big\{-\vartheta(t_1 + \frac{\lambda+\mu}{2\vartheta})^2\Big\} \\ &+ \frac{\sqrt{\pi}(\lambda-\mu)}{2\sqrt{\vartheta}}\Big\{1 - \Phi\Big(\sqrt{2\vartheta}\Big(t_1 + \frac{\lambda+\mu}{2\vartheta}\Big)\Big)\Big\}\Big] \end{aligned}$$

and

$$\begin{aligned} S_2(t_2) = \exp\Big\{\frac{(\lambda+\mu)^2}{4\vartheta}\Big\}\Big[&\frac{1}{2}\exp\Big\{-\vartheta(t_2 + \frac{\lambda+\mu}{2\vartheta})^2\Big\} \\ &+ \frac{\sqrt{\pi}(\mu-\lambda)}{2\sqrt{\vartheta}}\Big\{1 - \Phi\Big(\sqrt{2\vartheta}\Big(t_2 + \frac{\lambda+\mu}{2\vartheta}\Big)\Big)\Big\}\Big], \end{aligned}$$

where $\Phi(\cdot)$ denotes the standard normal cumulative distribution function.

Using the theorems and formulas in Tsiatis (1975), the marginal survival functions of the corresponding independent time to failure variables T_{I1} and T_{I2} are

$$\begin{aligned} MS_{I1}(t_1) = \Pr(T_{I1} > t_1) &= \exp\Big\{\int_0^{t_1} \frac{S_1'(u)}{S_1(u) + S_2(u)} du\Big\} \\ &= \exp\Big(-\lambda t_1 - \frac{1}{2}\vartheta t_1^2\Big) \end{aligned}$$

and

$$\begin{aligned} MS_{I2}(t_2) = \Pr(T_{I2} > t_2) &= \exp\Big\{\int_0^{t_2} \frac{S_2'(u)}{S_1(u) + S_2(u)} du\Big\} \\ &= \exp\Big(-\mu t_2 - \frac{1}{2}\vartheta t_2^2\Big), \end{aligned}$$

where $S_j'(u) = dS_j(u)/du$, $j = 1, 2$. Letting $f_{Ij}(t)$ be the marginal density function of T_{Ij} at t, the joint density function of T_{I1} and T_{I2} is given by

$$\begin{aligned} f_I(t_1, t_2) &= f_{I1}(t_1) f_{I2}(t_2) \\ &= (\vartheta t_1 + \lambda)\exp\Big(-\lambda t_1 - \frac{1}{2}\vartheta t_1^2\Big)(\vartheta t_2 + \mu)\exp\Big(-\mu t_2 - \frac{1}{2}\vartheta t_2^2\Big). \end{aligned}$$

After some algebra, it can be shown that the sub-survival function of T_{Ij} is exactly the same as the sub-survival function of T_j for $j = 1, 2$. More general results on necessary and sufficient conditions for the existence of a set of independent random variables with the same joint distribution of the times to failure are given in Langberg et al. (1978).

9.2.2 Cause-specific hazards model

The cause-specific hazards model (CS model) was discussed in Gaynor et al. (1993). For $j = 1, \ldots, J$, the cause-specific hazard function for cause j is defined by

$$h_{Cj}(t) = \lim_{\Delta t \to 0}\Big\{\frac{\Pr(t \le T < t + \Delta t,\ \delta = j | T \ge t)}{\Delta t}\Big\}.$$

The overall survival function of $T > t$ is

$$S_T(t) = \Pr(T > t) = \exp\Big\{ - \sum_{j=1}^{J} \int_0^t h_{Cj}(u)du \Big\}.$$

The cumulative incidence function and the sub-survival function for cause j are given by $F_j(t) = \Pr(T \le t,\ \delta = j) = \int_0^t h_{Cj}(u)S(u)du$ and $S_j(t) = \Pr(T > t,\ \delta = j) = \int_t^\infty h_{Cj}(u)S(u)du$. The probability of failing from cause j is $F_j(\infty)$ for $j = 1, \ldots, J$.

Let $\boldsymbol{x}$ be the $p \times 1$ vector of covariates and also let $\boldsymbol{\beta}_j$ and $h_{Cj0}(t)$ be the vector of the regression coefficients without an intercept and the cause-specific baseline hazard function at t for cause j, respectively. Assume the Cox proportional hazards structure for $h_{Cj}(t)$, i.e.,

$$h_{Cj}(t|\boldsymbol{x}) = h_{Cj0}(t)\exp(\boldsymbol{x}'\boldsymbol{\beta}_j).$$

Suppose that there are n observations with the vector of observed time $\boldsymbol{t} = (t_1, \ldots, t_n)'$, the $n \times p$ matrix of covariates $X = (\boldsymbol{x}_1, \ldots, \boldsymbol{x}_n)'$, the vector of cause indicator $\boldsymbol{\delta} = (\delta_1, \ldots, \delta_n)'$, and δ_i takes a value between 0 and J with 0 denoting a censored observation. Let

$$D_{obs} = (\boldsymbol{t}, \boldsymbol{\delta}, X) \tag{9.2}$$

denote the observed data. Assuming that there are two causes of failure in total, i.e., $J = 2$, the likelihood function given D_{obs} is

$$\begin{aligned} &L_C(\boldsymbol{\beta}_1, \boldsymbol{\beta}_2, h_{C10}, h_{C20}|D_{obs}) \\ &= \prod_{i=1}^{n} \Big\{ h_{C10}(t_i)\exp(\boldsymbol{x}_i'\boldsymbol{\beta}_1) \Big\}^{1\{\delta_i=1\}} \exp\Big\{ - H_{C10}(t_i)\exp(\boldsymbol{x}_i'\boldsymbol{\beta}_1) \Big\} \\ &\times \Big\{ h_{C20}(t_i)\exp(\boldsymbol{x}_i'\boldsymbol{\beta}_2) \Big\}^{1\{\delta_i=2\}} \exp\Big\{ - H_{C20}(t_i)\exp(\boldsymbol{x}_i'\boldsymbol{\beta}_2) \Big\}, \end{aligned} \tag{9.3}$$

where $H_{Cj0}(t_i) = \int_0^{t_i} h_{Cj0}(u)du$ for $j = 1, 2$ and $1\{A\}$ is the indicator function such that $1\{A\} = 1$ if A is true and 0 if A is not true. It is observed that the regression coefficients for the two causes are separated symmetrically in (9.3), and that for estimating each $\boldsymbol{\beta}_j$, failed observations from the other cause play the same role as censored observations. Hence, inference for the cause-specific hazards model with the proportional hazards assumption can be made by applying the Cox model to each cause, respectively, treating deaths from other causes as censored.

9.2.3 Mixture model

The use of the mixture model (M model) to analyze the survival data with competing risks was discussed in Larson and Dinse (1985). Assume the total number of failure from a specific cause follows a multinomial distribution, and the probabilities of failing from the J possible causes satisfy $p_1 + p_2 + \cdots + p_J = 1$. Define the hazard function conditional on failure from cause j by

$$h_{Mj}(t) = \lim_{\Delta t \to 0} \left\{ \frac{\Pr(t \le T < t + \Delta t | \delta = j,\ T \ge t)}{\Delta t} \right\}. \tag{9.4}$$

The cumulative incidence function and the sub-survival function for cause j are given by $F_j(t) = p_j\Big(1 - \exp\Big\{ - \int_0^t h_{Mj}(u)du \Big\}\Big)$ and $S_j(t) = p_j \exp\Big\{ - \int_0^t h_{Mj}(u)du \Big\}$, and the overall survival function of $T > t$ is

$$S_T(t) = \sum_{j=1}^{J} p_j \exp\Big\{ - \int_0^t h_{Mj}(u)du \Big\}.$$

Let $h_{Mj0}(t)$ be the conditional baseline hazard function at t for cause j. Assume the Cox proportional hazards structure for $h_{Mj}(t)$ with the vector of covariates $\boldsymbol{x}$ and the vector of the regression coefficients $\boldsymbol{\beta}_j$ by

$$h_{Mj}(t|\boldsymbol{x}) = h_{Mj0}(t)\exp(\boldsymbol{x}'\boldsymbol{\beta}_j).$$

When $J = 2$, we obtain the likelihood function given D_{obs} in (9.2) under the M model as follows:

$$\begin{aligned}
& L_M(\boldsymbol{\beta}_1, \boldsymbol{\beta}_2, h_{M10}, h_{M20}, \boldsymbol{p}_1|D_{obs}) \\
&= \prod_{i=1}^{n} \Big[p_{1i} h_{M10}(t_i)\exp(\boldsymbol{x}_i'\boldsymbol{\beta}_1)\exp\Big\{-H_{M10}(t_i)\exp(\boldsymbol{x}_i'\boldsymbol{\beta}_1)\Big\}\Big]^{1\{\delta_i=1\}} \\
&\times \Big[(1-p_{1i})h_{M20}(t_i)\exp(\boldsymbol{x}_i'\boldsymbol{\beta}_2)\exp\Big\{-H_{M20}(t_i)\exp(\boldsymbol{x}_i'\boldsymbol{\beta}_2)\Big\}\Big]^{1\{\delta_i=2\}} \\
&\times \Big[p_{1i}\exp\Big\{-H_{M10}(t_i)\exp(\boldsymbol{x}_i'\boldsymbol{\beta}_1)\Big\} \\
&+ (1-p_{1i})\exp\Big\{-H_{M20}(t_i)\exp(\boldsymbol{x}_i'\boldsymbol{\beta}_2)\Big\}\Big]^{1\{\delta_i=0\}}, \qquad (9.5)
\end{aligned}$$

where $\boldsymbol{p}_1 = (p_{11}, \ldots, p_{1n})'$ and $H_{Mj0}(t_i) = \int_0^{t_i} h_{Mj0}(u)du$ for $j = 1, 2$. In practice, the logistic regression model is often used for p_{1i} in (9.5). Specifically, we assume

$$p_{1i} = p(\boldsymbol{\phi}|\boldsymbol{z}_i) = \frac{\exp(\boldsymbol{z}_i'\boldsymbol{\phi})}{1+\exp(\boldsymbol{z}_i'\boldsymbol{\phi})}, \qquad (9.6)$$

where $\boldsymbol{z}_i$ is the $q \times 1$ vector of covariates and $\boldsymbol{\phi}$ is the vector of regression coefficients including an intercept. Notice that (9.6) can be easily extended to the general case with J causes.

9.2.4 Subdistribution model

The concept of "subdistribution hazard" was introduced and the subdistribution model (S model) for the cause of interest only was developed in Gray (1988) and Fine and Gray (1999). Assume cause 1 is the cause of interest. The subdistribution hazard for cause 1 is defined by

$$\begin{aligned}
h_{S1}(t) &= \lim_{\Delta t \to 0}\left\{\frac{\Pr(t \le T \le t+\Delta t, \delta = 1|T \ge t \cup (T \le t \cap \delta \ne 1))}{\Delta t}\right\} \\
&= \frac{\partial F_1(t)/\partial t}{1 - F_1(t)}, \qquad (9.7)
\end{aligned}$$

where $F_1(t) = \Pr(T \le t,\ \delta = 1)$. As discussed in Fine and Gray (1999), a regression model of (9.7) is developed by assuming the Cox proportional hazards structure for $h_{S1}(t)$ with the vector of covariates $\boldsymbol{x}$, the vector of the regression coefficients $\boldsymbol{\beta}_1$, and the subdistribution baseline hazard function $h_{S10}(t)$ as

$$h_{S1}(t|\boldsymbol{x}) = h_{S10}(t)\exp(\boldsymbol{x}'\boldsymbol{\beta}_1) \qquad (9.8)$$

and

$$F_1(t|\boldsymbol{x}) = 1 - \exp\Big\{-\int_0^t h_{S10}(u)\exp(\boldsymbol{x}'\boldsymbol{\beta}_1)du\Big\}. \qquad (9.9)$$

Fine and Gray (1999) pointed out that the covariates' effect on the cumulative incidence function for the cause of interest can be directly assessed through the corresponding vector of regression coefficients in the subdistribution model due to the proportional hazards structure of (9.8).

For the observed data D_{obs} in (9.2), assume that $\delta_i \geq 1$ for $i = 1, \ldots, n$, i.e., all failure times are completely observed. Then, for $J = 2$, the partial likelihood of $\boldsymbol{\beta}_1$ for cause 1 (Fine and Gray, 1999) is given by

$$L_p(\boldsymbol{\beta}_1|D_{obs}) = \prod_{i=1}^{n} \left[\frac{\exp(\boldsymbol{x}_i'\boldsymbol{\beta}_1)}{\sum_{j \in R_i^*} \exp(\boldsymbol{x}_j'\boldsymbol{\beta}_1)} \right]^{1\{\delta_i = 1\}}, \tag{9.10}$$

where R_i^* is defined as a special risk set at failure time t_i by

$$R_i^* = \big\{ j : (t_j \geq t_i) \cup (t_j \leq t_i \cap \delta_j \neq 1) \big\}. \tag{9.11}$$

Note that the risk set R_i^* is quite different than the risk set of $R_i = \big\{ j : t_j \geq t_i \big\}$ defined in Cox's partial likelihood (Cox, 1972, 1975) because the patients who failed from cause 2 before t_i are included in R_i^* while not in R_i. Since the distribution of time to failure due to cause 2 is not specified in this model, the full likelihood function given the n complete observations cannot be constructed.

9.2.5 Connections between the CS, M, and S models

Although the hazard functions of the CS, M, and S models are quite different, there are certain connections between these models. We formally state the results in the following two propositions.

Proposition 1 *The CS model is equivalent to the M model if for cause j, the cause-specific hazard function in the CS model equals to the multiplication of the conditional hazard function and the probability of failing from cause j in the mixture model, i.e.,*

$$h_{Cj}(t|\boldsymbol{x}) = p_j(\boldsymbol{\phi}|\boldsymbol{z}) h_{Mj}(t|\boldsymbol{x}), \tag{9.12}$$

where $p_j(\boldsymbol{\phi}|\boldsymbol{z}) = \exp(\boldsymbol{z}'\boldsymbol{\phi}_j)/\{1 + \sum_{k=1}^{J-1} \exp(\boldsymbol{z}'\boldsymbol{\phi}_k)\}$, *for* $j = 1, \ldots, J$, *with* $\boldsymbol{\phi} = (\boldsymbol{\phi}_1', \ldots, \boldsymbol{\phi}_{J-1}')'$ *and* $\boldsymbol{\phi}_J = \boldsymbol{0}$.

Proposition 2 *Assume cause 1 is the cause of interest. Then for cause 1, the S model is equivalent to the CS model if*

$$h_{S1}(t|\boldsymbol{x}) = \frac{h_{C1}(t|\boldsymbol{x}) \exp\Big\{ -\sum_{j=1}^{J} H_{Cj}(t|\boldsymbol{x}) \Big\}}{1 - \int_0^t h_{C1}(u|\boldsymbol{x}) \exp\Big\{ -\sum_{j=1}^{J} H_{Cj}(u|\boldsymbol{x}) \Big\} du}. \tag{9.13}$$

The proofs of the above two propositions follow from some straightforward algebra and the details of the proofs are omitted here for brevity. From these propositions, we see that (i) if (9.12) is assumed for $h_{Cj}(t|\boldsymbol{x})$, then the C model does not have a proportional hazards structure for the cause-specific hazard function; and (ii) if (9.13) is assumed, then the proportional hazards structure is no longer to hold for $h_{S1}(t|\boldsymbol{x})$ under the S model. In other words, if one of these three models is "true," the other two models do not have the proportional hazards structure.

9.2.6 Fully specified subdistribution model

The fully specified subdistribution model was developed in Ge and Chen (2012). For the sake of simplicity, assume there are two causes of failure. The extension for more than two causes can be found in Ge and Chen (2012). Assume cause 1 is the cause of interest. Let $T_j^* = T_j \times 1\{\delta = j\} + \infty \times 1\{\delta \neq j\}$, $j = 1, 2$, and let $T^* = \min\{T_1^*, T_2^*\}$, which is essentially the time to failure. The cumulative incidence functions of T^* for the two causes are

$$F_1(t) = \Pr(T^* \leq t, \delta = 1) = \Pr(T_1 \leq t, \delta = 1) \tag{9.14}$$

and

$$F_2(t) = \Pr(T^* \leq t, \delta = 2) = M_2(t)\Pr(\delta = 2), \tag{9.15}$$

where $M_2(t)$ is the cumulative incidence function conditional on failure from cause 2 defined by $M_2(t) = \Pr(T_2 \leq t|\delta = 2)$. Note that in (9.14) and (9.15), the correlation between T_1 and T_2 is not directly modeled, which is not identifiable as shown in Tsiatis (1975). Instead, $F_1(t)$ and $F_2(t)$ are related to each other naturally via

$$\Pr(\delta = 2) = 1 - \Pr(\delta = 1) = 1 - F_1(\infty). \tag{9.16}$$

Using the subdistribution hazard in (9.7) for cause 1, we have $F_1(t) = 1 - \exp\Big\{ - \int_0^t h_1(u)du\Big\}$. Notice that the subdistribution cumulative hazard function $H_1(t) = \int_0^t h_1(u)du$ is not a proper cumulative hazard function since $\lim_{t\to\infty} H_1(t) = -\log\big[1 - \lim_{t\to\infty} F_1(t)\big] < \infty$. Using the conditional hazard in (9.4) for cause 2, we have $M_2(t) = 1 - \exp\Big\{ - \int_0^t h_2(u)du\Big\}$. From (9.15), we see that

$$\lim_{t\to\infty} M_2(t) = \frac{\lim_{t\to\infty} \Pr(T^* \leq t, \delta = 2)}{\Pr(\delta = 2)} = 1,$$

which implies that $M_2(t)$ is a proper cumulative distribution function. Consequently, we have $\int_0^\infty h_2(u)du = \infty$, which indicates that $\int_0^t h_2(u)du$ is a proper cumulative hazard function.

Let $h_{10}(t)$ be the subdistribution baseline hazard function with the corresponding subdistribution cumulative baseline hazard $H_{10}(t) = \int_0^t h_{10}(u)du$ satisfying $\lim_{t\to\infty} H_{10}(t) < \infty$, and let $h_{20}(t)$ be the conditional baseline hazard function with the corresponding conditional cumulative baseline hazard $H_{20}(t) = \int_0^t h_{20}(u)du$. Assume the Cox proportional hazards structure for $h_1(t)$ and $h_2(t)$ with the vector of covariates $\boldsymbol{x}$ and the vector of the regression coefficients $\boldsymbol{\beta}_1$ and $\boldsymbol{\beta}_2$ by

$$h_1(t|\boldsymbol{x}) = h_{10}(t)\exp(\boldsymbol{x}'\boldsymbol{\beta}_1) \text{ and } h_2(t|\boldsymbol{x}) = h_{20}(t)\exp(\boldsymbol{x}'\boldsymbol{\beta}_2).$$

Hence, we have

$$F_1(t|\boldsymbol{x}) = 1 - \exp\Big\{ - H_{10}(t)\exp(\boldsymbol{x}'\boldsymbol{\beta}_1)\Big\} \tag{9.17}$$

and

$$M_2(t|\boldsymbol{x}) = 1 - \exp\Big\{ - H_{20}(t)\exp(\boldsymbol{x}'\boldsymbol{\beta}_2)\Big\}. \tag{9.18}$$

The model defined by (9.17) and (9.18) is called the "fully specified subdistribution (FS)" model (Ge and Chen, 2012). Under the FS model, the likelihood function given D_{obs} in (9.2) can be written as follows:

$$\begin{aligned}
&L_{FS}(\boldsymbol{\beta}_1, \boldsymbol{\beta}_2, h_{10}, h_{20}|D_{obs}) \\
&= \prod_{i=1}^{n} \Big[h_{10}(t_i) \exp(\boldsymbol{x}'\boldsymbol{\beta}_1) \exp\Big\{ - H_{10}(t_i) \exp(\boldsymbol{x}'\boldsymbol{\beta}_1) \Big\} \Big]^{1\{\delta_i=1\}} \\
&\quad \times \Big[h_{20}(t_i) \exp(\boldsymbol{x}'\boldsymbol{\beta}_2) \exp\big\{ - H_{20}(t_i) \exp(\boldsymbol{x}'\boldsymbol{\beta}_2) - H_{10}(\infty) \exp(\boldsymbol{x}'\boldsymbol{\beta}_1) \big\} \Big]^{1\{\delta_i=2\}} \\
&\quad \times \Big[\exp\big\{ - H_{10}(t_i) \exp(\boldsymbol{x}'\boldsymbol{\beta}_1) \big\} \\
&\quad - \big(1 - \exp\big\{ - H_{20}(t_i) \exp(\boldsymbol{x}'\boldsymbol{\beta}_2) \big\}\big) \exp\big\{ - H_{10}(\infty) \exp(\boldsymbol{x}'\boldsymbol{\beta}_1) \big\} \Big]^{1\{\delta_i=0\}}. \qquad (9.19)
\end{aligned}$$

The FS model is not only a natural extension of the subdistribution model of Fine and Gray (1999) but also provides justifications of Fine and Gray's partial likelihood in (9.10) and (9.11) under certain conditions. Assume that all n failure times are completely observed. Let $y_i = t_i$ when $\delta_i = 1$, and $y_i = \infty$ when $\delta_i \neq 1$. Write $0 = y_{(0)} < y_{(1)} < \cdots < y_{(D_1)} < y_{(D_1+1)} = \cdots = y_{(n)} = \infty$, where D_1 is the number of distinct failure times due to cause 1. Also the part of the likelihood function in (9.19) involving $\boldsymbol{\beta}_1$ is denoted by

$$L_{FS1}(\boldsymbol{\beta}_1, h_{10}|D_{obs}) = \prod_{i=1}^{n} \Big[h_{10}(y_{(i)}) \exp(\boldsymbol{x}'\boldsymbol{\beta}_1) \Big] \exp\Big\{ - H_{10}(y_{(i)}) \exp(\boldsymbol{x}'\boldsymbol{\beta}_1) \Big\}. \qquad (9.20)$$

With the n completely observed failure times, Ge and Chen (2012) showed that (i) assuming that the subdistribution baseline hazard rate h_{10} is zero after the last failure time due to cause 1, the partial likelihood function in (9.10) and (9.11) can be attained by the profile likelihood approach, i.e., plugging in the profile maximum likelihood estimator of h_{10} in the likelihood function in (9.20); and (ii) assuming that $h_{10}(t)$ is zero after the last failure time due to cause 1 and the prior of $h_{10}(t)$ is degenerate at 0 everywhere except at y_i's when $\delta_i = 1$, the partial likelihood function in (9.10) and (9.11) is proportional to the marginal posterior density of $\boldsymbol{\beta}_1$ under the independent Jeffreys type priors for $h_{10}(y_{(1)}), \ldots, h_{10}(y_{(D_1)})$. In addition, with the development of the FS model, formal model comparisons between the CS model, the M model, and the FS model can be carried out via Bayesian deviance information criterion (DIC) and logarithm of the pseudomarginal likelihood (LPML). Furthermore, the FS model also facilitates an efficient implementation of the Gibbs sampling algorithm. These two issues are further discussed in the subsequent section.

9.3 Bayesian inference

In this section, we first specify the models for baseline hazard functions and priors for all model parameters in order to carry out Bayesian inference for the CS, M, and FS models. We then present the posterior distributions for these three models, develop Markov chain Monte Carlo (MCMC) sampling algorithms to sample from the respective posterior distributions, and discuss Bayesian model comparison criteria. To ease the exposition, we assume $J = 2$ throughout the rest of the chapter.

9.3.1 Priors and posteriors

We assume piecewise constant hazard models for the baseline hazard functions $h_{Cj0}(t)$ (CS model), $h_{Mj0}(t)$ (M model), and h_{j0} (FS model) for $j = 1, 2$. We first construct $K_j + 1$ partitions of the time axis as follows: $0 = s_{j0} < s_{j1} < \cdots < s_{jK_j} < s_{j,K_j+1} = \infty$ for $j = 1, 2$. We choose s_{jK_j} to be sufficiently large so that there are no failure times beyond s_{jK_j}. For the CS model, we assume that the baseline hazard functions have the following form

$$\begin{aligned} h_{Cj0}(t) &= \lambda_{Cjk}, \; s_{j,k-1} < t \le s_{jk}, \; k = 1, \ldots, K_j, \\ h_{Cj0}(t) &= \lambda_{CjK_j}, \; t > s_{jK_j} \end{aligned} \tag{9.21}$$

for $j = 1, 2$. Similarly, for the M model, we assume

$$\begin{aligned} h_{Mj0}(t) &= \lambda_{Mjk}, \; s_{j,k-1} < t \le s_{jk}, \; k = 1, \ldots, K_j, \\ h_{Mj0}(t) &= \lambda_{MjK_j}, \; t > s_{jK_j} \end{aligned} \tag{9.22}$$

for $j = 1, 2$. For the FS model, the baseline hazard functions are assumed to take the form

$$\begin{aligned} h_{10}(t) &= \lambda_{1k}, \; s_{1,k-1} < t \le s_{1k}, \; k = 1, 2, \ldots, K_1, \\ h_{10}(t) &= \lambda_{1,K_1+1} \exp\{-(t - s_{1K_1})\}, \; t > s_{1K_1}; \\ h_{20}(t) &= \lambda_{2k}, \; s_{2,k-1} < t \le s_{2k}, \; k = 1, 2, \ldots, K_2, \\ h_{20}(t) &= \lambda_{2K_2}, \; t > s_{2K_2}. \end{aligned} \tag{9.23}$$

Note that the hazard functions $h_{Cj0}(t)$ and $h_{Mj0}(t)$ defined by (9.21) and (9.22) yield the proper cumulative hazard functions, respectively, while the hazard functions $h_{10}(t)$ and $h_{20}(t)$ in (9.23) lead to an improper cumulative hazard function for cause 1 and a proper cumulative hazard function for cause 2.

Let $\boldsymbol{\theta}_C = (\boldsymbol{\beta}_1', \boldsymbol{\beta}_2', \boldsymbol{\lambda}_C')'$, where $\boldsymbol{\lambda}_C = (\lambda_{C11}, \ldots, \lambda_{C1K_1}, \lambda_{C21}, \ldots, \lambda_{C2K_2})'$, $\boldsymbol{\theta}_M = (\boldsymbol{\beta}_1', \boldsymbol{\beta}_2', \boldsymbol{\phi}, \boldsymbol{\lambda}_M')'$, where $\boldsymbol{\lambda}_M = (\lambda_{M11}, \ldots, \lambda_{M1K_1}, \lambda_{M21}, \ldots, \lambda_{M2K_2})'$, and $\boldsymbol{\theta}_{FS} = (\boldsymbol{\beta}_1', \boldsymbol{\beta}_2', \boldsymbol{\lambda}_{FS}')'$, where $\boldsymbol{\lambda}_{FS} = (\lambda_{11}, \ldots, \lambda_{1,K_1}, \lambda_{1,K_1+1}, \lambda_{21}, \ldots, \lambda_{2K_2})'$. We assume independent improper uniform priors for $\boldsymbol{\beta}_1$, $\boldsymbol{\beta}_2$, and $\boldsymbol{\phi}$ and Jeffreys's priors for each of λ_{Cjk}, λ_{Mjk}, λ_{1k}, and λ_{2k} except for λ_{1,K_1+1}. For λ_{1,K_1+1}, we assume a Gamma prior given by

$$\pi(\lambda_{1,K_1+1}) \propto \lambda_{1,K_1+1}^{a-1} \exp(-b\lambda_{1,K_1+1}), \tag{9.24}$$

where $a > 0$ and $b \ge 0$ are prespecified hyperparameters. Thus, we have

$$\begin{aligned} \pi(\boldsymbol{\theta}_C) &\propto \textstyle\prod_{j=1}^2 \prod_{k=1}^{K_j} \frac{1}{\lambda_{Cjk}}, \\ \pi(\boldsymbol{\theta}_M) &\propto \textstyle\prod_{j=1}^2 \prod_{k=1}^{K_j} \frac{1}{\lambda_{Mjk}}, \\ \pi(\boldsymbol{\theta}_{FS}) &\propto \left[\textstyle\prod_{j=1}^2 \prod_{k=1}^{K_j} \frac{1}{\lambda_{jk}}\right] \pi(\lambda_{1,K_1+1}), \end{aligned} \tag{9.25}$$

where $\pi(\lambda_{1,K_1+1})$ is given by (9.24). Using (9.25), the posterior distributions of $\boldsymbol{\theta}_C$, $\boldsymbol{\theta}_M$, and $\boldsymbol{\theta}_{FS}$ can be written as

$$\begin{aligned} \pi(\boldsymbol{\theta}_C|D_{obs}) &\propto L_C(\boldsymbol{\beta}_1, \boldsymbol{\beta}_2, h_{C10}, h_{C20}|D_{obs}) \prod_{j=1}^{2} \prod_{k=1}^{K_j} \frac{1}{\lambda_{Cjk}}, \\ \pi(\boldsymbol{\theta}_M|D_{obs}) &\propto L_M(\boldsymbol{\beta}_1, \boldsymbol{\beta}_2, h_{M10}, h_{M20}, \boldsymbol{p}_1|D_{obs}) \prod_{j=1}^{2} \prod_{k=1}^{K_j} \frac{1}{\lambda_{Mjk}}, \\ \pi(\boldsymbol{\theta}_{FS}|D_{obs}) &\propto L_{FS}(\boldsymbol{\beta}_1, \boldsymbol{\beta}_2, h_{10}, h_{20}|D_{obs}) \Big[\prod_{j=1}^{2} \prod_{k=1}^{K_j} \frac{1}{\lambda_{jk}}\Big] \pi(\lambda_{1,K_1+1}), \end{aligned} \tag{9.26}$$

where $L_C(\boldsymbol{\beta}_1, \boldsymbol{\beta}_2, h_{C10}, h_{C20}|D_{obs})$, $L_M(\boldsymbol{\beta}_1, \boldsymbol{\beta}_2, h_{M10}, h_{M20}, \boldsymbol{p}_1|D_{obs})$, and $L_{FS}(\boldsymbol{\beta}_1, \boldsymbol{\beta}_2, h_{10}, h_{20}|D_{obs})$ are given in (9.3), 9.5), and (9.19), respectively.

Although the improper priors are specified for $\boldsymbol{\theta}_C$, $\boldsymbol{\theta}_M$, and $\boldsymbol{\theta}_{FS}$, the posterior distributions in (9.26) can still be proper under some mild conditions. Let $\nu_{jik} = 1$ if the ith subject failed or was censored in the kth interval $(s_{j,k-1}, s_{jk}]$, and 0 otherwise for $k = 1, 2, \ldots, K_j+1$, and $i = 1, 2, \ldots, n$, where $s_{j,K_j+1} = \infty$, for $j = 1, 2$. Also let X_j be a matrix with its ith row equal to $1\{\delta_i = j\}(\nu_{ji1}, \ldots, \nu_{jiK_j}, \boldsymbol{x}_i')$ for $j = 1, 2$ and $i = 1, \ldots, n$ and let X_M be a matrix with its ith row equal to $(1 - 2 \times 1\{\delta_i = 1\})\boldsymbol{x}_i'$ for $i = 1, \ldots, n$. Following the proof of Theorem 4 in Ge and Chen (2012), we can show that for the CS model and the FS model, the posterior distributions $\pi(\boldsymbol{\theta}_C|D_{obs})$ and $\pi(\boldsymbol{\theta}_{FS}|D_{obs})$ in (9.26) are proper if (i) when $\delta_i > 0$, $t_i > 0$ and (ii) X_1 and X_2 are of full rank. For the M model, again following the proof of Theorem 4 in Ge and Chen (2012) and using the results established in Chen and Shao (2001), we can show that the posterior distribution $\pi(\boldsymbol{\theta}_M|D_{obs})$ is proper if (i) when $\delta_i > 0$, $t_i > 0$, (ii) X_1 and X_2 are of full rank, (iii) X_M is of full rank, and (iv) there exists a positive vector $\boldsymbol{a} = (a_1, \ldots, a_n)' \in R^n$, i.e., each component $a_i > 0$, such that $X_M'\boldsymbol{a} = 0$.

9.3.2 Computational development

Due to the complexity of (9.26), it is not possible to carry out an analytical evaluation of these posterior distributions. Thus, we develop the Gibbs sampling algorithm for each of these three posterior distributions. Let $[A|B]$ denote the conditional distribution of A given B. For the CS model, the implementation of Gibbs sampling is straightforward. We sample from $[\boldsymbol{\beta}_1, \boldsymbol{\beta}_2|\boldsymbol{\lambda}_C, D_{obs}]$ and $[\boldsymbol{\lambda}_C|\boldsymbol{\beta}_1', \boldsymbol{\beta}_2', D_{obs}]$ in turns. It is easy to show that the conditional density of $(\boldsymbol{\beta}_1, \boldsymbol{\beta}_2)$ is log-concave in each regression coefficient. Thus, we can use the adaptive rejection algorithm of Gilks and Wild (1992) to sample $\boldsymbol{\beta}_1$ and $\boldsymbol{\beta}_2$. The conditional distribution of λ_{Cjk} follows a gamma distribution and sampling λ_{Cjk} is straightforward.

For the M model, when $\delta_i = 0$, we need to introduce a latent variable δ_i^*, which follows a Bernoulli distribution (since we consider $J = 2$) with the probability

$$
\begin{aligned}
&\Pr(\delta_i^* = 1|\boldsymbol{\beta}_1, \boldsymbol{\beta}_2, \boldsymbol{\phi}, \boldsymbol{\lambda}_M, \boldsymbol{x}_i)\\
=&p(\boldsymbol{\phi}|\boldsymbol{x}_i)\exp\Big\{-H_{M10}(t_i)\exp(\boldsymbol{x}_i'\boldsymbol{\beta}_1)\Big\}\Big[p(\boldsymbol{\phi}|\boldsymbol{x}_i)\exp\Big\{-H_{M10}(t_i)\exp(\boldsymbol{x}_i'\boldsymbol{\beta}_1)\Big\}\\
&+\{1 - p(\boldsymbol{\phi}|\boldsymbol{x}_i)\}\exp\Big\{-H_{M20}(t_i)\exp(\boldsymbol{x}_i'\boldsymbol{\beta}_2)\Big\}\Big]^{-1}, \qquad (9.27)
\end{aligned}
$$

where $p(\boldsymbol{\phi}|\boldsymbol{x}_i)$ is defined by (9.6). Let $\boldsymbol{\delta}^* = (\delta_i^* : \delta_i = 0, 1 \le i \le n)$. The Gibbs sampling algorithm requires to sample from the following distributions in turns: (i) $[\boldsymbol{\beta}_1, \boldsymbol{\beta}_2, \boldsymbol{\delta}^*|\boldsymbol{\phi}, \boldsymbol{\lambda}_M, D_{obs}]$, (ii) $[\boldsymbol{\phi}|\boldsymbol{\beta}_1, \boldsymbol{\beta}_2, \boldsymbol{\delta}^*, \boldsymbol{\lambda}_M, D_{obs}]$, and (iii) $[\boldsymbol{\lambda}_M|\boldsymbol{\beta}_1, \boldsymbol{\beta}_2, \boldsymbol{\delta}^*, \boldsymbol{\phi}, D_{obs}]$. For (ii), it is easy to show that the conditional density of $\boldsymbol{\phi}$ is log-concave in each component of $\boldsymbol{\phi}$ and thus we can use the adaptive rejection algorithm (Gilks and Wild, 1992) to sample from this conditional distribution. For (iii), the conditional distribution of λ_{Mjk} follows a gamma distribution and hence, sampling λ_{Mjk} is straightforward. For (i), observe that

$$
[\boldsymbol{\beta}_1, \boldsymbol{\beta}_2, \boldsymbol{\delta}^*|\boldsymbol{\phi}, \boldsymbol{\lambda}_M, D_{obs}] = [\boldsymbol{\beta}_1, \boldsymbol{\beta}_2|\boldsymbol{\phi}, \boldsymbol{\lambda}_M, D_{obs}][\boldsymbol{\delta}^*|\boldsymbol{\beta}_1, \boldsymbol{\beta}_2, \boldsymbol{\phi}, \boldsymbol{\lambda}_M, D_{obs}]. \qquad (9.28)
$$

In (9.28), we collapse out $\boldsymbol{\delta}^*$ in the conditional distribution $[\boldsymbol{\beta}_1, \boldsymbol{\beta}_2, \boldsymbol{\delta}^*|\boldsymbol{\phi}, \boldsymbol{\lambda}_M, D_{obs}]$. We can show that the conditional density of $[\boldsymbol{\beta}_1, \boldsymbol{\beta}_2|\boldsymbol{\phi}, \boldsymbol{\lambda}_M, D_{obs}]$ is log-concave, which can be sampled via the adaptive rejection algorithm (Gilks and Wild, 1992). Again, sampling $\boldsymbol{\delta}^*$ from (9.27) is straightforward. This approach is called the "collapsed Gibbs sampler" (Liu, 1994; Chen et al., 2000), which yields a much more efficient sampling algorithm. For the FS model, an efficient Gibbs sampling algorithm via the introduction of two sets of latent variables and the collapsed Gibbs method was developed in Ge and Chen (2012). Thus, the detail is omitted here for brevity.

9.3.3 Bayesian model comparison

To determine which of the CS, M, and FS models fit the data better and the values of K_j's, we consider Deviance Information Criterion (DIC) (Spiegelhalter et al., 2002) and Logarithm of the Pseudomarginal Likelihood (LPML) (Ibrahim et al., 2001). Let $\boldsymbol{\theta}$ denote the collection of model parameters. DIC is defined as

$$\text{DIC} = \text{Dev}(\hat{\boldsymbol{\theta}}) + 2p_D, \tag{9.29}$$

where $\text{Dev}(\boldsymbol{\theta})$ is a deviance function, $p_D = \overline{\text{Dev}} - \text{Dev}(\hat{\boldsymbol{\theta}})$, and $\overline{\text{Dev}}$ and $\hat{\boldsymbol{\theta}}$ are the posterior means of $\text{Dev}(\boldsymbol{\theta})$ and $\boldsymbol{\theta}$. The DIC in (9.29) is a Bayesian measure of predictive model performance, decomposed into a measure of fit and a measure of model complexity (p_D). The smaller the DIC value the better the model will predict new observations generated in the same way as the data. For the CS model, $\boldsymbol{\theta} = \boldsymbol{\theta}_C$ and $\text{Dev}(\boldsymbol{\theta}_C) = -2\log L_C(\boldsymbol{\beta}_1, \boldsymbol{\beta}_2, h_{C10}, h_{C20}|D_{obs})$. From (9.3), we have

$$\begin{aligned}
\text{Dev}(\boldsymbol{\theta}_C) =& -2\log L_C(\boldsymbol{\beta}_1, \boldsymbol{\beta}_2, h_{C10}, h_{C20}|D_{obs}) \\
=& -2\sum_{i=1}^{n}\Big[\log\Big\{h_{C10}(t_i)\exp(\boldsymbol{x}_i'\boldsymbol{\beta}_1)\Big\}^{1\{\delta_i=1\}} - H_{C10}(t_i)\exp(\boldsymbol{x}_i'\boldsymbol{\beta}_1)\Big] \\
& -2\sum_{i=1}^{n}\Big[\log\Big\{h_{C20}(t_i)\exp(\boldsymbol{x}_i'\boldsymbol{\beta}_2)\Big\}^{1\{\delta_i=2\}} - H_{C20}(t_i)\exp(\boldsymbol{x}_i'\boldsymbol{\beta}_2)\Big] \\
\equiv& \text{Dev}_1(\boldsymbol{\theta}_C) + \text{Dev}_2(\boldsymbol{\theta}_C).
\end{aligned} \tag{9.30}$$

Using (9.30), we see that under the CS model, DIC=DIC$_1$+DIC$_2$, where DIC$_j$ is the DIC defined for the Cox regression model for the survival data with $1\{\delta_i = j\}$ as the death indicator while treating other causes of death as censored. For the M and FS models, we simply let $\boldsymbol{\theta} = \boldsymbol{\theta}_M$ and $\boldsymbol{\theta} = \boldsymbol{\theta}_{FS}$ and define the deviance as $\text{Dev}(\boldsymbol{\theta}_M) = -2\log L_M(\boldsymbol{\beta}_1, \boldsymbol{\beta}_2, h_{M10}, h_{M20}, \boldsymbol{p}_1|D_{obs})$ and $\text{Dev}(\boldsymbol{\theta}_{FS}) = -2\log L_{FS}(\boldsymbol{\beta}_1, \boldsymbol{\beta}_2, h_{10}, h_{20}|D_{obs})$.

The LPML is given by

$$\text{LPML} = \sum_{i=1}^{n}\log(\text{CPO}_i), \tag{9.31}$$

where the Conditional Predictive Ordinate (CPO),

$$\text{CPO}_i = f(t_i|\boldsymbol{x}_i, \delta_i, D^{(i)}) = \int f(t_i|\boldsymbol{\theta}, \boldsymbol{x}_i, \delta_i)\pi(\boldsymbol{\theta}|D^{(i)})d\boldsymbol{\theta}, \tag{9.32}$$

$f(t_i|\boldsymbol{\theta}, \boldsymbol{x}_i, \delta_i)$ denotes the density or the survival probability based on the value of δ_i, $D^{(i)}$ is the data with the ith observation deleted, and $\pi(\boldsymbol{\theta}|D^{(i)})$ is the posterior distribution based on the data $D^{(i)}$. The larger the LPML value the better the model fits the data. According to Gelfand and Dey (1994), LPML implicitly includes a similar dimensional penalty as AIC asymptotically. For the CS model, $\boldsymbol{\theta} = \boldsymbol{\theta}_C$ and $f(t_i|\boldsymbol{\theta}_C, \boldsymbol{x}_i, \delta_i) = \{h_{C10}(t_i)\exp(\boldsymbol{x}_i'\boldsymbol{\beta}_1)\}^{1\{\delta_i=1\}}\exp\{-H_{C10}(t_i)\exp(\boldsymbol{x}_i'\boldsymbol{\beta}_1)\}\{h_{C20}(t_i)\exp(\boldsymbol{x}_i'\boldsymbol{\beta}_2)\}^{1\{\delta_i=2\}}\exp\{-H_{C20}(t_i)\exp(\boldsymbol{x}_i'\boldsymbol{\beta}_2)\}$. Similarly to DIC, when the independent priors are specified for $(\boldsymbol{\beta}_1, \lambda_{C11}, \ldots, \lambda_{C1K_1})$ and $(\boldsymbol{\beta}_2, \lambda_{C21}, \ldots, \lambda_{C2K_2})$, we have LPML = LPML$_1$ + LPML$_2$, where LPML$_j$ is the LPML defined for the Cox regression model for the survival data with $1\{\delta_i = j\}$ as the death indicator while treating other causes of death as censored. For the M and FS models, $f(t_i|\boldsymbol{\theta}, \boldsymbol{x}_i, \delta_i)$ is defined in the same way as for the CS model and the detail is omitted here for brevity.

9.4 Application to an AIDS study

A dataset with 329 homosexual men from the Amsterdam Cohort Studies on HIV infection and AIDS is analyzed. In the process of HIV infection, syncytium inducing (SI) HIV phenotype appears in many individuals. It is known that the appearance of this SI phenotype impairs AIDS prognosis. In the analysis of the time from HIV infection to SI appearance before AIDS diagnosis, AIDS plays a role of a competing event. The dataset is in the public domain and is available in the R package mstate (AIDSSI dataset) (de Wreede et al., 2011). A detailed description of this dataset can be found in Putter et al. (2007). Reduced susceptibility to HIV infection and delayed AIDS progression is related to the deletion of the CCR5 genotype. Subjects without the deletion are coded as WW (reference category) and subjects having the deletion on one of the chromosomes are coded as WM. In this dataset, there were no subjects with deletion on both chromosomes. We investigate whether this deletion has a protective effect on AIDS and SI appearance. Five subjects with unknown CCR5 genotype were excluded from the analysis. The numbers of patients coded as WW and WM are 259 and 65, respectively. There were 113 subjects with AIDS diagnosis and 107 subjects with SI appearance. The median time from HIV infection to AIDS diagnosis was 6.199 years, ranging from 1.44 to 13.36 years. The median time from HIV infection to SI appearance was 5.224 years, ranging from 0.112 to 13.94 years.

Since we have just one covariate (x), $\boldsymbol{\beta}_1$ and $\boldsymbol{\beta}_2$ reduce to β_{11} and β_{21}, respectively. In (9.6) we take $\boldsymbol{z} = (1, x)'$, so that $\boldsymbol{\phi} = (\phi_1, \phi_2)'$. Improper uniform priors were taken for β_{11}, β_{21}, and $\boldsymbol{\phi}$. In (9.21), (9.22), and (9.23), we took intervals with approximately the same number of events. In (9.24), $a = 1$ and $b = 0$ were specified. Different values of K_1 and K_2 were tried out for optimizing the model fitting. The values of DIC, p_D, and LPML for some combinations of K_1 and K_2 under all three models are reported in Table 9.1. We consider two FS models: one with AIDS diagnosis as the primary cause and another with SI appearance as the primary cause. The values of DIC and LPML range from 1704.4 to 1764.3 and from -881.2 to -853.6, respectively. These ranges highlight the role of the number of partitions of the time axis in model fitting. We see that $(K_1, K_2) = (10, 10)$ is the optimum combination of (K_1, K_2) for all models. From Table 9.1, we see that the M and FS models fit the data slightly better than the CS model although the differences among the models are not so remarkably large.

The posterior means (estimates), posterior standard deviations (SDs), and 95% highest posterior density (HPD) intervals of the parameters for the scenario of $(K_1, K_2) = (10, 10)$ are shown in Table 9.2. For the CS model (M model), β_{11} and β_{21} measure the effect of the CCR5 genotype on the cause-specific (conditional) hazard function of the times from HIV infection to AIDS diagnosis and SI appearance, respectively. For the FS model, β_{11} (β_{21}) is the coefficient of the subdistribution (conditional) baseline hazard corresponding to the primary (secondary) cause of interest. Therefore, it is not surprising that the estimates in Table 9.2 are different. For the time from infection to AIDS diagnosis the estimates of β_{11} are negative and significant, which means a protective effect of the deletion on one of the chromosomes (WM level of the CCR5 genotype) on AIDS. On the other hand, the effect of CCR5 on SI appearance was not significant whichever the model in Table 9.2. The subdistribution model (the S model) (Fine and Gray, 1999) was also fitted to the data. When the time to AIDS diagnosis is the primary cause, the estimate, standard error and 95% confidence intervals of β_{11} are -1.004, 0.295, and (-1.583, -0.426), respectively. When the time to SI appearance is the primary cause, we obtained 0.024, 0.227, and (-0.421, 0.468). These results are similar to the posterior summaries from the FS model.

The effect of the CCR5 genotype effect was further investigated by comparing the pos-

TABLE 9.1
The values of DIC, dimension penalty, and LPML for the AIDSSI data.

Model	K_1	K_2	DIC	p_D	LPML
CS	1	1	1762.6	4.0	-881.1
	5	5	1707.7	12.2	-854.0
	10	10	1706.8	22.4	-854.1
	15	15	1713.4	32.8	-858.2
M	1	1	1764.3	5.9	-881.2
	5	5	1707.5	12.6	-854.3
	10	10	**1704.4***	22.2	-853.7
	15	15	1711.2	31.5	-858.7
FS	1	1	1736.9	4.3	-868.2
(AIDS)	5	5	1709.5	12.5	-854.8
	10	10	1708.2	22.7	-854.7
	15	15	1714.9	33.1	-858.9
FS	1	1	1751.3	4.1	-875.4
(SI)	5	5	1707.3	12.8	-853.7
	10	10	1705.8	23.1	**-853.6***
	15	15	1712.9	33.1	-858.1

*The bold values indicate the best values of DIC or LPML corresponding to the best model.

TABLE 9.2
Posterior estimates of parameters for the models with $(K_1, K_2) = (10, 10)$.

Model	Coefficient	Estimate	SD	95% HPD interval
CS (AIDS)	β_{11}	-1.273	0.316	(-1.911, -0.678)
(SI)	β_{21}	-0.288	0.243	(-0.782, 0.167)
M (AIDS)	β_{11}	-1.055	0.540	(-2.081, -0.023)
(SI)	β_{21}	-0.506	0.339	(-1.143, 0.229)
	ϕ_1	-0.013	0.215	(-0.398, 0.450)
	ϕ_2	-0.767	0.538	(-1.757, 0.341)
FS	β_{11}	-1.044	0.311	(-1.662, -0.453)
(AIDS)	β_{21}	-0.892	0.314	(-1.556, -0.328)
FS	β_{11}	0.015	0.239	(-0.450, 0.483)
(SI)	β_{21}	-1.651	0.417	(-2.436, -0.813)

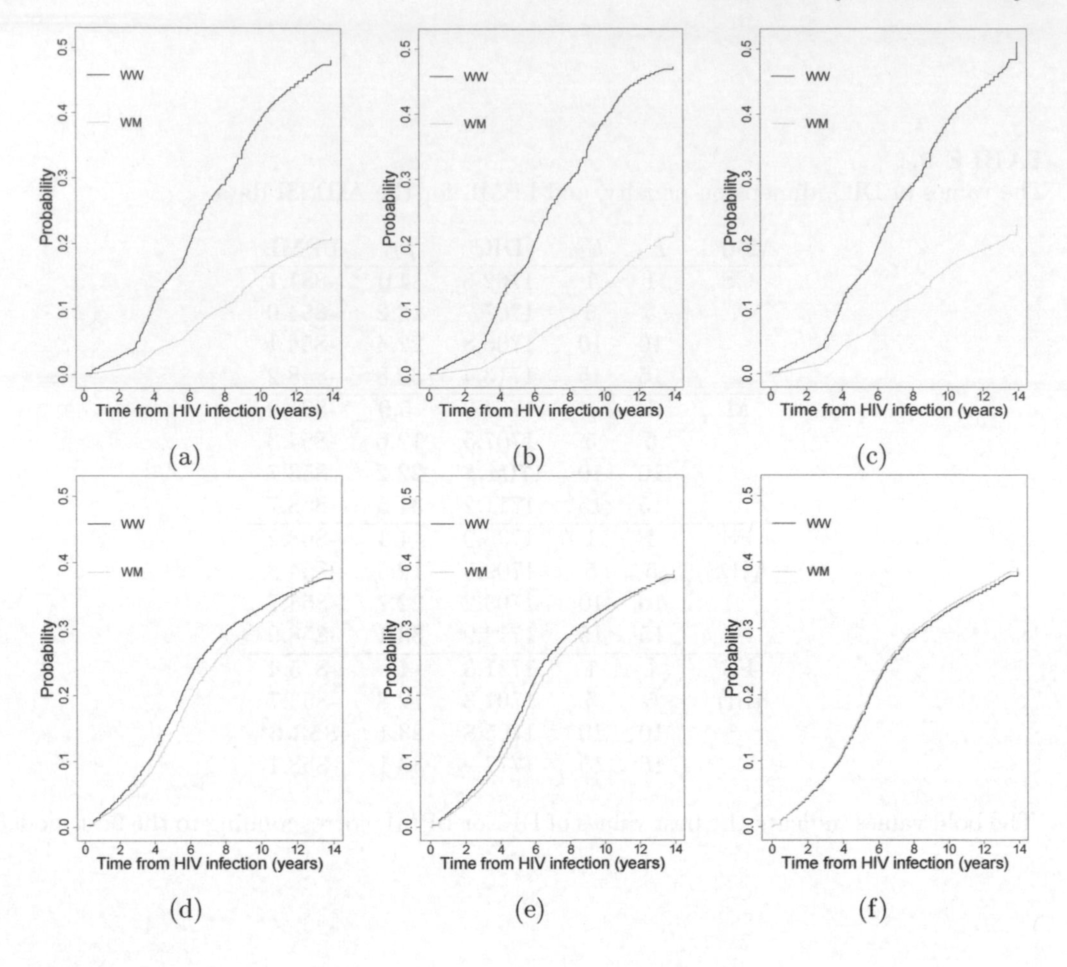

FIGURE 9.3
Plots of the cumulative incidence functions for AIDS and SI under the CS model ((a) and (d)), the M model ((b) and (e)), and the FS model ((c) and (f)), respectively.

terior means of the cumulative incidence functions at different times. The plots in Figure 9.3 show that when the primary cause is AIDS diagnosis, the estimates from the three models are very similar. When the primary cause is SI appearance, the nonsignificant effect of CCR5 is more noticeable in the estimates from the FS model.

After discarding the first 2,000 iterations of the Gibbs sampler, from the following 100,000 iterations with spacing of size 10 we obtain 10,000 samples for each parameter. The computational code was written in the FORTRAN language with the IMSL library. Convergence of the chains was monitored using graphical displays and the test statistic in Geweke (1992). Figure 9.4 shows the trace plots for the four parameters in the FS model when the primary cause is AIDS. These plots reveal a good mixing of the chains. Similar plots were drawn for the other parameters in the CS, M, and FS models.

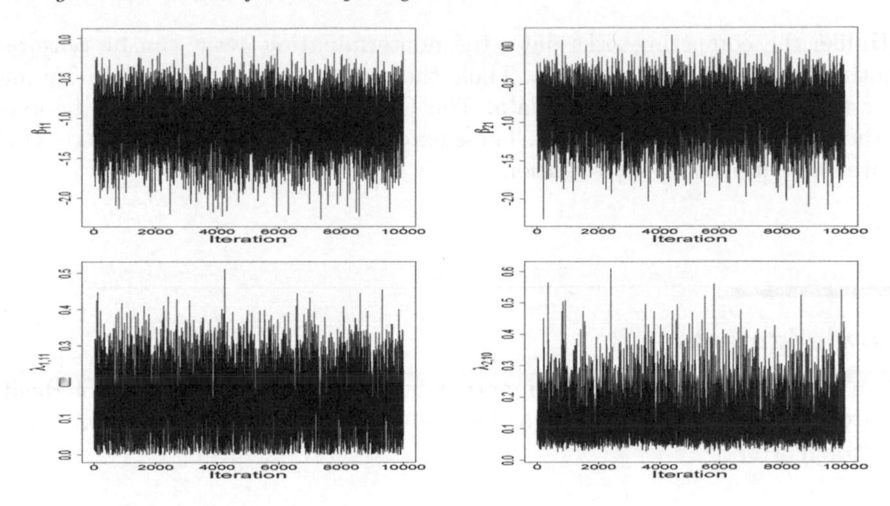

FIGURE 9.4
Trace plots of β_{11}, β_{21}, $\lambda_{1,11}$, and $\lambda_{2,10}$ in the FS model with $(K_1, K_2) = (10, 10)$ intervals when the primary cause is AIDS.

9.5 Discussion

The CS, M, and S models discussed in Section 9.2 are commonly used in analyzing survival data with competing risks. Recently, the S model becomes popular particularly in the medical literature, which may be partially due to the availability of the R package cmprsk in the public domain (http://cran.r-project.org/). The FS model is a natural extension of the S model. The FS model allows us to carry out a formal comparison between these three types of models. Based on the empirical results shown in Table 9.1 and Ge and Chen (2012), the DIC and LPML values indicate that the M and FS models outperform the CS model. The FS model has an attractive proportional hazards interpretation of regression coefficients in predicting the cause-specific mortality for the primary cause. In addition, the FS model is more parsimonious than the M model since under the FS model, $\Pr(\delta = 1|\boldsymbol{x}) = 1 - \exp\Big\{ - H_{10}(\infty)\exp(\boldsymbol{x}'\boldsymbol{\beta}_1)\Big\}$ while $\Pr(\delta = 1|\boldsymbol{x})$ needs to be modeled via (9.6) under the M model. Therefore, the FS model is more identifiable than the M model especially when the dimension of $\boldsymbol{x}$ is high. On the other hand, the M model is symmetric between the primary cause of death and the other cause of death while the FS model is asymmetric. Therefore, it is important to determine the primary cause before fitting the FS model. In most medical applications, the primary cause is known based on the goal of clinical investigation.

As we have emphasized, the regression coefficients are related to different hazard functions (subdistribution, cause-specific or conditional). Therefore, their interpretation should be cautious. The combined effect of all coefficients can be assessed through the cumulative incidence function (see Figure 9.3).

In this chapter, we assume the piecewise constant hazard models given by (9.21), (9.22), and (9.23) for the baseline hazard functions. The exponential model can be extended to the gamma process prior model (see, for example, Ibrahim et al. (2001)) or the autoregressive prior model proposed by Kim et al. (2007) for the baseline hazard function. As discussed in Section 9.1, another closely related type of survival data is the semi-competing risks data. Unlike the competing risks data, the nonterminating event can be censored by the terminating event but not vice versa. Thus, the semi-competing risks data are much more complex than the competing risks data. The CS, M, and FS models can be extended to model the semi-competing risks data. These extensions deserve to be future research topics, which are currently under investigation.

Acknowledgments

Dr. Chen's research was partially supported by U.S. National Institutes of Health (NIH) grants #GM 70335 and #CA 74015. Dr. de Castro acknowledges partial support from CNPq, Brazil.

Bibliography

Chen, M.-H. and Shao, Q.-M. (2001), 'Propriety of posterior distribution for dichotomous quantal response models with general link functions', *Proceedings of the American Mathematical Society* **129**, 293–302.

Chen, M.-H., Shao, Q.-M. and Ibrahim, J. G. (2000), *Monte Carlo Methods in Bayesian Computation*, New York: Springer-Verlag.

Cox, D. R. (1972), 'Regression models and life tables (with discussion)', *Journal of the Royal Statistical Society, Series B* **34**, 187–220.

Cox, D. R. (1975), 'Partial likelihood', *Biometrika* **62**, 269–276.

de Wreede, L. C., Fiocco, M. and Putter, H. (2011), 'mstate: An R package for the analysis of competing risks and multi-state models', *Journal of Statistical Software* **38**, 1–30.

Elashoff, R. M., Li, G. and Li, N. (2007), 'An approach to joint analysis of longitudinal measurements and competing risks failure time data', *Statistics in Medicine* **26**, 2813–2835.

Elashoff, R. M., Li, G. and Li, N. (2008), 'A joint model for longitudinal measurements and survival data in the presence of multiple failure types', *Biometrics* **64**, 762–771.

Fan, X. (2008), *Bayesian Nonparametric Inference for Competing Risks Data*, Unpublished Ph.D. Dissertation, Division of Biostatistics, Medical College of Wisconsin.

Fine, J. P. and Gray, R. J. (1999), 'A proportional hazards model for the subdistribution of a competing risk', *Journal of the American Statistical Association* **94**, 496–509.

Gail, M. (1975), 'A review and critique on some models used in competing risk analysis', *Biometrics* **31**, 209–222.

Gaynor, J. J., Feuer, E. J., Tan, C. C., Wu, D. H., Little, C. R., Strauss, D. J., Clarkson, B. D. and Brennan, M. F. (1993), 'On the use of cause-specific failure and conditional failure probabilities: examples from clinical oncology data', *Journal of the American Statistical Association* **88**, 400–409.

Ge, M. and Chen, M.-H. (2012), 'Bayesian inference of the fully specified subdistribution model for survival data with competing risks', *Lifetime Data Analysis* **18**, 339–363.

Gelfand, A. E. and Dey, D. K. (1994), 'Bayesian model choice: asymptotics and exact calculations', *Journal of the Royal Statistical Society, Series B* **56**, 501–514.

Geweke, J. (1992), Evaluating the accuracy of sampling-based approaches to the calculation of posterior moments, *in* J. M. Bernardo, J. O. Berger, A. P. Dawid and A. F. M. Smith, eds., 'Proceedings of the Fourth Valencia International Meeting, Peñíscola, 1991', Clarendon Press, Oxford, pp. 169–193.

Gilks, W. R. and Wild, P. (1992), 'Adaptive rejection sampling for Gibbs sampling', *Applied Statistics* **41**, 337–348.

Gray, R. J. (1988), 'A class of k-sample tests for comparing the cumulative incidence of a competing risk', *Annals of Statistics* **16**, 1141–1154.

Hu, W., Li, G. and Li, N. (2009), 'A Bayesian approach to joint analysis of longitudinal measurements and competing risks failure time data', *Statistics in Medicine* **28**, 1601–1619.

Huang, X., Li, G., Elashoff, R. M. and Pan, J. (2011), 'A general joint model for longitudinal measurements and competing risks survival data with heterogeneous random effects', *Lifetime Data Analysis* **17**, 80–100.

Ibrahim, J. G., Chen, M.-H. and Sinha, D. (2001), *Bayesian Survival Analysis*, New York: Springer-Verlag.

Jiang, H., Chappell, R. and Fine, J. P. (2003), 'Estimating the distribution of nonterminal event time in the presence of mortality or informative dropout', *Controlled Clinical Trials* **24**, 135–146.

Kim, S., Chen, M.-H., Dey, D. K. and Gamerman, D. (2007), 'Bayesian dynamic models for survival data with a cure fraction', *Lifetime Data Analysis* **13**, 17–35.

Langberg, N., Proschan, F. and Quinzi, A. J. (1978), 'Converting dependent models into independent ones, preserving essential features', *Annals of Probability* **6**, 174–181.

Larson, M. G. and Dinse, G. E. (1985), 'A mixture model for the regression analysis of competing risks data', *Applied Statistics* **34**, 201–211.

Liu, J. S. (1994), 'The collapsed Gibbs sampler in Bayesian computations with applications to a gene regulation problem', *Journal of the American Statistical Association* **89**, 958–966.

Prentice, R. L., Kalbfleisch, J. D., Peterson, A., Flournoy, N., Farewell, V. and Breslow, N. (1978), 'The analysis of failure times in the presence of competing risks', *Biometrics* **34**, 541–554.

Putter, H., Fiocco, M. and Geskus, R. (2007), 'Tutorial in biostatistics: Competing risks and multi-state models', *Statistics in Medicine* **26**, 2389–2430.

Spiegelhalter, D. J., Best, N. G., Carlin, B. P. and van der Linde, A. (2002), 'Bayesian measures of model complexity and fit (with discussion)', *Journal of the Royal Statistical Society, Series B* **64**, 583–639.

Tsiatis, A. (1975), 'A nonidentifiability aspect of the problem of competing risks', *Proceedings of the National Academy of Sciences of the United States of America* **72**, 20–22.

Zhang, Y., Chen, M.-H., Ibrahim, J. G., Zeng, D., Chen, Q., Pan, Z. and Xue, X. (2012), 'Bayesian gamma frailty models for survival data with semi-competing risks and treatment switching', *Technical Report #12-26, Department of Statistics, University of Connecticut* .

10

Pseudo-Value Regression Models

Brent R. Logan and Tao Wang

Division of Biostatistics, Medical College of Wisconsin

CONTENTS

10.1 Introduction 199
10.2 Applications 201
10.2.1 Survival data 201
10.2.2 Cumulative incidence for competing risks 202
10.2.3 Multi-state models 204
10.2.4 Quality adjusted survival 206
10.3 Generalized linear models based on pseudo-values 207
10.3.1 Estimation 207
10.3.2 Assumptions and formal justification 208
10.3.3 Covariate-dependent censoring 208
10.3.4 Clustered data 209
10.4 Model diagnosis 209
10.4.1 Graphical assessment 210
10.4.2 Tests of model fit 211
10.5 Software 211
10.6 Examples 212
10.6.1 Example 1: Survival and cumulative incidence 212
10.6.2 Example 2: Multi-state model 215
10.7 Conclusions 217
Bibliography 218

10.1 Introduction

Much of the survival analysis literature is focused on inference in the presence of missing data typically due to right censoring. Often these survival models are formulated in terms of hazard regression models. In this article we review an alternative approach to inference with incomplete survival data, based on pseudo-values or pseudo-observations obtained from a jackknife statistic constructed from non-parametric estimators for the quantity of interest. These pseudo-values are then used as outcome variables in a generalized linear model and model parameters are estimated using generalized estimating equations (GEE) (Liang and Zeger, 1986). This approach was first proposed by Andersen et al. (2003) for direct modeling of state probabilities in a multi-state model. This simple approach can be applied to regression models for any mean value parameter. In particular, the general approach allows for regression models to be extended to a number of non-standard settings, including survival probabilities at a fixed point in time (Klein et al., 2007), cumulative incidence (Klein and Andersen, 2005; Klein et al., 2008), restricted mean survival (Andersen et al., 2004), quality adjusted survival (Andrei and Murray, 2007; Tunes-da Silva and Klein, 2009), multi-state models (Andersen and Klein, 2007) , and clustered time to event data (Logan et al., 2011). These pseudo-values have also been used as outcome variables in a scatter plot or to compute pseudo-residuals for a regression model in order to facilitate goodness of fit assessment (Perme and Andersen, 2008; Andersen and Perme, 2010). The main advantage of pseudo-value regression is that it provides a simple and generalizable method of modeling complex time to event data which is often not easily modeled using standard techniques. Furthermore, pseudo-value regression is easily implemented using existing software packages once the pseudo-values have been obtained.

As an illustration of the technique, suppose that the survival time X were fully observed for all patients. Standard methods for quantitative data could then be applied to model parameters of the survival distribution (e.g., the mean). Alternatively, the patients' survival status at any time point could be determined by dichotomizing X as $I(X > t)$, and the probability of being alive at time t could be modeled using standard binary data approaches. Models for repeated binary data could also be performed by considering the vector of indicators over a set of time points. In general, suppose that we were interested in estimating $\theta = E[f(X)]$, for some f. This $f(X)$ may be a scalar such as the survival indicator $I(X > t)$ at a given time point t, a vector such as a collection of survival indicators at a set of pre-specified time points, or it may be function valued. If $X_1, ..., X_n$ based on a random sample were fully observed, then θ could be estimated by the average $\hat{\theta} = \sum_i f(X_i)/n$. For example, the survival probability at time t could be estimated by $\sum_i I(X_i > t)/n$. In the presence of right censoring or other kinds of incomplete data, such a simple estimator is not feasible. However, in many cases an estimator $\hat{\theta}$ is available which accounts for the incomplete data, such as the Kaplan-Meier estimator for the survival probability. Pseudo-values are defined based on this estimator as

$$Y_i = n\hat{\theta} - (n-1)\hat{\theta}^{-i}, i = 1, ..., n$$

where $\hat{\theta}^{-i}$ is the leave-one-out estimator of θ based on the sample of size $n-1$ with the ith observation deleted. Note that when $\hat{\theta}$ is unbiased, we have $E[Y_i] = \theta$. For complete data, $Y_i = f(X_i)$ so that the approach is equivalent to using the raw data. Therefore, the basic idea is to replace the incompletely observed data $f(X_i)$ with the pseudo-value Y_i, and then treat it as if it were raw data to be analyzed. For example, the pseudo-value can be used as an outcome variable in a generalized linear regression model for the conditional mean given the covariate $\theta(Z) = E[f(X)|Z]$ with link function $g(\cdot)$ given by

$$g(E[f(X_i)|Z_i]) = \alpha_0 + \beta Z_i.$$

Another important use of the Y_i's is to compute residuals to assess model fit.

In this chapter, we review the pseudo-value regression method and its application to a variety of settings. In Section 10.2, we discuss how pseudo-observations can be obtained in a variety of complex models. In Section 10.3, we describe the details of how pseudo-values can be used to estimate parameters of a generalized linear model (GLM) for the effect of covariates on a mean parameter of interest. We also explain the properties of the pseudo-values, the theoretical foundation of the regression procedure, the necessary assumptions of the model, and extensions to clustered data. Another important application of the pseudo-value approach is in model diagnostics, which is covered in Section 10.4. In Section 10.5, we describe available statistical software for implementing the techniques. We use two examples in Section 10.6 to illustrate the use of pseudo-values in the survival/competing risks setting as well as in a multi-state model setting. Finally, in Section 10.7, we give concluding remarks about the method.

10.2 Applications

10.2.1 Survival data

Pseudo-value regression has been used to model survival probabilities at a fixed time point (Klein et al., 2007), survival probabilities at late time points in the presence of crossing hazards (Logan et al., 2008), and restricted mean survival (Andersen et al., 2004). For modeling survival at a fixed time t, the parameter of interest is $\theta = S(t)$. Let $T_i = \min(X_i, C_i)$, where X_i is the survival time and C_i is the censoring time for observation i, and let $\delta_i = I(X_i < C_i)$ be the event indicator. Let $N(t) = \sum_i I(T_i \leq t, \delta_i = 1)$ be the counting process for the number of failures, and let $R(t) = \sum_i I(T_i \geq t)$ be the number at risk at time t. The Kaplan-Meier estimator is given by

$$\hat{S}(t) = \prod_{u \leq t}(1 - dN(u)/R(u)).$$

The Kaplan-Meier estimator is used to construct the pseudo-value as

$$Y_i = n\hat{S}(t) - (n-1)\hat{S}^{-i}(t), \tag{10.1}$$

where $\hat{S}^{-i}(t)$ is the Kaplan-Meier estimator with the ith observation omitted. Note that if there is no censoring prior to the time point t, Y_i simplifies to a simple binary indicator of whether the patient is alive at time t. Klein et al. (2007) proposed to model the survival probability given the covariate vector Z, $\theta(Z) = S(t|Z)$, at a single time point. They used these pseudo-values as responses to fit a GLM model $g(\theta(Z)) = \alpha + \beta Z$. Note that several link functions have been considered. The "logit" link function $g(u) = \log(u/(1-u))$ corresponds directly to logistic regression for the probability of survival at time t when there is no censoring prior to t, since the pseudo-values simplify to binary indicators of whether the patient is alive at time t. The complementary log-log link function $g(u) = \log(-\log(u))$ results in a GLM

$$\log(-\log(S(t|Z))) = \alpha(t) + \beta Z.$$

This is equivalent to the Cox proportional hazards model

$$\lambda(t|Z) = \lambda_0(t)\exp\{\beta Z\},$$

with cumulative baseline hazard function $\Lambda_0(t) = \int_0^t \lambda_0(u)du$, where $\alpha(t) = -\log(\Lambda_0(t))$. Note that this approach can be extended to multiple time points, in which case Y_i is a vector $(Y_i(t_1), \ldots, Y_i(t_k))$, with $Y_i(t_j) = n\hat{S}(t_j) - (n-1)\hat{S}^{-i}(t_j)$. In this vector case, the complementary log-log link function corresponds to a joint proportional hazards model across all the time points,

$$\log(-\log(S(t_j|Z))) = \log(\Lambda_0(t_j)) + \beta Z, \quad j = 1, \cdots, k \tag{10.2}$$

and the $\exp(\beta)$ parameters can be interpreted directly in terms of hazard ratios.

Several other applications of pseudo-values to survival have been proposed. Logan et al.

(2008) considered pseudo-value regression modeling of survival after a fixed time point t_0 in order to address whether there are late differences in survival. This approach is proposed for the setting where crossing hazards are anticipated and the researcher is focused on inference on late survival differences. Here the pseudo-values are computed at a set of times which all occur after a pre-specified time point t_0 of interest, so that the parameters of the model only reflect effects on survival after t_0.

An alternative parameter of interest in the survival setting is the restricted mean survival time, denoted by $\theta = \mu_\tau = E(X \wedge \tau) = \int_0^\tau S(t)dt$, where $\tau > 0$ is the upper limit. Andersen et al. (2004) proposed the use of pseudo-values to model restricted mean survival as an alternative to modeling of the hazard function. The restricted mean survival can be estimated by

$$\hat{\mu}_\tau = \int_0^\tau \hat{S}(u)du.$$

Then the ith pseudo-observation is given by

$$Y_{\tau i} = n \int_0^\tau \hat{S}(t)dt - (n-1) \int_0^\tau \hat{S}^{-i}(t)dt = \int_0^\tau Y_i(t).$$

Andersen et al. (2004) considered a linear model with identity link function $g(u) = u$ as well as a log-linear model with "log" link function $g(u) = \log(u)$. For pseudo-values of the τ-restricted mean, the identity link gives

$$\theta(Z) = E(X \wedge \tau | Z) = \alpha + \beta Z,$$

where β_k is interpreted as the difference between the τ-restricted mean for an observation with z_k=1 vs. one with $z_k = 0$, adjusting for the other covariates. With the "log" link, however,

$$\log \theta(Z) = \log(E(X \wedge \tau | Z)) = \tilde{\alpha} + \tilde{\beta} Z.$$

In this case, $e^{\tilde{\beta}_k}$ can be interpreted as the ratio of the two τ-restricted means. They also consider the use of pseudo-values for modeling the unrestricted mean survival, but found that other methods had better performance, possibly because of the difficulties in dealing with the unobserved tail of the survival distribution.

10.2.2 Cumulative incidence for competing risks

In the competing risks setting, a failure from one type of event precludes the observation of failure of another type. This can be expressed as a multi-state model using two causes of failure without loss of generality, where the states at time t represent alive, failure of type 1, or failure of type 2 as in Figure 10.1.

A common method for modeling this type of data is to apply Cox regression models to each cause-specific hazard function $\lambda_j(u)$

$$\lambda_j(u) = \lim_{\Delta t \to 0} \frac{1}{\Delta t} P(t \le X \le t + \Delta t, \epsilon = j | X \ge t), \quad j = 1, 2$$

where X is the time to failure from any cause and $\epsilon = 1, 2$ denotes the type of failure. However, it is complicated to piece these models together and interpret the covariate effects on the cumulative incidence function, defined as the probability of failing from cause 1 by time t. This is because the cumulative incidence function $F_j(t) = P(X \le t, \epsilon = j) = \int_0^t S(u-)\lambda_j(u)du$ depends on the hazard functions for both causes. Several researchers have proposed methods for directly modeling the covariate effects on the cumulative incidence

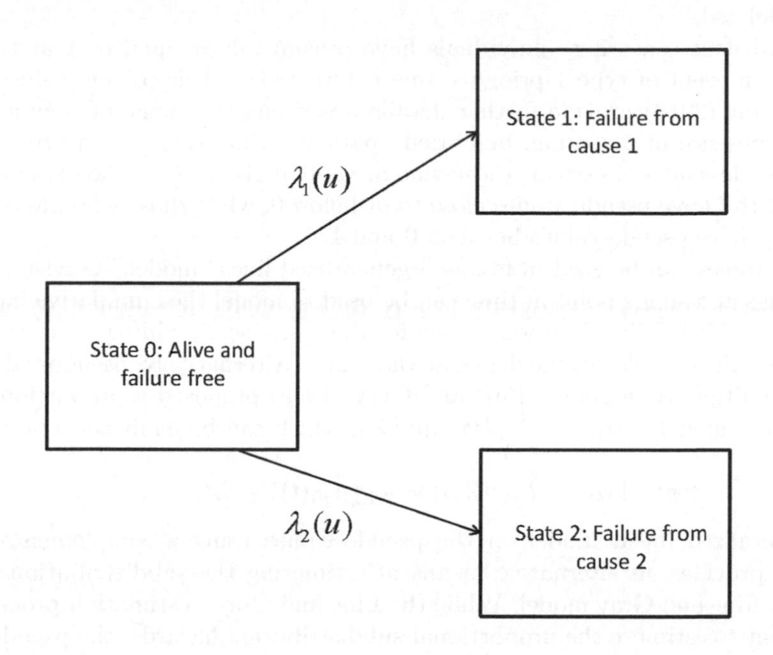

FIGURE 10.1
Competing risks model.

function. Fine and Gray (1999) proposed using a Cox type model to model the so-called subdistribution hazard $\tilde{\lambda}_j(t) = -d\log(1 - F_j(t))/dt$. Because $F_j(t) = 1 - \exp\{-\int_0^t \tilde{\lambda}_j(s)ds\}$ has a one-to-one relationship with the subdistribution hazard, the effects of covariates on $F_j(t)$ could therefore be evaluated via $\tilde{\lambda}_j(t)$. Fine (2001) further extended this approach to other link functions. Alternatively, Scheike et al. (2008) proposed to model the cumulative incidence function directly using a binomial regression model. Model fitting is based on the inverse probability of censoring weighting (IPCW) technique. Scheike and Zhang (2011) also described fitting this type of model using the R package "timereg."

Here we focus on the proposal by Klein and Andersen (2005) to use pseudo-value regression to model the cumulative incidence functions directly. We focus on the cumulative incidence for cause 1 without loss of generality. Note that the cumulative incidence function can be written as an expected value $F_1(t) = E(I(X \leq t, \epsilon = 1)$. Let T_i be the on-study time for patient i given by $T_i = (X_i \wedge C_i)$, where X_i is the event time and C_i is the censoring time. Also, $\delta_i = I(X_i \leq C_i)$ is the event indicator and $\epsilon_i \in \{1, 2\}$ denotes the failure type, observable if an event has occurred prior to C_i. Define $N_{1i}(t) = I(T_i \leq t)I(\epsilon_i = 1)\delta_i$, and let $N_1(t) = \sum_i N_{1i}(t)$ be the number of type 1 failures observed up to time t. Denote the number at risk at time t by $R(t)$. The cumulative incidence estimate is defined as

$$\widehat{F}_1(t) = \int_0^t \widehat{S}(u^-)R^{-1}(u)dN_1(u).$$

For each subject $i = 1, \ldots, n$, we calculate a pseudo-observation for the cause 1 cumulative incidence at each of several time points $t = t_1, \ldots, t_k$ as

$$Y_i(t_r) = n\widehat{F}_j(t_r) - (n-1)\widehat{F}_j^{-i}(t_r), \quad r = 1, \cdots, k \tag{10.3}$$

where $F_j^{-i}(t)$ is the cumulative incidence function based on the sample of size $n-1$ with the i-th subject deleted.

In a dataset without censoring, individuals have pseudo-values equal to 1 at time t if they experienced an event of type 1 prior to time t. Otherwise, their pseudo-values are 0. Andersen and Perme (2010) provide further details describing the range of pseudo-values over time in the presence of censoring, but briefly patients who experience a type 1 event prior to t have pseudo-values close to 1, those who are still at risk at t or who experienced a type 2 event prior to t have pseudo-values close to or below 0, while those who are censored prior to t typically have pseudo-values between 0 and 1.

These pseudo-values can be used in fitting a generalized linear model. As with survival data, pseudo-values at a single point in time can be used to model the cumulative incidence at that time. Often a "logit" link function is used for this purpose, providing an analog to logistic regression for the cumulative incidence at that time. Alternatively, pseudo-values can be computed at multiple time points. Fine and Gray (1999) proposed a proportional subdistribution hazards model $\tilde{\lambda}_j(t|Z) = \tilde{\lambda}_{j0}(t)\exp(\beta Z)$, which can be equivalently expressed as

$$\log(-\log(1 - F_j(t|Z))) = \log(\tilde{\Lambda}_{j0}(t)) + \beta Z.$$

Therefore, a generalized linear model on the pseudo-values using a complementary log-log link function provides an alternative means of estimating the subdistribution hazard parameters of the Fine and Gray model. While the Fine and Gray's estimation procedure is likely more efficient to estimate the proportional subdistribution hazards, the pseudo-value regression model provides much more flexibility in considering alternative models through the use of different link functions or select time points. This may be especially useful given the difficulty in interpreting a subdistribution hazards ratio clinically.

10.2.3 Multi-state models

In situations like bone marrow transplantation (BMT), a transplant patient could experience one or more intermediate (transient) events such as acute or chronic graft versus host disease (GVHD) as well as ultimate events such as relapse or death as they are being followed over time. Multi-state models are a useful tool for modeling such complicated time-to-event data. It can also be applied to analyze competing risks, as the latter is a special case of the multi-state models. A multi-state model (MSM) can be described by a multi-state stochastic process $X(t), t \in \Im$ with a finite state space $S = 0, 1, \cdots, M$. As an example, consider the MSM for outcomes after BMT and subsequent Donor Leukocyte Infusion (DLI) presented by Andersen and Klein (2007) and shown in Figure 10.2. Here patients can die in first remission (state 2) or experience disease relapse (state 1). Once they relapse, they may die with disease (state 3) or be treated with DLI and experience a disease remission (state 4). Even after a DLI-induced remission, they may die or experience a second disease relapse (state 5).

The transition probabilities in a MSM are defined as $P_{kh}(s,t) = P(X(t) = h|X(s) = k)$ for $k, h \in S$, and $s, t \in \Im, s \leq t$. The transition intensities are defined as

$$\lambda_{kh}(t) = \lim_{\Delta t \to 0} \frac{P_{kh}(t, t + \Delta t)}{\Delta t}.$$

In a MSM, the research interests often focus on estimation of transition intensities and transition probabilities, and the assessment of the effects of certain observed covariates on these probability quantities using regression models. Other quantities such as the state occupation probabilities and the probability of staying time in a state can be derived from

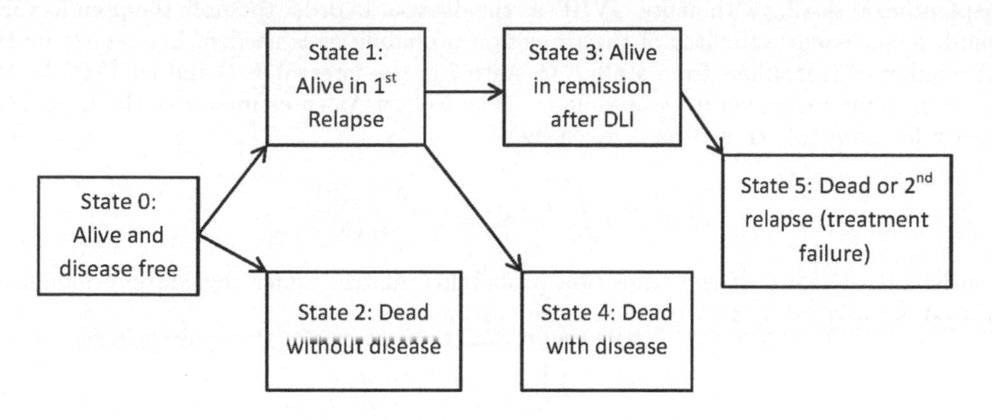

FIGURE 10.2
Multi-state model for current leukemia free survival.

the transition intensities or probabilities. For example, the state occupation probability

$$\pi_h(t) = P(X(t) = h) = \sum_{k \in S} \pi_k(0) P_{kh}(0, t).$$

When the initial distribution $\pi_k(0) = \delta_0$, i.e., degenerate at state 0, then $\pi_h(t) = P_{0h}(0, t)$. The probability of staying in a state h is $P_{hh}(s, t) = 1 - \sum_{k \neq h} P_{hk}(s, t)$. In general, an absorbing state is a state from which further transitions cannot occur, while a transient state is a state that is not absorbing. If h is an absorbing state, then $P_{hk}(s, t) = 0$ for any $k \in S, k \neq h$ and $t > s$. In the DLI example above, states 0, 1, and 3 are transient states while states 2, 4, and 5 are absorbing states. The current leukemia free survival (CLFS) is defined as the likelihood that a patient is alive and in remission at any time t, which can be written as $CLFS(t) = \pi_0(t) + \pi_3(t)$.

One straightforward approach for analyzing a MSM is to model all transition intensities based on the Cox proportional hazards model by left truncating on the time that a patient enters each state. This is often done under a Markov model assumption, where the transition intensity at time t only depends on the history through the state $X(t)$. We can then estimate the transition probabilities as product integrals from the transition intensities (Andersen et al., 1993),

$$P(s, t) = \prod\nolimits_{u \in (s,t]} (I + dA(u)),$$

where $P(s, t) = (P_{hj}(s, t))$, $A(t) = (A_{hj}(t))$ with $A_{hj}(t) = \int_0^t \lambda_{hj}(u)du$ being the integrated intensity, and $\prod$ is the product-integral of matrices. When transition probabilities are of main interest, however, plugging in the estimated transition intensities in the product integrals does not directly give estimates of the covariate effects on the transition probabilities, as the effects of covariates on the transition probabilities are highly complex nonlinear functions of the original covariate effects on the transition intensities (Andersen and Perme, 2008; Scheike and Zhang, 2007). This makes it difficult to assess the direct effects of covariates on transition probabilities. Scheike and Zhang (2007) proposed a binomial modeling approach to construct regression models directly on the transition probabilities, with model fitting based on the inverse probability of censoring weighting (IPCW) technique.

Alternatively, pseudo-values may be used for direct regression modeling of transition probabilities. Andersen et al. (2003) first proposed the use of pseudo-value regression models in this MSM context. They applied the method to an illness-death model in the bone marrow transplantation setting with acute GVHD as the disease. In order to apply the pseudo-value method, a consistent estimator of the transition probabilities is needed. Let $N_{hj}(t)$ be the total number of transitions from state h to state j in the interval $[0, t]$ and let $Y_h(t)$ be the number of individuals in state h at time $t-$. The Nelson-Aalen estimator of the integrated intensity for an h to j transition is given by

$$\hat{A}_{hj}(t) = \int_0^t I(Y_h(s) > 0) \frac{dN_{hj}(u)}{Y_h(u)}.$$

A consistent estimator of the transition probability matrix under the Markov model assumption is provided by the Aalen-Johansen estimator,

$$\hat{P}[s,t] = \prod\nolimits_{u \in (s,t]} [I + d\hat{A}(u)],$$

where $\hat{A}(t) = (\hat{A}_{hj}(t))$ and $\hat{A}_{hh}(t) = -\sum_{j \neq h} \hat{A}_{hj}(t)$. Andersen and Klein (2007) also suggested computing pseudo-values based on an alternative estimator for state probabilities of a transient state based on differences in Kaplan-Meier estimators. This approach was originally suggested by Pepe (1991) and is valid for non-Markov models as well. They illustrated this approach using models for the current leukemia free survival probability in the MSM in Figure 10.2. This approach will be illustrated later as well in a different MSM example. Note that additional estimators of transition probabilities are also available in the non-Markov setting (Meira-Machado et al., 2009). Once an estimator for a state probability or transition probability of interest is obtained, the pseudo-value can be computed as before. For example the pseudo-value for the transition probability from state 0 to state j by time t, which is equivalent to the state j probability at time t, is given by

$$Y_i(t) = n\hat{\pi}_j(t) - (n-1)\hat{\pi}_j^{-i}(t),$$

using the leave one out estimator $\hat{\pi}_j^{-i}(t)$. Note that in the absence of censoring by time t, the pseudo-value reduces to an indicator of whether the patient is in state j at time t, similar to what was seen with survival or competing risks data. These pseudo-values can then be used to model the state transition probability using the generalized linear model, $g(\theta(Z)) = g(\pi_j(t|Z)) = \alpha_0(t) + \beta Z$. Pseudo-value regression models for multi-state models have been described for simple multi-state models such as an illness-death model (Andersen et al., 2003), as well as more complicated multi-state models such as current leukemia free survival after a bone marrow transplant (Andersen and Klein, 2007; Liu et al., 2008).

10.2.4 Quality adjusted survival

Quality adjusted survival (QAS) attempts to account for both the length and quality of a patient's survival time. Here the health history of a patient is described by a process $X(t), t \leq 0$ taking values of $K + 1$ different health states. State 0 is the absorbing death state, while states $1, \ldots, K$ denote transient health states with different qualities associated with them through the specification of utility coefficients $Q(X(t))$. The QAS time up to τ is

$$\mu = \int_0^{T \wedge \tau} Q(X(u))du.$$

Tunes-da Silva and Klein (2009) proposed the use of pseudo-value regression to model the effect of covariates on QAS. They define a set of modes Ω based on distinct states entered and the number of times that state was entered, with utility q_j for mode j, and define X_j^E

and X_j^L as the times that a patient enters and leaves the mode, respectively. Then the mean QAS can be written as

$$\mu = \sum_{j\in\Omega} q_j E(X_j^L - X_j^E).$$

To construct pseudo-values, they propose to use the "event-marginal estimator" to estimate μ, given by

$$\hat{\mu} = \sum_{j\in\Omega} q_j \int_0^{\tau} (\hat{G}_j^L(u) - \hat{G}_j^E(u))du,$$

where $\hat{G}_j^k(t)$ is the Kaplan-Meier estimate of $G_j^k(t) = P(X_j^k > t)$. Then the pseudo-values are computed in the same way as before,

$$Y_i = n(\hat{\mu}) - (n-1)\hat{\mu}^{-i}.$$

They use an identity link function to provide direct inference on the mean QAS.

10.3 Generalized linear models based on pseudo-values

Once a pseudo-value (possibly a vector over multiple time points) for each patient is obtained, based on a consistent and approximately unbiased estimator of the parameter of interest, these pseudo-values are used to estimate the parameters of a generalized linear model. Let $Y_i = (Y_i(t_1), \ldots, Y_i(t_k))$ be a vector of pseudo-values for observation i over time points $t_1, \ldots, t_k$, with conditional expectation at time j of $\theta_j(Z_i) = E(Y(t_j)|Z_i)$, and suppose that we are interested in fitting the model $g(\theta_j|Z_i) = \alpha_j + \beta Z_i$, for $i = 1, \ldots, n; j = 1, \ldots, k$.

10.3.1 Estimation

Inference on the parameters may be performed using GEE. Let $\gamma = (\alpha_1, \ldots, \alpha_k, \beta)$ and $\tilde{Z}_{ij} = (I(t_l = t_j), l = 1, \ldots, k; Z_i)$ be the corresponding design matrix for observation i at time j. Define $g^{-1}(\gamma\tilde{Z}_i)$ to be a vector with elements $g^{-1}(\gamma\tilde{Z}_{ij})$, and let $dg^{-1}(\gamma\tilde{Z}_i)/d\gamma$ be the partial derivative matrix with elements $dg^{-1}(\gamma\tilde{Z}_{ij})/d\gamma_i$. Then the estimating equations to be solved are of the form

$$U(\gamma) = \sum_i \left(\frac{dg^{-1}(\gamma\tilde{Z}_i)}{d\gamma}\right)' V_i^{-1}\left(Y_i - g^{-1}(\gamma\tilde{Z}_i)\right) = \sum_i U_i(\gamma) = 0, \qquad (10.4)$$

where V_i is a working covariance matrix. Let $\hat{\gamma}$ be the solution to this equation. Based on the GEE results of Liang and Zeger (1986), under standard regulatory conditions, it follows that $\sqrt{n}(\hat{\gamma} - \gamma)$ is asymptotically multivariate normal with mean 0. The covariance matrix of $\hat{\gamma}$ can be estimated by the sandwich variance estimator

$$\hat{\Sigma}(\hat{\gamma}) = I(\gamma)^{-1}\left\{\sum_i U_i(\gamma)U_i(\gamma)'\right\} I(\gamma)^{-1},$$

where

$$I(\gamma) = \sum_i \left(\frac{dg^{-1}(\gamma\tilde{Z}_i)}{d\gamma}\right)' V_i^{-1}\left(\frac{dg^{-1}(\gamma\tilde{Z}_i)}{d\gamma}\right)$$

is the model-based equivalent of the information matrix. Estimation using a GEE approach requires selection of a working covariance matrix V_i. While an independent or identity matrix is often used for simplicity, choice of an appropriate working covariance matrix which more closely matches the true covariance may improve efficiency. Andersen and Klein (2007) conducted simulation studies of three possible working covariance matrices including an independent matrix, and found that the estimator based on an empirical working covariance matrix had slightly smaller mean squared error compared to the other two.

In practice, to implement the GEE algorithm one must select a set of time points on which to perform inference. Use of all event times is not practical for large datasets because the dimensions of the matrices in the GEE algorithm become cumbersome. If there is interest in a specific time point or set of time points, those could be used. Otherwise, most researchers have proposed to use 5-10 time points equally spaced on the event time scale to capture most of the information about the event time distribution, and simulations by Andersen and Perme (2010) and Klein and Andersen (2005) suggest that more than 5 time points equally spaced on the event time scale result in minimal improvements in efficiency.

10.3.2 Assumptions and formal justification

Application of the GEE results of Liang and Zeger (1986) require two key assumptions. First, the conditional expectation of the pseudo-values is equal to the conditional mean parameter of interest given the covariates, so that modeling the pseudo-value data does in fact correspond to a model for the conditional mean parameter of interest and the estimating equation has expectation 0. While it is straightforward to show that the marginal expectation of the pseudo-values is equal to the unconditional mean parameter θ, proof of the conditional mean is more challenging. Graw et al. (2009) showed using influence functions that the pseudo-values for the cumulative incidence have conditional expectation given Z which converges to the conditional cumulative incidence as the sample size increases, when right censoring is independent and does not depend on covariates. The other key assumption is that the pseudo-value observations are independent across i. In fact, the pseudo-value observations are dependent; however, Graw et al. (2009) established that the pseudo-values for the cumulative incidence are approximately independent as the sample size increases. By using a second order von Mises expansion, they showed that the solution to the generalized estimating equations based on pseudo-values for the cumulative incidence provide consistent estimators of the regression parameters. They also showed that these models are closely related to the weighted binomial regression approach (Scheike and Zhang, 2007; Scheike et al., 2008). Extension of this proof to survival data (both point-wise and restricted mean) is straightforward because the survival setting can be seen as a special case of competing risks. Extension of these results to more complex settings such as multi-state models is an area of ongoing research.

10.3.3 Covariate-dependent censoring

One of the main assumptions of the pseudo-value regression model is that the censoring is independent and does not depend on the covariates. Andersen and Perme (2010) showed that bias can be introduced when the censoring is dependent on covariates. They also proposed a way of modifying the calculation of the pseudo-values to avoid this bias. If censoring depends on a covariate W with values $1, 2, \ldots, m$, then one can replace the Kaplan-Meier estimator with a mixture estimator of survival

$$\hat{S}_M(t) = \sum_{w=1}^{m} p_w \hat{S}_w(t),$$

where p_w is the observed fraction of subjects with covariate $W = w$ and $\hat{S}_w(t)$ is the Kaplan-Meier estimate of survival in patients with $W = w$. Then the ith pseudo-observation, assuming it has covariate level $W = w$, simplifies to the ith pseudo-observation calculated using only patients with covariate $W = w$,

$$Y_i(t) = n_w \hat{S}_w(t) - (n_w - 1)\hat{S}_w^{-i}(t).$$

Andersen and Perme (2010) concluded that this mixture approach works well to correct the bias due to covariate-dependent censoring, but the variance of the pseudo-values based on the mixture estimate is slightly higher than the ones based on the standard Kaplan-Meier estimator.

10.3.4 Clustered data

The estimation techniques described earlier in this section were developed for independent observations. Logan et al. (2011) considered the problem of accounting for clustered data (i.e., center effects) when directly modeling cumulative incidence functions, and show that simple modifications to the estimating equations and sandwich variance estimates can be used to estimate a marginal model when the observations are clustered. Pseudo-values are computed for each patient using the leave-one-out estimate of the cumulative incidence function in (10.3), in the same way as for the independent observation setting. This works because the cumulative incidence estimator provides a consistent estimate of the marginal cumulative incidence function (over clusters). They showed following similar arguments as in Graw et al. (2009) that the pseudo-values have conditional mean given covariate Z equal to the (marginal over cluster) cumulative incidence given Z, and they are asymptotically independent across clusters. A marginal model approach is used to adjust the estimators and variances of the estimators for the within cluster correlation. The pseudo-values are then used in a generalized estimating equation set up to appropriately reflect the clustered data structure. This estimating equation simplifies to the usual one (10.4) when observations within a cluster are all independent or when an independent working covariance matrix is used. The sandwich variance estimators are set up to directly account for the clustered data structure, but the asymptotic convergence is driven by the number of clusters rather than the number of observations. The method can also be applied to model clustered survival data.

10.4 Model diagnosis

Statistical models often model the data under certain constraints or assumptions. The Cox model for example assumes that the hazard ratios of each covariate are constant over time. If a continuous covariate is used in the Cox model, we also need to check whether it has a linear relationship with $\log(-\log(S(t)))$. Otherwise, a non-linear transformation or categorization of the covariate might be needed. The Fine and Gray model for the cumulative incidence function has similar assumptions that the ratio of subdistribution hazards are constant over time. Pseudo-value regression models contain analogous assumptions which need to be assessed.

10.4.1 Graphical assessment

In survival analysis, model diagnosis is often hampered by the presence of censoring. Perme and Andersen (2008) proposed using the pseudo-values for diagnosis of the survival models. Since the pseudo-values can be used to calculate residuals for each individual at each time point regardless of censoring, one strategy to check the model assumptions such as proportional hazards is to look at the plots of the pseudo-value residuals versus the covariates in the same manner as checking the goodness-of-fit for other classical regression models.

For a Cox or additive survival model that we want to assess, we can calculate the predicted survival probabilities $\hat{S}(t|Z_i), i = 1, \cdots, n$, based on the model, and the pseudo-values $Y_i(t)$ of the survival probability from Equation (10.1). Perme and Andersen (2008) proposed to compare the pseudo-values with the predicted values using the pseudo-residuals defined as

$$\hat{\varepsilon}_i(t) = \frac{Y_i(t) - \hat{S}(t|Z_i)}{\hat{S}(t|Z_i)(1 - \hat{S}(t|Z_i))}.$$

Graphical plots are suggested for the diagnosis of model assumptions. For example, to assess the proportional hazards assumption for a continuous or categorical covariate in a Cox model, the residual plots should be centered around a mean of 0 and should not exhibit any trends when plotted against time or the covariates. In practice, a select number of time points are plotted, and the curves are smoothed to facilitate assessment of the residual mean function.

Another strategy proposed by Perme and Andersen (2008) is to construct pseudo-scatter plots and examine the relationship between the survival probability and the covariate Z directly for a given model. It is well known that a Cox model leads to a linear relationship between covariates and the complementary log-log transformation of the survival function

$$\log(-\log(S(t|Z))) = \log(\Lambda_0(t)) + \beta Z.$$

An estimate of $S(t|Z)$ can be obtained by smoothing the pseudo-values $Y_i(t)$ with respect to the covariate Z and time t. Then we can transform these smoothed estimates to build profile curves for $\log(-\log(S(t|Z)))$ versus Z at a select number of time points $t_1, \cdots, t_k$. If the Cox model assumptions hold, each of these profile curves should be approximately a straight line with the same slope β. For different time points, these profile curves should also be parallel with their intercepts $\log(\Lambda_0(t))$ being monotonically increasing as time t increases. When there is a nonlinear effect of Z, the profiles would deviate from a straight line with respect to Z. If the effect changes over time, then the profiles at different time points will have different slopes against Z. There are several advantages of this approach compared to traditional methods using Schoenfeld (Schoenfeld, 1982) or Martingale (Therneau et al., 1990) residuals. The pseudo-scatter plots allow for simultaneous assessment of the proportional hazards assumption and linearity of a covariate effect. It can also be adapted to diagnose the additive hazards model (Lin and Ying, 1994) by using a different transformation of the smoothed estimates of $S(t|Z)$, given by $-\log(S(t|Z))/t$. Finally, this graphical model diagnostic method can be extended to more complex model settings. For example, Andersen and Perme (2010) applied the pseudo-scatter plots approach to model diagnosis in fitting a Fine and Gray subdistribution hazards model. Under this model,

$$\log(-\log(1 - F_j(t|Z))) = \log(\tilde{\Lambda}_{j0}(t)) + \beta Z.$$

In order to check the model assumptions such as proportional subdistribution hazards or linearity of Z, we first estimate the pseudo-values $Y_i(t)$ for cumulative incidence as in Equation (10.3). These are smoothed with respect to time and the covariate Z, and transformed using the complementary log-log link function. The transformed values are used to construct profile plots against Z for several time points. If the proportional subdistribution hazards model holds and the functional form of the covariate Z is correctly specified, the profile plots should be parallel and approximately linear, as above in the survival setting.

Andersen and Perme (2010) pointed out that this type of diagnostic plots may be more informative than the previous residual plots because the scale of the plots makes it easier to determine the reason for the lack of fit. Another consideration is that this plot works only for models with a single covariate, or with independent covariates. When a covariate is associated with other covariates in the model, the profile curves do not have to be parallel and linear under the model assumptions. In this case, Perme and Andersen (2008) suggested to first use the pseudo-values of survival probabilities to construct smoothed curves with respect to a Z_1 covariate at different time points. Then they used the predicted survival probability for each individual calculated from the fitted model including all the covariates to construct smoothed curves for the predicted survival probabilities with respect to the Z_1 covariate at the same time points. Finally, at each time point, they transformed the smoothed pseudo-observations for the survival probabilities and the smoothed predicted survival probabilities and subtracted the two. The difference of the two smoothed profiles at a time point reflects the remaining effects (i.e., residuals) of Z_1 after the model fitting. If the model assumptions are correct, the residuals at each time point should not have a non-linear relationship with Z_1 even though they still could have a linear relationship with Z_1, which may represent a change of Z_1's effect due to the adjustment for other covariates.

10.4.2 Tests of model fit

Perme and Andersen (2008) also introduced several formal tests of model assumptions to augment the graphical diagnostics. For a GLM model formulated in Equation (10.2), they consider more flexible models either in terms of nonproportional hazards or non-linearity or both, and test for lack of fit of the GLM model relative to some more flexible models. For example, the model

$$\log(-\log E[S_i(t_r)|Z_i]) = \alpha(t_r) + \beta(t_r)g(Z_i), \quad i = 1, \cdots, n, \ r = 1, \cdots, k$$

allows flexibility from the proportional hazards assumption by using a time-varying parameter $\beta(t)$ and relaxes the linearity assumption by using a more flexible functional form $g(Z)$, such as a restricted cubic spline. Similar lack of fit tests can be constructed for other GLM models corresponding to additive hazards models or Fine and Gray competing risks models.

10.5 Software

SAS macros and R packages ("pseudo.r") for computing pseudo-values for standard settings such as survival, cumulative incidence and restricted mean survival have been described in Klein et al. (2008). Once the pseudo-values are computed, model fitting is straightforward as it is based simply on standard estimation routines for GLM and generalized estimating equations, both of which are available in software SAS (PROC GENMOD) or R ("geese" function in the "geepack" package). Perme and Andersen (2008) developed an R function "pseu.r" for getting the diagnostic plots on proportional hazards and linearity in fitting Cox and additive models. It can also perform some formal tests as we have mentioned in the previous subsection. More recently, Andersen and Perme (2010) developed a revised R function "pseucheck.r" which can provide diagnostic plots not only for the Cox model but also for the Fine and Gray model, linear regression model of restricted means, and GLM model using the pseudo-value approach.

10.6 Examples

10.6.1 Example 1: Survival and cumulative incidence

We illustrate the use of pseudo-value regression for survival and cumulative incidence as well as model diagnosis using a Center for International Blood and Marrow Transplant Research (CIBMTR) registry dataset of $n = 348$ adult MDS patients ages 45-65 undergoing bone marrow transplantation. Among them, 124 patients had a myeloablative conditioning intensity transplant (MCT) while 224 patients had a non-myeloablative/Reduced Intensity conditioning transplant (NST/RIC); 152 patients used 8/8 matched unrelated donors (MUD) and 196 patients had matched sibling donors (MSD); the number of patients who had early, advanced and other disease status prior to HCT are 90, 190 and 68, respectively.

We first look at overall survival, and consider diagnostic plots for 3 variables: age at transplant, interval from disease diagnosis to transplant (using a log transformation due to skewness), and the use of myeloablative conditioning regimen (1=Yes, 0=No). We ran "pseucheck.r" at the 10th to 90th percentiles of the death times to obtain the diagnostic plot in fitting the Cox proportional hazards model as shown in Figure 10.3(a), (c) and (e). In each plot, the lowest line corresponds to the 10th percentile and the highest line corresponds to the 90th percentile of the death times, and the lines are elevated from the lowest to the highest as time t increases. For patient age and logarithm of the time interval from disease diagnosis to transplant, we cut the edges off the plots as there were very few events at the edges. For conditioning intensity, the slopes of the curves are positive at early times and decrease as time increases to become negative at later times, which may indicate non-proportional hazards. This is confirmed by a formal test of proportional hazards using the supremum test based on cumulative sums of martingale residuals (p=0.038). Notice that the conditioning intensity is a binary covariate which only takes two possible values 0 for NST/RIC, and 1 for myeloablative; therefore, linearity is not an issue for this variable. For patient age, the lines are curved but parallel, which indicates that the proportional hazards assumption might be reasonable but the linearity of the age effect is questionable. For the log-transformed time from disease diagnosis to transplant, the curves are also approximately parallel but curved, suggesting a non-linear effect.

Because there was evidence of non-proportional hazards based on conditioning intensity, we fit two separate pseudo-value regression models for the mortality probabilities at day 100 and 2 years after transplantation, using the "logit" link function. Based on the residual plots, a natural breakpoint for dichotomizing age would be at approximately age 55, and we considered grouping logarithm of the time interval from disease diagnosis to transplant at ≤ 1.79, 1.79 to 2.48, and ≥ 2.48, which correspond approximately to 0-6, 6-12, and ≥ 12 months for the time from disease diagnosis to transplant. The two model details are shown in Table 10.1. Note that while the effect of myeloablative conditioning is not significant at either time point, the odds ratio (OR) for mortality switches direction between these two time points, which is consistent with the indication of our diagnostic plot.

Next, we look at relapse, where death in remission is considered a competing event. The diagnostic plots at the 10th to 90th percentiles of the relapse times are given in Figure 10.3 (b), (d) and (f) based on a complementary log-log transformation for the cumulative in-

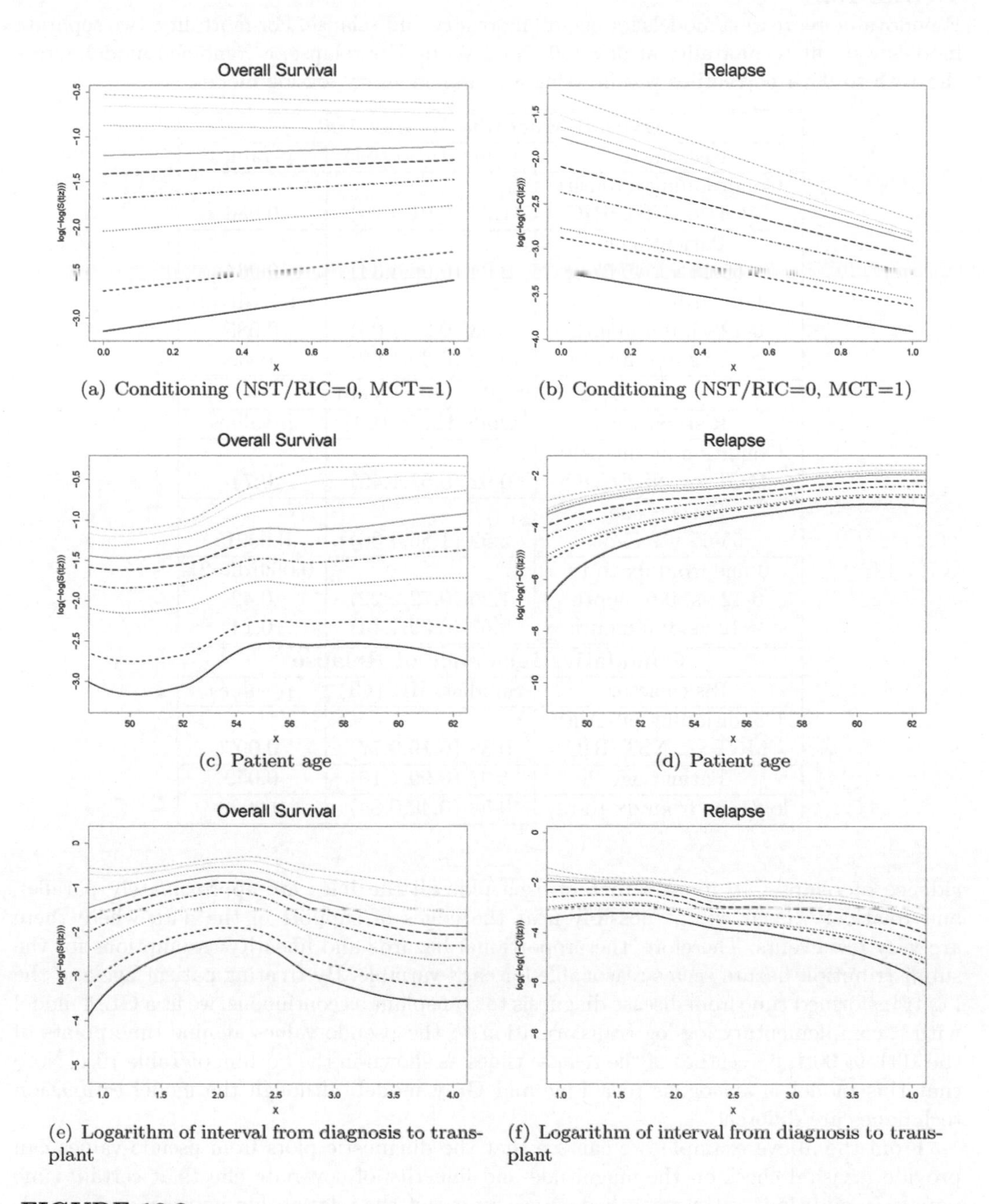

(a) Conditioning (NST/RIC=0, MCT=1)

(b) Conditioning (NST/RIC=0, MCT=1)

(c) Patient age

(d) Patient age

(e) Logarithm of interval from diagnosis to transplant

(f) Logarithm of interval from diagnosis to transplant

FIGURE 10.3
Goodness-of-fit in fitting a Cox proportional hazards model for overall survival and a Fine and Gray proportional subdistribution hazards model for relapse. Curves represent the smoothed scatterplots for the 10th to 90th percentiles of the event times, and they are plotted against 3 different covariates for each outcome.

TABLE 10.1
Pseudovalue regression models for overall mortality and relapse. For mortality, two separate models were fit to mortality at day 100 and 2 years. For relapse, a combined model across the 10th to 90th percentiles was fit using a complementary log-log model.

Overall Mortality at day 100		
Risk Factor	Odds Ratio (CI)	p-values
Conditioning intensity MCT vs. NST/RIC	1.95 (0.91,4.20)	0.086
Patient age 55-65 vs. 45-55	2.04 (0.96,4.34)	0.064
Time from dx to tx		0.047(df=2)
6-12 vs. 0-6 month	1.93 (0.91,4.09)	0.085
$\geq$ 12 vs. 0-6 month	0.72 (0.30,1.69)	0.45
Overall Mortality at 2 years		
Risk Factor	Odds Ratio (CI)	p-values
Conditioning intensity MCT vs. NST/RIC	0.92 (0.57,1.49)	0.74
Patient age 55-65 vs. 45-55	2.52 (1.56,4.06)	0.00015
Time from dx to tx		0.094(df=2)
6-12 vs. 0-6 month	1.26 (0.72,2.22)	0.42
$\geq$ 12 vs. 0-6 month	0.67 (0.39,1.14)	0.14
Cumulative Incidence of Relapse		
Risk Factor	Subdist. HR (CI)	p-values
Conditioning intensity MCT vs. NST/RIC	0.38 (0.19,0.77)	0.007
Patient age	1.07 (0.99,1.16)	0.059
log(time from dx to tx)	0.58 (0.40,0.84)	0.0038

cidence of relapse. It appears that in each plot all the lines are approximately parallel, and relatively linear except possibly near the edges of support of the data where there are very few events. Therefore, the proportional hazards and linearity assumptions for the subdistribution hazard seems reasonable for each variable. By treating patient age and the log-transformed time from disease diagnosis to transplant as continuous, we fit a GLM model with a complementary log-log transformation to the pseudo-values at nine time points of the 10th to 90th percentiles of the relapse times as shown in the bottom of Table 10.1. Note that this model is analogous to a Fine and Gray model, although the model estimation techniques are different.

From the above example, we can see that the diagnostic plots from pseudo-values can provide a visual check on the magnitude and linearity of covariate effects at certain time points. In addition, as the smoothed curves represent the relationship between each covariate and the "logit" or complementary log-log transformed survival or cumulative incidence functions at those time points, these diagnostic plots are also useful for choosing an appropriate nonlinear functional form or selecting some reasonable cut points for discretizing a continuous covariate when linearity of the covariate is violated.

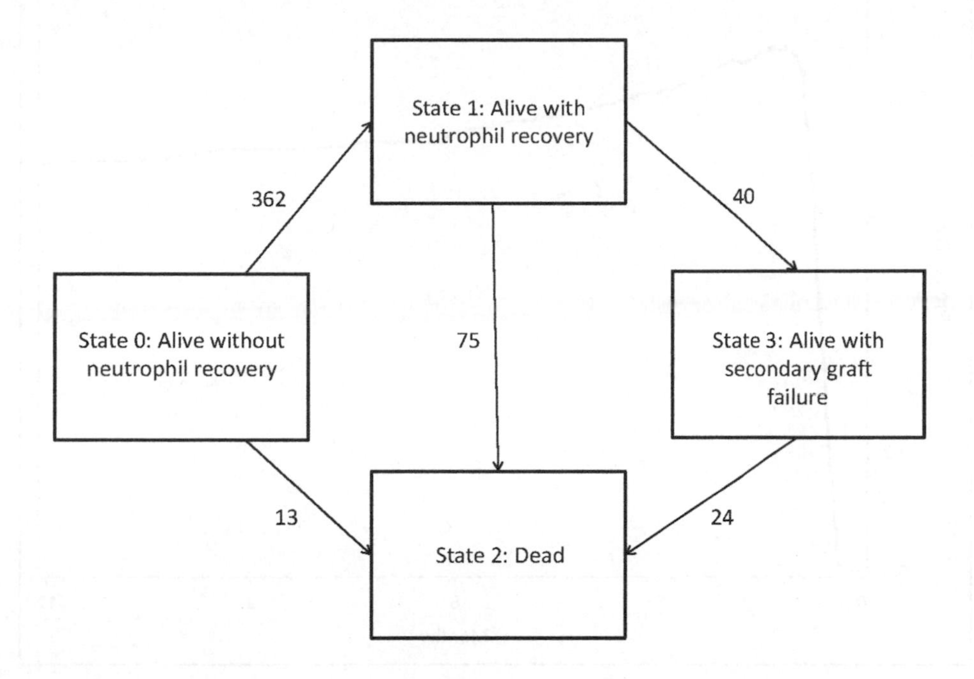

FIGURE 10.4
SAA Engraftment model

10.6.2 Example 2: Multi-state model

A second application involves outcomes after unrelated donor bone marrow transplantation for 375 patients with Severe Aplastic Anemia (SAA) using data reported to the CIBMTR. One concern post-transplant is that the donor cells will fail to engraft and repopulate the recipient's immune system; this can manifest as either lack of initial engraftment or recovery of neutrophils, or as secondary graft failure when the patient's neutrophil counts drop after initial recovery. This process can be shown as a multi-state model, where a patient starts post transplant in state 0 (alive without neutrophil recovery), and from there can either go to state 1 (alive with neutrophil recovery) or die prior to engraftment (state 2). Once they are in state 1, they can die or they can experience secondary graft failure (state 3), and from state 3 they can further progress to death. This multi-state model is summarized in Figure 10.4, which also shows the numbers of patients experiencing each transition as well as the number at risk.

One quantity that may be important to clinicians is the probability of being alive and engrafted (state 1) as a function of time; this combines both primary recovery and secondary graft failure into one summary endpoint describing a positive result. Note that the probability of being in state 1 can be written as the probability of being in state 0 or 1 minus the probability of being in state 0. Therefore, an estimate of the probability of being alive and engrafted is provided by a difference in the Kaplan-Meier estimates

$$\hat{P}_{01}(0,t) = \hat{S}_{2,3}(t) - \hat{S}_{1,2}(t),$$

where $\hat{S}_{s_1,\ldots,s_k}(t)$ is the Kaplan-Meier estimate treating transitions to states $s_1,\ldots,s_k$ as

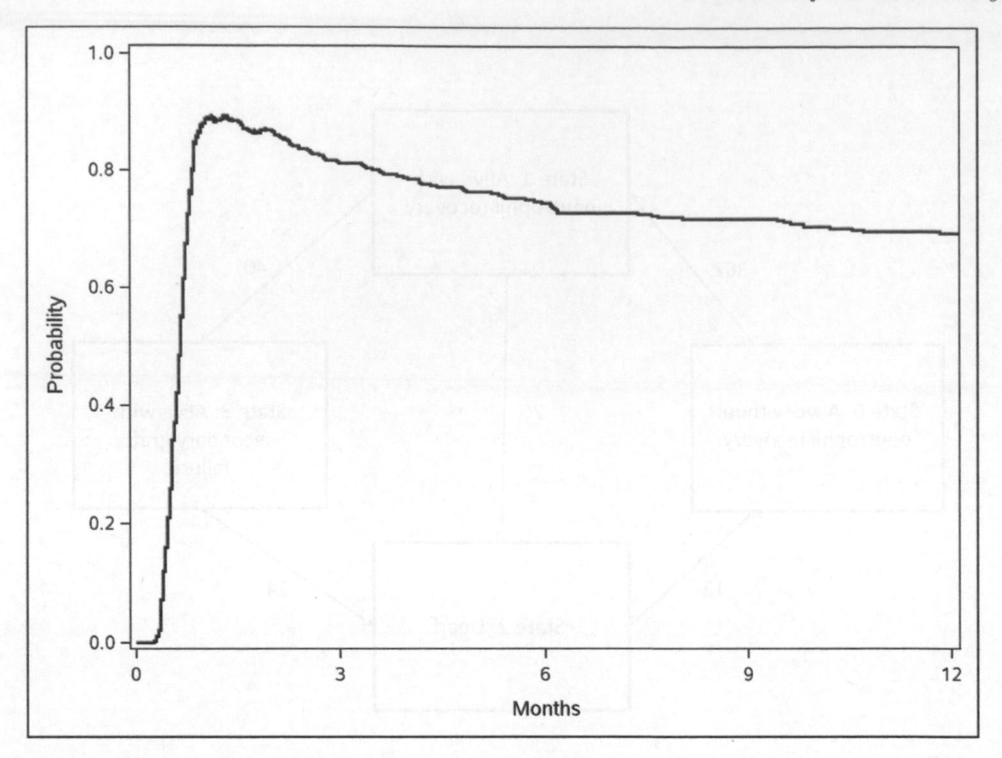

FIGURE 10.5
Probability of being alive and engrafted in SAA example.

events. An estimate of the marginal probability of being alive and engrafted is given in Figure 10.5.

To compute pseudo-values, we can take the difference in pseudo-values for each Kaplan-Meier estimate. Let

$$Y_i^{s_1,\ldots,s_k} = n\hat{S}_{s_1,\ldots,s_k}(t) - (n-1)\hat{S}^{-i}_{s_1,\ldots,s_k}(t)$$

be the pseudo-value based on $\hat{S}_{s_1,\ldots,s_k}(t)$. Then the pseudo-value for $P_{01}(0,t)$ is

$$Y_i(t) = Y_i^{2,3}(t) - Y_i^{1,2}(t).$$

We can use these pseudo-values to directly model the probability of being alive and engrafted as a function of covariates, including age at transplant, gender, karnofsky performance score (KPS), HLA matching of donor and recipient, and graft versus host disease (GVHD) prophylaxis. We use pseudo-values at 3 time points (3, 6, and 12 months), and a "logit" link function. The results of the regression model for survival with engraftment in terms of odds ratios for each of these covariates is given in Table 10.2. HLA mismatch is significantly associated with worse survival with engraftment, and patients age 21-40 have significantly better survival with engraftment compared to those ≤ 20 or > 40. Other risk factors are not significantly associated with the probability of being alive with neutrophil recovery.

TABLE 10.2
Results of pseudo-value regression model for the probability of being alive and engrafted.

Variable	Comparison	Odds Ratio (CI)	p-value
Age	21-40 vs. ≤ 20	2.06 (1.16,3.66)	0.014
	> 40 vs. ≤ 20	0.73 (0.34,1.56)	0.418
Gender	M vs. F	1.47 (0.91,2.38)	0.114
KPS	< 90 vs. ≥ 90	1.67 (0.96,2.92)	0.070
	Unknown vs. ≥ 90	1.45 (0.48,4.36)	0.506
HLA	Mismatched vs. Matched	0.50 (0.31,0.79)	0.003
GVHD	FK506 based vs. CSA+MTX	0.96 (0.55,1.69)	0.893
Proph.	Others vs. CSA+MTX	0.71 (0.36,1.39)	0.315

10.7 Conclusions

In this article, we reviewed the use of pseudo-values in survival and event history analysis. As pointed out by Perme and Andersen (2008), the main advantage of using the pseudo-value approach is that a pseudo-observation corresponding to an expectation of interest such as a survival probability can be estimated for each individual at any time point regardless of censoring. These pseudo-values can then be analyzed under a generalized linear model framework or be used for model diagnosis. Pseudo-value regression models are especially useful and straightforward to use when one is focusing inference on a single time point, as may be done either because of clinical interest or to avoid model assumptions such as proportional hazards. In this setting pseudo-value regression models function essentially as a censored data logistic regression model.

Pseudo-value regression works when the pseudo-values come from an approximately unbiased estimator of the mean parameter being modeled. This requirement is often satisfied when the censoring process is independent of the survival time. Although weighted estimators have been proposed to alleviate this concern when censoring is dependent on covariates, additional research work in this area is needed.

One drawback of the pseudo-value framework is that it may not be as efficient as other standard methods. For example, the pseudo-value regression parameter estimates using pseudo-values for survival data with a complementary log-log link function are not as efficient as the partial likelihood estimator of regression coefficients for a Cox survival model. Similarly, Klein and Andersen (2005) found a slight loss of efficiency of the pseudo-value regression parameter estimates using pseudo-values for cumulative incidence with a complementary log-log link function compared to the Fine and Gray model for competing risks data. However, while there is likely some loss of efficiency, the pseudo-value regression method is more flexible to handle the situation where the assumption of proportional hazard is violated, and it can be very easily applied to a broad class of multi-state models for which standard regression analysis is often not available.

Consideration of several issues may lead to improvements in efficiency for pseudo-value regression models. Selection of the number and position of the time points may affect efficiency, and while some simulations (Andersen and Perme, 2010; Klein and Andersen, 2005) have suggested that more than 5 time points equally spaced on the event time scale result in minimal improvements in efficiency, it would be desirable to use all time points. Several steps in that direction have already been taken. Liu et al. (2008) and Logan et al. (2008) proposed score tests in the two sample situations for current leukemia free survival and late survival, respectively, which utilize all the event times and have a closed-form expression. Andersen and Klein (2007) proposed an alternative estimating equation which includes estimation of the intercept $\alpha(t)$ based on work in recurrent events, as well as the addition of a penalty function to the GEE to smooth out $\alpha(t)$. Andersen and Perme (2010) proposed regression analysis of the cumulative incidence function using all time points with a smoothing spline. The choice of working covariance matrix V_i in the generalized estimating equation framework is another area which may affect efficiency. Simulations by Klein and Andersen (2005) for the competing risks model suggested that choosing V_i to approximately match the true covariance may improve the efficiency of the estimators. This may be especially important for the clustered data setting described in Section 10.3.4 when one needs to deal with two sources of correlation (within cluster and within individual). Finally in some settings multiple estimators are available as possible candidates from which to compute the pseudo-values, and it is unclear whether the choice of estimator may affect the efficiency of the pseudo-value regression parameter estimates.

Bibliography

Andersen, P. K., Borgan, Ø., Gill, R. D. and Keiding, N. (1993), *Statistical Models Based on Counting Processes*, Springer-Verlag, New York.

Andersen, P. K., Hansen, M. G. and Klein, J. P. (2004), 'Regression analysis of restricted mean survival time based on pseudo-observations', *Lifetime Data Anal* **10**(4), 335–350.

Andersen, P. K. and Klein, J. P. (2007), 'Regression analysis for multi-state models based on a pseudo-value approach with applications to bone marrow transplantation studies', *Scandinavian Journal of Statistics* **34**, 3–16.

Andersen, P. K., Klein, J. P. and Rosthøj, S. (2003), 'Generalised linear models for correlated pseudo-observations, with applications to multi-state models', *Biometrika* **90(1)**, 15–27.

Andersen, P. K. and Perme, M. P. (2008), 'Inference for outcome probabilities in multi-state models', *Lifetime Data Anal* **14(4)**, 405–431.

Andersen, P. K. and Perme, M. P. (2010), 'Pseudo-observations in survival analysis', *Stat Methods Med Res* **19**(1), 71–99.

Andrei, A.-C. and Murray, S. (2007), 'Regression models for the mean of the quality-of-life-adjusted restricted survival time using pseudo-observations', *Biometrics* **63**(2), 398–404.

Fine, J. P. (2001), 'Regression modeling of competing crude failure probabilities', *Biostatistics* **2**(1), 85–97.

Fine, J. P. and Gray, R. J. (1999), 'A proportional hazards model for the subdistribution of a competing risk', *Journal of the American Statistical Association* **94**, 496–509.

Graw, F., Gerds, T. A. and Schumacher, M. (2009), 'On pseudo-values for regression analysis in competing risks models', *Lifetime Data Anal* **15**(2), 241–255.

Klein, J. P. and Andersen, P. K. (2005), 'Regression modeling of competing risks data based on pseudovalues of the cumulative incidence function', *Biometrics* **61**(1), 223–229.

Klein, J. P., Gerster, M., Andersen, P. K., Tarima, S. and Perme, M. P. (2008), 'SAS and r functions to compute pseudo-values for censored data regression', *Comput Methods Programs Biomed* **89**(3), 289–300.

Klein, J. P., Logan, B., Harhoff, M. and Andersen, P. K. (2007), 'Analyzing survival curves at a fixed point in time', *Stat Med* **26**(24), 4505–4519.

Liang, K. Y. and Zeger, S. L. (1986), 'Longitudinal data analysis using generalized linear models', *Biometrika* **73**(1), 13–22.

Lin, D. Y. and Ying, Z. (1994), 'Semiparametric analysis of the additive risk model', *Biometrika* **81**(1), 61–71.

Liu, L.-Y., Logan, B. and Klein, J. P. (2008), 'Inference for current leukemia free survival', *Lifetime Data Anal* **14**(4), 432–446.

Logan, B. R., Klein, J. P. and Zhang, M.-J. (2008), 'Comparing treatments in the presence of crossing survival curves: an application to bone marrow transplantation', *Biometrics* **64**(3), 733–740.

Logan, B. R., Zhang, M.-J. and Klein, J. P. (2011), 'Marginal models for clustered time-to-event data with competing risks using pseudovalues', *Biometrics* **67**(1), 1–7.

Meira-Machado, L., de Ua-Alvarez, J., Cadarso-Surez, C. and Andersen, P. K. (2009), 'Multi-state models for the analysis of time-to-event data', *Stat Methods Med Res* **18**(2), 195–222.

Pepe, M. S. (1991), 'Inference for events with dependent risks in multiple end point studies', *Journal of American Statistical Association* **86(415)**, 770–778.

Perme, M. P. and Andersen, P. K. (2008), 'Checking hazard regression models using pseudo-observations', *Stat Med* **27**(25), 5309–5328.

Scheike, T. H. and Zhang, M.-J. (2007), 'Direct modeling of regression effects for transition probabilities in multistate models', *Scandinavian Journal of Statistics* **34(1)**, 17–32.

Scheike, T. H. and Zhang, M.-J. (2011), 'Analyzing competing risk data using the r timereg package', *J Stat Softw* **38**(2), 1–15.

Scheike, T. H., Zhang, M.-J. and Gerds, T. A. (2008), 'Predicting cumulative incidence probability by direct binomial regression', *Biometrika* **95**, 205–220.

Schoenfeld, D. (1982), 'Partial residuals for the proportional hazards regression model', *Biometrika* **69 (1)**, 239–241.

Therneau, T. M., Grambsch, P. M. and Fleming, T. R. (1990), 'Martingale-based residuals for survival models', *Biometrika* **77 (1)**, 147–160.

Tunes-da Silva, G. and Klein, J. P. (2009), 'Regression analysis of mean quality-adjusted survival time based on pseudo-observations', *Stat Med* **28**(7), 1054–1066.

11

Binomial Regression Models

Randi Grøn and Thomas A. Gerds

Department of Biostatistics, University of Copenhagen

CONTENTS

11.1 Introduction 222
11.1.1 Choice of time horizons 222
11.1.2 Modeling options 222
11.1.3 Right-censored data 223
11.1.4 Interval-censored data 223
11.1.5 Variance estimation 224
11.1.6 Time-varying covariates 224
11.1.7 Comparison with cause-specific modeling 224
11.2 Modeling 225
11.2.1 Logistic link 225
11.2.2 Log link 225
11.2.3 Complementary log-log link 226
11.2.4 Constant and time-varying regression coefficients 226
11.3 Estimation 227
11.3.1 Weighted response 227
11.3.2 Working censoring model 227
11.3.3 Weighted estimating equations 228
11.4 Variance estimation 228
11.4.1 Asymptotic variance estimate 228
11.4.2 Bootstrap confidence limits 229
Algorithm for estimating bootstrap confidence limits 230
11.5 Software implementation 230
11.6 Example 232
11.6.1 Melanoma data 232
11.6.2 Choice of link function 233
11.6.3 Effect of choice of time points 233
11.6.4 Compare confidence limits 234
11.7 Simulations 235
11.7.1 Competing risks model 235
11.7.2 Misspecified censoring model 235
11.7.3 Compare confidence limits 237
11.7.4 Effect of choice of time points 238
11.8 Final remarks 239
Bibliography 239

11.1 Introduction

Binomial regression is an estimation technique used for predicting a binary event status at future time points. In survival analysis the outcome is the time between a well-defined time origin (or) and the occurrence of an event. The key to using binomial regression for time-to-event outcome is the observation that at any time horizon after the time origin the event status is binary, taking the value 1 if the event has occurred, and 0 otherwise. Clearly, the event status at a single time horizon carries much less information than the time-to-event outcome. However, the time process $\{N_\epsilon(t) = \mathcal{I}\{T \leq t, \epsilon\} : t \in [0, \infty)\}$ represents the same information as (T, ϵ), where ϵ indicates the type of the event and T the event time. Indeed, there is a one-to-one correspondence between binomial regression models and (Doksum and Gasko, 1990) which holds also in more complex models for event history analysis (Jewell, 2005).

11.1.1 Choice of time horizons

In applications it is then possible and often useful to apply binomial regression to the event status at a sequence of time horizons $(t_1, \ldots, t_J)$. A sufficient choice is the sequence of all observed time points at which the event occurred in the current dataset. This is sufficient because the contrast between the event status of different patients is constant in the period between observed event times and this contrast is carrying the relevant information for regression. The process of binomial regression models applied at a sequence of time horizons may define time varying or time constant regression coefficients, or mixtures thereof (Martinussen and Scheike, 2006; Scheike et al., 2008). Simulations have shown that for estimating time constant effects it may be reasonable to only consider a selected set of 5 to 10 time horizons (Klein and Andersen, 2005). However, this may depend on the situation, and it is not totally clear how to optimally place the time horizons on the time scale. On the other hand it is clear that the choice of time horizons has a small sample effect on the numerical results of binomial regression analysis even if the regression coefficient is constant in time. Another challenge is to efficiently weight the sequence of binomial regression equations in order to combine information across time to obtain estimates of time constant regression coefficients. Furthermore, there may be problems with the convergence of the estimates at early time points where few subjects have experienced the event of interest, and at late time points where few subjects are at risk. We illustrate and discuss the choice of time horizons further in Sections 11.2.4, 11.3.1, 11.6, and 11.7.

11.1.2 Modeling options

The most commonly applied binomial regression model is the logistic regression model (Berkson, 1944), an attractive alternative is the log-binomial regression model (Wacholder, 1985; Blizzard and Hosmer, 2006; Marschner and Gillett, 2012). A good starting point for discussing modeling options is a generalized linear model for the absolute risk () of an event ϵ until time point t:

$$F_\epsilon(t|X) = \mathrm{E}(N_\epsilon(t)|X) = h(\beta_0(t), \beta(t), X). \tag{11.1}$$

The model describes the effect of a vector of predictor variables X measured at the time-origin or landmark by a possibly time-varying vector of regression coefficients $\beta(t)$, a link function h, and an intercept $\beta_0(t)$. The intercept characterizes the risk of an event for subjects with $X = 0$. Important special cases of model (11.1) are the proportional odds model and the proportional hazard model (Doksum and Gasko, 1990). Model (11.1) can also be regarded as a special case of the transformation model for survival analysis (Fine et al., 1998; Fine, 1999). We discuss different link functions and constraints of the regression coefficients in Section 11.2.

11.1.3 Right-censored data

A characteristic of event history analysis is the need to deal with (right-) censored data. If no event has occurred until time t, then the event status at time t is zero. However, if the patient was lost to follow up before time t, then the event status $N_\epsilon(t)$ is unknown, its value is right-censored. If the censoring mechanism is not informative, (Andersen et al., 1993), then the probability that the event has occurred to a patient whose status is censored may be estimated from the patients whose status was observed. There are two popular ways of approaching this. One works by weighting the observed status by so-called inverse of the probability of censoring weights (IPCW) (van der Laan and Robins, 2003). This approach was proposed by (Scheike et al., 2008) and is reviewed and further discussed in the present chapter (see in particular Sections 11.3 and 11.4). Another approach is to replace the possibly censored status by a jackknife pseudo-value. We refer to Andersen et al. (2003) and Chapter 10. Both approaches require a model for the censoring mechanism. The pseudo-approach in its basic form assumes that the censoring mechanism is independent of observed covariates, which is in contrast to the binomial regression approach. Specifically, the estimate of model (11.1) obtained with the IPCW approach depends explicitly on an estimate of the conditional probability function of not being censored given X, an infinite dimensional nuisance parameter. An example of a model for the conditional censoring distribution, which is attractive for its simplicity, is based on the assumption that the censoring probability does not depend on the covariates X. In this case the Kaplan-Meier estimate for the censoring times can be used to construct consistent estimates of the censoring weights. However, the censored times may depend on the covariates. And if this is the case, then the Kaplan-Meier model is misspecified. Moreover, the IPCW estimate of (11.1) based on the simple Kaplan-Meier model is always inefficient, as it ignores the covariate values that were collected for patients whose status was censored before time t. See van der Laan and Robins (2003) for a more technical explanation of this fact. Generally to increase efficiency and to reduce the risk of bias, one can specify a working regression model for the censoring times. We discuss this further in Section 11.3.2 and compare results obtained with different censoring models and investigate the effects of misspecification in Sections 11.6 and 11.7.

11.1.4 Interval-censored data

Binomial regression is also a very useful tool under more complex censoring schemes. This includes current status data and interval-censored data (Sun, 2006). In many applications it is only possible to observe the event time and status up to an interval. For example when the event status is monitored over time, e.g., at scheduled visits, exact event times are not observed but interval-censored. It is only known that the event occurred between two adjacent visits but the exact event time remains unknown. An extreme form of interval-censored data is called "current status data." Here only data are available from a single visit for each subject at which the event status is diagnosed. We do not detail the estimation techniques for current status and interval-censored data in this chapter, and instead refer to the respective literature (Rossini and Tsiatis, 1996; Jewell and van der Laan, 1997; Shiboski, 1998; Lin et al., 1998; van der Laan and Robins, 1998; Martinussen and Scheike, 2002; Jewell and van der Laan, 2004; Sun, 2006).

11.1.5 Variance estimation

Standard software for generalized linear models can be used to estimate the parameters of model (11.1). For this, the data have to be prepared in a special way which we describe in Section 11.5. This yields consistent estimates of the regression coefficients if both the model of interest for the absolute risk and the nuisance model for the censoring weights are correctly specified. But, the standard error obtained by software for generalized linear models does not automatically account for the variability of the estimate of the nuisance parameter. One approach to obtain unbiased estimates of the standard errors is to derive an explicit expression of the asymptotic distribution of the estimator for the regression coefficients. In Section 11.4.1 we review the approach of Scheike et al. (2008). An alternative approach is to apply bootstrap in a two-step procedure. In each bootstrap sample one first estimates the censoring weights and then solves an estimating equation using correspondingly weighted outcome (Section 11.4.2).

11.1.6 Time-varying covariates

A binomial regression model can incorporate time-varying covariates (van Houwelingen, 2006) by landmark analysis (see van Houwelingen and Putter (2012) and Chapter 21 for details). A landmark analysis starts by choosing a landmark time point. Then all individuals are included that are event-free at the landmark time point. At the landmark, the baseline information of covariates readily available at the time origin is augmented by information from time-dependent covariates which became available until the landmark. Also the state of the multi-state process at the landmark can be used as a covariate (van Houwelingen and Putter, 2008).

11.1.7 Comparison with cause-specific modeling

It is worth noting that there is an alternative way of modeling the cumulative incidence function in a multi-state model (Chapter 20). In this approach regression models are specified for all transition intensities and the results are combined. For example, Cox regression models can be fitted to all the cause-specific hazards of a competing risk model (Chapter 8) and the results subsequently combined into an estimate of the cumulative incidence function. The advantage of this approach is that one does not have to specify a model for the censoring mechanism. But the need to specify and estimate regression models for all the other transition intensities may turn out to be a disadvantage. This happens when there are many possible transitions, few observed transitions in the current dataset, and when it is not possible to observe transitions in continuous time. Furthermore, it may be difficult to summarize the total effect that a single predictor variable has on the cumulative incidence function. Here the direct binomial regression approach discussed in this chapter and the pseudo-value approach discussed in Chapter 10 have a clear advantage. In summary, the advantages of binomial regression are feasibility and direct interpretation of regression parameters. The disadvantages are that the censoring mechanism needs to be modeled and that the estimates of the regression coefficients depend on the estimate of the baseline risk (even in proportional hazards models).

11.2 Modeling

Consider a dataset that includes for each subject an event time T, a categorical marker ϵ, and a p-dimensional vector of covariates $X = (X_1, \ldots, X_p)$. The marker takes one of K different values indicating the event status at time T. It is assumed that it is coded such that $\epsilon = 1$ indicates that the event of interest has occurred at time T whereas $\epsilon > 1$ indicates a competing risk at time T. After a competing risk has occurred, the event of interest can either not occur anymore, for example when the subject has died, or it is not of interest any longer, for example when the subject has changed the treatment. All fatal events, such as death due to other causes after which the event of interest cannot occur, are competing risks. And it often makes sense to also define other events which change the risk of the event of interest as competing risks, and to study the time to whatever event comes first. Our set-up also covers more complex multi-state models with transient states which can be studied using a landmark approach. For a discussion of binomial regression in a multi-state model see Scheike and Zhang (2007).

The interpretation of the regression coefficients defined in (11.1) depends on the specific link function (Zhang and Fine, 2008; Ambrogi et al., 2008; Gerds et al., 2012) and on whether it is assumed that some or all regression coefficients are time constant. These modeling options are discussed in this section.

11.2.1 Logistic link

The choice $h(a, b, x) = \text{expit}(a + b^T x) = \exp(a + b^T x)/(1 + \exp(a + b^T x))$ defines a logistic regression model for the event status at time t in which the regression coefficients are log-odds ratios:

$$
\begin{aligned}
\text{OR}(t) &= \frac{F_\epsilon(t|X_1 + 1, X_2, \ldots, X_p)/(1 - F_\epsilon(t|X_1 + 1, X_2, \ldots, X_p))}{F_\epsilon(t|X_1, X_2, \ldots, X_p)/(1 - F_\epsilon(t|X_1, X_2, \ldots, X_p))} \\
&= \frac{\exp(\beta_0(t) + \beta_1(t)(X_1 + 1) + \beta_2(t)X_2 + \cdots + \beta_p(t)X_p)}{\exp(\beta_0(t) + \beta_1(t)X_1 + \beta_2(t)X_2 + \cdots + \beta_p(t)X_p)} \\
&= \exp(\beta_1(t)).
\end{aligned}
$$

In the special case where it is assumed that the regression coefficients do not depend on time, the model is also known as the "proportional odds model" (Bennett, 1983; Rossini and Tsiatis, 1996).

11.2.2 Log link

The choice $h(a, b, x) = \exp(a + b^T x)$ gives rise to a log-binomial regression model in which the regression coefficients are log-transformed absolute risk ratios.

$$
\begin{aligned}
\text{ARR}(t) &= \frac{F_\epsilon(t|X_1 + 1, X_2, \ldots, X_p)}{F_\epsilon(t|X_1, X_2, \ldots, X_p)} \\
&= \frac{\exp(\beta_0(t) + \beta_1(t)(X_1 + 1) + \beta_2(t)X_2 + \cdots + \beta_p(t)X_p)}{\exp(\beta_0(t) + \beta_1(t)(X_1) + \beta_2(t)X_2 + \cdots + \beta_p(t)X_p)} \\
&= \exp(\beta_1(t)).
\end{aligned}
$$

Absolute risk ratios are generally easier to understand than odds ratios. However, the log link comes at the cost of an increased risk of numerical instability when fitting the model with standard maximum likelihood procedures, in particular when the predictor variables are continuous. Furthermore, the model may predict risks below zero or above one (Blizzard and Hosmer, 2006; Gerds et al., 2012). To remedy these problems Marschner and Gillett (2012) derived an adapted EM algorithm.

11.2.3 Complementary log-log link

A third commonly used link function is $h(a, b, x) = 1 - \exp(-\exp(a + b^T x))$ which yields the Cox regression model (Cox, 1972) in absence of competing risks, and when there are competing risks the Fine and Gray model (Fine and Gray, 1999). In absence of competing risks the regression coefficients are hazard ratios:

$$\begin{aligned} \mathrm{HR}(t) &= \frac{\mathrm{d}\log(1 - F_\epsilon(t|X_1 + 1, X_2, \ldots, X_p))}{\mathrm{d}\log(1 - F_\epsilon(t|X_1, X_2, \ldots, X_p))} \\ &= \frac{\exp(\beta_0(t) + \beta_1(t)(X_1 + 1) + \beta_2(t)X_2 + \cdots + \beta_p(t)X_p)}{\exp(\beta_0(t) + \beta_1(t)(X_1) + \beta_2(t)X_2 + \cdots + \beta_p(t)X_p)} \\ &= \exp(\beta_1(t)). \end{aligned}$$

In the presence of competing risks, the regression coefficients have no direct interpretation (Fine and Gray, 1999; Ambrogi et al., 2008; Gerds et al., 2012).

11.2.4 Constant and time-varying regression coefficients

An important special case of model (11.1) is the following in which all covariates have time-constant effects:

$$F_\epsilon(t|X) = h(\beta_0(t) + \bar{\beta}_1 X_1 + \cdots \bar{\beta}_p X_p). \tag{11.2}$$

The vector of constant regression parameters $\bar{\beta} = (\bar{\beta}_1, \ldots, \bar{\beta}_p)$ can be estimated based on a sequence of time points (see Section 11.3). Interestingly, choosing a single time point is sufficient to identify $\bar{\beta}$. But, this will not be efficient since not all information is used. Combining results from multiple time horizons will increase the efficiency and the results can only be fully efficient if at least all time points are used at which the event of interest occurred in the current dataset. However, note that one may just as well choose time points at which no events were observed. For example, Klein and Andersen (2005) worked with a set of time points that were equally spaced on the event scale.

Another special case of (11.1) is the Cox-Aalen regression model (Scheike and Zhang, 2002, 2003; Scheike et al., 2008):

$$\begin{aligned} F_\epsilon(t|X) = 1 - \exp\{[\beta_0(t) + \beta_1(t)X_1 + \cdots + \beta_{k-1}(t)X_{k-1}] \\ \exp(\bar{\beta}_k^T X_k + \cdots + \bar{\beta}_p^T X_p)\} \end{aligned} \tag{11.3}$$

In this model the first $k - 1$ covariates are allowed to have time-varying additive effects as in Aalen's additive regression model (Aalen, 1989) and the remaining covariates have a time-constant multiplicative effect on the hazard, as in the Cox regression model (Cox, 1972, 1975). When fitting model (11.3) the choice of the time points affects the estimates of the time-constant parameters in a similar way as for (11.2). The choice of time points also defines the support of the estimates of the time-varying parameters. To reduce dimensionality of model (11.3) one can specify parametric or semiparametric constraints, such as $\beta(t) = \phi\log(t)$ (Bennett, 1983; Shiboski, 1998; Martinussen and Scheike, 2006).

11.3 Estimation

The estimators for the regression parameters in the models described in Section 11.2 are often called "maximum likelihood like estimators" (M-estimators) (Huber and Ronchetti, 2009). They really are zero estimators (Z-estimators) (Van der Vaart, 1998) defined as the solutions (zeros) to generalized estimating equations. In this section we describe so-called "inverse of the probability of censoring weighted" (IPCW) estimating equations.

11.3.1 Weighted response

Let C denote a right-censoring time and G its conditional survival function. We assume that the censoring event is non-informative in the sense that the risk of the event of interest does not change for individuals after they were lost to follow-up (Andersen et al., 1993). But we allow that the distribution of the censoring times depends on the covariates. Specifically, we assume that C is conditionally independent of (T, ϵ) given the observed predictors X and denote $G(t|X) = \mathrm{P}(C > t|X)$ for the conditional probability of not being lost to follow-up by time t given X. We also define the censoring indicator $\Delta = \mathcal{I}\{T \le C\}$, such that $\Delta = 1$ when the event time is observed, and $\Delta = 0$ when the event time is censored. The data observed for each subject are then $Y = (X, \tilde{T}, \Delta\epsilon)$, where $\tilde{T} = \min(T, C)$.

To avoid bias it is necessary to account for the possibility that the event may have occurred to subjects who were lost to follow-up (right censored) before the current time horizon. The idea is to replace the event status $N_\epsilon(t)$ by a weighted response $\tilde{N}_\epsilon(t) = \Delta N_\epsilon(t)/G(T-|X)$. Under the assumed conditional independence we have $\mathrm{E}(\Delta|X, T, \epsilon) = G(T-|X)$ and hence

$$E(\tilde{N}_\epsilon(t)) = E\left(E\left(\frac{\Delta N_\epsilon(t)}{G(T-|X)}\Big|T, \epsilon, X\right)\right) = E(N_\epsilon(t)|X) = F_\epsilon(t|X)\,. \tag{11.4}$$

This equality motivates the generalized estimating equations for the binomial regression model described below. Since, the numerator of the weighted response is zero when $\tilde{T} > t$ we can replace the denominator by $G(\min(t, T-)|X)$. To avoid problems with large weights it is sufficient to restrict the sequence of time horizons such that there exists $\eta > 0$ such that $\sup_x\{G(t|x)\} > \eta$. In practice it seems reasonable to choose $\eta > n/m(t)$ where $m(t)$ is the number of subjects who were right censored before time t and n is the sample size.

11.3.2 Working censoring model

To use the weighted response (11.4) in estimating equations we need to estimate the censoring distribution. The estimate $\hat{G}$ is based on a "working model" $\mathcal{G}$ which is a subset of all conditional survival functions. In the case were it is reasonable to assume that the censoring mechanism does not depend on the covariates a useful model consists of all marginal survival probability distributions. In this model the Kaplan-Meier estimator based on the censoring times consistently estimates G. However, as readily noted in the introduction this model does not lead to efficient estimating equations. To incorporate the covariate information of subjects censored before the current time horizon, one can specify a stratified survival model or a semiparametric regression model for the censoring times, for example a Cox regression model. However, a necessary condition for the estimating equations to yield consistent estimates of the regression coefficients is that the working model for the censoring distribution is correctly specified. More precisely, following Scheike et al. (2008) we assume that $\hat{G}$ is asymptotically regular, Gaussian linear with influence function IC_G

such that uniformly in $\mathcal{G}$

$$n^{1/2}(\hat{G} - G)(s, x) = n^{-1/2} \sum_{i=1}^{n} IC_G(s, x; Y_i) + o_P(1).$$

Note that recent work by Cheng and Huang (2010) indicates that it may be possible to relax the rate convergence.

The model $\mathcal{G}$ may be misspecified. In this case $G \notin \mathcal{G}$, and the best one can hope for is that $\hat{G}$ still converges, to $G^* \in \mathcal{G}$. For example, in the context of a Cox regression model, the estimate $\hat{G}$ converges to $G^* \in \mathcal{G}$ which has an interpretation as the least-false parameter (Hjort, 1992; Gerds and Schumacher, 2001). The effects of misspecification of the working model for G on the estimation of the binomial regression parameters are investigated in simulated data in Section 11.7.

11.3.3 Weighted estimating equations

Consider right-censored data from n independent individuals. At a fixed time horizon t, the Z-estimator $\hat{\beta}(t)$ solves the estimating equation, $U_n(\beta_0(t), \beta(t), \hat{G}) = 0$, corresponding to model (11.1) where

$$U_n(\beta_0(t), \beta(t), \hat{G}) = \sum_{i=1}^{n} \omega(t, X_i) \left\{ \frac{\Delta_i N_{\epsilon,i}(t)}{\hat{G}(T_i|X_i)} - h(\beta_0(t), \beta(t), X_i) \right\}. \tag{11.5}$$

The weights $\omega(t, X_i)$ may be p-dimensional and can depend on the model and on the link function h. For example Scheike et al. (2008) considered weights that depend on the directional derivatives $\partial h(\beta_0(t), \beta(t), X)/\partial \beta(t)$ of the current model.

Estimating equations for time-constant regression coefficients $\bar{\beta}$ are obtained relative to a sequence of time horizons $(t_1, \ldots, t_J)$. The estimates are solutions, $\bar{U}_n(\beta_0(t), \bar{\beta}, \hat{G}) = 0$ where

$$\bar{U}_n(\beta_0(t), \bar{\beta}, \hat{G}) = \sum_{i=1}^{n} \sum_{j=1}^{J} \omega(t_j, X_i) \left\{ \frac{\Delta_i N_{\epsilon,i}(t_j)}{\hat{G}(T_i|X_i)} - h(\beta_0(t_j), \bar{\beta}, X_i) \right\}. \tag{11.6}$$

Note that if the weights do not depend on the prediction horizon, then the information from all time points will contribute equally no matter for example the number of subjects at risk. This will not be efficient. However, it is difficult to derive explicit formulae for efficient weights. In Equation (11.6) the weights can also be used to model the correlation structure of data from the same subject obtained at the different time horizons.

11.4 Variance estimation

11.4.1 Asymptotic variance estimate

A natural way to estimate the variance of the Z-estimators of the regression coefficients is to estimate an asymptotic expression for the variance. To motivate this, it is instructive to first review the situation where all data are uncensored. In this case the data of subject i are $Y_i = (T_i, \epsilon_i, X_i)$. Denote P for the joint distribution of the vector Y_i, and $\psi(Y_i, \beta_0(t), \beta(t)) = \{N_{\epsilon,i}(t) - h(\beta_0(t), \beta(t), X_i)\}$ for the criterion function which defines the Z-estimator. The influence function of the Z-estimate $(\hat{\beta}_0, \hat{\beta}(t))$ at Y_i is given by (see page 47 in Huber and

Ronchetti, 2009):

$$IC_\beta(Y_i, \mathrm{P}, t) = \frac{\psi(Y_i, \beta_0(t), \beta(t))}{-\int \frac{\partial}{\partial \beta(t)} \psi(y, \beta_0(t), \beta(t)) \mathrm{P}(dy)}. \tag{11.7}$$

Furthermore, under the usual regularity conditions, which include differentiability of $(\beta_0(t), \beta(t)) \mapsto \psi(y, \beta_0(t), \beta(t))$ and finite second moments of $\psi(Y, \beta_0(t), \beta(t))$, the following von Mises expansion holds (Van der Vaart, 1998, Section 20):

$$\sqrt{n}(\hat{\beta}(t) - \beta(t)) = \frac{1}{\sqrt{n}} \sum_{i=1}^{n} IC_\beta(Y_i, \mathrm{P}, t) + o_{\mathrm{P}}(1). \tag{11.8}$$

This implies that $\sqrt{n}(\hat{\beta}(t) - \beta(t))$ is asymptotically normal with mean zero and asymptotic variance:

$$\mathrm{Var}(\hat{\beta}(t)) = \int IC_\beta^2(Y, \mathrm{P}, t)\mathrm{P}(dy).$$

Thus, based on an estimate $\widehat{IC}_\beta$ of the influence function the asymptotic variance of the Z-estimator $\hat{\beta}$ can then be estimated by

$$\widehat{\mathrm{Var}}(\hat{\beta}(t)) = \frac{1}{n} \sum_{i=1}^{n} \widehat{IC}_\beta^2(y, \mathrm{P}, t).$$

For right-censored data, under the assumption that the censoring model is correctly specified (see Section 11.3.2), and under further regularity conditions that are detailed in Appendix 1 of Scheike et al. (2008) one proves that the von Mises expansion (11.8) holds for the solution of (11.5). The most important condition which has to be ensured is that the probability of not being censored at the prediction horizon is uniformly bounded away from zero. This can be achieved by not using late time horizons where few subjects are at risk, (see Section 11.3.1). The influence function of the Z-estimator which solves (11.5) is given by

$$IC_\beta(\tilde{Y}_i, \tilde{\mathrm{P}}) = \frac{\tilde{\psi}(Y_i, \beta_0, \beta) - \int \frac{\int IC_G(s, x; y, \tilde{P}) \tilde{n}(y, s)}{G(s - |X_i)} \tilde{\mathrm{P}}(dy)}{-\int \frac{\partial}{\partial \beta} \tilde{\psi}(y, \beta_0, \beta) \tilde{\mathrm{P}}(dy)}. \tag{11.9}$$

Here the criterion function uses the weighted response

$$\tilde{\psi}(Y_i, \beta_0, \beta) = \left\{ \frac{\Delta_i N_{\epsilon,i}(t)}{G(T_i - |X_i)} - h(\beta_0(t), \beta, X_i) \right\},$$

and $\tilde{P}$ is the joint distribution of the right-censored observation $\tilde{Y}_i = (\tilde{T}_i, \Delta_i \epsilon_i, X_i)$.

11.4.2 Bootstrap confidence limits

Another way to take the uncertainty of the estimation of the censoring distribution into account is to use bootstrap. Here we consider two different types of confidence limits for the regression coefficients, percentile bootstrap confidence limits and Wald type confidence intervals based on bootstrap standard errors. For other and more sophisticated constructions, for practical advice and theory on bootstrap confidence limits we refer to Efron and Tibshirani (1993); DiCiccio and Efron (1996); Davison and Hinkley (1997); and Carpenter and Bithell (2000). The present situation seems to be covered by recent results on bootstrap consistency for M-estimation in semiparametric models (Cheng and Huang, 2010).

To construct confidence limits, repeat the following algorithm B times, where B is a large number, such as 10,000.

Algorithm for estimating bootstrap confidence limits

1. Draw a bootstrap sample from the observed data
2. Fix a set of time points $t_1 <, \ldots, < t_J$ for estimation based on the event times in the bootstrap sample
3. Estimate the censoring model based on the bootstrap sample
4. Calculate the weighted outcome based on the estimated weights for each subject i at each time point t_j (see Section 11.5 for details)
5. Compute the estimate $\hat{\beta}_b^*$ by solving the estimating equation in (11.5) applied to the bootstrap sample.

Based on the empirical distribution of the B bootstrap estimates: $(\hat{\beta}_1^*, \ldots, \hat{\beta}_B^*)$, a percentile bootstrap confidence set is defined as

$$\left(\tau^*_{n(\alpha/2)}, \tau^*_{n(1-\alpha/2)}\right),$$

where $\tau^*_{n(\alpha)}$ is the α percentile, in the empirical distribution of the bootstrap estimates. That is, α satisfies

$$P(\hat{\beta}^* \leq \tau^*_{n(\alpha)}) = \alpha .$$

Also a Wald test-based confidence interval is given by

$$(\hat{\beta} - z_q \cdot \widehat{\text{S.E}}(\hat{\beta}^*), \hat{\beta} + z_q \cdot \widehat{\text{S.E}}(\hat{\beta}^*)) , \tag{11.10}$$

where z_q is the q quantile of the standard normal distribution. The Wald confidence interval is calculated based on the empirical standard error of the bootstrap estimates:

$$\widehat{\text{S.E}}(\hat{\beta}^*) = \left(\frac{1}{B}\sum_{b=1}^{B}(\bar{\hat{\beta}}^* - \hat{\beta}^*)^2\right)^{1/2} , \tag{11.11}$$

where

$$\bar{\hat{\beta}}^* = \frac{1}{B}\sum_{b=1}^{B}\hat{\beta}^* .$$

11.5 Software implementation

The regression coefficients of the binomial regression model (11.1) can be estimated by solving the weighted estimating Equation (11.5). This can be implemented in most statistical software packages.

The first step is to transform the data from the format in Table 11.1 into the stacked format shown by the first 5 columns of Table 11.2. The second step is to specify and estimate a model for the conditional censoring distribution. To get individual weights, the estimate $\hat{G}$ needs to be evaluated "just before" the subject specific event time and possibly given covariates. In Table 11.1 this is illustrated for three time points t_1, t_2, t_3. Column 6 shows time point specific censoring weights evaluated at the subject specific event times and

TABLE 11.1
Unprepared competing risks data of n subjects.

id	Observed time	Event status	Covariates
1	$\tilde{T}_1$	$\tilde{\epsilon}_1$	x_1
2	$\tilde{T}_2$	$\tilde{\epsilon}_2$	x_2
$\vdots$	$\vdots$	$\vdots$	$\vdots$
n	$\tilde{T}_n$	$\tilde{\epsilon}_n$	x_n

TABLE 11.2
Stacked data format for estimation of binomial regression models.

id	Time grid	Observed time	Event status	Covariates	Weights	Weighted outcome
1	t_1	$\tilde{T}_1$	$\tilde{\epsilon}_1$	x_1	$\hat{G}_{11}$	$\tilde{N}_{\epsilon,1}(t_1)$
1	t_2	$\tilde{T}_1$	$\tilde{\epsilon}_1$	x_1	$\hat{G}_{21}$	$\tilde{N}_{\epsilon,1}(t_2)$
1	t_3	$\tilde{T}_1$	$\tilde{\epsilon}_1$	x_1	$\hat{G}_{31}$	$\tilde{N}_{\epsilon,1}(t_3)$
2	t_1	$\tilde{T}_2$	$\tilde{\epsilon}_2$	x_2	$\hat{G}_{12}$	$\tilde{N}_{\epsilon,2}(t_1)$
2	t_2	$\tilde{T}_2$	$\tilde{\epsilon}_2$	x_2	$\hat{G}_{22}$	$\tilde{N}_{\epsilon,2}(t_2)$
2	t_3	$\tilde{T}_2$	$\tilde{\epsilon}_2$	x_2	$\hat{G}_{32}$	$\tilde{N}_{\epsilon,2}(t_3)$
$\vdots$	$\vdots$	$\vdots$	$\vdots$	$\vdots$	$\vdots$	$\vdots$
n	t_1	$\tilde{T}_n$	$\tilde{\epsilon}_n$	x_n	$\hat{G}_{1n}$	$\tilde{N}_{\epsilon,n}(t_1)$
n	t_2	$\tilde{T}_n$	$\tilde{\epsilon}_n$	x_n	$\hat{G}_{2n}$	$\tilde{N}_{\epsilon,n}(t_2)$
n	t_3	$\tilde{T}_n$	$\tilde{\epsilon}_n$	x_n	$\hat{G}_{3n}$	$\tilde{N}_{\epsilon,n}(t_3)$

covariates $\hat{G}_{ji} = \hat{G}(\min(t_j, \tilde{T}_i-)|x_i)$. Finally, the weighted outcome $\tilde{N}_{\epsilon,i}$ for each subject at each of the time points (column 7) is a simple function of the other columns in the stacked dataset: $\tilde{N}_{\epsilon,i}(t_j) = \mathcal{I}\{\tilde{T}_i \le t_j, \tilde{\epsilon}_i = 1\}/\hat{G}_{ji}$.

To estimate time-varying regression coefficients based on the stacked data in Table 11.2, one includes an interaction term between the covariates and the vector of time points. To do this one specifies the time horizons as a factor or class variable with levels $(t_1, \ldots, t_J)$, in a model statement. Estimates of time constant regression coefficients are obtained by including the covariates and time points. For example, in R the estimation can be done using the function **geese** from the **geepack**-package.

```
geese(Weighted outcome~covariates+factor(time.grid),
      data=Table.8.2, id=Table.8.2$id,
      scale.fix=TRUE,
      family=gaussian,
      mean.link="cloglog",
      corstr="independence")
```

For example, to estimate the Cox model in survival or the Fine-Gray model in a competing risk model use the cloglog link function and choose `family=Gaussian` and `mean.link="cloglog"`. This produces estimating equations as the ones in Equation (11.6).

Time-varying coefficients can be obtained by replacing the formula with:

```
Weighted.outcome covariates*factor(time.grid).
```

In SAS the corresponding procedure is `proc genmod` where the model can be specified by a `model` statement. User defined link functions can be specified via the `fwdlink` and `invlink` statements as shown below where we again specify the cloglog link function.

```
proc genmod data=Table.8.2;
    class Time.grid id;
    ginv=1-exp(-exp(_xbeta_)); g=log(-log(1-_mean_));
    fwdlink link=g; invlink link=ginv;
    variance var=1;
    deviance dev= _resp_;
    model Weighted outcome = covariates Time.grid;
    repeated subject=id;
run;
```

As before time-varying coefficients can be obtained by replacing the model statement with:

```
Weighted.outcome=covariates*Time.grid
```

The standard errors obtained using **geese** or `proc genmod` do not reflect the statistical uncertainty incurred by the estimates of the censoring weights. To construct valid tests and confidence limits one possibility is to apply the bootstrap, see Section 11.4.2.

As far as we are aware the statistical software R (R Core Team, 2012) is the only package which provides routines to directly find the IPCW estimates of (11.5). This is implemented in the R packages **timereg** (Scheike and Zhang, 2011) and **riskRegression** (Gerds and Scheike, 2011). The function allows the user to specify a set of time points for estimation of time-constant coefficients, the link function and different ways to estimate the censoring weights. The electronic appendix of Gerds et al. (2012) describes the functionality and compares different link functions with respect to predictive performance.

11.6 Example

In this section we illustrate the binomial regression methods discussed in this chapter. For this purpose we use data which are described in Section 11.6.1 and freely available, e.g., from the online appendix of the book Andersen and Skovgaard (2010). We illustrate the effects of the link function and different choices of time points on the estimation of the regression coefficients and standard errors. We also compare the different bootstrap confidence intervals discussed in Section 11.4.2 with the Wald type interval which ignores the statistical uncertainty incurred by the censoring weights.

11.6.1 Melanoma data

The melanoma data include information on the survival time with malignant melanoma for 205 patients and were collected at Odense University Hospital by K.T. Drzewiecki. All the patients had their tumor removed by surgery in the period of 1962-1977 and were followed from day of surgery until death due to cancer or other causes, or end of study at December 31, 1977. At the end of study 57 patients had died from cancer and 14 patients

TABLE 11.3
Log absolute risk ratios in the melanoma data with three different confidence intervals.

	log(ARR)	95% confidence intervals		
	$\hat{\beta}$	Bootstrap percentile	Naive Wald	Bootstrap Wald
Sex: Male vs. Female	0.129	$(-0.326; 0.646)$	$(-0.283; 0.541)$	$(-0.360; 0.618)$
Ulceration: Present vs. Not present	0.785	$(0.237; 1.450)$	$(0.219; 1.350)$	$(0.166; 1.403)$
Log-thickness (1/100 mm)	0.323	$(0.092; 0.562)$	$(0.066; 0.580)$	$(0.083; 0.563)$

died from other causes. The remaining 134 patients were alive at the end of 1977 and their event time was right censored. Death due to cancer is the event of interest in the following.

11.6.2 Choice of link function

For the purpose of illustration we use a binomial regression model which includes only three variables into the linear predictor: sex, ulceration, (present/not present) and log-thickness, (operated tumor thickness in 1/100 mm). We first consider the log-link function. Table 11.3 shows corresponding estimates of time constant regression coefficients given as log-transformed absolute risk ratios (compare Section 11.2.2) with three different confidence intervals. The two bootstrap methods are described in Section 11.4.2. They take the variability of the estimates of the censoring distribution into account, as opposed to the naive Wald-type confidence limits which are based on the standard errors which assume the censoring weights are known. For the estimation we used all the event times in the dataset but discarded the first and last ten percent (155 time points).

For example the interpretation of the regression coefficient of ulceration is as follows. There is a 2.19 times ($\text{ARR}_{\text{Ulceration}} = \exp(\hat{\beta}_{\text{Ulceration}}) = \exp(0.785) = 2.19$) higher probability of dying from cancer during the next t days for a patient with ulceration than for a patient without ulceration, for fixed values of the other predictors.

When we change the link function to the logistic link, we obtain an odds ratio $\text{OR}_{\text{Ulceration}} = \exp(\hat{\beta}_{\text{Ulceration}}) = 2.92$. This means that the odds of dying of cancer is 2.92 times higher for patients with ulceration compared to patients without ulceration, keeping other variables fixed. Note in the competing risk model, the complementary probability of not experiencing the event includes both the risk of the competing causes and the chance of no event.

11.6.3 Effect of choice of time points

The regression coefficients in Table 11.3 were estimated based on all but the first and last ten percent of observed time points. Alternatively one could choose a smaller set of time points; for example, the set of event times at which the cause of interest occurred, deciles of the empirical distribution of the observed event times, or only a single time

TABLE 11.4
Effect of different choices of time horizons on the parameter estimates in the melanoma data.

	1 time point		3 time points		10 time points	
	$\hat{\beta}$	SE($\hat{\beta}$)	$\hat{\beta}$	SE($\hat{\beta}$)	$\hat{\beta}$	SE($\hat{\beta}$)
Sex: Male vs. Female	0.143	0.238	0.098	0.218	0.115	0.209
Ulceration: Present vs. Not present	0.845	0.348	0.733	0.295	0.797	0.291
Log-thickness (1/100 mm)	0.432	0.152	0.285	0.126	0.312	0.130

point. Andersen and Klein (2007) discusses how and how many time points are needed for estimation in a multi-state model using pseudo-values. They show by simulation that there is little advantage gained by introducing a large number of time points in the estimation. In their data example they use 10 time points equally spaced on the event time scale for the estimation.

Table 11.4 shows estimates of the log-absolute risk ratios based on 1, 3, and 10 time points, respectively, for the same model. The time points are chosen based on quantiles of the event time distribution in the melanoma dataset. Considered are three different sets of time points: the 50% quantile (1 time point), the 25, 50 and 75% quantiles (3 time points) and 10 quantiles equally spaced between the 10% and the 90% quantiles (10 time points). The results clearly demonstrate that there is a considerable small sample effect, as the estimates differ from each other and from the estimates based on the 155 time points in Table 11.3.

The estimated standard errors are also affected by the choice of time points. The standard errors generally become smaller as more time points are introduced in the estimation.

The selection of the set of time points should also account for the size of the estimated censoring weights $\hat{G}(\min(t, T_i-)|X_i)$. Too small weights even for only few subjects will have large effects on the estimates of the regression coefficients and lead to unstable results. This may be controlled by first locating the subjects for whom the observed event times return very small weights. Then the set of time horizons for estimation can be restricted such that the very small weights do not enter the estimating equations.

11.6.4 Compare confidence limits

Three different confidence limits are reported for each of the three estimates of the log absolute risk ratio in Table 11.3. We used 1,000 bootstrap replications and the Kaplan-Meier estimator in the estimation of the weights. To compare the naive Wald type confidence limits and the bootstrap based confidence limits we use all but the first and last ten percent of the unique observed event times in each bootstrap replication. The three methods give rise to the same overall conclusions of significance.

11.7 Simulations

The simulation study in Scheike et al. (2008) showed that the estimates of the regression coefficients had low small sample bias, when both the binomial regression model and the models for the censoring were correctly specified. Also the empirical variance of the $\hat{\beta}$ and the mean variance of the standard errors $\hat{\beta}$ were comparable to those obtained with the estimating technique of Fine and Gray (1999).

In this section we investigate the magnitude of the bias of the estimate of the regression parameter β and the coverage probability when the model for the censoring distribution is misspecified.

11.7.1 Competing risks model

Data are simulated from a Fine and Gray regression model based on the indirect simulation approach (see Fine, 1999; Beyersmann et al., 1993). It is assumed that the cumulative incidence function for cause 1 is given by

$$P(T_i \leq t, \epsilon_i = 1|X_i = (x_{i1}, x_{i2})) = 1 - (1 - p(1 - \exp(-t)))^{\exp(\beta x)} \tag{11.12}$$

For cause 2 the cumulative incidence function is assumed to be

$$P(T_i \leq t, \epsilon_i = 2|X_i = (x_{i1}, x_{i2})) = (1 - p)^{\exp(\beta x)}(1 - \exp(-t\exp(\beta x))\,. \tag{11.13}$$

We set $p = 0.66$ in all the simulations. The probability of a failure of cause 1 given the observed covariates is $p_1 = 1 - (1 - 0.66)^{\exp(\beta x)}$ and the cause of failure is determined by a coin toss with success probability p_1. The covariates $X = (X_1, X_2)$ are independent standard normal distributed and the relationship $\beta = (\beta_1, \beta_2)$ between X and T is constant over time.

11.7.2 Misspecified censoring model

We introduce conditionally independent right censoring as follows. Censoring times are simulated from a Cox-Weibull regression model by using the simulated values of the covariate matrix X. The shape parameter is set to 1 and the scale parameter to 1/2.

For the estimation of the censoring weights we consider two different ways of misspecifying the censoring distribution G:

1. Omitting covariate(s): The censoring time C depends on two covariates (X_1, X_2) which are also affecting the cumulative incidences. Estimation of the censoring weights is based either on the marginal Kaplan-Meier estimator which omits covariates, or on a misspecified Cox-model ($\text{Cox}_{\text{mis.}}$) which includes only X_1 as a predictor and omits X_2.
2. Wrong functional form: The censoring time C depends on (X_1, X_2, X_2^2), where (X_1, X_2) are also affecting the cumulative incidences. The estimation of the censoring weights is based either on the marginal Kaplan-Meier estimator, which omits covariates, or on a misspecified Cox-model ($\text{Cox}_{\text{mis.}}$) which includes only (X_1, X_2) and thus misspecifies the functional form for the association between the variable X_2 and the censoring hazard function.

The relationship between the cumulative incidences and X_2 is fixed by setting $\beta_2 = 0.5$ in all simulations. By η_{x_1}, η_{x_2} and η_{x^2} we denote the regression parameter values for the

TABLE 11.5
Summary of simulation results across 1,000 simulated datasets (n=500). Shown are Bias, MSE and Coverage Probabilities for estimates of β_1 and fixed $\beta_2 = 0.5$.

Parameters				Bias			MSE			Coverage Prob.		
				Estimates of β_1								
β_1	η_{x_1}	η_{x_2}	$\eta_{x_2^2}$	KM	Cox$_{mis.}$	Cox	KM	Cox$_{mis.}$	Cox	KM	Cox$_{mis.}$	Cox
0.3	0.0	0.0		0.007	0.008	0.008	0.008	0.007	0.007	0.945	0.958	0.957
0.5	0.0	0.0		0.008	0.008	0.009	0.009	0.009	0.009	0.948	0.953	0.952
0.3	0.1	0.1		−0.022	0.004	0.007	0.008	0.007	0.007	0.936	0.957	0.953
0.5	0.1	0.1		−0.026	0.000	0.004	0.009	0.008	0.008	0.930	0.947	0.949
0.3	0.3	0.1		−0.080	−0.001	0.005	0.014	0.008	0.008	0.815	0.953	0.956
0.5	0.3	0.1		−0.092	−0.003	0.004	0.016	0.008	0.008	0.800	0.963	0.965
0.3	0.1	0.3		−0.027	−0.003	0.006	0.008	0.007	0.007	0.921	0.962	0.960
0.5	0.1	0.3		−0.028	−0.004	0.009	0.009	0.008	0.009	0.927	0.949	0.950
0.3	0.3	0.3		−0.087	−0.011	0.005	0.015	0.008	0.008	0.813	0.951	0.959
0.5	0.3	0.3		−0.096	−0.015	0.005	0.017	0.008	0.008	0.792	0.959	0.969
0.3	0.1	0.1	0.3	−0.027	0.003	0.005	0.010	0.009	0.010	0.929	0.945	0.946
0.5	0.1	0.1	0.3	−0.029	0.004	0.006	0.010	0.010	0.010	0.936	0.955	0.956
0.3	0.3	0.1	0.3	−0.100	−0.003	−0.001	0.018	0.009	0.010	0.783	0.959	0.963
0.5	0.3	0.1	0.3	−0.105	0.002	0.003	0.019	0.009	0.011	0.770	0.969	0.973
0.3	0.1	0.3	0.3	−0.029	0.006	0.006	0.009	0.008	0.013	0.943	0.952	0.952
0.5	0.1	0.3	0.3	−0.035	0.006	0.008	0.011	0.010	0.012	0.904	0.953	0.957
0.3	0.3	0.3	0.3	−0.108	−0.007	−0.008	0.020	0.009	0.011	0.764	0.953	0.955
0.5	0.3	0.3	0.3	−0.116	0.001	0.000	0.022	0.010	0.012	0.716	0.961	0.958

effects on the censoring time hazard of the variables X_1, X_2 and X_2^2, respectively. Note that $\eta_{x_1} = 0$ means no correlation between C and X_1.

Table 11.5 summarizes the simulation results across 1,000 simulated datasets, each of size n=500. We report the bias, mean squared error (MSE) and coverage probabilities for the estimation of $\hat{\beta}_1$ subject to the two different misspecified models for the censoring distribution, as described above. Note that the model for the binomial regression model which describes the relationship between X and the cumulative incidence function is correctly specified. To estimate $\hat{\beta}$ we solved the estimating equations based on three time points: the 25, 50 and 75% quantiles of the event time distribution. For the coverage probabilities we applied the Wald-based confidence limits based on the naive standard error. The calculations are repeated for different choices of η_{x_1}, η_{x_2} and $\eta_{x_2^2}$.

The columns KM and Cox$_{\text{mis.}}$ in Table 11.5 refer to the Kaplan-Meier and misspecified Cox models for the estimation of the censoring weights, respectively. The columns denoted by Cox show results using the correctly specified model for the censoring weights. The first two rows in the table show results for ($\eta_{x_1} = \eta_{x_2} = 0$) corresponding to censoring times being independent of the covariates X_1 and X_2. Here we find that the coverage is better when the censoring weights are based on either of the Cox models as compared to the Kaplan-Meier estimator. This confirms that even though censoring is independent of the observed covariates the IPCW estimate which uses the Kaplan-Meier for the censoring model is not fully efficient. The upper half of Table 11.5 summarizes results from the setting where the misspecified censoring models omit one covariate ($\text{Cox}_{mis.}$) or all covariates (KM). The lower half of Table 11.5 summarizes results from the misspecified censoring model with a wrong functional form of the covariate X_2. Overall, the results in Table 11.5 indicate that the bias in $\hat{\beta}_1$ becomes smaller and the coverage probabilities higher, when the weights are based on the misspecified Cox models compared to when they are based on the marginal Kaplan-Meier. Similar results were obtained for the estimate $\hat{\beta}_2$ (results not shown).

11.7.3 Compare confidence limits

To compare the performance of the three different confidence intervals discussed in Section 11.4.2 we calculate the corresponding coverage probabilities in simulated data. For this purpose, we simulate data with different sample sizes $n = (250, 300, 500)$ and only one covariate X_1. The true relationship between X_1 and the cumulative incidence is denoted by $\beta_1 = 0.5$. We introduce independent right censoring based on the Cox-Weibull model with shape parameter 1 and scale parameter 1/2.

Based on 1,000 replications Table 11.6 shows the coverage probability for the bootstrap and the naive Wald type confidence limits for $\hat{\beta}_1$ for different values of β_1. For the estimation we used the following three time points: the 25%, 50% and 75% quantiles of the event time distribution in each bootstrap sample. The bootstrap confidence limits in Table 11.6 are estimated based on 1,000 bootstrap replications and the IPCW weights are estimated with the marginal Kaplan-Meier estimator.

Within the limitations of this simulation study we conclude that the bootstrap Wald confidence interval and the naive Wald confidence interval have satisfactory and comparable coverage probabilities. The coverage for the bootstrap quantile interval seems to be slightly

TABLE 11.6
Coverage probabilities for three different confidence intervals for the estimate of $\hat{\beta}$ (1,000 simulations).

Parameters		Coverage probabilities		
β	n	Wald Bootstrap	Bootstrap Quantile	Naive Wald
0.1	250	0.962	0.938	0.961
0.3	250	0.964	0.939	0.957
0.5	250	0.964	0.933	0.954
0.1	300	0.951	0.933	0.943
0.3	300	0.958	0.934	0.949
0.5	300	0.956	0.933	0.942
0.1	500	0.959	0.944	0.948
0.3	500	0.938	0.925	0.933
0.5	500	0.961	0.947	0.961

lower than that of the Wald-based intervals. This small difference between the quantile and Wald-based intervals may depend on the sample size.

11.7.4 Effect of choice of time points

We simulate data as described in 11.7.1 using two covariates $X = (X_1, X_2)$ varying log-hazard ratios $\beta = (\beta_{x_1}, \beta_{x_2})$ to see how the bias and MSE of $\hat{\beta}_1$ changes dependent of the selection of time points for estimation. η_{x_1} is the regression coefficient which characterizes the relationship between the X_1 and the censoring time. If $\eta_{x_1} = 0$ then there is no dependence between X_1 and C. For the censoring weights we use the marginal Kaplan-Meier estimator. Note that this model for the weights is misspecified in the case where $\eta_{x_1} \neq 0$.

For the estimation of the parameters β_{x_1} and β_{x_2} we use three different sets of time points based on the observed event time distribution in each of the simulated datasets:

1 The median of the event time distribution

10 10 quantiles from the event time distribution equally spaced between the 10% and the 90% quantile.

ALL All the observed times, discarding the first and last 10 percent.

Table 11.7 summarizes the bias and the mean squared error (MSE) of the estimates of β_{x_1} across 1,000 simulated datasets with sample size $n = 500$. There is a gain in efficiency if 10 time points are used compared to 1 time point when there is no misspecification in the IPCW. If we introduce dependence between censoring time C and covariate X_1 the bias

TABLE 11.7
Effect of the choice of time points on Bias and MSE for estimates of $\hat{\beta}_1$ (1,000 simulations).

Parameters			Bias			MSE		
			Number of time points					
			1	10	All	1	10	All
β_{x_1}	β_{x_2}	η_{x_1}		$\hat{\beta}_1$			$\hat{\beta}_1$	
0.3	0.1	0.0	0.001	0.001	0.001	0.030	0.026	0.026
0.5	0.1	0.0	0.005	0.004	0.004	0.093	0.089	0.089
0.3	0.3	0.0	0.009	0.003	0.004	0.010	0.006	0.006
0.5	0.3	0.0	0.009	0.004	0.004	0.032	0.027	0.027
0.3	0.1	0.1	−0.007	−0.030	−0.025	0.028	0.020	0.021
0.5	0.1	0.1	−0.011	−0.033	−0.028	0.085	0.076	0.077
0.3	0.3	0.1	−0.015	−0.032	−0.027	0.010	0.007	0.007
0.5	0.3	0.1	−0.005	−0.030	−0.025	0.028	0.021	0.022
0.3	0.1	0.3	−0.043	−0.107	−0.092	0.023	0.015	0.015
0.5	0.1	0.3	−0.045	−0.107	−0.090	0.075	0.057	0.060
0.3	0.3	0.3	−0.044	−0.106	−0.091	0.012	0.018	0.015
0.5	0.3	0.3	−0.057	−0.119	−0.103	0.020	0.016	0.016

seems to be higher if we use 10 or all time points compared to one time point. However in all settings the MSE was smaller for 10 or all time points compared to 1 time point.

11.8 Final remarks

Binomial regression is a flexible modeling approach which can be extended to complex multi-state models. Binomial regression yields direct models for covariate effects on the probability scale. Different interpretations can be obtained by changing the link function. A challenge is that one needs a (correctly) specified model for the censoring distribution. Also the choice of time points can influence the resulting estimates substantially, in particular if the β coefficient is in fact time varying. Then the estimate which combines information across different time points is expected to be variable and sensitive to the choice of time points.

Bibliography

Aalen, O. O. (1989), 'A linear regression model for the analysis of life times', *Statistics in Medicine* **8**(8), 907–925.

Ambrogi, F., Biganzoli, E. and Boracchi, P. (2008), 'Estimates of clinically useful measures in competing risks survival analysis', *Statistics in Medicine* **27**(30), 6407–6425.

Andersen, P. K., Borgan, Ø., Gill, R. D. and Keiding, N. (1993), *Statistical Models Based on Counting Processes*, Springer Series in Statistics, Springer, New York.

Andersen, P. K. and Klein, J. P. (2007), 'Regression analysis for multistate models based on a pseudo-value approach, with applications to bone marrow transplantation studies', *Scandinavian Journal of Statistics* **34**, 3–16.

Andersen, P. K. and Skovgaard, L. T. (2010), *Regression with Linear Predictors*, Statistics for Biology and Health, Springer, New York.

Andersen, P., Klein, J. and Rosthøj, S. (2003), 'Generalised linear models for correlated pseudo-observations, with applications to multi-state models', *Biometrika* **90**(1), 15–27.

Bennett, S. (1983), 'Analysis of survival data by the proportional odds model', *Statistics in Medicine* **2**(2), 273–277.

Berkson, J. (1944), 'Application of the logistic function to bio-assay', *Journal of the American Statistical Association* **39**, 357–365.

Beyersmann, J., Schumacher, M. and Allignol, A. (1993), *Competing Risks and Multistate Models with R*, Use R!, Springer, New York.

Blizzard, L. and Hosmer, D. (2006), 'Parameter estimation and goodness-of-fit in log binomial regression', *Biometrical Journal* **48**(1), 5–22.

Carpenter, J. and Bithell, J. (2000), 'Bootstrap confidence intervals: when, which, what? A practical guide for medical statisticians', *Statistics in Medicine* **19**(9), 1141–1164.

Cheng, G. and Huang, J. Z. (2010), 'Bootstrap consistency for general semiparametric m-estimation', *The Annals of Statistics* **38**, 2884–2915.

Cox, D. R. (1972), 'Regression models and life tables', *Journal of the Royal Statistical Society* **B 34**, 187–220.

Cox, D. R. (1975), 'Partial likelihood', *Biometrika* **62**, 269–276.

Davison, A. C. and Hinkley, D. V. (1997), *Bootstrap Methods and Their Applications*, Cambridge University Press, New York.

DiCiccio, T. and Efron, B. (1996), 'Bootstrap confidence intervals', *Statistical Science* pp. 189–212.

Doksum, K. A. and Gasko, M. (1990), 'On a correspondance between models in binary regression and in survival analysis', *International Statistical Review* **58**, 243–252.

Efron, B. and Tibshirani, R. J. (1993), *An Introduction to the Bootstrap*, Chapman & Hall/CRC, Florida.

Fine, J. (1999), 'Analysing competing risks data with transformation models', *J. R. Statist. Soc. B* **61**, 817–830.

Fine, J. and Gray, R. (1999), 'A proportional hazards model for the subdistribution of a competing risk', *Journal of the American Statictical Association* **94**, 496–509.

Fine, J., Ying, Z. and Wei, L. (1998), 'On the linear transformation model for censored data', *Biometrika* **85**, 980–986.

Gerds, T. A. and Scheike, T. H. (2011), *riskRegression: Risk regression for survival analysis.* R package version 0.0.5. **URL:** *http://CRAN.R-project.org/package=riskRegression*

Gerds, T. A., Scheike, T. H. and Andersen, P. K. (2012), 'Absolute risk regression for competing risks: interpretation, link functions, and prediction', *Statistics in Medicine* **31**(29), 3921–3930.

Gerds, T. and Schumacher, M. (2001), 'On functional misspecification of covariates in the cox regression model', *Biometrika* **88**, 572–580.

Hjort, N. L. (1992), 'On inference in parametric survival models', *International Statistical Review* **60**, 355–387.

Huber, P. J. and Ronchetti, E. M. (2009), *Robust Statistics*, Wiley Series in Probability and Statistics, New Jersey.

Jewell, N. P. (2005), 'Correspondance between regression models for complex binary outcome and those for structured multivariate survival analysis', *U.C. Berkeley Division of Biostatistics Working Paper Series* **195**.

Jewell, N. and van der Laan, M. (1997), Singly and doubly censored current status data with extensions to multi-state counting processes, *in* D.-Y. Lin, ed., 'Proceedings of First Seattle Conference in Biostatistics', Springer Verlag, pp. 171–84.

Jewell, N. and van der Laan, M. (2004), 'Current status data: review, recent developments and open problems', *Advances in Survival Analysis* **23**, 625–642.

Klein, J. P. and Andersen, P. K. (2005), 'Regression modeling of competing risks data based on pseudovalues of the cumulative incidence function', *Biometrics* **61**(1), 223–229.

Lin, D., Oakes, D. and Ying, Z. (1998), 'Additive hazards regression with current status data', *Biometrika* **85**(2), 289–298.

Marschner, I. and Gillett, A. (2012), 'Relative risk regression: reliable and flexible methods for log-binomial models', *Biostatistics* **13**(1), 179–192.

Martinussen, T. and Scheike, T. (2002), 'Efficient estimation in additive hazards regression with current status data', *Biometrika* **89**(3), 649–658.

Martinussen, T. and Scheike, T. (2006), *Dynamic Regression Models for Survival Data*, Springer.

R Core Team (2012), *R: A Language and Environment for Statistical Computing*, R Foundation for Statistical Computing, Vienna, Austria. ISBN 3-900051-07-0.
URL: *http://www.R-project.org/*

Rossini, A. and Tsiatis, A. (1996), 'A semiparametric proportional odds regression model for the analysis of current status data', *Journal of the American Statistical Association* **91**, 713–721.

Scheike, T. H. and Zhang, M.-J. (2002), 'An additive-multiplicative Cox-Aalen regression model', *Scandinavian Journal of Statistics* **29**(1), 75–88.

Scheike, T. H. and Zhang, M.-J. (2003), 'Extensions and applications of the Cox-Aalen survival model', *Biometrics* **59**(4), 1036–1045.

Scheike, T. H. and Zhang, M.-J. (2007), 'Direct modeling of regression effects for transition probabilities in multistate models', *Scandinavian Journal of Statistics* **34**, 17–32.

Scheike, T. H., Zhang, M.-J. and Gerds, T. A. (2008), 'Predicting cumulative incidence probability by direct binomial regression', *Biometrika* **95**, 205–220.

Scheike, T. and Zhang, M. (2011), 'Analyzing competing risk data using the R timereg package', *Journal of Statistical Software* **38**(2), 1–15.

Shiboski, S. C. (1998), 'Generalized additive models for current status data', *Lifetime Data Analysis* **4**, 29–50.

Sun, J. (2006), *The Statistical Analysis of Interval-Censored Failure Time Data*, Springer, New York.

van der Laan, M. J. and Robins, J. M. (2003), *Unified Methods for Censored Longitudinal Data and Causality*, Springer, New York.

van der Laan, M. and Robins, J. (1998), 'Locally efficient estimation with current status data and time-dependent covariates', *Journal of the American Statistical Association* **93**(442), 693–701.

Van der Vaart, A. W. (1998), *Asymptotic Statistics*, Cambridge University Press, New York.

van Houwelingen, H. (2006), 'Dynamic prediction by landmarking in event history analysis', *Scandinavian Journal of Statistics* **34**(1), 70–85.

van Houwelingen, H. and Putter, H. (2008), 'Dynamic predicting by landmarking as an alternative for multi-state modeling: an application to acute lymphoid leukemia data', *Lifetime Data Analysis* **14**(4), 447–463.

van Houwelingen, J. and Putter, H. (2012), *Dynamic Prediction in Clinical Survival Analysis*, Chapman & Hall/CRC, Florida.

Wacholder, S. (1985), 'Binomial regression in GLIM: estimating risk ratios and risk differences', *American Journal of Epidemiology* **123**, 174–184.

Zhang, M.-J. and Fine, J. (2008), 'Summarizing differences in cumulative incidence functions', *Stat Med* **27**(24), 4939–4949.

12

Regression Models in Bone Marrow Transplantation – A Case Study

Mei-Jie Zhang
Division of Biostatistics, Medical College of Wisconsin

Marcelo C. Pasquini
Division of Hematology and Oncology, Medical College of Wisconsin

Kwang Woo Ahn
Division of Biostatistics, Medical College of Wisconsin

CONTENTS

12.1 Introduction 243
12.2 Data 245
12.3 Survival analysis 245
12.3.1 Fitting Cox proportional hazards model 245
12.3.2 Adjusted survival curves based on a Cox regression model 248
12.4 Competing risks data analysis 252
12.4.1 Common approaches for analyzing competing risks data 252
12.4.2 Adjusted cumulative incidence curves based on a stratified Fine-Gray model 258
12.5 Summary 260
Bibliography 260

12.1 Introduction

Hematopoietic stem cell transplantation (HSCT) is a life-saving procedure for many cancer patients. It has been widely used for treating malignant and non-malignant diseases. Since the first successful transplantation using bone marrow from a human leukocyte antigen (HLA) identical sibling in 1968, more than 800,000 patients have received HSCT worldwide with an estimated annual number of transplantations around 60,000 currently (Bach et al., 1968; Gatti et al., 1968; Eapen and Rocha, 2008). The main reasons for the wide increase in HSCT are its demonstrated efficacy in many diseases, increased donor availability due in part to using stem cells from umbilical cord blood, increased use of peripheral blood stem cells, and improved transplant outcomes. However, HSCT also has severe side effects including graft failure and graft-versus-host disease complications. These complications are major causes for transplant-related death. Patients and transplant physicians are interested in knowing survival outcomes after HSCT and are interested in comparing outcomes between treatments. The main outcome events after HSCT are overall mortality and treatment failure which is defined as death or disease recurrence. Treatment failure is the complement of disease-free survival. Other outcomes of interest are engraftment, acute and chronic graft-versus-host-disease (GVHD), treatment-related mortality (TRM) which is defined as death without cancer relapse or progression, and cancer relapse or progression. Some of these events are competing risks, where a patient may fail due to one of the k causes, and the occurrence of one of these events precludes us from observing the other events. For example, death before developing acute GVHD precludes patients from getting acute GVHD. Thus, death pre-acute GVHD is considered a competing risk for acute GVHD. Another common competing risks in HSCT studies are TRM and cancer relapse or progression.

In HSCT studies researchers and patients often want to know the effect of treatment- and patient-related risk factors on the outcome, and to compare outcomes between treatment groups using existing non-clinical trial data. Regression techniques can be used to address these important questions. In addition, most HSCT studies involve censored observations, which results in observing only partial information on patients. The most common type of censoring is right censoring. Right censoring occurs when patients are lost to follow-up before the event occurred. Thus, it is unknown when the event will occur.

For right-censored survival data, one of the most widely used statistical regression models is the Cox proportional hazards regression model (Cox, 1972). The Cox model estimates the hazard rate as a function of risk factors. Cox proposed a partial likelihood-estimating approach. Andersen and Gill (1982) gave a detailed theoretical discussion of these techniques. Cox models can be fit using common statistical packages, such as SAS, SPSS, STATA, R and others. Recently, some alternative models have been developed, studied and applied to HSCT studies. Commonly used alternative models include Aalen's additive model (Aalen, 1989), partly parametric additive risk models (McKeague and Sasieni, 1994; Lin and Ying, 1995), and the flexible additive-multiplicative Cox-Aalen model (Scheike and Zhang, 2002, 2003). In this chapter we will focus on fitting a Cox model using a real HSCT dataset.

For analyzing competing risks data, the standard approach is to model cause-specific hazards for all causes. The Cox proportional hazards model has been most commonly used to model the hazard functions for all causes (Prentice et al., 1978; Cheng et al., 1998). A specific additive risk model has been considered by Shen and Cheng (1999), and a flexible Cox-Aalen model has been proposed and studied by Scheike and Zhang (2003). Since the cumulative incidence function (CIF) of a particular type of failure is a function of all cause-specific hazards, this modeling approach requires all cause-specific hazards to be modeled correctly, and it may be hard to evaluate the covariate effect on the cumulative incidence function directly and hard to identify which specific risk factor has an effect on the cumulative incidence function that changes over time. Recently, some new regression approaches have been proposed to model the CIF directly. Fine and Gray (1999) proposed a proportional regression model for the subdistribution hazard function which is based on earlier work by Gray (1988) and Pepe (1991). This approach is implemented in the **crr** function in the **cmprsk** package for R; see Gray (2013) for details. Since there is a direct relationship between the subdistribution hazard function and the CIF, one can directly interpret the covariate effect on the CIF based on the covariate effect on the subdistribution hazard function. Other flexible and more general models for the subdistribution hazard function have been proposed and studied (Sun et al., 2006). A second approach to direct modeling of the CIF is based on pseudo-values from a jackknife technique using nonparametric estimated CIF at some pre-fixed time points (Klein and Andersen, 2005; Klein, 2006). An R function and a SAS macro are also available for the pseudo-value approach (Klein et al., 2008). A third approach to direct modeling of the CIF is based on binomial regression modeling using inverse probability censoring weighting technique. A fully nonparametric regression model and a class of flexible and general semiparametric regression models have been proposed and studied (Scheike et al., 2008; Scheike and Zhang, 2008). An R package (**timereg**) has been developed for the direct binomial modeling approach (Scheike and Zhang, 2011).

In this chapter we will review some of the regression techniques with their applications to real HSCT data. The HSCT dataset is described in Section 12.2. The Cox model and its application are discussed in Section 12.3. Section 12.4 studies various competing risk modeling techniques.

12.2 Data

The data used for illustration came from a study comparing allogeneic HSCT versus autologous HSCT for diffuse large B cell lymphoma (DLBCL) (Lazarus et al., 2010). DLBCL is a type of non-Hodgkin lymphoma. The study included 916 adult DLBCL patients, between the ages of 18 and 60 years from 156 centers in 17 different countries, receiving autologous or matched sibling allogeneic HSCT reported to the Center for International Blood and Marrow transplantation (CIBMTR) from 1995 to 2003. The CIBMTR is comprised of clinical and basic scientists who confidentially share data on their blood and bone marrow transplants with the CIBMTR Data Collection Center located at the Medical College of Wisconsin. The CIBMTR is a repository of information about results of transplants at more than 450 transplant centers worldwide. For the purpose of illustration, we consider two outcomes: overall mortality, in which event is death from any cause, and treatment-related mortality (TRM), in which an event is defined as death without lymphoma progression. TRM and disease progression (defined as lymphoma recurrence or progressive lymphoma) are two competing risks. The main objective of this study is to determine the treatment effect of allogeneic HSCT (N=79) versus autologous HSCT (N=837). Comparing outcomes between treatment groups using registration data requires adjustment for the differences in baseline characteristics of patients and transplant-related risk factors. The variables considered in the CIBMTR study (Lazarus et al., 2010) include the main treatment effect, patient age, Karnofsky performance score at transplant, patient gender, disease stage at diagnosis, chemosensitive disease at transplant, B symptoms at diagnosis, time from diagnosis to transplant, extranodal disease, marrow involvement at diagnosis, source of stem cells, year of transplant, donor-patient gender match, GVHD prophylaxis, donor-patient CMV status, and purging; see Table 1 of Lazarus et al. (2010) for detail. In this example we will only consider the risk factors that were found to be significantly associated with overall mortality and TRM, which consists of the main treatment effect (allogeneic HSCT versus autologous HSCT), patient age at transplant ($18-30$ (N=120) versus $31-50$ (N=468) versus $50-60$ (N=325)), chemosensitive disease at transplant (sensitive (N=752) versus resistant (N=161)), year of transplant (1995–2000 (N=174) versus 2001–2003 (N=739)), and Karnofsky performance score at transplant ($<90\%$ (N=560) versus $90-100\%$ (N=331)).

12.3 Survival analysis

12.3.1 Fitting Cox proportional hazards model

Regression techniques are often used in cancer research to identify patient- and treatment-related risk factors which are associated with the outcomes, and to compare the treatment effect adjusting for potential imbalance of risk factors between treatment groups. This is commonly done by modeling the hazard function since there is a close relationship between survival probability, $S(t;\boldsymbol{Z}) = P(T > t|\boldsymbol{Z})$ and the hazard function,

$\lambda(t;\boldsymbol{Z}) = \lim_{\Delta t \to 0} P(T \leq t + \Delta t | T > t, \boldsymbol{Z})/\Delta t$: $S(t;\boldsymbol{Z}) = \exp\left\{-\int_0^t \lambda(u;\boldsymbol{Z})du\right\}$. Cox (1972) proposed the proportional hazards model,

$$\lambda(t|\boldsymbol{Z}) = \lambda_0(t)\exp(\boldsymbol{\beta}^T\boldsymbol{Z}), \tag{12.1}$$

where $\lambda_0(t) \geq 0$ is an unknown baseline hazard function and $\boldsymbol{Z} = (Z_1, \ldots, Z_p)^T$ is a p-dimensional vector of covariates. The proportionality of hazards in the Cox model provides a direct interpretation, where the exponential of the regression coefficient, $\exp(\beta_k)$, represents the relative risk of having an event for each one unit increment of a specific risk factor, Z_k. Cox (1972) suggested estimating the regression coefficients by maximizing the partial likelihood function and estimating the cumulative baseline hazard function, $\Lambda_0(t) = \int_0^t \lambda_0(u)du$ by a Nelson-Aalen type estimator. Both estimators are consistent and asymptotic variances can be consistently estimated. Statistical procedures for fitting a Cox model are available in most statistical packages. The use of a Cox model requires validating some statistical assumptions. The most important assumption that needs to be examined is the proportionality of hazards. When the proportionality assumptions are not valid, it indicates that the effects of covariates are not constant, which means the effects change over time. The proportional hazards assumption can be checked via various methods. Klein and Moeschberger (2003) gave a detailed review of various model diagnostic tests. One of the most popular methods is adding a time-dependent variable, $\tilde{Z}(t) = Z \times g(t)$ to the Cox model, for a monotone function $g(t)$ and testing whether the coefficient of $\tilde{Z}(t)$ is significant. The most commonly used functions for $g(t)$ are $\log(t)$ and t. If the covariate Z satisfies the proportional hazards assumption, the coefficient of $\tilde{Z}(t)$ should be zero. Another common approach is the graphical method. A survey of various graphical methods can be found in Chapter 11 of Klein and Moeschberger (2003). One of the simplest ways is comparing log-log survival curves. Assume that a covariate Z has two levels. We fit a Cox model to the other covariates stratified on Z, $\lambda_{0k}(t)\exp(\boldsymbol{\beta}^T\boldsymbol{X})$ where $\boldsymbol{X}$ are the other covariates and $k\ (= 1, 2)$ represents the level of Z. Let $\hat{S}(t|Z_k)$ be the estimator for $\exp\{-\int_0^t \lambda_{0k}(u)du\}$. Under the proportional hazards assumption, $-\log\{-\log \hat{S}(t|Z_1)\}$ and $-\log\{-\log \hat{S}(t|Z_2)\}$ should be parallel to each other. See Chapter 4 of Kleinbaum and Klein (2005) for mathematical details.

Application to example HSCT data

We analyze the overall mortality for the lymphoma transplant data using a Cox regression model. First, we check the proportionality assumption for the main treatment effect of allogeneic HSCT versus autologous HSCT by adding a time-dependent covariate. The test indicated that transplant type had a significant time-varying effect; see Table 12.1. Figure 12.1 plots the complementary log-log of the survival functions by transplant type and it confirms that the proportionality assumption does not hold for the treatment effect.

When the proportionality assumption does not hold for some risk factors, we need to make some adjustments to the modeling. Two approaches are commonly used: (i) fit a

TABLE 12.1
Checking the proportionality assumption.

Covariate	$\hat{\beta}(\hat{\sigma})$	HR (95% CI)	P
Z_1 : transplant type	1.46 (0.20)	4.33 (2.94 - 6.37)	< 0.0001
$Z_2 = Z_1 \times \log(t)$	-0.43 (0.11)	0.65 (0.52 - 0.81)	0.0001

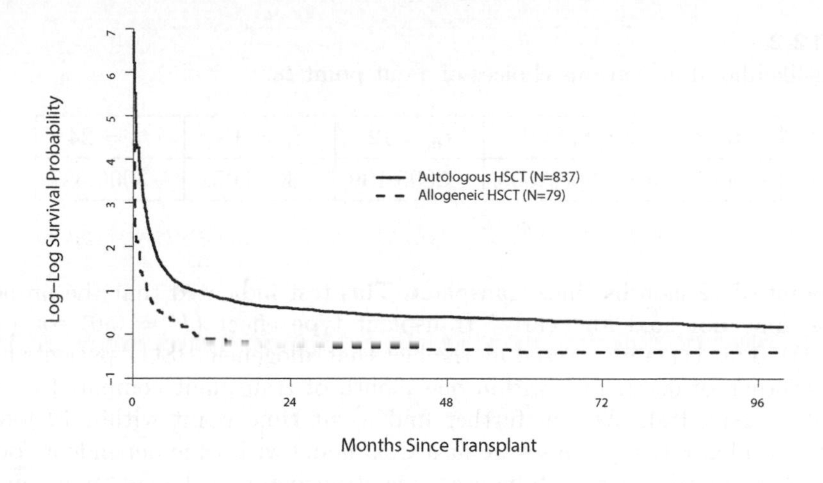

FIGURE 12.1
Log-log of survival plot.

stratified Cox model stratifying on these variables; and (ii) fit a time-varying effect Cox model, $\lambda_0(t)\exp\{\boldsymbol{\beta}(t)^T\boldsymbol{Z} + \boldsymbol{\gamma}^T\boldsymbol{X}\}$, where covariates $\boldsymbol{Z}$ have time-varying effects and $\boldsymbol{X}$ have constant effects.

To fit a stratified Cox model, the covariate having a time-varying effect should be categorical. If it is a continuous variable, it needs to be categorized. The stratification approach assumes that each stratum has a different baseline hazard rate, but the proportional hazards assumption is satisfied for the other covariates within each stratum. It also assumes that all strata share the same coefficient in general. For the HSCT data, to model the non-proportionality of the transplant type, we could fit a stratified Cox model,

$$\lambda_j(t;\boldsymbol{Z}) = \lambda_{j0}(t)\exp\{\boldsymbol{\gamma}^T\boldsymbol{X}\},$$

where $j = 1, 2$ for autologous HSCT and allogeneic HSCT, respectively, and $\boldsymbol{X}$ are the other significant covariates. Thus, the time-varying effect of transplant type is explained by two different baseline hazard rates $\lambda_{10}(t)$ and $\lambda_{20}(t)$. In practice, one of the advantages of using stratification is that no further proportionality conditions need to be examined. It has been implemented in most statistical packages. In SAS, we can use the "strata" statement in the Proc PHREG procedure.

Although stratification may be easier to implement in practice, its drawback is that the effect of the transplant type cannot be quantified directly. To model the effect of transplant type directly we model the main effect using time dependent covariates. We can fit a Cox model where the treatment has an early effect and a different late effect,

$$\lambda_0(t)\exp\left\{\beta_1 \times Z \times I(t \le t_0) + \beta_2 \times Z \times I(t > t_0) + \boldsymbol{\gamma}^T\boldsymbol{X}\right\},$$

where $Z = 0$ for autologous HSCT, $Z = 1$ for allogeneic HSCT, and $\boldsymbol{X}$ are the other significant covariates.

To determine the optimal cut point t_0 for the treatment effect we fit a series of models with different values of t_0. The cut point giving the largest partial likelihood is selected for the next step in the analysis. We use 6, 9, 12, 18, 24 months after transplant to choose the optimal cut point. Twelve months since transplant gave us the largest partial likelihood; see Table 12.2. This optimal cut point also can be verified in Figure 12.1. We need to further check the proportionality assumptions for the "early" and "late" transplant type effect

TABLE 12.2
Partial log-likelihood for various choices of a cut point t_0.

$t_0 = 6$	$t_0 = 9$	$t_0 = 12$	$t_0 = 18$	$t_0 = 24$
−3007.550	−3007.599	−3005.199	−3006.052	−3006.527

with cut point of 12 months since transplant. This test indicated that the proportionality assumption may not hold for "early" transplant type effect ($P = 0.01$ for $\{Z \times I(t \leq 12)\} \times \log(t)$). This is mainly caused by the fact that allogeneic HSCT patients had a much higher percentage of death rate within one month of transplant compared to autologous HSCT (12% versus 1%). We can further find a cut time point within 12 months since transplant. For illustrative purposes we fit a Cox model with time-dependent covariates for the main treatment effect using 12 months since transplant as the cut time point.

The transplant type is the main interest of the study (Lazarus et al., 2010) and the CIBMTR study has identified significant risk factors associated with overall mortality including patient's age, Karnofsky performance score, chemosensitive disease, and year of transplant. It is important to examine whether these risk factors had the same effect for allogeneic and autologous HSCT. Thus, we need to test for potential interaction between the main treatment effect and these risk factors, and further adjustments will be needed if the test indicates that any interaction exists. Here, potential interaction between six covariates and early and late treatment effects need to be tested. For the late treatment effect, there is no subject or no event occurred for some categories. Eight interaction variables ware tested and no strong interaction was observed; see Table 12.3.

Finally, we fit a Cox model with time-dependent covariates:

$$\lambda_0(t) \exp\left\{\beta_1 \times Z \times I(t \leq 12) + \beta_2 \times Z \times I(t > 12) + \gamma^T \boldsymbol{X}\right\}.$$

The results are given in Table 12.4. They indicate that the type of transplant had a time-varying effect on overall mortality: in the first 12 months within transplant, allogeneic HSCT had a higher mortality rate than autologous HSCT with relative risk of 2.75 (95% confidence interval: 2.03 - 3.72; $P < 0.0001$) for allogeneic HSCT versus autologous HSCT. For patients surviving 12 months after transplant the risk of death was similar for both types of transplants.

12.3.2 Adjusted survival curves based on a Cox regression model

In randomized clinical trials Kaplan-Meier survival curves are commonly used summary curves/statistics for the comparison between treatments. For non-randomized retrospective studies the distribution of some risk factors between the treatment groups is often different. The Kaplan-Meier curves for each treatment group represent a univariate unadjusted summary curve, which can be misleading when the distribution of covariates is unbalanced between the treatment groups. It has been proposed that for a non-randomized study adjusted survival curves should be computed. Such adjusted survival curves represent the survival experiences of the "average" patient in a given treatment group.

Based on a stratified Cox proportional hazards model, two methods of estimating the adjusted survival curves have been proposed. In the first method the adjusted survival probability for the ith treatment group is estimated by

$$\widetilde{S}_i(t) = \exp\left\{-\widehat{\Lambda}_{i0}(t) e^{\widehat{\boldsymbol{\beta}}^T \bar{\boldsymbol{Z}}}\right\},$$

TABLE 12.3
Interaction test.

Covariate	$\hat{\beta}(\hat{\sigma})$	P
Z_{1E} : "Early" HSCT effect	2.03 (0.51)	< 0.01
Z_{1L} : "Late" HSCT effect	1.06 (0.65)	0.10
Z_2 : Age 31-50 Years	0.20 (0.17)	0.26
Z_3 : Age 51-60 Years	0.60 (0.18)	< 0.01
Z_4 : Karnofsky: ≥ 90%	−0.23 (0.10)	0.02
Z_5 : Karnofsky: Missing	−0.18 (0.35)	0.60
Z_6 : Chemo-resistant	0.86 (0.12)	< 0.01
Z_7 : Year of transplant	0.53 (0.16)	< 0.01
$Z_8 = Z_2 \times Z_{1E}$	−0.78 (0.41)	0.06
$Z_9 = Z_3 \times Z_{1E}$	−0.38 (0.44)	0.39
$Z_{10} = Z_4 \times Z_{1E}$	−0.58 (0.31)	0.06
$Z_{11} = Z_5 \times Z_{1E}$	1.22 (0.83)	0.17
$Z_{12} = Z_6 \times Z_{1E}$	−0.04 (0.31)	0.89
$Z_{13} = Z_7 \times Z_{1E}$	−0.26 (0.36)	0.46
$Z_{14} = Z_6 \times Z_{1L}$	0.05 (0.79)	0.95
$Z_{15} = Z_7 \times Z_{1L}$	−1.59 (0.36)	0.05

where $\widehat{\Lambda}_{i0}(t)$ is the estimated cumulative baseline hazard function and $\bar{\boldsymbol{Z}}$ is the average value of $\boldsymbol{Z}$ from the pooled sample. Zhang et al. (2007) showed that this method is easy to implement in practice, but it has several drawbacks. First, the average value of a categorical variable could be quite meaningless. In addition, this method does not account for the sample variability in the risk factor from subject to subject.

The second method is based on the average of the estimated survival curves for each patient over the entire sample, i.e.,

$$\widehat{S}_i(t) = \frac{1}{n}\sum_{i=1}^{n} \exp\left\{-\widehat{\Lambda}_{i0}(t)e^{\widehat{\boldsymbol{\beta}}^T \boldsymbol{Z}_i}\right\}.$$

This is often called the "direct adjusted survival curve" and it produces a more representative curve. These direct adjusted survival curves estimate the survival probabilities in populations with similar prognostic factors.

Zhang et al. (2007) derived a variance estimate for $\widehat{S}_i(t)$ and for the difference of two adjusted survival curves between treatment groups. They also developed a SAS macro to implement the estimating procedure for the direct adjusted survival curves (the macro is available at http://www.mcw.edu/biostatistics).

To use the SAS macro, we need to load it into the current program:

```
%INCLUDE 'ADJSURV.sas';
```

The macro is invoked by running the following statement:

```
%ADJSURV(inputdata, time, event, group, covlist, model, outdata);
```

TABLE 12.4
Final proportional hazards model with time-dependent covariate for the type of HSCT.

Covariate	HR (95% CI)	P
Main effect:		$< 0.0001^a$
$\leq$ 12 months:		
Autologous	1.00	
Allogeneic	2.75 (2.03–3.72)	< 0.0001
> 12 months:		
Autologous	1.00	
Allogeneic	0.93 (0.43–1.99)	0.8427
Age:		$< 0.0001^a$
18–30	1.00	
31–50	1.10 (0.81–1.49)	0.5620
> 50	1.72 (1.26–2.35)	0.0006
Karnofsky score:		0.0059^a
< 90%	1.00	
90-100%	0.74 (0.61–0.89)	0.0014
Missing	0.91 (0.49–1.67)	0.7496
Chemotherapy-resistant disease:		
Sensitive	1.00	
Resistant	2.31 (1.86–2.86)	< 0.0001
Year of transplant:		
2001–2003	1.00	
1995–2000	1.55 (1.18–2.03)	0.0016

Note: [a]: 2 degree-of-freedom overall test.

where

- inputdata the input SAS dataset name;
- time the failure time variable;
- event the event indicator variable;
- group the treatment indicator variable;
- covlist list of all covariates (risk factors);
- model the option for model selection, which takes the value:
  1 for a stratified Cox model and
  2 for an unstratified Cox model;
- outdata the SAS output dataset name;

The group indicator variable group needs to be coded as $(1, \ldots, K)$, where K is the total number of treatment groups. The results are saved in the SAS output data file "outdata" which includes estimated direct adjusted survivals (surv1, . . . , survK), their standard errors (se1, . . . , seK), and estimated standard errors of the differences between any two adjusted survivals (se_12,..., se_(K-1)K). In practice we suggest estimating the direct adjusted survival probabilities based on a stratified Cox model which allows the treatment groups to have their own baseline hazards.

Application to example HSCT data

We estimate the direct adjusted survival probabilities for the large B-cell lymphoma (DLBCL) (Lazarus et al., 2010) data. The study indicated that patient age (Z_1=1, if age $31-50$; Z_2=1, if age > 50), Karnofsky performance score (Z_3=1, if 90-100%; Z_4=1, if Missing), chemotherapy-resistant disease (Z_5=1, if Resistant), and year of transplant (Z_6=1, if 1995 – 2000). The DLBCL dataset is saved in the SAS dataset "autoallo" which includes variables of time to death in months (time), death indicator variable (dead=1, if dead; =0, if alive), and treatment group variable (group=1, if autologous HSCT; =2, if allogeneic HSCT) were significant, where Z_i is an indicator variable for $1 =, \ldots, 6$. We use the following SAS statement to invoke the SAS macro:

```
%ADJSURV(autoallo, time, dead, group, Z1 Z2 Z3 Z4 Z5 Z6, 1, outSURV);
```

SAS micro computes pointwise survival probability with its standard error and standard error for the difference of two survival probabilities at each event time point. Based on the SAS output, we can compute the estimated adjusted survival probabilities with 95% pointwise confidence intervals by transplant type, the difference of the two adjusted survivals, and the P-value of pointwise tests of equal survival probability; see Table 12.5. From Cox model (see Table 12.4) we see a huge difference within 12 months of transplant ($RR = 2.75; p < 0.0001$). Since there is no difference between allogeneic HSCT and autologous HSCT 12 months after transplant ($RR = 0.93; p = 0.8427$), the differences in survival probability persisted after 12 months; see Table 12.5. To compare two survival curves over a given time period, one needs to compute the confidence band for the difference of two survival curves which is not available in SAS macro of %ADJSURV.

Figure 12.2 plots the estimated survival probabilities using the Kaplan-Meier method (Figure 12.2(a)) and the direct adjusted survival curves (Figure 12.2(b)). Both figures show a similar pattern in survival probabilities between allogeneic HSCT and autologous HSCT. However, the distribution of the chemotherapy-resistant disease is significantly different between two treatment groups, where only 128 (15%) autologous HSCT patients had resistant disease compared to 33 (42%) allogeneic HSCT patients ($P < 0.0001$), and the relative risk of mortality for resistant disease versus sensitive disease is 2.31 (95% CI: 1.86 – 2.86; $P < 0.0001$; see Table 12.4). Clearly, without any adjustment, the Kaplan-Meier method overestimates the survival difference between the two types of transplants.

TABLE 12.5
Estimated adjusted survival probabilities, by type of HSCT with pointwise CI and P-value.

	Autologous HSCT		Allogeneic HSCT			
Months	N_1	$\widehat{S}_1$(95% CI)	N_2	$\widehat{S}_2$(95% CI)	$\widehat{S}_1 - \widehat{S}_2$	P
24	420	$57(54-60)\%$	20	$32(22-42)\%$	25%	< 0.0001
48	279	$51(47-54)\%$	14	$29(20-39)\%$	22%	< 0.0001
72	163	$47(44-51)\%$	9	$27(18-37)\%$	20%	0.0001
96	78	$44(41-48)\%$	4	$27(18-37)\%$	17%	0.0010

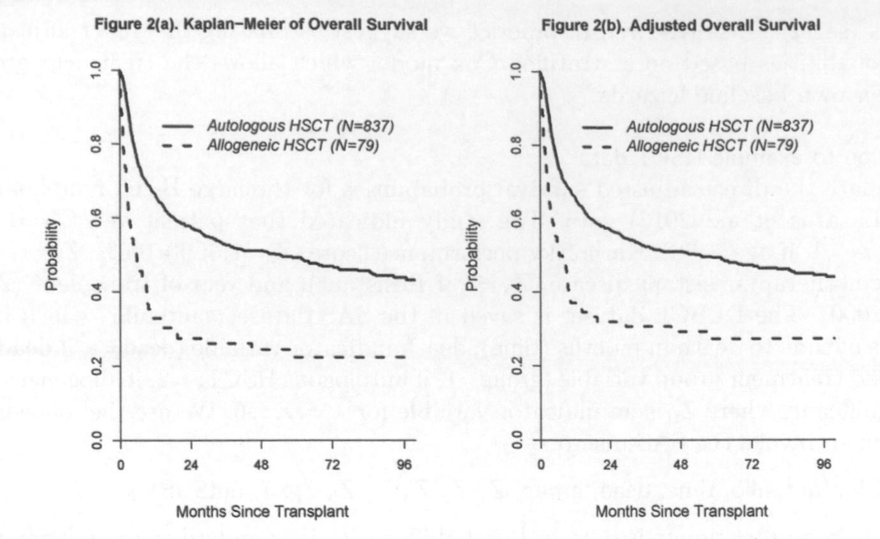

FIGURE 12.2
Direct adjusted survival curves.

This can be seen in Figure 12.2. For example, at 24 months after transplant, the difference $\widehat{S}_1(t) - \widehat{S}_2(t)$ for the Kaplan-Meier estimators and the direct adjusted survivals are $57.5\% - 27.5\% = 30.2\%$ and $57.1\% - 31.9\% = 25.2\%$, respectively. It shows that the Kaplan-Meier method overestimates the difference by about 5%.

12.4 Competing risks data analysis

12.4.1 Common approaches for analyzing competing risks data

For transplant studies, we often need to analyze competing risks data where a patient may fail due to one of K causes and the occurrence of one of these events precludes us from observing the other events. Common combinations of competing risks in HSCT studies include cancer relapse and treatment related mortality (TRM, defined as death without relapse), and GVHD and death without GVHD. When analyzing competing risks data we often wish to estimate and model the cumulative incidence function (CIF), i.e., the probability of failure from a specific cause. Assuming two type failures the CIF for cause 1 given covariates $\boldsymbol{z}$ is

$$F_1(t; \boldsymbol{z}) = P(T \leq t, \epsilon = 1 | \boldsymbol{z}),$$

where T is the failure time and ϵ indicates the cause of failure. One approach to analyzing competing risks data is to model cause-specific hazards for each cause, which is defined as

$$\lambda_k(t; \boldsymbol{z}) = \lim_{\Delta t \to 0} \frac{1}{\Delta t} P\left\{t \leq T \leq t + \Delta t, \epsilon = k | T \geq t, \boldsymbol{z}\right\}, \ k = 1, \ldots, K.$$

The cause-specific hazards for each cause need to be properly modeled since

$$F_1(t; \boldsymbol{z}) = \int_0^t \lambda_1(s; \boldsymbol{z}) \exp\left[- \int_0^s \{\lambda_1(u; \boldsymbol{z}) + \lambda_2(u; \boldsymbol{z})\} \, du \right] ds.$$

The commonly used Cox model, $\lambda_k(t; \boldsymbol{z}) = \lambda_{k0}(t)\exp\{\boldsymbol{\beta}_k^T \boldsymbol{z}\}$ has been proposed for modeling each cause, and for estimating the cumulative incidence functions (Cheng et al., 1998).

Application to example HSCT data

We apply the cause-specific hazard approach to the HSCT example data: 184 patients died without relapse (TRM) and 347 patients had cancer relapsed. We need to fit the Cox model for both TRM and relapse/progression even if we are only interested in modeling the CIF of TRM. First, testing proportionality assumptions by adding a time-dependent covariate, $Z \times \log(t)$ showed that transplant type (allogeneic versus autologous) had a time-varying effect for TRM ($P = 0.0026$) and a constant effect for relapse/progression ($P = 0.3690$). For illustrative purposes we fit a piecewise constant hazards Cox model with cut point of 12 months since transplant for both TRM and relapse/progression (Table 12.6).

Table 12.6 shows that TRM had a time-varying effect ($P = 0.0167$) and significant covariates include patient age, Karnofsky performance score, chemotherapy-resistant disease, and year of transplant, whereas relapse/progression had a non-significant constant effect and only patient age and chemotherapy-resistant disease were significant. Since cumulative incidence function of a specific cause is a function of all cause-specific hazards, modeling the CIF of a specific cause by cause-specific hazard approach needs to model cause-specific hazards correctly for all causes. It is therefore hard to evaluate the covariate effect on the CIF directly and hard to identify the time-varying effect on the CIF for a specific covariate.

Recently, some new regression techniques have been developed to model the CIF directly. The first approach is to directly model the subdistribution hazard function (Fine and Gray, 1999), where the subdistribution hazard of cause 1 is defined as

$$\lambda_1^*(t; \boldsymbol{z}) = -d\log\{1 - F_1(1; \boldsymbol{z})\}/dt.$$

The CIF of cause 1 can be expressed as $F_1(1; \boldsymbol{z}) = 1 - \exp\left\{-\int_0^t \lambda_1^*(u; \boldsymbol{z})du\right\}$. Thus, we can interpret the covariate effect on the CIF directly through the subdistribution hazard function. Fine and Gray (1999) proposed a Cox-type proportional subdistribution hazard model for a specific cause (Fine-Gray model),

$$\lambda_1^*(t; \boldsymbol{z}) = \lambda_{10}^*(t)\exp\{\boldsymbol{\beta}^T \boldsymbol{z}\}, \tag{12.2}$$

where $\lambda_{10}^*(t)$ is an unknown baseline subdistribution hazard function. Fine and Gray proposed using an inverse probability of censoring weighting (IPCW) technique to estimate $\boldsymbol{\beta}$ and the cumulative baseline subdistribution hazard function, $\Lambda_{10}^*(t) = \int_0^t \lambda_{10}^*(u)du$. This approach has been implemented in the **crr** function in the **cmprsk** R package which was developed by Gray (2013). Fine-Gray model assumes a constant proportional effect for each covariate. When the required constant effect assumption does not hold for a specific covariate, one may consider fitting a stratified subdistribution hazards model. This is not available in the **cmprsk** package.

The second approach is based on pseudo-values (Klein and Andersen, 2005) using a pre-selected grid of time points, $t_1, \ldots, t_M$. Five to ten time points with equal distance, or some fixed time points that are of interest to the researchers could be used. At each grid time point (t_j), one computes the nonparametric estimator for CIF, $\widehat{F}_1(t_j)$ based on the complete dataset and $\widehat{F}_i^{-i}(t_j)$ based on the sample obtained by deleting the ith observation. The pseudo-value of the ith subject at t_j is defined by $\hat{\theta}_{ij} = n\widehat{F}_1(t_j) - (n-1)\widehat{F}_i^{-i}(t_j)$. Then, we consider modeling the conditional CIF, $\theta_{ij} = F_1(t_j|\boldsymbol{Z}_i)$ by $\phi(\theta_{ij}) = \alpha_j + \boldsymbol{\beta}^T \boldsymbol{Z}_i$, where ϕ is a known link function. Some commonly used link functions, such as the logit link function, $\phi(\theta) = \log\{\theta/(1-\theta)\}$ and the complementary log-log link function, $\phi(\theta) = \log\{-\log(1-\theta)\}$

TABLE 12.6
Cox proportional hazards models for TRM and relapse/progression.

TRM	HR (95% CI)	*P*
Main effect:		$< 0.0001^a$
≤ 12 months: Autologous	1.00	0.0167^b
Allogeneic	4.89 (3.22–7.43)	< 0.0001
> 12 months: Autologous	1.00	
Allogeneic	1.09 (0.34–3.50)	0.8886
Age: 18–30	1.00	0.0006^c
31–50	0.96 (0.60–1.55)	0.8765
> 50	1.73 (1.07–2.80)	0.0250
Karnofsky score: < 90%	1.00	0.0010^c
90-100%	0.60 (0.45–0.81)	0.0008
Missing	1.41 (0.65–3.09)	0.3853
Chemotherapy-resistant disease: Sensitive	1.00	
Resistant	1.76 (1.22–2.53)	0.0024
Year of transplant: 2001–2003	1.00	
1995–2000	1.67 (1.08–2.59)	0.0022
Relapse/Progression	HR (95% CI)	*P*
Main effect:		0.7322^a
≤ 12 months: Autologous	1.00	0.5095^b
Allogeneic	1.14 (0.75–1.74)	0.5472
> 12 months: Autologous	1.00	
Allogeneic	0.69 (0.17–2.86)	0.6126
Age: 18–30	1.00	0.0005^c
31–50	1.56 (1.07–2.28)	0.0225
> 50	2.07 (1.40–3.05)	0.0003
Karnofsky score: < 90%	1.00	0.5323^c
90-100%	1.03 (0.82–1.28)	0.8298
Missing	0.62 (0.25–1.52)	0.2914
Chemotherapy-resistant disease: Sensitive	1.00	
Resistant	2.71 (2.10–3.49)	< 0.0001
Year of transplant: 2001–2003	1.00	
1995–2000	1.36 (0.99–1.84)	0.0510

Note: ^{a}: 2 degree-of-freedom overall test.

^{b}: Test early effect equals to late effect.

^{c}: 3 degree-of-freedom overall test.

TABLE 12.7
Regression analysis for TRM based on a subdistribution hazard approach and pseudo-vlaue approach.

	Subdistribution hazard		Pseudo-value	
Variable	exp($\hat{\beta}$) (95% CI)	P	exp($\hat{\beta}$) (95% CI)	P
Autologous	1.00		1.00	
Allogeneic	3.46 (2.26–5.30)	**<0.0001**	3.50 (2.33–5.24)	**<0.0001**
Age:				
18–30	1.00		1.00	
31–50	0.86 (0.53–1.40)	0.5452	0.75 (0.46–1.21)	0.2345
> 50	1.38 (0.85–2.25)	0.1974	1.28 (0.80–2.06)	0.3021
Karnofsky:				
$< 90\%$	1.00		1.00	
90-100%	0.65 (0.48–0.88)	**0.0051**	0.66 (0.48–0.91)	**0.0118**
Missing	1.67 (0.70–3.98)	0.2444	1.20 (0.48 –2.98)	0.7020
Sensitive	1.00		1.00	
Resistant	1.14 (0.78–1.68)	0.4971	1.13 (0.75–1.69)	0.5553
Year of TX:				
2001–2003	1.00		1.00	
1995–2000	1.77 (1.14–2.76)	**0.0117**	1.71 (1.08–2.71)	**0.0225**

have been suggested and used in biomedical studies. Klein et al. (2008) developed an R function and a SAS macro to compute the pseudo-values for right-censored competing risks data. The SAS GEE estimating procedure GENMOD has been applied to estimate the regression coefficients of α_j and β with pseudo-observation $\hat{\theta}_{ij}$. Gerds et al. (2009) showed that the coefficient estimates are asymptotically unbiased for right-censored competing risks data.

Application to example HSCT data

We applied the **crr** function to fit Fine-Gray model (12.2) for TRM in the HSCT example data. We also applied the pseudo-value approach to the HSCT example data for TRM using complementary log-log link function with 4 time points 12, 24, 48, and 96 months after transplantation. This fitted model is equivalent to the Fine-Gray model (12.2) at these time points. As expected both approaches give similar results; see Table 12.7. The table shows that in addition to transplant type only Karnofsky performance score and year of transplantation were significantly associated with TRM, and the effects of patient age and chemotherapy-resistant disease on TRM were not significant. However, patient age and chemotherapy-resistant disease were both significant for the cause-specific hazard function of TRM; see Table 12.6. It shows that it is hard to evaluate the covariate effect on the CIF directly based on cause-specific hazard functions. Figure 12.3 plots the univariate CIF of TRM by the transplant type. It shows that the transplant type had a time-varying effect on TRM. Thus, the Fine-Gray model is not an appropriate model for modeling the TRM.

The third approach to directly modeling the cumulative incidence function is based on

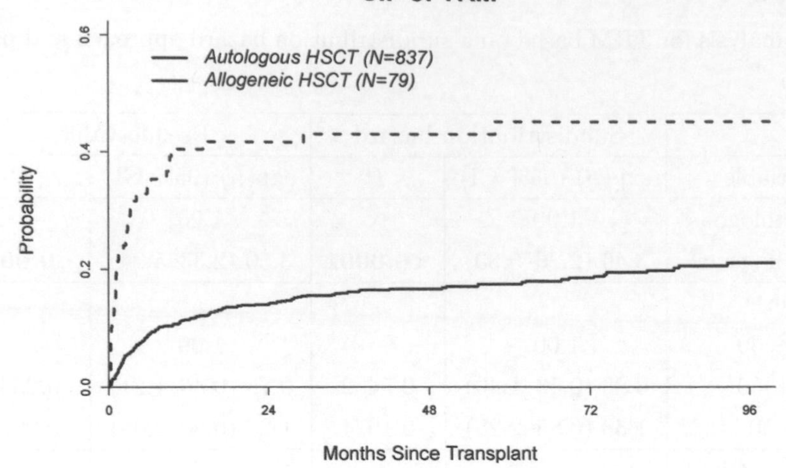

FIGURE 12.3
Univariate CIF of TRM for allogeneic HSCT versus autologous HSCT.

a direct binomial modeling approach. Scheike et al. (2008) and Scheike and Zhang (2008, 2011) considered a class of general models

$$h\{F_1(t;\boldsymbol{x},\boldsymbol{z})\} = \boldsymbol{\alpha}(t)^T\boldsymbol{x} + g(\boldsymbol{z},\boldsymbol{\gamma},t),$$

where h is a known link function, g is a known regression function, and $\boldsymbol{\alpha}(t)$ and $\boldsymbol{\gamma}$ are unknown regression coefficients. They focused on two classes of flexible models: proportional models

$$\text{cloglog}\{1 - F_1(t;\boldsymbol{x},\boldsymbol{z})\} = \boldsymbol{\alpha}(t)^T\boldsymbol{x} + \boldsymbol{\gamma}^T\boldsymbol{z} \tag{12.3}$$

and additive models

$$\log\{1 - F_1(t;\boldsymbol{x},\boldsymbol{z})\} = \boldsymbol{\alpha}(t)^T\boldsymbol{x} + (\boldsymbol{\gamma}^T\boldsymbol{z})t. \tag{12.4}$$

A direct binomial regression approach has been proposed to estimate the regression coefficients (Scheike et al., 2008), and the proposed estimating procedures have been implemented in an R function **comp.riks** which is available in the **timereg** R package (Scheike and Zhang, 2011). The **predict** function available in the **timereg** package also predicts the cumulative incidence probability for given values of covariates and constructs $(1-\alpha)100\%$ pointwise confidence intervals and confidence bands over a given time interval. Scheike and Zhang (2008) developed a goodness-of-fit test for testing whether a specific covariate has a constant effect on the CIF, which has been implemented in the **timereg** package.

Application to example HSCT data

We applied the **timereg** package to analyzing the HSCT example data. Let Z_0 be the transplant type indicator ($\mathsf{Z}_0 = 1$ for allogeneic transplant), and $\{\mathsf{Z}_1,\ldots,\mathsf{Z}_6\}$ be defined same as in Section (12.3.2). First, to check whether there is a time-varying effect on the TRM for each covariate we fit a flexible additive model

$$F_1(t;\boldsymbol{Z}) = 1 - \exp\{\boldsymbol{\alpha}(t)^T\boldsymbol{Z}\},$$

where $\boldsymbol{Z} = (1, Z_0, Z_1, Z_2, Z_3, Z_4, Z_5, Z_6)^T$. The result is given in Table 12.8. It shows that

TABLE 12.8
Testing of constant effect and non-significant effect for TRM.

	P-value	
Variable	Test constant effect	Test significant effect
Main effect: Autologous[a]	—	—
Allogeneic	<0.0001	0.0002
Age: 18 – 30[a]	—	—
31–50	0.5100	0.2750
> 50	0.5350	0.0514
Karnofsky score: < 90%[a]	—	—
90-100%	0.0270	0.0308
Missing	0.3710	0.4130
Chemo-resistant disease:		
Sensitive[a]	—	—
Resistant	0.3440	0.2250
Year of transplant:		
2001-2003[a]	—	—
1995–2000	0.0192	0.0008

Note: [a]: Baseline.

transplant type (allo vs auto) has a strong time-varying effect ($P < 0.0001$) and Karnofsky score and year of transplant have mild time-varying effects (P=0.0270 and P=0.0192, respectively), which indicates that the assumption of constant effect required in the Fine-Gray model is not valid. The test of non-significant effect indicates that patient age and chemotherapy resistant disease have no significant effect on TRM, which agrees with the conclusion from fitting a Fine-Gray model and the conclusion based on the pseudo-values approach; see Table 12.7.

Let $\boldsymbol{Z} = (Z_3, Z_4, Z_6)^T$. For illustrative and comparison purposes, we fit the Fine-Gray model (12.3)

$$F_1(t; \boldsymbol{Z}) = 1 - \exp\left\{-\exp\left(\alpha_0(t) + \beta_1 \mathrm{Z}_0 + \boldsymbol{\gamma}^T \boldsymbol{Z}\right)\right\} \tag{12.5}$$

and a proportional flexible model (12.4)

$$F_1(t; \boldsymbol{Z}) = 1 - \exp\left\{-\exp\left(\alpha_0(t) + \alpha_1(t) \mathrm{Z}_0 + \boldsymbol{\gamma}^T \boldsymbol{Z}\right)\right\}. \tag{12.6}$$

We compute the predicted cumulative incidence probabilities for given average covariate values for both models by transplant type. Figure 12.4(b) shows that for the allogeneic HSCT, the predicted CIF of TRM based on the flexible model (12.6) reaches 40% within about 9 months after HSCT and has very little additional TRM occurring afterwards, whereas in Figure 12.4(a) the predicted CIF of TRM based on Fine-Gray model (12.5) reaches 40% at about 24 months after HSCT, and it keeps increasing afterwards. The predicted cumulative incidence probabilities of TRM based on Fine-Gray model (12.5) are misleading since the underlying model assumption is not valid.

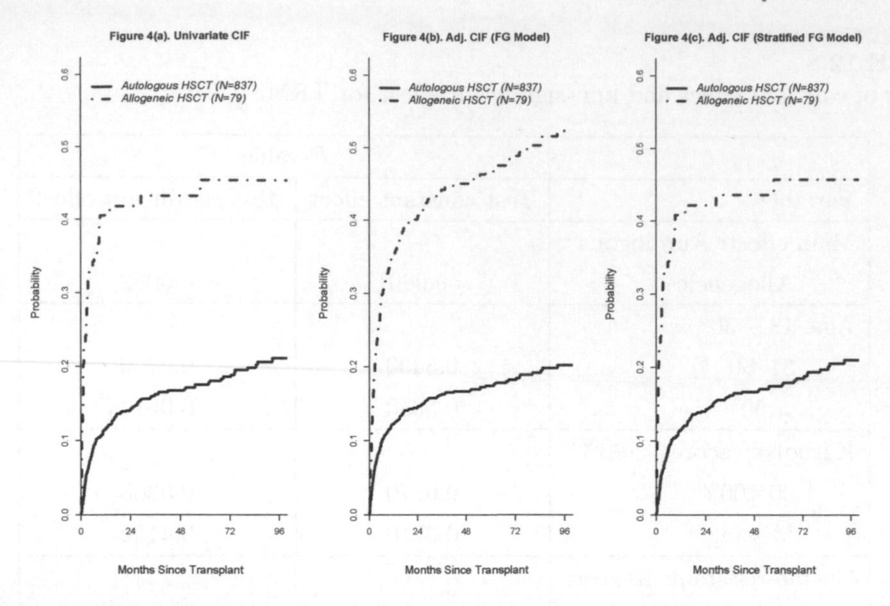

FIGURE 12.4
Cumulative incidence probabilities of TRM based on (a) Univariate, (b) Adjusted CIF based on the Fine-Gray model (12.5) assuming constant effect, (c) Adjusted CIF based a stratified Fine-Gray model (12.7).

12.4.2 Adjusted cumulative incidence curves based on a stratified Fine-Gray model

For non-randomized retrospective studies, the standard nonparametric estimated cumulative incidence curves for each treatment group can be misleading since the distribution of covariates is unbalanced between treatment groups. To adjust for the potentially imbalanced prognostic factors among treatment groups and for the potentially time-varying treatment effect, Zhang and Zhang (2011) considered fitting a stratified Fine-Gray model,

$$\lambda_1^*(t; l, \boldsymbol{z}) = \lambda_{10,l}^*(t) \exp\{\boldsymbol{\beta}^T \boldsymbol{z}\}, \tag{12.7}$$

where $\lambda_{10,l}^*(t)$ is the baseline subdistribution hazards for treatment $l = 1, \ldots, L$, and proposed to estimate the direct adjusted cumulative incidence probabilities by

$$\widehat{F}_{1l}(t) = \frac{1}{n} \sum_{i=1}^{n} \left[1 - \exp\left\{ -\widehat{\Lambda}_{10,l}^*(t) e^{\widehat{\boldsymbol{\beta}}^T \boldsymbol{Z}_i} \right\} \right],$$

where $\Lambda_{10,l}^*(t) = \int_0^t \lambda_{10,l}^*(u)du$ is the cumulative baseline subdistribution hazard for the lth treatment group. Zhang and Zhang (2011) derived variance estimates for $\widehat{F}_{1l}(t)$ and for the difference of two adjusted cumulative incidence curves between two treatment groups, and also developed a SAS macro to implement the proposed estimating procedures (SAS macro is available at http://www.mcw.edu/biostatistics).

Application to example HSCT data

We estimate the direct adjusted cumulative incidence probabilities of TRM using the HSCT example data. Model diagnostic tests indicate that only Karnofsky performance score and year of HSCT are significantly associated with TRM and transplant type had a time-varying

TABLE 12.9
Pointwise estimated cumulative incidence (CIF) probabilities of TRM by type of HSCT.

Univariate CIF						
	Autologous HSCT		Allogeneic HSCT			
Months	N_1	$\widehat{F}_{11}$(95% CI)	N_2	$\widehat{F}_{12}$(95% CI)	$\widehat{F}_{12} - \widehat{F}_{11}$	P
24	371	14(12 – 16)%	18	42(31 – 52)%	28%	< 0.0001
48	244	17(14 – 19)%	13	43(32 – 54)%	27%	< 0.0001
72	145	19(16 – 22)%	8	45(34 – 56)%	27%	< 0.0001
96	69	21(18 – 25)%	4	45(34 – 56)%	24%	< 0.0001
Adjusted CIF based on Fine-Gray Model (12.5)						
	Autologous HSCT		Allogeneic HSCT			
Months	N_1	$\widehat{F}_{11}$(95% CI)	N_2	$\widehat{F}_{12}$(95% CI)	$\widehat{F}_{12} - \widehat{F}_{11}$	P
24	371	14(12 – 17)%	18	40(29 – 51)%	26%	< 0.0001
48	244	17(14 – 19)%	13	45(33 – 57)%	29%	< 0.0001
72	145	18(16 – 21)%	8	49(36 – 60)%	30%	< 0.0001
96	69	21(18 – 25)%	4	54(40 – 66)%	33%	< 0.0001
Adjusted CIF based on Stratified Fine-Gray Model (12.7)						
	Autologous HSCT		Allogeneic HSCT			
Months	N_1	$\widehat{F}_{11}$(95% CI)	N_2	$\widehat{F}_{12}$(95% CI)	$\widehat{F}_{12} - \widehat{F}_{11}$	P
24	371	14(12 – 16)%	18	42(31 – 53)%	28%	< 0.0001
48	244	17(14 – 19)%	13	44(32 – 54)%	27%	< 0.0001
72	145	18(16 – 21)%	8	46(34 – 57)%	27%	< 0.0001
96	69	21(18 – 24)%	4	46(34 – 57)%	25%	< 0.0001

effect; see Table 12.8. For illustrative purposes we fit a stratified Fine-Gray model (12.7) stratified on transplant type. Based on the SAS output, we can estimate adjusted CIF with a 95% pointwise confidence interval, the difference of the two adjusted CIFs, and the P-value of pointwise tests of equal CIF; see Table 12.9. Figure 12.4(c) plots the estimated adjusted CIF of TRM based on a stratified Fine-Gray model (12.7) by transplant type. The adjusted risk factors are not strongly imbalanced (P=0.41 and P=0.02 of test equal distribution between transplant types for Karnofsky performance score and year of HSCT, respectively). As expected it is (Figure 12.4(c)) close to the univariate estimated CIF (Figure 12.4(a)). Fine-Gray model assumes constant effect for transplant type which is not true in the HSCT example data. Adjusted CIF of TRM based on Fine-Gray model overestimates the late difference from 25% to 33% at 8 years after transplant, see Figure 12.4(b) and Table 12.9. The plot of the adjusted cumulative incidence curves based on a stratified Fine-Gray model provides useful information to researchers and patients.

12.5 Summary

In this chapter we reviewed some basic regression techniques for analyzing the bone marrow transplant data. The Cox model can be applied for survival data and competing risks data. When the proportional hazards assumption is violated, a stratified Cox model or a Cox model with time–dependent variables can be used. For competing risks data, the Fine-Gray model, the pseudo-value approach, the direct binomial modeling approach can also be employed. Adjusted survival or cumulative incidence curves are recommended for graphical presentation. All methods have been successfully applied for the bone marrow transplant data. Discussion on other topics such as model diagnostics, multi-state modeling, and left truncation can be found in other chapters and Klein and Moeschberger (2003).

Bibliography

Aalen, O. O. (1989), 'A linear regression model for the analysis of life times', *Statistics in Medicine* **8**, 907–925.

Andersen, P. K. and Gill, R. D. (1982), 'Cox's regression model for counting processes: A large sample study', *The Annals of Statistics* **10**, 1100–1120.

Bach, F., Albertini, R., Joo, P., Anderson, J. and Bortin, M. (1968), 'Bone-marrow transplantation in a patient with the Wiskott-Aldrich syndrome', *Lancet* **2**, 1364–1366.

Cheng, S. C., Fine, J. and Wei, L. J. (1998), 'Prediction of cumulative incidence function under the proportional hazards model', *Biometrics* **54**, 219–228.

Cox, D. (1972), 'Regression models and life tables (with discussion)', *Journal of the Royal Statistical Society, Series B* **34**, 187–220.

Eapen, M. and Rocha, V. (2008), 'Principles and analysis of hematopoietic stem cell transplantation outcomes: the physicians perspective', *Lifetime Data Analysis* **14**, 379–388.

Fine, J. P. and Gray, R. J. (1999), 'A proportional hazards model for the subdistribution of a competing risk', *Journal of the American Statistical Association* **94**, 496–509.

Gatti, R., Meuwissen, H., H.D., A. and Hong, R. (1968), 'Immunological reconstitution of sex-linked lymphopenic immunological deficiency', *Lancet* **2**, 1366–1369.

Gerds, T., Graw, F. and Schumacher, M. (2009), 'On pseudo-values for regression analysis in competing risks models', *Lifetime Data Analysis* **15**, 241–255.

Gray, R. (2013), 'A package for subdistribution and analysis of competing risks', R package version 2.2-4, http://cran.r-project.org/package=cmprsk.

Gray, R. J. (1988), 'A class of k-sample tests for comparing the cumulative incidence of a competing risk', *The Annals of Statistics* **16**, 1141–1154.

Klein, J. (2006), 'Modeling competing risks in cancer studies', *Statistics in Medicine* **25**, 1015–1034.

Klein, J. and Andersen, P. (2005), 'Regression modeling of competing risks data based on pseudovalues of the cumulative incidence function', *Biometrics* **61**, 223–229.

Klein, J., Gerster, M., Andersen, P., Tarima, S. and Perme, M. (2008), 'SAS and r functions to compute pseudo-values for censored data regression', *Comput Methods Programs Biomed* **89**, 289–300.

Klein, J. and Moeschberger, M. (2003), *Survival Analysis: Techniques for Censored and Truncated Data*, second edn, Springer. New York, NY, USA.

Kleinbaum, D. G. and Klein, M. (2005), *Survival Analysis: A Self-Learning Text*, second edn, Springer. New York, NY, USA.

Lazarus, H., Zhang, M., Carreras, J., Hayes-Lattin, B., Ataergin, A., Bitran, J., Bolwell, B., Freytes, C., R.P., G., Goldstein, S., Hale, G., Inwards, D., Klumpp, T., Marks, D., Maziarz, R., McCarthy, P., Pavlovsky, S., Rizzo, J., Shea, T., Schouten, H., Slavin, S., Winter, J.N. snf van Besien, K., Vose, J. and Hari, P. (2010), 'A comparison of HLA-identical sibling allogeneic versus autologous transplantation for diffuse large B cell lymphoma: a report from the CIBMTR', *Biology of Blood and Marrow Transplantation* **16**(1), 35–45.

Lin, D. Y. and Ying, Z. (1995), 'Semiparametric analysis of general additive-multiplicative hazard models for counting processes', *The Annals of Statistics* **23**, 1712–1734.

McKeague, I. W. and Sasieni, P. D. (1994), 'A partly parametric additive risk model', *Biometrika* **81**, 501–514.

Pepe, M. S. (1991), 'Inference for events with dependent risks in multiple endpoint studies', *J. Amer. Statist. Assoc.* **86**, 770–778.

Prentice, R. L., Kalbfleisch, J. D., Peterson, A. V., Flournoy, N., Farewell, V. T. and Breslow, N. (1978), 'The analysis of failure time data in the presence of competing risks', *Biometrics* **34**, 541–554.

Scheike, T. H. and Zhang, M. (2003), 'Extensions and applications of the Cox-Aalen survival model', *Biometrics* **59**, 1033–1045.

Scheike, T. H., Zhang, M. and Gerds, T. (2008), 'Predicting cumulative incidence probability by direct binomial regression', *Biometrika* **95**, 205–220.

Scheike, T. and Zhang, M. (2002), 'An additive-multiplicative Cox-Aalen model', *Scandinavian Journal of Statistics* **28**, 75–88.

Scheike, T. and Zhang, M. (2008), 'Flexible competing risks regression modeling and goodness-of-fit', *Lifetime Data Anal* **14**, 464–483.

Scheike, T. and Zhang, M. (2011), 'Analyzing competing risk data using the r timereg package', *Journal of Statistical Software* **38**, 2.

Shen, Y. and Cheng, S. (1999), 'Confidence bands for cumulative incidence curves under the additive risk model', *Biometrics* **55**, 1093–1100.

Sun, L., Liu, J., Sun, J. and Zhang, M. (2006), 'Modeling the subdistribution of a competing risk', *Statistica Sinica* **16**, 1367–1385.

Zhang, X., Loberiza Jr., F., Klein, J. and Zhang, M. (2007), 'A SAS macro for estimation of direct adjusted survival curves based on a stratified cox regression model', *Computer Methods and Programs in Biomedicine* **88**, 95–101.

Zhang, X. and Zhang, M. (2011), 'SAS macros for estimation of direct adjusted cumulative incidence curves under proportional subdistribution hazards models', *Computer Methods and Programs in Biomedicine* **101**, 87–93.

Part III

Model Selection and Validation

In Part III we examine, in a series of chapters, two complementary problems in modeling survival data. The first is the body of techniques available for model selection and validation and the second is the robustness of the Cox regression model.

Model selection and model validation is an important problem in survival analysis. The most appropriate model depends in part on how the model will be used. Regression models can be used to test a particular hypothesis, perhaps adjusted for other covariates or the goal of the regression model can be to make a predictive model for the time to some event or the survival probability for an individual with known covariates.

In Chapter 13 Yong et al. examine classical techniques for selecting the best predictive model for survival based on a Cox proportional hazards model. The optimal model selection and validation procedures require the data to be split into three sub-samples with the first used to generate candidate regression models; the second to pick the best model using some measure of how well the model predicts outcome; and a third independent sub-sample which is used to validate the best model selected using sub-sample two. In some cases where data is limited, a cross-validation technique is used. There are a number of measures of how well a Cox model predicts outcome discussed in the chapter. These are measures of model fit such as the Akakie information criterion (AIC) and the Bayes information criterion; penalized partial likelihood methods such as the Lasso (Least Absolute Shrinkage and Selection Operator) selection; and measures based on the predictive ability of the model such as the C(Concordance)-statistic. Many of these measures are very similar to those used in the classical logistic regression problem. Yong et al. illustrate and compare all these methods using a well-known dataset from the Mayo Clinic Primary Biliary Cirrhosis (PBC) study.

In Chapter 14 Laud presents the Bayesian approach to model selection. Here Laud considers a quite general problem. Selection could be between two models or the selection problem could be to select between covariates, functional forms of covariates or alternative distributional assumptions in a Bayes analysis. In this approach Laud puts a prior on each model in a "model space," which is a collection of potential models. An element of this space is characterized by a model for all observable information in the experiment and includes, for example, the functional form of the relationship and a listing of the covariates which go into the model. Using this framework he examines the use of Bayes Factors to pick the best model; the Bayesian interpretation of the AIC and BIC as well as a number of Bayes-specific measures of fit. Given that the model space grows quite quickly he also considers model algorithms for this approach.

Meijer and Goeman in Chapter 15 examine model selection in a genetic framework. In this problem there are often a very large number of covariates to be tested in a survival model. These covariates come from, for example, individual genes or SNPs (single nucleotide polymorphisms), and possibly interactions between these genes. The goal is to select important covariates from a large set of candidate covariates. Here the authors present techniques based on prediction methods. In this approach, where there are typically more covariates than observations, one needs some special techniques other than the standard Cox model selection methods. These methods include penalized likelihood methods and screening using univariate models. A second general approach is to select variables by testing individual covariates and making adjustments to the α level of the individual tests to control the type I error rate.

In the final chapter of this part O'Quigley and Xu examine the robustness of the Cox model. They show that estimators that are robust to the censoring distribution can be found by a modification to the Cox model to have time-varying regression coefficients or by a modified estimating equation. They also present robust estimators of an average regression effect as well as an extensive set of simulations of these models.

13

Classical Model Selection

Florence H. Yong, Tianxi Cai, and L.J. Wei
Harvard School of Public Health

Lu Tian
Stanford University School of Medicine

CONTENTS

13.1 Introduction 265
13.2 Mayo Clinic primary biliary cirrhosis (PBC) data 266
13.3 Model building procedures and evaluation 267
  13.3.1 Variable selection methods 268
  13.3.2 Model evaluation based on prediction capability 269
13.4 Application of conventional model development and inferences 270
  13.4.1 Model building 270
  13.4.2 Selecting procedure using C-statistics 270
  13.4.3 Making statistical inferences for the selected model 270
13.5 Challenges and a proposal 272
  13.5.1 Over-fitting issue 273
  13.5.2 Noise variables become significant risk factors 273
  13.5.3 Utilizing cross-validation in model selection process 274
  13.5.4 3-in-1 dataset modeling proposal 275
13.6 Establishing a prediction model 275
  13.6.1 Evaluating model's generalizability 277
  13.6.2 Reducing over-fitting via 3-in-1 proposal 278
13.7 Remarks 279
Bibliography 281

13.1 Introduction

There are two major goals for conducting regression analysis: examining the association of the outcome variable and the covariates, and making prediction for future outcomes. The choice of the process for model building, selection and validation depends on the aim of the investigation. Generally it is difficult, if not impossible, that the selected model would be the correct one. On the other hand, a good approximation to the true model can be quite useful for making prediction. Model building and selection should not be a stand-alone procedure, we need valid inference for prediction with the final model.

To establish a prediction model, the same observations in a dataset are often used to build, select, and conduct inference. This traditional practice can lead to quite overly optimistic and unreliable model at the end. To tackle this issue, we present a model development strategy based on the well-known "machine learning" concept with additional inferential component. We recommend splitting the dataset randomly into two pieces. The first part is used for model building and selection via conventional cross-validation techniques, and the second part, often called the "holdout sample," is used for statistical inference based on the final selected model. Although conceptually this strategy is applicable to the general regression problem, our focus is on censored event time outcome variable.

The Cox proportional hazards (PH) model (Cox, 1972) is the most widely used model in analyzing event time data. Its statistical procedures for making inferences about the association between the event time and covariates are well developed and theoretically justified via martingale theory (Andersen and Gill, 1982). For prediction using Cox's model, van Houwelingen and Putter (2008) provide an excellent review of various prediction methods in clinical settings. Other recent development in this area involves high-dimensional data in which the number of observations is much smaller than the number of variables (van Houwelingen et al., 2006; Witten and Tibshirani, 2010). For typical study analyses, a vast majority utilizes all observations in the same dataset for model building and validation, despite a growing concern of the false positive findings (Ioannidis, 2005; Simmons et al., 2011).

The goal of our investigation is to establish a Cox prediction model and draw reliable inferences using such a model. We discuss in detail the model-building strategies from a prediction point of view. We will use a well-known dataset from a Mayo Clinic Primary Biliary Cirrhosis (PBC) study (Fleming and Harrington, 1991; Therneau and Grambsch, 2000) to guide us through each step of the process for conducting model building, selection and inference.

The article is organized as follows. In Section 13.2, we describe our study example. Section 13.3 summarizes various model building and selection procedures in the literature for the Cox model. Model evaluation based on predictive accuracy measures such as a censoring-adjusted C-statistic is introduced, which can be used to identify the optimal method among all candidate models to develop a final Cox model. Section 13.4 applies five candidate model selection methods to the PBC dataset to demonstrate the conventional model building, selection, and inference procedure. Section 13.5 presents some challenges using the conventional process. It shows that the conclusion of such inference can be quite misleading due to using an overly optimistic model building procedure. A prediction model development strategy that integrates cross-validation in the model building and validation, utilizes predictive measure to help identify the optimal model selection method, and conducts valid prediction inference on a holdout dataset is proposed. This 3-in-1-dataset modeling procedure is illustrated in detail with the PBC data in Section 13.6. We conclude with discussion of potential issues and interesting research problems on model selection in the Remarks section.

13.2 Mayo Clinic primary biliary cirrhosis (PBC) data

The Mayo clinical trial in PBC of the liver has been a benchmark dataset for illustration and comparison of different methodologies used in the analysis of event time outcome study (Fleming and Harrington, 1991; Therneau and Grambsch, 2000). The trial was conducted between January 1974 and May 1984 to evaluate the drug D-penicillamine versus placebo with respect to survival outcome. There were 424 patients who met the eligibility criteria for the trial, in which 312 cases participated in the double-blinded randomized placebo controlled trial and contained mostly complete information on the baseline covariates. Six patients were lost to followup soon after their initial clinic visit and excluded from Me study. The rest of the 106 cases did not participate in the randomized trial but were followed for survival with some basic measurements recorded.

Since there was no treatment difference with respect to the survival distributions at the end of study, the study investigators combined the data from the two treatment groups to establish models for predicting survival. In this article, we utilized data on all 418 patients to establish a prediction model for the patient's survival given their baseline covariates. The average follow-up time of these 418 patients was 5.25 years. Like other studies, there were missing covariate values among the patients ranging from 2 patients missing prothrombin time (protime) to patients missing triglyceride levels. For illustration, we imputed the missing values with their group sample median.

The outcome variable is the time to death ($time_i$). Censoring variable ($death_i$) for each case i has value 1 if the death date is available, or value 0 otherwise. The patient's baseline information consists of

- Demographic attributes: age in years, sex
- Clinical aspects: ascites (presence/absence), hepatomegaly (presence/absence), spiders (blood vessel malformations in the skin, presence/absence), edema (0 no edema and no diuretic therapy for edema, 0.5 edema untreated or successfully treated, 1 edema despite diuretic therapy)
- Biochemical aspects: serum bilirubin (mg/dl), albumin (g/dl), urine copper (μg/day), prothrombin time (standardised blood clotting time in seconds), platelet count n (number of platelets $\times$ 10^{-3} per mL^3), alkaline phosphotase (U/liter), ast (aspartate aminotransferase, once called SGOT (U/ml)), serum cholesterol (mg/dl), and triglyceride levels (mg/dl)
- Histologic stage of disease.

We applied logarithmic transformations to albumin, bilirubin, and protime in the process of model building, based on analyses of this dataset in Fleming and Harrington (1991).

To establish a prediction model, ideally one should have three similar but independent datasets, or split the dataset randomly into three subsets. Using the observations from the first subset, we fit the data with all model candidates of interest; using the data from the second piece, we evaluate those fitted models with intuitively interpretable, model-free criteria and choose a final model; and using the data from the third piece, we draw inferences about the selected model. In practice, if the sample size is not large, we may combine the first two steps with a cross-validation procedure.

We will use the PBC dataset to illustrate this model selection strategy in Section 13.5.4. First we review some classical algorithms for model selection and introduce some model-free, heuristically interpretable criteria for model evaluation.

13.3 Model building procedures and evaluation

Depending on the study question and subject matter knowledge, we may identify a set of potential explanatory variables which could be associated with the survival outcome in a Cox PH model, the hazard function at time t for m for an individual is:

$$\lambda(t|Z) = \lambda_0(t)\exp(\beta' Z),$$

where $\lambda_0(t)$ is an unknown baseline hazard function, $Z = (z_1, z_2, \ldots, z_p)'$ is the vector of explanatory variables of the individual, and $\beta = (\beta_1, \beta_2, \ldots, \beta_p)'$ is a $p \times 1$ vector of coefficients of the explanatory variables $Z_1, Z_2, \ldots, Z_p$. We estimate the parameter β by maximizing the partial likelihood:

$$L(\beta) = \prod_{r \in D} \frac{\exp(\beta' Z^{k_r})}{\sum_{k \in R_r} \exp(\beta' Z^k)},$$

where D is the set of indices of the failures, R_r is the set of indices of subjects at risk at time t_r, and k_r is the index of the failure at time t_r.

13.3.1 Variable selection methods

A classical variable selection method is the stepwise regression using $L(\beta)$ as the objective function and p-value as a criterion for inclusion or deletion of covariates. It combines forward selection and backward elimination methods, allowing variables to be added or dropped at various steps according to different pre-specified p-values for entry to or stay in the model. Variations of stepwise regression method have been proposed. For example, forward stepwise regression starts from a null model with intercept only, while backward stepwise regression starts from a full model. We use forward stepwise procedure, as backward stepwise selection may be more prone to the issues of collinearity.

To reduce overfitting (Harrell (2001); Section 13.5.1), we may introduce a penalty of complexity of the candidate models for the stepwise procedures using Akaike information criterion (AIC; Akaike (1974)) or Bayesian information criterion (BIC; Schwarz (1978)). Both AIC and BIC penalizes degrees of freedom (k) which is the number of nonzero covariates in regression setting, and their objective functions are:

AIC $= -2 * L(\beta) + 2 * k$; and

BIC $= -2 * L(\beta) + \log(\text{No. of Events}) * k$. The AIC's penalty for model complexity is less than that of BIC's. Hence, it may sometimes over-select covariates in order to describe the data more adequately; whereas BIC penalizes more and may under-select covariates (Acquah and Carlo, 2010). Note that we usually follow the principle of hierarchical models when building a model, in which interactions are included only when all the corresponding main effects are also included; however, this can be relaxed (Collett, 2003).

We can also select a model based on the maximization of a penalized partial likelihood (Verweij and Van Houwelingen, 2006) with different penalty functions including L_2 penalty, smoothing splines, and frailty models, which are studied extensively in the literature. Two commonly used methods are Lasso (Least Absolute Shrinkage and Selection Operator) selection (Tibshirani, 1996) and Ridge regression methods (Van Houwelingen, 2001).

Lasso Selection

Tibshirani (1996) proposed the Lasso variable selection procedures which was extended to the Cox model (Tibshirani, 1997). Instead of estimating β in the Cox model through maximization of the partial likelihood, we can find the β that minimizes the objective function {-log$L(\beta) + \lambda_1 ||\beta||_1$} (Park and Hastie, 2007), where

$$\hat{\beta} = arg\, min_{\beta}\{\text{-log}L(\beta) + \lambda_1 ||\beta||_1\}.$$

Lasso imposes an L_1 absolute value penalty, $\lambda_1 ||\beta||_1 = \lambda_1 \sum_{j=1}^{p} |\beta_j|$ to log $L(\beta)$, with $\lambda_1 \geq 0$. It does both continuous shrinkage and automatic variable selection simultaneously. Notice that Lasso penalizes all $\beta_j (j = 1, \ldots, p)$ the same way, and can be unstable with highly correlated predictors, which is common in high-dimensional data settings (Grave et al., 2011).

Different methods such as path following algorithm (Park and Hastie, 2007), coordinate descending algorithm (Wu and Lange, 2008), and gradient ascent optimization (Goeman, 2009) can be used to select variables and estimate the coefficients in Lasso models. Instead of using cross-validation to select the tuning parameters, we will consistently apply AIC or BIC to select models across various classical model selection methods for illustration.

Ridge regression

The Ridge penalty is a L_2 quadratic function, $\lambda_2||\beta||_2^2 = \lambda_2 \sum_{j=1}^{p} {\beta_j}^2$ in a general penalized regression setting. It achieves better prediction performance through a bias-variance trade-off. Of note, this method always keeps all predictors in the model and hence cannot produce a parsimonious model.

There are other penalized regression methods such as elastic net (Zou and Hastie, 2005) which combines both L_1 and L_2 penalty, the smoothly absolute clipped deviation (SCAD) penalty (Fan and Li, 2001, 2002), and various modification of the Lasso procedures. Other variable selection procedures and different combinations of model selection methods and algorithms to select the tuning parameter(s) have also been developed.

13.3.2 Model evaluation based on prediction capability

Many evaluation criteria can be used to select a model; however, if some covariates are difficult to obtain due to cost or invasiveness, a heuristically interpretable criterion is more informative than a purely mathematical one. Since it is desirable to examine the predictive adequacy of the Cox model for the entire study period, one of such criteria is the C(Concordance)-statistic (Pencina and D'Agostino, 2004).

C-statistics

To select a model with best predictive capability, C(Concordance)-statistics are routinely used to evaluate the discrimination ability and quantify the predictability of working models. Good predictions distinguish subjects with the event outcome from those without the outcome accurately and differentiate long-term survivors from the short-lived in survival context. The traditional C-statistic is a rank-order statistic for predictions against true outcomes (Harrell, 2001), and it has been generalized to quantify the capacity of the estimated risk score in discriminating subjects with different event times. Various forms of C-statistics are proposed in literature to provide a global assessment of a fitted survival model for the continuous event time. However, most of the C-statistics may depend on the study-specific censoring distribution.

Uno et al. (2011) proposed an unbiased estimation procedure to compute a modified C-statistic (C_τ) over a time interval $(0, \tau)$, which also has the same interpretation as Harrell's C-statistic for survival data, except that Uno's method is censoring-independent, and is given by (Uno et al., 2011) equations (5) and (6). This censoring-adjusted C-statistic is based on inverse-probability-of-censoring weights, which does not require a specific working model to be valid. The procedure is valid for both type I censoring without staggered entry, and random censoring independent of survival times and covariates (other conventional C-statistics may not be valid in this situation). A simulation study reported in Uno et al. (2011) did not find the procedure to be sensitive to violation of the covariate independent censoring assumption.

van Houwelingen and Putter (2008) and Steyerberg et al. (2010) provide a very helpful discussion of other assessments of predictive performance such as Brier score (Graf et al., 1999; Gerds and Schumacher, 2006). We show, as an example, the model-free, more recently developed censoring-adjusted C-statistic to evaluate the overall adequacy of the predictive model.

13.4 Application of conventional model development and inferences

The goal of this section is twofold: (1) to show the conventional way of analyzing time-to-event data, using the Mayo Clinic PBC dataset described in Section 13.2; and (2) to present some challenges and limitations, which lead us to propose an alternative strategy for selecting a model among several candidate model selection procedures and establishing a more reliable prediction model.

13.4.1 Model building

We apply five classical model selection algorithms to the PBC dataset for illustration. These candidate methods are: forward selection, backward elimination, stepwise regression, Lasso, and Ridge regression. For each of these methods, we build a model using AIC and BIC as model tuning criteria, respectively. For the Lasso and Ridge regression method, AIC (or BIC) as a function of the regularization parameter λ is plotted and evaluated to find the global minimum. Models are fitted using the λ at which the least AIC (or BIC) is achieved. The results are shown in Table 13.1, which consists of two parts. The first part summarizes all the resulting models via the aforementioned model building processes. The second part of the table summarizes how we obtained these models.

As a reference, all but two covariates (sex and alk.phos) contributed to a "significant" increase or decrease in risk ratio in univariate analysis ($p < 0.005$). Numerous studies in the literature have used Cox models to identify prognostic factors on event outcome. As shown in Table 13.1, the risk ratio estimates for each risk factor of interest can be very sensitive to what other covariates are put in the same model for evaluation.

13.4.2 Selecting procedure using C-statistics

Using the entire PBC dataset, Table 13.2 summarizes the censoring-adjusted C-statistics of the eight models presented in Table 13.1. The higher the measure, the better the model predicts throughout the course of study.

The two penalized regression methods yield slightly higher C-statistics. However, models derived from these two methods use more variables than the classical methods. The best single variable model, M4, has the lowest C-statistic. The predictive measures of M1, M2 and M3 are close to the models derived from the two penalized regression methods while using fewer variables. Inference for the difference in the C-statistic between models shows a difference between M1 and M4 using the method proposed by Uno et al. (2011). M1 appears as the most parsimonious model with reasonably good C-statistic of 0.790 among all these models.

13.4.3 Making statistical inferences for the selected model

Conventionally, once we find a desirable model, the risk score for this model can be estimated and used to differentiate the risk of the subjects in the cohort. These risk scores can be ranked to put subjects into different risk categories such as tertiles (or deciles if there are more data). We choose M1, and the Ridge BIC model which has the highest C-statistic in Table 13.2 for demonstration.

Table 13.3 presents the summary statistics of the difference in survival distributions depicted in Figure 13.1. The restricted mean survival time is computed as the area under

TABLE 13.1
Models derived from various classical model selection methods, using entire PBC dataset, hazard ratios $\exp(\hat{\beta})$ are presented.

Covariates	M1	M2	M3	M4	Lasso@		Ridge	
					AIC	BIC	AIC	BIC
logbili	2.334	2.312	2.279	2.688	2.213	2.137	2.016	1.651
edema	2.238	2.110	2.107		2.022	1.996	2.182	2.099
age	1.034	1.032	1.034		1.031	1.027	1.029	1.023
stage	1.386	1.394	1.412		1.366	1.326	1.369	1.284
lalb	0.120	0.119	0.128		0.168	1.180	0.166	0.199
lptime	8.164	7.267	8.004		6.535	5.513	7.715	6.898
ast			1.002		1.002	1.001	1.002	1.002
copper		1.002	1.001		1.001	1.001	1.001	1.002
ascites					1.320	1.291	1.407	1.498
trig		0.998	0.998		0.998	0.999	0.998	0.999
hepato					1.049		1.158	1.223
spiders							0.947	1.044
sex							1.057	1.063
chol							1.000	1.000
alk.phos							1.000	1.000
platelet							1.000	1.000

Note: All covariates were treated as continuous effects.

Model	Selection Method
M1	Several model building procedures using BIC as stopping criterion came up with this same model: a. Forward selection, BIC; b. Backward elimination, BIC; c. Stepwise, BIC
M2	Backward elimination, AIC
M3	a. Forward selection, AIC; b. Stepwise, AIC
M4	Best single variable model, logbili (log(bilirubin)) is the most significant variable (p < .00001)

the K-M survival curve, over the range from $[0, t_{max}]$, where t_{max} (= 12.5 years) is the maximum time for all K-M curves considered and serves as a common upper limit for the restricted mean calculation. The overall logrank test, and the logrank tests for the difference in survival distributions between any two risk categories all yield p-values < 0.00001. Both M1 and Ridge, BIC models produce similar results with little difference in C-Statistics, this further illustrates that M1 model is most preferable because it only takes 6 variables to achieve similar predictability.

TABLE 13.2
C-statistic of models using the full PBC dataset.

Model Selection Method	Model Size	C-Statistic
M4	1	0.748
M1	8	0.784
M3	6	0.790
M2	9	0.791
Lasso, AIC	11	0.794
Lasso, BIC	10	0.794
Ridge, AIC	16	0.796
Ridge, BIC	16	0.799

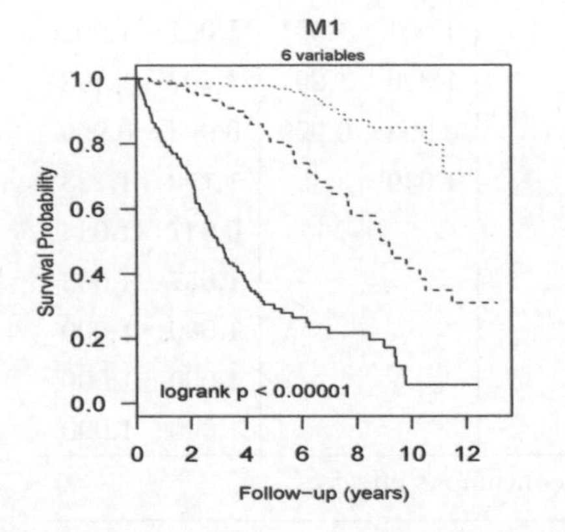

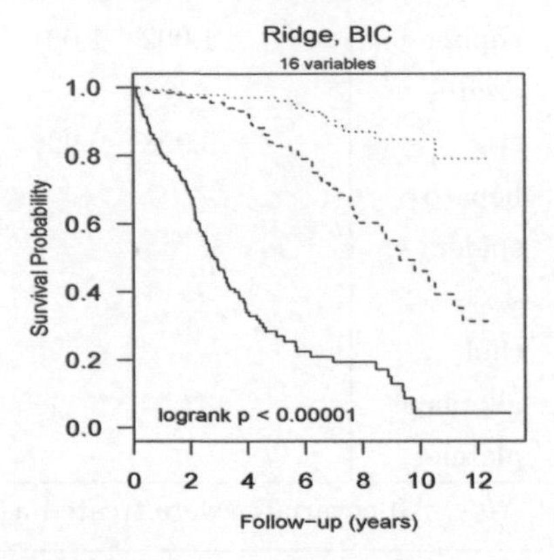

FIGURE 13.1
Kaplan-Meier curves of the survival time, stratified by tertiles of risk scores from two models: M1 - Six-variable model (left panel), and Ridge, BIC model (right panel).

TABLE 13.3
Summary statistics of the survival distributions by risk categories, scoring using the entire dataset.

Model Selection Method	Risk Categories	N Events/ Total	Restricted Mean (se) in years	Median (years)	(95% CI)
Stepwise, BIC	Low	14/140	11.31 (0.283)	NA	(NA, NA)
	Medium	45/139	8.66 (0.393)	9.19	(7.70, 11.47)
	High	102/139	4.21 (0.340)	2.97	(2.55, 3.71)
Ridge, BIC	Low	14/140	11.36 (0.278)	NA	(NA, NA)
	Medium	42/139	8.98 (0.381)	9.30	(7.79, NA)
	High	105/139	3.98 (0.320)	2.84	(2.44, 3.55)

13.5 Challenges and a proposal

The aforementioned process of using the same dataset for model building selection, and inference has been utilized in practice. This conventional process has potential of self-serving problem. In this section, we first summarize the reasons for the issue of over-fitting, then we will use the PBC dataset to demonstrate such an over-fitting problem with this conventional process.

13.5.1 Over-fitting issue

If we use the same dataset to construct a prediction rule and evaluate how well this rule predicts, the predictive performance can be overstated. Over-fitting occurs when a model describes the random variation of the observed data instead of the underlying relationship. Generally an overfit model indicates better fit and smaller prediction error than in reality because the model can be exceedingly complex to accommodate minor random fluctuation in observed data. This leads to the issue of over-optimism. We will demonstrate some covariates can be selected as statistically significant risk predictors of an event outcome even though there is no underlying relationship between them.

13.5.2 Noise variables become significant risk factors

Using the PBC dataset, we first randomly permutate the 418 survival time observations to break the ties between the observed or censored survival time and its covariate vector Z. Then we apply various traditional methods including forward selection, backward elimination, and stepwise regression to fit the data with newly permuted y' and 16 original covariates. These two steps are repeated 5,000 times.

Table 13.4 shows the median, interquartile range, and the range of the number of variables selected in 5,000 simulations. Using AIC as tuning criterion tends to over-select variables to achieve better model fit, while BIC tends to select fewer variables. Stepwise regression with BIC picked up at least one variable 25% of the time.

Consider one such realization, the risk ratio of the variable log(protime) is 11.5 (se=0.845, p=.0039). Using this single variable model to score the entire dataset, and stratify the risk scores into two strata, we have the left panel of the Kaplan-Meier (K-M) plot in Figure 13.2. It appears that this model is a reasonable prediction tool for survival.

If we randomly split the data into two parts, using the first half (called training data) to fit the model using Stepwise BIC approach, the upper part of the right panel shows some separation again; one may also pause here had we just given the training dataset. However, if we go one step further, using the training model to score the holdout sample (the other half) to evaluate the generalizability of the model, the lower right K-M plot shows no separation at all.

TABLE 13.4
Summary statistics of the number of variables selected in 5,000 runs.

Selection Procedure	Tuning Criterion	Median	(1st, 3rd) Quartile	(Min, Max)
Forward Selection	AIC	2	(1, 3)	(0, 9)
	BIC	0	(0, 1)	(0, 3)
Backward Elimination	AIC	3	(2, 4)	(0, 11)
	BIC	0	(0, 1)	(0, 6)
Stepwise	AIC	2	(1, 3)	(0, 10)
	BIC	0	(0, 1)	(0, 4)

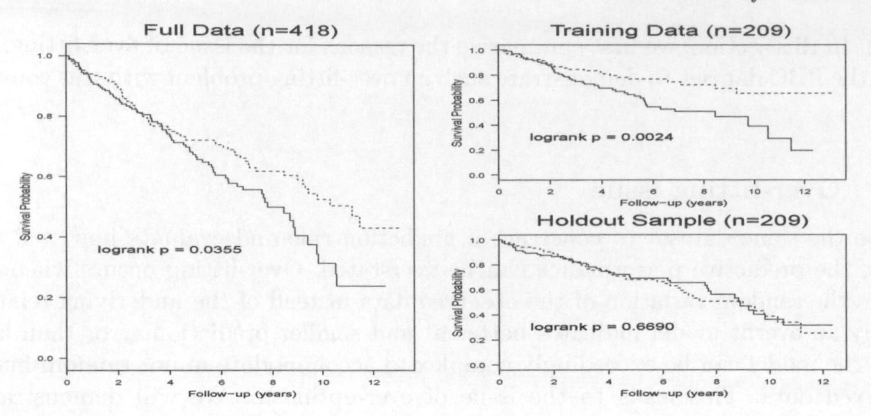

FIGURE 13.2
An example of using a holdout sample to show the over-fitting phenomena, using the entire dataset.

13.5.3 Utilizing cross-validation in model selection process

One way to address over-fitting is to use cross-validation (CV) techniques. It is preferred to evaluate the prediction error with independent data (validation data) separated from the data used for model building (training data). Two ways of conducting cross-validation are:

1. **K-fold cross-validation**
   - Randomly partition the entire original dataset into K groups
   - For $k = 1, \cdots, K$, do the following:
     - Retain a single k^{th} group as validation data
     - Use the rest $(K-1)$ groups as training data to estimate β and form prediction rule
     - Evaluate a predictive performance measure (e.g., C-Statistic) using the validation data
   - Compute the average of the K predictive performance measures

   All data are used for both training and validation, and each observation is used for validation exactly once.

2. **Monte-Carlo cross-validation**
   - Randomly subsample p percent of the entire dataset without replacement and retain it as validation data
   - Use the rest (1-p percent) data as training data to form prediction rule
   - Evaluate the prediction model using the validation data
   - Repeat the above steps M times
   - Average the M estimates to get the final estimate of the predictive accuracy measure

   In this schema, the results may vary if the analysis is repeated with different random splits; some observations may be selected more than once for training, while others may not be selected for validation. However, these can be resolved by increasing M, the number of times the CV is repeated.

CV is especially useful when we do not have enough observations to set aside for test set validation; it may address the over-fitting issue.

13.5.4 3-in-1 dataset modelling proposal

We present the following strategy to help us select a model with best predictive accuracy, utilizing CV in model selection process. First we randomly partition a given dataset into two parts, for example, $D_{train.val}$ has 50% of the data and $D_{hold.out}$ has the rest.

1. Model Building:
   For each candidate model selection method considered, use dataset $D_{train.val}$ to build a model and apply cross-validation to find the predictive accuracy measures of interest.

2. Model Selection:
   Identify the "optimal" model selection method(s) that gives us the most acceptable or highest predictive accuracy measures with a reasonable model size in Step 1. For the final model, we can either

   (a) Use the "optimal" method to refit the dataset $D_{train.val}$ to obtain the prediction equation for each subject; or
   (b) Apply the average model obtained from the training data portion of $D_{train.val}$ to $D_{hold.out}$ to report how good it is, and future data for application (e.g., identifying future study population for intervention).

   While the first approach can provide a simple scoring system (a linear combination of selected covariates in this example), the second approach as the "bagging" version, Breiman (1996), may have superior performance to the first one for "discrete" procedure such as stepwise regression and Lasso.

3. Statistical Inference:
   Dataset $D_{hold.out}$ is not involved in any training process; therefore, it is best suited for testing and reporting the predictive accuracy of the final model derived from $D_{train.val}$ using the optimal model selection method identified in Step 2 without over-fitting issue and biases. Additionally, using scores obtained from numerous training models developed during the cross-validation procedure in Step 1, model averaging can be applied to increase predictive accuracy in the holdout sample.

   We now use the PBC data to illustrate this proposal.

13.6 Establishing a prediction model

Conventional model building strategies using the entire dataset without external data validation may have limited application. For any given dataset, it will be ideal to be able to

(1) develop a predictive model with validation, and
(2) report how well the model performs externally.

Hereafter, we apply our proposal to the PBC dataset using Monte Carlo cross-validation to illustrate the idea.

1. First, retain a random 50% sample of data from the randomized trial portion of the dataset and another 50% of the follow-up portion of the data. This holdout dataset $D_{hold.out}$ consists of 209 observations and will be used to conduct inference, and examine the generalizability of our model developed by the other half of the dataset (called $D_{train.val}$).

TABLE 13.5
PBC Cross-validation data $D_{train.val}$: performance measures of different model selection methods, using random cross-validation with 2/3 of observations as training data and 1/3 of data as validation data.

Model Selection Method		C-	Median
Procedure	Tuning Criterion	Statistic	Model Size
Forward	AIC	0.737	6
Selection	BIC	0.733	4
Backward	AIC	0.742	7
Elimination	BIC	0.740	4
Stepwise	AIC	0.736	6
	BIC	0.744	4
Lasso	AIC	0.746	8
	BIC	0.746	6
Ridge	AIC	0.749	16
	BIC	0.757	16

2. Apply Monte Carlo CV to $D_{train.val}$ dataset, use 2/3 of the 209 observations as training data, and the rest 1/3 observations (70 in this case) as validation data.

3. In each CV run, evaluate the model selected by each candidate method using a predictive measure of interest: censoring-adjusted C-statistic proposed by Uno et al. (2011).

4. Repeat the model selection and computation of the above performance measures 200 times. The average over all 200 measures is presented in Table 13.5 using various model selection methods. The C-statistic measures how well the model predicts throughout the course of study.

In Table 13.5, the three traditional methods (forward selection, backward elimination, and stepwise regression), have comparable predictive performance in this dataset, with stepwise regression using BIC as stopping criterion yielding the highest censoring-adjusted C-statistic of 0.744. The Lasso and Ridge model selection methods yield slightly higher C-statistic than the other three methods with larger median model size. Ridge regression with BIC as stopping criterion using all 16 covariates yields the best predictive measure.

Stepwise regression using BIC as stopping criterion has comparable predictive performance as Lasso and Ridge regression on this particular dataset. However, the median model size of stepwise, BIC method is 4, compared with 6 in Lasso and 16 in Ridge regression method. The difference in the predictive measures between the traditional methods and the two penalized regression methods are not substantial in this dataset. Since CV-based estimator like all statistics is subject to some variability, the observed differences (if small) may be due to stochastic variation. If one decides that the slight difference in predictive accuracy does not outweigh the ease of implementation and smaller number of variables used by the traditional methods, one may choose the method that gives the highest predictive measures. In this case, stepwise regression using BIC as stopping criterion performs very well.

We should be aware that the CV-based prediction measures, such as the highest C-statistic of 0.757 derived from the Ridge BIC regression model, are optimistically biased because the models with the best estimated prediction accuracy measures are selected. Hence the true C-statistics may actually be lower.

13.6.1 Evaluating model's generalizability

To examine the generalizability of the model fit, we use the holdout sample for the evaluation and reporting. For the model selection method chosen by the CV procedure with best predictive measures, say, stepwise using BIC as stopping criterion, the following scoring algorithm is applied:

1. Use 2/3 of the $D_{train.val}$ dataset to build a model called $\mathcal{M}_1$, the rest 1/3 of this dataset is used to evaluate the predictive performance measures aforementioned.
2. Use β estimates from $\mathcal{M}_1$ to obtain a score for each subject in the holdout sample $D_{hold.out}$. This score $r_1 = \exp(\hat{\beta}Z)$, where Z is the covariate of each subject in dataset $D_{hold.out}$.
3. Repeat the above two steps 200 times, and we have $r_1, r_2, r_3, \ldots, r_{200}$ for each subject in the holdout sample derived from 200 training models $\mathcal{M}_1, \mathcal{M}_2, \mathcal{M}_3, \ldots, \mathcal{M}_{200}$.
4. Take the average of $r_1, r_2, r_3, \ldots, r_{200}$, which becomes the final risk score of the subjects in the holdout sample. The distribution of this summary risk score (r) can be obtained and used for risk profiling.

We stratify the summary risk scores (r) using tertiles of r. For the holdout sample, $D_{hold.out}$ dataset, Figure 13.3 displays the Kaplan-Meier curves of the survival time stratified by the risk scores using three risk categories. The left panel shows the results using the optimal model selection method stepwise BIC identified in Table 13.5. For reference, the right panel shows the scoring results using Lasso with BIC as stopping criterion, a penalized regression method that used fewer variables than Ridge regression.

Table 13.6 and Table 13.7 present the summary statistics of the difference in survival distributions depicted in Figure 13.3. All the reported log-rank test p-values, and confidence intervals based on the holdout sample are valid conditional on the scoring system derived from the training dataset. We also evaluate another schema by stratifying the risk scores

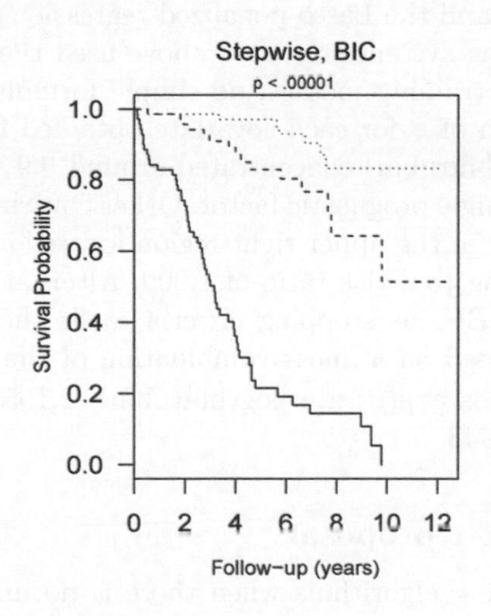

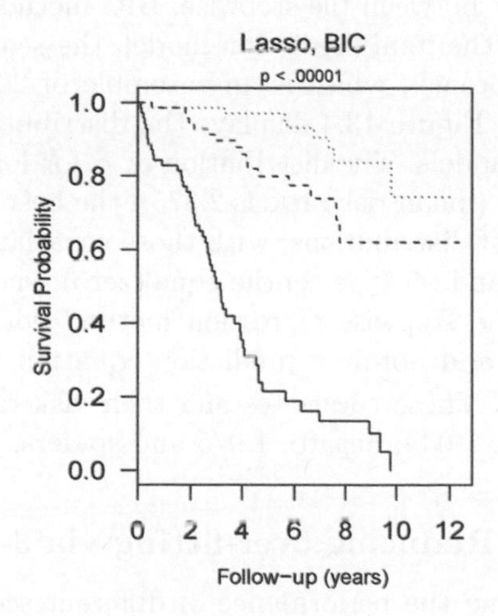

FIGURE 13.3
K-M curves of the survival time, stratified by tertile of risk scores of the holdout sample.

TABLE 13.6
Summary statistics of the survival distributions by risk categories, scoring using 200 training models on $D_{hold.out}$ dataset.

Model Selection Method	Risk Categories	N Events/ Total	Restricted Mean (se) in years	Median (years)	(95% CI)
Stepwise, BIC	Low	7/70	11.16 (0.370)	NA	(NA, NA)
	Medium	15/69	9.41 (0.606)	NA	(7.79, NA)
	High	55/70	3.76 (0.365)	3.15	(2.66, 4.05)
Lasso, BIC	Low	8/70	10.96 (0.406)	NA	(NA, NA)
	Medium	16/69	9.48 (0.586)	NA	(7.79, NA)
	High	53/70	3.74 (0.369)	3.15	(2.66, 4.05)

Note: Restricted mean with upper limit = 12.1 years.

TABLE 13.7
Logrank test p-values of the difference between the survival distributions by risk categories.

Survival Difference between	Model Selection Method	
Risk Categories	Stepwise, BIC	Lasso, BIC
Overall	$p < 10^{-7}, \chi^2_{(2)}$=124.63	$p < 10^{-7}, \chi^2_{(2)}$=119.47
Pairwise Comparison		
Low vs. Medium	0.0131	0.0151
Low vs. High	$< 10^{-7}$	$< 10^{-7}$
Medium vs. High	$< 10^{-7}$	$< 10^{-7}$

using four risk categories (low, medium low, medium high and high). The results (not shown) are similar between the stepwise, BIC method and the Lasso penalized regression method.

As for the final prediction model, the scoring system presented above used the average model approach, which is an ensemble of 200 training models, no simple formula can be expressed. Figure 13.4 displays the distribution of $\hat{\beta}$ for each covariate obtained from 200 training models. The distribution of $\hat{\beta}$ for log(bilirubin) concentrated around 0.9 with low variability (mean risk ratio is 2.375), the best single prognostic factor. Other covariates have a variety of distributions, with those covariates in the upper right region located closely at zero (5th and 95th percentile equal zero) leading to a risk ratio of 1.000. Alternatively, we can use the stepwise regression method with BIC as stopping criteria to fit the dataset $D_{train.val}$ and obtain a prediction equation based on a linear combination of the selected covariates. These covariates and their risk ratios $exp(\hat{\beta})$ are: log(bilirubin), 2.195; edema, 5.596; age, 1.044; hepato, 1.975 and spiders, 1.943.

13.6.2 Reducing over-fitting via 3-in-1 proposal

We examine the performance of different scoring algorithms when there is no underlying relationship between survival time observations and covariates in a high-dimensional setting using simulations on the PBC dataset as follows. We keep the original survival time observations $(time_i, death_i)$, simulate 50 independent binary random variables with event

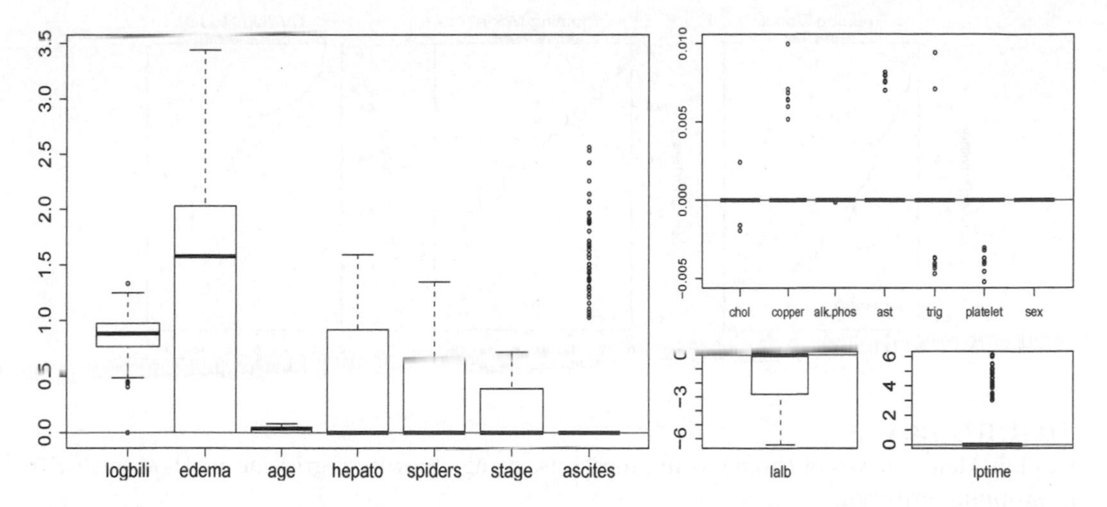

FIGURE 13.4
Distribution of $\hat{\beta}$ obtained from the 200 training models for each covariate.

rates ranging from 0.001 to 0.981 with 0.02 increment and 160 independent normal random variables with the same mean and standard deviation as the ten continuous covariates in PBC dataset (16 variables for each covariate distribution).

We randomly partition the dataset into two halves: $D_{train.val}$ and $D_{holdout}$, each has 209 observations ($n < p$). Applying stepwise regression with BIC as stopping criteria to the datasets, we present Figure 13.5 as follows:

- Leftmost panel: using the training data $D_{train.val}$ to build a model, and score on itself (conventional way)
- Middle panel: using the training data $D_{train.val}$ to build a model, but score on the holdout sample $D_{holdout}$
- Rightmost panel: apply our 3-in-1 modeling strategy similar to the procedure described in Section 13.6 with Stepwise BIC as the only candidate method considered, obtain an average model derived from 200 training models obtained during the CV process using $D_{train.val}$, score the holdout sample $D_{holdout}$ using this average model.

The leftmost panel shows over-fitting using conventional same dataset modeling way; the other two panels did not show separation of the K-M curves. Applying the average model derived from 200 training models in the spirit of bagging (i.e., bootstrap aggregate), all three K-M curves almost overlap each other, leading to the correct conclusion that there is no relationship between the survival time and covariates.

The example shows that a combination of cross-validation and holdout sample is useful in combating overfitting.

13.7 Remarks

It is important to consider model building, selection and inference processes simultaneously as a package. The usual practice of using the same dataset for implementing procedures

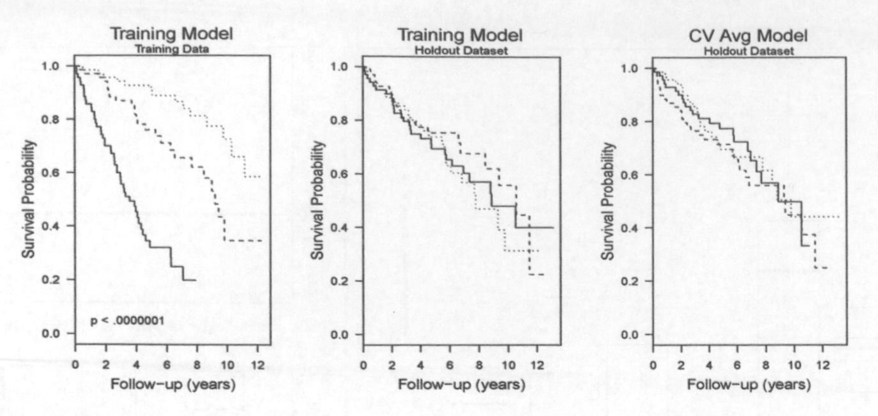

FIGURE 13.5
Kaplan-Meier curves of three scoring methods, using stepwise regression method with BIC as stopping criterion.

for these three steps may result in invalid inference as we demonstrated in this chapter. One may question about the efficiency issue for splitting the dataset for the final inference. However, with the conventional method, it is difficult to quantify the reliability of our claim at the end after an extensive, iterative model building process. This is probably why there are numerous false positive findings in all the scientific investigations. In fact, the sizes (or event rates) of most studies in practice may be too small for building reliable models for making valid inference.

We proposed a 3-in-1 dataset modeling strategy, achieving model building, selection, and holdout inference in one dataset. Cross-validation techniques are utilized to provide a sanity check for model fit to assess whether its predictive performance is acceptable. We can then select the optimal model building method to develop the final model and proceed with the inference part using the holdout sample, leading to more reproducible results and better application.

Needless to say, model building does not only depend on statistical grounds, knowledge of subject matter is absolutely essential in selecting the most appropriate model tailored to our needs. We focused on several classical methods and found that careful implementation of these methods could help us find a reasonably good predictive model. While censoring-adjusted C-statistic was used to evaluate predictive performance for illustration, other predictive measures or model evaluation methods can be considered. For example, Tian et al. (2007) proposed model evaluation based on the distribution of estimated absolute prediction error. Uno et al. (2011) looked at the incremental values of predictors. Furthermore, the candidate model selection methods considered were presented mainly under the framework that $n > p$. For the review on high-dimensional regression with survival outcomes, we refer to Chapter 5. They described some of the existing literature on dimension reduction, shrinkage estimation procedures with a range of penalty functions, and some hybrid procedures with univariate screening followed by shrinkage.

As mega datasets (genomic, data warehouse) become increasingly available, together with the ease of data storage and rapid development of data mining methodologies in censored data, these have enabled us to utilize more information for model development. It would be of interest to see how other datasets and predictive measures perform using our scoring algorithm. Moreover, what proportion of samples should we retain for holdout sample, other cross-validation techniques with different partition schema can also be considered to fine-tune our model building strategies. Additionally, we tend to develop methods separately for each step of the model building, selection and inference process. If one aims at making efficient and valid inference about a parameter, say, the restricted mean survival time, a more consistent and integrated process using criteria to increase the precision of the final inference procedure should be considered. These questions remain an area of active research.

With the advent of the information age and the vast growth in the availability of massive amount of data, the challenges presented a unique window of opportunity for us to re-examine our conventional model selection strategy. Alternative modeling strategies in the analysis of censored outcome data could be considered to utilize the data and increase the overall model predictability in this big data era and dawn of personalized medicine age.

Bibliography

Acquah, H. and Carlo, M. (2010), 'Comparison of Akaike information criterion (AIC) and Bayesian information criterion (BIC) in selection of an asymmetric price relationship', *Journal of Development and Agricultural Economics* **2**(1), 001–006.

Akaike, H. (1974), 'A new look at the statistical model identification', *IEEE Transactions on Automatic Control* **19**(6), 716–723.

Andersen, P. and Gill, R. (1982), 'Cox's regression model for counting processes: a large sample study', *The Annals of Statistics* **10**(4), 1100–1120.

Breiman, L. (1996), 'Bagging predictors', *Machine Learning* **24**(2), 123–140.

Collett, D. (2003), *Modeling Survival Data in Medical Research*, Chapman & Hall/CRC, Boca Raton, FL.

Cox, D. R. (1972), 'Regression models and life-tables', *Journal of the Royal Statistical Society - Series B* **34**, 187–220.

Fan, J. and Li, R. (2001), 'Variable selection via nonconcave penalized likelihood and its oracle properties', *Journal of the American Statistical Association* **96**(456), 1348–1360.

Fan, J. and Li, R. (2002), 'Variable selection for Cox's proportional hazards model and frailty model', *The Annals of Statistics* **30**(1), 74–99.

Fleming, T. R. and Harrington, D. P. (1991), *Counting Processes & Survival Analysis*, Applied probability and statistics, Wiley, New York.

Gerds, T. and Schumacher, M. (2006), 'Consistent estimation of the expected Brier score in general survival models with right-censored event times', *Biometrical Journal* **48**(6), 1029–1040.

Goeman, J. (2009), 'L_1 penalized estimation in the cox proportional hazards model', *Biometrical Journal* **52**(1), 70–84.

Graf, E., Schmoor, C., Sauerbrei, W. and Schumacher, M. (1999), 'Assessment and comparison of prognostic classification schemes for survival data', *Statistics in Medicine* **18**(17-18), 2529–2545.

Grave, E., Obozinski, G. and Bach, F. (2011), 'Trace lasso: a trace norm regularization for correlated designs', *arXiv:1109.1990.*

Harrell, F. (2001), *Regression modeling strategies: with applications to linear models, logistic regression, and survival analysis*, Springer, New York.

Ioannidis, J. (2005), 'Why most published research findings are false', *PLoS Medicine* **2**(8), e124.

Park, M. and Hastie, T. (2007), 'L_1-regularization path algorithm for generalized linear models', *Journal of the Royal Statistical Society: Series B (Statistical Methodology)* **69**(4), 659–677.

Pencina, M. and D'Agostino, R. (2004), 'Overall C as a measure of discrimination in survival analysis: model specific population value and confidence interval estimation', *Statistics in Medicine* **23**(13), 2109–2123.

Schwarz, G. (1978), 'Estimating the dimension of a model', *The Annals of Statistics* **6**(2), 461–464.

Simmons, J., Nelson, L. and Simonsohn, U. (2011), 'False-positive psychology undisclosed flexibility in data collection and analysis allows presenting anything as significant', *Psychological Science* **22**(11), 1359–1366.

Steyerberg, E., Vickers, A., Cook, N., Gerds, T., Gonen, M., Obuchowski, N., Pencina, M. and Kattan, M. (2010), 'Assessing the performance of prediction models: a framework for traditional and novel measures', *Epidemiology* **21**(1), 128–138.

Therneau, T. M. and Grambsch, P. M. (2000), *Modeling Survival Data: Extending the Cox Model*, Springer, New York.

Tian, L., Cai, T., Goetghebeur, E. and Wei, L. (2007), 'Model evaluation based on the sampling distribution of estimated absolute prediction error', *Biometrika* **94**(2), 297–311.

Tibshirani, R. (1996), 'Regression shrinkage and selection via the lasso', *Journal of the Royal Statistical Society. Series B (Methodological)* pp. 267–288.

Tibshirani, R. (1997), 'The lasso method for variable selection in the Cox model', *Statistics in Medicine* **16**(4), 385–395.

Uno, H., Cai, T., Pencina, M., D'Agostino, R. and Wei, L. (2011), 'On the C-statistics for evaluating overall adequacy of risk prediction procedures with censored survival data', *Statistics in Medicine* **30**(10), 1105–1117.

van Houwelingen, H. C., Bruinsma, T., Hart, A. A. M., van't Veer, L. J. and Wessels, L. F. A. (2006), 'Cross-validated Cox regression on microarray gene expression data', *Statistics in Medicine* **25**, 3201–3216.

van Houwelingen, H. C. and Putter, H. (2008), *Dynamic Prediction in Clinical Survival Analysis*, Monographs on statistics and applied probability, CRC Press, Boca Raton.

Van Houwelingen, J. (2001), 'Shrinkage and penalized likelihood as methods to improve predictive accuracy', *Statistica Neerlandica* **55**(1), 17–34.

Verweij, P. and Van Houwelingen, H. (2006), 'Penalized likelihood in Cox regression', *Statistics in Medicine* **13**(23-24), 2427–2436.

Witten, D. and Tibshirani, R. (2010), 'Survival analysis with high-dimensional covariates', *Statistical Methods in Medical Research* **19**(1), 29–51.

Wu, T. and Lange, K. (2008), 'Coordinate descent algorithms for lasso penalized regression', *The Annals of Applied Statistics* **2**(1), 224–244.

Zou, H. and Hastie, T. (2005), 'Regularization and variable selection via the elastic net', *Journal of the Royal Statistical Society: Series B (Statistical Methodology)* **67**(2), 301–320.

14

Bayesian Model Selection

Purushottam W. Laud

Division of Biostatistics, Medical College of Wisconsin

CONTENTS

14.1 Introduction 285
14.2 Posterior model probabilities and Bayes factor 286
14.3 Criterion-based model selection 287
14.3.1 Information criteria 287
14.3.1.1 BIC 288
14.3.1.2 DIC 288
14.3.2 Predictive criteria 289
14.3.2.1 Cross-valid prediction 289
14.3.2.2 Replicate experiment prediction 290
14.4 Search-based variable selection 291
14.4.1 Stochastic search variable selection 292
14.4.2 Reversible jump MCMC 293
14.5 Discussion 294
Bibliography 295

Model selection methods from the Bayesian perspective rely on full probability specification for each model under consideration. This includes the likelihood and the prior for each model. Beyond this, some methods employ the notions of prior and posterior probabilities on the models. Others omit this and rely on optimizing a criterion defined for each model. When the number of models under consideration is moderate, complete calculations of posterior probabilities or the selection criteria are feasible. For a large set of models, as is often the case in variable selection, various forms of stochastic searches can be quite effective. In this chapter we describe these three types of methods based on: (i) model probabilities, (ii) defined criterion, and (iii) stochastic search, and their use in survival analysis.

The plan for the chapter begins with an introduction with some needed notation in Section 26.1. We next describe the three types of methods in successive sections. A discussion section concludes the chapter.

14.1 Introduction

We consider model selection in a broad sense here. Firstly, it can be selection between two alternative models or several competing models. One common example of the latter case is when one wishes to select explanatory variables, from a potential set of these, that have an effect on the outcome in a regression model such as the Cox proportional hazards model. The former case is simpler to consider as it resembles hypothesis testing.

Secondly, model selection can involve exclusive ranges for parameter values, alternative distributional assumptions, or even some other aspects of models such as functional forms. The following three example scenarios are intended to convey this breadth of considerations to follow. Given an exchangeable (conditionally i.i.d.) sample from a population we may wish to determine whether the median survival time in the population is greater than a specified number θ_0, or not. Again with exchangeable observations of survival times, we may wish to select between the Weibull and the lognormal distributions for the population. Given survival data suitable for a regression on some covariates, we may wish to select between the Cox model and the proportional odds model; or between a model with frailties and one without.

From the Bayesian perspective, there is a unified approach to all of these problems. To describe it, let us establish some notation. We will denote a **model** by M or m, sometimes with a subscript to emphasize that more than one model is under consideration. When clear from the context, we will drop the subscript for brevity even though, strictly speaking, it should be used. The collection of all models under consideration will be denoted by $\mathcal{M}$ and often called the **model space**.

By a model we mean a complete probability specification for all the observables. Typically, this involves the likelihood (or the sampling distribution for the observables conditioned on the parameters of the model), and the prior distribution for the parameters. Generically, we will denote the **observables** by y and the **parameters** by θ. As parameters belong to specific models, we need a subscript m as in θ_m to be accurate. Again, we will omit this subscript if the context provides sufficient clarity. In some conditional models (typically regressions), we will denote the **fixed observables** by x and these will be considered constants throughout the context. Only y and θ will be given distributions.

The **sampling distribution** will be denoted by $p(y|\theta)$ mostly omitting, as is usual, the conditioning on any x variables present. Looking at this function (or any multiple of it) as a **likelihood** in θ, we will denote it by $L(\theta; y)$ thus making it clear that this is not a conditional distribution. The **prior** distribution for θ will be denoted $\pi(\theta)$. When considering several (countable) models in a model space $\mathcal{M}$, a full Bayesian specification requires probabilities for each model in the space which we will denote $p(m), m \in \mathcal{M}$ such that $p(m) \geq 0$, $\sum_{m \in \mathcal{M}} p(m) = 1$.

14.2 Posterior model probabilities and Bayes factor

With a full Bayesian specification, Bayes theorem yields the posterior probability of any model m^* in $\mathcal{M}$ as

$$p(m^*|y) = \frac{p(y|m^*)\, p(m^*)}{\sum_{m \in \mathcal{M}} p(y|m)\, p(m)} \tag{14.1}$$

where

$$p(y|m) = \int_{\Theta_m} p(y|\theta_m, m)\pi(\theta_m) d\theta_m, \tag{14.2}$$

Θ_m being the parameter space for model m. We note here that $p(y|m)$ is the marginal distribution of the data under model m, and is sometimes called – perhaps misleadingly – the "marginal likelihood."

While these expressions are rather simple, they have many interesting consequences.

One is the automatic control for the multiplicity of decisions, inherent in model selection, attained by good choices of prior model probabilities $p(m)$. This is discussed by Scott and Berger (2010). Another, quite distant from the first, yields a measure of evidence – called the Bayes Factor – provided by the data in favor of one model versus another, regardless of the prior model probabilities. Yet another is to note that the data marginal $p(y|m)$ may not exist if the prior $\pi(\cdot)$ is not a proper density. In Bayesian analysis, there are good reasons sometimes to use an improper density, i.e., one that does not assign finite total mass to the parameter space.

To introduce the Bayes Factor (BF hereafter), let $\mathcal{M}$ contain only two models, say M_1 and M_2. Thus $p(M_2) = 1 - p(M_1)$ so that the prior odds in favor of M_1 are $p(M_1)/p(M_2)$. Now, by (14.1), the posterior odds equal $p(M_1|y)/p(M_2|y) = p(y|M_1)p(M_1)/\{p(y|M_2)p(M_2)\}$. The ratio of posterior to prior odds is $p(y|M_1)/p(y|M_2)$, free of the prior model probabilities. This ratio was defined by Jeffreys (1998) in the early 1930s as the **Bayes Factor**. We thus have

$$BF_{1,2} = \frac{\text{Posterior odds of } M_1}{\text{Prior odds of } M_1} = \frac{p(y|M_1)}{p(y|M_2)}. \tag{14.3}$$

The subscript on BF identifies the numerator and the denominator. Clearly, $BF_{2,1} = 1/BF_{1,2}$. An excellent article detailing the use and interpretation of the BF is by Kass and Raftery (1995).

When $\mathcal{M}$ contains more than two models, the ratio of data marginals with two models yields the BF assuming a model space with just those two models. This can be seen by conditioning on the event $M_1 \cup M_2$ a priori, i.e., when other models are not considered. A ratio of posterior to prior odds of a particular model versus all others is possible to compute, but then it depends on prior probabilities of individual models in addition to the data marginals.

Computation of the BF is not necessarily a simple matter. In many situations, the data marginal in (14.2) is a high-dimensional integral. One often needs to resort to simulation techniques (Monte Carlo or Markov chain Monte Carlo). Many have discussed such methods, including Newton and Raftery (1994), Gelfand and Dey (1994), Chib (1995), Chib and Jeliazkov (2001), Chen and Shao (1997*a*,*b*). See also Chen, Shao and Ibrahim (2000). For survival analysis, Ibrahim et al. (1999) and Ibrahim and Chen (1998) provide a method for computing posterior model probabilities for proportional hazards models.

14.3 Criterion-based model selection

If we wish to avoid assigning prior probabilities to models, model selection can be based on the values of a criterion computed for each model. The criterion should have some appeal and interpretation suitable for the purpose. Here we view the criteria in current use as falling into two categories: those measuring the information content of models in some sense and following the lead of **Akaiki's Information Criterion (AIC)**, and those deriving from predictive considerations.

14.3.1 Information criteria

The very popular AIC is based on comparing models by the maximized likelihood under each with a penalty for the "size" of the model. The size is measured by the number of parameters and it works well in many cases. Formally, we can write it as

$$AIC_m = -2\log\left(\sup_{\theta_m \in \Theta_m} p(y|\theta_m)\right) + 2k_m \tag{14.4}$$

where k_m is the number of parameters in model m. The model with the smallest value for this criterion is selected since the criterion has an interpretation as the loss of information (measured by the Kullback-Leibler divergence) with respect to a true model. It can also be seen as an estimate of the expected predictive log-likelihood for replicated experiments in a frequentist setting. It should be noted, however, that k_m is difficult to specify in many situations including models with random effects and semi-parametric models such as the Cox model.

14.3.1.1 BIC

Schwarz (1978) introduced what is called the **Bayesian Information Criterion (BIC)**. It also takes the form for the maximized log-likelihood with a penalty. But the penalty involves the sample size n. We can define it as

$$BIC_m = -2\log\left(\sup_{\theta_m \in \Theta_m} p(y|\theta_m)\right) + k_m \log(n)\ . \tag{14.5}$$

The fixed multiple of the number of parameters in the AIC, namely 2, is replaced by $\log(n)$ in the penalty term. With this modification, the difference in BIC of two models approximates twice the log of BF under mild conditions in the following sense (Kass and Raftery, 1995):

$$\frac{\{BIC_{m_2}(n) - BIC_{m_1}(n)\}/2 - \log(BF_{1,2}(n))}{\log(BF_{1,2}(n))} \to 0 \text{ as } n \to \infty\ . \tag{14.6}$$

This allows the relative error of $e^{\{BIC_{m_2}(n) - BIC_{m_1}(n)\}/2}$ in approximating $B_{1,2}$ to be $O(1)$ with some priors. Kass and Wasserman (1995) show that BIC can provide a better approximation, to order $O(n^{-1/2})$, with the use of a reference prior called the unit information prior in the case of nested models. These approximations connecting this criterion to BF give it a Bayesian flavor and perhaps justifies its name.

Implicit in the definition of BIC is the notion that the number of parameters k_m as well as the sample size n are easily determined without any ambiguity. Volinsky and Raftery (2000) show clearly how, in the censored data case so common in survival analysis, using the number of uncensored observations is to be preferred to using the total sample size. This is essentially because censored and uncensored samples contribute different amounts of information to the likelihood. While they consider the cases of i.i.d. exponential data and regression data using Cox's partial likelihood, they comment that their method of deriving the appropriate choice of "n" via consideration of the unit information prior applies more broadly.

14.3.1.2 DIC

How to count the number of parameters in a model is a vexing problem, especially in models with random effects and hierarchical specifications. Such situations arise routinely in survival analysis, for example in frailty models and latent class models. Both AIC and BIC require this number k_m. To address this difficulty and with the intention of providing an AIC-like criterion suitable in a Bayesian analysis, Spiegelhalter et al. (2002) proposed to measure the complexity of a model in such a way that this measure of complexity would reduce to the number of parameters in the simple cases where the latter are easy to count.

Using this as the penalty and measuring goodness of fit via the deviance at the posterior mean of the parameter vector, they defined the **Deviance Information Criterion (DIC)**.

Let $D(\theta) = -\log\ L(\theta; y)$ denote the deviance and $\bar{D} = E_{\theta|y}(D(\theta; y))$ its posterior expectation. Then the measure of complexity is defined as $p_D = \bar{D} - D(E_{\theta|y}(\theta))$. It equals the posterior mean of the deviance minus the deviance at the posterior mean. It can be shown (Spiegelhalter et al., 2002) that, as the sample size grows, the expectation of p_D approximately equals the number of parameters p in a fixed effects models. Now, with $\bar{\theta} = E_{\theta|y}(\theta)$, define

$$DIC = D(\bar{\theta}) + 2p_D = 2\bar{D} - D(\bar{\theta})\ . \tag{14.7}$$

Given samples from the posterior distribution of θ, DIC can be calculated quickly and easily. Many Bayesian software packages now routinely make this available along with inferences for linear and generalized linear mixed models.

In survival analysis, Zhou et al. (2008) utilize the DIC in their joint spatial analysis of age at diagnosis of prostate cancer and survival times conditioned on this age. For both outcomes, they consider models without and with spatial correlation which is induced in the survival model via frailties. They conclude, through the use of DIC, that the frailty terms substantially improve the model fit. The best-fitting model shows correlation within counties for the survival outcome but not for the age at diagnosis. Kim et al. (2009) use the DIC to infer the number of distinct groups in their newly proposed latent cure rate marker model. They analyze data on prostate cancer recurrence and, using DIC (as well as the LPML; see below), conclude that there are three latent risk category groups. Ibrahim et al. (2008) derive expressions for and utilize the DIC in the context of variable selection in Cox regression with missing data. They illustrate the methodology using a bone marrow transplantation dataset with ten covariates.

14.3.2 Predictive criteria

While BIC and DIC follow the notion of a model's information content, other criteria have been proposed that focus more on a model's ability to predict future observations. These criteria can be seen to fall in two categories: one that uses a cross-validation approach and another that relies on the concept of a future replicate of the current experiment. The former was proposed in, and the latter inspired by, the work of Seymore Geisser, a strong proponent of the view that prediction is the primary goal of statistical methods.

14.3.2.1 Cross-valid prediction

Suppose the data y are composed of observations $y_1, \ldots, y_n$. With y_{-i} denoting all but the ith observation, one can compute the predictive distribution for a single observation given a fully specified model and data y_{-i}. How well y_i is predicted by this distribution can then be measured by the magnitude of this predictive distribution evaluated at y_i. The total predictive performance of the model could then be judged by combining all such measures for $i = 1, \ldots, n$. Geisser and Eddy (1979) introduced this idea and formalized it via what they termed pseudo-marginal likelihood. Using $f(\cdot)$ to denote a predictive density and the notation of Section 14.2 for a full Bayesian specification of a model, we have the following definition of the **Log Pseudo-Marginal Likelihood (LPML)**:

$$LPML = \sum_{i=1}^{n} \log\ f(y_i|y_{-i}) \tag{14.8}$$

where

$$f(y_i|y_{-i}) = \int_{\Theta} p(y_i|\theta)\pi(\theta|y_{-i})d\theta\ . \tag{14.9}$$

This $f(y_i|y_{-i})$ is often called the **Conditional Predictive Ordinate (CPO$_i$)**; see Geisser (1993). Plots of all n values of it have been proposed as a tool for model criticism by Gelfand et al. (1992). In addition, plots of ratios under two models can be used for model comparison as in Ibrahim et al. (2001*a*). Some computational aspects of CPO$_i$ are discussed in Chapter 10 of Chen, Shao and Ibrahim (2000).

Gelfand and Mallick (1995) use CPO$_i$ in the context of right-censored survival data models, using (14.9) for exact observations and the predictive survival probability $\int_{y_i}^{\infty} f(z|y_{-i})dz$ if the observation is right censored at y_i. For recurrent events models, Sinha et al. (2008) use plots of CPO$_i$ for model diagnostic purposes, while Ryu et al. (2007) employ CPO$_i$ and LPML for model diagnostics and choice in analysis of longitudinal studies with outcome-dependent follow-up. The LPML criterion has also been used by Hanson and coauthors in several articles: Jara et al. (2011); Zhao and Hanson (2011); Zhao et al. (2009); Hanson et al. (2009). The exponentiated difference of LPML's for two models has been termed the **Pseudo Bayes Factor** by Hanson (2006) and Hanson and Yang (2007) and interpreted in the manner of a predictive version of BF. It is well defined even under some improper priors, has computational stability when computed from moderate sized MCMC samples, and is reported by Hanson (2006) to be less sensitive to prior choice than is BF.

14.3.2.2 Replicate experiment prediction

A different approach that also uses predictive considerations as means to generate inferential procedures was advocated by Laud and Ibrahim (1995). They rely on a notion of an imaginary replicate of the (current) experiment that resulted in the data at hand. With Z denoting the future data in the replicate, they focus on the density of Z given the current data y and call it the **Predictive Density of a Replicate Experiment (PDRE)**, defining it as

$$f_{Z|y}(z) = \int_{\Theta} p(z|\theta)\pi(\theta|y)d\theta \; . \tag{14.10}$$

In a wide variety of situations including various regression models, the notion of a replicate experiment renders y and Z directly comparable, in fact, exchangeable a priori. Laud and Ibrahim (1995) then define three criteria – K, L and M – for model selection based on Kullback-Leibler information, Euclidean distance, and density ordinate measures of the discrepancy (or agreement) between the observed data and the PDRE. They also go on to propose calibration measures for L and M.

Among the three, the L measure has received the most attention as it is the simplest to interpret and can be decomposed into a component involving predictive variances and another that can be seen as a predictive bias. We only describe this measure and begin with its definition:

$$L_m^2 = E_m\{(Z-y)'(Z-y)|y\} \; . \tag{14.11}$$

It is easy to see that $L^2 = \sum_{i=1}^{n}\{Var(Z_i|y) + (E(Z_i|y) - y_i)^2\}$. The criterion was used in Ibrahim and Laud (1994) in the analysis of some designed experiments, for generalized linear models by Ibrahim and Chen (2000), Chen, Ibrahim and Shao (2000), and Meyer and Laud (2002). Gelfand and Ghosh (1998) used the replicate experiment formulation and considered many different loss function-based choices of measuring discrepancy between observed data and future replicates, obtaining the L criterion as a limiting case. They also reported that this is an interpretable criterion and that it performs as well as others considered. For generalized linear models, Chen et al. (2008) show relationships among various criteria and devise a strategy for computing them.

For survival analysis, and also for other models, Ibrahim et al. (2001*b*) give a detailed treatment of the L criterion. It utilizes the re-expression of the criterion as the sum of a predictive variance and a squared prediction error as

$$L^2(y) = \sum_{i=1}^{n} Var(Z_i|y) + \nu \sum_{i=1}^{n} (E(Z_i|y) - y_i)^2 \ .$$

Here ν is an additional tuning parameter which follows from a generalization and decision theoretic justification given by Gelfand and Ghosh (1998). Allowing a transformation of the data on time axis – typically a log transformation for survival analysis – the focus then is on computing the criterion when y contains possibly censored observations. The approach is to employ expectations of the censored observations with respect to their posterior predictive distribution. To see this more explicitly, let y_D denote data as observed, including exact observations for some individuals and censoring intervals for the rest. Let y^* denote predictions for the censored observations, restricted to their respectively observed censoring sets. Then define

$$L^2(y_D) = E_{y^*|y_D}\{L^2(y_{pc})\} = \int_{\Theta}\int_{y^*} L^2(y_{pc})p(y^*|\theta)\pi(\theta|y_D)dy^*d\theta$$

where y_{pc} denotes the predictively completed collection of all exact observations. Moreover, θ here denotes all parameters in the model, including the baseline hazard function that is typically modeled via a stochastic process prior. Calculation of the criterion is implemented in a relatively straightforward manner as part of the MCMC iterations for estimating the model.

Ibrahim et al. (2001*b*) also provide a useful construction of a calibration distribution for this criterion. Their illustrations include simulated data (parametric survival models), AIDS data (logistic regression) and breast cancer data (semi-parametric Cox regression). In joint modeling of longitudinal and survival data, Ibrahim et al. (2004) employ the L-measure to justify a quadratic trajectory model over one with a linear trajectory and to conclude that an important correlation between two antibody responses prior to relapse is nonzero in high-risk melanoma patients. Recently, Gu et al. (2011) have proposed a new measure that modifies the definition of the L-measure to suit censored data from a cure rate model while maintaining the predictive loss feature of the measure. They illustrate the use of the newly minted L-measure and compare it to DIC using breast cancer data.

14.4 Search-based variable selection

In the variable selection problem with p potential variables the number of models under consideration is 2^p. Calculating any of the criteria of the previous section, or posterior model probabilities, for each model is feasible only for p up to around 20 or so, even in relatively straightforward regression models. With the notation of Section 26.1 and Equations (14.1) and (14.2) in mind, it is natural to look for a sampling scheme that would visit various models in $\mathcal{M}$ with probabilities equal to the posterior probabilities. Indeed, it is possible to design Markov chains that have this as the stationary distribution on $\mathcal{M}$. It is also possible to design other stochastic search schemes that approximate this behavior in some sense. These developments originated with normal linear regression and spread to generalized linear, nonlinear and survival models. Early methods were provided by Carlin and Polson (1991), George and McCulloch (1993), Green (1995) and Carlin and Chib (1995). We give brief descriptions of these (more details are available, for example, in Chapter 9 of Chen, Shao and Ibrahim (2000)), and indicate their use in survival analysis. We also note that

O'Hara and Sillanpää (2009) provide a review of stochastic search methods and also include some performance comparisons.

14.4.1 Stochastic search variable selection

In order to conduct a stochastic search that automatically visits data-supported models more frequently, one must have a strategy that considers the distribution on the model space $\mathcal{M}$ as well as the likelihood for each model. This creates a difficulty as the model parameter θ_m has dimension varying with m. George and McCulloch (1993) circumvented this issue by constructing a prior for each regression coefficient that is a mixture of two normals, one with a small variance around zero and another with a much larger variance. This approach does not set a parameter value equal to zero, but then, if a posterior sample comes from the first component, the variable is taken to be omitted. Using normal linear regression for convenience, and following Chen, Shao and Ibrahim (2000), the method can be described as below.

Suppose there are p potential covariates forming the columns of X, and the regression coefficients make up the $p \times 1$ parameter θ so that $y|X, \theta, \sigma^2 \sim N(X\theta, \sigma^2 I)$. Now introduce latent variables $\gamma_j, j = 1, \ldots, p$ such that

$$\theta_j|\gamma_j \sim (1-\gamma_j)N(0, \tau_j^2) + \gamma_j N(0, c_j^2\tau_j^2) \tag{14.12}$$

where $P(\gamma_j = 1) = 1 - P(\gamma_j = 0) = p_j$. Setting τ_j small and c_j large essentially allows θ_j to be either near zero or away from zero with high probability. Assigning independent Bernoulli prior to $\gamma_j, j = 1.\ldots, p$ assigns prior model probabilities to all 2^p models in $\mathcal{M}$. Now, with an inverse-gamma conjugate specification for $\sigma^2|\gamma$ and an independent multivariate normal for $\theta|\gamma$, full conditionals needed for a Gibbs sampler are available in closed form. The computational implementation successively samples $\theta|y, \sigma^2, \gamma_j, j = 1, \ldots, p$, $\sigma^2|y, \theta, \gamma_j, j = 1, \ldots, p$, and each $\gamma_j|y, \theta, \sigma^2, \gamma_{-j}$ where γ_{-j} denotes all except the jth γ. It is interesting to note that, because of the hierarchical nature of the prior specification, the conditional for γ_j in the Gibbs cycle is free of the data y. It depends on it only through θ. In many cases, this sampler converges rapidly to the posterior distribution of the γ's, automatically visiting higher posterior probability models more frequently.

A different approach to the varying dimension problem was proposed by Carlin and Chib (1995) via the use of pseudo-priors. These are specified for $\theta_m, m \neq m^*|m^*$, so that, in a Gibbs cycle, conditioned on $m = M^*$ and the data, $\theta_{m^*}|data$ can be simulated from posterior using the data, the likelihood in $p(y|\theta_{m^*})$ and the prior $\pi(\theta_{m^*})$ while $\theta_m, m \neq m^*|data$ are generated from the pseudo-prior. This is justified by the assumption that y is independent of $\theta_m, m \neq m^*$ given m^*. We thus have the full set of parameters at each iteration. Kuo and Mallick (1998) follow a similar prescription for linear and generalized linear models.

In the survival analysis context, Lee and Mallick (2004) consider Cox's proportional hazards model

$$h(t_i|h_0, X_i) = h_0(t_i)e^{W_i}, \; W_i = X_i'\beta + \epsilon_i, \; \epsilon_i \sim N(0, \sigma^2), \; i = 1, \ldots, n.$$

With a gamma process prior on the cumulative hazard H_0 with mean H_0^* and weight a representing confidence in the mean function, they analytically integrate out the hazard to obtain a marginalized likelihood of the form

$$L(W|data) = e^{-\sum_{i=1}^n aB_iH_0^*(t_i)} \prod_{i=1}^n \{ah_0^*(t_i)B_i\}^{\delta_i} \; .$$

Here h_0^* is the hazard function corresponding to H_0^*, $B_i = -log\{1 - e^{W_i}/(a + A_i)\}$, $A_i =$

$\sum_{l \in R(t_i)} e^{W_l}$, $R(t_i)$ is the set of individuals at risk just before time t_i and δ_i is the usual non-censoring indicator. For variable selection, they define a vector γ with elements γ_j indicating $\beta_j = 0$ and β_γ as the vector of all nonzero elements of β. Then, with priors

$$\sigma^2 \sim InverseGamma(a_0, b_0/2),\ \gamma_i \overset{ind}{\sim} Bernoulli(\pi_i),$$
$$\beta_\gamma|\sigma^2 \sim N(0, c\sigma^2(X'_\gamma X_\gamma)^{-1}),\ W|\beta_\gamma \sim N(X_\gamma\beta_\gamma, \sigma^2 I),$$

they proceed to derive Gibbs conditionals justifying the following sequence of draws:

1. W from density proportional to

$$e^{-\sum_{i=1}^n aB_iH_0^*(t_i)} \left\{ \prod_{i=1}^n \{ah_0^*(t_i)B_i\}^{\delta_i} \right\} e^{-(W-X_\gamma\beta_\gamma)'(W-X_\gamma\beta_\gamma)/(2\sigma^2)}$$

2. γ from density proportional to

$$e^{-S(\gamma)/(2\sigma^2)} \prod_{i=1}^n \pi_i^{\gamma_i}(1-\pi)^{1-\gamma_i}$$

   where $S(\gamma) = W'W - \frac{c}{1+c}W'X_\gamma(X'_\gamma X_\gamma)^{-1}X'_\gamma W$

3. β_γ from multivariate normal with mean $V_\gamma X'_\gamma W$ and covariance matrix $V_\gamma = \sigma^2\frac{c}{1+c}(X'_\gamma X_\gamma)^{-1}$

4. σ^2 from $InverseGamma\left(\frac{n}{2}+a, \frac{1}{2}[(W - X_\gamma\beta_\gamma)'(W - X_\gamma\beta_\gamma) + b_0]\right)$.

It should be noted that the components of γ can be generated one at a time using conditional distributions as provided by Lee and Mallick. The authors illustrate their method with lymphoma and breast cancer datasets.

14.4.2 Reversible jump MCMC

Green (1995) tackled the varying dimension problem directly by showing that a Metropolis-Hastings (M-H) chain can be designed on the space in which the pair (m, θ_m) takes values. To take a glimpse at the idea, let us follow Green's notation and consider a space $\mathcal{X}$ on which we have a target distribution $\pi(x)$ of interest. We wish to construct a Markov transition kernel $P(x, dx')$ that is aperiodic, irreducible and satisfies

$$\int_A\int_B \pi(dx)P(x, dx') = \int_B\int_A \pi(dx')P(x', dx)\ . \tag{14.13}$$

Such a Markov chain is said to satisfy detailed balance. It specifies a certain reversibility of the chain, going from A to B and from B to A, for any reasonable subsets A, B of $\mathcal{X}$. We can then simulate this chain and obtain approximate samples from $\pi(dx)$. The M-H algorithm uses a proposal distribution $q(x'; x)$ to generate a next value x' from the current value x. The proposed value is accepted with probability

$$\alpha(x, x') = min\left\{1, \frac{\pi(x')q(x; x')}{\pi(x)q(x', x)}\right\}\ ; \tag{14.14}$$

otherwise, the chain remains at the current value x.

In the model selection case, the target distribution is $p(m|y)\pi(\theta_m|y, m)$. We need transitions from (m, θ_m) to $(m', \theta_{m'})$, where θ_m and $\theta_{m'}$ may be of different dimension, that

satisfy detailed balance. After explaining the intuition behind it via a simple example, Green (1995) provides the **Reversible Jump MCMC** algorithm. He defines the transition mechanism in the M-H manner: a proposal distribution and an acceptance probability. We put it in our context and notation as follows:
Let $j(m'|m)$ be the probability of proposing a new model m' from the current state (m, θ_m). Generate a random vector u of dimension d from a proposal density $q_{m,m'}(u|\theta_m)$. Associated with the reverse move, from $(m', \theta_{m'})$ to (m, θ_m), there is a random vector u' of dimension d' to be generated from $q_{m',m}(u'|\theta_{m'})$ such that $dimension(\theta_m) + d = dimension(\theta_{m'}) + d'$. Once u is generated, compute $(\theta_{m'}, u') = g_{m,m'}(\theta_m, u)$ deterministically through a bijection $g_{m,m'}$. Now use the acceptance probability

$$min\left\{1, \frac{p(m', \theta_{m'}|y)j(m|m')q_{m',m}(u'|\theta_{m'})}{p(m, \theta|y)j(m'|m)q_{m,m'}(u|\theta_m)}\left|\frac{\partial g_{m,m'}(\theta_m, u)}{\partial(\theta_m, u)}\right|\right\}. \tag{14.15}$$

This achieves detailed balance and provides a general MCMC method for the variable/model selection problem. Note that $p(m, \theta|y) = p(m|y)\pi(\theta_m|y, m)$ and, while the expression for $p(m|y)$ given by (14.1) contains a sum in the denominator, it cancels out in the ratio above.

This algorithm has been used widely in various forms, and for a variety of applications. Dellaportas and coauthors in Dellaportas et al. (2002), Ntzoufras et al. (2003) and Papathomas et al. (2011) have pointed to special cases where particular simplifications lead to efficiency gains. Performance comparisons by O'Hara and Sillanpää (2009) show some advantages of this method. Godsill (2001) gives a good perspective on how the various methods discussed above relate to each other and can be seen to emerge from a general Metropolis-Hastings specification of a Markov chain on what he terms a composite model space. This is a modification of the Carlin and Chib (1995) formulation that avoids drawing samples from a very large number of pseudo-priors. He also shows that Green's reversible jump algorithm can be seen to follow from this general formulation.

Overall, the stochastic search methods of this section have been used in survival analysis by only a few authors. As discussed above, the key to the method of Lee and Mallick (2004) is the analytic marginalization made possible by a particular choice of the nonparametric prior on the cumulative hazard function. Chen et al. (2009) jointly model categorical and survival outcomes in colorectal cancer. Their model is parametric and they employ the usual normal linear regression approach following a log transformation. Lee et al. (2011) carry out variable selection in high-dimensional data in the presence of right censoring by employing what they term as an adaptive jumping rule in their Markov chain transitions. It appears that these stochastic search techniques, despite their popularity in many areas, have not been widely adopted in survival analysis applications.

14.5 Discussion

Although this chapter focuses on the topic of Bayesian model selection, most Bayesians would agree that, whenever possible in the context of the application, one should employ **Bayesian Model Averaging (BMA)** rather than select a single model. From the predictive as well as the decision theoretic viewpoint, one should fully account for any post-data model uncertainty by averaging with respect to posterior model probabilities. Practical limitations of many applications, however, often demand a model choice. Recognition of this demand has led to the developments reported in this chapter. BMA has been discussed by many authors: Madigan and Raftery (1994); Raftery et al. (1997); Hoeting et al. (1999) as well as Clyde et al. (1996); Clyde (1999). Dunson and Herring (2005) develop this for a survival analysis scenario.

Adaptive shrinkage estimation techniques, generated via appropriate choices of priors, can guide variable selection. Casella and coauthors in Park and Casella (2008) and Kyung et al. (2010) show, in a linear model, that a product Laplace prior on the regression coefficients results in posterior median estimates similar to the Lasso and ridge regression estimates. They term the model the **Bayesian Lasso**. They go on to compare Bayesian point and interval estimates with frequestist penalized likelihood methods, and also suggest that credible intervals for the regression parameters could be used to guide variable selection. MacLehose and Dunson (2010) advance this idea substantially by using flexible Bayesian nonparametric models to achieve shrinkage to multiple points that are automatically determined by the data. They illustrate, for Parkinson's disease, how the method resulted in identifying two SNP's from among 270 in their analysis. In survival analysis, Lee et al. (2011) appears to be the only article to use a prior somewhat similar to the Bayesian Lasso. Garcia et al. (2010) undertake variable selection in Cox regression with missing data but choose a non-Bayesian solution, citing some difficulties with the Bayesian approach in this case.

An emerging set of models suitable for variable selection in high dimensions employ **Bayesian Nonparametric (BNP)** techniques such as the above cited model of MacLehose and Dunson (2010). Mixtures of Dirichlet processes were used by Guindani et al. (2009), and recently by Shahbaba and Johnson (2012) to identify gene signals. Giudici et al. (2003) define mixtures of products of Dirichlet processes for variable selection. Product Partition Models (PPM), being developed for the purpose of clustering in various applications – see Müller et al. (2011) and references therein – also show some promise for variable selection.

The choice of priors – $p(m), m \in \mathcal{M}$ and $\pi(\theta_m)$ in (14.1) and (14.2) – is an important consideration in determining the performance of the model selection procedures discussed here. It turns out that certain priors have desirable consequences in automatically controlling the problem of mutiple simultaneous decisions that are inherent in model selection. Scott and Berger (2006, 2010) show that a suitable hierarchical prior for $p(m)$, rather than fixed numerical values, can account for **multiplicity adjustment** in the Bayesian sense. This is a very appealing aspect of Bayesian model selection where priors can be used, not so much to represent information external to data, but to achieve control of an otherwise vexing problem. There is also considerable literature on the choice of the priors $\pi(\theta_m)$; see Liang et al. (2008); Wang and George (2007) and the references therein. Mixtures of g-priors, which are enhancements of Zellner's g-priors, are recommended as suitable priors.

Bibliography

Carlin, B. P. and Chib, S. (1995), 'Bayesian model choice via markov chain monte carlo methods', *Journal of the Royal Statistical Society, Series B: Methodological* **57**, 473–484.

Carlin, B. P. and Polson, N. G. (1991), 'Inference for nonconjugate Bayesian models using the gibbs sampler', *The Canadian Journal of Statistics / La Revue Canadienne de Statistique* **19**, 399–405.

Chen, M.-H., Huang, L., Ibrahim, J. G. and Kim, S. (2008), 'Bayesian variable selection and computation for generalized linear models with conjugate priors', *Bayesian Analysis* **3**(3), 585–614.

Chen, M.-H., Ibrahim, J. G. and Shao, Q.-M. (2000), 'Power prior distributions for generalized linear models', *Journal of Statistical Planning and Inference* **84**(1-2), 121–137.

Chen, M.-H. and Shao, Q.-M. (1997*a*), 'Estimating ratios of normalizing constants for densities with different dimensions', *Statistica Sinica* **7**, 607–630.

Chen, M.-H. and Shao, Q.-M. (1997*b*), 'On Monte Carlo methods for estimating ratios of normalizing constants', *The Annals of Statistics* **25**(4), 1563–1594.

Chen, M.-H., Shao, Q.-M. and Ibrahim, J. G. (2000), *Monte Carlo Methods in Bayesian Computation*, Springer-Verlag Inc., New York.

Chen, W., Ghosh, D., Raghunathan, T. E. and Sargent, D. J. (2009), 'Bayesian variable selection with joint modeling of categorical and survival outcomes: An application to individualizing chemotherapy treatment in advanced colorectal cancer', *Biometrics* **65**(4), 1030–1040.

Chib, S. (1995), 'Marginal likelihood from the gibbs output', *Journal of the American Statistical Association* **90**, 1313–1321.

Chib, S. and Jeliazkov, I. (2001), 'Marginal likelihood from the metropolis-hastings output', *Journal of the American Statistical Association* **96**(453), 270–281.

Clyde, M. A. (1999), Bayesian model averaging and model search strategies, *in* J. M. Bernardo, J. O. Berger, A. P. Dawid and A. F. M. Smith, eds., 'Bayesian Statistics 6 – Proceedings of the Sixth Valencia International Meeting', Clarendon Press [Oxford University Press], pp. 157–185.

Clyde, M., DeSimone, H. and Parmigiani, G. (1996), 'Prediction via orthogonalized model mixing', *Journal of the American Statistical Association* **91**, 1197–1208.

Dellaportas, P., Forster, J. J. and Ntzoufras, I. (2002), 'On Bayesian model and variable selection using mcmc', *Statistics and Computing* **12**(1), 27–36.

Dunson, D. B. and Herring, A. H. (2005), 'Bayesian model selection and averaging in additive and proportional hazards models', *Lifetime Data Analysis* **11**(2), 213–232.

Garcia, R. I., Ibrahim, J. G. and Zhu, H. (2010), 'Variable selection in the cox regression model with covariates missing at random', *Biometrics* **66**(1), 97–104.

Geisser, S. (1993), *Predictive Inference: an Introduction*, Chapman & Hall Ltd., New York.

Geisser, S. and Eddy, W. F. (1979), 'A predictive approach to model selection (corr: V75 p765)', *Journal of the American Statistical Association* **74**, 153–160.

Gelfand, A. E. and Dey, D. K. (1994), 'Bayesian model choice: Asymptotics and exact calculations', *Journal of the Royal Statistical Society, Series B: Methodological* **56**, 501–514.

Gelfand, A. E., Dey, D. K. and Chang, H. (1992), Model determination using predictive distributions, with implementation via sampling-based methods (disc: P160-167), *in* J. M. Bernardo, J. O. Berger, A. P. Dawid and A. F. M. Smith, eds., 'Bayesian Statistics 4. Proceedings of the Fourth Valencia International Meeting', Clarendon Press [Oxford University Press], pp. 147–159.

Gelfand, A. E. and Ghosh, S. K. (1998), 'Model choice: A minimum posterior predictive loss approach', *Biometrika* **85**, 1–11.

Gelfand, A. E. and Mallick, B. K. (1995), 'Bayesian analysis of proportional hazards models built from monotone functions', *Biometrics* **51**(3), 843–852.

George, E. I. and McCulloch, R. E. (1993), 'Variable selection via gibbs sampling', *Journal of the American Statistical Association* **88**, 881–889.

Giudici, P., Mezzetti, M. and Muliere, P. (2003), 'Mixtures of products of dirichlet processes for variable selection in survival analysis', *Journal of Statistical Planning and Inference* **111**(1-2), 101–115.

Godsill, S. J. (2001), 'On the relationship between Markov chain Monte Carlo methods for model uncertainty', *Journal of Computational and Graphical Statistics* **10**(2), 230–248.

Green, P. J. (1995), 'Reversible jump Markov chain Monte Carlo computation and Bayesian model determination', *Biometrika* **82**, 711–732.

Gu, Y., Sinha, D. and Banerjee, S. (2011), 'Analysis of cure rate survival data under proportional odds model', *Lifetime Data Analysis* **17**, 123–134.

Guindani, M., Müller, P. and Zhang, S. (2009), 'A Bayesian discovery procedure', *Journal of the Royal Statistical Society, Series B: Statistical Methodology* **71**(5), 905–925.

Hanson, T. E. (2006), 'Inference for mixtures of finite polya tree models', *Journal of the American Statistical Association* **101**(476), 1548–1565.

Hanson, T. E., Johnson, W. O. and Laud, P. W. (2009), 'Semiparametric inference for survival models with step process covariates', *The Canadian Journal of Statistics / La Revue Canadienne de Statistique* **37**(1), 60–79.

Hanson, T. E. and Yang, M. (2007), 'Bayesian semiparametric proportional odds models', *Biometrics* **63**(1), 88–95.

Hoeting, J. A., Madigan, D., Raftery, A. E. and Volinsky, C. T. (1999), 'Bayesian model averaging: a tutorial (pkg: P382-417)', *Statistical Science* **14**(4), 382–401.

Ibrahim, J. G. and Chen, M.-H. (1998), 'Prior distributions and Bayesian computation for proportional hazards models', *Sankhyā Series B* **60**, 48–64.

Ibrahim, J. G. and Chen, M.-H. (2000), 'Power prior distributions for regression models', *Statistical Science* **15**(1), 46–60.

Ibrahim, J. G., Chen, M.-H. and Kim, S. (2008), 'Bayesian variable selection for the Cox regression model with missing covariates', *Lifetime Data Analysis* **14**(4), 496–520.

Ibrahim, J. G., Chen, M.-H. and MacEachern, S. N. (1999), 'Bayesian variable selection for proportional hazards models', *The Canadian Journal of Statistics / La Revue Canadienne de Statistique* **27**, 701–717.

Ibrahim, J. G., Chen, M.-H. and Sinha, D. (2001*a*), *Bayesian Survival Analysis*, Springer, New York.

Ibrahim, J. G., Chen, M.-H. and Sinha, D. (2001*b*), 'Criterion-based methods for Bayesian model assessment', *Statistica Sinica* **11**(2), 419–443.

Ibrahim, J. G., Chen, M.-H. and Sinha, D. (2004), 'Bayesian methods for joint modeling of longitudinal and survival data with applications to cancer vaccine trials', *Statistica Sinica* **14**(3), 863–883.

Ibrahim, J. G. and Laud, P. W. (1994), 'A predictive approach to the analysis of designed experiments', *Journal of the American Statistical Association* **89**, 309–319.

Jara, A., Hanson, T., Quintana, F., Müller, P. and Rosner, G. (2011), 'DPpackage: Bayesian semi- and nonparametric modeling in R', *Journal of Statistical Software* **40**(5), 1–30. **URL:** *http://www.jstatsoft.org/v40/i05/*

Jeffreys, H. (1998), *Theory of Probability*, Oxford University Press, Oxford, England.

Kass, R. E. and Raftery, A. E. (1995), 'Bayes factors', *Journal of the American Statistical Association* **90**, 773–795.

Kass, R. E. and Wasserman, L. (1995), 'A reference Bayesian test for nested hypotheses and its relationship to the schwarz criterion', *Journal of the American Statistical Association* **90**, 928–934.

Kim, S., Xi, Y. and Chen, M.-H. (2009), 'A new latent cure rate marker model for survival data', *The Annals of Applied Statistics* **3**(3), 1124–1146.

Kuo, L. and Mallick, B. (1998), 'Variable selection for regression models', *Sankhyā Series B* **60**, 65–81.

Kyung, M., Gill, J., Ghosh, M. and Casella, G. (2010), 'Penalized regression, standard errors, and Bayesian lassos', *Bayesian Analysis* **5**(2), 369–412.

Laud, P. W. and Ibrahim, J. G. (1995), 'Predictive model selection', *Journal of the Royal Statistical Society, Series B: Methodological* **57**, 247–262.

Lee, K. E. and Mallick, B. K. (2004), 'Bayesian methods for variable selection in survival models with application to DNA microarray data', *Sankhyā: The Indian Journal of Statistics* **66**(4), 756–778.

Lee, K. H., Chakraborty, S. and Sun, J. (2011), 'Bayesian variable selection in semiparametric proportional hazards model for high dimensional survival data', *The International Journal of Biostatistics* **7**(1), 1–32, ISSN(online) 1557-4679, DOI: 10.2202/1557.4679.1301.

Liang, F., Paulo, R., Molina, G., Clyde, M. A. and Berger, J. O. (2008), 'Mixtures of g priors for Bayesian variable selection', *Journal of the American Statistical Association* **103**(481), 410–423.

MacLehose, R. F. and Dunson, D. B. (2010), 'Bayesian semiparametric multiple shrinkage', *Biometrics* **66**(2), 455–462.

Madigan, D. and Raftery, A. E. (1994), 'Model selection and accounting for model uncertainty in graphical models using occam's window', *Journal of the American Statistical Association* **89**, 1535–1546.

Meyer, M. C. and Laud, P. W. (2002), 'Predictive variable selection in generalized linear models', *Journal of the American Statistical Association* **97**(459), 859–871.

Müller, P., Quintana, F. and Rosner, G. (2011), 'A product partition model with regression on covariates', *Journal of Computational and Graphical Statistics* **20**, 260–278.

Newton, M. A. and Raftery, A. E. (1994), 'Approximate Bayesian inference with the weighted likelihood bootstrap (disc: P26-48)', *Journal of the Royal Statistical Society, Series B: Methodological* **56**, 3–26.

Ntzoufras, I., Dellaportas, P. and Forster, J. J. (2003), 'Bayesian variable and link determination for generalised linear models', *Journal of Statistical Planning and Inference* **111**(1-2), 165–180.

O'Hara, R. B. and Sillanpää, M. J. (2009), 'A review of Bayesian variable selection methods: What, how and which', *Bayesian Analysis* **4**(1), 85–118.

Papathomas, M., Dellaportas, P. and Vasdekis, V. G. S. (2011), 'A novel reversible jump algorithm for generalized linear models', *Biometrika* **98**(1), 231–236.

Park, T. and Casella, G. (2008), 'The Bayesian lasso', *Journal of the American Statistical Association* **103**(482), 681–686.

Raftery, A. E., Madigan, D. and Hoeting, J. A. (1997), 'Bayesian model averaging for linear regression models', *Journal of the American Statistical Association* **92**, 179–191.

Ryu, D., Sinha, D., Mallick, B., Lipsitz, S. R. and Lipshultz, S. E. (2007), 'Longitudinal studies with outcome-dependent follow-up: Models and Bayesian regression', *Journal of the American Statistical Association* **102**(479), 952–961.

Schwarz, G. (1978), 'Estimating the dimension of a model', *The Annals of Statistics* **6**, 461–464.

Scott, J. G. and Berger, J. O. (2006), 'An exploration of aspects of Bayesian multiple testing', *Journal of Statistical Planning and Inference* **136**(7), 2144–2162.

Scott, J. G. and Berger, J. O. (2010), 'Bayes and empirical-Bayes multiplicity adjustment in the variable-selection problem', *The Annals of Statistics* **38**(5), 2587–2619.

Shahbaba, B. and Johnson, W. (2012), 'Bayesian nonparametric variable selection as an exploratory tool for finding genes that matter', *Statistics in Medicine* **22**, DOI: 10.1002/51m.5680 (E pub ahead of print).

Sinha, D., Maiti, T., Ibrahim, J. G. and Ouyang, B. (2008), 'Current methods for recurrent events data with dependent termination: A Bayesian perspective', *Journal of the American Statistical Association* **103**(482), 866–878.

Spiegelhalter, D. J., Best, N. G., Carlin, B. P. and van der Linde, A. (2002), 'Bayesian measures of model complexity and fit (pkg: P583-639)', *Journal of the Royal Statistical Society, Series B: Statistical Methodology* **64**(4), 583–616.

Volinsky, C. T. and Raftery, A. E. (2000), 'Bayesian information criterion for censored survival models', *Biometrics* **56**(1), 256–262.

Wang, X. and George, E. I. (2007), 'Adaptive Bayesian criteria in variable selection for generalized linear models', *Statistica Sinica* **17**(2), 667–690.

Zhao, L. and Hanson, T. E. (2011), 'Spatially dependent polya tree modeling for survival data', *Biometrics* **67**(2), 391–403.

Zhao, L., Hanson, T. E. and Carlin, B. P. (2009), 'Mixtures of polya trees for flexible spatial frailty survival modeling', *Biometrika* **96**(2), 263–276.

Zhou, H., Lawson, A. B., Hebert, J. R., Slate, E. H. and Hill, E. G. (2008), 'Joint spatial survival modeling for the age at diagnosis and the vital outcome of prostate cancer', *Statistics in Medicine* **27**(18), 3612–3628.

15

Model Selection for High-Dimensional Models

Rosa J. Meijer and Jelle J. Goeman

Leiden University Medical Center

CONTENTS

15.1	Introduction	301
15.2	Selecting variables by fitting a prediction model	302
	15.2.1 Screening by penalized methods	303
	15.2.2 Screening by univariate selection	306
	15.2.3 Practical usefulness of methods possessing screening properties	307
	15.2.4 The Van de Vijver dataset	309
15.3	Selecting variables by testing individual covariates	310
	15.3.1 Methods for FWER control	312
	15.3.2 Methods for FDR control	315
	15.3.3 Confidence intervals for the number of true discoveries	316
15.4	Reducing the number of variables beforehand and incorporating background knowledge	318
15.5	Discussion	319
	Bibliography	319

15.1 Introduction

In recent years, quick developments in high-throughput biotechnology have enabled researchers to generate thousands of potentially interesting measurements per subject. Especially in the field of survival analysis, these measurements are extremely valuable because knowledge of the human genome could greatly enhance our understanding of many diseases and could lead to more accurate survival prediction models. However, the advent of gene expression data and other types of high-dimensional genomic data did not only give rise to numerous new opportunities but also brought new computational and methodological challenges. It is no longer possible to use standard survival prediction methods, such as multivariate Cox regression, directly, when the number of covariates greatly exceeds the number of subjects. Identifying influential covariates becomes more complicated because thousands of hypotheses have to be tested simultaneously. To control the number of false discoveries (i.e., the number of covariates that are believed to be influential while in fact they are not), proper adjustments for the number of tests performed are needed, the so-called multiple testing corrections.

In this chapter, the focus will be on answering the research question: "How to select important covariates from a large set of candidates?" These covariates can for example be genes, SNPs (single nucleotide polymorphisms), probes or proteins. In the remainder of this chapter we will often refer to them as genes, since our example datasets will be gene expression microarray datasets.

Before we can address this question, we first have to make more precise what we mean by "important." Suppose we have gene expression measurements and survival outcomes for patients diagnosed with a certain disease. Statistical methods are now essentially used for two reasons: either to predict survival times for individual patients based on their gene expression levels, or to identify genes whose expression levels differ between patients with good and bad prognosis, to gain more insight in the nature of the disease. Although it might seem plausible that genes that are differentially expressed between good and bad prognosis patients will end up in a prediction model and, likewise, that the genes entered in such a model are on itself predictive, this is not necessarily true. If several differentially expressed genes are highly correlated, for example, it is not unlikely that only one of these will be included in the prediction model, while all of them separately could be associated with the outcome variable. Similarly, a covariate which is not related to the response might be included in the prediction model because it can account for some variation in a second predictor which is related to the response. So although intertwined, the two questions of finding the right variables for prediction or finding the right variables for understanding underlying phenomena ask for different methods, as was already point out by Cox and Snell (1974). Essentially the differences boil down to the difference between univariate and multivariate regression.

In the remainder of this chapter we will discuss selection of important genes in a high-dimensional setting both from the variable selection perspective and from the multiple hypothesis testing perspective.

15.2 Selecting variables by fitting a prediction model

Suppose we have been given the, possibly censored, survival times for n individuals for which we also have gene expression information on m genes. Our aim is to construct a *parsimonious* model that accurately predicts survival time based on the gene expression measurements. Assuming that the number of genes exceeds the number of individuals ($m \gg n$), we cannot proceed by simply fitting the Cox proportional hazards model (Cox, 1972), but we first have to tackle the high-dimensionality problem by some form of dimension reduction.

Much research has been done on finding methods that not only have good prediction error, but also estimate the true "sparsity pattern," that is, the set of covariates with nonzero regression coefficients (Wasserman and Roeder, 2009). The ability to (at least asymptotically) find the underlying true model, is often referred to as the "oracle property." If a method possesses the oracle property, it will asymptotically select exactly those variables that are present in the true model and the corresponding parameter estimates will be asymptotically unbiased. Thus, such a method will asymptotically perform as well as an "oracle" that already knows the true set of relevant variables. Even though unbiasedness is a highly desirable property of an estimator, selecting the right predictors, regardless of the precise coefficient values, is of main importance in the context of variable selection. For that reason, in the remainder of this section we will primarily look at a method's ability to correctly select all influential variables.

Suppose the true model is a Cox model of the form:

$$h_i(t) = h_0(t) \exp(\boldsymbol{x}_i^T \boldsymbol{\beta}),$$

where $h_i(t)$ is the hazard function for individual i, having covariate vector $\boldsymbol{x}_i$ and $h_0(t)$ is the baseline hazard. A method that correctly selects all influential variables is said to be model selection consistent (Benner et al., 2010) which can be made precise as follows:

$$\lim_{n\to\infty} P(\hat{M}_n = M) = 1 \tag{15.1}$$

where $M = \{j : \beta_j \neq 0\}$ is the set of indices of all variables present in the true model, and $\hat{M}_n = \{j : \hat{\beta}_j \neq 0\}$ is the set of indices with parameter estimates unequal to zero, based on the specific prediction method and a dataset of size n.

In this section we will discuss several methods known to be model selection consistent or even known to possess the just described oracle property. The properties of these methods are promising in theory, but less is known about their exact behavior in practical settings, even though the methods are often used. For that reason the, sometimes very restrictive, assumptions underlying these methods will not only be stated but also discussed from a practical point of view, in order to clarify in which situations these methods can be expected to give reliable results and in which situations more caution has to be taken when interpreting the final model.

15.2.1 Screening by penalized methods

Estimation methods that are often associated with oracle properties are the so-called *penalized likelihood methods*. In the high-dimensional setting, fitting a model by maximizing the corresponding likelihood will result in severe overfitting. In order to still be able to use the likelihood, methods that maximize the likelihood but at the same time put a penalty on the parameters were developed. The function to maximize will have the form

$$l(\boldsymbol{\beta}) - p_\lambda(\boldsymbol{\beta}),$$

where $l(\boldsymbol{\beta})$ is the likelihood and $p_\lambda(\boldsymbol{\beta})$ is some penalty function that depends on a regularization parameter λ. The exact specification of the penalty function will determine the behavior of the final prediction rule. Fan and Li (2001) claim that a good penalty function should result in an estimator with three properties, namely *unbiasedness*, *sparsity* and *continuity*. Again, for variable selection, unbiasedness is not the first priority, but sparsity and continuity are. To fulfill these last two conditions, Fan and Li argue that only penalty functions that are non-differentiable in the origin qualify.

A penalty function satisfying this condition is the well-known l_1-norm penalty. The idea to use this penalty in combination with a regression model was first published by Tibshirani and resulted in a method that enabled estimation and variable selection simultaneously. For this reason, the method was named the *lasso*, which stands for "least absolute shrinkage and selection operator" (Tibshirani, 1996, 1997). The motivation for the lasso came from a similar method invented by Breiman; the *non-negative garotte* (Breiman, 1995). The garotte starts (in the linear model) by estimating the ordinary least squares (OLS) estimates and subsequently shrinks them by non-negative factors whose sum is constrained. However, to obtain the OLS estimates, the number of covariates cannot exceed the number of observations and these estimates can therefore not be calculated in high-dimensional problems. The lasso on the other hand combines shrinkage and estimation into one calculation step. For that reason it can easily be used with high-dimensional data.

The lasso can either be viewed as a penalized likelihood method, in which the penalty function is given by

$$p_\lambda(\boldsymbol{\beta}) = \lambda \sum_{j=1}^{m} |\beta_j|$$

but also as a constrained likelihood optimization procedure, where the regression coefficients result from the following optimization problem:

$$\hat{\boldsymbol{\beta}} = \text{argmax } l(\boldsymbol{\beta}), \text{ subject to } \sum_{j=1}^{m} |\beta_j| \leq s.$$

Both s and λ are user-specified parameters. Given the likelihood function, the two definitions are equivalent. Although first presented as a constrained optimization problem (as was the non-negative garotte), the penalized likelihood representation is the most common at present.

Because of the specific form of the penalty function, the lasso will shrink some coefficients exactly to zero and will result in a sparse model. This selection property has been a major reason for the method's popularity, even though only recently progress has been made in understanding the exact selection behaviour (Tibshirani, 2011). A very interesting question is whether the lasso is model selection consistent according to the definition as given in (15.1). In order for this criterium to hold, a necessary condition is that the true model is sparse as well. The lasso will never select more parameters in the final model than there are observations, and given that all true parameters are among those selected, the number of truly relevant variables cannot exceed n. Let $d = |M|$ be the number of variables in the true model. In order for the lasso to be model selection consistent, we know we should at least have $d \leq n$. This bound is usually not tight enough, however. Generally, consistency will require a *sparsity assumption* of the form

$$d \leq c_1 \sqrt{\frac{n}{\log(m)}}, \tag{15.2}$$

where c_1 is some constant factor (Bunea et al., 2007). This implies $d \ll n$. Given that the sparsity assumption holds, it can be shown that the lasso is model selection consistent, but only under rather restrictive conditions. If we let $m \gg n \to \infty$, the true model will be recovered with probability tending to 1 if the following two conditions are met:

1. The *neighbourhood stability condition* for the design matrix $\boldsymbol{X}$ (Meinshausen and Bühlmann, 2006) or the equivalent *irrepresentable condition* (Zhao and Yu, 2006; Zou, 2006), which basically says that the variables present in the true model M can neither be too strongly correlated with each other nor with the noise variables (i.e., the variables not in M) (Benner et al., 2010), and

2. The *'beta-min'-condition* (Bühlmann and van de Geer, 2011, p. 24) which states that all non-zero coefficients in the true model are sufficiently large. The importance of this condition is also illustrated by Leeb and Poetscher (2008).

Unfortunately, in genomics data it is rather unlikely for both assumptions to hold. Since the first condition is not only a *sufficient* but also an (essentially) *necessary* condition for model selection consistency, in general we cannot expect the selected set as retrieved by the lasso to be the true set of variables. The condition already gives an indication when to doubt the variables selected, namely in the situation where we have a strongly correlated design. Strong correlations between gene expressions or other types of genomic data are however the rule rather than the exception.

Retrieving all the relevant variables and none of the noise variables has proven to be a (too) difficult task for the lasso. However, retrieving all variables from the true model, whether or not accompanied by some noise variables is a desirable property in itself. We will refer to this as the *variable screening property* (Bühlmann and van de Geer, 2011) which can be made formal as follows:

$$\lim_{n\to\infty} P(\hat{M}_n \supseteq M) = 1$$

where M and $\hat{M}_n$ are as before the index sets of the variables in the true and estimated model, respectively. For this property to hold we again need the sparsity assumption and the 'beta-min'-condition, but the irrepresentable condition can be relaxed, leaving us with a less strong assumption on the design, namely the restricted eigenvalue condition (Bühlmann and van de Geer, 2011) which is technical but not overly restrictive in sparse problems. So in a sparse setting, as given by the sparsity assumption in Equation (15.2), where the true variables have corresponding coefficients above some detection limit (the 'beta-min'-assumption), the lasso has the ability to select them all. Even if there are some variables in the true model with coefficients that are too small to detect, one could still argue that the lasso is able to find the influential and for that reason *most relevant* variables.

Until now, we did not discuss the way to choose the specific value of the penalty parameter λ which heavily influences the lasso's screening behavior. Optimization of this parameter in terms of prediction accuracy is often done by maximizing the cross-validated log partial likelihood, as introduced by Verweij and Van Houwelingen (1993). Although the choice of λ in this way is not motivated by the lasso's screening property, it turns out that the lasso based on this optimal λ value (which we will denote by $\hat{\lambda}_{cv}$) often does possess the screening property:

$$\hat{M}(\hat{\lambda}_{cv}) \supseteq M.$$

This result is not only established empirically but can, at least in the linear model context, also be supported by theory (Bühlmann and van de Geer, 2011, p. 17).

One remark regarding the oracle and screening properties of the lasso that has to be made here is that almost all work has been done for (generalized) linear regression models. Much research on extending oracle results to censored survival data, especially in the $m \gg n$ case, remains to be done, even though the theory for the Cox and the linear model will probably show many similarities. A recent article on this topic is written by Bradic et al. (2011). Of course, model specific assumptions such as the assumption that the censoring times are conditionally independent of the survival times given the covariates have to be made in addition to the assumptions on sparsity, correlation structure and the minimal size of coefficients. However, most of these assumptions are quite general.

In the situation that the assumptions of the screening property hold, the lasso provides us with a selected group of variables, including the true but also noise variables. Is there a way to remove the false positives from the selection? Valuable methods that attempt to do so are the adaptive lasso (Zou, 2006), the relaxed lasso (Meinshausen, 2007) and smoothly clipped absolute deviation (SCAD) variants (Fan and Li, 2001; Zou and Li, 2008). The motivation behind the development of these methods was mainly two-fold. Firstly, it was observed that the lasso in combination with a penalty parameter chosen by cross-validation often resulted in a rather wide selection, potentially including many noise variables. Secondly, the model's coefficients as generated by the lasso are known to be biased towards zero. The aim of reducing this bias and diminishing the number of noise variables in the final selection resulted in the aforementioned methods. In the high-dimensional setting, these methods generally use the regular lasso as a first pre-selection step. Subsequently, a penalized likelihood which is only based on the remaining covariates is maximized. Not only will this often result in a much sparser model, because there is a high probability that some of the remaining coefficients will still get shrunk to zero, but the nonzero coefficients may also be less biased, because the penalty can be less severe. Most noise variables are indeed already eliminated by applying the lasso as a first step. For the SCAD and the adaptive lasso, especially the larger coefficients (i.e., those that correspond to the most influential variables) will remain practically unbiased.

15.2.2 Screening by univariate selection

Although the just discussed sparse regression techniques have proved useful for dealing with high-dimensional feature spaces, theoretical knowledge about their properties in so-called "ultra-high dimensional settings," where m grows at a non-polynomial rate with n, is still in its infancy (Gorst-Rasmussen and Scheike, 2012). Besides, the computational costs associated with these methods can be substantial in those situations. An ad-hoc approach to deal with this type of ultra-high dimensionality is to use an initial univariate screening step to reduce the number of covariates under consideration. Even though this procedure has been employed very often, it was only recently that Fan and Lv (2008) showed that this screening practice, which they named Sure Independence Screening (SIS), has desirable theoretical properties in the linear model setting. Recent work on SIS in combination with right-censored data has been conducted by Zhao and Li (2012) and Gorst-Rasmussen and Scheike (2012).

Using SIS in the Cox framework as proposed by Zhao and Li (2012) essentially comes down to fitting marginal Cox regressions for each covariate. Subsequently the final screened model will equal

$$\hat{M}_n = \{j : \boldsymbol{I}_j(\hat{\beta}_j)^{\frac{1}{2}}|\hat{\beta}_j| \geq \gamma_n\}$$

where $\boldsymbol{I}_j(\hat{\beta}_j)$ defines the information matrix at $\hat{\beta}_j$ and γ_n is some pre-specified cutoff, that depends on n. This procedure is termed a principled Cox sure independence screening procedure, abbreviated PSIS (Zhao and Li, 2012). To make the procedure less model-specific, Gorst-Rasmussen and Scheike (2012) propose to use a model-free statistic instead, which they call the "Feature Aberration at Survival Times" (FAST) and call their procedure FAST-SIS.

Under specific assumptions, both PSIS and FAST-SIS have been shown to possess the variable screening property. Moreover, the threshold γ_n can be chosen in such a way that the false selection rate becomes asymptotically negligible. The first assumption concerns the censoring mechanism. In order for the screening property to hold, the censoring times can not depend on the relevant variables nor the survival times (Gorst-Rasmussen and Scheike, 2012). Although restrictive, this assumption is still rather general. However, as before strong assumptions on sparsity and correlation structure are needed. The number of truly relevant variables should typically lie in the order $n/\log(n)$ and the covariates present in the true model M have to be independent of the irrelevant covariates. Furthermore, the validity of the proposed procedures hinges on whether the marginal Cox regressions or the FAST-statistics can reflect the importance of the corresponding covariates in the *joint model*. For this to hold, the design has to be close to orthogonal, which is an assumption that is easily violated. This is already known from the literature on *univariate gene selection* where genes are incorporated in a final multivariate model based on their marginal correspondence with the outcome variable. In practice this often leads to a model, in which many of the selected genes are mutually correlated and have an insignificant multivariate p-value (Van Wieringen et al., 2009).

To at least partly account for possible correlation between variables, as an alternative to univariate selection, *stepwise variable selection* procedures have been developed. See for example (Wiegand, 2010) for an overview on classical methods. A similar strategy has been suggested in the SIS literature as well. Fan and Lv (2008) propose to use an iterative SIS procedure (ISIS) in the situation wherein the assumptions on the correlation structure underlying regular SIS fail. Essentially, the first step of the ISIS procedure is just applying the SIS methodology. A small number of covariates is selected and within this subset, a (multivariate) variable selection procedure such as the lasso is used to even reduce the size of this initial selection. Secondly, the (univariate) relevance of the unselected covariates is reassessed, this time adjusted for the already selected covariates. Subsequently a small number of the most relevant features among them can be added to the selection and the last two steps are repeated until some stopping criterion is reached (Fan and Lv, 2008; Gorst-Rasmussen and Scheike, 2012). A similar iterated FAST-SIS procedure has already been developed for right-censored data (Gorst-Rasmussen and Scheike, 2012). Although heuristically appealing, theoretical support for the iterated SIS methods has not yet been developed. Since the comparable stepwise variable selection procedures have often been related to instability issues (Breiman, 1996), future research has to show the practical relevance of this procedure.

15.2.3 Practical usefulness of methods possessing screening properties

To evaluate the practical use of the lasso and its successors for analyzing high-dimensional data with the Cox proportional hazards model, Benner et al. (2010) conducted a study where the SCAD and the adaptive lasso were compared to the "standard" applications such as ridge regression, the lasso and the elastic net. The methods' performances were analyzed in various settings.

In the study of Benner et al. (2010), sparse methods are thus compared to methods not possessing oracle properties. While the emphasis in all previous described procedures lay on selecting as few variables as possible, using *ridge regression* or the *elastic net* will on the other hand usually result in a less sparse model. *Ridge regression* (Hoerl and Kennard, 1970; Van Houwelingen et al., 2006) is a shrinkage method with a quadratic penalty function. In contrast to the lasso, ridge regression does not perform variable selection, but only results in downward biased parameter estimates. Despite the fact that the final model incorporates all covariates, ridge regression is often found to be an effective prediction method in high-dimensional genomic applications (Bøvelstad et al., 2007; Van Wieringen et al., 2009). The *elastic net* was introduced as a method that combines the properties of both ridge and lasso regression. As mentioned before, the lasso will never select more than n parameters in the final model. Moreover, if a number of covariates is highly correlated the lasso tends to select only one of them. Zou and Hastie (2005) argued that highly correlated covariates could often be considered a group (e.g., a biological pathway) and that a variable selection method should ideally include the whole group once one variable among them is selected. To be able to select, if needed, more than n covariates and to select groups of variables instead of just one of them, the elastic net penalty is a combination of the lasso and the ridge penalty. Because of the lasso penalty, the elastic net is also a variable selection method and because of the ridge penalty the elastic net tends to select more variables than the lasso, especially when they are highly correlated. An interesting question is whether the methods connected to oracle properties will outperform methods that do not have these characteristics.

Benner et al. (2010) simulated high-dimensional datasets where the true solution was either very sparse ($d = 5$) or moderately sparse ($d = 30$). In the very sparse setting, the SCAD variants and the adaptive lasso came very close to selecting the underlying true model, while the lasso and the elastic net selected (as expected) too many variables. In the moderately sparse scenario, however, the two-stage procedures still selected very sparse models, only this time most true predictors were not included. The lasso and the elastic net on the other hand, selected, in addition to noise variables, also most of the true variables.

Benner et al. (2010) conclude that the performance of the SCAD and the adaptive lasso is highly dependent on the preselection procedure, for which the lasso is the natural choice.

When the starting solution is not close enough to the true model, there seems to be a considerable risk for SCAD and the adaptive lasso to break down completely, eventually resulting in models which are far too small or even include no covariates at all. Since there is no solution to this problem yet, the regular lasso and the elastic net are still the recommended methods for variable selection and prediction in actual data applications. At the same time it is pointed out that *"in nonsparse and even moderately sparse situations, the task of simultaneously selecting the correct variables and estimating the parameters consistently, remains an ongoing challenge."* This conclusion matches earlier findings from Bøvelstad et al. (2007) and Van Wieringen et al. (2009) who found ridge regression to have best predictive performance in high-dimensional gene expression data, even though the resulting prediction model was far from sparse. Although the relaxed lasso was not included in the comparison from the original paper (Meinshausen, 2007), it is already apparent that this method can also be expected to perform best when the true model is very sparse.

We should keep in mind that the validity of the screening and oracle properties of all lasso-type and SIS methods are based on a sparsity assumption. The question rises whether this assumption is reasonable in the genomics area.

Before we can try to answer this question, we should first understand the precise meaning of the sparsity assumption. The assumption that the underlying true model is sparse can easily be confused with the assumption that there is a limited number of variables associated with the response. Although this second assumption could be perceived as "univariate sparseness" which basically is the sparsity type demanded by SIS methods, the sparsity assumed by the lasso and its successors should be seen as "multivariate sparseness." Multivariate sparseness dictates the number of non-zero regression coefficients in the true model to be small, but not all selected variables have to be predictive in itself. Non-predictive variables accounting for variation in truly predictive regressors will be part of the true model as well. The issue of marginally uncorrelated, but jointly correlated variables, has been described earlier as the phenomenon of *unfaithfulness* in the causality literature (Wasserman and Roeder, 2009). When covariates are highly correlated, which is often the case in the high-dimensional setting (e.g., in gene and protein expression data), it can be argued that univariate as well as multivariate sparseness is not very plausible. If we consider a gene that is associated with survival and look into the genes it is correlated with, a similar association with survival can probably be demonstrated for those genes as well, contradicting univariate sparsity. Multivariate sparseness is even less likely, because even if only one gene among a group of correlated genes is associated with the response, the residual variation in this gene will partly be explainable by the correlated genes, which as a consequence will also be included in the true model, this time contradicting multivariate sparseness. Following this reasoning, the sparsity assumption underlying the lasso's oracle properties is unlikely to be met when working with gene expression data or other highly correlated designs. Note however that some data types, SNP data for example, can have a less strong correlation structure, which may result in a situation where univariate and multivariate sparseness coincide and could even be considered probable. In these situations, SIS methods would also be very well applicable.

Besides this theoretical argumentation which indicates that assuming true sparseness in genomics applications might not always be reasonable, this theory also seems to be supported by data analysis. Ein-Dor et al. (2005) showed that for a single breast cancer survival dataset, many equally predictive list of 70 genes could be constructed. So even though a single set of covariates, as for example generated by the lasso, performs well in terms of prediction, this does not necessarily mean that this set is unique or comes close to the "true" model. From this perspective, it is even questionable whether the notion of a true model is a useful one. Another indication that a true sparse model does not exist comes from the observation that using a ridge penalty in real data settings often results in a good prediction model. Since a ridge model can be considered the opposite of a sparse model, its good predictive ability could indicate that the underlying model is not sparse either.

15.2.4 The Van de Vijver dataset

In order to further illustrate these findings, we made prediction models based on either a lasso or a ridge penalty for a gene expression microarray dataset, with a survival outcome. The dataset of Van De Vijver et al. (2002) consists of gene expression profiles of 4,919 gene expression probes for 295 breast cancer patients. To measure the predictive performance of both the lasso and the ridge models, we used the *cross-validated partial log-likelihood (cvpl)*, a measure suggested by Van Houwelingen et al. (2006). The specific values for the penalty parameters were found by optimizing this same *cvpl* using Brent's algorithm for optimization. All calculations were made using the R-package `penalized` (Goeman, 2010). Although common practice, we chose not to standardize the (zero-centered) covariates. It is well known that the lasso has the disposition to firstly include variables that have a large variance and for that reason it is often advised to normalize the variances beforehand. However, if the measurements are made on the same scale, as is the case with gene expression measurements, larger variances are mostly a sign of biological variation and this variation might indeed point to better predictive ability, contradicting the need to standardize.

The following approach was taken; first a ridge and a lasso model were fitted based on the full dataset. The corresponding *cvpl*-values were given by -476.2 and -479.5, respectively, indicating that the ridge model predicts better than the lasso model. However, the ridge model included all 4,919 covariates, while the lasso model only included 16 predictors. To test whether most of the predictive information was indeed captured within these 16 predictors, we fitted a new lasso model, but this time based on all but the 16 just chosen predictors. The new model was found to fit worse than the first one, but still predicted considerably better than the null-model. We repeated the procedure of fitting a new model after removing the old predictors and could conclude that the model created after the fifth iteration (i.e., after removing approximately 60 predictors from the data) was hardly better than the null-model. After having removed 112 covariates, the lasso was no longer able to select any predictor. Fitting a ridge model on this stripped dataset instead, still resulted in a model with a *cvpl*-value of -480.7. Although worse than the original ridge model, this model still predicts almost as good as the first lasso model, indicating that the reduced dataset is still a source of information. Clearly, there are many variables weakly associated with the response, but their associations are individually too weak to get detected by the lasso. There seems strong reason to doubt the sparsity assumption, at least in this particular dataset.

Because it will be difficult to justify the sparsity assumption in many other real life settings, we should wonder how to interpret the parsimonious models resulting from the penalized or univariate screening methods. The answer could be that the resulting set of variables is indeed very valuable, because, based on these predictors, reliable prognostic tools can be produced for which only a restricted number of measurements have to be obtained. In the Van de Vijver dataset for example, the cross-validated 5-year survival probability calculated by either the lasso or a ridge model are quite similar, as illustrated by Figure 15.1.

Being able to derive sparse prediction models that still predict (almost) as well as models relying on far more measurements can be greatly beneficial in terms of time and money. However, although all screening methods discussed, under certain conditions, guarantee to provide us with all truly relevant variables, this claim will usually not be realistic. For that reason, one must be aware of the fact that, although the final models derived in this way are often useful for prediction purposes, biologically important factors can be missed and

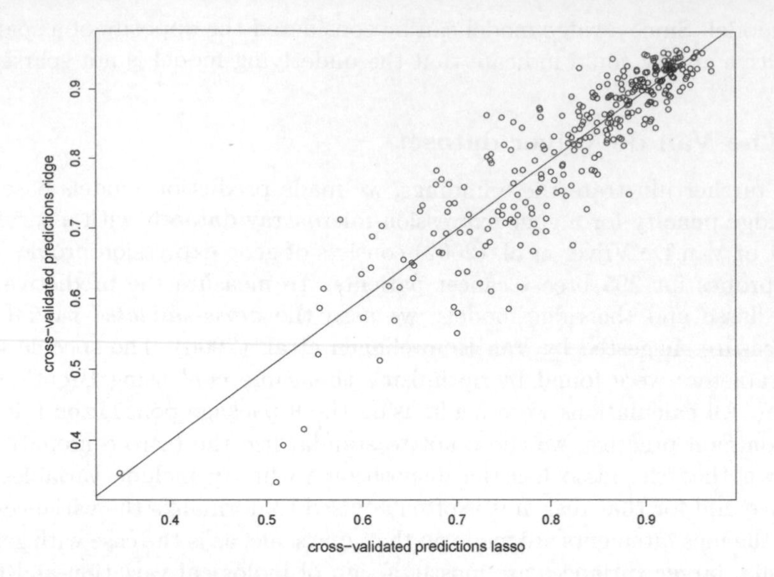

FIGURE 15.1
Cross-validated survival prediction at 5 years of diagnosis, based on a lasso or ridge model.

membership in the final prognostic list is not necessarily indicative of the variable's biological importance. To cite the conclusion of Ein-Dor et al. (2005) regarding the generated gene lists in the breast cancer study: *"Rather, in order to study the potential targets for treatment, one must scan the entire, wide list of survival-related genes."* This will be the subject of the next section.

15.3 Selecting variables by testing individual covariates

Suppose our data consists as before of n, potentially censored, survival times and a covariate matrix of size n by m. Where we previously wanted to select important variables based on their presence in *multivariate* prediction models, we will now approach the problem of selecting interesting covariates a little more directly, by formulating and testing m null-hypotheses. The collection of null-hypotheses will be denoted by $\mathcal{H} = \{H_1, \ldots, H_m\}$, and the individual hypotheses of no association between covariate i and the response are given by

$$H_i : \beta_i = 0$$

where β_i is the regression coefficient in a *univariate* Cox model where only the ith covariate is included.

There are three tests that are commonly used to test this type of null-hypothesis, namely the *Wald test*, the *score test* and the *likelihood ratio test*, each using a different test statistic. Under the null-hypothesis, the asymptotic distributions of the test statistics are known, which enables us to calculate a corresponding p-value p_i for every value of the test statistic T_i. Small p-values give reason to doubt the null-hypothesis which means that it is unlikely for the true regression coefficient to equal zero, which in turn indicates that the corresponding covariate is associated with the response and therefore important.

TABLE 15.1
Rejection/acceptation versus truth/falsehood of hypotheses.

	number not rejected	number rejected	
True null-hypotheses	U	V	m_0
False null-hypotheses	T	S	m_1
	$m - R$	R	m

Given our set of hypotheses, an unknown subset $\mathcal{T} \subset \mathcal{H}$ of size m_0 corresponds to the true hypotheses, while the remaining set $\mathcal{F} = \mathcal{H} \setminus \mathcal{T}$ of size $m_1 = m - m_0$ contains the false hypotheses. Note that the terminology is reversed with respect to the previous section. A false hypothesis corresponds to a truly relevant variable, whereas a true hypothesis corresponds to a variable which does not (significantly) influence the survival probability. Our goal of selecting influential covariates thus coincides with choosing a subset $\mathcal{R} \subseteq \mathcal{H}$ of hypotheses to reject. Ideally, this rejected set $\mathcal{R}$ is completely equivalent to the set $\mathcal{F}$. Two types of mistakes can be made, however: true hypotheses can get rejected, a so-called *type I error* or a *false positive* and on the contrary, a false hypothesis might not get detected and be accepted as a true one. This type of mistake is usually referred to as a *type II error* or a *false negative.* Rejecting a hypothesis is considered a discovery and a false rejection is called a "false positive." The situation can be summarized by Table 15.1, where V represents the number of type I errors and T the number of type II errors. The only quantities that are observable are those in the last row.

Although being unable to discover an important variable is regrettable, in hypothesis testing, identifying an effect that in reality is not present is usually considered more problematic. When there is only one hypothesis, the probability of making a type I error is bounded by the significance level α, which is most often chosen to be 0.05. By choosing α in this way, we can be reasonably confident that when we find an effect, this finding is not due to chance. Problems arise, however, when we want to perform more than one test. Because there is a risk of committing a type I error with every test, the chance of actually making one or more mistakes grows with the number of tests. More precisely, even when all m variables are in reality noise variables, the expected number of rejections will equal $m\alpha$. When m is a large number, which it will be in a high-dimensional data setting, the number of false positives can thus be very high.

To still retrieve reliable results, the focus in multiple testing problems is on keeping small either the number V of type I errors or the proportion of false rejections among all rejections, known as *the false discovery proportion* Q, where Q is defined as

$$Q = \frac{V}{\max(R, 1)} = \begin{cases} V/R & \text{if } R > 0 \\ 0 & \text{otherwise.} \end{cases}$$

Since both V and Q are random variables, we cannot control them directly, but we can control relevant aspects of their distributions. Different error rates focus on different distributional aspects (Shaffer, 1995), but the most popular ones are the *Family-wise Error Rate* (FWER), given by

$$\text{FWER} = \text{P}(V > 0) = \text{P}(Q > 0),$$

and the *False Discovery Rate* (FDR) (Benjamini and Hochberg, 1995), given by

$$\text{FDR} = \text{E}(Q).$$

By controlling the FWER or the FDR on a pre-specified α-level, an upper bound is set respectively on the probability of making any error or on the expected proportion of errors among the rejections. FWER control will automatically result in FDR control, but the reverse does not hold. Usually, controlling the FDR will result in more rejections than controlling the FWER.

The number of false positives we are willing to tolerate depends on the specific research setting. Genomics experiments are in an early stage often highly exploratory in nature. When using micro-arrays to measure gene expression levels, for example, the first step will usually consist of testing every probe for differential expression, even though many probes are not expected to be important for the specific disease under investigation. The desire to test them all merely stems from their availability and to a lesser extent from the fear of missing something important. The purpose of the experiment is often to come up with a list of promising candidates, to be further investigated in later validation experiments.

Controlling the FWER is especially important in these validation experiments. Because a validation experiment is usually the last step in a research project, the exploratory phase has already ended and the project is in the confirmatory phase. In order to be able to draw the right conclusions about which genes are truly associated with a particular disease, FWER control is necessary. In the exploratory phase, however, one could argue that a few false rejections are not that problematic, because these findings will probably be falsified in a subsequent validation experiment. Of course, even in exploratory research, some form of multiple testing correction should be used, because validation experiments are often costly in terms of time and money and it would be wasteful to proceed with many findings which will eventually turn out to be uninformative.

Controlling the FWER has as an advantage that, if you choose to follow up only on a subset of the rejected set, maybe because of biological information or limited capacity, the FWER is also controlled for this subgroup. This so-called *subsetting property* is not guaranteed in an FDR setting, in which it can happen that the proportion of false positives in a chosen subset exceeds the chosen α-level. However, when there is the opportunity of validating almost every finding, FWER control might be too restrictive and controlling the FDR can be more useful. The question of how to control the FWER and the FDR will be addressed in the next subsections.

15.3.1 Methods for FWER control

Probably the most well known method to control the FWER is the method of *Bonferroni*. When there are m null-hypotheses, using the Bonferroni method means testing the hypotheses simultaneously on level α/m instead of on level α. It can be easily shown that Bonferroni's method controls the FWER without any further assumption on the underlying number of true hypotheses (known as strong control) or on the dependency structure of the individual p-values. This method can for that reason be used in any situation. However, a drawback of the method is its conservativeness.

Holm's method (Holm, 1979) also strongly controls the FWER and has the same advantage as Bonferroni's method, namely its applicability in every possible setting. Additionally, this method will always reject at least as many hypotheses as Bonferroni's method. The gain in power lies in the *sequential* nature of Holm's method as opposed to the *single step* Bonferroni procedure. Holm's method first places the p-values in ascending order $p_{(1)}, \ldots, p_{(m)}$ and subsequently compares each p-value $p_{(i)}$ to its corresponding critical value $\alpha/(m-i+1)$. Holm's method finds the smallest j such that $p_{(j)}$ exceeds $\alpha/(m-j+1)$ and thereafter rejects all $j-1$ hypotheses with a p-value at most $\alpha/(m-j)$. The smallest p-value is thus compared to α/m, as would be the case when using a Bonferroni correction. If this p-value is smaller than α/m, the corresponding hypothesis gets rejected and the next p-value is tested on level $\alpha/(m-1)$, etc. Because the overall α-level is not distributed among all hypotheses, but only among the *remaining* hypotheses, Holm's method is uniformly more powerful than Bonferroni's method and as it is valid under the same assumptions, Holm's method should always be preferred to Bonferroni's method.

Again, we will use the Van de Vijver data to illustrate the just described multiple testing methods. The number of covariates in this dataset equals 4919, so this is the number of null-hypotheses we would like to test. Using the R-function `coxph` (from the `survival`-package), we can calculate the p-values of respectively the Wald, the score and the likelihood-ratio test statistic for all 4,919 hypotheses. We can subsequently use the R-function `p.adjust` to find out how many of those p-values can be rejected using either a Bonferroni or a Holm correction on level $\alpha = 0.05$. Using the likelihood-ratio test, we can reject 203 hypotheses according to the Bonferroni method and 206 hypotheses if we use Holm's method. Similar numbers of rejections are found by using the Wald test (230 and 232) or the score test (221 and 226). Although the differences in the number of rejections between Holm's and Bonferroni's method are quite small, they are clearly present. Despite the fact that the three test are asymptotically equivalent, it is clear that their finite sample performance can differ. The likelihood-ratio test is generally believed to be the most reliable test (see also a simulation study by Li et al. (1996)) and we will for that reason use this test throughout the rest of this chapter.

Bonferroni's and Holm's methods make no assumptions on the dependency structure of the p-values, which makes them generally applicable but sometimes unnecessarily conservative. If we are willing to make assumptions on the joint distribution of the p-values, we are often able to reject more hypotheses. One such assumption could be that the *Simes inequality* (Simes, 1986) holds for the subset of true hypotheses. The Simes inequality says that, for the ordered p-values of the m_0 *true* null-hypotheses $p_{(1)}, \dots, p_{(m_0)}$, the following holds

$$P\left(\bigcup_{i=1}^{m_0}\left\{p_{(i)} \leq \frac{i\alpha}{m_0}\right\}\right) \leq \alpha.$$

This inequality is not true in general but is shown to be valid in common situations as for example the situation in which the p-values arise from normal- or t-distributed one-sided test statistics with non-negative correlations, or for two-sided tests under more general correlation structures (Sarkar, 2008). Under the assumption that the inequality indeed holds, more powerful FWER controlling methods can be used. Note that the inequality is not about single hypotheses, but rather is a statement about a whole group of hypotheses, namely the group of all true hypotheses. Still, it can be used to determine which individual hypotheses should be rejected as shown independently by both Hommel (1988) and Hochberg (1988) who for this purpose combined the Simes inequality with the *closed testing procedure* of Marcus et al. (1976).

For m individual hypotheses $H_1, \dots, H_m$, the closed testing procedure first defines intersection hypotheses

$$H_I = \bigcap_{i \in I} H_i,$$

for all index sets $I \subseteq \{1, \dots, m\}$, with $I \neq \emptyset$. An intersection hypothesis is true if and only if all intersected hypotheses are true. Note that every individual hypothesis is an intersection hypothesis in itself. After defining all intersection hypotheses, all can be tested on level α, but only in a very specific order. A hypothesis H_I can only be tested if all hypotheses H_J with $I \subset J$ are already rejected. To understand why the FWER will be controlled in this way, even though every test is performed on level α, it is sufficient to realize that the intersection of exactly all m_0 *true* hypotheses, which will be denoted as H_T, is among the intersection hypotheses. Since this hypothesis corresponds to the *largest* group of non-important covariates, all other true hypotheses, corresponding to smaller groups, can by the imposed testing order only be rejected after H_T is rejected. If H_T does not get rejected, which will happen with probability at least $1 - \alpha$, this thus means that none of the rejections is a false positive, which implies that the FWER is indeed controlled on level α. If we assume the Simes inequality to be true, we can subsequently test every H_I based on this inequality, rejecting H_I if, for some j, the jth ordered p-value among p_i, $i \in I$, is smaller than $j\alpha/|I|$, knowing that this test will have the right α-level for all true intersection hypotheses. Because the individual hypotheses $H_1, \ldots, H_m$ are tested in the final steps of the closed testing procedure, we will know which of them can be rejected.

When there are m hypotheses of interest, the number of intersection hypotheses will equal $2^m - 1$ which can be an extremely large number. For large values of m, testing all intersection hypotheses will be impossible. Fortunately, the specific form of the Simes inequality allows us to formulate *shortcuts*, which are methods that can be used to determine which (individual) hypotheses can be rejected, without having to calculate all hypothesis tests. Using a shortcut should never result in more rejections than those that would be derived from carrying out the complete closed testing procedure, but still some shortcuts are more powerful than others. This is exactly the situation when Hommel's method (Hommel, 1988) is compared to Hochberg's method (Hochberg, 1988); although based on the same principle, the exact shortcut used by Hommel results in a more powerful procedure than the approximate one used by Hochberg. Although the formulation of Hommel's final procedure is quite complicated, Hochberg's method is easily explained. Each ordered p-value $p_{(i)}$ is compared to a critical value $\alpha/(m - i + 1)$, as was done in Holm's procedure. This time the largest j is found such that $p_{(j)}$ is smaller than or equal to $\alpha/(m - j + 1)$ and subsequently this hypothesis and all hypotheses corresponding to smaller p-values will be rejected. Compared to Holm's method, it is clear that Hochberg's method will result in at least as much rejections. Since Hommel's method is even more powerful, the same holds for this method. Remember, however, that this gain in power comes at the price of an extra assumption on the dependency structure of the p-values.

Applied to the Van de Vijver dataset, Hochberg's method finds 206 rejections, using the likelihood-ratio test. This number is identical to the number of rejections found with Holm's method. Although Hochberg's method is more powerful, the actual hypotheses and their corresponding p-values will determine if this power difference will also result in more rejections. Likewise, Hommel's method rejects 209 hypotheses. The advantage of using Hommel's method instead of Hochberg's method is comparable to the earlier discussed advantage of using Holm's method as opposed to the Bonferroni procedure. Table 15.2 summarizes the just described methods, their underlying assumptions and their corresponding outcome when applied to the Van de Vijver dataset.

All four methods discussed either made a general assumption or no assumption at all on the dependency structure of the p-values. However, it is also possible to adapt the procedure to the dependency structure that is *observed* in the data, by replacing the unknown true null-distribution with a permutation null-distribution. Westfall and Young (1993) propose a method that brings this idea into practice. For certain underlying dependency structures, permutation based methods can be a very powerful alternative to using Holm's or Hommel's method. Nevertheless, not every problem is eligible for a permutation-based method.

A last remark on FWER controlling methods is that they can provide, in addition to the information about which hypotheses to reject, a level of certainty about these rejections being correct rejections, as given by the p-value in case of one single test. The direct analogue of this p-value in the context of multiple testing is the *adjusted p-value*, which is defined as the smallest α-level at which the multiple testing procedure would reject this specific hypothesis. Adjusted p-values for the methods of Bonferroni, Holm, Hommel and Hochberg are easily computable by the R-function `p.adjust`.

TABLE 15.2
Overview of the different methods controlling the FWER, their underlying assumptions and their corresponding number of rejections when applied to the Van de Vijver dataset, with α-level 0.05.

FWER controlling methods

method	underlying assumptions	number of rejections
Bonferroni	none	203
Holm	none	206
Hochberg	Simes inequality holds	206
Hommel	Simes inequality holds	209

15.3.2 Methods for FDR control

As with the FWER-based methods, methods that control the FDR can also be divided into methods that are generally valid, methods that are valid under the assumption that the Simes inequality holds and methods based on permutations. Methods that control the FDR (or the FWER) only under the assumption of *independent* p-values have been developed as well, but those will not be discussed here, because this assumption is not tenable for applications in genomics research.

The first and still most widely used method to control the FDR was developed by Benjamini and Hochberg (1995). In contrast to the first methods to control the FWER, which do not assume a special p-value dependency structure and are for that reason sometimes quite conservative, the method of Benjamini and Hochberg is valid under assumptions that are virtually identical to those required for Simes inequality (Sarkar, 2008). The method uses the critical values suggested by the Simes inequality to test the individual hypotheses directly. Each ordered p-value $p_{(i)}$ is compared with the critical value $i\alpha/m$. Then the largest j such that $p_{(j)} \leq j\alpha/m$ is found and subsequently the hypotheses corresponding to the smallest j p-values are rejected. This method is the exact analogue of Hochberg's FWER controlling method, only the critical values are larger.

That higher critical values will normally lead to many more rejections can be illustrated by applying the method to the Van De Vijver dataset again. Where Hochberg's FWER controlling method only rejected 206 hypotheses using the likelihood-ratio test at level 0.05, Benjamini & Hochberg's FDR controlling method rejects 1,340 hypotheses at this same α-level. FDR controlling methods like the Benjamini and Hochberg method will especially have higher power compared to FWER-based methods, when the number of rejections is already high. In that case the denominator of the false discovery proportion Q gets large, leaving more room for extra rejections while still controlling the FDR criterium.

Although the method of Benjamini and Hochberg is already very powerful, possible improvements have been examined. It has been shown that the procedure actually controls the FDR not at level α, but at level $\pi_0\alpha$, where π_0 is the proportion of true hypotheses. If that proportion would be known to be smaller than one, the method's performance could be sharpened. In reality, however, this proportion will never be known, but will have to be estimated. Methods that try to find more rejections by using such an estimator of π_0 are called *adaptive procedures* (Benjamini et al., 2006). The usefulness of these methods will nonetheless heavily rely on the chosen estimation procedure, and it can even happen that instead of more rejections less rejections can be made.

A different approach to controlling the FDR has been taken by Benjamini and Yekutieli (2001). They propose a method that is valid under general p-value dependency structures

TABLE 15.3
Overview of the different methods controlling the FDR, their underlying assumptions and their corresponding number of rejections when applied to the Van de Vijver dataset, with α-level 0.05.

FDR controlling methods		
method	underlying assumptions	number of rejections
Benjamini & Yekutieli	none	614
Benjamini & Hochberg	Simes inequality holds	1340

and which thus does not depend on the validity of the Simes inequality. Each ordered p-value $p_{(i)}$ is this time compared to the critical value $i\alpha/(m\sum_{k=1}^{m} 1/k)$. As before, the largest j is found such that $p_{(j)}$ is smaller or equal to its corresponding critical value and all j hypotheses with the j smallest p-values are subsequently rejected.

The loss in power relatively to the Benjamini & Hochberg procedure is immediately visible when we look at Table 15.3 in which the behavior of the two methods on the Van de Vijver dataset is summarized. The Benjamini & Yekutieli procedure results in 614 rejections, compared to 1,340 found by the Benjamini & Hochberg procedure. Where the Benjamini & Hochberg procedure is strictly more powerful than Hochberg's FWER-based method, the same cannot be said for the Benjamini & Yekutieli procedure compared to its FWER-controlling counterpart, Holm's procedure. Although the Benjamini & Yekutieli method will usually result in more rejections, as is the case in our example dataset, in datasets with very few differentially expressed genes, it can happen that Holm's procedure rejects more. Because a FWER controlling method by definition also controls the FDR, in such extreme situations, Holm's method might be the preferred method to control the FDR, if no assumptions can be made on the p-value dependency structure.

Several authors have worked on FDR control by permutation or other types of resampling such as the bootstrap (Yekutieli and Benjamini, 1999; Romano et al., 2008). However, all methods proposed so far can only provide asymptotic control, are usually based on the bootstrap rather than on permutations, and often require substantial additional assumptions. It seems that FDR control by permutations is much more difficult than permutation-based FWER control. Although powerful permutation-based methods with exact finite sample FDR control would be highly desirable, they are not yet available.

Just as with the FWER-based methods, FDR-based methods also allow for the calculation of adjusted p-values. For the Benjamini & Hochberg and the Benjamini & Yekutieli methods these are also obtainable via the `p.adjust`-function. Interpreting the adjusted p-values is nevertheless less natural than in the FWER-based setting. Where a FWER-based adjusted p-value can be interpreted as a property of the hypothesis itself, this does not hold for FDR-based adjusted p-values. FDR control only tells us something about sets of hypotheses and not of individual hypotheses. In the same way, the adjusted p-values tell us something about the whole rejection set up to this single hypothesis, not of this hypothesis itself.

15.3.3 Confidence intervals for the number of true discoveries

Although the FDR is important in exploratory research, in the end it is only a statement about an average over the full set of rejections. Unfortunately, it is impossible to determine how many true or false rejections will (on average) be in an arbitrary subset of this final rejection set $\mathcal{R}$. However, this might be exactly what exploratory research would ideally be about: the *post hoc* choosing of hypotheses to continue with in a future validation experiment. Goeman and Solari (2011) propose a multiple testing method that does allow for such post hoc inference. They show that the earlier described closed testing procedure can be used to construct exact *simultaneous* confidence sets for the number of false rejections incurred when rejecting any specific set of hypotheses. What this means is that the researcher, *after having seen the results*, can choose his favorite set of hypotheses for further study, based on a confidence set for the number of false rejections or, if preferred, for the number of true discoveries. Because the confidence sets are simultaneous over all possible sets of rejected hypotheses, the researcher is free to optimize the selected set of hypotheses based on these confidence sets.

The possibility of deriving confidence sets simultaneously stems from the fact that they are derived from a single application of the closed testing procedure. Since all rejections of intersection hypotheses are simultaneously valid with probability $1 - \alpha$, the same holds for all confidence sets derived from these rejections.

As mentioned before, using a closed testing procedure can lead to computational difficulties when the number of individual hypotheses is large. However, using tests that allow for shortcuts, as for example the test based on the Simes inequality as used by Hochberg and Hommel, makes the procedure usable for a very large number of hypotheses as well. With the `R`-package `cherry` it is possible to use the Simes inequality to make confidence statements about the number of true discoveries for large sets of hypotheses. If we take for example all 4,919 hypotheses of the Van de Vijver dataset, we find that the 95% confidence interval for the number of true discoveries (i.e., false null-hypothesis) within these 4,919 probes is given by $[640, 4919]$, which means that there are at least 640 relevant probes among them, even though this does not imply that we will ever be able to pinpoint these probes.

Although it is informative to know how many relevant probes are present in the full dataset, it might be more interesting to make statements about the number of relevant probes in specifically defined subsets. Imagine for example that we would like to do a follow-up experiment on 500 probes. To increase the likelihood of finding truly relevant probes, we choose the set of probes that have the smallest p-values, based on a univariate Cox model and the likelihood-ratio test. The `cherry`-package again provides us with a 95% confidence interval for the number of true discoveries in this set, which is this time given by $[476, 500]$. So, assuming that the Simes inequality holds, we know that in the set of probes with the smallest 500 univariate p-values, at least 476 are truly related to the response, at a confidence level of 0.95.

Another subset we might be interested in is the subset of probes that the lasso selects to be in the final model. In the Van de Vijver dataset, we saw that the lasso constructs a model consisting of 16 predictors. Suppose we choose to follow up on these predictors in a validation experiment, what can we expect about the number of true findings in this set? Based on the simultaneous confidence set for the number of true findings, given by $[10, 16]$, we know that at least 10 probes selected by the lasso are also relevant from a univariate point of view. In Figure 15.2, we can even see that in order to choose a set containing at least 10 relevant probes, we only have to select the first 11 probes selected by the lasso, where the probes are sorted on increasing univariate p-value.

Compared to FWER or FDR controlling methods that only allow the user to specify the quality criterium and the α-level, resulting in a fixed set of rejected hypotheses, this new method gives the researcher freedom to compose his or her own preferred set of hypotheses, providing information about the risk of following up on this particular set.

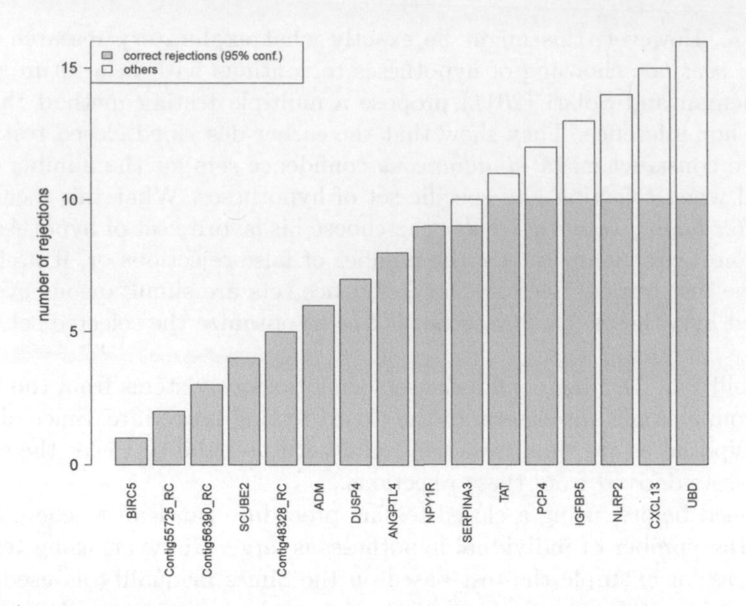

FIGURE 15.2
Number of correct rejections versus number of included probes for the 16 probes selected by the lasso. In case of an annotated probe, the x-label corresponds to the gene.

15.4 Reducing the number of variables beforehand and incorporating background knowledge

Until now, we only discussed how to obtain relevant information from an available gene expression dataset, without considering the possibility of first adding a few well-established prognostic factors to our covariates. Very often, clinical factors or molecular markers are already known to be prognostically relevant for the patient population under investigation (Benner et al., 2010; Binder et al., 2011). Many of those factors are not only easily available, e.g., age or gender, but can also be very influential and should for that reason not be ignored. Obtaining gene signatures, for example, that are actually predictive of age are, of course, not the findings we are hoping for. To prevent discoveries that will turn out to be just expensive ways to measure more easily accessible information, these "background" variables or confounders should both be included in our penalized regression model, as well as added to the null-hypotheses in the multiple testing framework. All penalized regression models can easily be extended to incorporate additional unpenalized covariates. By not adding an extra penalty term to these additional terms they will always be included in the final model. From a testing perspective, adjustment of the test results for possible confounders is only a manner of specifying these variables as covariates under the null-hypothesis.

Another point that has to be made here is that using the complete gene expression dataset is not necessarily the most promising strategy. Especially in the multiple testing framework, eliminating hypotheses beforehand, based on biological knowledge for example, can lower the multiple testing penalty considerably. FWER controlling methods will always be more powerful when fewer hypotheses are being tested. For FDR-based methods, this property does not hold in general, but if the discarded hypotheses were indeed true hypotheses, the power is expected to increase as well. Fortunately, especially in the gene expression context, there are often many "uninteresting" hypotheses that could be excluded in advance. Non-annotated probes for example are rarely considered interesting, since the name of the associated gene is yet unknown. Still, these non-annotated probes can make up for a considerable proportion of the complete set of measurements. In the Van de Vijver data for example, 35% of the probes is of this type. In a regression setting as well, it should be the rule rather than the exception to only include variables in which you would at least be interested when they were to appear in the final model, even though the exact number of variables to start with is less important in this situation.

A different approach to lower the multiple testing correction is by aggregating the data in a reasonable way. This type of dimension reduction will not always be possible, but in some situations it comes very naturally. Rather than testing every probe, for example, aggregated tests can be performed at the gene level or even at the level of a larger chromosomal region. Sometimes, the gene level will in fact be the preferred level of inference and in that case, conducting fewer tests on the gene level as opposed to more tests on the probe level is recommended. Choosing a test for aggregated hypotheses should be done carefully, however. Binder et al. (2011) discuss ways of data aggregation for lasso-type methods as well.

15.5 Discussion

The most important message of this chapter is that the way to retrieve information from a high-dimensional dataset should depend on the specific research question. If the aim is to find a relatively sparse prediction model, a multivariate approach, such as using a lasso model, is an appropriate choice. If, on the other hand, the aim is to find potential targets for future treatment, a univariate approach is a more sensible option.

Under rather strict assumptions on the correlation structure between the covariates and the sparsity of the underlying true model, the lasso approach will also lead to a set of predictors that incorporates most of the (biologically) relevant variables. This set could be compared to a set of predictors resulting from a multiple testing procedure. It is interesting to note that, whereas the emphasis in multiple testing procedures lies mainly on preventing type I errors, condoning the resulting type II errors, the lasso mostly focuses on including every potential "true" predictor, taking some additional noise variables more or less for granted. Again from a different perspective, this observation nicely illustrates the differences between the two approaches; for prediction purposes, in view of minimizing bias, it is better to include all true predictors with the risk of adding some extra noise variables whereas actually *proving* association asks for a more rigorous procedure.

Bibliography

Benjamini, Y. and Hochberg, Y. (1995), 'Controlling the false discovery rate - a practical and powerful approach to multiple testing', *Journal of the Royal Statistical Society - Series B* **57**(1), 289–300.

Benjamini, Y., Krieger, A. M. and Yekutieli, D. (2006), 'Adaptive linear step-up procedures that control the false discovery rate', *Biometrika* **93**(3), 491–507.

Benjamini, Y. and Yekutieli, D. (2001), 'The control of the false discovery rate in multiple testing under dependency', *Annals of Statistics* **29**(4), 1165–1188.

Benner, A., Zucknick, M., Hielscher, T., Ittrich, C. and Mansmann, U. (2010), 'High-dimensional Cox models: the choice of penalty as part of the model building process', *Biometrical Journal* **52**(1, SI), 50–69.

Binder, H., Porzelius, C. and Schumacher, M. (2011), 'An overview of techniques for linking high-dimensional molecular data to time-to-event endpoints by risk prediction models', *Biometrical Journal* **53**(2), 170–189.

Bøvelstad, H., Nygård, S., Størvold, H., Aldrin, M., Borgan, Ø., Frigessi, A. and Lingjærde, O. (2007), 'Predicting survival from microarray data - a comparative study', *Bioinformatics* **23**(16), 2080–2087.

Bradic, J., Fan, J. and Jiang, J. (2011), 'Regularization for Cox's proportional hazards model with NP-dimensionality', *Annals of Statistics* **39**(6), 3092–3120.

Breiman, L. (1995), 'Better subset regression using the nonnegative garrote', *Technometrics* **37**(4), pp. 373–384.

Breiman, L. (1996), 'Heuristics of instability and stabilization in model selection', *The Annals of Statistics* **24**(6), pp. 2350–2383.

Bühlmann, P. and van de Geer, S. (2011), *Statistics for High-Dimensional Data: Methods, Theory and Applications*, Springer, New York.

Bunea, F., Tsybakov, A. and Wegkamp, M. (2007), 'Sparsity oracle inequalities for the Lasso', *Electronic Journal of Statistics* **1**, 169–194.

Cox, D. (1972), 'Regression models and life-tables', *Journal of the Royal Statistical Society - Series B* **34**(2), 187–220.

Cox, D. R. and Snell, E. J. (1974), 'The choice of variables in observational studies', *Journal of the Royal Statistical Society - Series C* **23**(1), pp. 51–59.

Ein-Dor, L., Kela, I., Getz, G., Givol, D. and Domany, E. (2005), 'Outcome signature genes in breast cancer: is there a unique set?', *Bioinformatics* **21**(2), 171–178.

Fan, J. and Li, R. (2001), 'Variable selection via nonconcave penalized likelihood and its oracle properties', *Journal of the American Statistical Association* **96**(456), 1348–1360.

Fan, J. and Lv, J. (2008), 'Sure independence screening for ultrahigh dimensional feature space', *Journal of the Royal Statistical Society - Series B* **70**, 849–883.

Goeman, J. J. (2010), 'L-1 Penalized Estimation in the Cox Proportional Hazards Model', *Biometrical Journal* **52**(1, SI), 70–84.

Goeman, J. J. and Solari, A. (2011), 'Multiple Testing for Exploratory Research', *Statistical Science* **26**(4), 584–597.

Gorst-Rasmussen, A. and Scheike, T. (2012), 'Independent screening for single-index hazard rate models with ultrahigh dimensional features', *Journal of the Royal Statistical Society - Series B* **75**(2), 217–245.

Hochberg, Y. (1988), 'A sharper Bonferroni procedure for multiple tests of significance', *Biometrika* **75**(4), 800–802.

Hoerl, A. and Kennard, R. (1970), 'Ridge regression - biased estimation for nonorthogonal problems', *Technometrics* **12**(1), 69–82.

Holm, S. (1979), 'A simple sequantially rejective multiple test procedure', *Scandinavian Journal of Statistics* **6**(2), 65–70.

Hommel, G. (1988), 'A stagewise rejective multiple test procedure based on a modified Bonferroni test', *Biometrika* **75**(2), 383–386.

Leeb, H. and Poetscher, B. M. (2008), 'Sparse estimators and the oracle property, or the return of Hodges' estimator', *Journal of Econometrics* **142**(1), 201–211.

Li, Y.H., Klein, J.P. and Moeschberger, M.L. (1996), 'Effects of model misspecification in estimating covariate effects in survival analysis for small sample sizes', *Computational Statistics & Data Analysis* **22**(2), 177–192.

Marcus, R., Peritz, E. and Gabriel, K. (1976), 'Closed testing procedures with special reference to ordered analysis of variance', *Biometrika* **63**(3), 655–660.

Meinshausen, N. (2007), 'Relaxed lasso', *Computational Statistics & Data Analysis* **52**(1), 374–393.

Meinshausen, N. and Bühlmann, P. (2006), 'High-dimensional graphs and variable selection with the Lasso', *Annals of Statistics* **34**(3), 1436–1462.

Romano, J. P., Shaikh, A. M. and Wolf, M. (2008), 'Control of the false discovery rate under dependence using the bootstrap and subsampling', *Test* **17**(3), 417–442.

Sarkar, S. K. (2008), 'On the Simes inequality and its generalization', *Beyond Parametrics in Interdisciplinary Research: Festschrift in Honor of Professor Pranab K. Sen* **1**, 231–242.

Shaffer, J. (1995), 'Multiple hypothesis-testing', *Annual Review of Psychology* **46**, 561–584.

Simes, R. (1986), 'An improved Bonferroni procedure for multiple tests of significance', *Biometrika* **73**(3), 751–754.

Tibshirani, R. (1996), 'Regression shrinkage and selection via the lasso', *Journal of the Royal Statistical Society - Series B* **58**(1), 267–288.

Tibshirani, R. (1997), 'The lasso method for variable selection in the Cox model', *Statistics in Medicine* **16**(4), 385–395.

Tibshirani, R. (2011), 'Regression shrinkage and selection via the lasso: a retrospective', *Journal of the Royal Statistical Society - Series B* **73**(Part 3), 273–282.

Van De Vijver, M., He, Y., Dai, H., Hart, A., Voskuil, D., Schreiber, G., Peterse, J., Roberts, C., Marton, M., Parrish, M. et al. (2002), 'A gene-expression signature as a predictor of survival in breast cancer', *New England Journal of Medicine* **347**(25), 1999–2009.

van Houwelingen, H., Bruinsma, T., Hart, A., Van't Veer, L. and Wessels, L. (2006), 'Cross-validated Cox regression on microarray gene expression data', *Statistics in Medicine* **25**(18), 3201–3216.

Van Wieringen, W., Kun, D., Hampel, R. and Boulesteix, A. (2009), 'Survival prediction using gene expression data: a review and comparison', *Computational Statistics & Data Analysis* **53**(5), 1590–1603.

Verweij, P. and Van Houwelingen, H. (1993), 'Cross-validation in survival analysis', *Statistics in Medicine* **12**(24), 2305–2314.

Wasserman, L. and Roeder, K. (2009), 'High-dimensional variable selection', *Annals of Statistics* **37**(5A), 2178–2201.

Westfall, P. and Young, S. (1993), *Resampling-based multiple testing: Examples and methods for p-value adjustment*, Vol. 279, Wiley-Interscience, New York.

Wiegand, R. E. (2010), 'Performance of using multiple stepwise algorithms for variable selection', *Statistics in Medicine* **29**(15), 1647–1659.

Yekutieli, D. and Benjamini, Y. (1999), 'Resampling-based false discovery rate controlling multiple test procedures for correlated test statistics', *Journal of Statistical Planning and Inference* **82**(1-2), 171–196.

Zhao, P. and Yu, B. (2006), 'On model selection consistency of Lasso', *Journal of Machine Learning Research* **7**, 2541–2563.

Zhao, S. D. and Li, Y. (2012), 'Principled sure independence screening for Cox models with ultra-high-dimensional covariates', *Journal of Multivariate Analysis* **105**(1), 397–411.

Zou, H. (2006), 'The adaptive lasso and its oracle properties', *Journal of the American Statistical Association* **101**(476), 1418–1429.

Zou, H. and Hastie, T. (2005), 'Regularization and variable selection via the elastic net', *Journal of the Royal Statistical Society - Series B* **67**(2), 301–320.

Zou, H. and Li, R. (2008), 'One-step sparse estimates in nonconcave penalized likelihood models', *Annals of Statistics* **36**(4), 1509–1533.

16

Robustness of Proportional Hazards Regression

John O'Quigley
Université Pierre et Marie Curie - Paris VI

Ronghui Xu
University of California

CONTENTS

16.1 Introduction 323
16.2 Impact of censoring on estimating equation 325
16.2.1 Model-based expectations 325
16.2.2 More robust estimating equations 326
16.3 Robust estimator of average regression effect 329
16.3.1 The robust estimator 329
16.3.2 Interpretation of average regression effect 330
16.3.3 Simulations 331
16.4 When some covariates have non-PH 333
16.5 Proportional hazards regression for correlated data 335
Bibliography 337

16.1 Introduction

The usual censoring mechanism considered for survival models is one where the censoring is independent of the failure time given the covariate value. Sometimes a stronger assumption is needed in which we require independence between the censoring and failure times. Under the weaker assumption, the partial likelihood estimator for the regression coefficient β in the proportional hazards model was first shown to be consistent by Cox (1975) and later, more formally, by Tsiatis (1981). Consistency can be established using a martingale approach (Andersen and Gill, 1982). The two ways in which the main model assumptions may fail are that where the covariate specification is incorrect or that where the constancy of regression effect, β, is not respected by the observations. Either of these will result in a dependency of the usual estimates on the censoring mechanism. In certain simplified cases, say a simple binary grouping variable, the only thing we need consider is the time dependency. This is because any transformation of the grouping variable that still allows us to distinguish the groups will not change the model essentially; the only inadequacy we need consider is then that of the constancy of regression effect. Again, this argument extends immediately, to p groups, represented by $p-1$ binary indicator variables. Moving away from this situation to that of ordinal variables, or even continuous variables, then the covariate representation can be crucial. Some transformation, for example using normal order statistics, or using a uniform transformation of the scale, can produce a model that may be more robust to small departures from assumptions. If so, we would expect the impact of the censoring to be weak if not almost absent.

The estimating equation derives from the model. If we modify our model by transforming the covariate specification, then we will also change the estimating equation. That is fairly obvious. What may be less obvious is that we can still maintain our model but modify the estimating equation directly. If we are concerned that high covariate values may have an undue influence on inference, in particular if only a few are represented in the data, we could, for example, bound or transform the value in the equation itself without making any such transformation in the actual model. Another example would be where we use the actual covariate values in the model but where, for the purposes of robustness, we work with the ranks of the covariates in the estimating equation. Under time dependent effects, expressed as $\beta(t)$, the partial likelihood estimator converges to some population parameter depending in a complex way on the underlying censoring mechanism. In both cases we would like to have procedures that reduce the impact of censoring on inference.

One reason for this is that the practical application of the estimator is otherwise limited. When effects depend on time, although the partial likelihood estimator is consistent under the proportional hazards model, it is not consistent for any meaningful parameter, i.e., one that does not involve censoring, under broader models in which the regression effect $\beta(t)$ is not constant through time. Indeed, the effect of censoring on the partial likelihood estimate can be considerable (Xu and O'Quigley, 2000), whereas, for the estimator described here, not only does censoring not impact the population parameter to which we converge but also the parameter can be given a concrete interpretation as "average effect." The estimator can be seen to be consistent in the more usual situation in which the data are generated by a mechanism that is not exactly equal to but only approximated by the working model in which hazard ratios are taken to be constant.

Lancaster and Nickell (1980), Gail et al. (1984), Struthers and Kalbfleisch (1986), Bretagnolle and Huber-Carol (1988), O'Quigley and Pessione (1989, 1991), Anderson and Fleming (1995), and Ford et al. (1995) have all studied the fit of the model in the presence of time dependency of regression effect. A general formulation of this, and a useful starting point, is to write the model in the following way:

$$\lambda(t|Z(t)) = \lambda_0(t)\exp\{\beta(t)Z(t)\}, \tag{16.1}$$

where $Z(t)$ is a possibly time-dependent covariate, λ is the conditional hazard function, λ_0 is the baseline hazard and $\beta(t)$ is the time-varying regression effect. For simplicity of notation we first assume covariates of dimension one. Particular cases of multiple covariates are considered later. Under this model, an estimator of average regression effect can be a more suitable summary measure of effect, i.e., a quantity consistent for $\int \beta(t)dF(t)$, where $F(t)$ is a distribution function, and in the following we take it to be that of the failure time. This would be robust to censoring since it does not involve the censoring distribution. With reasonably large datasets, relative to the number of studied covariates, it is possible to estimate $\beta(t)$ under model (16.1), as in Murphy and Sen (1991), Gray (1992), and Xu and Adak (2002), among others. However, the interpretation of an average effect is usually much more straightforward in applications, and also estimation of an average effect can be used in a preliminary analysis of a dataset with time-varying regression effects.

16.2 Impact of censoring on estimating equation

In order to fix the notation we take T_i, C_i and $Z_i(\cdot)$, $i = 1, ..., n$, to be a random sample from the distribution of T, C and $Z(\cdot)$ which satisfies model (16.1). Here T is the failure time random variable and $Z(\cdot)$ the covariate as described in Section 16.1, C is a censoring time random variable. The time-dependent covariate $Z(\cdot)$ is assumed to be a predictable process and, for notational simplicity, assumed to be of dimension one. For each subject i we observe $X_i = \min(T_i, C_i)$, and $\delta_i = I(T_i \leq C_i)$. Define also $Y_i(t) = I(X_i \geq t)$, $N_i(t) = I\{T_i \leq t, T_i \leq C_i\}$ and $\bar{N}(t) = \sum_1^n N_i(t)$. The useful concept of risk set is defined via the set $\mathcal{R}(t) = \{i\,; Y_i(t) = 1\}$. The size of the risk set at time t is denoted $n(t) = \sum_{i=1}^n Y_i(t)$. We will first assume C to be independent of T and $Z(\cdot)$, and will refer to this assumption as independent censorship. This assumption has been used under non-proportional hazards models by Cheng et al. (1995) and Ying et al. (1995). For a dataset of n subjects, it is useful to note the number of distinct failures k, so that $k = \sum_{i=1}^n \delta_i$. The ordered failures, from smallest to largest, are denoted $T_{(i)}$, $i = 1, ..., k$ and only when the original times are ordered, and there is no censoring, do we have that $T_i = T_{(i)}$, otherwise they differ and the index i has different ranges.

Suppose that Model (16.1) generates the observations but that we use the more restricted proportional hazards model of which (16.1) is a generalization, for the purposes of analysis. We might imagine that the resulting estimate would be interpretable as one of average effect and, indeed, this is true when there is no censoring. Unfortunately though, the partial likelihood estimator is not robust to censoring in this situation and we need consider something else if we are to recover robustness, i.e., an estimator that is unaffected asymptotically by an independent censorship. For this purpose the relevant distribution is not, as we might guess, the distribution of T given the covariate but, in fact, it is the conditional distribution of the covariate given the failure time T. This conditional distribution enables us to view the score equation from the partial likelihood as an estimating equation for the average effect and, as a result, suggests a way to obtain robustness to censoring.

16.2.1 Model-based expectations

Time plays two roles in model (16.1). First, $Z(\cdot)$ is a stochastic process with respect to time, so that $Z(t)$ is a random variable at any fixed t and may have different distributions at different time points t. Secondly, the failure time variable T is a non-negative random variable denoting time. While it is immediate to understand the distribution of T given the covariates, at any fixed time t there are two different conditional distributions of $Z(t)$ on T that are of interest to us. First, the conditional distribution of $Z(t)$ given $T \geq t$, which can be estimated by the empirical distribution of $Z(t)$ in the risk set at time t under the independent censorship assumption. The other conditional distribution is that of $Z(t)$ given that $T = t$, which can be interpreted as the distribution of $Z(t)$ among individuals who fail at time t in the population. Under the assumption that T has a continuous distribution we usually observe only one failure at a time and it is difficult to estimate this latter conditional distribution. We can, however, obtain a consistent estimate by using Model (16.1), as is described in Theorem 1. Define

$$\pi_i(\beta, t) = \frac{Y_i(t)\exp\{\beta Z_i(t)\}}{\sum_{j=1}^n Y_j(t)\exp\{\beta Z_j(t)\}}. \tag{16.2}$$

The product of the π's over the observed failure times gives the partial likelihood (Cox, 1972, 1975) under a proportional hazards model. When $\beta = 0$, $\{\pi_i(0, t)\}_i$ is the empirical distribution that assigns equal weight to each sample subject in the risk set. The following theorem (Xu, 1996; Xu and O'Quigley, 1999) states that $\{\pi_i(\beta(t), t)\}_i$ provides a consistent estimate of the conditional distribution of $Z(t)$ given $T = t$ under (16.1).

Theorem 1 *Under Model (16.1) and an independent censorship, assuming $\beta(t)$ is known, the conditional distribution function of $Z(t)$ given $T = t$ is consistently estimated by*

$$\hat{P}(Z(t) \leq z | T = t) = \sum_{\{j: Z_j(t) \leq z\}} \pi_j(\beta(t), t).$$

Theorem 1 is mainly of theoretical interest here, as $\beta(t)$ is not known in practice. In addition, if we assume a general form of the relative risk $r(t; Z)$ and define $\{\pi_i\}_i$ through $r(t; Z)$ in place of $\exp(\beta Z)$, the proof of Theorem 1 can be easily modified to show that the same result holds for general $r(t; Z)$. This is of interest in its own right and useful for the discussion of our estimator under other non-proportional hazards models. Expectations with respect to $\pi_j(\beta, t), j = 1, ..., n$, are written,

$$\mathcal{E}(Z : \beta, t) = \sum_{j=1}^{n} Z_j(t)\pi_j(\beta, t) = \frac{\sum_{j=1}^{n} Y_j(t) Z_j(t) \exp\{\beta Z_j(t)\}}{\sum_{j=1}^{n} Y_j(t) \exp\{\beta Z_j(t)\}}. \tag{16.3}$$

In accordance with Theorem 1, $\mathcal{E}(Z : \beta(t), t)$ converges in probability to $E\{Z(t)|T = t\}$ under the model.

16.2.2 More robust estimating equations

We find the estimate $\hat{\beta}$ for β from the score equation $U(\hat{\beta}) = 0$ which in the case of no censoring and time-invariant covariates, is given by,

$$U(\beta) = \sum_{i=1}^{n} \{Z_i - \mathcal{E}(\beta, X_i)\} = 0. \tag{16.4}$$

Dividing both sides of (16.4) by n, $\sum Z_i/n$ then converges in probability to the marginal expectation of Z. The second term on the left-hand side, if we replace β by $\beta(t)$, would be $\sum \mathcal{E}(Z : \beta(t), X_i)/n = \int \mathcal{E}(Z : \beta(t), t) dF_n(t)$ where $F_n(t)$ is the empirical distribution function of T. This is a double (empirical) expectation, and, since $\mathcal{E}(\beta(t), t)$ consistently estimates $E\{Z|T = t\}$ in this case, it again gives a consistent estimate of the marginal expectation of Z under Model (16.1). Therefore (16.4) can be viewed as an estimating equation (Godambe and Kale, 1991). In the presence of censoring and for time-dependent covariates in general, we weight the summands in (16.4) (i.e., the Schoenfeld residuals) by the increments of a consistent estimate of the marginal failure time distribution $F(t)$, such as the Kaplan and Meier (1958) estimate. Thus (16.4) is generalized to

$$\sum_{i=1}^{n} \delta_i W(X_i)\{Z_i(X_i) - \mathcal{E}(Z : \beta, X_i)\} = \sum_{i=1}^{n} \int_0^\infty W(t)\{Z_i(t) - \mathcal{E}(Z : \beta, t)\} dN_i(t) = 0, \tag{16.5}$$

where $W(t) = \hat{S}(t) / \sum_1^n Y_i(t)$, and $\hat{S}(t)$ is the left continuous version of the Kaplan-Meier estimate of the marginal survivorship function $S(t) = 1 - F(t)$. Assuming no ties it can be verified that $W(X_i)$ is the jump of the Kaplan-Meier curve at an observed failure time X_i. In practice ties may be split randomly, or some other approaches can be adopted (Peto, 1972; Breslow, 1974). We denote the solution to (16.5) as $\tilde{\beta}$.

Typically $Z_i(t)$ is defined as a function of measurements on the ith subject but we can consider more general expressions and in particular those that may involve measurements taken on other subjects. This would be the case, for example, when, in the interest of robustness, we replace the Z by its rank in the risk set. The value of the rank can only be determined by making use of the values of all subjects in the risk set at that time. If, instead of being the observed covariate value, we redefine the quantity $Z_i(t)$ to be a function of this value and any other information contained in $\mathcal{F}_t$, the collective failure, censoring and covariate information prior to time t on the entire study group, we can write; $Z_i(t) = \psi_i(t, \mathcal{F}_t)$ for some bounded function ψ. To keep things uncluttered we drop use of the notation $\mathcal{F}_t$ in the definition of ψ but keep in mind that when we write $\psi_i(t)$, however defined, we only ever allow ourselves to make use of information contained in the set $\mathcal{F}_t$.

$$\sum_{i=1}^{n} \delta_i W(X_i)\{\psi_i(X_i) - \mathcal{E}(\psi : \beta, X_i)\} = \sum_{i=1}^{n} \int_0^\infty W(t)\{\psi_i(t) - \mathcal{E}(\psi : \beta, t)\}dN_i(t) = 0, \tag{16.6}$$

Note that this is not the same as simply making the transform ψ on Z and then fitting a proportional hazards model to the resulting observations. Under such a transform the model is no longer the same and there is no reason to suppose that the regression coefficient(s), β, would remain the same. They would not even necessarily have the same sign. In this above expression, the relevant probabilities are still those given by Equation (16.2) and, for any bounded ψ, the estimated $\hat{\beta}$ converge (a.s.) to the same population value β. Recall that we denote the distinct failures by; $T_{(1)} < T_{(2)} \ldots < T_{(k)} (< t^*)$, leaving $n - k$ censored observations. Let $Z_{i\ell}$ denote the covariate value for the ℓth subject of the $n(T_{(i)})$ at risk at $T_{(i)}$, of which Z_{ii} denotes this value for the actual subject failing at $T_{(i)}$. Denote the rank of $Z_{i\ell}$ among these $n(T_{(i)})$ subjects as $r(Z_{i\ell})$. Let

$$p_{i\ell} = r(Z_{i\ell})/n(T_{(i)}) - 1/2n(T_{(i)}) \ , \qquad y_{i\ell} = \text{logit}\ p_{i\ell}$$

and

$$v_i = n^{-1}(T_{(i)}) \sum_j \log\left(\frac{2n(T_{(i)})}{2j-1} - 1\right), \quad i = 1, \ldots, n,$$

where the index j runs from 1 to $n(T_{(i)})$. A robust test can be based upon comparing p_{ii} $i = 1, \ldots, k$ with its expected value under the null hypothesis of 0.5. This corresponds to O'Brien (1978) suggestion as a way to reduce the impact of potential outliers and, thereby indirectly, to reduce the impact of censoring on the test. O'Brien also suggested that for reasons of efficiency and computational tractability it may be preferable to compare y_{ii} $i = 1, \ldots, k$ with the value zero, basing a test on

$$w = \sum_{i=1}^{k} y_{ii} \Big/ \left(\sum_{i=1}^{k} v_i\right)^{\frac{1}{2}}.$$

In the interests of robustness to scale of the covariate, and indirectly robustness to the conditional independence, also known as the covariate dependent, censoring assumption, O'Brien (1978) proposed the use of these tests. O'Quigley and Prentice (1991) showed how such tests could be placed within the general framework of proportional hazards type models. This enables rank-based tests that are often only valid under an independent censoring assumption to remain valid under these broader assumptions. Letting

$$I(\beta) = \sum_{i=1}^{n} \int_0^\infty W(t)\{\psi_i(t) - \mathcal{E}(\psi : \beta, t)\}^2 dN_i(t) = 0, \tag{16.7}$$

A score test based on the proportional hazards model but with covariate specification $Z_\ell(t) = \psi_\ell(t, \mathcal{F}_t) = y_{i\ell}$ when $T_{(i-1)} < t \leq T_{(i)}$ for $i = 1, \ldots, k$, and is equal to $y_{k\ell}$ when $t > T_{(k)}$ where $T_{(0)} = 0$ and $y_{i\ell} = 0$ at times at which the ℓth subject is not at risk, leads exactly to the logit-rank test described by O'Brien (1978). Simple calcuations show that $U(0) = \sum_{i=1}^k y_{ii}$, and that the information can be written under the null as,

$$I(0) = \sum_{i=1}^{k} \sum_{j=1}^{n_i} n_i^{-1} \left\{ \text{logit } (jn_i^{-1} - (2n_i)^{-1}) \right\}^2 = \sum_{i=1}^{k} v_i.$$

Similarly, O'Quigley and Prentice (1991) showed how O'Brien's original suggestion corresponds to making use of an estimating equation arising under the model, when we obtain the term $\sum_{i=1}^k p_{ii}$ as a result of the coding; $\psi_\ell(t, \mathcal{F}_t) = p_{i\ell}$ when $T_{(i-1)} < t \leq T_{(i)}$ for $i = 1, \ldots, k$ and, otherwise is equal to $p_{k\ell}$ when $t > T_{(k)}$, and where $p_{i\ell} = 0$ at times at which the ℓth subject is not at risk. Since $\sum_{j=1}^m j = m(m+1)/2$ and $\sum_{j=1}^m j^2 = m(m+1)(2m+1)/6$ there is a simple expression for the variance of $\sum_{i=1}^n p_{ii}$ from the above information equation as

$$\text{var} \left(\sum_{i=1}^{k} p_{ii} \right) = \sum_{i=1}^{k} (4n_i + 1)(n_i - 1)/12n_i.$$

Note that the mean of $\sum_{i=1}^k p_{ii}$ under H_0 is $k/2$ so that an easily evaluable approximate test obtains by referring

$$\frac{36(2\sum_{i=1}^k p_{ii} - k)^2}{\sum_{i=1}^k (4n_i + 1)(n_i - 1)/n_i}$$

to chi-square tables on one degree of freedom. In all of this we have taken the weights to not vary with time and so, implicitly, we have $W(X_i) = 1/k$. However, we may gain yet further in robustness to censoring in cases where we may not be that close to the null and where there are time trends in the effect. It may then pay to use an inverse probability weighting rather than a constant one, in particular to use $W(t) = \hat{S}(t)/\sum_1^n Y_i(t)$ where $\hat{S}(t)$ is the left continuous version of the Kaplan-Meier estimate of the marginal survival $S(t) = 1 - F(t)$. This becomes more important when the question is one of estimation of an effect that changes through time rather than the situation of small local departures from a null hypothesis of no effect. This particular problem is developed in more detail below. For tests based on different estimating equations note that there are a very large number of possibilities of which the above are only the most immediate ones. One situation studied by O'Quigley and Prentice (1991) was to consider $\psi_\ell(t, \mathcal{F}_t)$ to not depend on information on subjects $i \neq \ell$, in which case scores chosen at the start of the study remain unchanged throughout the study. This is easier to do in practice and turns out to be fully efficient. Tests analogous to the logit-rank tests can be obtained by simply transforming the marginal empirical distribution of the Z_i to be some given distribution, e.g., the uniform over $(0, 1)$. The use of normal order statistics in place of the original measurements amounts to a transformation to normality conditional on the risk sets or unconditionally if applied only once at the start of the study. Given this interpretation of the effect of different kinds of rankings, an another approach, although not rank invariant, would be to replace the original z_i measurements by $G^{-1}\{\tilde{F}(Z_i)\}$ where $\tilde{F}(\cdot)$ is some consistent estimate of the cumulative distribution function for Z and $G(\cdot)$ is the cumulative distribution function we are transforming to, for example the standard normal. In order to compare different possible transformations we can use the result of O'Quigley and Prentice (1991). Specifically, suppose that the mechanism generating the observations is given by the model with $\psi_i(t, \mathcal{F}_t)$ but that the test is based on using $\psi_i^*(t, \mathcal{F}_t)$. Then the asymptotic efficiency can be written

$$e(\psi, \psi^*) = \frac{(\int \phi[g\{Z(t)\}, g^*\{Z(t)\}]\lambda_0(t)dt)^2}{\int \phi[g\{Z(t)\}, g\{Z(t)\}]\lambda_0(t)dt \int \phi[g^*\{Z(t)\}, g^*\{Z(t)\}]\lambda_0(t)dt},$$

where the integrals are over $(0, \infty)$,

$$\phi(a(t), b(t)) = E\{Y(t)a(t)b(t)\} - E\{Y(t)a(t)\}E\{Y(t)b(t)\}/E\{Y(t)\},$$

and the expectations are taken with respect to the distribution of $\mathcal{F}_t$ and where

$$\lim_{n\to\infty} \psi(t, \mathcal{F}_t) = g\{Z(t)\}, \quad \lim_{n\to\infty} \psi^*(t, \mathcal{F}_t) = g^*\{Z(t)\}$$

for continuous monotonic functions $g(\cdot)$ and $g^*(\cdot)$. Applying the above expression for $e(\psi, \psi^*)$, to time-independent covariates, we find that the asymptotic relative efficiency is equal to one for the logit-rank procedure and the simplified logit-rank procedure (where simplified means using a fixed non-time-dependent covariate equal to the initial ranking at the beginning of the study). The same applies to other pairs of procedures, the normal-rank and simplified normal rank for instance. Using normal scores instead of logit scores and vice versa leads to an asymptotic relative efficiency of 0.97 in the absence of censoring.

16.3 Robust estimator of average regression effect

Define

$$S^{(r)}(\beta, t) = n^{-1}\sum_{i=1}^{n} Y_i(t)e^{\beta Z_i(t)}Z_i(t)^r, \quad s^{(r)}(\beta, t) = ES^{(r)}(\beta, t),$$

for $r = 0, 1, 2$, where the expectations are taken with respect to the true distribution of $(T, C, Z(\cdot))$. Then $\mathcal{E}(Z : \beta, t) = S^{(1)}(\beta, t)/S^{(0)}(\beta, t)$. From Theorem 1, $s^{(1)}(\beta(t), t)/s^{(0)}(\beta(t), t) = E\{Z(t)|T = t\}$, and $s^{(1)}(\beta, t)/s^{(0)}(\beta, t)$ is what we get when we impose a constant β through time in place of $\beta(t)$. Define also

$$V(\beta, t) = \frac{S^{(2)}(\beta, t)}{S^{(0)}(\beta, t)} - \frac{S^{(1)}(\beta, t)^2}{S^{(0)}(\beta, t)^2}, \quad v(\beta, t) = \frac{s^{(2)}(\beta, t)}{s^{(0)}(\beta, t)} - \frac{s^{(1)}(\beta, t)^2}{s^{(0)}(\beta, t)^2}. \tag{16.8}$$

16.3.1 The robust estimator

When there are time trends and the true mechanism generating the observations involves $\beta(t)$, not a constant, then the usual partial likelihood estimator is not consistent. It is also very non-robust to an independent censoring mechanism and the population quantity to which the estimator converges depends very strongly on the level of censoring. We can again consider the estimating equation and it turns out to be easy to not only obtain an estimator of average effect - when $\beta(t)$ depends on t - but the estimator itself is very robust to an independent censoring mechansim. Using the weights $W(t) = \hat{S}(t)/\sum_1^n Y_i(t)$ in the estimating equation leads to the estimator $\tilde{\beta}$ rather than the usual partial likelihood estimate $\hat{\beta}$. Xu (1996) showed the following theorem.

Theorem 2 *Under Model (16.1) the estimator $\tilde{\beta}$ converges in probability to a constant β^*, where β^* is the unique solution to the equation*

$$\int_0^{\infty} \left\{ \frac{s^{(1)}(\beta(t), t)}{s^{(0)}(\beta(t), t)} - \frac{s^{(1)}(\beta, t)}{s^{(0)}(\beta, t)} \right\} dF(t) = 0, \tag{16.9}$$

provided that $\int_0^\infty v(\beta^*, t)dF(t) > 0$.

The equation does not involve censoring, and thus neither does the limit β^*. In this sense it can be considered to be robust to an independent censoring mechanism. In contrast the maximum partial likelihood estimator $\hat{\beta}_{PL}$ was shown by Struthers and Kalbfleisch (1986) to converge to the solution of the equation

$$\int_0^\infty \left\{ \frac{s^{(1)}(\beta(t),t)}{s^{(0)}(\beta(t),t)} - \frac{s^{(1)}(\beta,t)}{s^{(0)}(\beta,t)} \right\} s^{(0)}(\beta(t),t)\lambda_0(t)dt = 0. \tag{16.10}$$

In general this solution is not robust to censoring, even when independent and the unknown censoring mechanism impacts the limit through the factor $s^{(0)}(\beta(t),t)$. It is not then easy to obtain any kind of useful interpretation under non-proportional hazards. The dependence of $\hat{\beta}_{PL}$ on censoring is also clear from simulation results and we will recall some of those later.

Theorem 3 *Under Model (16.1)* $\sqrt{n}(\tilde{\beta} - \beta^*)$ *is asymptotically normal with mean zero and variance* σ^2 *given in Xu and O'Quigley (2000).*

A variance estimator based on empirical influence function was developed by Xu and Harrington (2001).

The fact that β^* is free of censoring requires that the censoring distribution be independent of the covariates. For two-group comparisons Boyd et al. (2009) and Hattori and Henmi (2012) further incorporated weights in the definitions of $S^{(1)}$ and $S^{(0)}$, so that the resulting population parameter is not affected by a censoring mechanism that is independent of the failure time distribution conditional on the covariates. Hattori and Henmi (2012) also considered improving efficiency of the estimator by applying the covariate adjustment method of Lu and Tsiatis (2008) based on the semiparametric theory.

16.3.2 Interpretation of average regression effect

The solution β^* to Equation (16.9) can be viewed as an average regression effect. In the equation $s^{(1)}(\beta(t),t)/s^{(0)}(\beta(t),t) = E\{Z(t)|T = t\}$ from Theorem 1, and $s^{(1)}(\beta^*,t)/s^{(0)}(\beta^*,t)$ results when $\beta(t)$ is restricted to be a constant; the difference between these two is zero when integrated out with respect to the marginal distribution of failure time. Suppose, for instance, that $\beta(t)$ decreases over time, then earlier on $\beta(t) > \beta^*$ and $s^{(1)}(\beta(t),t)/s^{(0)}(\beta(t),t) > s^{(1)}(\beta^*,t)/s^{(0)}(\beta^*,t)$; whereas later $\beta(t) < \beta^*$ and $s^{(1)}(\beta(t),t)/s^{(0)}(\beta(t),t) < s^{(1)}(\beta^*,t)/s^{(0)}(\beta^*,t)$. Applying a first-order Taylor series approximation to the integrand of (16.9), we have

$$\int_0^\infty v(t)\{\beta(t) - \beta^*\}dF(t) \approx 0, \tag{16.11}$$

where $v(t) = v(\beta(t),t) = \text{Var}\{Z(t)|T = t\}$ according to Theorem 1. Therefore

$$\beta^* \approx \frac{\int_0^\infty v(t)\beta(t)dF(t)}{\int_0^\infty v(t)dF(t)} \tag{16.12}$$

is a weighted average of $\beta(t)$ over time. According to (16.12) more weights are given to those $\beta(t)$'s where the marginal distribution of T is concentrated; and more weights are given to those $\beta(t)$'s where the conditional distribution of $Z(t)$ has larger variance. In (16.12) if $v(t)$, the conditional variance of $Z(t)$, changes relatively little with time apart from for large t, when the size of the risk sets becomes very small, we can make the approximation $v(t) \equiv c$ and it follows that

$$\beta^* \approx \int_0^\infty \beta(t)dF(t) = E\{\beta(T)\}. \tag{16.13}$$

In general when $\beta(t)$ is close to zero, we know that the distribution of $Z(t)$ does not change much over time because there is very little seletive elimination from the risk set due to the covariate effect (Prentice, 1982). The approximate constancy of this conditional variance is also used in the sample size calculation for two-group comparisons (Kim and Tsiatis, 1990). In practice $v(t)$ will often be approximately constant, an observation supported by our own practical experience as well as with simulated datasets. For a comparison of two groups coded as 0 and 1, the conditional variance is of the form $p(1-p)$ for some $0 < p < 1$, and this changes relatively little provided that, throughout the study, p and $1-p$ are not too close to zero. In fact we only require the weaker condition that $\text{Cov}(v(T), \beta(T)) = 0$ to obtain (16.13), a constant $v(t)$ being a particular example. Even when the condition does not hold we would still anticipate the approximation as being valuable and the best way to see this is via simulations.

Xu and Harrington (2001) showed that (16.13) holds exactly for two-group log-logistic (i.e., proportional odds) models with equal group memberships. In particular, for $\rho \geq 0$, define survival function

$$\begin{aligned} H_0(t) &= \exp(-e^t), \ \ \rho = 0; \\ H_\rho(t) &= (1 + \rho e^t)^{-1/\rho}, \ \ \rho > 0. \end{aligned}$$

This is the G^ρ family distribution of Harrington and Fleming (1982). Consider a linear transformation model of the form

$$g(T) = \alpha Z + \epsilon, \tag{16.14}$$

where $g(\cdot)$ is an unspecified strictly increasing function, and ϵ is from the G^ρ family. For a binary group indicator Z, (16.14) is a special case of model (16.1), since for any two-group case we simply have $\beta(t) = \log\{\lambda_1(t)/\lambda_0(t)\}$. It is well-known that when $\rho = 0$, (16.14) is the proportional hazards model with $\beta = -\alpha$; when $\rho = 1$, (16.14) is the log-logistic model, which is also called the "proportional odds model." Model (16.14) is also related to the gamma frailty model for $\rho > 0$: it is equivalent to

$$\lambda(t|z) = \lambda_0(t)\exp(-\alpha Z)\omega,$$

where ω is the unobserved gamma frailty with mean one and variance ρ. In general (16.14) provides a much broader class of models than the proportional hazards. Xu and Harrington (2001) showed that under model (16.14), $\beta^* = -\alpha/(\rho + 1)$; and for $\rho = 1$ and $P(Z = 0) = P(Z = 1) = 1/2$, this further equals to $\int_0^\infty \beta(t)dF(t)$. This way the interpretation of the average regression effect over time is directly equivalent to the treatment effect under this broad class of models.

16.3.3 Simulations

In Xu and O'Quigley (2000) simulations were used to study the finite sample behavior of the estimator $\tilde{\beta}$ with the partial likelihood estimator $\hat{\beta}_{PL}$, as well as to investigate the accuracy of the approximation of $\int \beta(t)dF(t)$. Data were generated from a simple two-step time-varying regression coefficients model with $\beta(t) = \beta_1$ when $t < t_0$ and β_2 otherwise. The details are described in Xu and O'Quigley (2000) and, here, we simply reproduce Table

TABLE 16.1
Comparison of $\hat{\beta}_{PL}$, $\tilde{\beta}$, β^* and $\int \beta(t)dF(t)$.

β_1	β_2	t_0	% censored	$\hat{\beta}_{\mathrm{PL}}$	$\tilde{\beta}$	β^*	$\int \beta(t)dF(t)$
1	0	0.1	0%	0.155 (0.089)	0.155 (0.089)	0.156	0.157
			17%	0.189 (0.099)	0.158 (0.099)	0.156	0.157
			34%	0.239 (0.111)	0.160 (0.111)	0.156	0.157
			50%	0.309 (0.130)	0.148 (0.140)	0.156	0.157
			67%	0.475 (0.161)	0.148 (0.186)	0.156	0.157
			76%	0.654 (0.188)	0.161 (0.265)	0.156	0.157
3	0	0.05	0%	0.716 (0.097)	0.716 (0.097)	0.721	0.750
			15%	0.844 (0.107)	0.720 (0.106)	0.721	0.750
			30%	1.025 (0.119)	0.725 (0.117)	0.721	0.750
			45%	1.294 (0.133)	0.716 (0.139)	0.721	0.750
			60%	1.789 (0.168)	0.716 (0.181)	0.721	0.750
			67%	2.247 (0.195)	0.739 (0.255)	0.721	0.750

Note: $\lambda_0(t) = 1$, $\beta(t) = \beta_1$ when $t < t_0$ and β_2 otherwise, $Z \sim$ U(0,1), point censoring at t_0. In the $(\cdot)$ are standard errors. Sample size 1,600 with 200 simulations each.

1 from that paper: 200 simulations were carried out with sample size 1,600 for each set of the results. Clearly, $\hat{\beta}_{PL}$ is very non-robust under independent censoring, the value to which it converges changing substantially as the rate of censoring increases. This underlines the difficulty in the interpretation of the partial likelihood estimate under non-proportional hazards, a fact that has been alluded to in the literature. The estimate $\tilde{\beta}$, on the other hand, consistently estimates the population average β^* regardless of the censoring. The bracketed figures in Table 16.1 give the standard errors of the estimates from the 200 simulations. An important observation is that between $\tilde{\beta}$ and $\hat{\beta}_{\mathrm{PL}}$, for the cases studied, any gains in efficiency of the partial likelihood estimate are very quickly lost to the potentially large biases caused by the censoring.

Next we consider a more gradually changing $\beta(t)$. We simulate data from a two-sample log-logistic model (16.14). This model has an attenuating hazard ratio, i.e., a monotone $\beta(t)$ that tend to zero as $t \to \infty$. The censoring distribution is uniform $(0, \tau)$. For various combinations of group differences $\alpha = 1$, 2 and 3 and censoring percentages, we see from Table 16.2 that $\tilde{\beta}$ always consistently estimates β^*, while the bias of $\hat{\beta}_{PL}$ is typically one standard deviation or larger. Notice that for this model, as mentioned earlier, it was shown in Xu and Harrington (2001) that $\beta^* = \int_0^\infty \beta(t)dF(t)$, and in fact they both equal to $-\alpha/2$. In the table β^*_τ reflects the finite follow-up time under the uniform $(0, \tau)$ censoring. Table 16.2 reflects the types of non-proportionality and censoring distributions that are likely to arise in practice, and it is important to be aware of the behavior of the partial likelihood estimator under these situations.

TABLE 16.2
Comparison of $\hat{\beta}_{PL}$, $\tilde{\beta}$, β^*_τ and $\int \beta(t)dF(t)$ — log-logistic model.

α	% censored	$\hat{\beta}_{\text{PL}}$	$\tilde{\beta}$	β^*_τ	$\int \beta(t)dF(t)$
1	0%	-0.505 (0.048)	-0.505 (0.048)	-0.500	-0.500
	35%	-0.645 (0.067)	-0.580 (0.081)	-0.578	-0.584
	57%	-0.746 (0.076)	-0.670 (0.111)	-0.676	-0.684
2	0%	-1.000 (0.060)	-1.000 (0.060)	-1.000	-1.000
	35%	-1.258 (0.064)	-1.141 (0.082)	-1.143	-1.195
	56%	-1.480 (0.084)	-1.339 (0.106)	-1.339	-1.409
3	0%	-1.508 (0.077)	-1.508 (0.077)	-1.500	-1.500
	35%	-1.843 (0.079)	-1.694 (0.091)	-1.684	-1.838
	53%	-2.124 (0.089)	-1.940 (0.117)	-1.941	-2.151

Note: $\log(T) = \alpha Z + \epsilon$, $\epsilon \sim$ Logistic, $P(Z = 0) = P(Z = 1) = 0.5$, uniform $(0, \tau)$ censoring. Sample size 1,600 with 200 simulations each.

16.4 When some covariates have non-PH

The non-robustness issues considered in this chapter arise when the proportional hazards (PH) assumption is violated. These issues have been so well recognized in the literature that the top medical journals today require verifying the proportional hazards assumption whenever the model is used. It is also known, however, unlike the linear regression models, the proportional hazards regression models are typically not nested; dropping or adding a covariate to an existing PH model can result in non-proportional hazards (Ford et al., 1995). The only known exception is when the covariate under consideration follows a positive stable distribution.

Proportional hazards regression models are often applied in biomedical studies to compare treatments in randomized clinical trials, or to study exposure effects in observational studies. In the former it is always more efficient asymptotically to adjust for covariates (Schoenfeld, 1983; Lagakos and Schoenfeld, 1984), while in the latter it is crucial to adjust for uncontrolled confounders for the purpose of causal inference.

Strandberg et al. (2012) considered the case where the main treatment or exposure effect follows the proportional hazards assumption, while the effects of additional covariates do not necessarily. More specifically, they consider

$$\lambda(t|Z_1, Z_2) = \lambda_0(t) \exp\{\beta_1 Z_1 + \beta_2(t) Z_2\}. \tag{16.15}$$

The above, of course, can be seen as a special case of Model (16.1) with multiple covariates, and the general theory under Model (16.1) as described earlier extends in a straightforward fashion. In particular, (16.10) is now a system of equations whose solution, β_{PL}, is the population limit of the partial likelihood estimator as $n \to \infty$. An approximate equation similar to that of (16.11) can be written down:

$$\int_0^\infty v(t)\{\beta(t) - \beta_{PL}\} s^{(0)}(\beta(t), t)\lambda_0(t)dt \approx 0, \tag{16.16}$$

where $v(t)$ is the same as before. Since $s^{(0)}$ and λ_0 are scalars, if $v(t)$ is a diagonal matrix, then components of β_{PL} are weighted averages of the corresponding components of $\beta(t)$ over time. In particular, under Model (16.15), since β_1 is constant, then $\beta_{PL}^{(1)} = \beta_1$ if $v(t)$ is diagonal.

Strandberg et al. (2012) carried out extensive simulations to study the commonly used partial likelihood estimator of β_1 under various shapes of $\beta_2(t)$ and $\lambda_0(t)$. It is perhaps not surprising that the shape of $\lambda_0(t)$ has very little impact on the estimated β_1. When Z_1 and Z_2 are independent like in randomized clinical trials, the bias of $\hat{\beta}_1$ is typically no more than 15%. This can be explained by the fact that $v(t)$ should be close to diagonal for most of the t values, when Z_1 and Z_2 are independent. But when Z_1 and Z_2 are dependent, which is typically the case in observational studies between exposure and potential confounders, the bias can be very severe. In Table 16.3 we give some results that are consistent with the findings of Strandberg et al. (2012), but were nonetheless not included in the paper due to space limitation.

For the table, data were generated as piecewise constant to approximate different shapes of $\beta_2(t)$: (a) increasing, (b) decreasing, (c) increasing then decreasing, and (d) decreasing then increasing. Ten change points were used resulting in 11 steps for $\beta_2(t)$. The change points in each case were chosen so that approximately equal numbers of events fell in each interval. The baseline hazard function $\lambda_0(t)$ was constant. The covariate Z_1 was binary with equal probabilities of group membership, and Z_2 was continuous: $Z_2 \sim$ U(-2,0) if $Z_1 = 0$, and $Z_2 \sim$ U(0,2) if $Z_1 = 1$. Censoring was generated using U(0,τ), where τ was chosen so there was approximately 20-40% censoring in each case. Each simulation was run with 1,000 repetitions. In each repetition the proportional hazards model with constant β_1 and β_2 was fitted. The average and standard deviation over the repetitions of the partial likelihood estimate $\hat{\beta}_1$ were tabulated, together with the coverage probabilities (CP) of the 95% confidence intervals for β_1 based on $\hat{\beta}_1$ and its estimated standard error. '% Bias' was calculated using $(\hat{\beta}_1 - \beta_1)/\beta_1 \times 100$ when $\beta_1 \neq 0$ and averaged over the 1,000 repetitions.

In Table 16.3 different strengths of the main effect were considered: $\beta_1 = 2$, 0.5 and 0. When $\beta_1 = 2$, the percentage of bias ranged from -43% to 17%, and the 95% confidence interval coverage probability varied from 45% to 96%, with the most severe bias and the lowest coverage probabilities occurring for case (d). For $\beta_1 = 0.5$, the percent bias ranged greatly from -94% to 51%, and the coverage probabilities ranged from 75% to 95%. Although there were relatively good coverage probabilities even when the bias was as much as 50%, they were probably reflections of over-estimated standard errors. When $\beta_1 = 0$, the coverage probability was between 83% and 97%, indicating an often inflated type I error rate for testing $\beta_1 = 0$. Although censoring has some tendency to lessen the time-varying effect of $\beta_2(t)$ by making the later observations unobservable, it can worsen the bias and the coverage probability of the 95% confidence intervals in some cases.

Anderson and Fleming (1995) showed that in randomized clinical trials when $\beta_1 = 0$ under Model (16.15), the asymptotic bias of the partial likelihood estimator $\hat{\beta}_1$ is zero. This is no longer the case when Z_1 and Z_2 are dependent. The above findings have important implications in observational studies when the proportional hazards model is used to adjust for potential confounders. Since the proportional hazards assumption is often likely violated in practice, an estimator that is robust and interpretable becomes crucial.

TABLE 16.3
Effect of non-PH on estimation of β_1 when $Z^{(1)}$ and $Z^{(2)}$ are strongly dependent, $n = 100$.

		No Censoring			Censoring		
	β_1	$\hat{\beta}_1$ (SD)	CP	% Bias	$\hat{\beta}_1$ (SD)	CP	% Bias
(a)	2	1.42 (0.36)	0.79	-29%	1.82 (0.42)	0.96	-9%
(b)		2.11 (0.84)	0.72	5%	1.76 (0.76)	0.79	-12%
(c)		2.33 (0.52)	0.88	17%	2.05 (0.54)	0.96	3%
(d)		1.13 (0.38)	0.45	-43%	1.29 (0.47)	0.70	-35%
(a)	0.5	0.61 (0.38)	0.95	22%	0.76 (0.48)	0.91	51%
(b)		0.05 (0.44)	0.75	-89%	0.03 (0.48)	0.77	-94%
(c)		0.42 (0.41)	0.94	-16%	0.45 (0.46)	0.95	-9%
(d)		0.25 (0.37)	0.93	-50%	0.26 (0.47)	0.92	-48%
(a)	0	0.34 (0.39)	0.86	-	0.40 (0.50)	0.86	-
(b)		-0.35 (0.39)	0.86	-	-0.43 (0.47)	0.83	-
(c)		0.07 (0.40)	0.95	-	0.07 (0.45)	0.94	-
(d)		-0.05 (0.37)	0.97	-	-0.08 (0.45)	0.95	-

16.5 Proportional hazards regression for correlated data

The proportional hazards regression model has been extended to correlated survival data, first as the frailty model which contains a random intercept on the baseline hazard function, then more generally, as the mixed effects model which allows random effects on arbitrary covariates. The proportional hazards mixed-effects model (PHMM) is parallel to the linear, non-linear and generalized linear mixed-effects models (LMM, NLMM, and GLMM). For clustered data let $i = 1, ..., n$ denote the clusters, and $j = 1, ..., n_i$ denote the observations from a cluster. The observed data for individual ij is $(X_{ij}, \delta_{ij}, Z_{ij})$, where $X_{ij} = \min(T_{ij}, C_{ij})$, $\delta_{ij} = I(T_{ij} \leq C_{ij})$, and the notation are otherwise the same as earlier for independent and identically distributed (*i.i.d.*) data. The proportional hazards mixed-effects model can be written as

$$\lambda_{ij}(t) = \lambda_0(t) \exp(\beta' Z_{ij} + b_i' W_{ij}), \tag{16.17}$$

where compared to Model (16.1), we have the addition of the random effects b_i for cluster i, and W_{ij} is usually a sub-vector of Z_{ij} corresponding to those covariates that have random effects. W may also include a '1' if there is a random effect on the baseline hazard itself. When $W = 1$, (16.17) becomes the shared frailty model. The inclusion of the $b'W$ term in the model represents cluster by covariate interactions, such as a treatment by center interaction in multi-center clinical trials. The random effects b_i's are assumed to be *i.i.d.* according to some distribution; in this way the estimation of the random effects is different from the fixed effects when cluster itself is treated as a categorical variable. This is often referred to as "borrowing strength," since the estimation of b_i makes use of the whole dataset, and not just the data from cluster i. Typically the b_i's are assumed to be from $N(0, \Sigma)$.

Inference under Model (16.17) can be carried out using nonparametric maximum likelihood (Vaida and Xu, 2000, NPMLE) or penalized partial likelihood (Ripatti and Palmgren, 2000, PPL). Gamst et al. (2009) established the consistency and asymptotic normality of the NPMLE, and carried out extensive comparisons of the NPMLE and the PPL using simulations. The PPL is faster to compute, but is less accurate when the sample size is small, and also its variance estimation is not straightforward. The NPMLE is numerically stable and accurate, but is more computationally intensive. The two estimators are implemented in R packages **phmm** and **coxme**, respectively.

Under the special case of frailty model O'Quigley and Stare (2002) demonstrated the robustness in the fixed effects estimation against frailty distribution misspecifications. Xu and Gamst (2007) studied the effect of non-proportional hazards on the estimation of the parameters. They considered data generated under

$$\lambda_{ij}(t) = \lambda_0(t)\exp\{\beta(t)'Z_{ij} + b_i'W_{ij}\}, \tag{16.18}$$

where the fixed regression effect β was allowed to vary with time t. Parallel to the developments in Section 16.3, denote

$$S^{(r)}(\beta; t, \theta) = \frac{1}{N}\sum_{i=1}^{n}\sum_{j=1}^{n_i} Y_{ij}(t)\exp(\beta' Z_{ij}) Z_{ij}^{\otimes r} E_\theta(e^{b_i' W_{ij}} | y_i), \tag{16.19}$$

for $r = 0, 1, 2$, where $N = \sum_{i=1}^n n_i$, $\theta = (\beta, \Sigma, \lambda_0)$ contains all the parameters under the PHMM (16.17), and $a^{\otimes 0} = 1$, $a^{\otimes 1} = a$ and $a^{\otimes 2} = aa'$ for a vector a. Consider the NPMLE $\hat{\theta} = (\hat{\beta}, \hat{\Sigma}, \hat{\lambda}_0)$, which can be estimated using a Monte Carlo *EM* algorithm (Vaida and Xu, 2000). It can be verified that at the convergence of the *EM* algorithm, $\hat{\beta}$ satisfies

$$\sum_{i=1}^{n}\sum_{j=1}^{n_i}\int_0^\infty \left\{ Z_{ij} - \frac{S^{(1)}(\beta; t, \theta)}{S^{(0)}(\beta; t, \theta)} \right\} dN_{ij}(t) = 0. \tag{16.20}$$

The above equation can be seen directly from the equations that are solved at the *M*-steps; it can also be derived as the score equation via direct differentiation of the log likelihood using one-dimensional submodels, as in Murphy (1995).

Denote $s^{(r)}(\beta; t, \theta) = E\{S^{(r)}(\beta; t, \theta)\}$, $r = 0, 1, 2$, and $v(\beta; t, \theta) = s^{(2)}(\beta; t, \theta)/s^{(0)}(\beta; t, \theta) - \{s^{(1)}(\beta; t, \theta)/s^{(0)}(\beta; t, \theta)\}^{\otimes 2}$ is the derivative of $s^{(1)}/s^{(0)}$ with respect to the first argument β.

Theorem 4 *Under regularity conditions and assuming that* $\int v(\beta; t, \theta)s^{(0)}(\beta(t); t, \theta)\lambda_0(t)dt$ *is positive definite, as* $n \to \infty$ *the NPMLE* $\hat{\beta}$ *converges in probability to* β^**, which is the unique zero of the following equation:*

$$\int_0^\infty \left\{ \frac{s^{(1)}(\beta(t); t, \theta(t))}{s^{(0)}(\beta(t); t, \theta(t))} - \frac{s^{(1)}(\beta; t, \theta)}{s^{(0)}(\beta; t, \theta)} \right\} s^{(0)}(\beta(t); t, \theta(t))\lambda_0(t)dt = 0, \tag{16.21}$$

where $\theta(t) = (\beta(t), \Sigma, \lambda_0)$.

The asymptotic normality of $\hat{\beta}$ under Model (16.18) was established in Dupuy (2009).

Using a first-order Taylor expansion of $\{\cdot\}$ in (16.21), we have

$$\int_0^\infty \{\beta(t) - \beta^*\} v(\tilde{\beta}(t); t, \tilde{\theta}(t)) s^{(0)}(\beta(t); t, \theta(t))\lambda_0(t)dt = 0, \tag{16.22}$$

where $\tilde{\beta}(t)$ is between $\beta(t)$ and β^*, and $\tilde{\theta}(t) = (\tilde{\beta}(t), \Sigma, \lambda_0)$. Solving (16.22) for β^* we see that the population value β^* underlying $\hat{\beta}$ is a weighted average of $\beta(t)$ over time. Meanwhile from (16.20) we see that when there is censoring, the censored observations lose their contribution to the estimating equation; therefore, the equation gives insufficient weights to the later censored observations. This results in an average β value that is biased towards the earlier values of $\beta(t)$, as compared to the uncensored case.

The above findings are consistent with the fixed-effects-only case discussed earlier in the chapter. What is perhaps more curious is how the estimates of the variance components in Σ are affected by non-proportionality. Xu and Gamst (2007) gave analytic results for a single covariate with both non-proportional and random effects:

$$\lambda_{ij}(t) = \lambda_0(t) \exp[\{\beta(t) + b_i\} Z_{ij}]. \tag{16.23}$$

Under the PHMM we fit $\lambda_{ij}(t) = \lambda_0^*(t) \exp\{(\beta^* + b_i^*) Z_{ij}\}$. Earlier results of this chapter imply that in the absence of censoring the least-false parameter value

$$\beta^* + b_i^* \approx \int \{\beta(t) + b_i\} dF_i(t) = \int \beta(t) dF_i(t) + b_i, \tag{16.24}$$

where $F_i(t)$ is the marginal distribution function of the failure times in cluster i. If we assume for the moment that the covariate values are non-negative, then a larger b_i implies higher relative risks and shorter failure times in cluster i, i.e., $F_i(\cdot)$ puts more weight on earlier times. If $\beta(t)$ is decreasing, this shows that b_i and $\int \beta(t) dF_i(t)$ are positively correlated, therefore $\Sigma^* = \text{Var}(b_i^*) = \text{Var}(\beta^* + b_i^*) > \text{Var}(b_i) = \Sigma$. When allowing censoring in general, $F_i(\cdot)$ is replaced by the intensity of the counting process, which also puts more weight on the earlier values of $\beta(t)$ under larger b_i. The above argument does not rely on the positivity of the covariate values, and it shows that $\Sigma^* > \Sigma$ if $\beta(t)$ is decreasing. A similar argument shows that $\Sigma^* < \Sigma$ if $\beta(t)$ is increasing.

When there are more than one covariates the effect of non-proportional hazards on the estimation of Σ is more complex. Xu and Gamst (2007) provided simulation results to illustrate various scenarios, as well as a method for checking the proportional hazards assumption in Model (16.17).

Bibliography

Andersen, P. and Gill, R. (1982), 'Cox's regression model for counting processes: a large sample study', *Annals of Statistics* **10**, 1100–1120.

Anderson, G. L. and Fleming, T. R. (1995), 'Model misspecification in proportional hazards regression', *Biometrika* **82**, 527–541.

Boyd, A. P., Kittelson, J. M. and Gillen, D. L. (2009), 'Estimation of treatment effect under nonproportional hazards and covariate dependent censoring', *Biostatistics* **8**, 1–15.

Breslow, N. (1974), 'Covariance analysis of censored survival data', *Biometrics* **30**, 89–99.

Bretagnolle, J. and Huber-Carol, C. (1988), 'Effects of omitting covariates in Cox's model for survival data', *Scandinavian Journal of Statistics* **15**, 125–138.

Cheng, S. C., Wei, L. J. and Ying, Z. (1995), 'Analysis of transformation models with censored data', *Biometrika* **82**, 835–845.

Cox, D. R. (1972), 'Regression models and life tables (with discussion)', *Journal of the Royal Statistical Society, Series B* **34**, 187–220.

Cox, D. R. (1975), 'Partial likelihood', *Biometrika* **62**, 269–276.

Dupuy, J. (2009), 'On the random effects Cox model with time-varying regression parameter', *Journal of Statistical Theory and Practice* **3**, 763–776.

Ford, I., Norrie, J. and Ahmadi, S. (1995), 'Model inconsistency, illustrated by the Cox proportional hazards model', *Statistics in Medicine* **14**, 735–746.

Gail, M. H., Wieand, S. and Piantadosi, S. (1984), 'Biased estimates of treatment effect in randomized experiments with nonlinear regressions and omitted covariates', *Biometrika* **71**, 431–444.

Gamst, A., Donohue, M. and Xu, R. (2009), 'Asymptotic properties and empirical evaluation of the *npmle* in the proportional hazards mixed-effects model', *Statistica Sinica* **19**, 997–1011.

Godambe, V. and Kale, B. (1991), Estimating functions: an overview. In *Estimating Functions*, Clarendon Press, Oxford.

Gray, R. J. (1992), 'Flexible methods for analyzing survival data using splines, wich applications to breast cancer prognosis', *Journal of the American Statistical Association* **87**, 942–951.

Harrington, D. P. and Fleming, T. R. (1982), 'A class of rank test procedures for censored survival data', *Biometrika* **69**, 553–566.

Hattori, S. and Henmi, M. (2012), 'Estimation of treatment effects based on possibly misspecified cox regression', *Lifetime Data Analysis* **18**, 408–433.

Kaplan, E. and Meier, P. (1958), 'Non-parametric estimation from incomplete observations', *Journal of the American Statistical Association* **53**, 457–481.

Kim, K. and Tsiatis, A. A. (1990), 'Study duration for clinical trials with survival response and early stopping rule', *Biometrics* **46**, 81–92.

Lagakos, S. W. and Schoenfeld, D. A. (1984), 'Properties of proportional-hazards score tests under misspecified regression models', *Biometrics* **40**, 1037–1048.

Lancaster, T. and Nickell, S. (1980), 'The analysis of re-employment probabilities for the unemployed', *Journal of the Royal Statistical Society, Series A* **143**, 141–165.

Lu, X. and Tsiatis, A. A. (2008), 'Improving the efficiency of the log-rank test using auxiliary covariates', *Biometrika* **95**, 679–694.

Murphy, S. (1995), 'Asymptotic theory for the frailty model', *Annals of Statistics* **23**(1), 182–198.

Murphy, S. A. and Sen, P. K. (1991), 'Time-dependent coefficients in a cox-type regression model', *Stochastic Processes and Their Applications* **39**, 153–180.

O'Brien, P. (1978), 'A non-parametric test for association with censored data', *Biometrics* **34**, 243–250.

O'Quigley, J. and Pessione, F. (1989), 'Score tests for homogeneity of regression effect in the proportional hazards model', *Biometrics* **45**, 135–144.

O'Quigley, J. and Pessione, F. (1991), 'The problem of a covariate-time qualitative interaction in a survival study', *Biometrics* **47**, 101–115.

O'Quigley, J. and Prentice, R. (1991), 'Nonparametric tests of association between survival time and continuously measured covariates: the logit-rank and associated procedures', *Biometrics* **47**, 117–127.

O'Quigley, J. and Stare, J. (2002), 'Proportional hazards models with frailties and random effects', *Statistics in Medicine* **21**, 3219–33.

Peto, R. (1972), 'Contribution to the discussion of paper by D.R. Cox', *Journal of the Royal Statistical Society, Series B* **34**, 205–207.

Prentice, R. (1982), 'Covariate measurement errors and parameter estimation in a failure time regression model', *Biometrika* **69**, 331–342.

Ripatti, S. and Palmgren, J. (2000), 'Estimation of multivariate frailty models using penalized partial likelihood', *Biometrics* **56**, 1016–1022.

Schoenfeld, D. A. (1983), 'Sample size formula for the proportional-hazards regression model', *Biometrics* **39**, 499–503.

Strandberg, E., Lin, X. and Xu, R. (2012), 'Estimation of main effects when covariates have non-proportional hazards', *Communications in Statistics - Simulation and Computation* **accepted**.

Struthers, C. A. and Kalbfleisch, J. D. (1986), 'Misspecified proportional hazards model', *Biometrika* **73**, 363–369.

Tsiatis, A. A. (1981), 'A large sample study of Cox's regression model', *Annals of Statistics* **9**, 93–108.

Vaida, F. and Xu, R. (2000), 'Proportional hazards model with random effects', *Statistics in Medicine* **19**, 3309–3324.

Xu, R. (1996), *Inference for the Proportional Hazards Model*, Ph.D. thesis of the University of California, San Diego.

Xu, R. and Adak, S. (2002), 'Survival analysis with time-varying regression effects using a tree-based approach', *Biometrics* **58**, 305–315.

Xu, R. and Gamst, A. (2007), 'On proportional hazards assumption under the random effects models', *Lifetime Data Analysis* **13**, 317–332.

Xu, R. and Harrington, D. P. (2001), 'A semiparametric estimate of treatment effects with censored data', *Biometrics* **57**, 875–885.

Xu, R. and O'Quigley, J. (1999), 'A r^2 type measure of dependence for proportional hazards models', *Journal of Nonparametric Statistics* **12**, 83–107.

Xu, R. and O'Quigley, J. (2000), 'Estimating average regression effect under non-proportional hazards', *Biostatistics* **1**, 423–439.

Ying, Z., Jung, S. H. and Wei, L. J. (1995), 'Survival analysis with median regression models', *Journal of the American Statistical Association* **90**, 178–184.

Part IV

Other Censoring Schemes

Part IV deals with estimation for models with more complex censoring and sampling schemes than simple right censoring. Many of these censoring schemes have problems in deriving inference techniques since the counting process techniques used in the study of inference techniques for right-censored data no longer hold true. This is particularly true for interval (Chapter 18) and current status data (Chapter 19).

In Chapter 17 by Borgan and Samuelsen the focus is on estimation based on one of two sampling schemes. In these two sampling schemes there are cases that have failed and controls that are still alive. Often for all patients we need to confirm diagnosis, collect additional data or examine some biological sample to check on the assignment of patients to treatment or control and make comparisons. This confirmatory data analysis may be quite expensive and often uses up individual patient biological material which may be useful in other experiments. Clearly in such a situation a design that uses up the least amount of data is the best. In particular this is a concern when we have a large cost of verifying some biological covariate.

There are two common cost effective designs for rare diseases, the nested case-control study and the case-cohort study. For both designs complete covariate information is collected on all failures (cases) and on a sample of survivors (controls). In a nested case-control study complete information is collected on each case and on a small sample of survivors at the time of the case's failure. For a case-cohort design a subcohort is randomly selected from the full cohort and these are used as controls at all event times at which they are at risk. Chapter 17 looks at classical versions of the two designs using simple random sampling.

In the remaining two chapters of this part we look at two other censoring schemes. In Chapter 18 by Sun and Li the problem of interval censoring is examined. Interval censoring arises when for some individuals all we know is that the event of interest is in an interval $(L, R]$. This type of censoring arises in many studies when the event of interest can only be detected at a visit to the physician. Interval censoring is more general than right censoring or left censoring which are special cases of this scheme. In this type of censoring, since the counting process approach does not work, many of the estimators are obtained using an EM algorithm. The chapter studies the usual interval censored case as well as some modified interval censoring schemes and examines how one can estimate the survival function and regression coefficients in the Cox model.

In the final chapter of this part, Jewell and Emerson present a special case of interval censoring, namely current status data. This type of censoring occurs when the subject is observed only once for the occurrence of failure. At this single planned time we know only the observation time and the failure status. The failure time is either right censored (subject alive) or left censored (subject dead) at the observation time. They look at how one can estimate a survival curve or regression models for the survival probability using current status by studying a dataset of avalanche victims.

17

Nested Case-Control and Case-Cohort Studies

Ørnulf Borgan and Sven Ove Samuelsen
University of Oslo

CONTENTS

17.1	Introduction	343
17.2	Cox regression for cohort data	344
17.3	Nested case-control studies	345
	17.3.1 Sampling of controls	345
	17.3.2 Estimation and relative efficiency	346
	17.3.3 Example: radiation and breast cancer	347
	17.3.4 A note on additional matching	348
17.4	Case-cohort studies	348
	17.4.1 Sampling of the subcohort	348
	17.4.2 Prentice's estimator	349
	17.4.3 IPW estimators	350
	17.4.4 Stratified sampling of the subcohort	351
	17.4.5 Example: radiation and breast cancer	351
	17.4.6 Post-stratification and calibration	352
17.5	Comparison of the cohort sampling designs	353
	17.5.1 Statistical efficiency and analysis	353
	17.5.2 Study workflow and multiple endpoints	353
	17.5.3 Simple or stratified sampling	354
17.6	Re-use of controls in nested case-control studies	354
17.7	Theoretical considerations	355
	17.7.1 Nested case-control data	355
	17.7.2 Case-cohort data	357
17.8	Nested case-control: stratified sampling and absolute risk estimation	358
	17.8.1 Counter-matching	358
	17.8.2 Estimation of absolute risk	359
17.9	Case-cohort and IPW-estimators: absolute risk and alternative models	360
17.10	Maximum likelihood estimation	361
17.11	Closing remarks	362
	Bibliography	364

17.1 Introduction

In cohort studies, regression methods are commonly applied to assess the influence of risk factors and other covariates on mortality or morbidity; in particular Cox-regression is much used. Estimation in Cox's model is based on a partial likelihood, which at each observed death or disease occurrence ("failure") compares the covariate values of the failing individual to those of all individuals at risk. Thus Cox regression requires collection of covariate information for all individuals in the cohort, even when only a small fraction of these actually get diseased or die. This may be very expensive, or even logistically impossible. Further when covariate measurements are based on biological material stored in biobanks, it will imply a waste of valuable material that one may want to save for future studies.

Cohort studies are considered to be the most reliable study design in epidemiology, while traditional case-control studies are easier and quicker to implement, but statistically less efficient. Further the modeling framework of traditional case-control studies (contingency tables, logistic regression) does not consider the time aspect of the development of a disease. The cohort sampling methods considered in this chapter are developed to provide study designs that, like cohort studies, take the time aspect into account and at the same time maintain the cost-effectiveness of traditional case-control studies.

There are two main types of cohort sampling designs: nested case-control studies and case-cohort studies. For both types of designs, covariate information is collected for all failing individuals ("cases"), but only for a sample of the individuals who do not fail ("controls"). This may save valuable biological material and drastically reduce the workload of data collection and error checking compared to a full cohort study. Further, as most of the statistical information in a rare disease situation will be contained in the relatively few cases (and the controls), such studies may still be sufficient to give reliable answers to the questions of interest.

The two types of cohort sampling designs differ in the way controls are selected. For nested case-control sampling, one selects for each case a small number of controls from those at risk at the case's failure time, and a new sample of controls is selected for each case. For the case-cohort design a subcohort is selected from the full cohort, and the individuals in the subcohort are used as controls at all failure times when they are at risk. In their original forms, the designs use simple random sampling without replacement for the selection of controls and subcohort (Thomas, 1977; Prentice, 1986). Later the designs have been modified to allow for stratified random sampling (Langholz and Borgan, 1995; Borgan et al., 2000).

The purpose of the chapter is to review and discuss the classical versions of the nested case-control and case-cohort designs using simple random sampling and their modifications with stratified sampling. Our main focus is on estimation of relative risks using partial likelihoods and pseudo-likelihoods or weighted likelihoods that resemble the full cohort partial likelihood. But we also consider other topics like estimation of absolute risk and maximum likelihood estimation for the entire cohort, treating unobserved covariates as missing data.

17.2 Cox regression for cohort data

We first review Cox regression for cohort data. Consider a cohort $\mathcal{C} = \{1, \ldots, n\}$ of n independent individuals, and let $h_i(t)$ be the hazard rate for the ith individual with vector of covariates $\mathbf{x}_i = (x_{i1}, \ldots, x_{ip})'$. The covariates may be time-fixed or time-dependent, but we have suppressed the time-dependency from the notation. We assume that the covariates of individual i are related to its hazard rate by Cox's regression model:

$$h_i(t) = h(t \,|\, \mathbf{x}_i) = h_0(t) \exp(\boldsymbol{\beta}' \mathbf{x}_i). \tag{17.1}$$

Here $\boldsymbol{\beta} = (\beta_1, \ldots, \beta_p)'$ is a vector of regression coefficients describing the effects of the covariates, while the baseline hazard rate $h_0(t)$ corresponds to the hazard rate of an individual with all covariates equal to zero. In particular we may interpret e^{β_j} as a *hazard ratio* (or more loosely a *relative risk*). We will focus on model (26.4), which adopts the exponential relative risk function $r(\boldsymbol{\beta}, \mathbf{x}_i) = \exp(\boldsymbol{\beta}'\mathbf{x}_i)$. But it should be noted that most results for Cox regression carry over to other relative risk functions, one example being the excess relative risk function $r(\boldsymbol{\beta}, \mathbf{x}_i) = \prod_{j=1}^{p} \{1 + \beta_j x_{ij}\}$.

The individuals in the cohort may be followed over different periods of time, from an entry time to an exit time corresponding to failure or censoring. The risk set $\mathcal{R}(t)$ is the collection of all individuals who are under observation just before time t, and $n(t) = \#\mathcal{R}(t)$ is the number at risk at that time. We denote by $t_1 < t_2 < \cdots < t_d$ the times when failures are observed and, assuming no tied failure times, we let i_j denote the individual who fails at t_j.

We assume throughout that late entries and censorings are independent in the sense that the additional knowledge of which individuals have entered the study or have been censored before time t does not carry information on the risks of failure at t, cf. Kalbfleisch and Prentice (2002, Sections 1.3 and 6.2). Then the vector of regression coefficients in (26.4) is estimated by $\widehat{\boldsymbol{\beta}}$, the value of $\boldsymbol{\beta}$ maximizing Cox's partial likelihood

$$L(\boldsymbol{\beta}) = \prod_{j=1}^{d} \frac{\exp(\boldsymbol{\beta}'\mathbf{x}_{i_j})}{\sum_{k \in \mathcal{R}(t_j)} \exp(\boldsymbol{\beta}'\mathbf{x}_k)}. \tag{17.2}$$

It is well known that $\widehat{\boldsymbol{\beta}}$ can be treated as an ordinary maximum likelihood estimator (Andersen and Gill, 1982). In particular, $\widehat{\boldsymbol{\beta}}$ is approximately multivariate normally distributed around the true value of $\boldsymbol{\beta}$ with a covariance matrix that may be estimated by the inverse information matrix.

17.3 Nested case-control studies

In this section we consider the original form of the nested case-control design, where the controls are selected by simple random sampling. A modification of the design using stratified sampling is discussed in Section 17.8.1.

17.3.1 Sampling of controls

The nested case-control design was originally suggested by Thomas (1977), but see also Prentice and Breslow (1978). For this design, if a case occurs at time t, one selects $m - 1$ controls by simple random sampling from the $n(t) - 1$ non-failing individuals in the risk set $\mathcal{R}(t)$. The set consisting of the case and these $m - 1$ controls is called a sampled risk set and denoted $\widetilde{\mathcal{R}}(t)$. Covariate values are ascertained for the individuals in the sampled risk sets, but are not needed for the remaining individuals in the cohort. Figure 17.1 illustrates the basic features of a nested case-control study for a hypothetical cohort of seven individuals when one control is selected per case (i.e., when $m = 2$).

Note that the selection of controls is done independently at the different failure times. Thus subjects may serve as controls for multiple cases, and a case may serve as control for other cases that failed when the case was at risk. For example, the case at time t_4 in the figure had been selected as control at the earlier time t_1. A basic assumption for

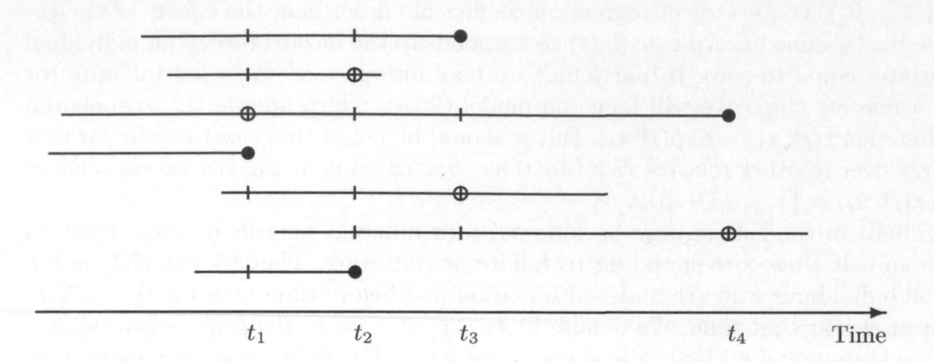

FIGURE 17.1
Nested case-control sampling, with one control per case, from a hypothetical cohort of seven individuals. Each individual is represented by a line starting at an entry time and ending at an exit time corresponding failure or censoring. Failure times are indicated by dots (•), potential controls are indicated by bars (|), and the sampled controls are indicated by circles (○).

valid inference is that not only delayed entries and censorings, but also the sampling of controls are independent in the sense that the additional knowledge of which individuals have entered the study have been censored or have been selected as controls before time t does not carry information on the risks of failure at t. This assumption will be violated if, e.g., in a prevention trial, individuals selected as controls change their behavior in such a way that their risk of failure is different from similar individuals who have not been selected as controls.

17.3.2 Estimation and relative efficiency

For nested case-control data the vector of regression coefficients in (26.4) may be estimated by $\widehat{\boldsymbol{\beta}}_{\text{ncc}}$, the value of $\boldsymbol{\beta}$ maximizing the partial likelihood

$$L_{\text{ncc}}(\boldsymbol{\beta}) = \prod_{j=1}^{d} \frac{\exp(\boldsymbol{\beta}'\mathbf{x}_{i_j})}{\sum_{k \in \widetilde{\mathcal{R}}(t_j)} \exp(\boldsymbol{\beta}'\mathbf{x}_k)}, \tag{17.3}$$

cf. Thomas (1977), Oakes (1981) and Section 17.7.1 below. Note that (17.3) is similar to the full cohort partial likelihood (17.2), except that the sum in the denominator is only over subjects in the sampled risk set. Also note that (17.3) coincides with a conditional likelihood for matched case-control data under a logistic regression model (Breslow, 1996).

Inference concerning the regression coefficients, using usual large sample likelihood methods, can be based on the partial likelihood (17.3). More specifically, under weak regularity conditions, one may prove that (Goldstein and Langholz, 1992; Borgan et al., 1995)

$$\sqrt{n}\left(\widehat{\boldsymbol{\beta}}_{\text{ncc}} - \boldsymbol{\beta}\right) \xrightarrow{d} \text{N}(\mathbf{0}, \boldsymbol{\Sigma}_{\text{ncc}}^{-1}), \tag{17.4}$$

and that $\boldsymbol{\Sigma}_{\text{ncc}}$ may be estimated consistently by $n^{-1}\mathbf{I}_{\text{ncc}}(\widehat{\boldsymbol{\beta}}_{\text{ncc}})$, where $\mathbf{I}_{\text{ncc}}(\boldsymbol{\beta}) = -\partial^2 \log L_{\text{ncc}}(\boldsymbol{\beta})/\partial\boldsymbol{\beta}'\partial\boldsymbol{\beta}$ is the observed information matrix. An outline of the main steps in the proof is given in Section 17.7.1. Furthermore nested models may be compared by likelihood ratio tests. For computing one may use standard software for Cox regression (like coxph in the **survival** package in R), formally treating the label of the sampled risk sets as a stratification variable in the Cox regression, or software for conditional logistic regression.

The relative efficiency of a nested case-control study compared to a full cohort study is the ratio of the variance of the estimator for full cohort data to the variance of the estimator based on nested case-control data. If there is only one covariate in the model, and its regression coefficient equals zero, the relative efficiency of the nested case-control design compared to a full cohort study is $(m-1)/m$, independent of censoring and covariate distributions (Goldstein and Langholz, 1992). When the regression coefficient differs from zero, and when more than one regression coefficient has to be estimated, the efficiency may be lower (Borgan and Olsen, 1999).

17.3.3 Example: radiation and breast cancer

For illustration we will use data from a cohort of female patients who were discharged from two tuberculosis sanatoria in Massachusetts between 1930 and 1956 to investigate breast cancer risk of radiation exposure due to fluoroscopy (Hrubec et al., 1989). Radiation doses have been estimated for 1,022 women who received radiation exposure to the chest from X-ray fluoroscopy lung examinations. The remaining 698 women in the cohort received treatments that did not require fluoroscopic monitoring and were radiation unexposed. The patients had been followed up until the end of 1980, by which time 75 breast cancer cases were observed.

For this cohort radiation data have been collected for all 1,720 women. But the workload of exposure data collection would have been reduced if the investigators had used a cohort sampling design. To illustrate the methods of this chapter, we will here and in Section 17.4.5 select nested case-control and case-cohort samples from the cohort and analyse the sampled data.

To model the effect of radiation dose on breast cancer risk, it is common to adopt an excess relative risk model (e.g., Preston et al., 2002). However, for our illustrative purpose, we choose to stay within the framework of Cox's regression model, and we will use $x = \log_2(\text{dose} + 1)$, with dose measured in grays (Gy), as covariate. Table 17.1 gives parameter estimates for cohort data and nested case-control data with two controls per case (i.e., $m = 3$) when age is used as time scale in (26.4). The cohort estimate is based on radiation data for all 1,720 women, while we only need radiation information for 211 women (75 cases and 136 controls) for the nested case-control data. (The number of women selected as controls is less than 150, since 14 women were members of two sampled risk sets.) Nevertheless, the parameter estimates are fairly similar for the two designs, and the standard

TABLE 17.1
Estimates of the effect of the covariate $x = \log_2(\text{dose} + 1)$ on breast cancer risk for cohort data and nested case-control data with two controls per case.

Study design	Parameter estimate	Standard error	Wald test statistic	P-value
Cohort data	0.491	0.162	3.04	0.002
Nested case-control data	0.539	0.231	2.34	0.019

error of the nested case-control estimate is only 40% larger than the standard error of the cohort estimate.

17.3.4 A note on additional matching

In order to keep the presentation simple, we have so far considered the proportional hazards model (26.4), where the baseline hazard rate is the same for all individuals. Often this will not be reasonable. To control for the effect of one or more confounding factors, one may adopt a stratified version of (26.4), where the baseline hazard differs between population strata generated by the confounders. For instance, for the breast cancer example of Section 17.3.3, one may control for calendar time and age at first treatment by allowing the baseline to depend on these covariates. The regression coefficients are, however, assumed to be the same across population strata. Thus the hazard rate of an individual i from population stratum c is assumed to take the form

$$h_i(t) = h_{0c}(t)\exp(\boldsymbol{\beta}'\mathbf{x}_i). \tag{17.5}$$

When the stratified proportional hazards model (17.5) applies, the sampling of controls in a nested case-control study should be restricted to those at risk in the same population stratum as the case. We say that the controls are matched by the stratification variable(s). In particular if an individual in population stratum c fails at time t, one selects at random $m-1$ controls from the $n_c(t)-1$ non-failing individuals at risk in this population stratum. Then the partial likelihood (17.3) still applies, and the estimation of the vector of regression coefficients is carried out as described in Section 17.3.2.

17.4 Case-cohort studies

Case-cohort studies are considered in this section. We start out with the original version of the design, where the subcohort is selected by simple random sampling, and discuss two ways of analysing the case-cohort data. Then the modification with stratified sampling of the subcohort is considered, and finally some comments on post-stratification are given.

17.4.1 Sampling of the subcohort

The case-cohort design was originally suggested by Prentice (1986), although related designs without taking a time-perspective into consideration had previously been considered (Kupper et al., 1975; Miettinen, 1982). For the case-cohort design one selects a subcohort $\widetilde{\mathcal{C}}$ of size $\widetilde{m}$ from the full cohort by simple random sampling, and the individuals in the subcohort are used as controls at all failure times when they are at risk. Covariate values are ascertained for the individuals in $\widetilde{\mathcal{C}}$ as well as for the cases occurring outside the subcohort, but they are not needed for the non-failures outside the subcohort. Similarly to nested case-control sampling, an assumption for valid inference is that individuals sampled to the subcohort do not change their behavior in such a way that their risk of failure is different from (similar) individuals who have not been selected to the subcohort. Figure 17.2 illustrates a case-cohort study for the hypothetical cohort of Figure 17.1 with subcohort size $\widetilde{m} = 3$.

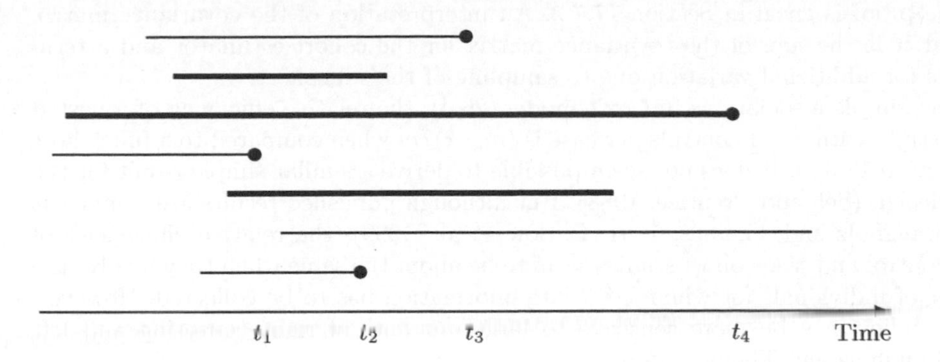

FIGURE 17.2
Case-cohort sampling, with subcohort size $\widetilde{m} = 3$, from the hypothetical cohort of Figure 17.1. Failure times are indicated by dots (•), and the individuals in the subcohort are indicated by thick lines.

17.4.2 Prentice's estimator

Different methods have been suggested for estimating the regression coefficients in (26.4) from case-cohort data. The original suggestion of Prentice (1986) consists of maximizing what is referred to as a pseudo-likelihood:

$$L_{\mathrm{P}}(\boldsymbol{\beta}) = \prod_{j=1}^{d} \frac{\exp(\boldsymbol{\beta}'\mathbf{x}_{i_j})}{\sum_{k\in\mathcal{S}(t_j)} \exp(\boldsymbol{\beta}'\mathbf{x}_k)}. \tag{17.6}$$

Here the sum in the denominator is over the set $\mathcal{S}(t_j)$ consisting of the subcohort individuals at risk with the case i_j added when it occurs outside the subcohort. A modification of Prentice's pseudo-likelihood, where a case outside the subcohort is not included in $\mathcal{S}(t_j)$ was considered by Self and Prentice (1988). However, their main aim was to study large sample properties of the case-cohort estimator, not to provide an alternative to (17.6) for practical use.

Each factor of the product in (17.6) is of the same form as a factor of the product in the partial likelihood (17.3). In (17.6), however, controls from the subcohort are used over again for each case and thus the factors are dependent. This has the consequence that (17.6) is not a partial likelihood (Langholz and Thomas, 1991). Thus standard errors cannot be computed directly from the information matrix of (17.6) and likelihood ratio statistics will not follow chi-square distributions. But (17.6) provides unbiased estimating equations (Prentice, 1986), and one may show that the maximum pseudo-likelihood estimator $\widehat{\boldsymbol{\beta}}_{\mathrm{P}}$ is approximately normally distributed (Self and Prentice, 1988).

More specifically, for the information matrix $\mathbf{I}_{\mathrm{P}}(\boldsymbol{\beta}) = -\partial^2 \log L_{\mathrm{P}}(\boldsymbol{\beta})/\partial\boldsymbol{\beta}'\partial\boldsymbol{\beta}$ we have under standard assumptions that $n^{-1}\mathbf{I}_{\mathrm{P}}(\widehat{\boldsymbol{\beta}}_{\mathrm{P}}) \to \boldsymbol{\Sigma}$ in probability, where $\boldsymbol{\Sigma}$ is the same limit as for the cohort information matrix. For the case-cohort estimator we then have that

$$\sqrt{n}(\widehat{\boldsymbol{\beta}}_{\mathrm{P}} - \boldsymbol{\beta}) \xrightarrow{d} \mathrm{N}(\mathbf{0}, \boldsymbol{\Sigma}^{-1} + \frac{1-p}{p}\boldsymbol{\Sigma}^{-1}\boldsymbol{\Delta}\boldsymbol{\Sigma}^{-1}), \tag{17.7}$$

where p is the (limiting) proportion of the cohort that is sampled to the subcohort, and $\boldsymbol{\Delta}$ is the limit in probability of the covariance matrix of the individual score-contributions. An outline of the proof is given in Section 17.7.2. An interpretation of the covariance matrix for $\widehat{\boldsymbol{\beta}}_{\mathrm{P}}$ is that it is the sum of the covariance matrix for the cohort estimator and a term that accounts for additional variation due to sampling of the subcohort.

For testing simple associations in Cox's model (26.4), the relative efficiency of a nested case-control study with $m-1$ controls per case is $(m-1)/m$ when compared to a full cohort study (cf. Section 17.3.2). It does not seem possible to derive a similar simple result for the case-cohort design (Self and Prentice, 1988). But although published results are somewhat conflicting (Langholz and Thomas, 1991; Barlow et al., 1999), the relative efficiencies of nested case-control and case-cohort studies seem to be about the same when they involve the same number of individuals for whom covariate information has to be collected. However, case-cohort studies may be more sensitive to large amounts of right censoring and left truncation (Langholz and Thomas, 1991).

The covariance matrix of the case-cohort estimator can be estimated by a straightforward plug-in procedure using (17.7). The estimate of $\boldsymbol{\Sigma}$ is obtained directly from $\mathbf{I}_{\mathrm{P}}(\widehat{\boldsymbol{\beta}}_{\mathrm{P}})$ and in programs that allow for calculation of score-contributions and *dfbetas* the covariance matrix $\boldsymbol{\Delta}$ is just replaced by the empirical counterpart; see Therneau and Li (1999) for details. An alternative to this model based estimator is simply to use a robust sandwich type estimator (Lin and Ying, 1993; Barlow, 1994).

17.4.3 IPW estimators

Prentice's pseudo-likelihood (17.6) can be calculated for time-dependent covariates also when covariate information for cases outside the subcohort are ascertained only at their failure times. This may be useful in some situations. However, with fixed covariates (or when the full covariate paths of time-dependent covariates are easily retrieved) it would seem that information may be lost by this estimating procedure. Another proposal for case-cohort studies, first suggested by Kalbfleisch and Lawless (1988), is to maximize the weighted likelihood (or, more precisely, weighted pseudo-likelihood)

$$L_{\mathrm{W}}(\boldsymbol{\beta}) = \prod_{j=1}^{d} \frac{\exp(\boldsymbol{\beta}'\mathbf{x}_{i_j})}{\sum_{k\in\widetilde{\mathcal{S}}(t_j)} \exp(\boldsymbol{\beta}'\mathbf{x}_k) w_k}, \tag{17.8}$$

where $\widetilde{\mathcal{S}}(t_j)$ is the set consisting of the subcohort individuals at risk at time t_j together with all cases that are at risk at that time. The weights are $w_k = 1$ for cases (whether in the subcohort or not) and $w_k = 1/p_k$ for non-failures in the subcohort, where the p_k's are appropriate inclusion probabilities. Note that we have assumed that cases are sampled with probability one, so inverse probability weighting (IPW) is used. The estimator is thus an application of the Horvitz-Thompson method.

Kalbfleisch and Lawless (1988) assumed that the inclusion probabilities p_k were known. Later Borgan et al. (2000), in the context of stratified case-cohort studies (Section 17.4.4), suggested to set p_k equal to the proportion of non-failures in the subcohort compared to all non-failures. We denote the estimator thus obtained by $\widehat{\boldsymbol{\beta}}_{\mathrm{W}}$. For a standard case-cohort study with simple random sampling of the subcohort, this estimator coincides with a suggestion by Chen and Lo (1999). We may show that

$$\sqrt{n}(\widehat{\boldsymbol{\beta}}_{\mathrm{W}} - \boldsymbol{\beta}) \xrightarrow{d} \mathrm{N}(\mathbf{0}, \boldsymbol{\Sigma}^{-1} + \frac{q(1-p)}{p}\boldsymbol{\Sigma}^{-1}\boldsymbol{\Delta}_0\boldsymbol{\Sigma}^{-1}), \tag{17.9}$$

where $\boldsymbol{\Sigma}$ and p are defined in connection with (17.7), q is the (limiting) proportion of non-failures in the cohort, and $\boldsymbol{\Delta}_0$ is the limit in probability of the covariance matrix of the individual score-contributions among non-failures. The result is a special case of (17.10) below. Since $q < 1$ and the variance of score-contributions over non-failures is smaller than over the full cohort we find, by comparing (17.7) and (17.9), that the variance of $\widehat{\beta}_{\mathrm{W}}$ is smaller than the variance of the original Prentice estimator $\widehat{\beta}_{\mathrm{P}}$. However, in practice this matters little unless there is a fairly large proportion of cases, and in such situations the case-cohort design will typically not be chosen.

Estimation of the covariance matrix of the IPW estimator may be performed by a slight modification of the plug-in procedure described at the end of Section 17.4.2; see e.g., Langholz and Jiao (2007). We may also use the robust estimator, although now this estimator is theoretically conservative.

17.4.4 Stratified sampling of the subcohort

In the presentation of case-cohort sampling in Section 17.4.1, we assume that covariate information is collected only for the cases and the non-failures in the subcohort. However, since a case-cohort study is performed within a well-defined cohort, there will typically be additional background data that are available for all cohort members. For instance a surrogate measure of an exposure of interest may be available for everyone. Such background data may be used to classify the cohort individuals into S distinct strata. With n_s individuals in stratum s, one then selects a random sample of $\widetilde{m}_s$ individuals to the subcohort $\widetilde{\mathcal{C}}$ from each stratum s; $s = 1, 2, \ldots, S$. By selecting the subcohort by stratified sampling, one may increase the variation in the subcohort of a covariate of main interst, and thereby achieve a more efficient estimation of the effect of this covariate (Samuelsen, 1989; Borgan et al., 2000).

As for the simple case-cohort design, there are different options for analysing stratified case-cohort data. Borgan et al. (2000) discussed three estimators. We will here restrict attention to their Estimator II, denoted $\widehat{\beta}_{\mathrm{II}}$, which is a generalization of the IPW estimator of Section 17.4.3. We then use the weighted likelihood (17.8) with weights $w_k = 1$ for cases and $w_k = n_s^0/\widetilde{m}_s^0$ for non-failing subcohort members from stratum s. Here n_s^0 and $\widetilde{m}_s^0$ are the number of non-failures in the cohort and subcohort, respectively, who belong to stratum s.

For Estimator II we have (Borgan et al., 2000; Samuelsen et al., 2007)

$$\sqrt{n}(\widehat{\beta}_{\mathrm{II}} - \boldsymbol{\beta}) \xrightarrow{d} \mathrm{N}(\mathbf{0}, \boldsymbol{\Sigma}^{-1} + \sum_{s=1}^{S} \frac{q_s(1-p_s)}{p_s} \boldsymbol{\Sigma}^{-1} \boldsymbol{\Delta}_{0s} \boldsymbol{\Sigma}^{-1}). \tag{17.10}$$

Here q_s is the (limiting) proportion of non-failures who belong to stratum s and p_s is the (limiting) proportion of non-failures in this stratum who are sampled to the subcohort. Furthermore $\boldsymbol{\Delta}_{0s}$ is the limit in probability of the covariance matrix of the individual score-contributions among non-failures in stratum s. We may estimate the covariance matrix of Estimator II by a plug-in procedure using (17.10); details are provided in Langholz and Jiao (2007) and Samuelsen et al. (2007). For stratified case-cohort data, one should avoid using the robust covariance estimator as this tends to give variance estimates that are quite a bit too large.

17.4.5 Example: radiation and breast cancer

We consider the data from Section 17.3.3 on radiation and breast cancer. To illustrate the case-cohort methodology, we first select at random a subcohort of 150 individuals from the full cohort. There were 7 cases in the selected subcohort, so for a case-cohort analysis radiation data are needed for 218 women. Table 17.2 gives Prentice's estimate and the IPW estimate obtained by maximizing (17.6) and (17.8). The estimates were computed using the

TABLE 17.2
Estimates of the effect of the covariate $x = \log_2(\text{dose} + 1)$ on breast cancer risk for cohort data and case-cohort data where the subcohort is selected by simple random sampling (for Prentice's estimator and the IPW estimator) or stratified random sampling (for Estimator II). See the text for details.

Estimator	Parameter estimate	Standard error	Wald test statistic	P-value
Cohort estimator	0.491	0.162	3.04	0.002
Prentice's estimator	0.519	0.215	2.41	0.016
IPW estimator	0.524	0.211	2.48	0.013
Estimator II	0.509	0.184	2.75	0.006

cch command in the **survival** package in R (version 2.13.0) with options method="Prentice" and method="LinYing", respectively. The two case-cohort estimates are very close, and so are their standard errors. Comparing the case-cohort estimates with the nested case-control estimate of Table 17.1, we note that the estimates and their standard errors are fairly similar. This is a common observation for situations where case-cohort data and nested case-control data contain about the same number of individuals for whom covariate information is ascertained.

For the breast cancer cohort, information on the number of fluoroscopic examinations is available for each woman. The number of examinations may be used as a surrogate for the radiation exposure in situations where the latter is costly to obtain. So to illustrate a stratified case-cohort study, we stratify the cohort into three strata: (i) the 698 women with no fluoroscopic examinations (i.e., the unexposed women), (ii) the 765 women with 1-149 examinations, and (iii) the 257 women with 150 examinations or more. From each stratum we select a random sample of 50 women to the subcohort, thereby selecting a higher proportion of the women with many fluoroscopic examinations. From the stratified case-cohort data we obtain Estimator II of Section 17.4.4 by the cch command with option method="II.Borgan". The estimate is given in the last line of Table 17.2. We see that the estimate is close to the cohort estimate and that its standard error is clearly smaller than the standard errors for the estimates from the non-stratified case-cohort data.

17.4.6 Post-stratification and calibration

When describing the stratified case-cohort design in Section 17.4.4 we assume, for the ease of presentation, that the cohort is stratified according to some background information (like a surrogate measure for exposure) that is available for everyone. However, all information that is recorded for every cohort member may be used for stratification, including information on entry and follow-up times and whether an individual is a case or not. When follow-up information is used to stratify the cohort, it is common to say that the cohort is post-stratified.

Note that we may use stratified sampling to select the subcohort at the outset of the study, and later redefine the strata by post-stratification. In fact, this is what we do in Section 17.4.4 when we define the cases as a separate stratum. Another possibility is to post-stratify on follow-up time (appropriately partioned). As shown by Samuelsen et al. (2007), this corresponds to the "local averaging" estimator of Chen (2001*b*).

The idea of post-stratification can also be applied to background variables known for the entire cohort. Suppose a simple or stratified case-cohort sample has been selected at the outset of a study. At the analysis stage, one may then post-stratify according to known background variables, thereby modifying the sampling fractions. Such an approach may lead to improved efficiency when the background variables are strongly related to the covariates of the Cox model. In particular, if the background variables are included as covariates in (26.4), the efficiency of the corresponding regression coefficients will improve greatly. But obviously such an approach will break down with very fine post-stratification. Then one option is to modify the weights using the double weighting method of Kulich and Lin (2004) or the calibration technique of Breslow et al. (2009*a*,*b*) so that they better reflect the full cohort information.

17.5 Comparison of the cohort sampling designs

If one wants to apply a cohort sampling design, a choice between a nested case-control and a case-cohort study has to be made. The choice between the two designs depends on a number of issues, and it has to be made on a case-by-case basis. We here discuss some issues that should be considered to arrive at a useful design for a particular study.

17.5.1 Statistical efficiency and analysis

As mentioned in Section 17.4.2, the statistical efficiencies of the nested case-control and case-cohort designs seem to be about the same when they involve comparable numbers of individuals for whom covariate information has to be ascertained. Thus efficiency considerations are usually not important for design decisions when studying a single disease endpoint. If multiple endpoints are of interest, the situation may be different; cf. Section 17.5.2.

The statistical analysis of nested case-control data by Cox's regression model (26.4) parallels the analysis of cohort data, and it may be performed by means of the usual likelihood based methods and standard Cox regression software (or by software for conditional logistic regression). For case-cohort data likelihood methods do not apply, and even though standard Cox regression software may be "tricked" to do the analysis (Therneau and Li, 1999; Langholz and Jiao, 2007), this has made inference for case-cohort data more cumbersome. But with the development of specialized computer software for case-cohort data (like cch in the **survival** package in R), this drawback has become less important.

In a nested case-control study, the controls are sampled from those at risk at the cases' failure times. Therefore one has to decide the time scale to be used in the analysis (e.g., age or time since employment) before the controls are sampled. This does not apply to a case-cohort study, where the subcohort is selected without consideration of at risk status. Moreover, while other failure time models than (26.4) may be used to analyze case-cohort data (cf. Section 17.9), the analysis options for nested case-control data are more restricted. Thus case-cohort data allow for more flexibility in model choice at the analysis stage (see, however, Section 17.6).

17.5.2 Study workflow and multiple endpoints

Cohort sampling is useful both for prospective studies, like disease prevention trials, and for retrospective studies, where the events have already happened, but covariate information is costly to retrieve (e.g., from paper files). For the former case, the workflow can be made more predictable with a case-cohort design. Since the subcohort is sampled at the outset of the study, more efforts can be used in early phases on processing subcohort information, while the information on the cases may be processed later. For a nested case-control study, however, control selection and ascertainment of covariate values for the controls have to wait until the cases occur.

In a nested case-control study, as described in Section 17.3, the controls are matched to their cases. So if one wants to study more than one type of endpoint (e.g., more than one disease), new controls have to be selected for each endpoint. Here a case-cohort design may give large savings by allowing the subcohort individuals to be used as controls for multiple endpoints. However, in Section 17.6 we describe a method that also allows the controls in a nested case-control study to be used for more than one endpoint.

Cohort sampling designs are often used in studies of biomarkers from cohorts with stored biological samples. For such studies one should be aware of possible effects of analytic batch and long-term storage. If these effects are substantial, a case-cohort study may give biased results, and it is advisable to use a nested case-control design with matching on storage time with the cases and their controls analyzed in the same batch. Otherwise a case-cohort design may be the preferred approach, since it allows us to re-use the biomarkers for other endpoints (Rundle et al., 2005).

17.5.3 Simple or stratified sampling

In Section 17.4.4 we discuss the possibility of selecting the subcohort for a case-cohort study by stratified random sampling, and in Section 17.8.1 we consider a stratified version of the nested case-control design. Stratified sampling may be a useful option when stratification can be based on a surrogate measure of an exposure of main interest. One should be aware, however, that there is "no free lunch" so the efficiency gain for the exposure of main interest will often be accompanied by a loss in efficiency for other covariates. Thus stratified sampling may be a useful option for studies with a focused research question, but less so for a subcohort that is assembled to serve as controls for multiple endpoints.

17.6 Re-use of controls in nested case-control studies

In the partial likelihood (17.3), a case and its controls are included only at the failure time of the case. When the covariate information obtained for cases and controls is time-fixed (or the full trajectories of time-dependent covariates can be obtained), one may consider to break the matching between a case and its controls and analyse the nested case-control data as if they were case-cohort data (with a non-standard sampling scheme for the subcohort). In this way the covariate information for cases and controls may be used whenever the individuals are at risk. Such re-use of controls may lead to more efficient estimators and counter some of the limitations of nested case-control studies discussed in Section 17.5. One should be aware, however, that in some situations the matching is needed to avoid bias (Section 17.5.2), and then one should avoid breaking the matching between a case and its controls.

One way one may re-use the controls is by estimating the probability that an individual is ever sampled as control, and then apply the weighted likelihood (17.8). For instance Samuelsen (1997), see also Suissa et al. (1998), suggested using the weights $w_k = 1/p_k$, where $p_k = 1$ for cases and

$$p_k = 1 - \prod \left(1 - \frac{m-1}{n(t_j)-1}\right) \tag{17.11}$$

for control individual k. The product in (17.11) is over failure times t_j when individual k is at risk. As for case-cohort data, the weighted likelihood does not possess likelihood properties, so variance estimation requires special attention. Samuelsen (1997) developed a variance estimator with the weights (17.11) which implies that the robust variance tends to be conservative. However, other authors have found the robust variance to be adequate in most situations (Samuelsen et al., 2007; Saarela et al., 2008; Støer and Samuelsen, 2012).

We have applied the weighted likelihood (17.8) with weights (17.11) for the nested case-control data of Section 17.3.3. We then obtain the estimate 0.474 with standard error 0.207 (using Samuelsen's variance estimator), corresponding to a Wald test statistic of 2.30.

The inclusion probability (17.11) may be modified in several ways. Additional matching (Section 17.3.4) can be accounted for by replacing the number at risk $n(t_j)$ by the number at risk who satisfy the matching criteria and by also restricting the product to be over failure times satisfying the matching criteria (Salim et al., 2009; Cai and Zheng, 2012). Furthermore nested case-control studies from partly overlapping cohorts can be combined (Salim et al., 2009), and with controls sampled for multiple endpoints, one may calculate an overall inclusion probability by taking the product in (17.11) to be over event times for all endpoints (Saarela et al., 2008).

Alternative weights or inclusion probabilities have been suggested. Chen (2001*b*) considered "local averaging" based on partitioning the follow-up time into disjoint intervals. The weights are the inverse sampling fractions among non-failures in each of the intervals. It has been argued (Samuelsen et al., 2007) that Chen's estimator can be seen as post-stratification (Section 17.4.6) on the censoring interval, and that variances can be obtained from (17.10). It has also been suggested to estimate the inclusion probabilities by logistic regression using the indicator of being sampled among non-cases as response and the right-censoring time as covariate (Robins et al., 1994; Samuelsen et al., 2007; Saarela et al., 2008). The local averaging approach can be seen as a special case of this approach using censoring interval as a categorical covariate. If there is a strong dependency between an exposure and censoring, these estimators can give efficiency improvements compared to weights using (17.11); see Chen (2001*b*) and Samuelsen et al. (2007).

17.7 Theoretical considerations

In Section 17.7.1 we derive the partial likelihood (17.3) for nested case-control data and sketch the main steps in the derivation of the large sample properties of the maximum partial likelihood estimator of $\boldsymbol{\beta}$. Further in Section 17.7.2 we give an outline of the proof of the large sample properties of the estimator for $\boldsymbol{\beta}$ based on pseudo-likelihoods for case-cohort data.

17.7.1 Nested case-control data

For nested case-control data we adopt the counting process formulation of Borgan et al. (1995). To this end we introduce, for all individuals i and all possible sampled risk sets $\mathbf{r}$, the counting process

$$N_{i,\mathbf{r}}(t) = \sum_{j \geq 1} I\{t_j \leq t,\, i_j = i,\, \widetilde{\mathcal{R}}(t_j) = \mathbf{r}\}. \tag{17.12}$$

This process has a jump at time t if the ith individual fails at that time *and* the set $\mathbf{r}$ is selected as the sampled risk set. Therefore, assuming Cox's model (26.4), the intensity process of (17.12) takes the form

$$\lambda_{i,\mathbf{r}}(t) = Y_i(t) h_i(t) \pi(\mathbf{r} \,|\, t, i) = Y_i(t) h_0(t) \exp(\boldsymbol{\beta}'\mathbf{x}_i) \pi(\mathbf{r} \,|\, t, i). \tag{17.13}$$

Here $Y_i(t)$ is an at risk indicator for the ith individual and $\pi(\mathbf{r} \,|\, t, i)$ is the conditional probability that $\mathbf{r}$ is selected as the sampled risk set, given "the past" and given that individual i fails at time t. When the controls are selected by simple random sampling without replacement

$$\pi(\mathbf{r} \,|\, t, i) = \frac{1}{\binom{n(t)-1}{m-1}} \tag{17.14}$$

for all subsets $\mathbf{r}$ of the risk set $\mathcal{R}(t) = \{i \,|\, Y_i(t) = 1\}$ that contain i and are of size m. (For all other subsets, $\pi(\mathbf{r} \,|\, t, i) = 0$.)

To derive the partial likelihood, we first note that

$$\pi(i \,|\, t, \mathbf{r}) = \frac{\lambda_{i,\mathbf{r}}(t)}{\sum_{k \in \mathbf{r}} \lambda_{k,\mathbf{r}}(t)} \tag{17.15}$$

is the conditional probability that individual i fails at time t given "the past" *and* given that a failure occurs for an individual in $\mathbf{r}$ at that time. When the controls are selected by simple random sampling, we obtain from (17.13) and (17.14) that the conditional probability takes the form

$$\begin{aligned} \pi(i \,|\, t, \mathbf{r}) &= \frac{Y_i(t) \exp(\boldsymbol{\beta}'\mathbf{x}_i) \pi(\mathbf{r} \,|\, t, i)}{\sum_{k \in \mathbf{r}} Y_k(t) \exp(\boldsymbol{\beta}'\mathbf{x}_k) \pi(\mathbf{r} \,|\, t, k)} \\ &= \frac{Y_i(t) \exp(\boldsymbol{\beta}'\mathbf{x}_i)}{\sum_{k \in \mathbf{r}} Y_k(t) \exp(\boldsymbol{\beta}'\mathbf{x}_k)}. \end{aligned} \tag{17.16}$$

The partial likelihood is obtained by multiplying together such conditional probabilities over all failure times t_j, cases i_j, and sampled risk sets $\widetilde{\mathcal{R}}(t_j)$:

$$L_{\text{ncc}}(\boldsymbol{\beta}) = \prod_{j=1}^{d} \pi(i_j \,|\, t_j, \widetilde{\mathcal{R}}(t_j)) = \prod_{j=1}^{d} \frac{\exp(\boldsymbol{\beta}'\mathbf{x}_{i_j})}{\sum_{k \in \widetilde{\mathcal{R}}(t_j)} \exp(\boldsymbol{\beta}'\mathbf{x}_k)}. \tag{17.17}$$

The at risk indicators may be omitted from (17.17) since the case and the sampled controls are all at risk. This gives a justification of (17.3).

The maximum partial likelihood estimator $\widehat{\boldsymbol{\beta}}_{\text{ncc}}$ solves $\mathbf{U}_{\text{ncc}}(\boldsymbol{\beta}) = 0$, where $\mathbf{U}_{\text{ncc}}(\boldsymbol{\beta}) = \partial \log L_{\text{ncc}}(\boldsymbol{\beta}) / \partial \boldsymbol{\beta}$ is the score function. Using (17.12), we may write

$$\mathbf{U}_{\text{ncc}}(\boldsymbol{\beta}) = \sum_{\mathbf{r}} \int_0^{\tau} \sum_{i \in \mathbf{r}} \left\{ \mathbf{x}_i - \frac{\sum_{k \in \mathbf{r}} Y_k(t) \mathbf{x}_k \exp(\boldsymbol{\beta}'\mathbf{x}_k)}{\sum_{k \in \mathbf{r}} Y_k(t) \exp(\boldsymbol{\beta}'\mathbf{x}_k)} \right\} dN_{i,\mathbf{r}}(t), \tag{17.18}$$

where τ is the upper time limit of the study. For the true value of $\boldsymbol{\beta}$, standard counting process theory (e.g., Aalen et al., 2008) gives the decomposition $dN_{i,\mathbf{r}}(t) = \lambda_{i,\mathbf{r}}(t)dt +$

$dM_{i,\mathbf{r}}(t)$, where the $M_{i,\mathbf{r}}(t)$ are orthogonal martingales. Using this decomposition, (17.13) and (17.14), we find that the score function takes the form

$$\mathbf{U}_{\mathrm{ncc}}(\boldsymbol{\beta}) = \int_0^\tau \sum_{\mathbf{r}} \sum_{i \in \mathbf{r}} \left\{ \mathbf{x}_i - \frac{\sum_{k\in\mathbf{r}} Y_k(t)\mathbf{x}_k \exp(\boldsymbol{\beta}'\mathbf{x}_k)}{\sum_{k\in\mathbf{r}} Y_k(t)\exp(\boldsymbol{\beta}'\mathbf{x}_k)} \right\} dM_{i,\mathbf{r}}(t) \tag{17.19}$$

when evaluated at the true value of $\boldsymbol{\beta}$. Thus the score (17.19) is a sum of stochastic integrals, and hence a mean zero martingale. By the martingale central limit theorem, one may then prove that $n^{-1/2}\mathbf{U}_{\mathrm{ncc}}(\boldsymbol{\beta}) \xrightarrow{d} \mathrm{N}(\mathbf{0}, \boldsymbol{\Sigma}_{\mathrm{ncc}})$ where $\boldsymbol{\Sigma}_{\mathrm{ncc}}$ is the limit in probability of the predictable variation process of $n^{-1/2}\mathbf{U}_{\mathrm{ncc}}(\boldsymbol{\beta})$. Further one may prove that $1/n$ times the observed information matrix $\mathbf{I}_{\mathrm{ncc}}(\boldsymbol{\beta}) = -\partial^2 \log L_{\mathrm{ncc}}(\boldsymbol{\beta})/\partial\boldsymbol{\beta}'\partial\boldsymbol{\beta}$ converges in probability to $\boldsymbol{\Sigma}_{\mathrm{ncc}}$. These results are key to prove (17.4). The formal proof is similar to the one of Andersen and Gill (1982) for cohort data; see Borgan et al. (1995) for details.

17.7.2 Case-cohort data

We will now indicate how one may derive the large sample properties of case-cohort estimators. For ease of presentation, we focus on the estimator of Self and Prentice (1988), which is asymptotically equivalent to the Prentice estimator of Section 17.4.2. The Self and Prentice estimator $\widehat{\boldsymbol{\beta}}_{\mathrm{SP}}$ is obtained by maximizing the pseudo-likelihood $L_{\mathrm{SP}}(\boldsymbol{\beta})$ obtained from (17.8) when cases outside the subcohort $\widetilde{C}$ are not included in the sums in the denominator. The large sample properties of the IPW estimator of Section 17.4.3 and the stratified estimator of Section 17.4.4 may be derived along similar lines (Borgan et al., 2000; Samuelsen et al., 2007).

To derive the large sample properties of $\widehat{\boldsymbol{\beta}}_{\mathrm{SP}}$, we note that the normalized score $\mathbf{U}_{\mathrm{SP}}(\boldsymbol{\beta}) = \partial \log L_{\mathrm{SP}}(\boldsymbol{\beta})/\partial\boldsymbol{\beta}$ of the Self and Prentice pseudo-likelihood may be decomposed as

$$n^{-1/2}\mathbf{U}_{\mathrm{SP}}(\boldsymbol{\beta}) = n^{-1/2}\mathbf{U}(\boldsymbol{\beta}) + n^{-1/2}\sum_{i=1}^{n}\left(1 - \frac{n}{\widetilde{m}}V_i\right)\mathbf{Z}_i, \tag{17.20}$$

where $\mathbf{U}(\boldsymbol{\beta})$ is the score of the full cohort partial likelihood (17.2), V_i is an indicator that individual i is selected to the subcohort, and

$$\mathbf{Z}_i = \sum_{j=1}^{d} Y_i(t_j)\left\{\mathbf{x}_i - \frac{\sum_{k=1}^{n} Y_k(t_j)\mathbf{x}_k\exp(\boldsymbol{\beta}'\mathbf{x}_k)}{\sum_{k=1}^{n} Y_k(t_j)\exp(\boldsymbol{\beta}'\mathbf{x}_k)}\right\}\frac{\exp(\boldsymbol{\beta}'\mathbf{x}_i)}{S_{\widetilde{C}}^{(0)}(\boldsymbol{\beta}, t_j)}. \tag{17.21}$$

In (17.21), $Y_i(t_j)$ is an indicator that individual i is at risk at time t_j, and

$$S_{\widetilde{C}}^{(0)}(\boldsymbol{\beta}, t_j) = \frac{n}{\widetilde{m}}\sum_{k\in\widetilde{C}} Y_k(t_j)\exp(\boldsymbol{\beta}'\mathbf{x}_k).$$

The leading term on the right-hand side of (17.20) is the normalized score of the full cohort partial likelihood, and converges weakly to a mean zero multivariate normal distribution with covariance matrix $\boldsymbol{\Sigma}$ (Andersen and Gill, 1982). For the second term we may, conditional on the complete cohort history and after approximating (17.21) by a quantity $\mathbf{Z}_i^*$ that only depends on observations from individual i, apply a central limit theorem for finite populations (e.g., page 353 in Lehmann, 1975). The result is that the second term on the right-hand side of (17.20) converges weakly to a mean zero multivariate normal distribution with covariance matrix $\{(1-p)/p\}\boldsymbol{\Delta}$, where p is the limit of $\widetilde{m}/n$ and $\boldsymbol{\Delta}$ is the limit in probability of the finite population covariance matrix of the $\mathbf{Z}_i^*$'s. Further the two terms in (17.20) are asymptotically independent. It follows that $n^{-1/2}\mathbf{U}_{\mathrm{SP}}(\boldsymbol{\beta}) \xrightarrow{d} \mathrm{N}(\mathbf{0}, \boldsymbol{\Sigma} + \{(1-p)/p\}\boldsymbol{\Delta})$.

Further one may prove that $1/n$ times $\mathbf{I}_{\mathrm{SP}}(\boldsymbol{\beta}) = -\partial^2 \log L_{\mathrm{SP}}(\boldsymbol{\beta})/\partial\boldsymbol{\beta}'\partial\boldsymbol{\beta}$ converges in probability to $\boldsymbol{\Sigma}$. These results are key to prove that the normalized Self and Prentice estimator converges weakly to the limiting multivariate normal distribution of (17.7). Further details are given in Borgan et al. (2000), while Self and Prentice (1988) provide a formal proof.

17.8 Nested case-control: stratified sampling and absolute risk estimation

In Section 17.3 we discuss the classical nested case-control design, where the controls are selected by simple random sampling, and show how we may estimate relative risks. Here we will consider stratified (or counter-matched) sampling of the controls and describe estimation of absolute risks.

17.8.1 Counter-matching

In Section 17.4.4 we discuss stratified sampling for the case-cohort design. In a similar manner one may adopt a stratified version of nested case-control sampling (Langholz and Borgan, 1995). For this design, called counter-matching, one applies information available for all cohort subjects to classify each individual at risk into one of S distinct strata. We denote by $\mathcal{R}_s(t)$ the subset of the risk set $\mathcal{R}(t)$ that belongs to stratum s, and let $n_s(t) = \#\mathcal{R}_s(t)$ be the number at risk in this stratum just before time t. If a failure occurs at time t, we want to sample our controls such that the sampled risk set will contain a specified number m_s of individuals from each stratum $s = 1, \ldots, S$. This is obtained as follows. Assume that an individual i who belongs to stratum $s(i)$ fails at t. Then for $s \neq s(i)$ one samples randomly without replacement m_s controls from $\mathcal{R}_s(t)$. From the case's stratum $s(i)$ only $m_{s(i)} - 1$ controls are sampled. The failing individual is, however, included in the sampled risk set $\widetilde{\mathcal{R}}(t)$, so this contains a total of m_s from each stratum. Even though it is not made explicit in the notation, we note that the classification into strata may be time-dependent; e.g., one may stratify according to the quartiles of a time-dependent surrogate measure of an exposure of main interest. It is crucial, however, that the information on which the stratification is based is known just before time t.

Inference for counter-matched nested case-control data may be based on a partial likelihood similar to (17.3); however, weights have to be inserted in the partial likelihood in order to reflect the different sampling probabilities. To see how this should be done, we take the approach of Section 17.7.1. To this end we note that, in probabilistic terms, the counter-matched design may be desribed as follows. Consider a set $\mathbf{r} \subset \mathcal{R}(t)$ with $i \in \mathbf{r}$ and assume that $\mathbf{r} \cap \mathcal{R}_s(t)$ is of size m_s for $s = 1, \ldots, S$. Then, if individual i fails at time t, the probability that $\mathbf{r}$ is selected as the sampled risk set becomes

$$\pi(\mathbf{r} \mid t, i) = \left\{ \binom{n_{s(i)}(t) - 1}{m_{s(i)} - 1} \prod_{s \neq s(i)} \binom{n_s(t)}{m_s} \right\}^{-1} = \frac{n_{s(i)}(t)}{m_{s(i)}} \left\{ \prod_{s=1}^{S} \binom{n_s(t)}{m_s} \right\}^{-1}.$$

When inserting these sampling probabilities in (17.16), we obtain after cancellation of common factors

$$\pi(i \mid t, \mathbf{r}) = \frac{Y_i(t) \exp(\boldsymbol{\beta}'\mathbf{x}_i) w_i(t)}{\sum_{k \in \mathbf{r}} Y_k(t) \exp(\boldsymbol{\beta}'\mathbf{x}_k) w_k(t)}, \tag{17.22}$$

where $w_k(t) = n_{s(k)}(t)/m_{s(k)}$. This yields the partial likelihood

$$L_{\rm cm}(\boldsymbol{\beta}) = \prod_{j=1}^{d} \pi(i_j \mid t_j, \widetilde{\mathcal{R}}(t_j)) = \prod_{j=1}^{d} \frac{\exp(\boldsymbol{\beta}'\mathbf{x}_{i_j}) w_{i_j}(t_j)}{\sum_{k \in \widetilde{\mathcal{R}}(t_j)} \exp(\boldsymbol{\beta}'\mathbf{x}_k) w_k(t_j)}. \tag{17.23}$$

Inference concerning the regression coefficients, using usual large sample likelihood methods, can be based on this weighted partial likelihood (Borgan et al., 1995; Langholz and Borgan, 1995). Moreover, software for Cox regression can be used to fit the model provided the software allows us to specify the logarithm of the weights as "offsets."

17.8.2 Estimation of absolute risk

In Sections 17.3.2 and 17.8.1 we discuss how to estimate the regression coefficients in (26.4) from simple and counter-matched nested case-control data. From these we immediately get estimates of the hazard ratios (relative risks) e^{β_j}. We will here indicate how we may estimate *absolute risks.*

Consider an individual with vector of covariates $\mathbf{x}$ who has not failed by time s. We here assume that the covariates are fixed, but note that the results may be generalized to external time-varying covariates (Langholz and Borgan, 1997). Assuming Cox's model $h(t \mid \mathbf{x}) = h_0(t)\exp(\boldsymbol{\beta}'\mathbf{x})$, the absolute risk (i.e., probability) that the individual will fail before time $t > s$ is given by

$$p(s, t \mid \mathbf{x}) = 1 - \exp\left\{-\exp(\boldsymbol{\beta}'\mathbf{x}) \int_s^t h_0(u)du\right\}. \tag{17.24}$$

To estimate the absolute risk, we need to estimate both the vector of regression coefficients $\boldsymbol{\beta}$ and the cumulative baseline hazard $H_0(t) = \int_0^t h_0(u)du$. The latter may be estimated by the Breslow type estimator (Borgan and Langholz, 1993; Borgan et al., 1995)

$$\widehat{H}_0(t) = \sum_{t_j \le t} \frac{1}{\sum_{k \in \widetilde{\mathcal{R}}(t_j)} \exp(\widehat{\boldsymbol{\beta}}'\mathbf{x}_k) w_k(t_j)}. \tag{17.25}$$

Here the weights $w_k(t_j)$ depend on the sampling scheme for controls. When controls are selected by simple random sampling we use the weights $w_k(t_j) = n(t_j)/m$, while for the counter-matched design the weights are given just below (17.22). Then (17.24) may be estimated by the Kaplan-Meier type estimator

$$\widehat{p}(s, t \mid \mathbf{x}) = 1 - \prod_{t_j \le t} \left\{1 - \exp(\widehat{\boldsymbol{\beta}}'\mathbf{x})\Delta\widehat{H}_0(t_j)\right\}, \tag{17.26}$$

where $\Delta\widehat{H}_0(t_j)$ is the increment of (17.25) at time t_j. Alternatively we may use the asymptotically equivalent estimator

$$\widetilde{p}(s, t \mid \mathbf{x}) = 1 - \exp\left\{\exp(\widehat{\boldsymbol{\beta}}'\mathbf{x})\left(\widehat{H}_0(t) - \widehat{H}_0(s)\right)\right\}. \tag{17.27}$$

Variance estimation and large sample properties of (17.26) are studied in Borgan and Langholz (1993) and Borgan et al. (1995). Generalizations of the absolute risk estimator (17.26) to competing risks and Markov chain models are studied in Langholz and Borgan (1997) and Borgan (2002), respectively. The estimator (17.27) may not be generalized in a similar way.

17.9 Case-cohort and IPW-estimators: absolute risk and alternative models

Absolute risk can be estimated from case-cohort data using an IPW or Horvitz-Thompson approach. Thus Prentice (1986) suggested to estimate the cumulative baseline hazard function $H_0(t)$ by the Breslow type estimator

$$\widetilde{H}_0(t) = \sum_{t_j \le t} \frac{1}{\sum_{k \in \widetilde{\mathcal{C}}(t_j)} \exp(\widehat{\boldsymbol{\beta}}'\mathbf{x}_k)(n/\widetilde{m})},$$

where $\widetilde{\mathcal{C}}(t_j)$ is the set of subcohort members at risk at time t_j and n and $\widetilde{m}$ are the sizes of the cohort and subcohort, respectively. The individual contributions to the denominator are thus weighted by the inverse of the subcohort sampling fraction $\widetilde{m}/n$. Large sample result for $\widetilde{H}_0(t)$ was developed by Self and Prentice (1988).

Alternatively one may base estimation on the full case-cohort sample consisting of all cases and the subcohort using

$$\widehat{H}_0(t) = \sum_{t_j \le t} \frac{1}{\sum_{k \in \widetilde{\mathcal{S}}(t_j)} \exp(\widehat{\boldsymbol{\beta}}'\mathbf{x}_k) w_k}$$

with $\widetilde{\mathcal{S}}(t_j)$ and w_k as in (17.8). This formula immediately generalizes to stratified case-cohort studies (Section 17.4.4) and to nested case-control studies with estimated weights $w_k = 1/p_k$ (Section 17.6). Large sample theory for such estimators for stratified case-cohort studies was provided by Kulich and Lin (2004) and Breslow and Wellner (2007). One may then estimate the absolute risk (17.24) by plug-in rules corresponding to (17.26) or (17.27).

The Horvitz-Thompson approach can quite generally be applied when fitting failure time models other than the proportional hazards model. For instance Kalbfleisch and Lawless (1988) worked with fully parametrically specified failure time models with individual likelihood contributions $l_k(\boldsymbol{\theta})$ (with model parameters $\boldsymbol{\theta}$) suggesting to maximize a weighted likelihood

$$\widetilde{l}(\boldsymbol{\theta}) = \sum_{k \in \mathcal{S}} l_k(\boldsymbol{\theta}) w_k$$

under case-cohort and other sampling plans, where $\mathcal{S}$ is the set of individuals (cases and controls) sampled. Kalbfleisch and Lawless (1988) also suggested large sample properties based on $\widetilde{l}(\boldsymbol{\theta})$ similar to (17.10) and with a derivation along the lines of Section 17.7.2.

Kulich and Lin (2000) considered an additive hazard model $h_i(t) = h_0(t) + \boldsymbol{\beta}'\mathbf{x}_i$ and developed a weighted version of the Lin-Ying estimator (Lin and Ying, 1994) under a stratified case-cohort design. They thereby developed parallel results to those presented in Borgan et al. (2000) for Cox regression.

Several other models have been addressed with various sampling plans and similar IPW approaches, such as proportional odds models (Chen, 2001*a*), accelerated failure time models (Kong and Cai, 2009), semi-parametric transformation models (e.g., Kong et al., 2004; Lu and Tsiatis, 2006; Chen et al., 2012) and correlated failure time models (Lu and Shih, 2006; Kang and Cai, 2009). When we weight the cases and controls with IPW, the underlying idea is to reconstruct the full cohort by letting each control represent a number of the individuals not sampled. This has the implication that methods developed for cohort data can usually be modified to case-cohort data and other types of case-control data where inclusion probabilities can be calculated. Often large sample results follow in the same vein as for case-cohort data (see Section 17.7.2), but special care may have to be taken to account for the structure of the estimating equations or sampling plan.

17.10 Maximum likelihood estimation

In the previous sections we have estimated the regression coefficients using partial likelihoods and weighted pseudo-likelihoods; see, e.g., (17.3), (17.6), (17.8), and (17.23). Then we only use data for the cases and the sampled controls/subcohort. Another approach is a full maximum likelihood solution, where the entire cohort is used in the estimation and unobserved covariates are treated as missing data. We here give an outline of this approach and discuss some of its strengths and limitations.

For ease of presentation we restrict attention right-censored survival data, see Saarela et al. (2008) and Støer and Samuelsen (2012) for a discussion on how the results may be extended to cover left-truncation as well. The situation is as follows. We consider a cohort $\mathcal{C} = \{1, \ldots, n\}$ of n independent individuals. Each individual is observed until failure or censoring. For individual $i \in \mathcal{C}$ we denote by v_i the time of failure or censoring, and let δ_i be an indicator taking the value 1 if the individual is observed to fail at v_i and the value 0 if the individual is censored. We now assume that some covariates are observed for all individuals in the cohort, and we denote by $\mathbf{x}_i^{\mathrm{o}}$ the vector of these covariates for the ith individual. Additional covariate information is collected for all cases and for a number of controls. The controls may be selected by nested case-control or case-cohort sampling. We let O_i be a case-control indicator that is 1 if individual i is a case or a control and 0 otherwise. Then $\mathcal{O} = \{i \in \mathcal{C} : O_i = 1\}$ is the set of individuals for whom additional covariate data are obtained. We denote by $\mathbf{x}_i^{\mathrm{u}}$ the vector of these additional covariates for the ith individual. Thus the available data for all cohort members (i.e., for $i \in \mathcal{C}$) are $(v_i, \delta_i, \mathbf{x}_i^{\mathrm{o}})$, while for all cases and controls (i.e., for $i \in \mathcal{O}$) we additionally observe $\mathbf{x}_i^{\mathrm{u}}$.

In order to derive the likelihood for the data, we assume that all covariates are time-fixed and make the following assumptions (Saarela et al., 2008):

(i) The random vectors $(v_i, \delta_i, \mathbf{x}_i^{\mathrm{o}}, \mathbf{x}_i^{\mathrm{u}})$; $i \in \mathcal{C}$; are independent.

(ii) The conditional distribution of the vector $(O_1, \ldots, O_n)$ of case-control indicators depends only on data observed for all $i \in \mathcal{C}$.

Assumption (i) is common in the survival analysis literature, but stronger than the independent censoring assumption imposed in the previous sections. Assumption (ii) ensures that the $\mathbf{x}_i^{\mathrm{u}}$ are missing at random for $i \in \mathcal{C} \setminus \mathcal{O}$, and the assumption is fulfilled for the nested case-control and case-cohort designs. Assuming (i) and (ii) the likelihood takes the same form regardless of what kind of sampling scheme has been used to determine the set $\mathcal{O}$; see Saarela et al. (2008) for a detailed argument. The likelihood becomes (conditional on the covariates $\mathbf{x}_i^{\mathrm{o}}$):

$$
\begin{aligned}
L = & \prod_{i \in \mathcal{O}} p\left(v_i, \delta_i \,|\, \mathbf{x}_i^{\mathrm{u}}, \mathbf{x}_i^{\mathrm{o}}\right) dG\left(\mathbf{x}_i^{\mathrm{u}} \,|\, \mathbf{x}_i^{\mathrm{o}}\right) \\
& \times \prod_{i \in \mathcal{C} \setminus \mathcal{O}} \int_{\mathcal{X}_{\mathrm{u}}} p\left(v_i, \delta_i \,|\, \mathbf{x}_i^{\mathrm{u}}, \mathbf{x}_i^{\mathrm{o}}\right) dG\left(\mathbf{x}_i^{\mathrm{u}} \,|\, \mathbf{x}_i^{\mathrm{o}}\right).
\end{aligned} \tag{17.28}
$$

Here $p\left(v_i, \delta_i \,|\, \mathbf{x}_i^{\mathrm{u}}, \mathbf{x}_i^{\mathrm{o}}\right)$ is the conditional density of (v_i, δ_i) given $(\mathbf{x}_i^{\mathrm{u}}, \mathbf{x}_i^{\mathrm{o}})$, and $G\left(\mathbf{x}_i^{\mathrm{u}} \,|\, \mathbf{x}_i^{\mathrm{o}}\right)$ is the conditional distribution of $\mathbf{x}_i^{\mathrm{u}}$ given $\mathbf{x}_i^{\mathrm{o}}$. Further the integral in (17.28) is over the space $\mathcal{X}_{\mathrm{u}}$ of all possible values of the covariate vector $\mathbf{x}_i^{\mathrm{u}}$.

To achieve a full maximum likelihood solution, we need to specify the conditional distributions in (17.28). We assume a Cox model for the hazard of the ith individual, i.e.,

$$h_i(t) = h_0(t) \exp\left(\beta_{\mathrm{o}}' \mathbf{x}_i^{\mathrm{o}} + \beta_{\mathrm{u}}' \mathbf{x}_i^{\mathrm{u}}\right). \tag{17.29}$$

Then the conditional density of (v_i, δ_i) given $(\mathbf{x}_i^{\mathrm{u}}, \mathbf{x}_i^{\mathrm{o}})$ takes the form

$$p\,(v_i, \delta_i \,|\, \mathbf{x}_i^{\mathrm{u}}, \mathbf{x}_i^{\mathrm{o}}) = \{h_0(v_i) \exp(\beta_{\mathrm{o}}' \mathbf{x}_i^{\mathrm{o}} + \beta_{\mathrm{u}}' \mathbf{x}_i^{\mathrm{u}})\}^{\delta_i} \tag{17.30}$$

$$\times \exp\left\{ - \exp(\beta_{\mathrm{o}}' \mathbf{x}_i^{\mathrm{o}} + \beta_{\mathrm{u}}' \mathbf{x}_i^{\mathrm{u}}) \int_0^{v_i} h_0(u) du \right\}.$$

In (17.29) we may choose between a non-parametric specification of the baseline hazard (as in the ordinary Cox regression model) or a parametric specification, e.g., assuming a Weibull baseline $h_0(t) = \alpha\kappa(\alpha t)^{\kappa-1}$. For a full likelihood solution, we also need to specify the conditional distribution of $\mathbf{x}_i^{\mathrm{u}}$ given $\mathbf{x}_i^{\mathrm{o}}$. Again we may adopt a non-parametric or a parametric specification.

Inspired by Kulathinal and Arjas (2006), Saarela et al. (2008) take a fully parametric approach by assuming a parametric baseline hazard and a parametric specification of the conditional distribution of $\mathbf{x}_i^{\mathrm{u}}$ given $\mathbf{x}_i^{\mathrm{o}}$. The logarithm of the likelihood (17.28) may then be maximized by a suitable optimization routine. The optimization may be very time consuming, however, since the integrals in (17.28) typically will have to be evaluated numerically or by Monte Carlo integration. But substantial computational savings may be obtained by grouping individuals in $\mathcal{C} \setminus \mathcal{O}$ with similar values of v_i and $\mathbf{x}_i^{\mathrm{o}}$ (Støer and Samuelsen, 2012).

Scheike and Juul (2004) consider a semi-parametric model for nested case-control data, where the baseline hazard in (17.29) and the distribution of $\mathbf{x}_i^{\mathrm{u}}$ given $\mathbf{x}_i^{\mathrm{o}}$ are given non-parametric specifications. When maximizing the likelihood using the EM algorithm, it is assumed that the cumulative baseline hazard is a step function with jumps at the observed failure times and that the distribution of $\mathbf{x}_i^{\mathrm{u}}$ given $\mathbf{x}_i^{\mathrm{o}}$ has point masses at the observed covariate values. A similar approach for case-cohort data is considered by Scheike and Martinussen (2004). We should also mention that Zeng and Lin (2007) have developed a general framework for maximum likelihood estimation in semi-parametric regression models for censored data, which may be adopted to nested case-control and case-cohort data (Zeng et al., 2006).

Maximum likelihood estimation may give an efficiency gain compared to estimation based on partial likelihoods and weighted pseudo-likelihoods. The gain may be substantial for estimation of the effect of $\mathbf{x}_i^{\mathrm{o}}$. When $\mathbf{x}_i^{\mathrm{o}}$ and $\mathbf{x}_i^{\mathrm{u}}$ are correlated, one may also get an improved estimation of the effect of $\mathbf{x}_i^{\mathrm{u}}$; see e.g., Støer and Samuelsen (2012). However, the increased efficiency comes at a cost. The computations may be quite extensive, and the maximum likelihood approach is vulnerable to misspecification of the conditional distribution of $\mathbf{x}_i^{\mathrm{u}}$ given $\mathbf{x}_i^{\mathrm{o}}$ (Støer and Samuelsen, 2012). At the time of writing of this chapter, there is limited practical experience with the full maximum likelihood approach for nested case-control and case-cohort data. So even though the methodology shows promise, it is too early to say if it will be of importance for epidemiological practice in the future.

17.11 Closing remarks

Nested case-control and case-cohort designs are increasingly being used in epidemiology and biomarker studies. In this chapter we have surveyed methods for analysing nested case-control and case-cohort data. In our survey we have chosen to focus on partial likelihood and weighted pseudo likelihood methods for Cox regression when the controls/subcohort are selected by simple or stratified random sampling. This choice is motivated by our belief that these are the methods that are most useful for practitioners. But some other material, like absolute risk estimation, IPW estimation for general failure time models, and maximum likelihood estimation for sampled cohort data, have also been discussed. However, many important topics are not discussed in our survey, and in this final section we mention some of them.

For nested case-control studies, we have discussed simple and counter-matched sampling of controls. However, the counting process framework of Section 17.7.1 allows for quite general control sampling schemes. Two examples are quota sampling and counter-matched sampling with additional randomly sampled controls (Borgan et al., 1995). The partial likelihood for a specific sampling scheme may be derived in a similar manner as (17.23). All that is needed is to use the appropriate sampling probabilities in the partial likelihood (Langholz and Goldstein, 1996; Langholz, 2007).

Instead of a case-cohort design with a sampled subcohort, one can sometimes take a sample only of the individuals who do not fail. This has been referred to as a "classical case-control" design (Chen and Lo, 1999; Chen, 2001*b*). Such data can be analysed with an IPW-approach with the limiting distribution given by (17.9) under the proportional hazards model (26.4).

When studying a rare disease, one will commonly include all cases in a nested case-control or case-cohort study, and we have made this assumption throughout. However, if the disease is more common, it could be useful not to include all cases. If we in a case-cohort study (with a randomly selected subcohort) have a random sample of the cases, the pseudo likelihood (17.6) may be adopted without specifying the sampling fraction of cases. The weighted likelihood (17.8) can also be extended to allow for a random sample of cases. But then the sampling fraction (or an estimate of it) needs to be inserted in (17.8). An additional term also needs to be added to the large sample covariance matrix and its estimate (Chen and Lo, 1999; Breslow and Wellner, 2007; Gray, 2009). For nested case-control studies, Langholz and Borgan (1995) provide a brief discussion of case sampling.

The weights in a counter-matched nested case-control study will change over time according to the numbers at risk in the various strata. In a similar manner, and in the spirit of Barlow (1994) and Lin and Ying (1994), Borgan et al. (2000) proposed time-dependent weighting schemes for stratified case-cohort sampling. So at time t the weights $w_k = n_s^0/\widetilde{m}_s^0$ in (17.8) are replaced by the number of non-case cohort members in stratum s who are *at risk* at t divided by the number of non-case subcohort members in the stratum who are *at risk* at that time. Kulich and Lin (2004) also considered time-dependent weights and developed large sample results for these also incorporating auxiliary cohort information. However, unless incidence or censoring is strongly dependent on covariates, this modification will not increase efficiency much.

The estimators for case-cohort data based on (17.6) and (17.8) are not fully efficient, and it is then a question of how much one may lose by using these estimators compared to fully efficient estimators. This question has been addressed by Nan et al. (2004) and Nan (2004). They consider the situation where the only covariates in (26.4) are the ones observed for the cases and the subcohort members, and compare the asymptotic variances of the case-cohort estimators with the asymptotic information bound and with an efficient estimator for case-cohort data, respectively. Their results indicate that the efficiency loss is modest when the proportion of cases in the cohort is fairly small (not more than 10%, say) and the sampling fraction is at least as large as the proportion of cases. We are not aware of corresponding studies for nested case-control data, but conjecture that similar results hold here as well.

In our discussion of estimation of regression coefficients and relative risks, we have assumed that the regression model (26.4) is correctly specified. However, any model is an approximation to reality and, even when the model does not fully capture the relation between the covariates and the hazard, it may still provide a useful framework for summarizing covariate effects. But this makes it important to understand the behaviour of parameter estimates when the presumed model is misspecified. Struthers and Kalbfleisch (1986) and Hjort (1992) have studied the behaviour of the maximum partial likelihood estimator for cohort data, and they show that when the model is misspecified $\widehat{\boldsymbol{\beta}}$ will converge in probability to a "least false parameter" $\boldsymbol{\beta}^*$. Their arguments go through with only minor modifications for the case-cohort estimators of Section 17.4 and the IPW estimator for nested case-control data of Section 17.6, so for a misspecified model these estimators will also converge to $\boldsymbol{\beta}^*$. However, the traditional estimator $\widehat{\boldsymbol{\beta}}_{\text{ncc}}$ for nested case-control data obtained from (17.3), will converge to a "least false parameter" $\boldsymbol{\beta}^*_{\text{ncc}}$ that differs from $\boldsymbol{\beta}^*$ (Xiang and Langholz, 1999). As a consequence, the traditional approach to nested case-control data may give estimates that differ systematically from those of the corresponding cohort study. However, the magnitude of the misspecification must be quite large in order to produce bias of practical importance (Xiang and Langholz, 1999).

Bibliography

Aalen, O. O., Borgan, Ø. and Gjessing, H. K. (2008), *Survival and Event History Analysis: A Process Point of View*, Springer-Verlag, New York.

Andersen, P. K. and Gill, R. D. (1982), 'Cox's regression model for counting processes: A large sample study', *Annals of Statistics* **10**, 1100–1120.

Barlow, W. E. (1994), 'Robust variance estimation for the case-cohort design', *Biometrics* **50**, 1064–1072.

Barlow, W. E., Ichikawa, L., Rosner, D. and Izumi, S. (1999), 'Analysis of case-cohort designs', *Journal of Clinical Epidemiology* **52**, 1165–1172.

Borgan, Ø. (2002), 'Estimation of covariate-dependent Markov transition probabilities from nested case-control data', *Statistical Methods in Medical Research* **11**, 183–202. Correction **12**:124.

Borgan, Ø., Goldstein, L. and Langholz, B. (1995), 'Methods for the analysis of sampled cohort data in the Cox proportional hazards model', *Annals of Statistics* **23**, 1749–1778.

Borgan, Ø. and Langholz, B. (1993), 'Non-parametric estimation of relative mortality from nested case-control studies', *Biometrics* **49**, 593–602.

Borgan, Ø., Langholz, B., Samuelsen, S. O., Goldstein, L. and Pogoda, J. (2000), 'Exposure stratified case-cohort designs', *Lifetime Data Analysis* **6**, 39–58.

Borgan, Ø. and Olsen, E. F. (1999), 'The efficiency of simple and counter-matched nested case-control sampling', *Scandinavian Journal of Statistics* **26**, 493–509.

Breslow, N. E. (1996), 'Statistics in epidemiology: The case-control study', *Journal of the American Statistical Association* **91**, 14–28.

Breslow, N. E., Lumley, T., Ballantyne, C. M., Chambless, L. E., and Kulich, M. (2009*a*), 'Using the whole cohort in the analysis of case-cohort data', *American Journal of Epidemiology* **169**(11), 1398–1405.

Breslow, N. E., Lumley, T., Ballantyne, C. M., Chambless, L. E. and Kulich, M. (2009*b*), 'Improved Horvitz-Thompson estimation of model parameters from two-phase stratified samples: Applications in epidemiology', *Statistics in Biosciences* **1**, 32–49.

Breslow, N. E. and Wellner, J. A. (2007), 'Weighted likelihood for semiparametric models and two-phase stratified samples, with application to Cox regression', *Scandinavian Journal of Statistics* **34**, 86–102. Correction **35**:186-192.

Cai, T. and Zheng, Y. (2012), 'Evaluating prognostic accuracy of biomarkers in nested case-control studies', *Biostatistics* **13**, 89–100.

Chen, H. Y. (2001*a*), 'Weighted semiparametric likelihood methods for fitting a proportional odds regression model to data from the case-cohort design', *Journal of the American Statistical Association* **96**, 1446–1457.

Chen, K. (2001*b*), 'Generalized case-cohort estimation', *Journal of the Royal Statistical Society: Series B (Statistical Methodology)* **63**, 791–809.

Chen, K., Liuquan, S. and Xiangwei, T. (2012), 'Analysis of cohort survival data with transformation model', *Statistica Sinica* **22**, 489–508.

Chen, K. and Lo, S.-H. (1999), 'Case-cohort and case-control analysis with Cox's model', *Biometrika* **86**, 755–764.

Goldstein, L. and Langholz, B. (1992), 'Asymptotic theory for nested case-control sampling in the Cox regression model', *Annals of Statistics* **20**, 1903–1928.

Gray, R. J. (2009), 'Weighted analyses for cohort sampling designs', *Lifetime Data Analysis* **15**, 24–40.

Hjort, N. L. (1992), 'On inference in parametric survival data models', *International Statistical Review* **60**, 355–387.

Hrubec, Z., Boice, Jr., J. D., Monson, R. R. and Rosenstein, M. (1989), 'Breast cancer after multiple chest fluoroscopies: second follow-up of Massachusetts women with tuberculosis', *Cancer Research* **49**, 229–234.

Kalbfleisch, J. D. and Lawless, J. F. (1988), 'Likelihood analysis of multi-state models for disease incidence and mortality', *Statistics in Medicine* **7**, 149–160.

Kalbfleisch, J. D. and Prentice, R. L. (2002), *The Statistical Analysis of Failure Time Data*, 2nd edn, Wiley, Hoboken, New Jersey.

Kang, S. and Cai, J. (2009), 'Marginal hazards regression for retrospective studies within cohort with possibly correlated failure time data', *Biometrics* **65**, 405–414.

Kong, C. and Cai, J. (2009), 'Case-cohort analysis with accelerated failure time model', *Biometrics* **65**, 135–142.

Kong, C., Cai, J. and Sen, P. K. (2004), 'Fitting semiparametric transformation regression models to data from a modified case-cohort design', *Biometrika* **88**, 255–268.

Kulathinal, S. and Arjas, E. (2006), 'Bayesian inference from case-cohort data with multiple end-points', *Scandinavian Journal of Statistics* **33**, 25–36.

Kulich, M. and Lin, D. Y. (2000), 'Additive hazards regression for case-cohort studies', *Biometrika* **87**, 73–87.

Kulich, M. and Lin, D. Y. (2004), 'Improving the efficiency of relative-risk estimation in case-cohort studies', *Journal of the American Statistical Association* **99**, 832–844.

Kupper, L. L., McMichael, A. J. and Spirtas, R. (1975), 'A hybrid epidemiologic study design useful in estimating relative risk', *Journal of the American Statistical Association* **70**, 524–528.

Langholz, B. (2007), 'Use of cohort information in the design and analysis of case-control studies', *Scandinavian Journal of Statistics* **34**, 120–136.

Langholz, B. and Borgan, Ø. (1995), 'Counter-matching: A stratified nested case-control sampling method', *Biometrika* **82**, 69–79.

Langholz, B. and Borgan, Ø. (1997), 'Estimation of absolute risk from nested case-control data', *Biometrics* **53**, 767–774. Correction **59**:451.

Langholz, B. and Goldstein, L. (1996), 'Risk set sampling in epidemiologic cohort studies', *Statistical Science* **11**, 35–53.

Langholz, B. and Jiao, J. (2007), 'Computational methods for case-cohort studies', *Biometrics* **51**, 3737–3748.

Langholz, B. and Thomas, D. C. (1991), 'Efficiency of cohort sampling designs: some surprising results', *Biometrics* **47**, 1563–1571.

Lehmann, E. (1975), *Nonparametrics*, Holden-Day, San Francisco.

Lin, D. Y. and Ying, Z. (1993), 'Cox regression with incomplete covariate measurements', *Journal of the American Statistical Association* **88**, 1341–1349.

Lin, D. Y. and Ying, Z. (1994), 'Semiparametric analysis of the additive risk model', *Biometrika* **81**, 61–71.

Lu, S. and Shih, J. H. (2006), 'Case-cohort design and analysis for clustered failure time data', *Biometrics* **62**, 1138–1148.

Lu, W. B. and Tsiatis, A. A. (2006), 'Semiparametric transformation models for the case-cohort study', *Biometrika* **93**, 207–214.

Miettinen, O. (1982), 'Design options in epidemiologic research: An update', *Scandinavian Journal of Work, Environment & Health* **8** (supplement 1), 7–14.

Nan, B. (2004), 'Efficient estimation for case-cohort data', *Canadian Journal of Statistics* **32**, 403–419.

Nan, B., Emond, M. J. and Wellner, J. A. (2004), 'Information bounds for Cox regression models with missing data', *Annals of Statistics* **32**, 723–753.

Oakes, D. (1981), 'Survival times: Aspects of partial likelihood (with discussion)', *International Statistical Review* **49**, 235–264.

Prentice, R. L. (1986), 'A case-cohort design for epidemiologic cohort studies and disease prevention trials', *Biometrika* **73**, 1–11.

Prentice, R. L. and Breslow, N. E. (1978), 'Retrospective studies and failure time models', *Biometrika* **65**, 153–158.

Preston, D. L., Mattsson, A., Holmberg, E., Shore, R. Hildrethe, N. G. and Boice, Jr., J. D. (2002), 'Radiation effects on breast cancer risk: A pooled analysis of eight cohorts', *Radiation Research* **158**, 220–235.

Robins, J. M., Rotnitzky, A. and Zhao, L. P. (1994), 'Estimation of regression-coefficients when some regressors are not always observed', *Journal of the American Statistical Association* **89**, 846–866.

Rundle, A. G., Vineis, P. and Ahsan, H. (2005), 'Design options for molecular epidemiology research within cohort studies', *Cancer Epidemiology, Biomarkers & Prevention* **14**, 1899–1907.

Saarela, O., Kulathinal, S., Arjas, E. and Läärä, E. (2008), 'Nested case-control data utilized for multiple outcomes: A likelihood approach and alternatives', *Statistics in Medicine* **27**, 5991–6008.

Salim, A., Hultman, C., Sparén, P. and Reilly, M. (2009), 'Combining data from 2 nested case-control studies of overlapping cohorts to improve efficiency', *Biostatistics* **10**, 70–79.

Samuelsen, S. O. (1989), Two incompleted data problems in life-history analysis: Double censoring and the case-cohort design, PhD thesis, University of Oslo.

Samuelsen, S. O. (1997), 'A pseudolikelihood approach to analysis of nested case-control studies', *Biometrika* **84**, 379–394.

Samuelsen, S. O., Ånestad, H. and Skrondal, A. (2007), 'Stratified case-cohort analysis of general cohort sampling designs', *Scandinavian Journal of Statistics* **34**, 103–119.

Scheike, T. H. and Juul, A. (2004), 'Maximum likelihood estimation for Cox's regression model under nested case-control sampling', *Biostatistics* **5**, 193–206.

Scheike, T. H. and Martinussen, T. (2004), 'Maximum likelihood estimation for Cox's regression model under case-cohort sampling', *Scandinavian Journal of Statistics* **31**, 283–293.

Self, S. G. and Prentice, R. L. (1988), 'Asymptotic distribution theory and efficiency results for case-cohort studies', *Annals of Statistics* **16**, 64–81.

Støer, N. and Samuelsen, S. O. (2012), 'Comparison of estimators in nested case-control studies with multiple outcomes', *Lifetime Data Analysis* **18**, 261–283.

Struthers, C. A. and Kalbfleisch, J. D. (1986), 'Misspecified proportional hazard models', *Biometrika* **73**, 363–369.

Suissa, S., Edwardes, M. D. D. and Boivin, J. F. (1998), 'External comparisons from nested case-control designs', *Epidemiology* **9**, 72–78.

Therneau, T. M. and Li, H. (1999), 'Computing the Cox model for case-cohort designs', *Lifetime Data Analysis* **5**, 99–112.

Thomas, D. C. (1977), 'Addendum to: "Methods of cohort analysis: appraisal by application to asbestos mining," by F. D. K. Liddell, J. C. McDonald and D. C. Thomas', *Journal of the Royal Statistical Society: Series A (General)* **140**, 469–491.

Xiang, A. H. and Langholz, B. (1999), 'Comparison of case-control to full cohort analysis under model misspecification', *Biometrika* **86**, 221–226.

Zeng, D. and Lin, D. Y. (2007), 'Maximum likelihood estimation in semiparametric regression models with censored data (with discussion)', *Journal of the Royal Statistical Society: Series B (Statistical Methodology)* **69**, 507–564.

Zeng, D., Lin, D. Y., Avery, C. L., North, K. E. and Bray, M. S. (2006), 'Efficient semiparametric estimation of haplotype-disease associations in case-cohort and nested case-control studies', *Biostatistics* **7**, 486–502.

18

Interval Censoring

Jianguo Sun
Department of Statistics, University of Missouri

Junlong Li
Department of Biostatistics, Harvard University

CONTENTS

18.1	Introduction	369
18.2	Likelihood function and an example	371
18.3	Current status data	373
18.4	Univariate interval-censored data	374
18.5	Multivariate interval-censored data	378
18.6	Competing risks interval-censored data	380
18.7	Informatively interval-censored data	381
18.8	Other types of interval-censored data	382
18.9	Software and concluding remarks	383
	Bibliography	383

The literature on the statistical analysis of interval-censored failure time data has grown rapidly in last twenty years or so and among others, one relatively complete review is given by Sun (2006), one of only two books currently available on interval-censored data. The other book is the edited volume given by Chen et al. (2012). Interval-censored data includc the usual right-censored failure time data as a special case, but have much more complex structure and provide less relevant information than the right-censored data. We will discuss several types of interval-censored data including univariate interval-censored data, multivariate interval-censored data and competing risks interval-censored data. For each topic, the focus will be on the discussion of some basic concepts and issues that commonly occur in the analysis of such data and the review of some recent advances or literature, mainly after Sun (2006).

18.1 Introduction

Interval-censored failure time data are a special type of failure time data, which involve interval censoring and have drawn more and more attention during last 20 years or so. It is well known that one essential and special feature of failure time data is censoring and there exist different types of censoring. Among them, the type that has been studied most in the literature is right censoring and an extensive literature including many textbooks have been established for the analysis of right-censored failure time data (Kalbfleisch and

Prentice, 2002; Klein and Moeschberger, 2002). In contrast, the literature on the analysis of interval-censored failure time data is quite limited and in particular, there currently exist only two books on it: one is Sun (2006), which gives a relatively complete review of the literature, and the other is Chen et al. (2012), an edited volume. In addition, a couple of review papers have been published including Gómez et al. (2009) and Zhang and Sun (2010*b*). In the context of failure time data, interval censoring means that the failure time variable of interest is observed or known only to lie within some intervals or windows instead of being observed exactly (Finkelstein, 1986; Kalbfleisch and Prentice, 2002; Sun, 2006). If the interval includes only or reduces to a single time point, one obtains the exact failure time.

One field that often produces interval-censored failure time data is medical or health studies that entail periodic follow-ups. In this situation, an individual due for the prescheduled observations for a clinically observable change in disease or health status may miss some observations and return with a changed status. Accordingly, we only know that the true event time is greater than the last observation time at which the change has not occurred and less than or equal to the first observation time at which the change has been observed to occur, thus giving an interval which contains the real (but unobserved) time of occurrence of the change.

A more specific example of interval-censored data arises in the acquired immune deficiency syndrome (AIDS) trials (De Gruttola and Lagakos, 1989) that, for example, are interested in times to AIDS for human immunodeficiency virus (HIV) infected subjects. In these cases, the determination of AIDS onset is usually based on blood testing, which can be performed obviously only periodically but not continuously. In consequence, only interval-censored data may be available for AIDS diagnosis times. A similar case is for studies on HIV infection times. If a patient is HIV positive at the beginning of a study, then the HIV infection time is usually determined by a retrospective analysis of his or her medical history. Therefore, we are only able to obtain an interval given by the last HIV negative test date and the first HIV positive test date for the HIV infection time.

In reality, interval censoring can occur in different forms and each form represents one type of interval-censored failure time data. Among them, an important type of interval-censored failure time data is the so-called current status data (Jewell and van der Laan, 1995; Sun and Kalbfleisch, 1993). This type of interval censoring means that each subject is observed only once for the status of the occurrence of the failure event of interest. In other words, one does not directly observe the occurrence of the failure event of interest, but instead only knows the observation time and whether or not the event has occurred at the time. In consequence, the failure time is either left- or right-censored. One type of studies that usually produce current status data is cross-sectional studies on failure events (Keiding, 1991). Another type is tumorigenicity studies and in this situation, the time to tumor onset is usually of interest, but not directly observable (Dinse and Lagakos, 1983). In these cases, one only knows or observes the exact value of the observation time, which is usually the death or sacrifice time of the subject. Note that for the first example, current status data occur due to the study design, while for the second case, they are often observed because of the inability of measuring the failure variable directly and exactly. Sometimes we also refer current status data to as case I interval-censored data and the general case as case II interval-censored data (Groeneboom and Wellner, 1992).

Another type of interval-censored data is the so-called doubly censored data (De Gruttola and Lagakos, 1989; Sun, 2002). By this, we mean that the failure time of interest is defined as or represents the time between two related events and the observed data on the times to the occurrences of both events are interval-censored. In contrast, the interval-censored data discussed above can be regarded as a special case of such doubly censored data in which one observes the times to the first event exactly and thus can treat them being zero for simplicity. An example of doubly censored data is provided by the AIDS studies discussed above when the variable of interest is AIDS incubation time (De Gruttola and Lagakos, 1989), the time from HIV infection to AIDS diagnosis, with both the HIV infection time and the AIDS diagnosis time being right- or interval-censored.

Grouped failure time data, which will not be discussed here, is another special case of interval-censored data and often arise in, for example, large animal studies. By grouped failure time data, we usually mean that the intervals for any two subjects either are completely identical or have no overlapping. It is easy to see that for the analysis of such data, one could readily employ the methods available for right-censored data and no new methods are needed in theory. In other words, the statistical inference about grouped failure time data is relatively straightforward. In the following, the focus will be on general interval-censored data in which the observed intervals for the failure times of interest may overlap in any way.

The remainder of this chapter is organized as follows. We will begin in Section 18.2 with describing the commonly used likelihood function and the fundamental and important assumption behind it, noninformative interval censoring. It means that the censoring mechanism does not contribute to the likelihood function. A specific example of interval-censored data is then provided to illustrate what was discussed. Sections 18.3-18.8 will be organized according to the types or structures of the data. For completeness, we will first briefly discuss in Section 18.3 the analysis of univariate current status data, as the next chapter will provide more details on it. Section 18.4 will consider the analysis of general or case II univariate interval-censored data followed by the analysis of multivariate interval-censored failure time data such as bivariate failure time data in Section 18.5. Section 18.6 will investigate the analysis of competing risks interval-censored failure time data and the analysis of informatively interval-censored data will be the focus of Section 18.7. Section 18.8 will briefly cover a few other types of interval-censored data that are not touched on above. These include doubly censored data, interval-censored data from multi-state models and interval-censored data with missing or mismeasured covariates. In all of the sections, the discussion will be mainly on the existing literature on three basic topics in analyzing failure time data: nonparametric estimation of a survival function, nonparametric comparison of survival functions and regression analysis, with the focus on some of the recent advances primarily after Sun (2006). To conclude, we will briefly review some software packages, especially *R* packages, available for the analysis of interval-censored data and give some general remarks in Section 18.9.

18.2 Likelihood function and an example

To describe the likelihood function, we will first define some notation. In the following, we will use T to denote the failure time of interest. By saying that T is interval-censored, we mean that only available information for T is an interval denoted by $I = (L, R]$ such that $T \in I$. Using this notation, we see that current status data correspond to the situation where either $L = 0$ or $R = \infty$. Interval-censored data reduce to right-censored data if $L = R$ or $R = \infty$ for all subjects in the study. Note that a more general way to describe interval-censored data is to assume that the observation on T is given by a group of intervals (Turnbull, 1976). However, we will not discuss this general representation as the resulting likelihood functions in both cases have essentially the same structure.

Now suppose that there is a failure time study consisting of n independent subjects and let T_i and $I_i = (L_i, R_i]$ be defined as above but associated with subject i. Define $F(t) = P(T \leq t)$, the cumulative distribution function of T, and $S(t) = 1 - F(t)$, the survival function of T. Let $0 = t_0 < t_1 < ... < t_m < t_{m+1} = \infty$ denote the unique ordered elements of $\{0, \{L_i\}_{i=1}^n, \{R_i\}_{i=1}^n, \infty\}$, α_{ij} the indicator of the event $(t_{j-1}, t_j] \subseteq I_i$, and $p_j = S(t_{j-1}) - S(t_j)$. Then for inference about S or for $\mathbf{p} = (p_1, ..., p_{m+1})'$, the likelihood function that is commonly used has the form

$$L_S(\mathbf{p}) = \prod_{i=1}^{n} \left[S(L_i) - S(R_i) \right] = \prod_{i=1}^{n} \sum_{j=1}^{m+1} \alpha_{ij}\, p_j \,. \tag{18.1}$$

In reality, there may exist some covariates denoted by Z and in this case, the likelihood function above becomes

$$L_S(\mathbf{p} \,|\, Z_i's) = \prod_{i=1}^{n} \left[S(L_i|Z_i) \;-\; S(R_i \,|\, Z_i) \right] = \prod_{i=1}^{n} \sum_{j=1}^{m+1} \alpha_{ij}\, p_j(Z_i) \,. \tag{18.2}$$

Of course, the likelihood functions given above come with some assumptions. The most fundamental and important one is perhaps the so-called "noninformative interval censoring," which can be described by the following equality Sun (2006)

$$P(T \le t | L = l, R = r, L < T \le R) = P(T \le t | l < T \le r) \,. \tag{18.3}$$

The assumption above essentially says that, except for the fact that T lies between l and r which are the realizations of L and R, the interval $(L, R]$ (or equivalently its endpoints L and R) does not provide any extra information for T. In other words, the probabilistic behavior of T remains the same except that the original sample space $T \ge 0$ is now reduced to $l = L < T \le R = r$. In the existence of covariates, the assumption (18.3) becomes

$$P(T \le t | L = l, R = r, L < T \le R, Z = z) \;=\; P(T \le t | l < T \le r, Z = z) \,. \tag{18.4}$$

One could also employ different ways to characterize the noninformative interval censoring assumption. For example, one can use a stochastic process to describe the underlying interval censoring mechanism by assuming that there exists a sequence of observation times or an observation process. Then the noninformative assumption means that the process is independent of the failure time or process of interest (Groeneboom and Wellner, 1992; Lawless and Babineau, 2006). In practice, one question of interest is the conditions under which the assumption (18.3) or (18.4) holds and for this, the readers are referred to the discussion given in Oller et al. (2007) among others. In the following, all discussion will be based on the assumption (18.3) or (18.4) unless specified otherwise.

It is worth noting that in the case of right-censored failure time data, the noninformative censoring can be described in a much simpler format. In this case, it means that the censoring time or variable is independent of the failure time of interest completely or conditionally given covariates. It is clear that the two censoring mechanisms are quite different as only one variable is involved or needed with respect to right censoring. In the case of interval censoring, two variables L and R are needed and furthermore, they together with T have a natural relationship $L < T \le R$.

To help understand and illustrate the concepts and discussion above, we now consider a specific example of interval-censored data arising from an AIDS clinical trial, AIDS Clinical Trial Group 181, on HIV-infected individuals. The study is a natural history substudy of a comparative clinical trial of three anti-pneumocystis drugs and concerns the opportunistic infection cytomegalovirus (CMV). During the study, among other activities, blood and urine samples were collected from the patients at their clinical visits and tested for the presence of CMV, which is also commonly referred to as "shedding" of the virus. These samples and tests provide observed information on the two variables of interest, the times to CMV shedding in blood and in urine.

The observed dataset is given in dataset I of Appendix A in Sun (2006) and contains the observed intervals for the times to CMV shedding in blood and urine from 204 patients who provided at least one urine and blood sample during the study. More specifically, for the two failure times of interest, only interval-censored data are available with the intervals given by the last clinical visit time at which the shedding had not happened and the first clinical visit time at which the shedding had already occurred. Note that in this case, we actually have bivariate interval-censored data. If it is reasonable to assume that the clinical visit times of the patients have nothing to do with their disease status, then we would have noninformative interval censoring. Otherwise, one may have to consider the existence of informative interval censoring and this latter situation could be the case if, for example, the patients paid clinical visits because they felt their disease got worse. Among others, Goggins and Finkelstein (2000) discussed this dataset.

18.3 Current status data

In this section, we will briefly discuss statistical analysis of current status data. Let T, $F(t)$ and $S(t)$ be defined as above and suppose that one observes only current status data on T, which are usually denoted by $\{C_i, \delta_i = I(T_i \le C_i)\}_{i=1}^n$, where C_i represents the observation time on subject i. Note that in this case, the noninformative censoring means that T_i and C_i are independent completely or given covariates. For the situation, it is easy to see that the likelihood function given in (1) reduces to

$$L_S(\mathbf{p}) = \prod_{i=1}^{n} \Big[1 - S(C_i)\Big]^{\delta_i} \Big[S(C_i)\Big]^{1-\delta_i} .$$

To find the nonparametric maximum likelihood estimator (NPMLE) of S or maximize the likelihood function above, let the t_j's denote the ordered observation times as defined above and Q_j the set of subjects who are observed at t_j, $j = 1, ..., m$. Define $d_j = \sum_{i \in Q_j} 1(T_i \le t_j)$ and let n_j denote the number of elements in Q_j. Then the NPMLE of S can be shown (Sun, 2006) to be equal to the isotonic regression of $\{d_1/n_1, ..., d_m/n_m\}$ with weights $\{n_1, ..., n_m\}$. Using the max-min formula for the isotonic regression Barlow et al. (1972), one can derive the NPMLE of S as

$$\hat{S}(t_j) = 1 - \max_{u \le j} \min_{v \ge j} (\sum_{l=u}^{v} d_j / \sum_{l=u}^{v} n_j) .$$

One question of both practical and theoretical interest about the NPMLE of S is its asymptotic properties such as the convergence rate. For this, many studies have been performed including a sequence of papers by Groeneboom and his collaborators (Groeneboom and Wellner, 1992, Groeneboom et al., 2010; Groeneboom et al., 2012). For more complete discussion and recent references on this, the readers are referred to Banerjee (2012) and Chapter 19 of this book.

With respect to nonparametric comparison of survival functions based on current status data, several procedures have been proposed including the ones given in references (Andersen and Ronn, 1995; Groeneboom (2012): Sun, 1999; Sun and Kalbfleisch, 1993). Here we remark that most of the existing procedures including the ones for general interval-censored data assume that the censoring mechanism is the same for different treatments. More specifically for current status data, this means that the observation times C_i's follow the same distributions for subjects in different arms. One exception is the test procedure given in Sun (1999), which allows the distributions of the C_i's to depend on the treatment arms.

One needs to perform regression analysis if there exist covariates and one is interested in, for example, quantifying the effect of some covariates on the failure time of interest or predicting survival probabilities for new individuals (Zhang, 2009). For this, the first step is usually to specify an appropriate regression model. For failure time data, the most commonly used model is perhaps the proportional hazards model (Cox, 1972) given by

$$\lambda(t|Z=z) = \lambda_0(t)\, e^{\beta' z} \tag{18.5}$$

with respect to the hazard function of T given covariates $Z = z$. Here $\lambda_0(t)$ denotes the unknown baseline hazard function and β the vector of unknown regression parameters. To fit current status data to the model, Huang (1996) provided an ICM-type algorithm for estimation of unknown parameters in addition to investigating the properties of the resulting estimates of the unknown parameters. For the same problem, a Newton-Raphson algorithm could also be used (Sun, 2006) and a more recent discussion on it is given by Zhang (2012).

In addition to the proportional hazards model (18.5), of course, there exist many other models that may be used. For example, another attractive semiparametric regression model often used in practice is the additive hazards model given by

$$\lambda(t|Z=z) = \lambda_0(t) + \beta' z \tag{18.6}$$

again with respect to the hazard function of T given $Z = z$. It specifies that the effects of the covariates are additive rather than multiplicative as in model (18.5). For inference about this model based on current status data, among others, Lin et al. (1998) and Martinussen and Scheike (2002) developed some estimating equation approaches and Chen and Sun (2008) gave a multiple imputation procedure. More recently, Wang and Dunson (2011) considered the fitting of the proportional odds model described below to current status data in the Bayesian framework. Zhang (2012) also discussed the fitting of the proportional odds model as well as the linear transformation model described below to current status data. We remark that unlike most methods developed for right-censored data, estimating regression parameters under interval censoring usually involves estimation of both the parametric and the nonparametric parts. In other words, for interval-censored data, one has to deal with estimation of some unknown baseline functions in order to estimate regression parameters.

There are many other issues in analyzing failure time data. One is that in dealing with failure time data, a basic assumption that is usually not explicitly described is that the failure is always assumed to occur. In some situations, this may not be the case and the so-called cure model is often employed. Among others, Ma (Ma, 2009; Ma, 2011) recently studied the fitting of the cure model to current status data. Another issue of practical interest is the misclassification problem and, among others, McKeown and Jewell (2010) discussed it in the context of current status data.

18.4 Univariate interval-censored data

Now we will consider statistical analysis of general univariate interval-censored failure time data. As mentioned above, we will mainly discuss the three basic topics: nonparametric

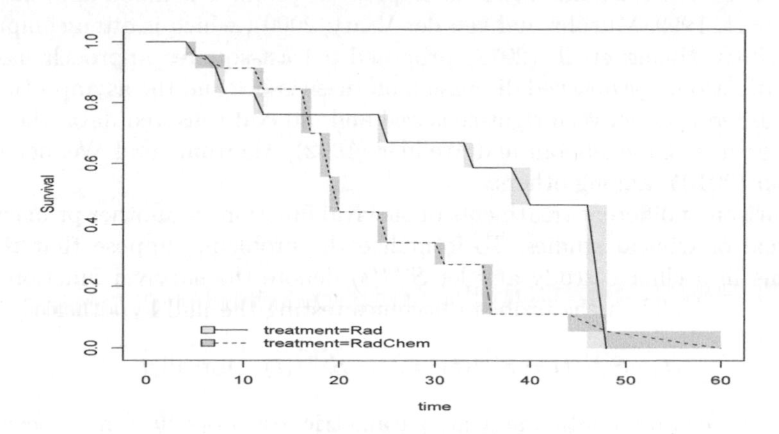

FIGURE 18.1
Nonparametric estimation of survival functions for breast cancer data: Rad=Radiation; RadChem=Radiation plus Chemotherapy.

estimation of a survival function, nonparametric comparison of survival functions and regression analysis, with the focus on some of the recent advances.

For nonparametric estimation, one can easily see that given general interval-censored data, the problem of finding the NPMLE of S becomes that of maximizing $L_S(\mathbf{p})$ given in (18.1) under the constraint that $\sum_{j=1}^{m+1} p_j = 1$ and $p_j \geq 0$, where $j = 1, ..., m+1$ (Li et al., 1997). Obviously, the likelihood function L_S depends on S only through the values $\{S(t_j)\}_{j=1}^{m}$. Thus the NPMLE $\hat{S}$ of S can be uniquely determined only over the observed intervals $(t_{j-1}, t_j]$ and the behavior of S within these intervals will be unknown. Several methods have been proposed for maximizing $L_S(\mathbf{p})$ with respect to $\mathbf{p}$. The first and simplest one is perhaps the self-consistency algorithm given by Turnbull (1976). One drawback of the algorithm is that the convergence can be slow. Corresponding to this, Groeneboom and Wellner (1992) developed an iterative convex minorant (ICM) algorithm, which can be seen as an optimized version of the well-known pool-adjacent-violator algorithm for the isotonic regression (Robertson et al., 1998). Another faster algorithm, a hybrid one that combines the two approaches above, is the EM-ICM algorithm given in Wellner and Zhan (1997). The other authors who recently investigated the same problem include Yavuza and Lambert (2010) and Dehghan and Duchesne (2011). The former employed Bayesian penalized B-splines to obtain the smooth estimation, while the latter generalized the self-consistency algorithm to the case where there exists a continuous covariate. To give a graphical idea about the nonparametric estimation based on interval-censored data, Figure 18.1 displays the estimated survival functions given by the self-consistency algorithm on a well-known set of the interval-censored data arising from a breast cancer study (Finkelstein and Wolfe, 1985). The study consists of 94 patients given either radiation therapy alone or radiation therapy plus adjuvant chemotherapy.

It is worth to emphasize that all algorithms mentioned above are iterative as for general interval-censored data, there is no closed form for the NPMLE of S unlike with current status data. Also it should be noted that for general interval-censored data, the NPMLE may not be unique and the solutions derived from the algorithms mentioned above may not be the NPMLE. One sufficient condition for the uniqueness of the NPMLE is that the log likelihood is strictly concave. Another important but difficult issue related the NPMLE of S is the consistent variance estimation as the standard Fisher information matrix approach does not work here. For this, one way is to employ the profile likelihood approach (Murphy and van der Vaart, 1999; Murphy and van der Vaart, 2000), which is often computationally intensive. Recently Huang et al. (2012) proposed a least-squares approach based on the efficient score function. For detailed discussion on these issues and the asymptotic properties as well as the differences between right-censored and interval-censored data, the readers are referred to references Groeneboom and Wellner (1992), Maathuis and Wellner (2008) and Zhang and Sun (2010*b*) among others.

The comparison of different treatments or survival functions is another primary objective in most medical or clinical studies. To formalize the problem, suppose that there are K treatment arms in a clinical study and let $S^{(k)}(t)$ denote the survival function of the kth arm with $k = 1, ..., K$. Then the problem becomes testing the null hypothesis

$$H_0 : S^{(1)}(t) = S^{(2)}(t) = ... = S^{(K)}(t) \quad \text{for all } t\,.$$

In the case of right-censored data, many nonparametric test procedures have been developed and most of them can be classified into two categories: rank-based tests and survival-based tests. The fundamental difference between them is that the former relies on the differences between the estimated hazard functions, while the latter bases the comparison on the differences between the estimated survival functions. Among them, the log-rank test is perhaps the most widely used method and a few of them have been generalized to the case of interval-censored data (Sun, 2006; Zhu, 2008*a*). For example, recently Fay and Shih (2012) and Oller and Gómez (2012) proposed some generalized rank-based and survival-based test procedures, respectively, for interval-censored data. To give a representative of such procedures, in the following, we describe the one proposed by Zhao et al. (2008).

Consider a survival study consisting of n independent subjects with n_k subjects from treatment arm k, $k = 1, ..., K$. Let the T_i's and I_i's be defined as before and D_{k1} and D_{k2} denote the sets of indices of the subjects in treatment arm k whose failure times are observed exactly and interval-censored, respectively. Define $n_{k1} = |D_{k1}|$, $n_{k2} = |D_{k2}|$, $n_1 = \sum_{k=1}^{K} n_{k1}$ and $n_2 = \sum_{k=1}^{K} n_{k2}$, and let $\hat{S}$ denote the NPMLE of the common survival function under the hypothesis H_0. To test the hypothesis H_0, Zhao et al. (2008) proposed to use the test statistic $U_\xi = (U_\xi^{(1)}, \cdots, U_\xi^{(K)})^T$, where

$$U_\xi^{(k)} = \frac{n_1}{n_{k1}} \sum_{i \in D_{k1}} \frac{\xi\{\hat{S}(T_i-)\} - \xi\{\hat{S}(T_i)\}}{\hat{S}(T_i-) - \hat{S}(T_i)} + \frac{n_2}{n_{k2}} \sum_{i \in D_{k2}} \frac{\xi\{\hat{S}(L_i)\} - \xi\{\hat{S}(R_i)\}}{\hat{S}(L_i) - \hat{S}(R_i)}$$

and ξ is a positive known function over $(0, 1)$. In practice, different ξ can be used and will yield different test statistics. The statistics above were motivated by the log-rank test given in Peto and Peto (1972), which has a similar form with $\xi(x) = x\log(x)$, for right-censored data. Under some regularity conditions and H_0, Zhao et al. (2008) showed that as $n \to \infty$, $n_{k1}/n \to p_{k1}$, $n_{k2}/n \to p_{k2}$ with $0 < p_{kj} < 1$, $U_\xi/\sqrt{n}$ has an asymptotic normal distribution with mean zero. They also gave a consistent estimate of the asymptotic covariance matrix. For an illustration, the application of the test procedure described above with $\xi(x) = x\log(x)$ to the breast cancer data used in Figure 16.1 gives a $p-$value being 0.007 for comparing the two treatment groups. The result suggests that the patients given the radiation therapy alone had a higher survival rate than the other patients.

With respect to regression analysis of general interval-censored data, as with other types of failure time data, the proportional hazards model (18.5) and the additive hazards model (18.6) are commonly used. Among recent work on this, Zhang et al. (2010) and Heller

(2011) investigated the fitting of model (18.5) to interval-censored data. The former studied the spline-based maximum likelihood approach in which monotone B-spline were used to approximate the baseline cumulative hazard function, while the latter gave a weighted estimating equation method. To fit model (18.6) to interval-censored data, Chen and Sun (2010*a*) and Wang et al. (2010) developed a multiple imputation procedure and an estimating equation approach, respectively.

Many other semiparametric models have been employed for regression analysis of interval-censored data. One such model is the proportional odds model (Sun, 2006; Sun et al., 2007) expressed as

$$\log\left\{\frac{F(t|z)}{1-F(t|z)}\right\} = h(t) + \beta' z \tag{18.7}$$

with respect to the CDF $F(t|z)$ of the failure time T of interest given $Z = z$. Here $h(t)$ is an unknown monotone-increasing function, also referred to as the "baseline log odds," and β represents a vector of regression parameters as in models (18.5) and (18.6). The accelerated failure time model (Betensky et al., 2001; Li and Pu, 2003) has also been considered for analyzing interval-censored data and it assumes that T and Z have the following relationship

$$\log(T) = \beta' Z + \varepsilon\,. \tag{18.8}$$

In the above, β is defined as before and ε is an error term whose distribution is usually unspecified.

It is easy to see that the four semiparametric models (18.5–18.8) are all specific models in terms of the functional form of the effects of covariates. Sometimes one may prefer a model that gives more flexibility. One such model that has been investigated for interval-censored data is the linear transformation model that specifies the relationship between the failure time T and the covariate Z as

$$h(T) = \beta' Z + \varepsilon\,. \tag{18.9}$$

Here $h : \mathcal{R}^+ \to \mathcal{R}$ ($\mathcal{R}$ denotes the real line and $\mathcal{R}^+$ the positive half real line) is an unknown strictly increasing function and the distribution of ε is assumed to be known. It is apparent that model (18.9) gives different models depending on the specification of the distribution of ε and especially, it includes models (18.5) and (18.7) as special cases. Among others, Zhang et al. (2005) and Chen and Sun (2010*b*) considered the fitting of model (18.9) to interval-censored data and proposed an estimating equation-based procedure and a multiple imputation approach, respectively. Applying the estimating equation procedure to the breast cancer data discussed above again, we obtained the estimate of the group effect parameter β being -0.697 with the estimated standard error of 0.251 by specifying model (18.5). If assuming model (18.7), one would get the estimate of -1.04 with the estimated standard error being 0.372. Both results indicate that the adjuvant chemotherapy significantly increased the hazard rate. These results are similar to those given by others such as Finkelstein (1986) and Chen and Sun (2010*b*).

For a practical problem, of course, one may also employ some parametric models such as piecewise exponential models (Zhang and Sun, 2010*b*) if there exists some prior information about the appropriateness of the parametric model. A main advantage of adopting a parametric model is that one can readily apply the maximum likelihood approach for inference. On the other hand, it is well known that parametric models usually may be too difficult to be verified and are less flexible than semiparametric models.

Other recent work on regression analysis of interval-censored data includes Kim and Jhun (2008) and Liu and Shen (2009), who considered the fitting of a cure model to interval-censored data, and Wang et al. (2012*b*) and Wang et al. (2012*c*), who discussed the application of Bayesian approaches to the problem. In addition, Zhu et al. (2008*b*) proposed a transformation approach; Lin and Wang (2010) studied the fitting of a probit model to interval-censored data; Lee et al. (2011) examined the use of three imputation approaches for the problem. Also Sun et al. (2007) developed a model checking procedure for model (18.7) based on interval-censored data, Zhang and Davidian (2008) proposed a group of smooth semiparametric regression models, and Zhang (2009) discussed the survival prediction under model (18.9).

18.5 Multivariate interval-censored data

Multivariate interval-censored data arise if a failure time study involves several related failure time variables of interest and some of them suffer interval censoring. It is apparent that the analysis would be straightforward if the variables are independent and when they are not independent of each other as usually the case, one needs and should employ the inference procedures that can take into account the correlation among the failure time variables. Another difference between univariate and multivariate interval-censored data is that for the latter, a new and unique issue that does not exist for the former is to make inference about the association between the failure time variables. In the following, we will first discuss a couple of basic issues related to the analysis of multivariate interval-censored data and then some recent advances with the focus on bivariate interval-censored data.

As with univariate interval-censored data, nonparametric estimation of a survival function is also one of the primary objectives of interest in practice for the analysis of multivariate interval-censored data. To discuss this, consider a survival study that involves n independent subjects from a homogeneous population with each subject giving rise to two failure times denoted by T_{1i} and T_{2i}, $i = 1, ..., n$. Let $F(t_1, t_2) = P(T_{1i} \leq t_1, T_{2i} \leq t_2)$ denote their joint cumulative distribution function and suppose that only interval-censored failure time data in the form

$$\{ U_i = (L_{1i}, R_{1i}] \times (L_{2i}, R_{2i}] \, , \, i = 1, \cdot \cdot, n \}$$

are available, where $(L_{1i}, R_{1i}]$ and $(L_{2i}, R_{2i}]$ represent the intervals to which T_{1i} and T_{2i} belong, respectively. It is easy to see that the observation on each subject could be a point, line segment or rectangle. If one treats points as rectangles that are degenerate in both dimensions and line segments as rectangles that are degenerate in one dimension, then the observed data consist of a collection of n rectangles.

For the determination of the NPMLE of F, note that as with univariate data, it will be a step function. Let

$$H = \{ H_j = (r_{1j}, s_{1j}] \times (r_{2j}, s_{2j}] \, , \, j = 1, ..., m \}$$

denote the disjoint rectangles that constitute the regions of possible support of the NPMLE of F, and define

$$\alpha_{ij} = I(H_j \subseteq (L_{1i}, R_{1i}] \times (L_{2i}, R_{2i}])$$

and

$$p_j = F(H_j) = F(s_{1j}, s_{2j}) - F(r_{1j}, s_{2j}) - F(s_{1j}, r_{2j}) + F(r_{1j}, r_{2,j}) \, ,$$

$i = 1, ..., n$, $j = 1, ..., m$. Then the likelihood function has the form

$$L_F(\mathbf{p}) = \prod_{i=1}^{n} F(U_i) = \prod_{i=1}^{n} \sum_{j=1}^{m} \alpha_{ij} \, p_j \tag{18.10}$$

with $\mathbf{p} = (p_1, ..., p_m)'$, and the NPMLE of F can be determined by maximizing (18.10) over the p_j's subject to $p_j \geq 0$ and $\sum_{j=1}^{m} p_j = 1$. One can easily see that the likelihood functions given (18.1) and (18.10) actually have the same structure and thus the algorithms developed for the maximization of (18.1) could be employed here for the maximization of (18.10) assuming that H is known. However, for a given dataset, the determination of H is actually difficult or not straightforward and for this, some algorithms have been developed and need to be used (Sun, 2006). It is apparent that the same holds for general multivariate interval-censored data.

As mentioned above, the estimation of the association between related failure time variables is often of interest for multivariate failure time data. To discuss this in the context of bivariate data, let $S_1(t)$ and $S_2(t)$ denote the marginal survival functions of possibly related T_1 and T_2, respectively, and $S(t_1, t_2) = P(T_1 > t_1, T_2 > t_2)$ their joint survival function. For the problem, one common way is to assume that $S(t_1, t_2)$ can be expressed by a copula model as $S(t_1, t_2) = C_\alpha(S_1(t_1), S_2(t_2))$, where C_α is a distribution function on the unit square and $\alpha \in R$ is a global association parameter. One attractive feature of the above expression is its flexibility as it includes as special cases many useful bivariate failure time models such as the Archimedean copula family. Another attractive feature is that under this expression, the marginal distributions do not depend on the choice of the association structure and thus one can model the marginal distributions and the association separately. Among others, Wang and Ding (2000) and Sun et al. (2006) discussed this approach for bivariate interval-censored data. Of course, by using higher order copula models instead of the bivariate model, one can apply the same approach to general multivariate interval-censored data.

The copula model approach discussed above can be employed for regression analysis of multivariate interval-censored data too. For example, Wang et al. (2008) and Zhang et al. (2008) recently applied it to the fitting of models (18.5) and (18.7), respectively, to bivariate current status data and developed efficient estimates of regression parameters. To describe the relationship among correlated failure time variables, instead of using the copula model, another commonly used approach is to employ frailty or latent variable models in which some latent variables are used to characterize the correlation. Among others, Hens et al. (2009) and Jonker et al. (2009) considered this approach for regression analysis of bivariate interval-censored data, and Chen et al. (2009*a*) and Nielsena and Parner (2010) applied it to regression analysis of general multivariate interval-censored data. A third approach that is often adopted for multivariate failure time data is the marginal approach that leaves the correlation arbitrary. Among others, Chen et al. (2007) and Tong et al. (2008) developed such approaches for fitting models (18.7) and (18.6), respectively, to general multivariate interval-censored data. For all three approaches, of course, one could apply them in the Bayesian framework (Gómez et al., 2009; Komàrek and Lesaffre, 2007).

It is well known that multivariate failure time data can be seen as a special case of clustered failure time data and one key feature of the latter is that the cluster size, the number of correlated failure time variables, can differ from cluster to cluster. Thus the existing methods for multivariate interval-censored data cannot be directly applied to clustered interval-censored data. Some recent references on the statistical analysis of clustered interval-censored data include Chen et al. (2009*b*), Zhang and Sun (2010*a*), Xiang et al. (2011) and Li et al. (2012). In particular, Chen et al. (2009*b*) and Zhang and Sun (2010*a*) discussed the fitting of the proportional hazards model (18.5) to them; Xiang et al. (2011) considered the inference about a cure model; Li et al. (2012) proposed regression analysis of data of this kind with the additive models. Compared to other types of interval-censored data discussed above, however, there exists very limited literature on clustered interval-censored data.

18.6 Competing risks interval-censored data

Competing risks failure time data arise when there exist several related failure causes or types. The underlying structure behind them is actually similar to that behind multivariate failure time data. A key difference between the two types of data, also a key feature of the former, is that one observes only one failure type or one of several related underlying failure time variables. For a patient with several diseases, for example, one can only observe the death time caused by one of them, but not the death times due to other diseases. In other words, other disease death times are censored by the death time of the disease that occurs the first. In the following, we will first discuss the analysis of competing risks interval-censored data and then informatively interval-censored data with the focus on nonparametric estimation and regression analysis.

Consider a competing risks study that involves m different types of failures. Let T denote the real or observable failure time and $J \in \{1, \cdots, m\}$ the cause or type of failure. In this case, one function that is usually of interest is the cumulative incidence function defined as $F_j(t) = P(T \le t; J = j)$ for failure type j. For its estimation, one of early work was given by Jewell et al. (2003), who gave several estimates. Recently Groeneboom et al. (Groeneboom et al., 2008*a*; Groeneboom et al., 2008*b*) derived the nonparametric maximum likelihood estimates of the function and established their consistency and local limiting distributions. Maathuis and Hudens (2011) also investigated the same problem. Note that all four references mentioned above are for competing risks current status data only and there does not seem to exist similar work for general competing risks interval-censored data.

In addition to the notation above, suppose that there also exists a vector of covariates Z and one is interested in regression analysis. In this case, one can write the conditional cumulative incidence function $F_j(t|Z) = P(T \le t; J = j|Z)$ as

$$F_j(t|Z) = P(T \le t|J = j, Z)\, P(J = j|Z),$$

$j = 1, \cdots, m$. To conduct regression analysis, one way is to directly model $F_j(t|Z)$ or the conditional distribution function $P(T \le t|J = j, Z)$ along with $P(J = j|Z)$. Note that the regression parameters in these two different modeling approaches will have different meanings and one may base the selection on the regression parameter that is preferred. Assume that one would like to take the second approach and T is continuous. For this, a common approach is to assume that the cause-specific hazard function corresponding to $P(T \le t|J = j, Z)$ and $P(J = j|Z)$ follow, for example, the proportional hazards model $\lambda_j(t|Z) = \lambda_{0j}(t)\, e^{\beta_j^T Z}$ and a parametric model $g_j(\gamma, Z)$, respectively. Here $\lambda_{0j}(t)$ is an unknown baseline hazard function, β_j and γ are unknown parameters, and g_j is a known and positive function satisfying $\sum_{j=1}^m g_j(\gamma, Z) = 1$. For each j, let $\Lambda_j(t) = \int_0^t \lambda_{0j}(t)$. Then we have

$$F_j(t|Z) = g_j(\gamma, Z)\Big[1 - \exp\big\{ - \Lambda_j(t) \exp(\beta_j^T Z)\big\}\Big].$$

For inference, suppose that one observes only the current status data discussed in Section 18.3. For each j, define $\delta_j = I\{T \le C, J = j\}$, $j = 1, \cdots, m$, $\delta_{m+1} = I\{T > C\}$ and $\Delta = (\delta_1, \cdots, \delta_m, \delta_{m+1})$. Let $\theta = (\beta_1^T, \cdots, \beta_m^T, \gamma^T)^T$ and $\Lambda = (\Lambda_1, \cdots, \Lambda_m)^T$ and suppose that the joint distribution function of (C, Z) does not involve θ and Λ. Then the likelihood contribution from a single observation $X = (C, \Delta, Z)$ has the form

$$L(\theta, \Lambda; X) = \prod_{j=1}^m F_j(c|Z)^{\delta_j} \{1 - \sum_{k=1}^m F_k(c|Z)\}^{\delta_{m+1}}$$

$$= \prod_{j=1}^{m} \left[g_j(\gamma, Z) \left\{ 1 - \exp\left(- \Lambda_j(c) \exp(\beta_j^T Z) \right) \right\} \right]^{\delta_j}$$

$$\times \left[\sum_{k=1}^{m} g_k(\gamma, Z) \exp\left\{ - \Lambda_k(c) \exp(\beta_k^T Z) \right\} \right]^{\delta_{m+1}}$$

and the likelihood function from n i.i.d. copies of X is given by the product of $L(\theta, \Lambda; X)$. Sun and Shen (2009) investigated this and established the asymptotic properties of the resulting maximum likelihood estimates.

18.7 Informatively interval-censored data

Now we discuss the analysis of informatively interval-censored data and for this, we will first consider current status data. As before, let T and C denote the failure time of interest and the observation time, respectively. By informative censoring, it means that T and C are correlated and thus the underlying probability space is similar to that for $m = 2$ competing risks problems without censoring. On the other hand, the data structures behind them are different as for current status data, C is always observable but not T, while both T and C could be observed for exact or right-censored competing risks data. As with informatively right-censored failure time data, the survival function of T is generally unidentifiable based on informatively current status data without some assumptions. Wang et al. (2012*a*) recently discussed this and proposed two estimates of the survival function under the copula model framework.

As mentioned above, one area that often produces current status data is tumorigenicity experiments and in this case, T and C represent the tumor onset time and the death time, respectively. In addition, sacrifice is often used in these studies and in this situation, the death will not be observed. In other words, there may exist right censoring on the observation time C. Also often used in tumorigenicity studies, one way to analyze informatively current status data with right censoring is to formulate the problem using the illness-death or three-state model consisting of health, illness (tumor in the tumorigenicity study) and death states. Here the illness and death correspond to the failure event of interest and the observation event, respectively, and an extensive literature has been established on this in the context of tumorigenicity studies. Recent references on this for informative current status data include Frydman and Szarek (2009) and Kim et al. (n.d.). The former developed the nonparametric maximum likelihood approach and the latter discussed the regression analysis problem under the proportional hazards model.

For the analysis of general informatively interval-censored data, first note that as discussed above, the censoring mechanism behind interval censoring is usually much more complicated than that behind both right censoring and current status data case. In consequence, it is usually difficult or impossible to generalize the methods developed for informatively right-censored data or informative current status data to them. To further see this, note that one can write the likelihood contribution from a single interval-censored observation as

$$Pr(L \leq T \leq R) = Pr(l \leq T \leq r | L = l, R = r) \, Pr(L = l, R = r) \, .$$

This indicates that to conduct regression analysis, one would have to specify some joint models for $Pr(L \leq T \leq R)$ or for both $Pr(l \leq T \leq r | L = l, R = r)$ and $Pr(L = l, R = r)$. This is quite different from the case of right-censored data or current status data, and usually not easy. The authors who recently discussed this joint modeling approach include

Zhang et al. (2007) and Wang et al. (2010). They considered the cases where T follows models (18.5) and (18.6), respectively, marginally and in both cases, model (18.5) was also used to model the censoring variables.

18.8 Other types of interval-censored data

In the previous sections, we have discussed several types of interval-censored failure time data that one commonly faces in practice. A few other types of interval-censored data that were not touched above may occur too in failure time studies. They include doubly censored data, interval-censored data from multi-state models and interval-censored data with missing covariates. In this section, we will briefly discuss them along with some others.

By failure time variable or interval-censored failure time variable, one generally means a variable measuring the time from zero to the occurrence of an event of interest. A generalization of this is the variable that measures the elapse time between two successive events such as the onset of a disease and death due to the disease. Given such a variable of interest plus interval censoring, one will have a doubly censored data analysis problem as discussed above. In other words, the interval-censored data discussed above can be regarded as a special case of doubly censored data. Sun (2006) devoted one chapter for the analysis of doubly censored data. Recent references on doubly censored data include Komàrek and Lesaffre (2008) and Deng et al. (2009), which discussed the Bayesian approach and the nonparametric estimation problem, respectively.

Note that one way to formulate doubly censored data is to employ a three-state model similar to the one discussed in the previous section. Corresponding to this, a natural question of interest will be the analysis of interval-censored data arising from a multi-state model, naturally and commonly used in, for example, epidemiological or disease progression studies. Several authors actually have recently considered this problem including Barrett et al. (2011), Chen et al. (2010), Cook et al. (2008), Joly et al. (2012) and Yang and Nair (2011). In particular, Chen et al. (2010) investigated the maximum likelihood approach, and Yang and Nair (2011) compared the multi-state-based analysis and the simple failure time data analysis in the presence of interval censoring.

Missing covariates can occur in any regression analysis as well as covariate mismeasurement and other related problems. The same can happen in failure time studies with interval censoring. For example, Wen and Lin (2011) gave a set of current status data with missing covariates arising from the 2005 Taiwan National Health Survey and developed a semiparametric maximum likelihood estimation procedure under the proportional hazards model (18.5). Wen (2012) and Wen et al. (2011) discussed regression analysis of interval-censored data under model (18.5) in the presence of measurement errors on covariates. Also Li and Nan (2011) considered current status data arising from case-cohort studies. In all discussion so far, it has been supposed that interval censoring occurs on the failure time variable of interest and in reality, it could occur on covariates too. In other words, covariates may have missing values in the form of censoring and among others, Schoenfeld et al. (2011) provided some discussion on such interval-censored data.

18.9 Software and concluding remarks

The software development is always important for both the development and promotion of new statistical methodology. Although there does not seem to exist a single commercially available statistical software yet that provides a relatively complete and extensive coverage for the analysis of interval-censored data, many individual functions or packages have been developed and are available online. For example, in SPLUS , the function *kaplanMeier* can be used to compute the Turnbull estimator (Turnbull, 1976). In SAS, the procedure *LIFEREG* allows one to fit the parametric accelerated failure time model to interval-censored data. In the following, we will describe several R packages that can be easily implemented.

Actually a number of R packages have recently been developed for the analysis of interval-censored data and can be found in the survival analysis part of R websites. For the determination of the NPMLE of a survival function, for example, one can employ the **Icens** or **MLEcens** packages, developed by Gentleman and Vandal (2013) and Maathuis (2013), respectively. The two packages apply to both univariate and bivariate interval-censored data. For doubly censored data, one may apply the **dblcens** developed by Zhou et al. (2012). To conduct nonparametric comparison of survival functions, the available R packages include the **glrt** and **interval** packages developed by Zhao and Sun (2010) and Fray (2011), respectively. The former include three generalized log-rank tests and the score test, while the latter also allows one to use some generalized log-rank tests as well as Wilcoxon type tests.

To perform regression analysis of interval-censored data, one may apply the **intcox** package, developed by Henschel et al. (2013), if the proportional hazards model (18.5) is of interest. One could also use the *Icens* function in package **Epi**, developed by Carstensen et. al. (2013), to fit the multiplicative relative risk and additive excess risk models to interval-censored data. Furthermore, the **dynsurv** and **survBayes** packages, developed by Wang et al. (2012*d*) and Henschel et al. (2012), allow one to fit the time-varying coefficient model and the proportional hazards model in the Bayesian framework to interval-censored data.

Finally we remark that methodologically, there are still many open questions in the analysis of interval-censored data. Examples include but are not limited to model checking techniques and joint modeling of longitudinal and interval-censored data. Also many of the existing procedures need proper theoretical justification. One major difficulty with interval-censored data is that there lacks basic tools as simple and elegant as the partial likelihood and the martingale theory for right-censored data. Instead, one often has to rely on the empirical process theory as well as the optimization theory (Groeneboom and Wellner, 1992; Huang and Wellner, 1997).

Acknowledgment
The authors wish to thank Professor Thomas H. Scheike and other volume editors, Professors John P. Klein, Joseph G. Ibrahim and Hans C. van Houwelingen for the invitation of writing this chapter, and also the editors and two associate editors for their helpful commments and suggestions.

Bibliography

Andersen, P. and Ronn, B. (1995), 'A nonparametric test for comparing two samples where all observations are either left- or right-censored', *Biometrics* **51**, 323–329.

Banerjee, M. (2012), 'Current status data in the twenty-first century: Some interesting developments', *Interval-Censored Time-to-Event Data: Methods and Applications*, Chen, D.G. and Sun, J. and Peace, K.E., eds. CRC Press Taylor & Francis Group pp. 45–90.

Barlow, R., Bartholomew, D., Bremner, J. and Brunk, H. (1972), *Statistical inference under order restrictions*, New York: Wiley.

Barrett, J., Siannis, F. and Farewell, V. (2011), 'A semi-competing risks model for data with interval-censoring and informative observation: An application to the MRC cognitive function and ageing study', *Statistics in Medicine* **30**, 1–10.

Betensky, R., Rabinowitz, D. and Tsiatis, A. (2001), 'Computationally simple accelerated failure time regression for interval censored data', *Biometrika* **88**, 703–711.

Carstensen B., Plummer, M., Hills, M. and Laara, E. (2013), R package Epi: A package for statistical analysis in epidemiology. http://cran.r-project.org/web/packages/Epi/index.html.

Chen, B., Yi, G. and Cook, R. (2010), 'Analysis of interval-censored disease progression data via multi-state models under a nonignorable inspection process', *Statistics in Medicine* **29**, 1175–1189.

Chen, D., Sun, J. and Peace, K. (2012), *Interval-Censored Time-to-Event Data: Methods and Applications*, CRC Press Taylor & Francis Group.

Chen, L. and Sun, J. (2008), 'A multiple imputation approach to the analysis of current status data with the additive hazards model', *Communications in Statistics: Theory and Methods* **38**, 1009–1018.

Chen, L. and Sun, J. (2010*a*), 'A Multiple Imputation Approach to the Analysis of Interval-censored Failure Time Data with the Additive Hazards Model', *Computational Statistics and Data Analysis* **54**, 1109–1116.

Chen, L. and Sun, J. (2010*b*), 'Multiple imputation to regression analysis of interval-censored failure time data with linear transformation models', *Far East Journal of Theoretical Statistics* **33**, 41–55.

Chen, M., Tong, X. and J., S. (2007), 'The proportional odds model for multivariate interval-censored failure time data', *Statistics in Medicine* **26**, 5147–5161.

Chen, M., Tong, X. and Sun, J. (2009*a*), 'A frailty model approach for regression analysis of multivariate current status data', *Statistics in Medicine* **28**, 3424–3436.

Chen, P., Shen, J. and Sun, J. (2009*b*), 'Statistical Analysis of Clustered Current Status Data', *International Journal of Intelligent Technologies and Applied Statistics* **2**, 21–31.

Cook, R., Zeng, L. and Lee, K.-A. (2008), 'A Multistate Model for Bivariate Interval-Censored Failure Time Data', *Biometrics* **64**, 1100–1109.

Cox, D. (1972), 'Regression models and life-tables (with discussion)', *Journal of the Royal Statistical Society: Series B* **34**, 187–220.

De Gruttola, V. and Lagakos, S. (1989), 'Analysis of doubly-censored survival data, with application to AIDS', *Biometrics* **45**, 1–11.

Dehghan, M. and Duchesne, T. (2011), 'A generalization of Turnbull's estimator for nonparametric estimation of the conditional survival function with interval-censored data', *Lifetime Data Analysis* **17**, 234–255.

Deng, D., Fang, H. and Sun, J. (2009), 'Nonparametric estimation for doubly censored failure time data', *Journal of Nonparametric Statistics* **21**, 801–814.

Dinse, G. and Lagakos, S. (1983), 'Regression analysis of tumor prevalence data', *Applied Statistics* **32**, 236–248.

Fay, M. and Shih, J. (2012), 'Weighted logrank tests for interval censored data when assessment times depend on treatment', *Statistics in Medicine* **31**, 3760–3772.

Fray, M. (2011), R package interval: Weighted logrank tests and NPMLE for interval censored data. http://cran.r-project.org/web/packages/interval/index.html.

Finkelstein, D. (1986), 'A proportional hazards model for interval-censored failure time data', *Biometrics* **42**, 845–854.

Finkelstein, D. and Wolfe, R. (1985), 'A semiparametric model for regression analysis of interval-censored failure time data', *Biometrics* **41**, 933–945.

Frydman, H. and Szarek, M. (2009), 'Nonparametric estimation in a Markov illness-death process from interval censored observations with missing intermediate transition status', *Biometrics* **65**, 143–151.

Gentleman, R. and Vandal A. (2013), R package ICENS: NPMLE for censored and truncated data Version 1.30.0. http://bioconductor.org/packages/2.11/bioc/man/Icens.pdf

Goggins, W. and Finkelstein, D. (2000), 'A proportional hazards model for multivariate interval-censored failure time data', *Biometrics* **56**, 940–943.

Gómez, G., Oller, R., Calle, M. and Langohr, K. (2009), 'Tutorial on methods for interval-censored data and their implementation in R', *Statistical Modeling* **9**, 299–319.

Groeneboom, P. (2012), 'Likelihood ratio type two-sample tests for current status data', *Scandinavian Journal of Statistics* **39**, 645–662.

Groeneboom, P., Jongbloed, G. and Witte, B. (2010), 'Maximum smoothed likelihood estimation and smoothed maximum likelihood estimation in the current status model', *The Annals of Statistics* **38**, 352–387.

Groeneboom, P., Jongbloed, G. and Witte, B. (2012), 'A maximum smoothed likelihood estimator in in the current status continuous mark model', *Journal of Nonparametric Statistics* **24**, 85–101 .

Groeneboom, P., Maathuis, M. and Wellner, J. (2008*a*), 'Current status data with competing risks: Consistency and rates of convergence of the MLE', *The Annals of Statistics* **36**, 1031–1063.

Groeneboom, P., Maathuis, M. and Wellner, J. (2008*b*), 'Current status data with competing risks: Limiting distribution of the MLE', *The Annals of Statistics* **36**, 1064–1089.

Groeneboom, P. and Wellner, J. (1992), *Information Bounds and Nonparametric Maximum Likelihood Estimation*, DMV Seminar, Band 19, New York: Birkhauser.

Heller, G. (2011), 'Proportional hazards regression with interval censored data using an inverse probability weight', *Lifetime Data Analysis* **17**, 373–385.

Hens, N., Wienke, A., Aerts, M. and Molenberghs, G. (2009), 'The correlated and shared gamma frailty model for bivariate current status data: An illustration for cross-sectional serological data', *Statistics in Medicine* **28**, 2785–2800.

Henschel, V., Heiss, C. and Mansmann, U. (2012), R package survBayes: Fits a proportional hazards model to time to event data by a Bayesian approach. http://cran.rproject.org/web/packages/survBayes/index.html

Henschel, V., Mansmann, U. and Heiss, C. (2013), R package intcox: Iterated convex minorant algorithm for interval censored event data. http://cran.r-project.org/web/packages/intcox/index.html

Huang, J. (1996), 'Efficient estimation for the proportional hazards model with interval censoring', *The Annals of Statistics* **24**, 540–568.

Huang, J. and Wellner, J. (1997), Interval censored survival data: a review of recent progress. *Proceedings of the First Seattle Symposium in Biostatistics: Survival Analysis*, Lin, D., Fleming, T., eds. New York: Springer-Verlag.

Huang, J., Zhang, Y. and Hua, L. (2012), 'Consistent variance estimation in interval-censored data', *Interval-Censored Time-to-Event Data: Methods and Applications*, Chen, D.G. and Sun, J. and Peace, K.E., eds. CRC Press Taylor & Francis Group pp. 233–268.

Jewell, N. and van der Laan, M. (1995), 'Generalizations of current status data with applications', *Lifetime Data Analysis* **1**, 101–110.

Jewell, N., van der Laan, M. J. and Hennemean, T. (2003), 'Nonparametric estimation from current status data with competing risks', *Biometrika* **90**, 183–197.

Joly, P., Gerds, T., Qvist, V., Commenges, D. and Keiding, N. (2012), 'Estimating survival of dental fillings on the basis of interval-censored data and multi-state models', *Statistics in Medicine* **31**, 1139–1149.

Jonker, M., Bhulai, S., Boomsma, D., Ligthart, R. and Vander Vaart, A. (2009), 'Gamma frailty model for linkage analysis with application to interval-censored migraine data', *Biometrics* **10**, 187–200.

Kalbfleisch, J. and Prentice, R. (2002), *The Statistical Analysis of Failure Time Data*, New York: Wiley.

Keiding, N. (1991), 'Age-specific incidence and prevalence: A statistical perspective (with discussion)', *Journal of the Royal Statistical Society: Series A* **154**, 371–412.

Kim, Y. and Jhun, M. (2008), 'Cure rate model with interval censored data', *Statistics in Medicine* **27**, 3–14.

Kim, Y.-J., Kim, J., Nam, C. and Kim, Y.-N. (2012), 'Statistical analysis of dependent current status data, in *Interval-Censored Time-to-Event Data: Methods and Applications*, Chen, D.G., Sun, J., and Peace, K.E., eds. CRC Press Taylor & Francis Group pp. 113–148.

Klein, J. and Moeschberger, M. (2002), *Survival Analysis*, New York: Springer.

Komàrek, A. and Lesaffre, E. (2007), 'Bayesian accelerated failure time model for correlated interval-censored data with a normal mixture as error distribution', *Statistica Sinica* **17**, 549–569.

Komàrek, A. and Lesaffre, E. (2008), 'Bayesian accelerated failure time model with multivariate doubly interval-censored data and flexible distributional assumptions', *Journal of the American Statistical Association* **103**, 523–533.

Lawless, J. and Babineau, D. (2006), 'Models for interval censoring and simulation-based inference for lifetime distributions', *Biometrika* **93**, 671–686.

Lee, T., Zeng, L., Darby, J. and Dean, C. (2011), 'Comparison of imputation methods for interval censored time-to-event data in joint modeling of tree growth and mortality', *Canadian Journal of Statistics* **39**, 438–457.

Li, J., Wang, C. and Sun, J. (2012), 'Regression analysis of clustered interval-censored failure time data with the additive hazards model', *Journal of Nonparametric Statistics* **24**, 1041–1050.

Li, L. and Pu, Z. (2003), 'Rank estimation of log-linear regression with interval-censored data', *Lifetime Data Analysis* **9**, 57–70.

Li, L., Watkins, T. and Yu, Q. (1997), 'An EM algorithm for estimating survival functions with interval-censored data', *Scandinavian Journal of Statistics* **24**, 531–542.

Li, Z. and Nan, B. (2011), 'Relative risk regression for current status data in case-cohort studies', *Canadian Journal of Statistics* **39**, 557–577.

Lin, D., Oakes, D. and Ying, Z. (1998), 'Additive hazards regression with current status data', *Biometrika* **85**, 289–298.

Lin, X. and Wang, L. (2010), 'A semiparametric probit model for case 2 interval-censored failure time data', *Statistics in Medicine* **29**, 972–981.

Liu, H. and Shen, Y. (2009), 'A semiparametric regression cure model for interval-censored data', *Journal of the American Statistical Association* **104**, 1168–1178.

Ma, S. (2009), 'Cure model with current status data', *Statistica Sinica* **19**, 233–249.

Ma, S. (2011), 'Additive risk model for current status data with a cured subgroup', *Annals of the Institute of Statistical Mathematics* **63**, 117–134.

Maathuis, M. (2013), R package MLEcens: Computation of the MLE for bivariate (interval) censored data. http://cran.r-project.org/web/packages/MLEcens/MLEcens.pdf

Maathuis, M. and Hudens, M. (2011), 'Nonparametric inference for competing risks current status data with continuous, discrete or grouped observation times', *Biometrika* **98**, 325–340.

Maathuis, M. and Wellner, J. (2008), 'Inconsistency of the mle for the joint distribtion of interval-censored survival times and continuous marks', *Scandinavian Journal of Statistics* **35**, 83–103.

Martinussen, T. and Scheike, T. (2002), 'Efficient estimation in additive hazards regression with current status data', *Biometrika* **89**, 649–658.

McKeown, K. and Jewell, N. (2010), 'Misclassification of current status data', *Lifetime Data Analysis* **16**, 215–230.

Murphy, S. and van der Vaart, A. (1999), 'Observed information in semiparametric models', *Bernoulli* **5**, 381–412.

Murphy, S. and van der Vaart, A. (2000), 'On profile likelihood', *Journal of the American Statistical Association* **95**, 449–465.

Nielsena, J. and Parner, E. (2010), 'Analyzing multivariate survival data using composite likelihood and flexible parametric modeling of the hazard functions', *Statistics in Medicine* **29**, 2126–2136.

Oller, R. and Gómez, G. (2012), 'A generalized Fleming and Harrington's class of tests for interval-censored data', *Canadian Journal of Statistics* **40**, 1–16.

Oller, R., Gómez, G. and Calle, M. (2007), 'Interval censoring: identifiability and the constant-sum property', *Biometrika* **94**, 61–70.

Peto, R. and Peto, J. (1972), 'Asymptotically efficient rank invariant test procedures', *Journal of the Royal Statistical Society: Series A* **135**, 185–207.

Robertson, T., Wright, F. and Dykstra, R. (1998), *Order Restrict Statistical Inference*, New York: John Wiley.

Schoenfeld, D., Rajicic, N., Ficociello, L. and Finkelstein, D. (2011), 'A test for the relationship between a time-varying marker and both recovery and progression with missing data', *Statistics in Medicine* **30**, 659–799.

Sun, J. (1999), 'A nonparametric test for current status data with unequal censoring', *Journal of the Royal Statistical Society: Series B* **61**, 243–250.

Sun, J. (2002), Statistical analysis of doubly interval-censored failure time data, *Handbook of Statistics: Survival Analysis*, Balakrishnan, N. and Rao, C.R., eds. Elsevier 105–122.

Sun, J. (2006), *The Statistical Analysis of Interval-Censored Failure Time Data*, New York: Springer.

Sun, J. and Kalbfleisch, J. (1993), 'The analysis of current status data on point processes', *Journal of the American Statistical Association* **88**, 1449–1454.

Sun, J. and Shen, J. (2009), 'Efficient estimation for the proportional hazards model with competing risks and current status data', *Canadian Journal of Statistics* **37**, 592–606.

Sun, J., Sun, L. and Zhu, C. (2007), 'Testing the proportional odds model for interval-censored data', *Lifetime Data Analysis* **13**, 37–50.

Sun, L., Wang, L. and Sun, J. (2006), 'Estimation of the association for bivariate interval-censored failure time data', *Scandinavian Journal of Statistics* **33**, 637–649.

Tong, X., Chen, M. and Sun, J. (2008), 'Regression analysis of multivariate interval-censored failure time data with application to tumorigenicity experiments', *Biometrical Journal* **50**, 364–374.

Turnbull, B. (1976), 'The empirical distribution with arbitrarily grouped censored and truncated data', *Journal of the Royal Statistical Society: Series B* **38**, 290–295.

Wang, C., Sun, J., Sun, L., Zhou, J. and Wang, D. (2012*a*), 'Nonparametric estimation of current status data with dependent censoring', *Lifetime Data Analysis* **18**, 434–445.

Wang, L. and Dunson, D. (2011), 'Semiparametric Bayes' proportional odds models for current status data with underreporting', *Biometrics* **67**, 1111–1118.

Wang, L., Lin, X. and Cai, B. (2012*b*), 'Bayesian semiparametric regression analysis of interval-censored data with monotone splines', *Interval-Censored Time-to-Event Data: Methods and Applications*, Chen, D.G. and Sun, J. and Peace, K.E., eds. CRC Press Taylor & Francis Group pp. 149–166.

Wang, L., Sun, J. and Tong, X. (2008), 'Efficient estimation for the proportional hazards model with bivariate current status data', *Lifetime Data Analysis* **14**, 134–153.

Wang, L., Sun, J. and Tong, X. (2010), 'Regression Analysis of Case II Interval-censored Failure Time Data with the Additive Hazards Model', *Statistica Sinica* **20**, 1709–1723.

Wang, W. and Ding, A. (2000), 'On assessing the association for bivariate current status data', *Biometrika* **87**, 879–893.

Wang, X., Sinha, A., Yan, J. and Chen, M. (2012*c*), 'Bayesian inference for interval-censored survival data', *Interval-Censored Time-to-Event Data: Methods and Applications*, Chen, D.G. and Sun, J. and Peace, K.E., eds. CRC Press Taylor & Francis Group pp. 167–198.

Wang, X., Yan, J. and Chen, M-H. (2012*d*), R package dynsurv: Dynamic models for survival data. http://cran.r-project.org/web/packages/dynsurv/index.html

Wellner, J. and Zhan, Y. (1997), 'A hybrid algorithm for computation of the nonparametric maximum likelihood estimator from censored data', *Journal of the American Statistical Association* **92**, 945–959.

Wen, C. (2012), 'Cox regression for mixed case interval-censored data with covariate errors', *Lifetime Data Analysis* **18**, 321–338.

Wen, C., Huang, Y. and Chen, Y. (2011), 'Cox regression for current status data with mismeasured covariates', *Canadian Journal of Statistics* **39**, 73–88.

Wen, C. and Lin, C. (2011), 'Analysis of current status data with missing covariates', *Biometrics* **67**, 760–769.

Xiang, L., Ma, X. and Yau, K. (2011), 'Mixture cure model with random effects for clustered interval-censored survival data', *Statistics in Medicine* **30**, 995–1006.

Yang, Y. and Nair, V. (2011), 'Parametric inference for time-to-failure in multi-state semi-Markov models: A comparison of marginal and process approaches', *Canadian Journal of Statistics* **39**, 537–555.

Yavuza, A. and Lambert, P. (2010), 'Smooth estimation of survival functions and hazard ratios from interval-censored data using Bayesian penalized B-splines', *Statistics in Medicine* **30**, 75–90.

Zhang, B. (2012), 'Regression analysis of for current status data', *Interval-Censored Time-to-Event Data: Methods and Applications*, Chen, D.G. and Sun, J. and Peace, K.E., eds. CRC Press Taylor & Francis Group pp. 91–112.

Zhang, B., Tong, X. and Sun, J. (2008), 'Efficient estimation for the proportional odds model with bivariate current status data', *Far East Journal of Theoretical Statistics* **27**, 113–132.

Zhang, M. and Davidian, M. (2008), "Smooth" semiparametric regression analysis for arbitrarily censored time-to-event data", *Biometrics* **64**, 567–576.

Zhang, X. and Sun, J. (2010*a*), 'Regression analysis of clustered interval-censored failure time data with informative cluster size', *Computational Statistics & Data Analysis* **54**, 1817–1823.

Zhang, Y., Hua, L. and Huang, J. (2010), 'A spline-based semiparametric maximum likelihood estimation method for the Cox model with interval-censored data', *Scandinavian Journal of Statistics* **37**, 338–354.

Zhang, Z. (2009), 'Linear transformation models for interval-censored data: prediction of survival probability and model checking', *Statistical Modeling* **9**, 321–343.

Zhang, Z. and Sun, J. (2010*b*), 'Interval censoring', *Statistical Methods in Medical Research* **19**, 53–70.

Zhang, Z., Sun, L., Sun, J. and Finkelstein, D. (2007), 'Regression analysis of failure time data with informative interval censoring', *Statistics in Medicine* **26**, 2533–2546.

Zhang, Z., Sun, L., Zhao, X. and Sun, J. (2005), 'Regression analysis of interval censored failure time data with linear transformation models', *The Canadian Journal of Statistics* **33**, 61–70.

Zhao, Q. and Sun, J. (2010) R Package glrt: Generalized logrank tests for interval-censored failure time data. http://cran.r-project.org/web/packages/glrt/index.html

Zhao, X., Zhao, Q., Sun, J. and Kim, J. (2008), 'Generalized log-rank tests for partly interval-censored failure time data', *Biometrical Journal* **50**, 375–385.

Zhou, M., Lee, L. and Chen, K. (2012), R package dblcens: Compute the NPMLE from doubly censored data. http://cran.r-project.org/web/packages/dblcens/

Zhu, C., Yues, K., Sun, J. and Zhao, X. (2008*a*), 'A nonparametric test for interval-censored failure time data with unequal censoring', *Communications in Statistics: Theory and Methods* **37**, 1895–1904.

Zhu, L., Tong, X. and Sun, J. (2008*b*), 'A transformation approach for the analysis of interval-censored failure time data', *Lifetime Data Analysis* **14**, 167–178.

19

Current Status Data: An Illustration with Data on Avalanche Victims

Nicholas P. Jewell and Ruth Emerson

University of California

CONTENTS

19.1 Introduction 391
19.2 Estimation of a single distribution function 393
19.2.1 Inference 395
19.3 Regression methods 398
19.4 Competing risks 404
19.5 Sampling and measurement issues 406
19.6 Other topics 406
19.7 Discussion 407
Acknowledgments 408
Bibliography 408

Current status data provides information on the survival status of individuals at single screening times rather than standard observation, possibly right-censored, of failure times. It thus represents an extreme form of interval censoring where censoring intervals are either $[0, C]$ or (C, ∞) where C represents the screening time on an appropriate timescale. Considerable attention has been given to estimation of a survival function based on such data, often supplemented by estimation of regression coefficients from a variety of standard models in the context where observed covariates influence survival characteristics. We here review these methods using a particularly accessible example on mortality of avalanche victims to motivate and illustrate methods, assumptions and results.

19.1 Introduction

Current status data provides information on the survival status of individuals at single screening times rather than standard observation, possibly right-censored, of failure times. It thus represents an extreme form of interval censoring where censoring intervals are either $[0, C]$ or (C, ∞), where C represents the time at screening on an appropriate timescale. Considerable attention has been given to estimation of a survival function based on such data, and estimation of regression coefficients from a variety of standard models where covariates influence survival characteristics. Earliest work was motivated by applications in demography (Diamond et al., 1986) and epidemiology (Becker, 1989), followed by carcinogenicity studies, partner studies of Human Immunodeficiency Virus (HIV) transmission (Jewell and Shiboski, 1990; Shiboski and Jewell 1992), and age-incidence estimation (Keiding, 1991).

Here, we motivate relevant estimation problems, and outline both parametric and non-parametric approaches to such, using data from people buried in avalanches in both Switzerland and Canada (Haegeli et al., 2011). This dataset contains information on 1,247 individuals who suffered a complete avalanche burial (coverage of the person's head and chest) between October 1, 1980 and September 30, 2005 in situations where there was complete information on the duration of burial and mortality outcome; of these, 946 data points are from Switzerland and 301 from Canada. We note here that the records do not distinguish between people who were dead upon discovery or who died subsequent to their rescue (that is, their injuries were so great, or their breathing sufficiently impaired, that they did not recover). The data was collected from databases of the Canadian Avalanche Centre and the WSL Institute for Snow and Avalanche Research SLF in Switzerland. Two thirds of the Canadian data were from the later half of the relevant time period, that is after October 1, 1992, whereas the Swiss data was more uniformly spread over the entire 25-year interval. Covariate information included the date and location of the avalanche, the type of outdoor activity involved (for example, backcountry skiing, snowmobiling, mountaineering and ice climbing, etc.), and the depth of burial (this variable was missing for 10% of the individuals), in addition to the crucial censoring variable C (the time from avalanche to discovery, or duration of burial). Some of the Canadian data also provided some additional information on the cause of death (asphyxia, hypothermia, and trauma) when fatality occurred, and on the snow climate at the location of the avalanche (maritime, continental, and transitional). Specifically, cause of death was recorded for 88% of the 162 Canadian fatalities, and snow climate for 89% of the 301 Canadian observations. We note that while the climate characterizations are related to snow density, this important characteristic was not measured. Figure 19.1 uses a variation of a binned scatter plot to show the data distribution of deaths and survivors, with regard to their burial times, for both countries.

The description of the survival curve, where time is measured chronologically from the time of avalanche (the time 'origin') until death, has formed the basis for international recommendations for rescue and resuscitation as well as improvements in safety and rescue devices. This has traditionally been based solely on Swiss data (Falk et al., 1994). For further discussion of the data, and its detailed interpretation and implications, we refer to Haegeli et al. (2011).

It is immediately apparent that the available data is exactly in current status format.

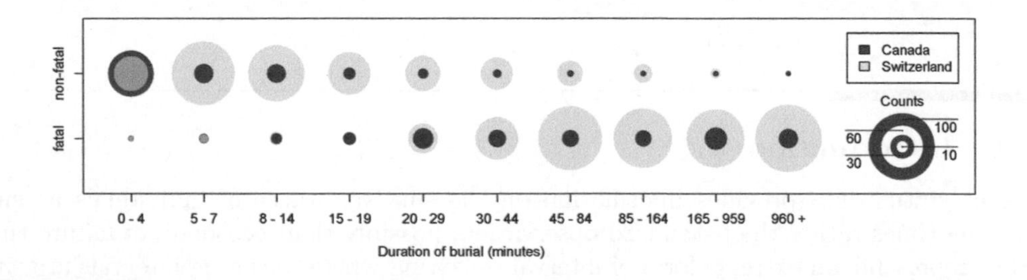

FIGURE 19.1
The distribution of burial durations for avalanche deaths and survivors in Canada and Switzerland.

That is, at the time of discovery (referred to as the "screening time" in many applications), the survival status of the individual is observed and known (with the slight proviso noted earlier about the definition of mortality); however, the actual time of death, where this is observed on discovery, is not known. Thus, if the random variable T denotes the time from avalanche until death (with distribution function F), and C is the time from avalanche until discovery, then the observed outcome data is simply n observations of the random variable (Y, C) where $Y = I(T \leq C)$. In the application, we assume that the discovery time C is also random, following a distribution function G; most current status techniques do not depend on whether C is assumed fixed (as in some applications) or random.

Here, we also assume that the n observations on (Y, C) are i.i.d.; this would, of course be violated, in cases where a specific avalanche buried multiple individuals (this is unknown to us regarding the data source). There is considerable recent research on the generalization of methods for current status data to cover multivariate or clustered current status observations where the i.i.d assumption is violated. In the bivariate case, where clusters involve at most two individuals, see Wang and Ding (2000) and Jewell et al. (2005). In general, and for more recent work, see, for example, Chen et al. (2009), Cook and Tolusso (2009), and Wen and Chen (2011).

As current status data is a special case of interval censored data, many of the statistical estimation and inference problems can be tackled using general interval censoring ideas (Sun, 2006). However, given the very specialized nature of current status data, there are often insights and more efficient algorithms available for this special case. For more extensive reviews of statistical methods for current status data, we also refer the reader to Jewell and van der Laan (2004b) and Sun (2006, Chapter 5). Banerjee (2012) provides an excellent review of theoretical developments and future areas of research.

19.2 Estimation of a single distribution function

Assuming an i.i.d. random sample of the population as described above, the likelihood of the data is given by:

$$L = \prod_{i=1}^{n} F(c_i)^{y_i}(1 - F(c_i))^{1-y_i} \mathrm{d}G(c_i), \tag{19.1}$$

where the critical assumption of *independence of T and C* is exploited. In the avalanche example, this is a plausible assumption, akin to claiming, for example, that rescues are not performed more quickly in cases where there is some information available on the survival status of the individual (e.g., imagine a scenario where skiers wear a sensor that would transmit vital signs to rescuers). See for example van der Laan and Robins (1998) for a discussion of theory in the case where the screening time, C, and failure time, T, are dependent through observable (time-dependent) covariates (e.g., the measure of transmitted vital signs in our hypothetical example).

With independent screening, estimation of F can be simply based on the conditional likelihood of Y, given C, given by

$$CL = \prod_{i=1}^{n} F(c_i)^{y_i}(1 - F(c_i))^{1-y_i}. \tag{19.2}$$

Parametric estimation of F is then straightforward using standard maximum likelihood techniques. A very simple illustration occurs when F is assumed to be Exponential, with constant hazard λ, with all screening times the same, that is $c_i = c$ for all i. In this case, the maximum likelihood estimator of λ is just

$$\hat{\lambda} = \big[-\log\big(\frac{n-r}{n}\big)\big]/c, \tag{19.3}$$

where $r = \sum_{i=1}^{n} y_i$, the total number of failures observed by time c. In such a simple case, it is possible to calculate the amount of (asymptotic) information lost in terms of estimation of λ, the constant hazard, when only current status information is available at a sole screening time c, as compared to standard observation of exact failure times with right censoring at c. By computing the (expected) information, it is straightforward to show that the asymptotic relative efficiency (ARE) of the current status data estimator, as compared to the right-censored data estimator, is simply

$$ARE = \frac{(1-p)\big(\log(1-p)\big)^2}{p^2} \tag{19.4}$$

where $p = 1 - e^{-\lambda c}$ is the probability of failure by time c. Thus, for example, if the expected fraction of uncensored failures is 0.5, the ARE of having only current status information is 96.1%. If $p = 0.1$, the ARE rises to 99.9%. On the other hand, if 90% of observations are expected to fail by time c, the ARE drops to 65.5%. The high efficiency of the limited current status information is perhaps counter-intuitive, but largely depends on the strong parametric assumption as is evident when considering nonparametric approaches. On the other hand, in situations where the relative efficiency is high, use of current status data may be preferred if measurement of exact failure times is subject to substantial amounts of error. With the avalanche data, there were no attempts to record the actual time to death for fatalities, and so only current status information was available.

The nonparametric maximum likelihood estimator (NPMLE) of F, based on (19.2), can be estimated algorithmically using the pool-adjacent-violators-algorithm of Ayer et al. (1955). In R, it can be calculated using the *gpava* function within the package *isotope* (de Leeuw, 2009). It is also possible to use the EM method to calculate the NPMLE, but the resulting algorithm is considerably slower. For the avalanche data, Figure 19.2 displays the NPMLE of both the Swiss and Canadian data separately as well as combined. From these estimates it is apparent that, at first glance, survival prospects are much worse for Canadian avalanche victims than in Switzerland. Of course, it remains to be seen the extent to which the observed survival differences can be attributed to chance variation or due to differences in the nature of the avalanches and victims. We will investigate both of these issues in further detail below.

Figure 19.3 reproduces the same NPMLE for Switzerland and, for comparison, the survival curve of a Weibull parametric model first to the same data. While the Weibull model is reasonable for the left-hand tail (the first 100 minutes of burial), it does not fit quite so well after that time period where it overstates survival probabilities up to almost 300 minutes and then understates them thereafter. On the other hand, for the same data, the likelihood ratio test unequivocally rejects the null hypothesis of a constant hazard (i.e., an Exponential distribution) with a p-value $< 10^{-8}$; as Figure 19.3 shows indirectly, the hazard function for the fitted Weibull distribution for Switzerland decreases over time buried. It remains to be seen the extent to which this is due to the risk of varying causes of death changing differentially over time, or whether covariates (such as depth of burial) are distributed differently for early versus late screening times. We follow up on both of these issues below.

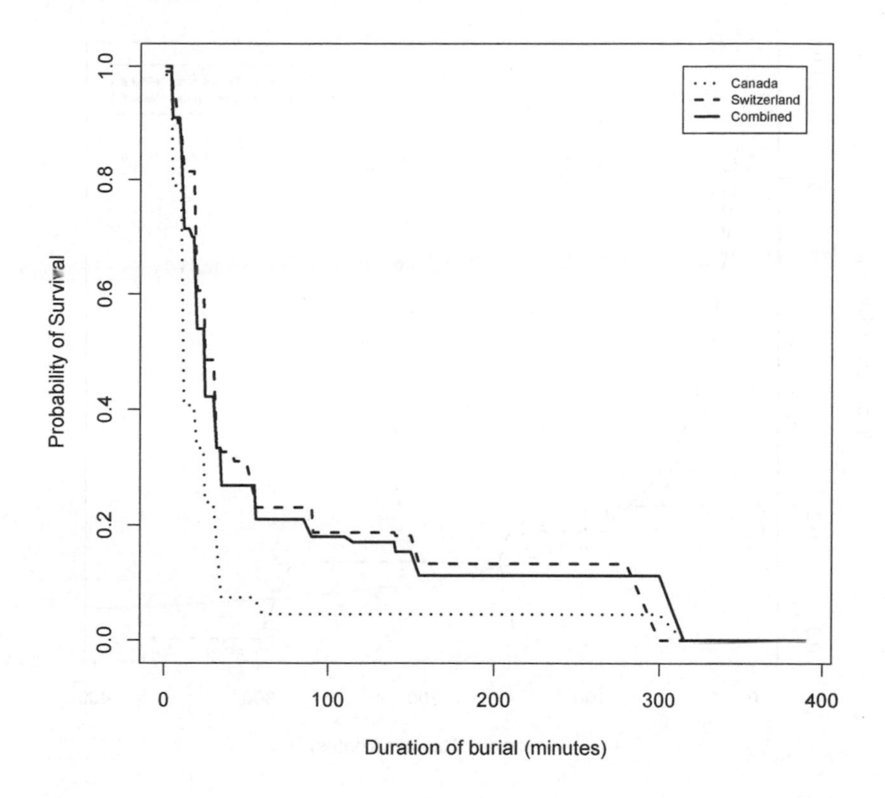

FIGURE 19.2
Plot of the NPMLE estimates of the survival distribution, S, for Canada, Switzerland, and the two countries combined.

19.2.1 Inference

In principal, it is straightforward, at least for large samples, to compute confidence bands (either pointwise or for the entire survival curve) associated with parametric estimation, using (asymptotic) Fisher information calculations. However, for nonparametric estimation, assessment of uncertainty in estimation of the survival distribution is much more complex. First, we note that it is not possible to carry out simple efficiency comparisons of the NPMLE and the Kaplan-Meier curve (based on right-censored data observations), even at a single point in time using variance comparisons. Even pointwise confidence intervals associated with the NPMLE are much more complicated when nonparametric methods are used. Difficulties are immediate because (i) the convergence for the NPMLE is now at rate $n^{1/3}$ (as opposed to the familiar $\sqrt{n}$ rate for the Kaplan-Meier estimator), and (ii) the limit distribution of the estimator of F at a single point point t_0 is not Gaussian (Groenboom and Wellner, 1992). Thus it is not appropriate to focus on the (asymptotic) variance of the NPMLE based on any form of current status data as a step towards confidence interval construction. The slow rate of convergence illustrates the substantial loss of information associated with current status observations, as compared to standard right censoring, when nothing is known *a priori* regarding the shape of the survival curve; note, however, that

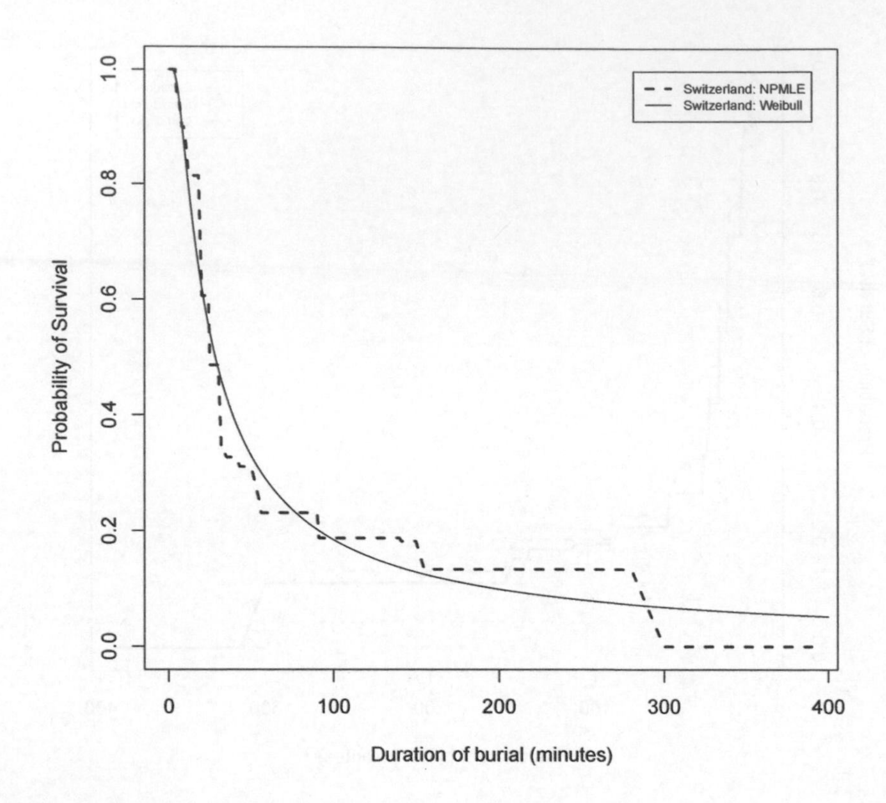

FIGURE 19.3
Plot of the NPMLE estimate of the survival distribution, S, for Switzerland together with an estimate based on a Weibull distribution assumption.

comparisons can still be made with regard to estimation of *smooth functionals* of F where rates of convergence for both methods are equivalent.

Various approaches have been developed for pointwise confidence intervals for F, based on the NPMLE for standard current status data, and are reviewed in Banerjee (2012). In particular, it is possible to compute Wald-type confidence intervals based on the limiting distribution of the NPMLE of $F(t_0)$. However, this approach requires estimation of the *density function* associated with the distribution of the status screening variable, denoted above by G. Even more challenging is that the method also requires estimation of the density function f, associated with F, at t_0. This is much harder, with only current status information on the survival times available, as compared to estimating the density associated with G given uncensored screening times (C) are observed for all individuals.

These difficulties motivated research into alternative methods that avoid such complex estimation problems along the way. The bootstrap procedure suggests a potentially simpler approach. In general, however, the standard bootstrap procedure yields inconsistent estimates of pointwise confidence intervals, whether data is sampled with replacement from the original data or generated from the NPMLE estimator (Sen, Banerjee and Woodroofe, 2010). As a modification, a smoothed version of the bootstrap is possible, as is the m out of n bootstrap (Politis et al. 1999). For practical implementation, the latter procedure necessarily involves choice of the block size m. Asymptotically, m must be chosen so that $m \to \infty$ and $m/n \to 0$ as $n \to \infty$, although these requirements provide little guidance for a finite sample size. Banerjee and Wellner (2005) suggest an intricate procedure for selecting m, based itself on bootstrapping. The method can be adapted to provide symmetric confidence intervals as these often perform better in finite samples. Banerjee and Wellner (2005) provide further implementation details although by now it should be clear that we have lost much of the simplicity that motivated the bootstrap method as an appealing procedure.

The most attractive method is thus based on the likelihood-ratio (LR) method that does not require estimation of nuisance parameters. This approach is developed in detail in Banerjee and Wellner (2001; 2005), and constructs confidence limits for $F(t_0)$ by considering tests of the null hypothesis, $H_0 : F(t_0) = a$, against its complement. These methods (and their asymptotics) depend on the screening time distribution G being continuous; that is, there can be no tied screening times. This is, however, not true, for the avalanche data, at least as currently available. (It is apparent that there is, in fact, a fair amount of rounding, or digit preference, with many burial times given to the nearest 5 minutes.) This raises an interesting question (originally posed by Kalbfleisch and Jewell) regarding the appropriate asymptotics when there are only a few screening times with multiple observations sharing the same screening value. In the most trivial case, when there is a single screening time, C_0, a suitable confidence interval, associated with the nonparametric estimate of $F(C_0)$, can be based on standard $\sqrt{n}$ asymptotics and the familiar χ^2 distribution. This approach can readily be generalized to a small number of screening times for each of which there are multiple observations associated. However, at some point, as the number of screening times increases (so that the number of observations at each becomes small), we might expect the asymptotics to approach the continuous G case described above. In practice, of course, we do not know how the number of screening times will increase as the sample size grows asymptotically large. Generally speaking, this suggests that if the number of screening times is much less than the total sample size, standard normal approximations to the distribution of the NPMLE should be reasonable. On the other hand, if the number of screening times is essentially the same as the sample size, the LR method due to Banerjee and Wellner (2001; 2005) should be used.

This leaves the question of what should be done when the dataset sits between these two extremes (as it does with the avalanche data), where there are a large number of screening times but also multiple observations at many of the observed screening times, a scenario where neither of the two "extreme" methods may be effective. This problem has been studied intensively by Tang et al. (2012) who provide a method that reaffirms the two "extreme" approaches when appropriate, and identifies a third boundary possibility that establishes a connection between the two extremes, the three situations distinguished by how fast the number of screening times grows as compared to the sample size. Since, as noted, we cannot tell from a single dataset which scenario is appropriate, Tang et al. (2012) suggest an adaptive inference scheme, using the boundary result to construct Wald- or LR-type confidence intervals, without assuming knowledge of the rate of growth of the number of monitoring times as the sample size increases. Figure 19.4 illustrates the Tang et al. (2012) estimates of Wald-type 95% pointwise confidence intervals, associated with the NPMLE, for the Swiss avalanche data. With the avalanche data, there is an additional complication in that it is necessary to calculate local averages of the relative density of the screening observations per unit time from the data. Here, the density of screening times clearly varies substantially across the entire scale of burial times as the empirical distribution function estimate for G is skewed. To partially alleviate this issue, and to focus on the more informative data associated with shorter burial times, we ignored burial times greater than 240 minutes (4 hours); for the Swiss data, 150 (of the original 946) data points had burial durations greater than 240 minutes and all but two of these individuals had perished on recovery. Further,

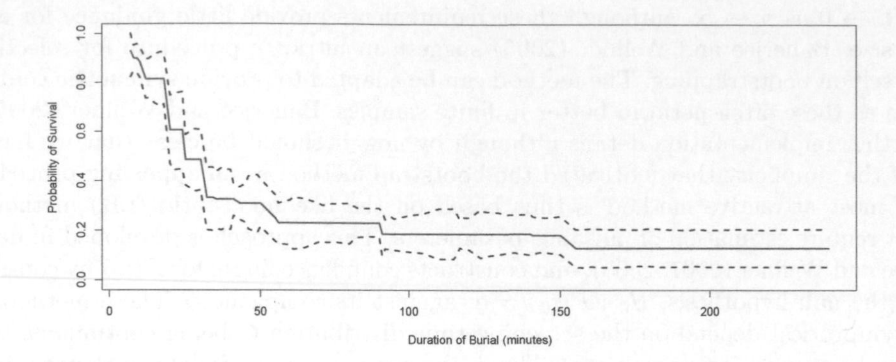

FIGURE 19.4
The NPMLE estimate of the survival distribution, S, for the Swiss avalanche data (solid line), together with Wald-type pointwise 95% confidence intervals (dashed lines) based on the adaptive method of Tang et al. (2012).

for burial times ≤ 150 minutes, the bandwidth used to estimate the density of screening times is determined pointwise, whereas a single bandwidth is used to achieve robustness for times between 150 and 240 minutes where the number of observations is relatively much more sporadic. Figure 19.4 shows that the pointwise limits place the Canadian NPMLE on the lower edge of the Swiss confidence intervals (after 30 minutes), suggesting that the differences between the two countries regarding mortality patterns are unlikely to be due solely to chance. We will examine this comparison more formally in Section 19.3 below. Finally, we note that the method of Tang et al. (2012) is not invariant under a monotone transformation of the time axis; for very uneven patterns of screening times this therefore suggests using a pre-analysis monotone transformation of the time measurements to make the observation screening times correspond more closely to an evenly spaced grid, with a subsequent transformation back to the original time scale after the confidence limits have been calculated.

19.3 Regression methods

For the avalanche data, we are interested in comparison of the survival experiences for Swiss and Canadian victims. Further, there is interest in whether, and the extent to which, survival patterns have evolved over time, and whether the risk of death is influenced by the nature of the activity (prior to the avalanche) or the snow climate, and whether such factors contribute to the noted survival differences between the two countries. These questions can all be tackled using various regression models that exploit different assumptions about the covariates influence on survival.

We first note the correspondence between regression models that link T and Y to an underlying k-dimensional covariate vector $\mathbf{Z}$ (Doksum and Gasko, 1990). This is particularly useful since estimates of parameters from the regression model for the observed Y can then be interpreted in terms of the parameters in the regression model for the unobserved T. We illustrate this simple idea for three familiar regression models.

For example, suppose that the survival times follow a proportional hazards model (Cox, 1972) of the form

$$S(t|\mathbf{Z} = \mathbf{z}) = [S_0(t)]^{e^{\beta \mathbf{z}}} \tag{19.5}$$

where S_0 is the baseline survival function for the sub-population where $\mathbf{Z} = \mathbf{0}$, and β is a k-dimensional vector of regression coefficients. Each component of β gives the relative hazard associated with a unit increase in the corresponding component of $\mathbf{Z}$, holding all other covariates fixed. On the other hand, this model can be equivalently written in terms of $p(\mathbf{z}|c) = E(Y|C = c, \mathbf{Z} = \mathbf{z})$, as

$$\log - \log(1 - p(\mathbf{z}|c)) = \log - \log[S_0(c)] + \beta \mathbf{z}. \tag{19.6}$$

This is thus a generalized linear model for Y with complementary log-log link and offset given by $\log - \log[S_0(c)]$. In the special case where the baseline survival curve is itself parametrically described, this offset may be simplified further. For example, if the baseline survival curve follows a Weibull distribution with hazard function $e^a bt^{b-1}$, then

$$\log - \log(1 - p(\mathbf{z}|c)) = a + b \log c + \beta \mathbf{z}. \tag{19.7}$$

As an aside, this allows easy estimation of the Weibull model without covariates, illustrated in Figure 19.2 for Switzerland, using any generalized linear model program that allows for the complementary log-log link with binomial data.

In the familiar semi-parametric Cox model, where the baseline survival curve is unspecified, the offset curve in (19.6) is then given by an arbitrary increasing function of the "covariate" C. It is important that additivity of the effects of C and the covariates $\mathbf{Z}$ are preserved in the generalized linear model, but specialized estimation algorithms are now required that account for the nonparametric shape of $\log - \log[S_0(c)]$ while exploiting its monotonicity (Shiboski, 1998). Huang (1996) discusses the asymptotic properties of the maximum likelihood estimators for the semi-parametric model.

As an alternative approach, assume that T follows the proportional odds model (Bennett, 1983), defined by

$$1 - S(t|\mathbf{Z} = \mathbf{z}) = \frac{1}{1 + e^{-\alpha(t) - \beta \mathbf{z}}}, \tag{19.8}$$

where $S_0(t) = \frac{1}{1+e^{\alpha(t)}}$. In this case, Y is associated with $\mathbf{Z}$ via the logit link:

$$\log \frac{p(\mathbf{z}|c)}{1 - p(\mathbf{z}|c)} = \alpha(c) + \mathbf{z}, \tag{19.9}$$

Finally, suppose T follows the accelerated failure time regression model,

$$\log T = \alpha + \beta \mathbf{z} + e, \tag{19.10}$$

at a given level of the covariates $\mathbf{Z}$, where the random variable e is independent of $\mathbf{Z}$ and follows a distribution H. Then, Y follows a generalized linear model in $\mathbf{Z}$ using the link function H^{-1}, with fixed offset given by $\log C$, where the intercept is now $-\alpha$ and slope $-\beta$. Technically, this is not a standard generalized linear model since the link function is often unspecified in the accelerated failure time model, and estimation in such a case would require methods to estimate the regression coefficients with an unknown link. For methods to approach this problem generally, see Weisberg and Welsh (1994), Mallick and Gelfand (1994), and Young and He (1997), for example.

As for the proportional hazards model, estimation of the coefficients in parametric versions of these regression models follows simply from estimation of the appropriate coefficients in the GLM. Again, semi-parametric estimation where the baseline survival function is left unspecified is more complex. Fully efficient semi-parametric estimators have been studied in each of these cases but involves sophisticated asymptotic theory (Huang, 1995; Rossini and Tsiatis, 1996; Rabinowitz et al., 1995; Murphy et al., 1999; Shen, 2000; Tian and Cai, 2006).

It is important to note that none of these approaches can easily accommodate time-dependent covariates as the above derivations have all tacitly assumed fixed covariates. This is not the case, however, for additive hazards models, at least in simple cases with suitable assumptions. To see this, first consider the basic additive hazards model

$$\lambda(t|\mathbf{Z}=\mathbf{z}) = \lambda_0(t) + \beta\mathbf{z}, \tag{19.11}$$

where λ_0 is the hazard function when $\mathbf{Z}=\mathbf{0}$, and β is a k-dimensional vector of regression coefficients as before. We now make the assumption for the moment that the screening time *C is independent of the covariates* $\mathbf{Z}$ in addition to the failure time T. In order to estimate β, Lin et al. (1998) focused on the survival processes associated with the screening time rather than on T itself. Specifically, the trick is to consider events where failure is associated with the observation of the monitoring time (i.e., duration of burial in the avalanche example), but consider observed Cs where $T > C$ (or $Y=1$) as "failures" and those for whom $T \leq C$ (or $Y=0$) as censored. For an individual with covariates $\mathbf{z}$, the hazard for such "failures" is then simply the hazard function associated with the screening time distribution G (λ_G), times $Pr(T > C|\mathbf{Z}=\mathbf{z})$, that is

$$\lambda_G(c)S(c|\mathbf{Z}=\mathbf{z}) = \lambda_G(c)\exp(-\Lambda_0(c))\exp(-\beta\mathbf{z}c) \tag{19.12}$$

from (11), where $\Lambda_0(c) = \int_0^c \lambda_0(u)du$. Of course, this is just the proportional hazards model for this filtered screening time process with time-dependent covariates $\mathbf{z}c$. Thus, standard software for the proportional hazards model can directly be used with, for the ith individual, observation time c_i, covariates $\mathbf{z}_i c_i$ with $y_i = 0$ corresponding to failures, and $y_i = 1$ to censored observations. This is easily extended to time-dependent $\mathbf{z}$ by simply replacing $\mathbf{z}c$ in (12) by $\int_0^c \mathbf{z}(u)du$, of course assuming that the *entire* covariate process $\{\mathbf{z}(u) : o \leq u \leq c\}$ is observed and not just $\mathbf{z}(c)$. This approach, while easy to implement, is not fully (semi-parametrically) efficient as the "censoring" distribution associated with the filtered monitoring time process also depends on β so that the censoring times are informative.

The assumption that the screening time process is independent of the covariates is necessarily very restrictive and unlikely to be true in practice. For example, there is no a priori reason to believe that the burial time distribution in Switzerland is the same as in Canada, for example. However, if it is assumed that the monitoring times depend on the covariates through proportional hazards models themselves, the above derivation can be generalized in a relatively straightforward way (Lin et al. 1998). Thus, if $\lambda_G(c|\mathbf{Z}=\mathbf{z}) = \lambda_{G0}(c)\exp(\gamma\mathbf{z})$, where λ_{G0} is the baseline hazard function for the screening time distribution (that is, when $\mathbf{z}=0$) and γ is a vector of regression coefficients. In this case, the hazard function for the filtered screening time process, (19.12), is modified to

$$\lambda_G(c)S(c|\mathbf{Z}=\mathbf{z}) = \lambda_G(c)\exp(-\Lambda_0(c))\exp(\gamma\mathbf{z})\exp(-\beta\mathbf{z}c). \tag{19.13}$$

This is again a proportional hazards model that can be used to directly estimate γ and, more importantly, our original parameter of interest β. However, it is more efficient to first estimate γ directly from the proportional hazards model applied directly to the observations of the screening times C_i and the covariates $\mathbf{z}_i$. The estimate $\hat{\gamma}$ can then be used in (19.13) to supply a fixed offset, leaving only β to be estimated. Although this is straightforward with proportional hazards software allowing offsets, the naive standard error provided from such will not be correct as it does not allow for the estimation of γ. Use of the bootstrap would be a possibility here to avoid the more complex computation of an estimate of the (asymptotic) standard error provided in Lin et al. (1998).

Martinussen and Scheike (2002) tackle both the issue of semi-parametric efficiency and allow for a general dependence structure between the screening times and covariates rather than restricting to a proportional hazards relationship. However, their approach now requires estimation of the baseline hazard function for T, namely Λ_0, making this approach more akin computationally to those used for other regression models. Lu and Song (2012) provide a more stable computational approach to obtaining a semi-parametric efficient estimator but again restricted to the assumption of a proportional hazards model for the screening times and covariates. Their approach also requires an estimate of Λ_0.

It is instructive to apply these ideas to the avalanche data. For example, the two group comparison between the Swiss and Canadian survival curves (whose NPMLEs are illustrated in Figure 19.2) can be immediately obtained using a simple indicator variable for the country location in the model given by (19.6) or (19.7) based on a proportional hazards assumption. If the baseline survival distribution is assumed to be Weibull, as in (19.7), then the Relative Hazard of death, comparing Canada to Switzerland, is estimated to be 1.9 with a 95% confidence interval of (1.5, 2.5), so that the mortality hazard rate in Canada is about double what is observed in Switzerland. This is, of course, a strikingly significant difference with a p-value $< 10^{-7}$. On the other hand, if we make no underlying distributional assumption and fit the semi-parametric Cox model, the estimate of the same Relative Hazard is even larger at 2.2 with a 95% confidence interval of (1.8, 2.8), obtained using the algorithm of Shiboski (1998). This semi-parametric estimate therefore reflects that there is probably a small amount of bias induced by assuming a Weibull survival model for both countries. Necessarily, the semi-parametric estimate is somewhat less precise; here, estimates of the standard error of estimators are based on the ideas of Huang (1996), although the finite sample properties of this approach have never been fully explored. Since the estimators of the regression coefficients enjoy $n^{1/2}$ asymptotics, the bootstrap provides a viable alternative although we do not pursue this further here.

Note that both of these analyses directly account for variation in the burial time distributions across the two countries which are substantially shorter in Canada than in Switzerland on average (median burial times of 18 and 35 minutes in Canada and Switzerland, respectively). An interaction term included in the Weibull model allows for differing scale parameters in the two countries, but the associated coefficient is not significant (($p = 0.6$), and so this complication was deemed unnecessary.

We illustrate also the additive hazards model here in this simple setting with a single binary covariate. Assuming equivalence of the distributions of duration of burial time between the two countries yields an estimated additive shift of the hazard of 0.0087 (per minute), or 0.52 (per hour), with Canada displaying the higher hazard as we have already seen from our proportional hazards analysis. The effect is statistically significant ($p = 0.03$), although much less so than under proportional hazards. However, this analysis is certainly distorted by the fact that the screening time distribution varies across countries as noted above. In fact, fitting a proportional hazards model to the fully observed burial times and country indicates a Relative Hazard of 1.40, comparing Canada to Switzerland, with a 95% confidence interval of (1.23, 1.59). Using the estimate of γ, arising from this analysis, as a fixed coefficient of country in the model (19.13) yields an estimate of β, the additive hazard shift between countries, of 0.017 (per minute), almost double what our prior biased analysis suggested. However, this new estimate is itself subject to residual bias since there is some evidence that the screening time hazard functions for the two countries are not proportional. In fact, fitting a simple interaction term (for country with time of follow-up on the screening time scale) shows that the Relative Hazard comparing Canada to Switzerland increases with time ($p = 0.03$). Thus, to pursue the additive model further, even in this simple case, would require the more complex evaluation of the semi-parametric efficient estimator of Martinussen and Scheike (2002), or consideration of a more complicated model for the relationship between the screening time distributions and country that does satisfy the proportional hazards model.

Returning to the proportional hazards regression setting, we now investigate the effects of other covariates on survival, in part to determine whether they can explain the differences between the two countries. For example, the median burial depth in the Canadian data is 100 centimeters, whereas for Switzerland it is less at 80 cm. Could this, at least partially, explain different mortality patterns over time? That is, when individuals are recovered at the same time is the risk of death greater for those who are buried more deeply? When the burial depth variable is added to the Weibull model in a logarithmic form, it indeed shows a significant effect with an estimated Relative Hazard of 1.33, for every tenfold increase in depth, with a 95% confidence interval of (1.16, 1.53) and associated p-value of $6 \times 10M^{-5}$ (a similar estimate of 1.38, with 95% confidence interval of (1.22, 1.57) is obtained from the semi-parametric version). While this is not surprising, adjusting for burial depth in the Weibull model actually *increases* the estimated Relative Hazard, comparing Canada to Switzerland, to 2.1, showing that, at least naively, burial depth does not explain any of the country difference in survival curves. We note here that this was a complete case analysis, ignoring 65 and 53 individuals in Switzerland and Canada, respectively, who have missing data on depth. There is little evidence of interaction between depth and country ($p = 0.5$).

Continuing to adjust for country and depth, a further Weibull proportional hazards analysis of activity (prior to burial) showed no difference between survival comparing out-of-bounds and backcountry skiing; there were slight (statistically insignificant) differences for other recreational activity, and snowmobiling, as compared to backcountry skiing with a 30% reduction in hazard, and 42% increase, respectively. Similarly, there was no significant difference in risk comparing the activities mountaineering (including ice climbing) and mechanized skiing. For simplicity we therefore combined the activities backcountry skiing, other recreational, and snowmobiling into one activity class (the reference category), mountaineering and mechanized skiing into a second class, leaving non-recreational as a separate activity class. Table 19.1 shows the Weibull regression model accommodating these three activity groups through indicator variables, supplemented by the year of occurrence of the avalanche (measured in single year increments); in this table the variable Country = 1 refers to Canada. For comparison, we show the regression coefficient estimates from the semi-parametric proportional hazards model with an unspecified baseline hazard function; in this model, standard error estimates are based on the procedure suggested by Huang (1996). We now observe a slight reduction in the country effect, with an estimated Relative Hazard of 1.8, controlling for the main effects of the other covariates in the Weibull model. There is a substantially increased Relative Hazard of 1.7 associated with avalanche victims who were pursuing mountaineering, ice climbing and mechanized skiing as compared to the expanded backcountry skiing activity category. On the other hand, the hazard is reduced by 55% for those buried while involved in non-recreational activities. Note that these effects already control for the length of time buried and the depth of the burial. Finally, there is a slight trend in increasing risk over chronological time with roughly a 1% hazard increase per year over the observed quarter century; however, this trend is not statistically significant ($p = 0.2$). Qualitatively, the coefficient estimates from fitting the semi-parametric proportional hazards model are very similar although they surprisingly enjoy slightly lower standard errors. In addition, in this version of the model, the effect of chronological time is elimitated entirely. Since the semi-parametric model is more general, it is likely that these

TABLE 19.1
Regression coefficient (log Relative Hazards) estimates (B), and associated estimated standard errors (SE), for both a Weibull proportional hazards regression model and a semi-parametric version that leaves the baseline hazard function unspecified ($n = 1127$).

	Weibull Model		Unspecified Baseline Hazard Model	
Variable	B	SE	B	SE
Country	0.567	0.152	0.430	0.121
$\log_{10}$(Burial Depth in cm.)	0.285	0.072	0.358	0.069
Mountaineering, Ice Climbing & Mechanized Skiing	0.544	0.188	0.482	0.151
Non-Recreational	-0.791	0.290	-0.521	0.263
Year	0.010	0.007	-0.001	0.007

results suffer from less bias (assuming the model is correct), and are subsequently more reliable.

For Canada, we can expand the analysis to include snow climate information using a single indicator variable (with transitional and continental climates as the reference category as these showed almost identical survival patterns). The results are shown in Table 19.2. The increased risks associated with avalanches in maritime snow climates is large with an estimated Relative Hazard of 2.8 with a 95% confidence interval of (1.2, 6.8) based on the Weibull model, the loss of precision reflecting the much smaller amount of Canadian data. The effect is slightly smaller in the semi-parametric model with an estimated Relative Hazard of 1.8. Of interest, the Canadian analysis, using the Weibull model, suggests some interactive effects in that burial depth has no influence on mortality (as compared to the Swiss data implicitly), and the chronological time trend is much stronger here than we noted for the combined data (in Table 19.1) although the precision for estimating this effect is low. However, neither of these observations are supported by the semi-parametric model estimates which is likely more reliable. These subtle changes nevertheless illustrate the influence of model assumptions that must therefore be considered carefully (let alone whether the proportional hazards model is adequate, an issue not explored further here). For the Weibull model the coefficient on log(Burial Time), as in (19.7), now shows a (non-significantly) increasing hazard as the burial time increases (in contrast to what we saw before for the Swiss data in Figure 19.2); this estimate is not displayed in Table 19.2. Haegeli et al. (2011) argue that the effect of a maritime snow climate explains the early differences between the survival curves for Switzerland and Canada (Figure 19.2) as the Swiss snow climate is described as "transitional to partly continental," although this interpretation is clearly tentative (see Roggia, 2011). While our goal has not been to implement a definitive regression analysis and interpretation, the dataset serves as an excellent introduction of how current status techniques can attack central survival analysis questions using this extreme form of interval censored data.

Finally, we make some brief remarks regarding estimation of regression coefficients for survival models applied to current status data when some covariate data is missing. For parametric models, the analogue to generalized models that is described above allows methods for missing data for such models to be employed directly. On the other hand, for estimation

TABLE 19.2
Regression coefficient (log Relative Hazards) estimates (B), and associated estimated standard errors (SE), for both a Weibull proportional hazards regression model and a semi-parametric version that leaves the baseline hazard function unspecified ($n = 218$).

	Weibull Model		Unspecified Baseline Hazard Model	
Variable	B	SE	B	SE
$\log_{10}$(Burial Depth in cm.)	-0.035	0.214	0.296	0.168
Mountaineering, Ice Climbing & Mechanized Skiing	0.700	0.316	0.518	0.268
Non-Recreational	-0.339	1.072	-0.383	0.913
Year	0.043	0.029	-0.001	0.022
Maritime Snow Climate	1.041	0.448	0.574	0.312

of the semi-parametric proportional hazards model with current status data, see Wen and Lin (2011).

19.4 Competing risks

As noted in the Introduction, some of the Canadian avalanche data has information on the cause of death when mortality was observed. Of the 162 deaths in the Canadian data, 116 were attributed to asphyxia, 27 to trauma, and 19 recorded as cause unknown. It is possible that the risks associated with different causes evolve quite differently as the amount of time buried increases. For example, one might think that trauma deaths likely occur quickly (due to perhaps a head injury as the avalanche occurs), whereas asphyxia deaths occur over longer time periods. Jewell et al. (2003) discuss the analysis of current status data in such a competing risk environment.

If we denote the cause of death by a random variable J, interest then focuses on estimation of the cumulative incidence functions associated with each cause

$$F_j(t) = pr(T \leq t, J = j). \tag{19.14}$$

In the avalanche example, J takes on three values, say 1, 2, 3, and the overall survival function is given by $S(t) = 1 - F(t) = 1 - F_1(t) - F_2(t) - F_3(t)$. Jewell et al. (2003) consider the NPMLEs of F_j for $j = 1,,3$ (and simpler-to-compute alternative estimators). The NPMLEs are a special case of competing risk data estimation for data subject to general interval censoring, considered in Hudgens et al. (2001).

While the EM algorithm could again be used to compute the NPMLE estimators, Jewell and Kalbfleisch (2004) describe a much faster iterative algorithm that generalizes pool-adjacent-violators, and establish convergence. The R-package *MLEcens* provides an alternate estimation procedure based on the height map algorithm of Maathuis (2005) supplemented by convex optimization. Groeneboom et al. (2008a, b) establish the asymptotic properties of the NPMLEs of the cumulative incidence functions F_j using methods that require significant extensions of the ideas used for the single cause situation.

In implementing the NPMLEs for competing risks there is a slight additional issue of identifiability that rises in certain scenarios: when one of the screening times has all observed failures of a single type, any possible jump in the cumulative incidence curves for the other types is not identifiable since the mass for such types does not appear in the likelihood function. This occurs at several burial time points for the Canadian avalanche data. Jewell and Kalbfleisch (2004) used the convention that such unidentifiable jumps were zero; an alternative convention, at the other extreme, raises the estimate of the relevant sub-distribution function to the highest value possible compatible with estimates at higher screening times preserving monotonicity (essentially adding all mass possible). Figure 19.5 shows the NPMLE estimates of F_j for $j = 1, 2, 3$ for the Canadian data, with both conventions displayed when there is such non-identifiability. As the raw data suggest, the estimates reflect that death is much more likely due to asphyxia than trauma and that the cumulative risk for asphyxia-related deaths continues to grow at least through the first 40 minutes of burial.

Regression models for competing risks current status data have been less studied than in the single cause case and any generalized linear model for the observed data (survival or death plus cause) has a more complex outcome than binary. Jewell (2007) shows the polytomous logistic regression model for the observed outcomes with current status data directly corresponds to a generalization of the proportional odds survival model (for the full unobserved failure times and causes) to the competing risks situation (in much the same way as occurs in the single cause setting as described in Section 19.3, but only in the special

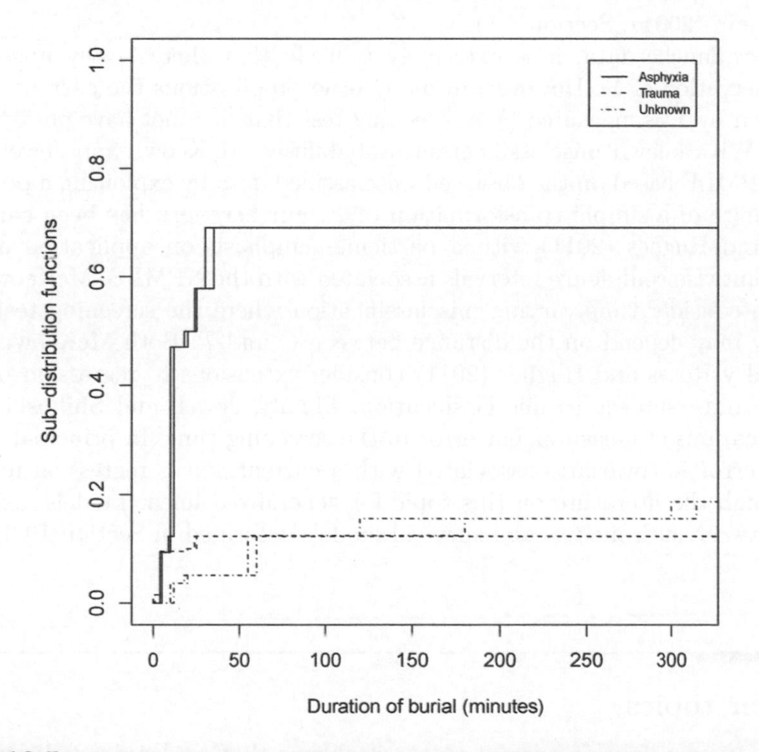

FIGURE 19.5
The NPMLE estimates of the cumulative incidence functions, F_1, F_2, F_3, for the Canadian avalanche data, for three causes of deaths: asphyxia, trauma and unknown.

case where the underlying cause-specific hazards are proportional. Sun and Shen (2009) describe estimation for the proportional hazards model with competing risks and current status data but have to approach estimation of the underlying parameters (and baseline cumulative incidence functions) from first principals rather than relying on existing software for generalized linear models. Unfortunately, this algorithm is not yet widely accessible in available software.

19.5 Sampling and measurement issues

To this point we have assumed complete ascertainment of all avalanche victims, or at least a random sample of the universe. However, it is plausible that data on avalanche survivors may be underreported in comparison to cases where a death occurs. That is, the sampling frequency of "cases" (i.e., where $Y_i = 1$ or $T \leq C$) differs from that for "controls" ($Y = 0$ or $T > C$). Assuming the availability of separate simple random samples of both cases and controls, Jewell and van der Laan (2004*a*) show that such data only (nonparametrically) identifies the odds function associated with F, that is $\log\left[\frac{F(t)}{1-F(t)}\right]$, up to a constant. If the two sampling frequencies are known, then simple weighting of the observations (inversely proportional to the probability of selection) prior to application of the pool-adjacent-violators algorithm produces the NPMLE. For parametric models, the situation is somewhat more complex, although an iterative algorithm due to Scott and Wild (1997) can be used directly to obtain the maximum likelihood estimator of F. See Jewell and van der Laan (2004*a*, Section 4.1).

With the avalanche data, it is extremely unlikely that there is any misclassification present in observation of Y. However, in many other applications the current status of an individual given by Y is measured by a screening test that may not have perfect sensitivity or specificity. With known misclassification probabilities, McKeown and Jewell (2010) determine the NPMLE based on the observed misclassified data by exploiting a pool-adjacent-violators estimate of a simple transformation of F. Further work has been carried out by Sal y Rosas and Hughes (2011) with a particular emphasis on application of likelihood ratio based pointwise confidence intervals associated with the NPMLE. McKeown and Jewell (2010) also consider time-varying misclassification where the screening test sensitivity and specificity may depend on the distance between C and T. Both McKeown and Jewell (2010) and Sal y Rosas and Hughes (2011) consider extensions to regression problems for current status data subject to misclassification. Finally, Jewell and Shiboski (1990) discuss the implications of measurement error in the screening time. In principal, the issue of measurement error in covariates associated with a current status regression model can be explored through the literature on this topic for generalized linear models, exploiting the connection between such models and survival models described in Section 19.3.

19.6 Other topics

There are many extensions to current status problems that we have not discussed in this introduction to simple analyses of current status data, in large part because they are not relevant to the motivating avalanche data. We simply note several of these here to encourage further reading. First, in some applications it is possible to screen individuals more than once with regard to their survival status. This, of course, leads to the more general case of interval-censored data; the book by Sun (2006) provides an excellent introduction to this topic. Of course, more frequent screening is advantageous as it provides more granulated information about the relevant failure time; however, it is apparent that this is not ethical or practical in the case of avalanche victims!

There has been very little attention given to the design of current status observation schemes, in part because many applications simply do not permit investigator choices regarding the screening time distribution G (as is the case in the avalanche example). Optimal choice of G, when F is described parametrically, is closely connected to optimal design of dose levels in binary response experiments. In the nonparametric case, the optimal design depends on what aspect of F is of most interest, the mean or variance for example. Different functionals of interest lead to differing optimal designs. In addition, even if a single functional is chosen, the optimal design still depends on the form of the unknown F. Thus, one can choose an optimal design for a specific functional based on an assumed "target" F with the knowledge that the choice will not be quite optimal if the true F is somewhat different. These issues, and the description of optimal choices of G, are discussed in Jewell et al. (2006). This work contains a brief description of the extension of optimal designs to include the situation where predictive covariates are available a priori, under the weaker assumption that the failure time T and screening time C are now independent conditional on the observed covariates. Particularly with time-dependent covariates, this approach suggests the possibility of choosing screening times closer to the likely failure time by exploiting information on the time-dependent covariate. van der Laan and Robins (1998) develop estimators for this situation where the screening time C is associated with T through such covariates.

A second area of intense interest has been the extension of the ideas to current status observation of multi-state processes, the simplest being the familiar illness-death model where there are two failure times of interest (both measured from the same time origin), for example, T_1 being the time until the first event ("illness"), and T_2 the time until the second event ("death"), where necessarily $T_2 > T_1$. Particular interest has focused on cases where T_2 is observed exactly, or subject to right censoring, whereas there is only current status observation of T_1. The one-sample case of such data structures is studied in van der Laan et al. (1997). A regression version of the problem is studied in detail in Young et al. (2008), following work by Dunson and Baird (2001). The version of the same multi-state process where only current status information on *both* failure times, T_1 and T_2, at a single screening time C is available is discussed in McKeown and Jewell (2011) with application to estimation of recent HIV incidence rates based on simultaneous accurate and diluted assays. It is important to note that, in either observational setting, van der Laan and Jewell (2003) showed that you cannot improve on the naive NPMLE estimator of T_2, say, that ignores the additional information of whether the first event has occurred by the screening time ($T_1 < C$), or not, in cases where it is known that $T_2 > C$, in regard to nonparametric estimation of smooth functionals of the distribution of T_2. (Of course, if $T_2 \leq C$, necessarily $T_1 \leq C$ also).

19.7 Discussion

Motivated by avalanche data, we have briefly discussed various parametric and nonparametric approaches to the analysis of current status data. The last twenty years have seen an explosion of interest in theoretical, computational and applied problems associated with current status data and, more generally, interval censoring. We have necessarily omitted much of this work by our selective choice of topics. For example, some work focuses on cases where the failure time distribution F has a possible mass at infinity, that is a "cured," or "immortal" subpopulation. Various regression models for current status data in this setting have been investigated by Lam and Xue (2005) and Ma (2009; 2011). There is also considerable work in the application of Bayesian methods to current status data problems; see for example, Dunson and Dinse (2002), Wang and Dunson (2010), and Cai et al. (2011), for example.

A serious impediment to the practical implementation of the methods we have discussed (and those we have not) is the lack of an easily accessible open-source suite of software applications. Only some of the methods we have discussed can currently be tackled with existing software tools. The development of additional widely available programs would make the ideas and methods accessible to a much wider set of investigators, and allow the methods to be applied to a broader set of important scientific problems.

Acknowledgments

The authors wish to thank Pascal Haegeli for provision of, and permission to use, the avalanche data from Haegeli et al. (2011). The first author appreciates the comments of James Hanley who showed him an earlier version of this data that appeared in Falk et al. (1994). Thanks also to Marco Carone for help in producing Figure 19.1, Runlong Tang for help in producing Figure 19.4, Steve Shiboski for the semi-parametric regression model estimates of Tables 19.1 and 19.2, and Marloes Maathuis for Figure 19.5. We also acknowledge support for this research from the National Institute of Allergy and Infectious Diseases through grant #2R56AI070043-05A1.

Bibliography

Ayer, M., Brunk, H.D., Ewing, G.M., Reid, W.T. and Silverman, E. (1955). An empirical distribution function for sampling with incomplete information. *Ann. Math. Statist.* 26: 641-647.

Banerjee, M. (2012). Current status data in the 21st century: Some interesting developments. In *Interval-Censored Time-to-Event Data: Methods and Applications*, D-G. Chen, J. Sun, K.E. Peace: editors. Boca Raton, Florida: Chapman & Hall/CRC Press.

Banerjee, M. and Wellner, J.A. (2001). Likelihood ratio tests for monotone functions. *Ann. Statist.* 29: 1699-1731.

Banerjee, M. and Wellner, J.A. (2005). Confidence intervals for current status data. *Scand. J. Statist.* 32: 405-424.

Becker, N.G. (1989). *Analysis of Infectious Disease Data.* Chapman & Hall: New York, N.Y.

Bennett, S. (1983). Analysis of survival data by the proportional hazards model. *Stat. in Med.* 2: 273-277.

Cai, B., Lin, X. and Wang, L. (2011). Bayesian proportional hazards model for current status data with monotone splines. *Comp. Stat. & Data Anal.* 55: 2644-2651.

Chen, P., Shen, J. and Sun, J. (2009). Statistical analysis of clustered current status data. *Int. J. of Intell. Tech. and App. Stat.* 2: 109-1190.

Cook, P. and Tolusso, D. (2009). Second-order estimating equations for the analysis of clustered current status data. *Biostatistics* 10: 756-772.

Cox, D.R. (1972). Regression models and life-tables. *J. Royal Stat. Soc. Series B* 34: 187-220.

de Leeuw, J., Hornik, K., Mair, P. (2009). Isotone Optimization in R: Pool-Adjacent-Violators Algorithm (PAVA) and Active Set Methods. *J. of Stat. Software* 32(5), 1-24. URL http://www.jstatsoft.org/v32/i05/.

Diamond, I.D, McDonald, J.W. and Shah, I.H. (1986). Proportional hazards models for current status data: application to the study of differentials in age at weaning in Pakistan. *Demography* 23: 607-620.

Doksum, K.A. and Gasko, M. (1990). On a correspondence between models in binary regression analysis and in survival analysis. *Int. Stat. Rev.* 58, 243–52.

Dunson, D.B. and Baird, D.D. (2001). A flexible parametric model for combining current status and age at first diagnosis data. *Biometrics* **57**, 396–403.

Dunson, D.B. and Dinse, G.E. (2002). Bayesian models for multivariate current status with informative censoring. *Biometrics* **58**, 79–88.

Falk, M., Brugger, H. and Adler-Kastner, L. (1994) Avalanche survival chances. *Nature* 368: 21.

Groeneboom, P., Maathuis, M.H. and Wellner, J.A. (2008*a*) Current status data with competing risks: Consistency and rates of convergence of the mle. *Ann. Statist.* 36: 1031-1063.

Groeneboom, P., Maathuis, M.H. and Wellner, J.A. (2008*b*) Current status data with competing risks: Limiting distribution of the mle. *Ann. Statist.* 36: 10641-1089.

Groeneboom, P. and Wellner, J.A. (1992) *Nonparametric Maximum Likelihood Estimators for Interval Censoring and Deconvolution.* Boston: Birkhäuser-Boston.

Haegeli, P., Falk, M., Brugger, H., Etter, H-J. and Boyd, J. (2011) Comparison of avalanche survival patterns in Canada and Switzerland. *Can. Med. Assoc. Journal* 183(7): 789-795.

Huang, J. (1995). Maximum likelihood estimation for proportional odds regression model with current status data. In *Analysis of Censored Data*, H.L. Koul and J.V. Deshpande: editors, Vol. 27, 129-145, Hayward, California: Institute of Mathematical Statistics.

Huang, J. (1996). Efficient estimation for the proportional hazards model with interval censoring. *Ann. Statist.* 24: 540-568.

Hudgens, M.G., Satten, G.A. and Longini, I.M. (2001) Nonparametric maximum likelihood estimation for competing risks survival data subject to interval censoring and truncation. *Biometrics* 57: 74-80.

Jewell, N.P. (2007). Correspondences between regression models for complex binary outcomes and those for structured multivariate survival analyses. In *Advances in Statistical Modeling and Inference*, V. Nair: editor, 45-64, Hackensack, New Jersey: World Scientific.

Jewell, N.P. and Kalbfleisch, J.D. (2004). Maximum likelihood estimation of ordered multinomial parameters. *Biostatistics* 5: 291-306.

Jewell, N.P. and Shiboski, S. (1990). Statistical analysis of HIV infectivity based on partner studies. *Biometrics* 46: 1133-1150.

Jewell, N.P. and van der Laan, M. (2004*a*). Case-control current status data. *Biometrika* 91: 529-541.

Jewell, N.P. and van der Laan, M. (2004*b*). Current status data: Review, recent developments and open problems. In Advances in *Survival Analysis, Handbook in Statistics* #23: 625-642, Amsterdam: Elsevier.

Jewell, N.P., van der Laan, M. and Henneman, T. (2003). Nonparametric estimation from current status data with competing risks. *Biometrika* 90: 183-197.

Jewell, N.P., van der Laan, M. and Lei, X. (2005). Bivariate current status data with univariate monitoring times. *Biometrika* 92: 847-862.

Jewell, N.P., van der Laan, M.J. and Shiboski, S. (2006). Choice of monitoring mechanism for optimal nonparametric functional estimation for binary data. *Int. J. of Biostat.* 2: Issue 1, Article 7.

Keiding, K. (1991). Age-specific incidence and prevalence: a statistical perspective. *J. Royal Stat. Soc. Series A* 154: 371-412.

Lam, K.F. and Xue, H. (2005). A semi parametric regression cure model with current status data. *Biometrika* 92: 573-586.

Lin, D.Y., Oakes, D. and Ying, Z. (1998). Additive hazards regression with current status data. *Biometrika* 85: 285-298.

Lu, X. and Song, P. X-K. (2012). Additive hazards regression with current status data. *Comp. Stat. & Data Anal.* 56: 2051-2058.

Ma, S. (2009). Cure model with current status data. *Stat. Sinica* 19: 233-249.

Ma, S. (2011). On efficient estimation in additive hazards regression with current status data. *Ann. Inst. Stat. Math.* 63: 117-134.

Maathuis, M.H. (2005). Reduction algorithm for the MLE for the distribution function of bivariate interval censored data. *J. Comp. Graph. Statist.* 14: 352-362.

Mallick, B.K. and Gelfand, A.E. (1994). Generalized linear models with unknown link functions. *Biometrika* 81: 237-245.

Martinussen, T. and Scheike, T.H. (2002). Efficient estimation in additive hazards regression with current status data. *Biometrika* 89: 649-658.

McKeown, K. and Jewell, N.P. (2010). Misclassification of current status data. *Lifetime Data Anal.* 16: 215-230.

McKeown, K. and Jewell, N.P. (2011). Current status observation of a three-state counting process with application to simultaneous accurate and diluted HIV test data. *Can. J. of Stat.* 39: 475-487.

Murphy, S.A., van der Voart, A.W. and Wellner, J.A. (1999). Current status regression. *Math. Meth. Statist.* 8: 407-425.

Neuhaus, J.M. (1999). Bias and efficiency loss due to misclassified responses in binary regression. *Biometrika* 86: 843-855.

Politis, D.N., Romano, J.P. and Wolf, M. (1999). *Subsampling.* New York: Springer.

Rabinowitz, D., Tsiatis, A. and Aragon, J. (1995). Regression with interval-censored data. *Biometrika* 82: 501-513.

Roggia, G. (2011). Researchers fail to dig out enough avalanche data. *Can. Med. Assoc. Journal* 183(8): 934.

Rossini, A.J. and Tsiatis, A.A. (1996). A semi parametric proportional odds regression model for the analysis of current status data. *J. Amer. Stat. Assoc.* 91: 713-721.

Sal y Rosas, V.G. and Hughes, J.P. (2011). Nonparametric and semi parametric analysis of current status data subject to outcome misclassification. *Stat. Comm. in Infectious Diseases* 3: Issue 1, Article 7.

Scott, A.J. and Wild, C.J. (1997). Fitting regression models to case-control data by maximum likelihood. *Biometrika* 84: 57-71.

Sen, B., Banerjee, M. and Woodroofe, M.B. (2010). Inconsistency of bootstrap: the Grenander estimator. *Ann. Statist.* 38: 1953-1977.

Shen, X. (2000). Linear regression with current status data. *J. Amer. Stat. Assoc.* 95: 842-852.

Shiboski, S.C. (1998). Generalized additive models for current status data. *Lifetime Data Anal.* 4: 29-50.

Shiboski, S.C. and Jewell, N.P. (1992). Statistical analysis of the time dependence of HIV infectivity based on partner study data. *J. Amer. Stat. Assoc.* 87: 360-372.

Sun, J. (2006). *The Statistical Analysis of Interval-Censored Failure Time Data.* New York: Springer.

Sun, J. and Shen, J. (2009). Efficient estimation for the proportional hazards model with competing risks and current status data. *Can. J. of Stat.* 37: 592-606.

Tang, R., Banerjee, M. and Kosorok, M.R. (2012). Likelihood based inference for current status data on a grid: A boundary phenomenon and an adaptive inference procedure. *Ann. Statist.* 40: 45-72.

Tian, L. and Cai, T. (2006). On the accelerated failure time model for current status and interval censored data. *Biometrika* 93: 329-342.

van der Laan M.J., Jewell, N.P. and Petersen, D.R. (1997). Efficient estimation of the lifetime and disease onset distribution. *Biometrika* 84: 539-554.

van der Laan M.J. and Jewell, N.P. (2003). Current status and right-censored data structures when observing a marker at the censoring time. *Ann. Statist.* 31: 512-535.

van der Laan M.J. and Robins, J.M. (1998). Locally efficient estimation with current status data and time-dependent covariates. *J. Amer. Stat. Assoc.* 93: 693-701.

Wang, W. and Ding, A.A. (2000). On assessing the association for bivariate current status data. *Biometrika* 87: 879-893.

Wang, L. and Dunson, D.B. (2010). Semiparametric Bayes' proportional odds models for current status data with underreporting. *Biometrics* 67: 1111-1118.

Weisberg, S. and Welsh, A.H. (1994). Adapting for the missing link. *Ann. Statist.* 22: 1674-1700.

Wen, C-C. and Chen, Y-H. (2011). Nonparametric maximum likelihood analysis of clustered current status data with the gamma-frailty Cox model. *Comp. Stat. & Data Anal.* 55: 1053-1060.

Wen, C-C. and Lin, C-T. (2011). Analysis of current status data with missing data. *Biometrics* 67: 760-769.

Young, J.G., Jewell, N.P. and Samuels, S.J. (2008). Regression analysis of a disease onset distribution using diagnosis data. *Biometrics* 64: 20-28.

Young, M.R. and He, X. (1997). Robust estimation of semi parametric regression models. Working Paper #9712-03, *University of Michigan Business School, Research Support.*

Part V

Multivariate/Multistate Models

Part V deals with two related areas of survival analysis. The first involves multistate models. These are models for the complete disease/recovery process of a patient. The second is the area of multivariate models for survival where we have some dependency between a set of event times. In this part we have right-censored and possibly left-truncated data.

Multistate models are used to study disease or treatment progression. In a multistate model there are number of health states $\{0, 1, \ldots, p\}$ that a patient may be in at time t. For example in the simple "illness-death" model a patient starts in a healthy disease-free state from which he/she can contract a disease and move to an "ill" state or he/she can die and move to the dead state. Patients in the ill state can move to the dead state or return to the healthy state. In this example we have two transient states "alive and disease free" and "alive with disease" and one absorbing state "dead." Events in the multistate model problem are the times of transition from one state to the next and the data on an individual is the set of event times and an indicator of what event occurred.

There are several parameters associated with these types of models. The primary parameter is the set of rates of transition from one state to the next. These are usually estimated assuming a Markov transition model where the rate depends only on what state the patient is currently in not how he/she got there. These rates look like Nelson-Aalen estimators discussed in Part I. Investigators are also interested in the "state occupation probability" which is the chance a patient is in a given state at a particular time point t. Regression models are typically constructed by fitting Cox regression models to each of the intensities. The set of covariates one obtains from these covariates are often hard to interpret when interest is in making inference about state occupation probabilities. Here one may use a more direct regression model based on the pseudo-observation approach presented here in Chapter 10. In Chapter 20 Andersen and Perme explore these techniques using classical methods.

Chapter 21 by Putter deals with a related problem called "landmarking." This type of analysis arises in two related types of problems. Consider a patient given a hematopoietic stem cell transplantation. A major problem encountered in the recovery process is acute graft-versus-host disease (agvhd). This disease occurs when the patient's remaining immune system attacks the graft causing problems in the gut, liver and skin. To confirm that this disease causes increased mortality we want to compare patients with and without agvhd. One approach that is often made by investigators is to define a fixed-time covariate based on occurrence or non-occurrence of agvhd. This method leads to highly biased estimates since patients who live longer are more likely to develop agvhd and those with shorter lives have less time to develop agvhd. To solve this problem we could use time-dependent covariates for the agvhd indicator or we could pick a "landmark" time at which we will start the clock running, say 6 months, and fit Cox models among the 6-month survivors with agvhd as a fixed covariate. Landmark analysis is also used when we want to look at long-term survival based on covariates collected at some time post-transplant rather than at transplant. Again we start the clock running at some landmark time where we would have access to these post-transplant covariates. Putter looks at these problems, how to analyze landmark data and the advantages and disadvantages of landmarking.

The next set of chapters deal with dependence problems in multistate modeling. Included here are three chapters on general theory of the most popular method of dealing with dependence in survival analysis followed by three chapters that apply these methods to special problems in survival analysis.

The most common way to generate multivariate survival data is by the use of a frailty model. For simplicity suppose we are given a set of twins. We assume that the hazard rates of the twins are $(w\lambda_1(t), w\lambda_2(t))$, respectively. The frailty, $w = W$, is a random, unobservable random variable often assumed to have a mean of 1 and a variance σ^2. W represents the genetic and environmental effects shared by two twins. When $w < 1$ then the hazard rate of a pair of twins is smaller than expected and the twins are more robust while if $w > 1$ both twins in a pair tend to fail at a faster rate and these twins are more frail than the average twin. The parameter σ^2 acts like a dependence parameter since when σ^2 is 0 we have independence (since the mean is assumed to be one). In most cases a parametric model is assumed for W. If the Laplace transform of this distribution, defined by

$$LP[s] = E_W[\exp\{-sW\}]$$

is known then the joint survival function for a pair of twins with event times (T_1, T_2) is given by

$$S(t_1, t_2) = LP\left(\sum_{j=1}^{2} \Lambda_j(t_j)\right)$$

where $\Lambda()$ is the conditional cumulative hazard function. The most common models for the frailty distribution are the gamma, normal, inverse Gaussian and the positive stable distribution.

Inference for frailty models often proceeds in one of two ways. In a conditional analysis estimation of both the frailty distribution parameters and the parameters of the conditional failure rates, $\lambda(\cdot)$ is performed. Two popular approaches are an EM algorithm technique or a penalized likelihood approach. An alternate approach is to fit models to the marginal hazards of (T_1, T_2) ignoring the frailty distribution using independence working models. These models are typically Cox regression models and these marginal hazard rates are of a different functional form from the conditional rates. Using the frailty distribution, robust variance estimators are then made to the marginal models to incorporate the association between individuals within a pair.

In Chapter 22 Hougaard summarizes the literature on frailty models and inference for these models using a classical approach. Gustafson looks at Bayesian estimation for frailty models in Chapter 23. Shih in Chapter 24 looks at the marginal modeling approach using a copula approach to the dependence between failure times. Copulas provide basic information on dependence between random variable by breaking the information in the joint distribution into information on the marginal distributions and information on how these margins are coupled together to obtain the joint distribution. The copula is a joint distribution on the unit cube $[0, 1]^k$ found by making probability integral transformations on each of the margins. That is the copula for $1(X_1, \ldots, X_k)$ is the joint distribution of $(F_1(X_1), \ldots, F_k(X_k))$. Assuming a particular copula in many cases is equivalent to assuming certain frailty distributions.

In the remaining three chapters of this part we examine related models and uses of multistate models. In Chapter 25 by Diao and Zeng the problem of clustered competing risks data is studied. Presented here are extensions of the frailty and marginal models to the competing risks problem. In Chapter 26 Ye and Yu consider models which examine both the development through time of some biomarker and the time to some event. For example, an investigator may be interested in modeling how measures of blood glucose levels in diabetics are changing over time as well as the time to development of coronary complications of the disease. Chapter 27 by Brandeen-Roche studies the use of frailty models in familial studies. In such studies we have a population-based sample of individuals with a specific characteristic or disease. For each participant, information is collected on all first-degree relatives. Of interest is not the risk factors for disease but rather a study of the association between relatives in hopes of understanding the strength and mechanics of the relationship between disease and genetic relationships within a family.

20

Multistate Models

Per Kragh Andersen

Department of Biostatistics, University of Copenhagen

Maja Pohar Perme

Department of Biostatistics and Medical Informatics, University of Ljubljana

CONTENTS

20.1	Introduction	417
20.2	Models and inference for transition intensities	418
	20.2.1 Models for homogeneous populations	419
	20.2.2 Regression models	419
	20.2.3 Inference for transition intensities	420
	20.2.4 Inference for marginal rate functions	423
	20.2.5 Example	424
20.3	Models for transition and state occupation probabilities	429
	20.3.1 Plug-in models based on intensities	429
	Markov processes	429
	A note on interval-censoring	431
	Non-Markov processes	431
	20.3.2 Direct models for probabilities	432
	Regression models	432
	20.3.3 Example	433
20.4	Comments	436
	Bibliography	437

20.1 Introduction

Event history analysis deals with data obtained by observing individuals over time focusing on events occurring for the individuals. Thus, typical outcome data consist of *times of occurrence* of events and of the *types of events* which occurred. Frequently, an event may be considered as a *transition* from one state to another and, therefore, *multistate models* (MSM) will often provide a relevant modeling framework for event history data. Previous summary papers on multistate models include Hougaard (1999), Commenges (1999), Andersen and Keiding (2002), Putter et al. (2007) and Meira-Machado et al. (2009).

A *multistate process* is a (continuous-time) stochastic process $(X(t), t \in \mathcal{T})$ with a finite *state space* $\mathcal{S} = \{0, 1, \ldots, p\}$ and with right-continuous sample paths: $X(t+) = X(t)$. Here, $\mathcal{T} = [0, \tau]$ or $[0, \tau)$ with $\tau \leq +\infty$. A multistate process $X(\cdot)$ generates a *history* $\mathcal{X}_t$ (a σ-algebra) consisting of the observation of the process in the interval $[0, t]$. Relative to this history we may define *transition probabilities* by

$$P_{hj}(s,t) = \text{Prob}(X(t) = j \mid X(s) = h, \mathcal{X}_{s-})$$

for $h, j \in \mathcal{S}, s, t \in \mathcal{T}, s \leq t$ and *transition intensities* (or *transition hazards*) by the derivatives

$$\alpha_{hj}(t) = \lim_{\Delta t \to 0} P_{hj}(t, t + \Delta t)/\Delta t$$

which we shall assume exist. Some transition intensities may be 0 for all t. The *state occupation probabilities* are $\pi_h(t) = \text{Prob}(X(t) = h), h \in \mathcal{S}$ and, in particular, the *initial distribution* is $\pi_h(0) = \text{Prob}(X(0) = h), h \in \mathcal{S}$. We may then write

$$\pi_h(t) = \sum_{j \in \mathcal{S}} \pi_j(0) P_{jh}(0,t).$$

Graphically, multistate models may be illustrated using diagrams with boxes representing the states and with arrows between the states representing the possible transitions, i.e., the non-zero transition intensities. We shall illustrate this in connection with the example below.

A state $h \in \mathcal{S}$ is *absorbing* if for all $t \in \mathcal{T}, j \in \mathcal{S}, j \neq h, \alpha_{hj}(t) = 0$, that is, no arrows in the diagram begin in h; otherwise h is *transient*. Notice that the $P_{hj}(\cdot,\cdot)$ and thereby the $\alpha_{hj}(\cdot)$ depend on both the probability measure (Prob) and on the history, though this dependence has been suppressed in the notation. If $\alpha_{hj}(t)$ only depends on the history via the state $h = X(t)$ occupied at time t then the process is *Markovian*.

Sometimes one is interested in considering an extended history which also includes observed *covariates*. If only time-fixed covariates Z are studied, then the observed history is $\mathcal{F}_t = \mathcal{X}_t \vee \mathcal{Z}_0$ whereas time-dependent covariates $Z(t)$ may give rise to an extended history of the form $\mathcal{F}_t = \mathcal{X}_t \vee \mathcal{Z}_t$ where $\mathcal{Z}_t$ is the history generated by the covariates in $[0,t]$. (Here, for σ-algebras $\mathcal{A}$ and $\mathcal{B}$, $\mathcal{A} \vee \mathcal{B}$ is the smallest σ-algebra containing both $\mathcal{A}$ and $\mathcal{B}$.)

Thus, parameters of interest in MSMs include transition intensities, transition probabilities, state occupation probabilities, and distributions of time spent in each state.

Throughout this work, the described methods shall be illustrated using the data on peritoneal dialysis patients described in Pajek (2012). The data include 132 patients who started peritoneal dialysis (PD) at Manchester Royal Infirmary and who were not eligible for transplant. The patients are followed up until mid-2011. The PD is considered as the dialysis option that allows for a better quality of life, but may lead to potentially fatal complications after long-term use. In the course of the study, 95 patients switched to hemodialysis (HD), either due to a complication (infection) or based on their own decision. The understanding of the impacts of switch to hemodialysis and the timing of this switch was the main goal of the study. Table 20.1 gives some dataset information.

20.2 Models and inference for transition intensities

An attractive feature of multistate models based on transition intensities is that all hazard-based models known from survival analysis apply, see, e.g., Andersen et al. (1993). This includes both estimation, testing and model checking. For ease of notation we first fix two states, $h, j \in \mathcal{S}$, and denote the $h \to j$ transition intensity $\alpha(t)$.

TABLE 20.1
Description of data from 132 patients treated with peritoneal dialysis.

Age (mean, SD)	66.6	14.3
Female sex	44	33.3%
Comorbidities present	105	79.5%
Creatinine, (μmol/l) (mean, SD)	593	230
HD switch	50	37.8%
HD switch due to peritonitis	24	48% of patients on HD
Death	82	62.1%
HD switch and death	32	39% of patients on HD

20.2.1 Models for homogeneous populations

We first study the case of no covariates where both non-parametric and parametric models are relevant. In the former, $\alpha(t)$ is left completely unspecified while, in the latter, the simplest possible model is the constant hazard model, $\alpha(t) = \alpha$. Since the constant hazard assumption is often too simple to apply in practice, a frequently used extension of this is the piecewise constant hazard model where, for some time intervals given by cut-points $0 = \tau_0 < \tau_1 < \cdots < \tau_K = \tau$, it is assumed that $\alpha(t) = \alpha_\ell$ for $\tau_{\ell-1} \leq t < \tau_\ell, \ell = 1, \ldots, K$. Another extension of the simple constant hazard model is the Weibull model with $\alpha(t) = \alpha t^\gamma, \gamma \geq 0$ which for $\gamma = 0$ reduces to the constant hazard model. Such models are all Markovian as the transition intensity at time t does not depend on other aspects of the past history, $\mathcal{X}_{t-}$ than the state (h) occupied at $t-$. This is an important class of models where, as we shall see in later sections, transition *probabilities* may be derived from the intensities. Another important class of models is *semi-Markov* models where the $h \to j$ transition intensity at time t also depends on the *duration* or *sojourn time* in state h, that is $t - T$ where $T(\leq t)$ is the time of entry into state h. The constant hazard model is, obviously, always Markov. A model where there is only duration dependence (and no direct dependence on the baseline time variable, "calendar time", t) is sometimes called *homogeneous semi-Markov* to distinguish it from a general semi-Markov model where $\alpha(\cdot)$ may depend on both calendar time, t and duration, $t - T$. We shall return to general semi-Markov models below when regression models are to be introduced. In multistate models with many states and many possible transitions, the transition intensities may depend on the past in more complicated ways, e.g., via the number of previous states visited or via the total time spent in certain states. However, we shall not cover examples of that kind in what follows.

20.2.2 Regression models

Most regression models involve a linear predictor, that is a linear function of the covariates for individual i, $i = 1, \ldots, n$, with some unknown regression coefficients, β_m, (or regression functions, $\beta_m(t)$):

$$\text{LP}_i(t) = \sum_{m=1}^{k} \beta_m(t) Z_{mi}(t)$$

where covariates are allowed to be time-dependent. We shall restrict attention to regression models with a linear predictor. This means that, when covariates are to be included into models for transition intensities, a choice of link function has to be made, i.e., one needs to specify how $\mathrm{LP}_i(t)$ relates to $\alpha(t \mid Z_i)$. Many hazard models are multiplicative, i.e., $\log(\alpha(t \mid Z_i))$ is linear in the covariates, but also additive hazard models are frequently studied. The main example of an additive hazard model is *Aalen's non-parametric model* given by

$$\alpha(t \mid Z_i) = \beta_0(t) + \mathrm{LP}_i(t),$$

i.e., all regression functions are left unspecified, though semi-parametric versions where some or all regression functions (except $\beta_0(t)$) are constant also exist, see, e.g., Martinussen and Scheike (2006).

In what follows we will only discuss multiplicative models which means that, when introducing covariates into the non-parametric model above, the semi-parametric Cox regression model

$$\alpha(t \mid Z_i) = \alpha_0(t)\exp(\mathrm{LP}_i(t)) \tag{20.1}$$

is obtained. Here, the baseline hazard $\alpha_0(t)$ is left completely unspecified while the regression coefficients, β_m are usually assumed constant leading to the basic *proportional hazards* assumption of this model. The covariates, however, may be time-dependent.

In a similar way, covariates may be introduced into parametric hazard models. This leads to fully parametric models for $\alpha(t \mid Z_i)$ of the same multiplicative form as the Cox Model (20.1) and with the baseline hazard, $\alpha_0(t)$ being, e.g., constant or piecewise constant.

As for the models for homogeneous populations, the regression models lead to Markov processes if the hazard at ("calendar") time t does not depend on other aspects of $\mathcal{F}_{t-}$ than the state (h) occupied at $t-$, that is, if the baseline hazard is a function of t and if no function of the past is included as a time-dependent covariate. Similarly, the process is homogeneous semi-Markov if the hazard at time t only depends on the duration, $t - T$, of the stay in the current state, h.

However, the allowance for time-dependent covariates provides a means for studying general semi-Markov processes where the hazard at time t depends on both time variables, t and $t - T$. In the framework of the semi-parametric Cox regression model this is done by choosing one of the time variables as the "baseline" time variable, i.e., $\alpha_0(\cdot)$ is a function of that time variable, while functions of the other time variable may be included as time-dependent covariates, e.g.,

$$\alpha(t \mid Z_i) = \alpha_0(t)\exp(\beta_0 f(t - T_i) + \mathrm{LP}_i(t)),$$

with $f(\cdot)$ being some pre-specified function of the current duration like the identity $f(d) = d$ or the indicator $f(d) = I(d \leq d_0)$ corresponding to some duration threshold, d_0. Such a model is frequently used to test the Markov null hypothesis: $\beta_0 = 0$. More recently, non-parametric methods for testing the Markov assumptions have been developed (Rodríguez Girondo and Uña Álvarez, 2011).

20.2.3 Inference for transition intensities

We will assume that independent multistate processes

$$(X_i(t), 0 \leq t \leq C_i; i = 1, \ldots, n)$$

are observed in continuous time, that is, times of transition are observed exactly, except for the fact that independent right censoring at $C_i(\leq \tau)$ is allowed for $X_i(\cdot)$. In the next section, interval-censoring will be touched upon. Left truncation may quite easily be incorporated, as well, but for sake of simplicity we have chosen not to do so since that would require specific assumptions concerning the information available at the time of left truncation. The data for individual i can then be represented as a multivariate counting process

$$N_{hji}(t), h, j \in \mathcal{S}, h \neq j, t \leq C_i$$

counting the number of direct $h \to j$ transitions observed for subject i in $[0, t]$ (where some h, j combinations may not be possible). The model is then specified by the transition intensities $\alpha_{hji}(t \mid Z_i(t))$ leading to the intensity process

$$\lambda_{hji}(t) = Y_{hi}(t)\alpha_{hji}(t \mid Z_i(t))$$

for $N_{hji}(t)$ with respect to the filtration $(\mathcal{F}_t)$ and the relevant probability measure. That is,

$$\mathrm{E}(N_{hji}(t) \mid \mathcal{F}_{t-}) = \int_0^t \lambda_{hji}(u)\mathrm{d}u. \tag{20.2}$$

Here,

$$Y_{hi}(t) = I(X_i(t-) = h)$$

is the indicator of $X_i(\cdot)$ being in state h at time $t-$. For a filtration of the form $\mathcal{F}_t = \mathcal{X}_t \vee \mathcal{Z}_0$, the likelihood for the model parameters, say $\boldsymbol{\theta}$, is obtained via Jacod's formula, see, e.g., Andersen et al. (1993), Chapter II:

$$L(\boldsymbol{\theta}) = \prod_i \prod_{h,j} \left(\prod_t \lambda_{hji}(t)^{\Delta N_{hji}(t)} \right) \exp(-\int_0^{C_i} \lambda_{hji}(u)\mathrm{d}u). \tag{20.3}$$

The filtration will have the form $\mathcal{X}_t \vee \mathcal{Z}_0$ when, either, only time-fixed covariates are included, or when time-dependent covariates only depend on the past history of the multi-state process because in these cases the covariates do not impose extra randomness for $t > 0$. Usually, (20.3) is referred to as a likelihood in this case even though potential randomness in the right-censoring times, C_i, and in the time-fixed covariates is not accounted for. However, in the case of an extended filtration $\mathcal{F}_t = \mathcal{X}_t \vee \mathcal{Z}_t$, that is, when more general time-dependent covariates are allowed, (20.3) is only a partial likelihood (see, e.g., Andersen et al. (1993), Chapters II–III, for further discussion).

At any rate, (20.3) may be used as the basis for inference on $\boldsymbol{\theta}$. For the purely non-parametric model: $\alpha_{hji}(t) = \alpha_{hj}(t)$, completely unspecified, this leads (with proper definition of non-parametric maximum likelihood estimation) to the *Nelson-Aalen* estimator

$$\widehat{A}_{hj}(t) = \int_0^t \frac{\mathrm{d}N_{hj}(u)}{Y_h(u)} \tag{20.4}$$

for the cumulative transition intensity

$$A_{hj}(t) = \int_0^t \alpha_{hj}(u)\mathrm{d}u.$$

In (20.4), $N_{hj}(t)$ and $Y_h(t)$ are the aggregated processes

$$N_{hj} = \sum_i N_{hji}, \quad Y_h = \sum_i Y_{hi}.$$

For the piecewise constant hazard model

$$\alpha_{hji}(t) = \alpha_{hj\ell} \text{ for } \tau_{hj,\ell-1} \leq t < \tau_{hj,\ell}, \ell = 1, \ldots, K_{hj},$$

(20.3) leads to the following "occurrence/exposure rate" estimators

$$\widehat{\alpha}_{hj\ell} = \frac{D_{hj\ell}}{T_{h\ell}}$$

with $D_{hj\ell} = N_{hj}(\tau_{hj,\ell}-) - N_{hj}(\tau_{hj,\ell-1})$ being the total number of $h \to j$ transitions in interval ℓ and

$$T_{h\ell} = \int_{\tau_{hj,\ell-1}}^{\tau_{hj,\ell}} Y_h(u)\mathrm{d}u$$

the total time at risk for such transitions in that interval.

Variance estimates are available from the second derivative of $-\log L(\boldsymbol{\theta})$, though robust standard errors may also be applied (e.g., Lin et al. (2000)).

For the Cox regression model

$$\alpha_{hji}(t \mid Z_i(t)) = \alpha_{hj0}(t)\exp(\mathrm{LP}_{hji}(t))$$

a number of options is available for specification of the linear predictor. The most obvious choice is to let both the covariates and the regression coefficients vary among transition types, i.e., to let

$$\mathrm{LP}_{hji}(t) = \sum_{m=1}^{k_{hj}} \beta_{hjm} Z_{hjmi}(t).$$

However, (see, e.g., Andersen et al. (1993), Chapter VII, for further discussion), working with a single ("long") vector, $(\beta_1, \ldots, \beta_k)$, of regression coefficients and defining, appropriately, type specific covariates $Z_{hjmi}(t), m = 1, \ldots, k$, provides added flexibility in that models where some coefficients are shared between several transition types may be studied. This specification of the linear predictor:

$$\mathrm{LP}_{hji}(t) = \sum_{m=1}^{k} \beta_m Z_{hjmi}(t)$$

leads to the following version of the *Cox partial likelihood*

$$\mathrm{CL}(\beta) = \prod_t \prod_{hji} \left(\frac{\exp(\mathrm{LP}_{hji}(t))}{\sum_r Y_{hr}(t)\exp(\mathrm{LP}_{hjr}(t))} \right)^{\Delta N_{hji}(t)} \tag{20.5}$$

as a profile likelihood from (20.3) obtained by partial maximization over $A_{hj0}(\cdot)$. When no regression coefficients are common to different transition intensities, (20.5) is a product over transitions types, h, j, with separate parameters for the different types, i.e., maximization may be performed for one type of transition at a time. In general, the whole Cox partial likelihood (20.5) must be maximized in one analysis (utilizing the concept of *stratified* Cox models, see, e.g., Andersen et al. (1993), Chapter VII, and Andersen and Keiding (2002)). Furthermore, the cumulative baseline hazard is estimated by the Breslow estimator

$$\widehat{A}_{hj0}(t) = \int_0^t \frac{\mathrm{d}N_{hj}(u)}{\sum_r Y_{hr}(u)\exp(\widehat{\mathrm{LP}}_{hjr}(u))}.$$

For the parametric regression model with piecewise constant transition intensities, (20.3) leads to so-called *Poisson* regression. When covariates are all discrete, that is, when the covariate vector takes values, v, in a finite set V, the data may be summarized as the "tables":

$$(D_{hj\ell,v}, T_{h\ell,v}, v \in V, \ell = 1, \ldots, K_{hj}).$$

Here, $D_{hj\ell,v}$ are the $h \to j$ transition counts and $T_{h\ell,v}$ the total time at risk in h in each cell obtained by a cross-classification of v and time-interval, ℓ and these tables are *sufficient*. Furthermore, the likelihood is proportional to that obtained by, formally, treating the $D_{hj\ell,v}$ as independent Poisson with means

$$\alpha_{hj\ell,v} T_{h\ell,v}$$

proportional to $T_{h\ell,v}$, i.e., in a multiplicative hazard model, $\log(T_{h\ell,v})$ is used as an *offset* in a generalized linear model with $D_{hj\ell,v}$ as outcome variable. That is, the $h \to j$ transition intensity $\alpha_{hji}(t)$ is a piecewise constant function of time and is linear in the covariates:

$$\log(\alpha_{hji}(t \mid Z_i(t) = v)) = \log(\alpha_{hj\ell,0}) + \sum_{v \in V} \beta_v \text{ for } \tau_{hj,\ell-1} \leq t < \tau_{hj,\ell}.$$

For large datasets, this sufficiency reduction of the follow-up data may lead to considerable simplification of the inference without seriously losing the flexibility of the non-parametric baseline hazard in the Cox model (though, of course, a choice of time-intervals needs to be made). Furthermore, Poisson regression has the advantages that non-proportional hazards is simply an interaction between time and covariates and, further, that non-homogeneous semi-Markov models are easily handled by splitting follow-up time according to both ("calendar") time t and duration, $t - T$. This means that, when analyzing such models, a choice of "baseline time variable" (calendar time or duration) is not needed since both appear in the log-hazard model on equal footings. Finally, the parameters for effects of such time variables will be part of the standard regression output from any computer package in contrast to the Breslow estimates from the Cox models which will usually appear in the output, in either tabular or graphical form, only when specifying the relevant options.

20.2.4 Inference for marginal rate functions

The basic counting processes, $N_{hji}(t)$, have intensity processes, $\lambda_{hji}(t) = Y_{hi}(t)\alpha_{hji}(t)$ directly depending on the transition intensities, $\alpha_{hji}(t)$. By (20.2), these are conditional expectations

$$\mathrm{E}(N_{hji}(t) \mid \mathcal{F}_{t-}) = \int_0^t \lambda_{hji}(u)\mathrm{d}u$$

given the entire history $\mathcal{F}_{t-}$. Sometimes, one may be interested in studying only certain marginal properties of the multistate process, i.e., conditioning on less information than $\mathcal{F}_{t-}$:

$$\mathrm{E}(N_{hji}(t) \mid \mathcal{G}_{t-}) = \int_0^t \widetilde{\lambda}_{hji}(u)\mathrm{d}u,$$

where $\mathcal{G}_t \subset \mathcal{F}_t$. Here, the derivatives $\widetilde{\lambda}_{hji}(t)$ of the mean functions are known as the (marginal) *rate* functions and may (by the Innovation Theorem, e.g., Andersen et al. (1993), Section II.4) be written as

$$\widetilde{\lambda}_{hji}(t) = \mathrm{E}(\lambda_{hji}(t) \mid \mathcal{G}_{t-}).$$

Typically, score equations derived from (20.3) will still be unbiased estimating equations for these marginal rates. Thus, in Cook and Lawless (2007), Chapter III, and Lin et al. (2000) the special case of recurrent events was studied conditioning only on time-fixed covariates, Z_i. Also marginal models for clustered survival data have been studied (e.g., Martinussen and Scheike (2006), Chapter 9) where conditioning is only done on past information for subject i. Since these estimating equations are no longer likelihood score equations, robust variance estimation is always needed. Furthermore, the right censoring times must be assumed marginally independent of the multistate process.

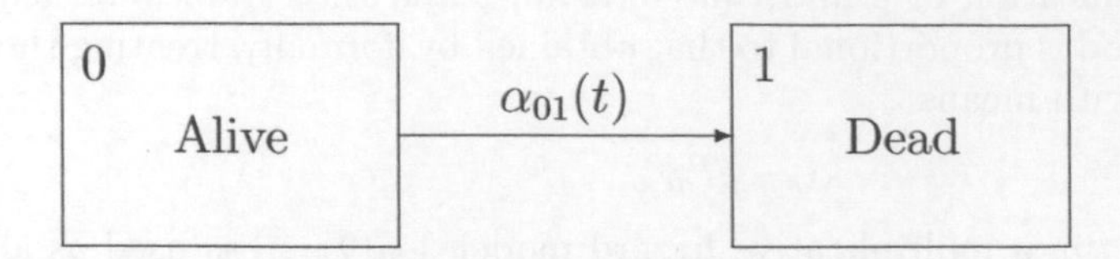

FIGURE 20.1
The two-state model for survival data.

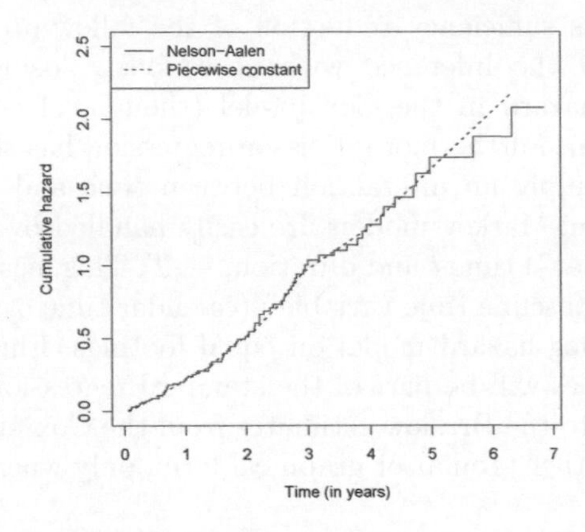

FIGURE 20.2
Nelson-Aalen estimate and the piecewise constant estimate of the cumulative hazard for the two-state model for survival data.

20.2.5 Example

We start by considering the simple model for survival data given in Figure 20.1, with the transient state "alive" (0) and the absorbing state "dead" (1). Figure 20.2 presents the Nelson-Aalen estimate of the cumulative hazard function for the transition $0 \rightarrow 1$. A piecewise constant estimate using 10 time intervals split at quantiles of event times is superimposed. The two estimators agree well, the cumulative hazard function is almost a straight line indicating that the hazard of dying is almost constant throughout the follow-up.

The association between survival time and covariates is explored using both the Cox model and the Poisson model assuming a piecewise constant hazard in the ten time intervals. The piecewise constant hazard assumption of the Poisson model seems reasonable (Figure 20.2) and the two models agree well; see Table 20.2.

The most important covariate seems to the presence of comorbidities, patients with comorbidities have a considerably increased hazard of dying. Age is marginally significant but creatinine levels do not seem to be associated with the hazard of dying. When including the switch to HD as a time-dependent covariate, age becomes a more important predictor while creatinine levels and the presence of comorbidities keep a similar value. The switch to HD is an important predictor, an individual who has already switched is at a higher risk than those still on PD.

Next, say we are interested in a *competing risks* model with two types of adverse events,

TABLE 20.2
Cox and Poisson models for transition intensity in the two-state model, with and without the time-dependent variable "switch to HD."

	Cox			Poisson		
	β	SE	p	β	SE	p
Age (per 10 years)	0.204	0.111	0.066	0.214	0.112	0.056
Comorbidities present	0.967	0.361	0.007	0.989	0.359	0.006
Creatinine (per 100 μmol/l)	0.001	0.063	0.989	0.007	0.062	0.912
Age (per 10 years)	0.276	0.111	0.013	0.292	0.112	0.009
Comorbidities present	0.943	0.361	0.009	0.964	0.360	0.007
Creatinine (per 100 μmol/l)	-0.054	0.066	0.416	-0.049	0.066	0.455
Switch to HD	0.919	0.271	0.001	0.956	0.271	< 0.001

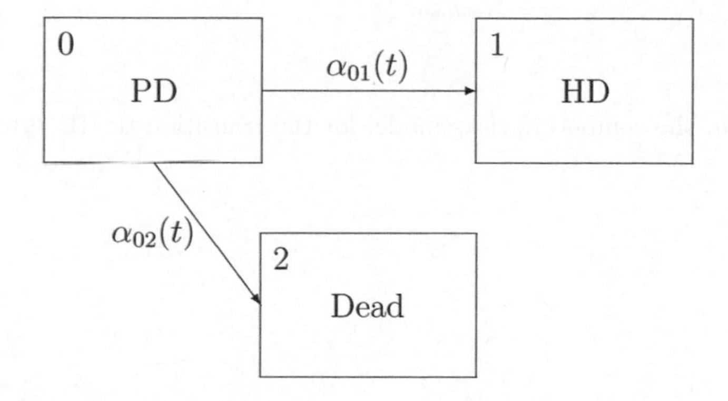

FIGURE 20.3
The competing risks model.

both considered as absorbing states: failure of the PD technique (switching to HD, denoted as 1) and death (2); see Figure 20.3.

The Nelson-Aalen estimates of the two cumulative hazard functions are presented in Figure 20.4. The hazards of dying and switching to HD both seem rather constant in time.

Table 20.3 reports the results of Cox models for transition intensities in the competing risks model. Since the likelihood in (20.3) factorizes into distinct terms that include only the parameters for one transition each, two separate Cox models are fitted, each considers one of the possible outcomes as the event of interest and censors the times to the other outcome. We can see that while high creatinine levels increase the hazard of switching to HD, creatinine is not associated with the hazard of dying from state 0. Conversely, the presence of comorbidities increases the hazard of dying but does not seem to be associated with the technique failure risk.

If transition from HD to death is also of interest, the competing risks model is extended to the illness-death model without recovery; see Figure 20.5.

Regarding the state of being on HD as a transient state does not affect the analysis of transition intensities out of state 0, hence Figure 20.4 and Table 20.3 still apply. The intensity for the $1 \to 2$ transition estimated using the Nelson-Aalen estimator is presented

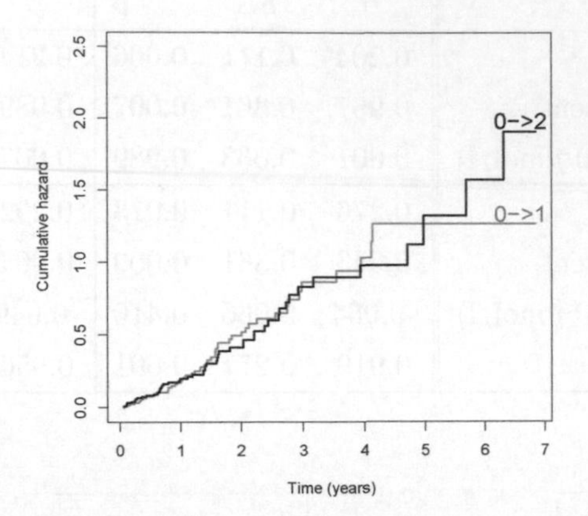

FIGURE 20.4
Nelson-Aalen estimates in the competing risks model for the transition to HD (grey) and death (black).

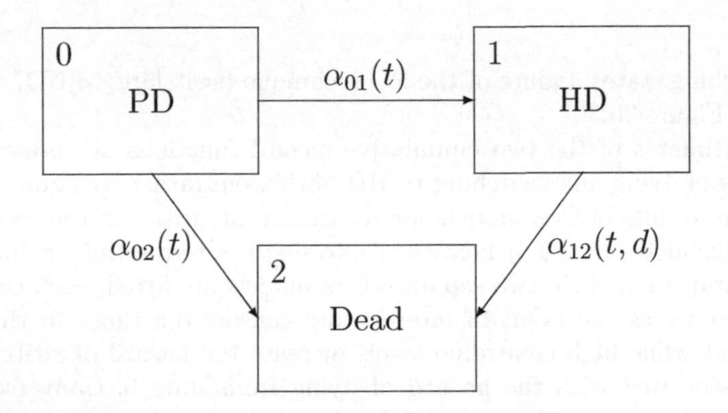

FIGURE 20.5
The illness-death model without recovery.

TABLE 20.3
Cox models for transition intensities in the competing risks model.

	$0 \to 1$			$0 \to 2$		
	β	SE	p	β	SE	p
Age (per 10 years)	-0.09	0.124	0.468	0.031	0.142	0.827
Comorbidities present	0.47	0.381	0.218	1.107	0.486	0.023
Creatinine (per 100 μmol/l)	0.17	0.064	0.008	0.02	0.084	0.815

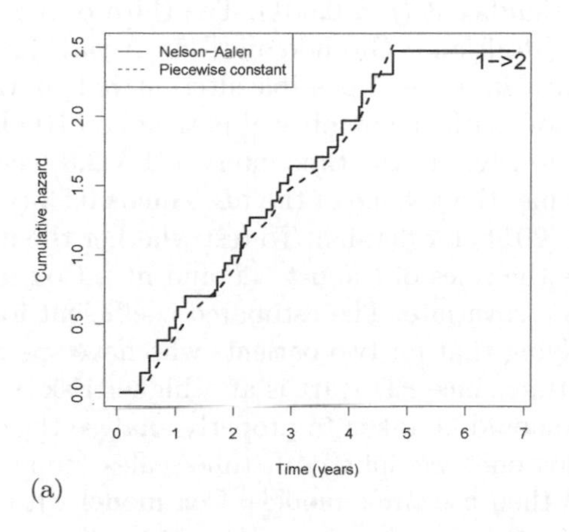

(a)

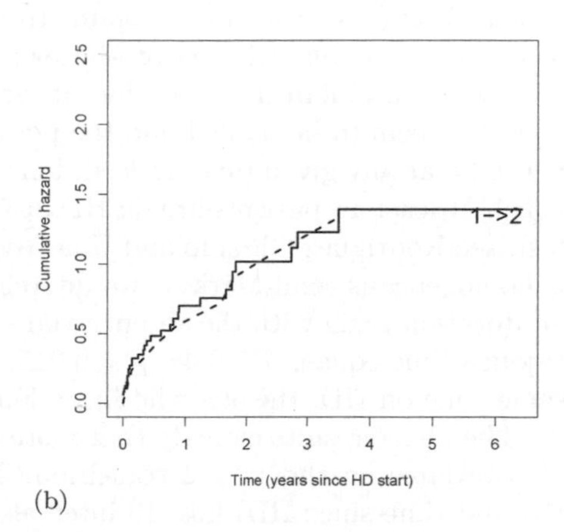

(b)

FIGURE 20.6
Nelson-Aalen estimates and estimates based on piecewise constant intensities for transitions out of state 1 (HD to death) as (a) a function of total time and (b) duration in state 1.

in Figure 20.6a. The curve is approximately linear but considerably steeper than those for transitions out of state 0. The estimate based on piecewise constant intensities agrees well, therefore no important differences between the Cox and the Poisson model using 10 intervals for the baseline function can be expected.

The results of the two models are given in Table 20.4. Only patients who ever reach state 1 are at risk for this transition and the smaller sample size (50) reflects in the wider confidence intervals. The results of both models are close, even if the effect of comorbidities is formally significant in the Poisson model but not in the Cox model. Age is found to be a strong risk factor of dying in both models.

TABLE 20.4
Cox and Poisson model for $1 \to 2$ transition intensity in the illness-death model.

	Cox			Poisson		
	β	SE	p	β	SE	p
Age (per 10 years)	0.534	0.173	0.002	0.592	0.173	0.001
Comorbidities present	0.996	0.558	0.074	1.128	0.552	0.041
Creatinine (per 100 μmol/l)	-0.229	0.132	0.083	-0.262	0.139	0.06

When considering the $1 \to 2$ transition, the time scale since HD start may also be of interest, Figure 20.6b presents the cumulative hazard out of state 1 as a function of duration in state 1. The cumulative hazard on this scale is not linear; the hazard of dying seems to be the highest just after the switch to HD and we therefore expect that the process is not Markov. The medical explanation here is that some patients had to switch to HD due to a potentially lethal infection and may have died shortly after the switch. The outcome of the infection should be known in a maximum of two months, so a test of the Markov assumption can be performed using the indicator function of surviving 2 months $I(t - T_i > 2 \text{ months})$ as a time-varying covariate in the Cox model. The estimated parameter β is strongly significant ($p < 0.001$) and negative (-2.11), indicating that the hazard lowers substantially after the critical phase. If a linear effect of duration $t - T_i$ is assumed instead, the estimated parameter equals -1.02 (per year) and is again strongly significant ($p < 0.001$). The third option, i.e., to use a non-parametric test of Markovianity (Rodríguez Girondo and Uña Álvarez, 2011), may be less useful in this case, since its generality in terms of possible alternative hypotheses does not seem to be needed and the power is low with the number of patients on HD being below 25 at any given time. Indeed, limiting ourselves to the time interval $[1.4, 2.8]$ years in which at least 15 patients are on HD at any time, the p value of the test equals 0.53 (global test, see Rodríguez Girondo and Uña Álvarez (2011) for details). To test whether the model is homogeneous semi-Markov, we interchange the roles of t and $t - T$ and fit a Cox model on duration scale with the sojourn time T as a covariate. The estimated coefficient for the sojourn time equals $\hat{\beta} = 0.42$, $p = 0.015$, implying that for two patients who have spent the same time on HD, the one who has a longer time since PD start is at a higher risk.

The above results indicate that more care should be taken to properly analyse the effect of covariates on the $1 \to 2$ transition. To this end, we split both time-scales (time since PD and time since HD) into 10 intervals and then use three models: Cox model with time since PD as the baseline time variable and adjusting for time since HD using a time-varying categorical variable; Cox model with time since HD as the baseline and time to HD as a categorical variable; and Poisson model with both times as categorical covariates. The estimated parameters for the three covariates are presented in Table 20.5, as expected, all three models yield similar results. The interpretation is very similar with all three models: the hazard ratio for age, for example, compares the hazards of two individuals that differ for 10 years in age, but have the same time since PD and the same time since HD. The models only differ in what they regard as equal times; in the Poisson model, time is categorized into intervals on both scales, while the Cox models regard one of the scales as time and the other as a covariate split into intervals. Comparing the results with the $1 \to 2$ transition fit in Table 20.4, we can see that the inclusion of the time since HD did not change importantly the parameter estimates in our case.

TABLE 20.5
Cox and Poisson models for transition $1 \to 2$ (HD to death) taking into account both time scales. X_1=Age (per 10 years), X_2=Comorbidities present, X_3=Creatinine (per 100 μmol/l).

	Cox (time since PD)			Cox (time since HD)			Poisson		
	β	SE	p	β	SE	p	β	SE	p
X_1	0.533	0.207	0.010	0.522	0.181	0.004	0.515	0.182	0.005
X_2	0.617	0.596	0.300	0.793	0.561	0.158	0.776	0.559	0.164
X_3	-0.062	0.118	0.599	-0.119	0.118	0.311	-0.126	0.121	0.298

20.3 Models for transition and state occupation probabilities

As illustrated in Section 20.2, intensity-based MSMs are rich and standard software offers the necessary means to perform the analysis. However, since the interpretation of probabilities is more simple than that of intensities, it is of considerable interest to extend the methods from the previous section with techniques for inference for MSM transition probabilities and state occupation probabilities. For certain MSMs, including Markov processes, explicit formulas relate such probabilities to transition intensities, thereby allowing for simple plug-in probability estimation once intensity models are established. We shall first (Section 20.3.1) review techniques of that kind.

For regression situations, however, plug-in methods do not provide us with simple parameters describing the association between covariates and outcome probabilities. This is because of the non-linearity of the relation between intensities and probabilities and, hence, even intensity models with a simple link function (such as the Cox model or the additive hazard model), lead to complicated relations between covariates and outcome probabilities. For these reasons, direct (marginal) regression models for outcome probabilities are of interest and we shall review a number of such techniques in Section 20.3.2, including methods based on pseudo-observations and direct binomial regression models.

20.3.1 Plug-in models based on intensities

Markov processes

Suppose that the multistate process $X(t), t \in \mathcal{T}$ is Markov and let $\mathbf{P}(s,t)$ be the $(p+1) \times (p+1)$ transition probability matrix, i.e.,

$$P_{hj}(s,t) = \text{Prob}(X(t) = j \mid X(s) = h), h, j \in \mathcal{S}, s \leq t.$$

Let $\mathbf{A}(t)$ be the corresponding $(p+1) \times (p+1)$ cumulative transition intensity matrix, i.e.,

$$\mathrm{d}A_{hj}(t) = \text{Prob}(X(t + \mathrm{d}t) = j \mid X(t) = h), h \neq j, h, j \in \mathcal{S}$$

and let $A_{hh}(t) = -\sum_{j \neq h} A_{hj}(t)$, that is $\sum_j A_{hj}(t) = 0$.

We assume that $\mathbf{A}$ is absolutely continuous, that is, $\mathrm{d}A_{hj}(t) = \alpha_{hj}(t)\mathrm{d}t$ where

$$\alpha_{hj}(t) = \lim_{\Delta t \to 0} \frac{1}{\Delta t} P_{hj}(t, t + \Delta t)$$

is the $h \to j$ transition intensity. For given $\mathbf{A}$, the transition probability matrix $\mathbf{P}$ is the unique solution to $\mathbf{P}(s,s) = \mathbf{I}$, the $(p+1) \times (p+1)$ identity matrix, and the Kolmogorov forward differential equations

$$\frac{\partial}{\partial t}\mathbf{P}(s,t) = \mathbf{P}(s,t)\boldsymbol{\alpha}(t)$$

or, written coordinate-wise

$$\frac{\partial}{\partial t} P_{hj}(s,t) = \sum_l P_{hl}(s,t)\alpha_{lj}(t).$$

One can show that the solution is the *matrix product-integral*

$$\mathbf{P}(s,t) = \prod_s^t (\mathbf{I} + \boldsymbol{\alpha}(u)\mathrm{d}u)$$

defined by

$$\lim_{\max|s_i - s_{i-1}| \to 0} \prod (\mathbf{I} + \mathbf{A}(s_i) - \mathbf{A}(s_{i-1})),$$

where $s = s_1 < \cdots < s_{i-1} < s_i < \cdots = t$ is a partition of the interval from s to t (e.g., Andersen et al. (1993), Section II.6).

This provides us with the solution to the non-parametric estimation problem for Markov transition probabilities: estimate $A_{hj}(t)$ by the Nelson-Aalen estimator $\widehat{A}_{hj}(t)$, cf. (20.4), and let $\widehat{A}_{hh} = -\sum_j \widehat{A}_{hj}$. Transition probabilities are then estimated by the *Aalen-Johansen* estimator, the finite matrix product obtained by plugging the Nelson-Aalen estimator into the product-integral

$$\widehat{\mathbf{P}}(s,t) = \prod_{(s,t]} (\mathbf{I} + \mathrm{d}\widehat{\mathbf{A}}(u)).$$

For the special case of the two-state model for survival data, the latter simply gives the Kaplan-Meier estimator for $S(t) = P_{00}(0,t)$.

Parametric models with constant transition intensities may also be handled quite easily. This is because, in this case, there exists a simple exponential representation of the product-integral and, thereby, a simple formula for $P_{hj}(s,t)$. For $d = t - s$ this is given by

$$\mathbf{P}(d) = \exp(d\mathbf{A}) = \mathbf{V}\mathrm{diag}(e^{\boldsymbol{\rho} d})\mathbf{V}^{-1} \tag{20.6}$$

where $\boldsymbol{\rho}$ are the eigenvalues for the intensity matrix, $\mathbf{A}$, and $\mathbf{V}$ the matrix of eigenvectors, that is, $\mathbf{A}$ is estimated by maximum likelihood and plugged into (20.6).

The model with piecewise constant intensities can also be handled though the expressions for transition probabilities become slightly more involved. Suppose we wish to estimate $P_{hj}(s,t)$ where s and t belong to adjacent intervals, $[\tau_{\ell-1}, \tau_\ell)$ and $[\tau_\ell, \tau_{\ell+1})$ in which the intensities are constant. Then we may use the exponential formulas from (20.6) for $P_{hm}(s,\tau_\ell)$ and $P_{mj}(\tau_\ell, t)$, $m \in \mathcal{S}$ and the *Chapman-Kolmogorov equations*:

$$P_{hj}(s,t) = \sum_{m \in \mathcal{S}} P_{hm}(s,\tau_\ell) P_{mj}(\tau_\ell, t)$$

to estimate $P_{hj}(s,t)$.

For the models with constant or piecewise constant intensities, standard errors for the transition probability estimates are derived from the likelihood-based estimated covariance matrix for $\widehat{\alpha}_{hj\ell}$ and the delta-method.

For Markov regression models with time-fixed covariates (both for the semi-parametric Cox Model (20.1) and for the similar Poisson model) transition probabilities for given covariates, Z_0,

$$P_{hj}(s,t \mid Z_0)$$

may be estimated completely analogously by plugging the estimated regression intensities into the product integral. Thereby, such probabilities may be predicted for given covariates and standard errors for the predictions may be obtained via the delta method (see Andersen et al. (1991) and Shu and Klein (2005) for the Cox model and the additive model, respectively, in general Markov processes, and Cheng et al. (1998), Shen and Cheng (1999) and Scheike and Zhang (2003) for the competing risks model with Cox hazards, additive hazards and more flexible hazards, respectively). However, as mentioned above, this does not lead to simple relations between covariates and transition probabilities. As an example we study the Markov illness-death model (Figure 20.5) with Cox type $0 \to 1, 0 \to 2$ and $1 \to 2$ transition intensities. In this model, the transition probability $P_{01}(0,t)$ is given by

$$P_{01}(0,t \mid Z) = \int_0^t P_{00}(0,u- \mid Z)\alpha_{01}(u \mid Z)P_{11}(u+,t \mid Z)\mathrm{d}u \tag{20.7}$$

with

$$P_{00}(0,u \mid Z) = \exp\left(-A_{01}(u \mid Z) - A_{02}(u \mid Z)\right)$$

and

$$P_{11}(u,t \mid Z) = \exp\left(-\int_u^t \alpha_{12}(x \mid Z)\mathrm{d}x\right).$$

Thus, for $\alpha_{hj}(u \mid Z_i) = \alpha_{hj0}(u)\exp(\mathrm{LP}_{hji})$ the way in which $P_{01}(0,t \mid Z)$ depends on Z is not described by simple parameters.

For the competing risks model (Figure 20.3), $P_{11}(\cdot,\cdot) = 1$ and (20.7) reduces to the *cumulative incidence function*

$$P_{01}(0,t \mid Z) = \int_0^t P_{00}(0,u- \mid Z)\alpha_{01}(u \mid Z)\mathrm{d}u.$$

A note on interval-censoring

In the derivation of the likelihood, (20.3), continuous observation of $X(\cdot)$ was assumed, i.e., times of transitions were observed exactly, except possibly for right censoring. For a Markov model with piecewise constant intensities the likelihood for interval-censored data may be written down in terms of intensities based on the explicit relation between transition intensities and probabilities. For a single individual, observed at times $s_0, s_1, \ldots, s_r$ to be in states $x_0 = X(s_0), x_1 = X(s_1), \ldots, x_r = X(s_r)$ the likelihood contribution is

$$\prod_{j=1}^{r} P_{x_{j-1}x_j}(s_{j-1}, s_j).$$

This means that Markov models with piecewise constant transition intensities (including regression models) may be handled quite easily, see, e.g., Kay (1986). Non-parametric inference based on interval-censored data have been studied in some special cases, including the 3-state Markov illness-death model without recovery (Frydman, 1992, 1995), see also Commenges (2002).

Non-Markov processes

For semi-Markov processes without loops, that is, when only a finite number of paths from any state $h \in \mathcal{S}$ to another state $j \in \mathcal{S}$ is possible, explicit expressions for transition probabilities like (20.7) are available. As a first example we consider the semi-Markov illness-death model without recovery and where the transition intensity from 1 to 2, $\alpha_{12}(t, t-T_1 \mid Z)$ depends on both "calendar" time, t, and duration $t - T_1$ in state 1, where T_1 is the time of transition from 0 to 1. Here, $P_{01}(0,t)$ is still given by (20.7) with P_{11} now specified as

$$P_{11}(u,t \mid Z, T_1) = P_{11}(u,t \mid Z, u) = \exp\left(-\int_u^t \alpha_{12}(x, x-u \mid Z)\mathrm{d}x\right).$$

Asymptotics for this model was studied by Shu et al. (2007) for the special case of $\alpha_{12}(x, x-u \mid Z) = \alpha_{12}(x-u \mid Z)$ (i.e., a homogeneous semi-Markov model as described in Section 20.2.1).

For general multi-state models, Datta and Satten (2001) studied estimation of the state occupation probabilities $\pi_h(t), h \in \mathcal{S}$. They showed that the product-integral estimator is consistent without the Markov assumption and related this fact to the estimation of marginal rates as discussed in Section 20.2.

20.3.2 Direct models for probabilities

In some models, transition probabilities may be estimated directly. Thus, for a general non-Markov illness-death process without recovery, Meira-Machado et al. (2006) derived estimators for the transition probabilities $P_{00}(s,t)$, $P_{01}(s,t)$, $P_{11}(s,t)$ based on the following representation. If T_0 and T_2 are the sojourn time spent in state 0 and the time to absorption in state 2, respectively, and if H is the survival function for T_0 then

$$P_{00}(s,t) = \frac{H(t)}{H(s)}, P_{01}(s,t) = \frac{\mathrm{E}(\phi_{st}(T_0,T_2))}{H(s)}, P_{11}(s,t) = \frac{\mathrm{E}(\widetilde{\phi}_{st}(T_0,T_2))}{\mathrm{E}(\widetilde{\phi}_{ss}(T_0,T_2))},$$

where $\phi_{st}(u,v) = I(s < u \leq t, v > t)$ and $\widetilde{\phi}_{st}(u,v) = I(u \leq s, v > t)$. Here, H can be estimated by the Kaplan-Meier estimator, $\widehat{H}$. The estimators for the expectations $\mathrm{E}(\phi_{st}(T_0,T_2))$ and $\mathrm{E}(\widetilde{\phi}_{st}(T_0,T_2))$ presented in Meira-Machado et al. (2006) have later been revised; see Meira-Machado et al. (2011).

Without right censoring, the estimator of $P_{hj}(s,t)$ reduces to the relative frequency of processes in state j at time t among those in state h at time $s < t$. Meira-Machado et al. (2006) derived large sample properties of these estimators which may be generalized to more complicated non-Markov processes. It should be noted, however, that consistency of these estimators only holds if the support of distribution of T_2 is contained in that of the censoring time.

Finally, for a transient state the state occupation probability may be estimated by "Kaplan-Meier differences" (Pepe, 1991). As a simple example, let us once more study the illness-death model without recovery where T_0 is the time spent in state 0 and T_2 the time to death with survival functions H and S, respectively. Then $H(t)$ is the probability that the process is in state 0 at time t and $S(t)$ is the probability that it is in either state 0 or state 1 at time t. With Kaplan-Meier estimators $\widehat{H}$ and $\widehat{S}$ for T_0 and T_2, respectively, the state occupation probability $\pi_1(t)$ for the transient state 1 can, therefore, be estimated without a Markov assumption by

$$\widehat{\pi}_1(t) = \widehat{S}(t) - \widehat{H}(t).$$

This approach may be used for transient states in more general MSMs.

Regression models

Without censoring, state occupation indicators $I(X_i(t) = h)$ would always be observed and could thereby be used as outcome variables in generalized linear models for

$$\pi_{hi}(t \mid Z_i) = \mathrm{E}(I(X_i(t) = h \mid Z_i)),$$

that is, a model of the form

$$g(\pi_{hi}(t)) = \mathrm{LP}_{hi}(t)$$

with link function g. Here, the estimating equations would be $\sum_i U_i(\beta,t) = 0$, for (all or) selected $t-$values, with

$$U_i(\beta,t) = \left(\frac{\partial}{\partial\beta} g^{-1}(\mathrm{LP}_{hi}(t))\right)^{\mathsf{T}} \mathbf{V}_i^{-1}(I(X_i(t) = h) - g^{-1}(\mathrm{LP}_{hi}(t))). \qquad (20.8)$$

In (20.8), $\mathbf{V}_i$ is a working covariance matrix, frequently chosen simply to be the identity.

With censoring, $I(X_i(t) = h)$ is not always observed and modifications to (20.8) are needed. One possibility is to replace $I(X_i(t) = h)$ by its pseudo-observation (e.g., Andersen and Klein (2007)) given by

$$\widehat{\pi}_{hi}(t) = n\widehat{\pi}_h(t) - (n-1)\widehat{\pi}_h^{-i}(t)$$

where $\widehat{\pi}_h(t)$ is a well-behaved estimator for $\pi_h(t)$ based on the entire sample of size n while $\widehat{\pi}_h^{-i}(t)$ is the same estimator applied to the sample of size $n-1$ obtained by eliminating subject i. For this approach to work, right censoring should be independent of both the multistate process $X_i(\cdot)$ and of the covariates, Z_i (though the latter assumption may be relaxed by basing the pseudo-observations on an inverse probability of censoring-weighted estimator; see Binder et al. (2012) for a study of the competing risks model).

A related method, also based on inverse probability of censoring weighting, was studied by Scheike and Zhang (2007). Here, the starting point is once more (20.8) where $I(X_i(t) = h)$ is now replaced by

$$\frac{I(X_i(t) = h)I(C_i > t)}{G_C(t)}$$

with G_C denoting (an estimate of) the censoring distribution, $G_C(t) = \text{Prob}(C_i > t)$. For the special case of the competing risks model this approach was shown by Graw et al. (2009) to be asymptotically equivalent to that based on pseudo-observations. For this model, both methods are also closely related to the regression techniques suggested by Fine and Gray (1999) for the clog-log link function and Fine (2001) for other links. Here, the cumulative incidence function, $P_{0h}(0,t)$ was studied via the sub-distribution hazard

$$\widetilde{\alpha}_h(t) = \frac{\partial}{\partial t}(-\log(1 - P_{0h}(0,t)))$$

and the Cox regression score equations were modified by inverse probability of censoring weights. We prefer to formulate the models directly in terms of $P_{0h}(0,t)$, mainly because of the awkward interpretation of the sub-distribution hazard, which is

$$\widetilde{\alpha}_h(t)\text{d}t = \text{Prob(failure from cause } h \text{ in } (t, t+\text{d}t) \mid$$

$$\textit{either} \text{ alive at time } t- \textit{ or} \text{ failure from a competing cause in } [0,t)).$$

20.3.3 Example

We start by analyzing the simple two-state survival model; see Figure 20.1. The survival probability is given in Figure 20.7a. The predicted curve for given values of covariates can be calculated using the estimated baseline hazard values and coefficients in Table 20.2. Because of the one-to-one relationship between the hazard and the survival function, the estimated coefficients directly reflect in the differences between the predicted curves at different covariate values; see Figure 20.7b.

The Aalen-Johansen estimates of the cumulative incidence functions $\widehat{P}_{01}(0,t)$ and $\widehat{P}_{02}(0,t)$ for the competing risks model described in Figure 20.3 are given in Figure 20.8. The two curves are very similar, both reaching almost 0.5 after 7 years, the probability of either dying or switching to HD is thus almost 1 at that time, the estimated event free survival probability equals $\widehat{P}_{00}(0,7) = 0.034$.

The two curves resulting from a piecewise constant model using 10 intervals (at quantiles of the time to exit from state 0) agree reasonably well with the non-parametric Aalen-Johansen estimates. Plug-in estimates can be calculated using the estimated baseline hazard

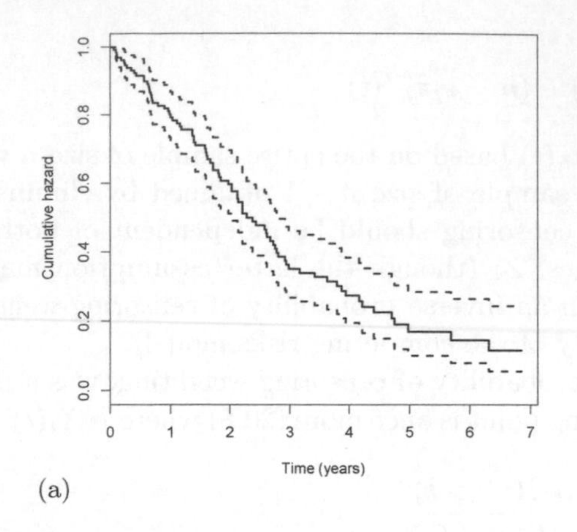

(a)

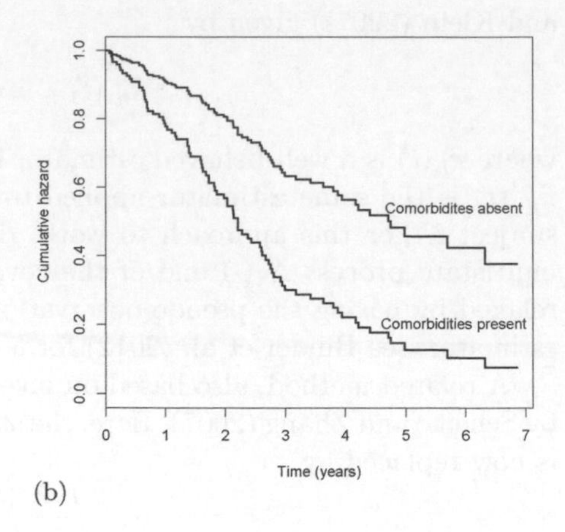

(b)

FIGURE 20.7
(a) Kaplan-Meier estimate with 95% confidence interval for the two-state model; (b) predicted curves for individuals with respect to the presence of comorbidities, evaluated at mean values of age and creatinine.

and coefficients of the Cox model; see Table 20.3 and Figure 20.9. Note that Cox models for both transitions are needed to estimate each of the transition probabilities and the correspondence between the transition intensity and probability is no longer simple. For example, though proportional hazards were assumed in the fitted Cox models, the association of creatinine with the probability of dying seems to change in time; see Figure 20.9b.

The association between the covariates and the cumulative incidence functions can be modeled directly using the Fine-Gray approach (Fine and Gray, 1999) or the method based on pseudo-observations (Andersen and Klein, 2007). Table 20.6 presents the results using the clog-log link with 10 time points for the pseudo-observations based method. Both models lead to the same qualitative conclusions, higher creatinine seems to be importantly associated with higher probability of switching to HD and is the only significant variable.

When considering the illness-death model (Figure 20.5), the Aalen-Johansen estimator can be used to estimate the transition probabilities. Note that unlike in the case of the cumulative hazard function, both $P_{01}(0,t)$ and $P_{02}(0,t)$ are affected by regarding the state 1 as transient; the estimated curves are given in Figure 20.10. The Aalen-Johansen estimator is based on the Markov assumption, which, as we already know, does not hold in our example. This is not important for the $P_{01}(0,t)$ estimation since it is in fact a state occupation probability $\pi_1(t)$ and the Aalen-Johansen estimator thus consistent regardless of Markov assumption. Figure 20.10a compares the different approaches to estimating $P_{01}(0,t)$. As expected, all the methods give very similar results with the Kaplan-Meier differences (Pepe, 1991) and the Meira-Machado estimate even completely overlapping. The Markov assumption is more important when estimating $P_{02}(s,t)$; Figure 20.10b shows that Aalen-Johansen is overestimating compared to the Meira-Machado approach.

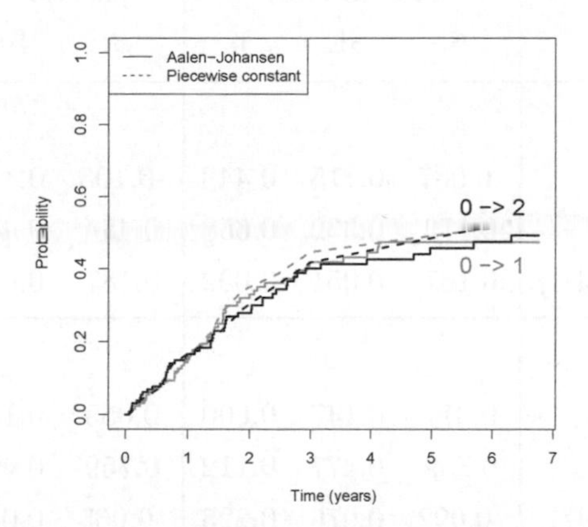

FIGURE 20.8
Cumulative incidence estimates (state occupation probabilities): Aalen-Johansen and piecewise constant hazards. Grey=HD, black=death.

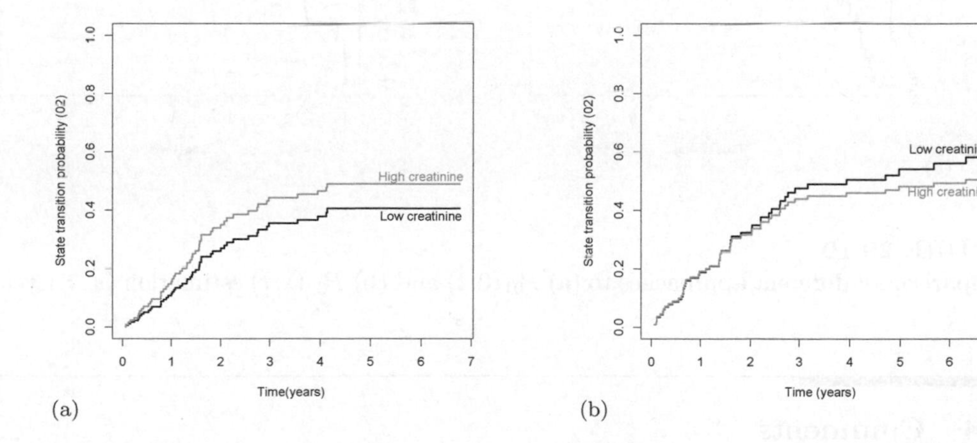

FIGURE 20.9
Competing risks model: (a) Plug-in estimate for $0 \to 1$ transition probability; (b) plug-in estimate for $0 \to 2$ transition probability. Evaluated at age 70, with present comorbidities and creatinine equal 450 or 650 μg/l.

TABLE 20.6
Estimates in direct regression models for cumulative incidences.

	Fine & Gray			pseudo - cloglog		
	β	SE	p	β	SE	p
$0 \to 1$						
Age (per 10 years)	-0.087	0.115	0.448	-0.105	0.110	0.341
Comorbidities present	0.173	0.392	0.658	0.484	0.444	0.276
Creatinine (per 100 μmol/l)	0.161	0.051	0.002	0.187	0.055	0.001
$0 \to 2$						
Age (per 10 years)	0.077	0.147	0.600	0.055	0.147	0.710
Comorbidities present	0.759	0.477	0.112	0.759	0.495	0.125
Creatinine (per 100 μmol/l)	-0.092	0.076	0.225	-0.065	0.085	0.442

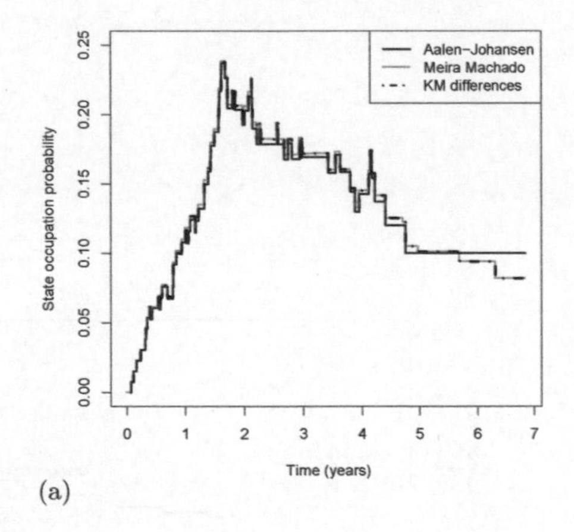

(a)

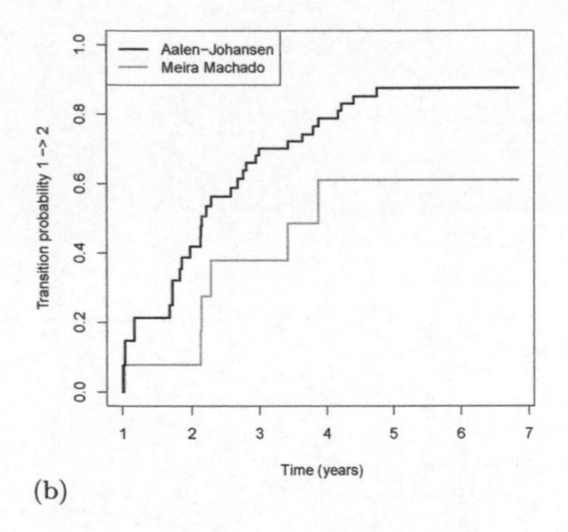

(b)

FIGURE 20.10
Comparison of different approaches to (a) $\widehat{P}_{01}(0,t)$ and (b) $\widehat{P}_{02}(s,t)$ estimation ($s = 1$ year).

20.4 Comments

We have presented a series of statistical methods for MSMs which may be useful in the analysis of follow-up data in a number of situations. One class of models was based on transition intensities. These are the most fundamental parameters in MSMs and they are the parameters which enter directly in the likelihood for continuously observed follow-data, cf. (20.3). We focused on two broad classes of intensity models: non- or semi-parametric models, and parametric models with piecewise constant transition intensities. One purpose of our illustrative example was to emphasize that these models, in fact, tend to provide very similar results. However, for the model with piecewise constant intensities the choice of intervals may affect the results.

Another purpose of our illustrative example was to demonstrate that the MSMs may provide insight into the data which may be overlooked if one instead analyzes the time to death using more simple survival analysis techniques. Thus, prognostic factors for survival may influence different transition intensities quite differently.

Another class of models focused on outcome probabilities, that is, state occupation probabilities and transition probabilities. One advantage of such models is the more direct interpretation of probabilities than intensities. While transition intensities provide a *local* (in time) description of the dynamics of the model, the probabilities give a global description which has been accumulated over time. For some intensity-based models (Markovian models and semi-Markov models without loops) transition probabilities were easily estimable by plug-in methods while the (marginal) state occupation probabilities could be estimated in more general classes of multistate models (Figures 20.7-20.10). When probabilities were estimated by plugging-in regression models for transition intensities no simple relationship between covariates and probabilities was obtained though predictions for given covariates were quite simple. In such situations, direct regression models for the outcome probabilities provided an alternative option which we exemplified using pseudo-observations; see Table 20.6.

Some limitations of the methods should be mentioned. First of all, we have only exemplified analysis of continuously observed data where all transition times were observed exactly. For interval-censored data, the only general approach (for Markovian models) seems to be models with piecewise constant intensities. In this connection it is reassuring to note the similarity between results from models with piecewise constant intensities and those from non- or semi-parametric inference. One should notice that the complexity of the inference increases with the number of possible transitions in the model, and we have only exemplified analyses with few states.

One final remark is that our example should be regarded as purely illustrative with the purpose of showing how the different models we have discussed may be handled in practice. Thus, our example was not intended to provide definitive analyses for the diabetes data. For this to be the case, much more attention must be paid to the goodness of fit of the models. For models based on intensities, techniques known from survival analysis may be applied while goodness of fit of models for pseudo-observations was discussed by Klein and Andersen (2005). A general method that can be used with all the described methods was described in Pohar Perme and Andersen (2008).

Bibliography

Andersen, P. K., Børgan, O., Gill, R. D. and Keiding, N. (1993), *Statistical models based on counting processes*, Springer-Verlag, New York.

Andersen, P. K., Hansen, L. S. and Keiding, N. (1991), 'Non- and semi-parametric estimation of transition probabilities from censored observation of a non-homogeneous Markov process', *Scandinavian Journal of Statistics* **18**, 153–167.

Andersen, P. K. and Keiding, N. (2002), 'Multi-state models for event history analysis', *Statistical Methods in Medical Research* **11**, 91–115.

Andersen, P. K. and Klein, J. P. (2007), 'Regression analysis for multistate models based on a pseudo-value approach, with applications to bone marrow transplantation studies', *Scandinavian Journal of Statistics* **34**, 3–16.

Binder, N., Gerds, T. A. and Andersen, P. K. (2012), 'Pseudo-observations for competing risks with covariate dependent censoring', *Research Report, Department of Biostatistics, University of Copenhagen* **6**.

Cheng, S. C., Fine, J. P. and Wei, L. J. (1998), 'Prediction of cumulative incidence function under the proportional hazards model', *Biometrics* **54**, 219–228.

Commenges, D. (1999), 'Multi-state models in epidemiology', *Lifetime Data Analysis* **5**, 315–327.

Commenges, D. (2002), 'Inference for multi-state models from interval-censored data', *Statistical methods in Medical Research* **11**, 167–182.

Cook, R. J. and Lawless, J. F. (2007), *The statistical analysis of recurrent events*, Springer-Verlag, New York.

Datta, S. and Satten, G. A. (2001), 'Validity of the Aalen-Johansen estimators of stage occupation probabilities and Nelson-Aalen estimators of integrated transition hazards for non–Markov models', *Statistics and Probability Letters* **55**, 403–411.

Fine, J. P. (2001), 'Regression modeling of competing crude failure probabilities', *Biostatistics* **2**, 85–97.

Fine, J. P. and Gray, R. J. (1999), 'A proportional hazards model for the subdistribution of a competing risk', *Journal of the American Statistical Association* **94**, 496–509.

Frydman, H. (1992), 'A non-parametric estimation procedure for a periodically observed three-state Markov process, with application to AIDS', *Journal of the Royal Statistical Society — Series B* **54**, 853–866.

Frydman, H. (1995), 'Nonparametric estimation of a Markov 'illness-death' process from interval- censored observations, with application to diabetes survival data', *Biometrika* **82**, 773–789.

Graw, F., Gerds, T. and Schumacher, M. (2009), 'On pseudo-values for regression analysis in multi-state models', *Lifetime Data Analysis* **15**, 241–255.

Hougaard, P. (1999), 'Multi-state models: a review', *Lifetime Data Analysis* **5**, 239–264.

Kay, R. A. (1986), 'Markov model for analysing cancer markers and disease states in survival models', *Biometrics* **42**, 855–865.

Klein, J. P. and Andersen, P. K. (2005), 'Regression modeling of competing risks data based on pseudovalues of the cumulative incidence function', *Biometrics* **61**, 223–229.

Lin, D. Y., Wei, L. J., Yang, I. and Ying, Z. L. (2000), 'Robust inferences for counting processes under Andersen-Gill model', *Journal of the Royal Statistical Society — Series B* **62**, 711–730.

Martinussen, T. and Scheike, T. H. (2006), *Dynamic regression models for survival data*, Springer-Verlag, New York.

Meira-Machado, L., Ũna-Álvarez, J. and Cadarso-Suárez, C. (2006), 'Nonparametric estimation of transition probabilities in a non-Markov illness death model', *Lifetime Data Analysis* **12**, 325–344.

Meira-Machado, L., Ũna-Álvarez, J. and Datta, S. (2011), 'Conditional transition probabilities in a non-markov illness-death model', *Discussion Papers in Statistics and Operation Research, University of Vigo* **3**.

Meira-Machado, L., Uña Álvarez, J., Cadarso-Suárez, C. and Andersen, P. K. (2009), 'Multistate models for the analysis of time-to-event data', *Statistical Methods in Medical Research* **18**, 195–222.

Pajek, J. (2012), 'The impact of technique failure on survival of patients treated with peritoneal dialysis', *Marn Pernat, A.(ed.). Final programme, invited lectures and abstracts. Ljubljana: Slovenian Society of Nephrology* pp. 58–60.

Pepe, M. S. (1991), 'Inference for events with dependent risks in multiple endpoint studies', *Journal of the American Statistical Association* **86**, 770–778.

Pohar Perme, M. and Andersen, P. K. (2008), 'Checking hazard regression models using pseudo-observations', *Statistics in Medicine* **27**, 5309–5328.

Putter, H., Fiocco, M. and Geskus, R. B. (2007), 'Tutorial in biostatistics: competing risks and multi-state models', *Statistics in Medicine* **26**, 2389–2430.

Rodríguez Girondo, M. and Uña Álvarez, J. (2011), 'A nonparametric test for markovianity in the illness-death model', *Discussion Papers in Statistics and Operation Research, University of Vigo* **4**.

Scheike, T. H. and Zhang, M. J. (2003), 'Extensions and applications of the Cox-Aalen survival model', *Biometrics* **59**, 1036–1045.

Scheike, T. H. and Zhang, M. J. (2007), 'Direct modeling of regression effects for transition probabilities in multistate models', *Scandinavian Journal of Statistics* **34**, 17–32.

Shen, Y. and Cheng, S. C. (1999), 'Confidence bands for cumulative incidence curves under the additive risk model', *Biometrics* **55**, 1093–1100.

Shu, Y. and Klein, J. P. (2005), 'Additive hazards Markov regression models illustrated with bone marrow transplant data', *Biometrika* **92**, 283–301.

Shu, Y., Klein, J. P. and Zhang, M. J. (2007), 'Asymptotic theory for the Cox semi-Markov illness-death model', *Lifetime Data Analysis* **13**, 91–117.

21

Landmarking

Hein Putter

Department of Medical Statistics and Bioinformatics, Leiden University Medical Center

CONTENTS

21.1	Landmarking	441
	21.1.1 Immortal time bias	442
	21.1.2 Landmarking	442
21.2	Landmarking and dynamic prediction	445
	21.2.1 Dynamic prediction	445
	21.2.2 The AHEAD data	446
	21.2.3 Dynamic prediction and landmarking	447
	21.2.4 Landmark super models	449
	21.2.5 Application to the AHEAD data	449
21.3	Discussion	453
	21.3.1 Implementation of landmarking	453
	21.3.2 When to use landmarking	454
	Bibliography	454

This chapter is about landmarking. Section 21.1 will discuss the origin of landmarking and its use for estimation of the effect of time-dependent covariates. The topic of Section 21.2 is the use of landmarking for dynamic prediction. Section 21.3 concludes with a discussion of practical issues.

21.1 Landmarking

The term "landmarking" originates from a debate on the effect of response to chemotherapy on survival in cancer patients (Anderson et al., 1983). Chemotherapy is given to cancer patients with the intention to destroy cancer cells, thereby shrinking the tumor; it is given either before or after surgical resection, or both. If the size of the tumor has decreased to a sufficient degree the patient is said to have responded to the chemotherapy (ignoring the distinction that is usually made between complete and partial response). Depending on the type of cancer, it typically takes in the order of 3 to 6 months to establish whether or not a patient has responded to chemotherapy. A comparison between chemotherapy responders and non-responders was often performed to support the claim of effectiveness of chemotherapy in the absence of a randomized clinical trial. The patients who responded to chemotherapy could be considered as the treated group, the non-responders as those who did not receive treatment. The typical way such an analysis was performed was by making two groups, one of "responders," the other of "non-responders," and by simply comparing survival between these two groups.

21.1.1 Immortal time bias

Apart from the fact that there could be factors that influence both the probability of a response and survival, which could invalidate the claim of a causal effect of response on survival, there is another fundamental problem with this approach. Consider a potential responder, a patient who would, given enough time, eventually respond to chemotherapy. If the patient died before his response to chemotherapy he would be considered a "non-responder" because his response was never observed. Thus, the patients in the response group are in some sense immortal until their time of response; if they had died before their response they would not have been in the response group. Such a requirement to be alive at the time of response is not present for the non-response group, thus giving the responders an unfair advantage. Epidemiologists call this phenomenon "immortal time bias." The same problem comes in many other disguises. One prominent example, which has been the subject of debate in cardiovascular research (Clark et al., 1971; Gail, 1972; Mantel and Byar, 1974), involves comparison of survival between patients with and without heart transplant. In the naïve approach described above, patients with heart transplant must have survived from diagnosis to time of heart transplant in order to be assigned to the heart transplant group. Other examples are the effect of recurrence on survival in cancer, the effect of transplant failure on survival in transplant studies, the effect of compliance to treatment on recurrence, the effect of drug-specific adverse events on recurrence. A final interesting example is a series of papers in the *Annals of Internal Medicine* on the effect of survival of winning an Oscar among U.S. Academy actors. A controversial paper by Redelmeier and Singh (2001) claimed to have shown that Oscar winners lived longer than their non-Oscar winning peers. The authors offered a number of explanations for this remarkable result, but of course the most likely explanation for their finding is immortal time bias. A subsequent paper using a correct analysis with time-dependent covariates found no evidence that Oscar winners lived longer (Sylvestre et al., 2006). In a nice review paper, Dafni (2011) gives a historical overview of inadequate comparisons between "responders" and "non-responders." By now, a growing number of researchers is aware of the problem of immortal time bias, but unfortunately the incorrect approach is still prevalent in medical journals.

The crucial issue is of course that the grouping "responder" versus "non-responder" is not yet known at baseline. When studying survival, an approach that makes groups based on something that will happen in the future is subject to bias. The paper of Anderson et al. (1983) proposed two alternatives for a correct statistical analysis of this type of problems. The first, well-known, approach is to consider response to chemotherapy as a time-dependent covariate, that is to let the value of the covariate in a Cox model change value from 0 (no Oscar winner) to 1 (Oscar winner) at the time of winning the Oscar. The second approach is the landmark approach which is discussed in the next subsection.

21.1.2 Landmarking

The idea is to fix a landmark time point in advance. For the example of the effect of response to chemotherapy on survival a good choice could be 6 months, because the majority of the chemotherapy responses will have occurred by then and not many patients will have died yet. The response status for each patient is then assessed at the landmark time point. That means that a patient with a response at 7 months will be considered to be part of the non-responder group at the landmark time point of 6 months. The subsequent survival analysis starts at the landmark time point, which implies that all follow-up before the landmark

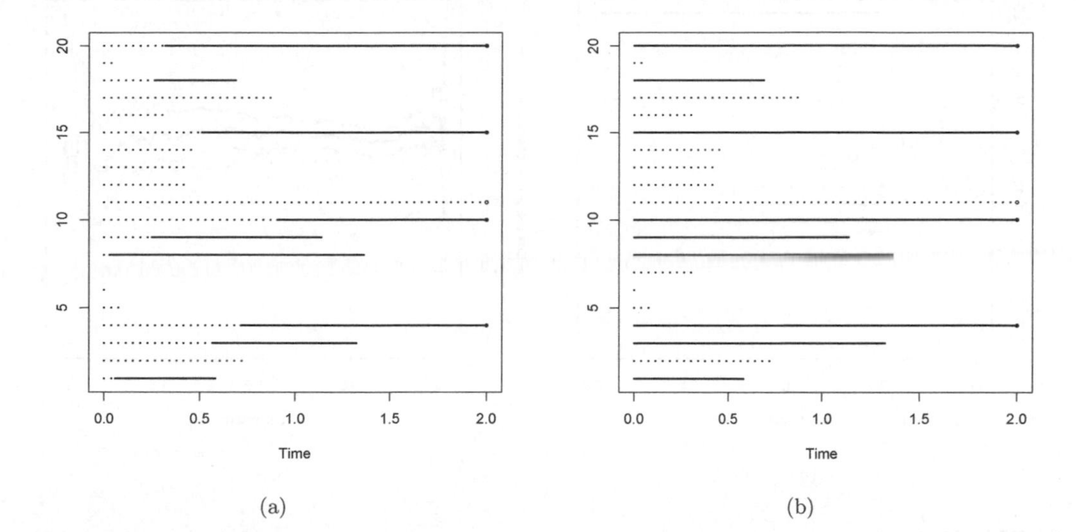

FIGURE 21.1
(a) Original data for the first 20 patients. (b) Groups made based on response status.

time point is discarded. All patients who died or were censored before the landmark time point are *excluded* from the landmark dataset.

An illustration based on a simulated dataset will serve to offset the different approaches. The set-up is very loosely based on the example of response to chemotherapy and its effect on survival. The simulated data consist of $n = 1,000$ patients. For each patient, time to response T_{resp} was generated uniformly on $(0, 1)$ with probability 0.5, while no response ($T_{\text{resp}} = \infty$) was generated with probability 0.5. Time to death T_{death} was exponential with mean 1, *independent of* T_{resp}. Death could happen before the response, in which case response is not observed. Censoring was applied at 2 (years). Figure 21.1(a) shows the original data for the first 20 patients in the left plot. Each line is one patient, the solid part of the lines corresponds to follow-up after response, while the dotted part of the lines corresponds to follow-up before (without response). The dots at 2 years correspond to censored observations, the rest are events. Figure 21.1(b) shows the set-up of the incorrect analysis for the same 20 patients based on making two groups as if they were known from baseline. As can be clearly seen, see for instance patients 1, 3 and 4, follow-up that should have been assigned to the non-response group is incorrectly assigned to the response group. Since events are correctly assigned to the two groups, hazard rates (events divided by follow-up time) are underestimated in the response group and overestimated in the non-response group, leading to a negative bias in the hazard ratio of response versus non-response for survival.

Indeed, Cox regression based on the analysis illustrated in Figure 21.1(b) gives an estimated regression coefficient (log hazard ratio) of -0.696 with a standard error of 0.076 ($p < 0.0001$), leading to a clear false positive result claiming that response to chemotherapy significantly improves survival. A time-dependent Cox regression using response status as time-dependent covariate results in an estimated coefficient of -0.078 with a standard error of 0.081 ($p = 0.34$). This result is in line with the true effect of 0 that was used in the data generation. The landmark analysis at 6 months created two groups, based on the response status at 6 months used as a time-fixed covariate in a Cox regression based on the landmark

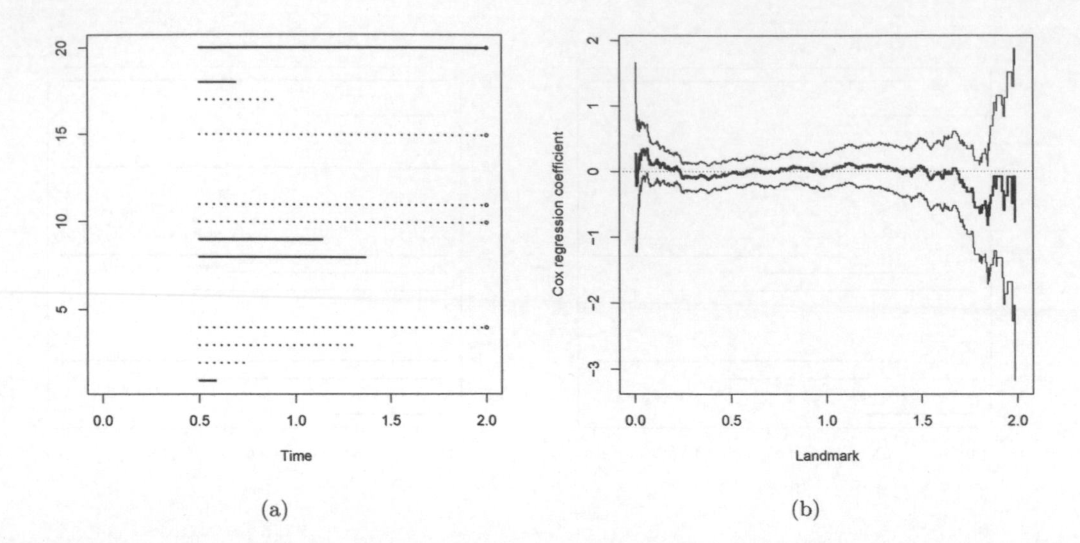

FIGURE 21.2
(a) The landmark dataset at 6 months for the 20 patients of Figure 21.1. (b) Regression coefficients and 95% confidence intervals for all possible landmark time points.

dataset obtained by removing everyone who was censored or who died before 6 months. The estimated regression coefficient was -0.025 with a standard error of 0.103, also not significantly different from 0. Figure 21.2(a) shows how groups are assigned to the patients. Patient 1, who responded to chemotherapy after 0.06 years, is assigned to the responder group, while patient 3, who responded to chemotherapy after 0.57 years (i.e., after the landmark time point), is assigned to the non-responder group. Patients 5–7, 12–14, 16 and 19 are excluded from the landmark dataset because they died before 6 months. Figure 21.2(b) shows the regression coefficients and 95% confidence intervals, obtained for the landmark analyses using as landmark time points all possible time points between 0 and 2 years. Clearly, estimates of the effect of response differ from landmark to landmark, but for each landmark time point the true effect of 0 is contained in the 95% confidence interval of the estimate. Note also that the precision of the estimate of the effect of chemotherapy varies appreciably. Initially confidence intervals of the landmark estimates are rather wide because the group sizes (responder vs. non-responder) are unbalanced. With increasing landmark time point the precision initially increases as more and more subjects become responders, until after one year the precision decreases again because the sample size of the landmark datasets (those still alive) becomes smaller.

This simple simulation study took the effect β of the time-dependent covariate $Z(t)$ ("response") in the Cox model

$$\lambda(t \mid Z(t)) = \lambda_0(t) \exp(\beta Z(t)) \tag{21.1}$$

to be equal to 0. Suppose that, as in the simulation study, $Z(t)$ is a 0/1 covariate that is initially 0 and then jumps to 1 if some event happens (a response, or start of treatment). If (21.1) is the correct model, then a time-dependent Cox model will yield a consistent estimate of β. In a landmark analysis at the landmark time point s the estimated effect will be attenuated towards 0, because the comparison is between subjects with $Z(s) = 1$ and subjects with $Z(s) = 0$, for which some will later on switch to $Z(t) = 1$. If interest is in an unbiased estimator of β, a time-dependent Cox regression is preferred (if that model is correct). The landmarking procedure will however yield a correct procedure for testing the null hypothesis $\beta = 0$.

In the present general context where interest is in the effect of an intermediate event like response to chemotherapy on an event of interest (death) a good choice of the landmark time point will take into consideration both when the intermediate event is likely to occur and when the event of interest is likely to occur. With respect to the former the landmark time point should be chosen late enough to allow a sufficiently large proportion of the population to have experienced the intermediate event. With respect to the latter the landmark time point should be early enough so that not too many events of interest have occurred. Typically such a choice can (and should) be made from prior knowledge, without looking at the data. Other choices can be used for sensitivity analyses, as in Dezentjé et al. (2010). Given the above, landmarking is most useful when the periods in which the intermediate event and the event of interest occur can be separated. If there is large overlap the landmark method will lose power either because insufficient events have occurred or because too many events of interest or censored observations had to be discarded. A final important point, discussed by Dafni (2011), is the fact that landmarking can only claim association, no causal relations, because the group membership is not determined on the basis of randomization, even if the data originate from a randomized clinical trial. This caveat is less of a problem when landmarking is used for prediction, as discussed in the next section.

21.2 Landmarking and dynamic prediction

Landmarking is now frequently used by biostatisticians if the principal interest is in *estimation* of the effect of a time-dependent covariate on survival. This aspect was discussed in the previous section. A novel use of landmarking is in *dynamic prediction*, which is the topic of the present section. First in Subsection 21.2.1 it will be clarified what is meant by dynamic prediction. Then Subsection 21.2.2 will introduce a dataset and a dynamic prediction problem that will be used as illustration in the rest of this section. After a brief discussion of the traditional approach in this kind of problem the general use of landmarking for dynamic prediction is examined in Subsection 21.2.3 and the idea of combining landmark models in Subsection 21.2.4 is introduced. Finally, Subsection 21.2.5 illustrates the use of landmarking for dynamic prediction for the data of Subsection 21.2.2.

21.2.1 Dynamic prediction

Among the many prediction models that have been developed in medicine, the overwhelming majority of these have been developed with the intention of providing predictions for patients with a certain disease from the moment of diagnosis or from the start of treatment for that disease. These models are clearly useful to inform patients about their prognosis and to guide clinicians in making treatment decisions. But they do not tell the whole story. A patient may want to know what the probability is that he/she is alive 5 years later at the start of treatment. But after successful primary treatment the patient is usually followed up at regular visits to the hospital, for instance each year. The question "what is the probability that I will still be alive after 5 years" is equally pressing two years later as it was at the start of treatment. In these two years new information has become available for this patient. This new information could be in the form of clinical events (for instance a local recurrence may have occurred one year after surgery) or in the form of measurements of biomarkers. Also the fact that no clinical events have occurred in the mean time is important information

TABLE 21.1
Demographics of Patients in AHEAD Study.

Covariate	N	(%)
Gender		
Male	1564	(39%)
Female	2468	(61%)
Education		
Less than high school	1736	(43%)
High school	1212	(30%)
Some college	1084	(27%)
BMI		
≤ 25	2244	(56%)
25–30	1388	(34%)
> 30	390	(10%)
Missing	10	
Smoking		
Never	1997	(50%)
Past	1683	(42%)
Current	324	(8%)
Missing	28	

that would typically improve prognosis for the patient. This new information, in the form of time-dependent covariates, should be incorporated in the prediction probabilities. This aspect of obtaining prediction probabilities not only from baseline but also at later points in time is called "dynamic prediction."

21.2.2 The AHEAD data

As illustration of the use of landmarking for dynamic prediction, data of the Asset and Health Dynamics Among the Oldest Old (AHEAD), now part of the wider U.S. Health and Retirement Study (HRS), will be used (Juster and Suzman, 1995). The AHEAD survey includes a nationally representative sample of initially non-institutionalized persons born before 1923, aged 70 and older in 1993. The present analysis uses only the non-Hispanic white subset. The outcome of interest is overall survival; the time scale is age. Table 21.1 shows the frequency in these data of the time-fixed covariates considered in the illustration (BMI and smoking status are assessed at entry into the study). The time-dependent covariate of interest is whether or not the subject is disabled according to the Basic Activities of Daily Living (ADL) scale by Katz et al. (1963), which includes items for walking, bathing, dressing, toileting and feeding. A subject is defined to be ADL disabled here if he/she responds "with difficulty" for at least one of the ADL items. In mathematical notation, the time-dependent covariate of interest is

$$Z_{\mathrm{ADL}}(t) = \begin{cases} 1, & \text{if subject is ADL disabled at age } t; \\ 0, & \text{otherwise.} \end{cases}$$

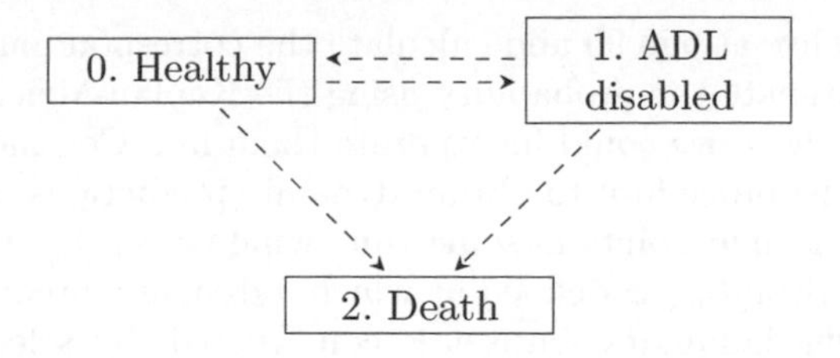

FIGURE 21.3
Graphical illustration of the reversible illness-death model for the HRS.

The general objective is dynamic prediction of survival of at least 10 years beyond age s, with given ADL disability status at age s and given covariates. If T denotes the random survival time of the individual, the objective is to estimate $P(T > s + 10 \,|\, T > s, Z_{\mathrm{ADL}}(s), Z^*)$, for given time-fixed covariates Z^*.

Before considering landmarking for solving this problem, it is instructive to examine the standard way of obtaining such dynamic prediction probabilities in this context. The standard approach is to cast the problem in the form of a multistate model, which may be formalized as a random process $X(t)$ in time. An obvious multistate model would have states (0) alive without ADL disability (healthy), (1) alive with ADL disability, and (2) dead. This multistate model is illustrated in Figure 21.3. The multistate model represented by Figure 21.3 is a reversible illness-death model. The illness state is state 1, ADL disabled, and the multistate model is reversible because recovery from ADL disability is possible (there is a transition from ADL disability to healthy). For a total of 4,032 subjects, 1,929 transitions from healthy to ADL disabled occurred and 679 recoveries (transitions from ADL disability to healthy). A total of 1,994 deaths were observed, 922 from the healthy state and 1,072 from ADL disability. The fact that the data are actually interval-censored is conveniently ignored for the purpose of illustration. In the notation just introduced the objective is to estimate $P(X(s+w) < 2 \,|\, X(s) = 0, Z^*)$ and $P(X(s+w) < 2 \,|\, X(s) = 1, Z^*)$ for given time-fixed covariates Z^*.

Dynamic prediction probabilities would typically be obtained by first estimating the transition hazards. In the presence of covariates, separate Cox models could be fitted for each of the four transitions. If the Markov assumption holds, there are no easy closed-form expressions for the prediction probabilities, because the multi-state model is reversible, but the Aalen-Johansen estimator (Aalen and Johansen, 1978) can be used to obtain estimates of prediction probabilities. This is quite laborious but it has been implemented in the R package `mstate` (de Wreede et al., 2010). If the Markov assumption does not hold, no direct ways are available to obtain the dynamic prediction probabilities, so one would have to resort to the use of (micro-)simulation. In the HRS data, there is in fact evidence that history of ADL disability increases the disability rate, which would indeed point to a violation of the Markov assumption. All this indicates that it is not straightforward to obtain dynamic prediction probabilities using the multistate approach.

21.2.3 Dynamic prediction and landmarking

The idea to use landmarking for dynamic prediction stems from van Houwelingen (2007). A similar approach was described by Zheng and Heagerty (2005) as partly conditional modeling. Suppose we want to estimate the probability, given alive at age 80, of survival until age 90 and suppose that we had an enormous database at our disposal. The basic idea is very simple: we would select a subset of the data, consisting of everyone alive at age 80. This would be a *landmark dataset*. Then in case of no censoring we would simply count how many individuals are alive at age 90 and calculate the corresponding proportion. If there is censoring, we could estimate the probability using the Kaplan-Meier estimator and if there are also covariates involved, we could incorporate them in a Cox model.

In general terms, the procedure to obtain dynamic predictions using landmarking is to select a set of landmark time points in some time window $[s_0, s_1]$. For each such landmark time point s a time horizon t_{hor} is defined at which a dynamic prediction is to be obtained. Then the corresponding landmark dataset is constructed, by selecting all individuals at risk at s. Define $Z(s)$ to be the vector of time-fixed and time-dependent covariates (for the time-dependent covariates the *current* value at s should be taken). We can then fit a simple Cox proportional hazards model

$$\lambda(t \mid Z(s), s) = \lambda_0(t \mid s) \exp(Z(s)^\top \beta(s))$$

for $s \leq t \leq t_{\text{hor}}$, enforcing administrative censoring at t_{hor}. This means that all events and follow-up after t_{hor} are ignored. After having obtained estimates $\hat{\beta}(s)$ and $\hat{\Lambda}_0(t \mid s)$ of the regression coefficient and the cumulative baseline hazard, respectively, the cumulative hazard specific to a subject with covariate values $Z^*(s)$ can be calculated as $\exp(Z^*(s)^\top \hat{\beta}(s))\hat{\Lambda}_0(t \mid s)$. An estimate of the prediction probability $P(T > t_{\text{hor}} \mid T > s, Z^*(s))$ is then given by $\exp(-\exp(Z^*(s)^\top \hat{\beta}(s))\hat{\Lambda}_0(t_{\text{hor}} \mid s))$.

It is important to note that for fixed s and t_{hor}, the Cox model

$$\lambda(t \mid Z(s), s) = \lambda_0(t \mid s) \exp(Z(s)^\top \beta(s))$$

uses $Z(s)$ as *time-fixed* covariates and $\beta(s)$ as *time-fixed* covariate effects. The regression coefficients and baseline hazards do depend on s, the landmark time point, but $Z(s)$ and $\beta(s)$ do not depend on t. There is an interesting and useful property of robustness of the landmark dynamic predictions. Using results of Xu and O'Quigley (2000) it was shown in van Houwelingen (2007) that *even if the effect of $Z(s)$ is time-varying*, the above model will give accurate dynamic predictions of $P(T > t_{\text{hor}} \mid T > s, Z^*(s))$, provided administrative censoring is enforced at t_{hor} during estimation of the Cox model, and that the prediction is only used at t_{hor}. If the true effect of the covariate is time-varying and equals $\beta(t)$, then

$$\begin{aligned} P(T > t_{\text{hor}} \mid T > s, Z^*(s)) &= \exp\Big(-\int_s^{t_{\text{hor}}} \lambda_0(u) \exp(Z^*(s)^\top \beta(u)) du\Big) \\ &= \exp\Big(-\Lambda_0(t_{\text{hor}} \mid s) \exp(Z^*(s)^\top \overline{\beta}(t_{\text{hor}} \mid s))\Big) \,. \end{aligned}$$

Here $\overline{\beta}(t_{\text{hor}} \mid s)$ is simply defined so that the last equation holds (more precisely it will also depend on $Z^*(s)$); it can be approximated as

$$\overline{\beta}(t_{\text{hor}} \mid s) \approx \frac{\int_s^{t_{\text{hor}}} \lambda_0(u)\beta(u)du}{\int_s^{t_{\text{hor}}} \lambda_0(u)du} \,,$$

that is, as a weighted time-average of $\beta(t)$ between the prediction time s and the horizon t_{hor}. Fitting a Cox proportional hazards model on the landmark dataset at s, ignoring the time-varying effect of Z and applying administrative censoring at t_{hor}, would also yield approximately a time-average of $\beta(t)$ between s and t_{hor}, but with subtly different weights (Struthers and Kalbfleisch, 1986; Xu and O'Quigley, 2000)

$$\overline{\beta}_{\text{Cox}}(t_{\text{hor}} \mid s) \approx \frac{\int_s^{t_{\text{hor}}} S(u|s)C(u|s)\lambda(u)\beta(u)du}{\int_s^{t_{\text{hor}}} S(u|s)C(u|s)\lambda(u)du} \,.$$

Here $S(u|s)$ and $C(u|s)$ are the conditional survival and censoring function, given alive at

s, and $\lambda(u)$ is the marginal hazard. If t_{hor} is not too far away from s, then $\overline{\beta}_{\text{Cox}}(t_{\text{hor}} \mid s) \approx \overline{\beta}(t_{\text{hor}} \mid s)$, and since it can be shown that also the baseline hazards agree, it follows that

$$\Lambda_{\text{Cox}}(t_{\text{hor}} \mid s, Z^*(s)) \approx \Lambda(t_{\text{hor}} \mid s, Z^*(s)) \ .$$

For more details the reader is referred to van Houwelingen (2007) and Section 7.1 of van Houwelingen and Putter (2012).

21.2.4 Landmark super models

The procedure of estimating parameters by fitting the simple Cox model

$$\lambda(t \mid Z(s), s) = \lambda_0(t \mid s) \exp(Z(s)^\top \beta(s))$$

for $s \leq t \leq t_{\text{hor}}$ can be done for each landmark point separately. But one would expect the coefficients $\beta(s)$ to depend on s in a smooth way. This can be exploited by defining models combining the models of different landmark time points. One could use splines or a parametric model, such as

$$\beta(s) = \beta_0 + \beta_1 s + \beta_2 s^2 \ .$$

Such a combined model can be fitted using standard software, by stacking the landmark datasets and performing a single Cox regression, stratified by landmark. The estimated coefficients will be correct, but for the standard errors a correction is needed to account for the fact that data of the same patient are used repeatedly. Sandwich estimators (Lin and Wei, 1989) can be used for this purpose. The baseline hazard can be estimated by the Breslow estimator; it will depend on s unless both $Z(s)$ and $\beta(s)$ are constant. The resulting model is called a "landmark super model."

We can go one step further by also combining the baseline hazards for the different landmark time points with the aim of adding more structure and to make it easier to interpret the models. We may assume a model

$$\lambda_0(t \mid s) = \lambda_0(t) \exp(\theta(s))$$

with $\theta(s_0) = 0$ for identifiability. In our application a quadratic function

$$\theta(s) = \theta_1 s + \theta_2 s^2 \ .$$

will be taken. This model can also be fitted directly by applying a simple Cox model to the stacked dataset, like the stratified landmark super model, only the landmark time s is not used as stratifying variable but as a covariate.

It should be noted that the landmark super model is not a comprehensive probability model, but a sequence of optimal models indexed by landmark time point s. This implies that the landmark super model does not satisfy the coherence condition of Jewell and Nielsen (1993).

21.2.5 Application to the AHEAD data

In applying the landmarking approach to the AHEAD data, the endpoint is defined as survival in a window of fixed width $w = 10$ years from the moment s of prediction, i.e., for each s, the time horizon for prediction is defined as $t_{\text{hor}} = s + 10$. The landmark time points used are 16 points, equally spaced, from age $s_0 = 75$ to age $s_1 = 90$. For each landmark (prediction) time point s, a landmark dataset was constructed, containing all relevant information needed for the prediction. Each dataset consisted of all patients at risk

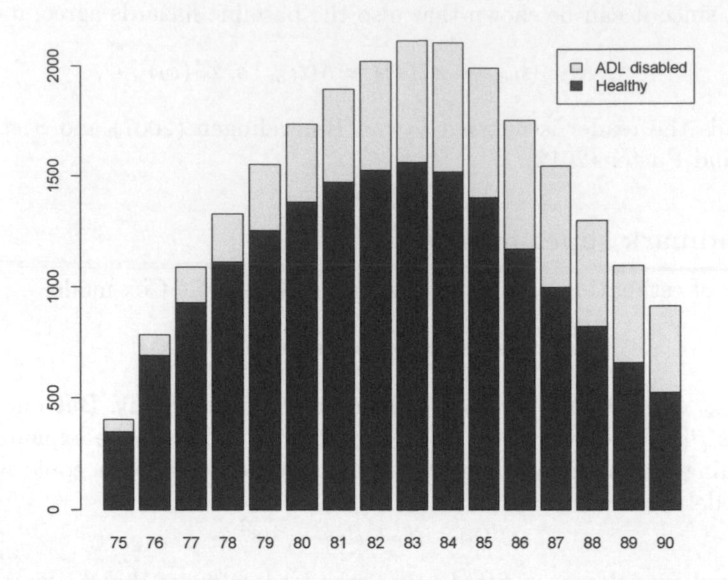

FIGURE 21.4
Frequencies of ADL disabled and healthy subjects for each of the landmark datasets.

(i.e., alive and under follow-up), the current value of ADL disability was computed and the horizon was set at $t_{\text{hor}} = s + 10$ years. At each landmark point a simple Cox model was fitted on $(s, s + 10)$ and used to obtain a prediction of survival at $s + 10$.

Figure 21.4 shows frequencies of ADL disabled and healthy subjects for each of the landmark datasets. The total height of the bars are the number of subjects at risk. Because of the delayed entry the total sample size of the landmark datasets is initially increasing, then the total sample size decreases again due to individuals dying. The proportion of subjects with ADL disability increases with age. Figure 21.5 shows regression coefficients with 95% confidence intervals for each of the covariates at each landmark time point. With the possible exception of smoking, the regression coefficients seem to be reasonably stable over (landmark) time. Figure 21.5 also shows smoother (mostly constant) regression coefficients obtained from the landmark super model. How these were obtained will be explained now.

The objective of a landmark super model is to obtain a parsimonious model combining possibly different effects over (landmark) time of covariates. For numeric stability, landmark time was standardized using $\overline{s} = (s - 75)/15$, which runs from 0 to 1. Table 21.2 shows the result of a backward selection procedure using Wald tests where, starting from a full model with linear and quadratic interactions of landmark time and covariates, the linear and quadratic terms of covariates were removed and replaced by a constant in case these linear and quadratic interactions were not significant at the 0.05 level. Only interactions of landmark time and smoking were retained, in accordance with the results of the separate landmark models of Figure 21.5. The smooth lines in Figure 21.5 overlay the regression coefficients of the separate landmark models. Table 21.2 also shows the coefficients θ_1 and θ_2 defining $\theta(s) = \theta_1 \overline{s} + \theta_2 \overline{s}^2$. Figure 21.6 shows the estimate of the cumulative baseline hazard $\Lambda_0(t)$ (left) and $\exp(\theta(s))$ (right), together defining the baseline cumulative hazard $\Lambda_0(t \mid s) = \Lambda_0(t) \exp(\theta(s))$ at s.

As mentioned before, dynamic predictions from the landmark super model at time s with covariate values $Z^*(s)$ are obtained from the estimates $\hat{\beta}(s)$, $\hat{\theta}$ and $\hat{\Lambda}_0(t)$ using $\hat{\Lambda}_0(t \mid s) =$

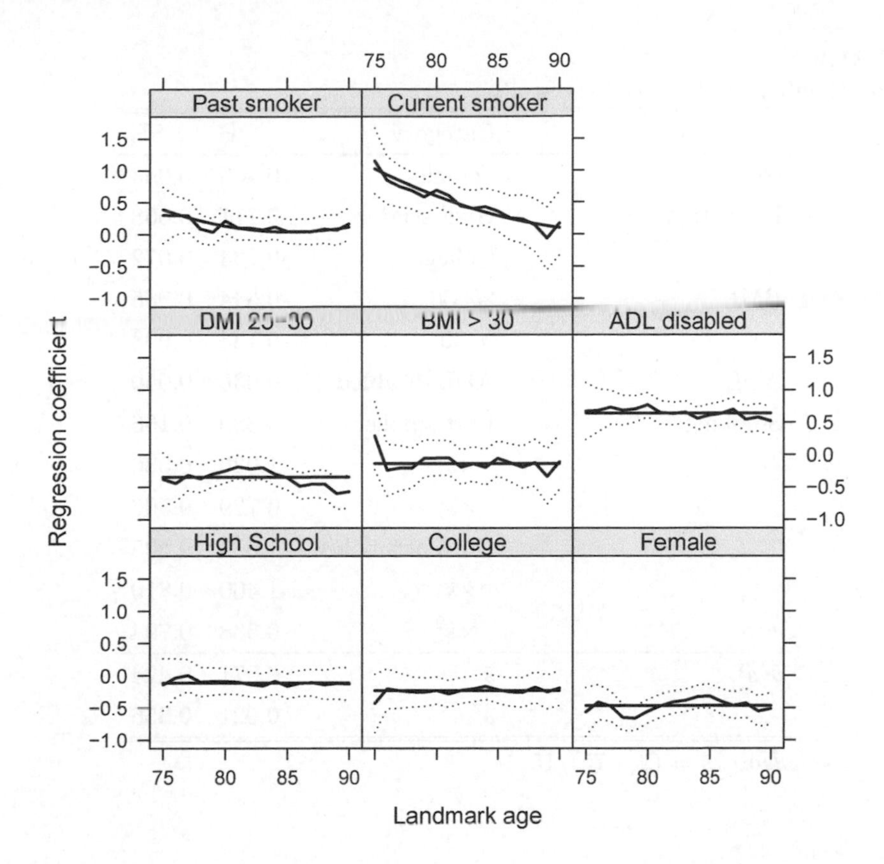

FIGURE 21.5
Regression coefficients with 95% confidence intervals for the separate landmark analyses. The smoothed lines show the regression coefficients implied by the landmark super model of Table 21.2.

$\hat{\Lambda}_0(t)\exp(\hat{\theta}(s))$ and

$$\widehat{P}(T > s + w \mid T > s, Z^*(s)) = \exp(-\exp(Z^*(s)^\top \hat{\beta}(s))(\hat{\Lambda}_0(s + w \mid s) - \hat{\Lambda}_0(s \mid s)) \ .$$

Figure 21.7 shows dynamic prediction probabilities of being alive 10 years after the prediction time point for all prediction time points between age 75 and 90, for all combinations of male/female, never/past/current smoker, and ADL disabled/healthy, as derived from the landmark super model. Comparing the top row (males) with the bottom row (females), it is clear that women live longer than men with comparable smoking and ADL disability status. Also, comparing the dashed lines for ADL disabled with the solid lines for healthy, clearly healthy people have a higher chance of living another 10 years than comparable ADL disabled people. Finally, an interesting observation can be made by comparing current smokers with never smokers. Comparing the healthy males, at age 75 never smokers have a probability of above 60% of living another 10 years, while for male current smokers this probability is less than 30%. In contrast, at age 90 there is no difference anymore between never smokers and current smokers, for healthy males the probability of living another 10 years is 25%, irrespective of smoking status. These same observations could of course already have been made from Table 21.2 and Figure 21.5, from the panel showing the effect of current smoking with respect to never smoking (top middle). The estimated regression coefficient is initially

TABLE 21.2
Regression Coefficients for AHEAD Study.

Covariate	Category	B	SE
Gender	Female	-0.465	0.063
Education	High school	-0.111	0.068
	College	-0.234	0.072
BMI	25–30	-0.344	0.065
	> 30	-0.135	0.098
ADL	ADL disabled	0.636	0.050
Smoking	Past smoker	0.389	0.166
	$\times\bar{s}$	-1.020	0.586
	$\times\bar{s}^2$	0.739	0.506
	Current smoker	1.024	0.205
	$\times\bar{s}$	-1.460	0.810
	$\times\bar{s}^2$	0.538	0.794
$\theta(s)$	$\bar{s}$	0.971	0.424
	$\bar{s}^2$	0.021	0.356

Note: $\bar{s} = (s - 75)/15$.

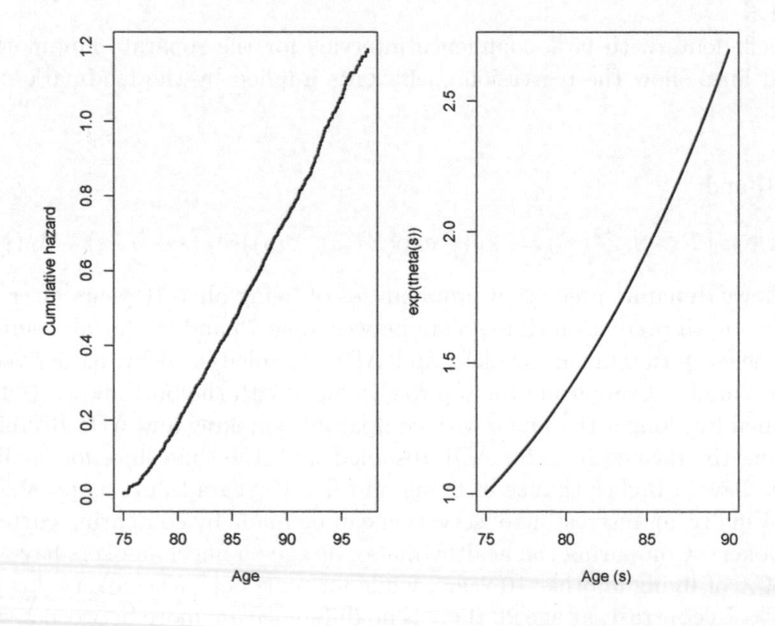

FIGURE 21.6
Estimates of the cumulative baseline hazard $\Lambda_0(t)$ (left) and of $\exp(\theta(s))$ (right) from the landmark super model.

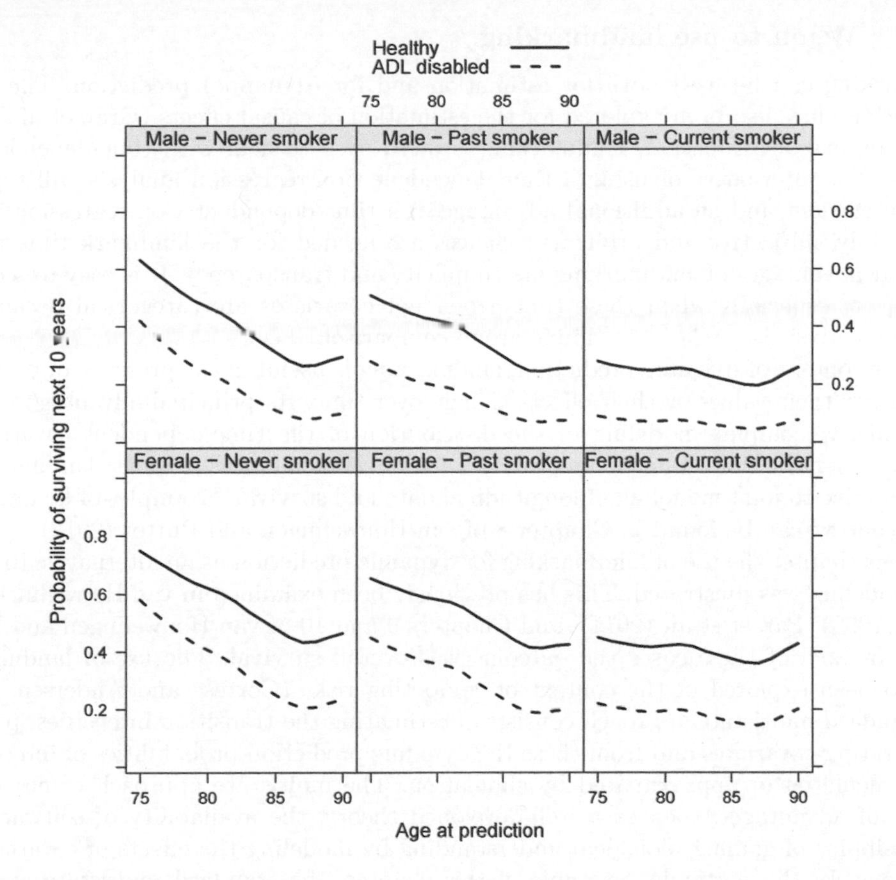

FIGURE 21.7
Dynamic prediction probabilities estimated from the landmark super model.

(at age 75) approximately 1, indicating a considerably lower probability of surviving 10 more years for current smokers, compared to never smokers. At age 90, however, the estimated regression coefficient is approximately 0, indicating no difference in the probability of surviving 10 more years between current and never smokers.

21.3 Discussion

21.3.1 Implementation of landmarking

In contrast to a multistate approach it is quite straightforward to implement landmarking for dynamic prediction. For R (R Development Core Team, 2012), the companion package dynpred of the book *Dynamic Prediction in Clinical Survival Analysis* (van Houwelingen and Putter, 2012) is available on CRAN (`cran.r-project.org`). The package dynpred contains functions to create landmark datasets, applying administrative censoring at a given horizon (*cutLM*), and to calculate dynamic "death within a window" prediction curves (*Fwindow*). On the book's website `http://www.msbi.nl/DynamicPrediction`, R code (using the dynpred package) of all the analyses in the book is available for download.

21.3.2 When to use landmarking

Landmarking can be used both for estimation and for (dynamic) prediction. The use of landmarking has also been explored for the estimation of causal effects (Gran et al., 2010). In the context of estimation, landmarking is often used to deal with time-dependent covariates. The alternative of using a time-dependent Cox regression analysis will typically be more efficient and an additional advantage of a time-dependent Cox regression is that no (possibly subjective and arbitrary) choices are needed for the landmark time points. The clear advantages of landmarking are simplicity and transparency. It is easy to see what is going on, especially when these time-dependent covariates are categorical because the resulting analysis is a relatively simple group comparison.

In the context of dynamic prediction, landmarking is useful in the presence of covariates when either their values or their effects change over time. Its principal advantage is again its simplicity: complex modeling of the distribution of the time-dependent covariates is avoided. When the time-dependent covariates are longitudinal measurements landmarking is an alternative to joint modeling of longitudinal data and survival. Examples of landmarking in this context can be found in Chapter 8 of van Houwelingen and Putter (2012).

In this chapter the use of landmarking for dynamic prediction as an alternative to multi-state modeling was illustrated. This has previously been examined in van Houwelingen and Putter (2008), Parast et al. (2011), and Chapters 9 and 10 of van Houwelingen and Putter (2012). In each of these cases the outcome was overall survival. The use of landmarking has also been explored in the context of competing risks (Cortese and Andersen, 2010). The standard multistate approach consists of estimating the transition intensities, possibly incorporating covariates and from these the dynamic prediction probabilities of interest are either calculated or approximated by simulation. The multistate approach comes with a number of advantages, such as a well-developed theory, the availability of software, and the possibility of gaining biological understanding by modeling the effects of covariates on transitions. In the particular example of this chapter, the standard multistate approach posed difficulties in obtaining dynamic prediction probabilities, especially since the Markov assumption was not satisfied. In such cases landmarking comes with a number of advantages. It is more directly targeted towards obtaining dynamic predictions; it bypasses models for the transition hazards and thereby uses sparser models. It is also robust against violations of the Cox proportional hazards assumptions; predictions obtained from a multistate model may be off the mark if assumptions are violated or if model fit is not good. There is no such thing as a free lunch: a possible disadvantage of landmarking is that predictions obtained by landmarking may be less efficient than those obtained from multistate models if the underlying assumptions of these models are correct. At present it is not sufficiently clear whether the loss of efficiency outweighs the gain in robustness and in what circumstances. More experience and research is needed to clarify this issue.

Bibliography

Aalen, O. O. and Johansen, S. (1978), 'An empirical transition matrix for nonhomogeneous Markov chains based on censored observations', *Scandinavian Journal of Statistics* **5**, 141–150.

Anderson, J. R., Cain, K. C. and Gelber, R. D. (1983), 'Analysis of survival by tumor response', *Journal of Clinical Oncology* **1**, 710–719.

Clark, D. A., Stinson, E. B., Griepp, R. B., Schroeder, J. S., Shumway, N. E. and Harrison, D. D. (1971), 'Cardiac transplantation in man, VI: prognosis of patients selected for cardiac transplantation', *Annals of Internal Medicine* **75**, 15–21.

Cortese, G. and Andersen, P. K. (2010), 'Competing risks and time-dependent covariates', *Biometrical Journal* **52**, 138–158.

Dafni, U. (2011), 'Landmark analysis at the 25-year landmark point', *Circulation - Cardiovascular Quality and Outcomes* **4**, 363–371.

de Wreede, L. C., Fiocco, M. and Putter, H. (2010), 'The `mstate` package for estimation and prediction in non- and semi-parametric multi-state and competing risks models', *Computer Methods and Programs in Biomedicine* **99**, 261–274.

Dezentjé, V. O., van Blijderveen, N. J. C., Gelderblom, H., Putter, H., van Herk-Sukel, M. P. P., Casparie, M. K., Egberts, A. C. G., Nortier, J. W. R. and Guchelaar, H.-J. (2010), 'Effect of concomitant CYP2D6 inhibitor use and tamoxifen adherence on breast cancer recurrence in early-stage breast cancer', *Journal of Clinical Oncology* **28**, 2423–2429.

Gail, M. H. (1972), 'Does cardiac transplantation prolong life? A reassessment', *Annals of Internal Medicine* **76**, 815–871.

Gran, J. M., Røysland, K., Wolbers, M., Didelez, V., Sterne, J. A. C., Ledergerber, B., Furrer, H., von Wyl, V. and Aalen, O. O. (2010), 'A sequential Cox approach for estimating the causal effect of treatment in the presence of time-dependent confounding applied to data from the Swiss HIV Cohort Study', *Statistics in Medicine* **29**, 2757–2768.

Jewell, N. P. and Nielsen, J. P. (1993), 'A framework for consistent prediction rules based on markers', *Biometrika* **80**, 153–164.

Juster, F. T. and Suzman, R. (1995), 'An overview of the Health and Retirement Study', *The Journal of Human Resources* **30**, 7–56.

Katz, S., Ford, A. B., Moskowitz, R. W., Jackson, B. A. and Jaffe, M. W. (1963), 'The index of ADL: a standardized measure of biological and psychosocial function', *Journal of the American Medical Association* **185**, 914–919.

Lin, D. Y. and Wei, L. J. (1989), 'The robust inference for the Cox proportional hazards model', *Journal of the American Statistical Association* **84**, 1074–1078.

Mantel, N. and Byar, D. P. (1974), 'Evaluation of response-time data involving transient states: an illustration using heart-transplant data', *Journal of the American Statistical Association* **69**, 81–86.

Parast, L., Cheng, S. and Cai, T. (2011), 'Incorporating short-term outcome information to predict long-term survival with discrete markers', *Biometrical Journal* **53**, 294–307.

R Development Core Team (2012), *R: A Language and Environment for Statistical Computing*, R Foundation for Statistical Computing, Vienna, Austria. URL: `http://www.R-project.org/`.

Redelmeier, D. A. and Singh, S. M. (2001), 'Survival in Academy Award-winning actors and actresses', *Annals of Internal Medicine* **134**, 955–962.

Struthers, C. A. and Kalbfleisch, J. D. (1986), 'Misspecified proportional hazards', *Biometrika* **73**, 363–369.

Sylvestre, M.-P., Huszti, E. and Hanley, J. A. (2006), 'Do Oscar winners live longer than less successful peers? A reanalysis of the evidence', *Annals of Internal Medicine* **145**, 361–363.

van Houwelingen, H. C. (2007), 'Dynamic prediction by landmarking in event history analysis', *Scandinavian Journal of Statistics* **34**, 70–85.

van Houwelingen, H. C. and Putter, H. (2008), 'Dynamic predicting by landmarking as an alternative for multi-state modeling: an application to acute lymphoid leukemia data', *Lifetime Data Analysis* **14**, 447–463.

van Houwelingen, H. C. and Putter, H. (2012), *Dynamic Predicion in Clinical Survival Analysis*, Chapman & Hall / CRC Press, Boca Raton.

Xu, R. and O'Quigley, J. (2000), 'Estimating average regression effect under non-proportional hazards', *Biostatistics* **1**, 423–439.

Zheng, Y. Y. and Heagerty, P. J. (2005), 'Partly conditional survival models for longitudinal data', *Biometrics* **61**, 379–391.

22

Frailty Models

Philip Hougaard

Biometric Division, Lundbeck

CONTENTS

22.1 Introduction 458
22.2 Purpose of a frailty model 459
22.2.1 Multivariate data examples where a frailty model is useful 459
22.2.2 Multivariate data examples where a frailty model is less useful 459
22.2.3 Univariate data examples 460
22.3 Models for univariate data 460
22.4 Shared frailty models for multivariate data 462
22.5 Frailty models for recurrent events data 463
22.6 Specific frailty distributions 463
22.6.1 Gamma 463
22.6.2 Positive stable 464
22.6.3 PVF 465
22.6.4 Lognormal 465
22.6.5 Differences between the models 465
22.7 Estimation 466
22.8 Asymptotics 466
22.8.1 The parametric case 467
22.8.2 The non-parametric case 467
22.8.3 The semi-parametric case 467
22.9 Extensions 468
22.10 Goodness-of-fit 468
22.10.1 Goodness-of-fit of models without frailty 469
22.10.2 Goodness-of-fit of models with frailty 469
22.10.3 Alternative models 469
22.11 Applications 470
22.12 Key aspects of using frailty models 471
22.13 Software 471
22.14 Literature 472
22.15 Summary 472
Bibliography 472

22.1 Introduction

A frailty model is a random effects model dedicated to survival data. This means that there is a model describing a basic random variation and on top of this there is an additional random variation (which is the one that the term *frailty* refers to). This random effect can be used for univariate (independent) data in order to obtain a more flexible model, or specifically describe overdispersion in relation to the basic random variation. It is, however, more interesting to use the random effect in the multivariate case as a means to model dependence between the observations. Four cases of multivariate survival data stand out as particularly relevant for the use of a frailty model: first, the case of related individuals, such as twins, family members or matched pairs; second, the case of similar organs (or other kinds of components) in an individual, for example, right and left eye or a set of teeth; third, the case of recurrent events, where multiple occurrences of similar events are happening for an individual, for example, epilectic seizures, or hypoglycaemic episodes in diabetes; fourth, times that come out from a designed experiment, where a single individual goes through multiple treatments, where in each case, the time to some event is recorded. The model will be introduced as covering survival times for one or more related individuals, but it can also be used in the other cases mentioned, as will be demonstrated later.

However, the frailty model is somehow less transparent than other random effects models, because the basic variation is described by a hazard function instead of a random variable. In popular terms, a steeply increasing hazard function describes a distribution with small variation, whereas a decreasing hazard function corresponds to a distribution with large variation. The frailty model is basically formulated as specifying that the hazard conditionally on the frailty, say Y, is of the form

$$Y\mu(t). \tag{22.1}$$

Here, $\mu(t)$ is the hazard function (or more precisely, the conditional hazard function). As Y is unknown, this is not immediately a useful model. But as for other random effects models, the way of handling the unknown values is to integrate the random components out. This gives a marginal distribution of the survival times, which are the observed quantities. The advantage of this formulation is that by letting Y being shared among several individuals, it can be used to create dependence between the times. Like in other random effects models, the random effects might reflect the effect of unobserved covariates.

Historically, frailty models have been used, at least since Greenwood & Yule (1920), who derived the negative binomial distribution of counts as a gamma distribution mixture of Poisson variables. More precisely, the number of events, say K, is assumed to be Poisson distributed with mean $Y\kappa$, conditionally on Y and using κ as a parameter (which in the above terminology should correspond to $\mu(t)$ integrated over the relevant time period). Assuming that Y follows a gamma distribution, it can be integrated out to give the marginal distribution of K, which then follows the negative binomial. In the present terminology this is a frailty model for recurrent events. Actually, κ is redundant in the expression, as the gamma distribution already includes a scale parameter. It is only included here to make it clearer that one can model the scale factor independently of the frailty distribution. For bivariate data, the model, still using a gamma distribution for the frailty variable, was introduced by Clayton (1978). The term *frailty* was suggested by Vaupel et al. (1979). They studied the effect of calendar time in long-term demographic studies of overall mortality, which in the present terminology is a kind of univariate data. Also that paper used the gamma distribution. There were other papers describing discrete random effects, whereas the use of other continuous distributions was made easier by using the general results for Laplace transforms, as suggested by Hougaard (1984).

This chapter is structured in the following way. Section 22.2 describes for which purposes a frailty model is useful. Models for univariate data are described in Section 22.3. Models for multivariate parallel data, such as times for related individuals, several organs or a designed experiment are described in Section 22.4. Section 22.5 describes models for recurrent events. Section 22.6 discusses the choice of distribution for the frailty. Section 22.7 describes estimation approaches and Section 22.8 asymptotic results. Section 22.9 discusses extensions to more complex models for multivariate data. Evaluation of goodness-of-fit is described in Section 22.10. Some applications are presented in Section 22.11. Section 22.12 is a kind of checklist for the application of frailty models. Software is described in Section 22.13. References to further general literature are given in Section 22.14. Finally, Section 22.15 is a short summary.

22.2 Purpose of a frailty model

The most important feature of a frailty model is that it can be used to model dependence between several time variables. This is obtained by assuming that the frailty is shared among several time variables. Basically, the time variables are conditionally independent given the frailty, but marginally (that is, when the frailty is integrated out), the observations are dependent. Thus the frailty approach is a way of generating dependence between time variables.

22.2.1 Multivariate data examples where a frailty model is useful

In this case, the frailty model is created by the conditional independence setup. It is, for example, useful to model the genetic effect in a twin. In most cases, the actual genes are not known, but it is known that identical twins have the same genes. The conditional independence assumption means that when the genes are accounted for, the survival times of twins are independent. As the genes are unknown, their effects have to be integrated out, and this implies that the times will be dependent. If some genes are known, they can be included as fixed effects. Inclusion of known covariates is therefore a key aspect of the model, and this turns the random effects model into a mixed effects model. In the twin case, it is the dependence as modelled by the frailty that is the interesting aspect of the model.

In other cases, the most interesting aspect in the study may be the effect of covariates, and the frailty is only included in order to account for the dependence. Still, the frailty model is useful and the frailty setup does the job of describing the dependence. This dependence is sometimes a nuisance, but in other cases, such as a cross-over experiment, the dependence is a design tool that reduces the unexplained random error and therefore allows for more precise evaluation of the effect of key covariates (typically, treatment).

22.2.2 Multivariate data examples where a frailty model is less useful

Several different types of events could be considered. One is the case of evaluating the relationship between myocardial infarctions and death, where the dependence is more direct. In this case, the study aims at examining whether the hazard of death increases after the occurrence of a myocardial infarction (and, of course, one would want to quantify this effect). The assumption of conditional dependence does not fit this kind of dependence. It is more reasonable to apply a multistate model. In practice an illness-death model with the effect of infarction (ill state) on death described directly by the transition intensity is needed. To set up a frailty model for such data would further require an assumption that the effect of the unknown frailty is quantitatively similar on the death hazard and the infarction hazard. Such an assumption would, in general, be unrealistic.

22.2.3 Univariate data examples

While the multivariate data are the real drivers of frailty models, they may have some use even in the case of univariate (independent) data. If $\mu(t)$ is a restricted parametric model, including a frailty on top of this can create a more flexible model. This could be interpreted as a model with overdispersion compared to the model given by $\mu(t)$ but alternatively, it could be used pragmatically just as a model with more parameters than the original model. As an example of creating a more flexible model, one can take the proportional hazards regression model. Depending on the choice of distribution for the frailty, this leads to a model with non-proportional hazards. This can be used not only to derive a test for hazard proportionality but also as a model in its own right, for use when the proportional hazards model is not fulfilled.

22.3 Models for univariate data

The frailty model will be presented in the general case, meaning that at this stage, the calculations apply to all distributions for the frailty, of course, satisfying that $Y \geq 0$.

The univariate model for the hazard is simply given by Equation (22.1), inserting either a parametric expression for $\mu(t)$, or extending this to a regression model of the Cox proportional hazards model form

$$Y \exp(\beta' z)\mu_0(t), \tag{22.2}$$

where z is a vector of covariates with corresponding regression coefficients β and $\mu_0(t)$ is the conditional hazard function corresponding to $z = 0$. The function $\mu_0(t)$ can be parametric or non-parametric. From this expression, one can derive the conditional survivor function, which in the absence of covariates is

$$S(t \mid Y) = \exp(-YM(t)), \tag{22.3}$$

where $M(t) = \int_0^t \mu(u)du$ is the integrated conditional hazard. As Y is unobserved and independent for all times, it has to be considered random and integrated out. This integration can be formulated on the probability level or the likelihood level. Starting on the probability level, integrating Y out gives the expression

$$S(t) = \int_0^\infty \exp(-yM(T))g(y)dy. \tag{22.4}$$

Strictly speaking, this expression is valid only when the frailty has a continuous distribution. If the distribution has discrete components, one should just integrate over that measure instead. The trick is that the integration above is the same integration as used in the Laplace transform for the distribution of Y. The Laplace transform is defined as $L(s) = E\{\exp(-sY)\}$. Using this, the survivor function becomes

$$S(t) = ES(t \mid Y) = L(M(t)). \tag{22.5}$$

Combining this expression with the previous makes it possible to derive the distribution of the frailty among the survivors at time t. As death has a preference for happening to subjects with high frailty (high risk), the distribution among survivors moves towards 0. Using the Laplace transform, the density among survivors becomes

$$\exp\{-yM(t)\}g(y)/L(M(t)).$$

By defining $\theta = M(t)$, this is of the form

$$\exp(-\theta y)g(y)/\int_0^\infty \exp(-\theta s)g(s)ds, \tag{22.6}$$

which is a member of the natural exponential family generated by $g(y)$. It follows directly that the mean in this distribution is

$$E(Y \mid T > t) = -L'(M(t))/L(M(t)). \tag{22.7}$$

From Equation (22.5), it is easy to derive the density of the time variable as

$$f(t) = -\mu(t)L'(M(t)). \tag{22.8}$$

Combining the above expressions gives the simple formula for the relation between the hazard in the marginal distribution and the corresponding one in the conditional distribution

$$\omega(t) = \mu(t)E(Y \mid T > t).$$

On the one hand, this expression describes the marginal value as a mean of the conditional value, but it is an ever changing mean because the population changes composition due to the gradual selective removal of high-risk individuals.

On the likelihood level, considering only a single individual with censored data, consisting of an observation time T and an event indicator D, the standard likelihood expression (hazard in the power of D times the survivor function) becomes

$$\int_0^\infty y^D\mu(T)^D \exp(-yM(T))g(y)dy, \tag{22.9}$$

where $g(y)$ is the density of the frailty distribution. Alternatively, one can use the Laplace transform calculations and this gives directly that the likelihood contribution is Equation (22.8) in the case of an event and Equation (22.5) in the case of a censoring and these expressions can be united as

$$(-1)^D\mu(T)^D L^{(D)}(M(T)), \tag{22.10}$$

where $L^{(p)}(s)$ is used to denote the pth derivative of $L(s)$. In the parametric case, this expression is ready to use, just by inserting the values of the relevant functions. In the non-parametric and semi-parametric cases, one has to extend the continuous hazards into discrete expressions with a hazard term for each observed event time, and this turns the integral $M(t)$ into a sum.

Depending on the model, some parameters may not be identifiable. For example, if $\mu(t)$ is a completely unspecified hazard function, including a random effect will not make the model larger, which implies that the various parameters cannot be identified. On the other hand, if $\mu(t)$ is a hazard function given by a few parameters, or if the model includes covariates, it is possible that the model with frailty (like Equation (22.5)) is more general than the corresponding model just given without frailty (like in Equation (22.3) without Y).

On a lower level, there are often identifiability issues as the frailty distribution may include a scale parameter but also the model for the hazard ($\mu(t)$) has a similar scale parameter. This is typically handled by introducing a scale restriction on the frailty distribution, in some cases $EY = 1$.

22.4 Shared frailty models for multivariate data

When the frailty is shared among several individuals, it leads to dependence between the times. To be more precise, conditionally on the frailty, the individuals are assumed to have independent times, modelled as described in Equation (22.1), but as the frailty is shared, the actual times are dependent. Thus, one can say that the frailty generates dependence between the times. For example, the frailty can describe the effect of shared genes among family members. In this setup, the dependence is necessarily positive.

So the hazard function model conditional on the frailty will have the form

$$Y\mu_j(t) \tag{22.11}$$

for the jth individual, where $\mu_j(t)$ can denote either of $\mu_j(t)$ (one hazard function per coordinate), $\mu(t)$ (symmetric) or $\exp(\beta' z_j)\mu_0(t)$ (proportional hazards). Marginal distributions can be evaluated using the formulas presented in Section 22.3, so here the focus is on bivariate calculations. The assumption of independence conditionally on the frailty implies that the likelihood in Equation (22.9) can be generalized to

$$\int_0^\infty y^{D_1+D_2}\mu_1(T_1)^{D_1}\mu_2(T_2)^{D_2}\exp(-y\{M_1(T_1)+M_2(T_2)\})g(y)dy \tag{22.12}$$

in the bivariate case. This formula, as well as the formulas below, can be extended to more than two individuals.

Again, using the Laplace transform allows for direct computation of the survivor function. This is based on the bivariate conditional survivor function being of the form

$$S(t_1, t_2 \mid Y) = \exp(-Y\{M_1(t_1) + M_2(t_2)\}). \tag{22.13}$$

The integration is essentially the same as in the univariate case, giving the bivariate survivor function as

$$S(t_1, t_2) = L\{M_1(t_1) + M_2(t_2)\}. \tag{22.14}$$

To handle possible censored data, this expression needs to be differentiated towards the coordinates, which correspond to actual events. This gives

$$(-1)^{D_1+D_2}\mu_1(T_1)^{D_1}\mu_2(T_2)^{D_2}L^{(D_1+D_2)}(M_1(T_1)+M_2(T_2)). \tag{22.15}$$

This extends Equation (22.10) to the bivariate case. Again, it is ready to use for parametric models, whereas the non-parametric and semi-parametric cases need to be extended to have a discrete component at each of the event times in the dataset.

The expressions above are given using the conditional hazard function $\mu(t)$ and this is the standard way of thinking in a random effects model. In a normal distribution repeated measurements model, this is known as a "subject-specific model." Alternatively, one might invert the expression in Equation (22.5) to give $M(t)$ as function of $S(t)$ for each coordinate and insert this in Equation (22.14). This gives the bivariate survivor function $S(t_1, t_2)$ as function of the univariate marginal survivor functions $S_1(t_1)$ and $S_2(t_2)$. Within survival data, this is known as a copula approach referring to separate modeling of dependence and marginal distributions. This would correspond to what in a normal distribution repeated measurements model is known as a "population-average model."

22.5 Frailty models for recurrent events data

In the case of recurrent events, the frailty is not shared among several individuals. Instead, it describes a random subject effect, and as such is assumed to be constant. This corresponds to a conditional Poisson model. That is, at any time the hazard of the subject to experience a new event is modelled by Equation (22.1). Conditionally on the frailty, events occur according to Poisson process with rate $Y\mu(t)$. In particular, it is noted that the time scale used in this model is time since starting the process. Still conditional on the frailty, the number of events over the interval from 0 to t, say K, is Poisson distributed with mean $YM(t)$. Inserting the corresponding terms and integrating out over the frailty (using the Laplace transform) gives the expression

$$Pr(K = k) = (-1)^k \{M(t)\}^k L^{(k)}(M(t))/k!. \qquad (22.16)$$

The presence of a frailty term leads to overdispersion compared to the Poisson model. A classical example is the negative binomial model, which occurs when assuming a gamma distributed frailty and that all subjects are followed for the same time period. In general, frailty models are much more flexible, by first allowing other distributions than the gamma and second by allowing each subject to have his own period of observation. The expression in Equation (22.16) describes only the distribution of the number of events in the interval. If the data also give the exact times of the events, the likelihood needs a factor $\mu(T_j)$ for each of the event times. In particular, these extra terms do not depend on the parameters of the frailty distribution. This implies that if all subjects are followed for the same period, the frailty parameters can be determined using only the event counts.

22.6 Specific frailty distributions

Outside of survival data (like in the linear normal model), random effects are typically chosen as being normally distributed, because this leads to simple marginal distributions and other nice properties. In other cases, a discrete mixture of, typically, two components is used, leading to a latent subgroup model. Distributional properties of the two-component model are not particularly simple, but the model has pedagogical advantages as it is very easy to explain.

In survival data, the normal distribution does not lead to simpler calculations than other models and it is therefore no longer particularly attractive. The discrete model is as above; it has pedagogical advantages, but has technical complexities that imply that it is not attractive overall. However, the setup in Equation (22.1) is indeed very convenient for many distributional choices because the necessary integrations can be performed simply and exact by means of the Laplace transform of the frailty distribution. Some convenient models are discussed below.

22.6.1 Gamma

The density of the gamma model is

$$g(y) = \theta^\delta y^{\delta-1} \exp(-\theta y)/\Gamma(\delta). \qquad (22.17)$$

It has two parameters (δ, θ), which both need to be positive. It turns out that some models cannot identify a scale parameter in the model, which means that it might be preferable to restrict the scale parameter in some way. For this model, the natural choice is to request $EY = 1$, and as this mean generally is δ/θ, this is obtained by setting $\theta = \delta$. For probability calculations, it is in some cases an advantage not to implement the restriction, so it will only be applied when explicitly said so. The limiting case of $\delta \to \infty$ corresponds to degenerate frailty, which again corresponds to independence between the time variables. The Laplace transform is $L(s) = E\exp(-sY) = \theta^\delta/(\theta+s)^\delta$. The gamma distributions turn out to have many interesting properties. The Laplace transform has rather simple derivatives and this makes it computationally simple to handle any number of time variables. The pth derivative of the Laplace transform is $L^{(p)}(s) = (-1)^p\theta^\delta(\theta+s)^{-(\delta+p)}\Gamma(\delta+p)/\Gamma(\delta)$. Based on observing survival times for a group of individuals, in which case one can easily experience both death times and censoring times, it is simple to calculate the conditional distribution of the frailty given the set of observations. Indeed, this conditional distribution is still a gamma distribution, where the δ value is increased by the number of events and the θ value is increased by the integral of the hazard function (that is, to $\theta + M(t)$). A by-product of this calculation is that the conditional distribution of the frailty among survivors in any given age is still a gamma distribution, and the δ parameter is unchanged; and the θ parameter is substituted by $\theta + M(t)$.

Integrating Y out in Equation (22.2) (using the restriction $EY = 1$) leads to the hazard function

$$\mu_0(t)\exp(\beta'z)/\{1 + M_0(t)\exp(\beta'z)/\theta\}. \tag{22.18}$$

This is a regression model with non-proportional hazards. The special case of $\theta = 1$ gives the proportional odds model. Indeed, the expression implies that the model parameters are identifiable even in the univariate case. This may be an advantage in that case, but in multivariate data models this may be a disadvantage, as it implies that non-proportionality may wrongly be interpreted as dependence between the times. This implies that by implementing an assumption of proportional hazards, say due to smoking, it is possible to collect a dataset on fathers and use this to identify the dependence between fathers and daughters. It is, of course, not sensible that it is possible to derive such a dependence without having combined data on fathers and daughters.

22.6.2 Positive stable

The positive stable distributions make up another distribution family, which is classical within probability theory. However, such distributions are less used in ordinary statistical models owing to infinite means. The most convenient definition of these distributions is indeed via the Laplace transform, that is, the distribution with parameter $\delta > 0$ and $0 < \alpha \le 1$ has Laplace transform $L(s) = E\exp(-sY) = \exp(-\delta s^\alpha/\alpha)$. This model includes a scale parameter, and this is most easily handled by setting $\delta = \alpha$.

In frailty models, the infinite mean does not lead to infinite survival, because the frailty influences the survival time through the hazard function. Thus the long tail of Y instead influences the hazard immediately after start of observation. Therefore, this distributional family works as an alternative to the gamma distribution and indeed has some other nice properties although it is more complicated to work with these distributions than the gamma. In particular, when combined with a Weibull model for the hazards conditionally on the frailty, this family simply implies that also the marginal distribution is in the Weibull family, but the Weibull shape parameter is reduced, corresponding to the distribution showing an increased variation. This model is also unique for the proportional hazards model, where if the conditional hazards are proportional, also the hazards in the marginal distributions are proportional, although the proportionality factors are closer to 1. The boundary case of

$\alpha = 1$ corresponds to degenerate frailty, which again corresponds to independence between the time variables.

This family was suggested for use as a frailty distribution in Hougaard (1986*a*) for the univariate case and in Hougaard (1986*b*) for the multivariate case.

22.6.3 PVF

The PVF (power variance function) is a three-parameter distribution family that includes both the gamma distributions (for $\alpha = 0$) and the positive stable distributions (for $\theta = 0$). It also includes the inverse Gaussian distributions (for $\alpha = 1/2$). Generally, the family has three parameters, $\alpha \leq 1$, $\delta > 0$ and $\theta \geq 0$, (with the further restriction that $\theta > 0$ in the case of $\alpha \leq 0$) but as one of them is a scale parameter, it is essentially a two-parameter model. This distribution family has a Laplace transform, which in the case of $\alpha \neq 0$ is

$$L(s) = \exp[-\delta\{(\theta + s)^{\alpha} - \theta^{\alpha}\}/\alpha]. \tag{22.19}$$

While it is more complicated than either of the gamma and positive stable formulas, it is still explicit and can be used as an extended model and thus as a goodness-of-fit test for either of the models.

In the case of $\alpha < 0$, the frailty distribution has a point mass at 0, which implies that some subjects will not experience the event. This is obviously not satisfied when lifetimes are modelled, but when the event is not certain to happen, it may be a useful feature.

This family was suggested for use as a frailty distribution in Hougaard (1986*a*).

22.6.4 Lognormal

The lognormal distribution (where the logarithm to the frailty follows a normal distribution, say $N(\theta, \sigma^2)$) is not suitable for exact calculations because the Laplace transform does not have a mathematically convenient expression, but as it is a standard distribution and many software packages have methods to integrate over the normal, it is sometimes used for this purpose. The natural restriction of the scale parameter is to select $\theta = 0$, which implies that the median frailty (and the geometric mean) is 1. The limit $\sigma^2 \to 0$ corresponds to independence.

22.6.5 Differences between the models

Standard random effects models (that is, outside of survival data) are almost purely based on the normal distribution. Therefore, it is a kind of luxury to have several families to choose between. This also means that we are not used to assess the differences between such models.

From a theoretical perspective, most frailty families (including the gamma distribution) will, combined with a regression model lead to distributions that identify the parameters even with univariate data. This is a convenient property in the study of univariate data, but in relation to multivariate data, this is not an advantage, because it implies that the model does not adequately discriminate between dependence and covariate influence. The positive stable model is an exception to this, and this implies that the model in the multivariate case purely describes dependence.

In terms of fit, one can consider the density of a bivariate survival time. To see the real difference between the models, it is relevant to normalise so the marginal distributions are controlled, most conveniently to a uniform distribution. In that frame, the gamma distribution will show a relative high density at late time points, whereas the positive stable will show a relatively high density at early time points.

Having several models for the dependence also creates a need for comparing the degree of dependence across models. As the correlation between the times is not adequate in a non-parametric model (and indeed, as the dependence is nonlinear), alternative measures are needed. Classical non-parametric measures of dependence are Kendall's τ and Spearmans ρ. Both of these depend only on the frailty distribution, meaning that they are invariant under nonlinear transformations of the time scale. Kendalls τ is the most simple to calculate and equals $1/(1+2\delta)$ for the gamma model and $1-\alpha$ for the positive stable.

From a computational point of view, many exact calculations are derivable from the Laplace transform, so having a sensible expression for this is a key feature. This is particularly simple for the gamma model, but exact evaluations are also possible for positive stable, PVF and inverse Gaussian distributions. On the other hand, the lognormal model does not allow this, but it is still preferred in some software packages.

22.7 Estimation

The frailty model estimates that have been suggested in the literature are mainly maximum likelihood estimates, although some of them are based on various approximations. Two main principles have been used. One is to use likelihood expressions, where the frailty directly enters, such as Equation (22.12). A classical example is the EM algorithm, where one of the steps is based on the so-called "full likelihood," which is the term under the integral sign (meaning it covers both the observations and the frailty values), which is then contrasted to the observed likelihood, which is the integrated version. The other approach is to integrate the frailty out analytically, that is, as described in Equation (22.15).

For a parametric hazard model, the expression for the hazard function can be inserted into these formulas and this typically works well. For a non-parametric hazard model, the model will be computationally more complex. Where the standard univariate tools either have explicit estimation of the non-parametric terms (like Kaplan-Meier and Nelson-Aalen) or have excluded the terms (like the Cox model), a frailty model cannot handle these terms in such simple ways. One approach is to include a hazard parameter for each time point with observed events. This leads to a multivariate estimation problem, potentially of high dimension, but with today's computers this is not really an issue as long as the number of different event times is below, maybe 2000. Alternatively, the hazard parameters can be handled approximately by being profiled out of the likelihood. The penalized likelihood approach uses such an approximation by estimating the hazard parameters using the predicted frailty values.

22.8 Asymptotics

Asymptotic methods are more complicated for frailty models than for standard survival time models, because the martingale methods are not directly applicable. However, some results are available. A specific issue is that the hypothesis of independence is a boundary case, corresponding to a degenerate frailty distribution. This boundary value may or may not correspond to a parameter value inside the natural parameter set, but in both cases, problems may occur.

Any censoring needs to happen independently of the frailty. However, in many cases, it is acceptable to allow censoring to depend on the events that have occurred within the cluster. For example, in the recurrent events case, censoring could be related to the number of events already occurred.

22.8.1 The parametric case

In the parametric case, the gamma frailty distribution is typically well behaved in asymptotical calculations. Of course, it is necessary to request identifiability (which is why the assumption of $EY = 1$ is typically implemented). The identifiability condition may further rule out some models in the univariate case, but it will not give problems in the multivariate case. Actually, the model can be extended to allow a slight negative dependence in the bivariate case, and this implies that the hypothesis of independence (which in other words could be formulated as degenerate frailty; or $\theta = \infty$). To let this work, one has to reparametrize the model, for example, using the frailty variance (say $\varphi = 1/\theta$) as parameter. In this case, independence is obtained for $\varphi = 0$ and the model is a valid model also for some negative values of φ. Thus, the hypothesis of independence is in the interior of the extended parameter set. This implies that there are no specific problems in testing independence.

For the positive stable frailty model, the model cannot be extended to negative dependence. The independence boundary is strict. The Fisher information at the boundary is infinite (both when the marginal distributions are known and when they follow the Weibull), as shown (for complete data) by Oakes & Manatunga (1992). Away from the independence boundary, the model does not show problems. For univariate data, there are identifiability problems in some models, whereas in the multivariate case, one may just need to fix the scale parameter.

The PVF model has two parameters to model the dependence and this implies that there will be identifiability issues near the independence boundary. This model has not been examined in detail, but away from the boundary the model is not expected to show problems. In the univariate case, identifiability needs to be checked.

The lognormal model has not been studied in detail, but the model is not expected to show problems.

22.8.2 The non-parametric case

For the non-parametric case, the data need to be multivariate, as the univariate case will show identifiability problems. Murphy (1995) showed a number of results for the gamma shared frailty model, covering both the multivariate and the recurrent events cases. She showed that one could apply maximum likelihood estimation and that tests for the frailty distribution parameter could be based on the profile likelihood, where the hazard function parameters have been profiled out.

22.8.3 The semi-parametric case

Parner (1998) extended results regarding the gamma frailty model from the non-parametric case to the semi-parametric case. With covariates, the frailty parameter can also be found based on univariate data. Kosorok et al. (2004) made further evaluations covering several one-parameter frailty models for univariate data.

Martinussen & Pipper (2005) evaluated the shared positive stable frailty model with covariates. They show consistency and large sample properties, but did not consider the boundary case of independence.

22.9 Extensions

The shared frailty model above is the standard model. In some cases, however, extensions are needed. For example, one would like to have observations that show different degrees of dependence in one case. For example, when studying twins, there are both monozygotic (identical) and dizygotic (fraternal) twins. A simple analysis technique is to analyse each type separately and this can be done using the standard shared frailty model. However, by making a combined model, one can implement common parameters, such as common marginal distributions or common effect of some covariates. This can be done by using a model that describes the differences in dependence for monozygotic and dizygotic twins. More generally, model extensions can be used to model more complex relationships, such as those that occur when studying complete pedigrees or in hierarchical models, where the various random effects are nested. The problem with these extensions is that the complexity of the model clearly increases.

One of the models suggested is the correlated gamma model, in which each individual has his own frailty value. In the bivariate case models the frailties are assumed to be of the form $(Y_1, Y_2) = (X_0 + X_1, X_0 + X_2)$, where X_0, X_1, X_2 are independent gamma variables with the same scale θ and with shape parameter δ_0 for X_0 and δ_1 for both X_1 and X_2, which gives a symmetric model. To set the mean to 1, one restricts $\delta_0 + \delta_1 = \theta$. This model was also considered by Parner (1998) and one can apply asymptotic methods in that case.

Another model suggested is the bivariate lognormal model, meaning that $(Y_1, Y_2) = (\exp(X_1), \exp(X_2))$, where (X_1, X_2) follow a bivariate normal distribution. The natural restriction to obtain identifiability is $EX_1 = EX_2 = 0$. This model has been suggested as a model for treatment-center interaction as described in Yamaguchi et al. (2002). In their model, center is the clustering unit, with Y_1 being the effect for treatment 1 and Y_2 the effect for treatment 2. There is one or more patients for each combination of center and treatment. There is, of course, also a fixed effect (a covariate) to describe the overall treatment effect.

Gorfine & Hsu (2011) discuss an extension of the model to describe competing risks for related individuals. The competing risks model has the added complexity that each subject has at most one event, which in the single individual case leads to unidentifiability of some quantities. This problem does not go away by having related individuals, so it needs special handling.

Another type of extension is to let go of the assumption of the frailty being constant. One can instead assume that the frailty is a stochastic process. A key paper on this is Gjessing et al. (2003). Clearly, the computational complexity increases dramatically.

Overall, these extensions add flexibility, but they also make the setup more complicated, and some models will show some undesired properties, such as identifiability with some datasets, where common sense would say that the relevant parameters should not be identifiable. This implies that when using these models, one should check that the results are adequate for the problem at hand.

22.10 Goodness-of-fit

Goodness-of-fit has two relations to frailty models. First, frailty models give new ways of testing goodness-of-fit of other models, and, second, frailty models also need to be checked in goodness-of-fit tests. Finally, alternative models can be considered.

22.10.1 Goodness-of-fit of models without frailty

Clearly, the frailty models have a role to play for testing the assumptions in standard survival time models, particularly in the univariate case.

For example, a one-sample parametric model, like the Weibull, can be tested by including a frailty term, and then testing whether this gives a better fit. If the frailty is assumed to be gamma distributed this makes a dramatic change of the functional form, in particular, if the Weibull shape parameter is larger than 1 (meaning that the conditional hazard increases with time). The hazard, when the frailty is integrated out, will then be first increasing and then decreasing.

An even more interesting application of this idea is to test the proportional hazards model. The original model is then a proportional hazards model, that is, of the form $\exp(\beta' z)\mu_0(t)$. By including also a gamma distributed frailty term as in Equation (22.2), and integrate it out, the resulting model becomes one of converging hazards, as described in Equation (22.18). The hazard ratio between two covariate values (which is constant in the proportional hazards model) becomes a monotone function of time, and it converges to 1 as time goes to infinity. There are, of course, many other ways of testing this assumption, but this approach has the special feature of modeling deviations in a special type of alternatives (which may or may not be an advantage, but it appears that this type of alternative is realistic in many cases) and it does this by adding only one extra parameter, and this potentially makes it rather powerful. Even in the case of many covariates, there is only one parameter to describe the non-proportionality.

22.10.2 Goodness-of-fit of models with frailty

Of course, also frailty models need tools for testing their goodness-of-fit. Any given frailty distribution can be evaluated by studying another frailty distribution. If the family of distributions is larger, this can be used for a statistical test, whereas otherwise one can only compare the results informally. In this way a gamma or positive stable frailty model can be tested by using the PVF model.

Shih & Louis (1995) suggest a goodness-of-fit test for the gamma frailty distribution in a parametric model. They suggested evaluating the average of the posterior expected frailties as function of time (t). Here "posterior" refers to including the event information on the cluster during the period $(0, t]$. If the gamma model is satisfied, the mean will fluctuate around 1. This was generalised to semi-parametric models by Glidden (1999).

22.10.3 Alternative models

An alternative approach to test a frailty model for bivariate or recurrent events data is to set up the corresponding multistate model. In these models, a frailty model implies a relation between the various transition hazards. This can be compared to the models one would set up for the multistate model, as described in Hougaard (2000, Chapter 6). Figure 22.1 illustrates the bivariate multistate model. The twins are considered symmetric, so only the hazard functions corresponding to death of twin 1 are inserted on the figure, and this is done in the most general setting; that is, without assuming that the process is Markov. Therefore, the second hazard function depends also on t_1, which is the time of the first death in the pair of twins. Based on the gamma frailty model, the hazard functions in the multistate model become $\lambda_1(t) = \mu(t)\delta/(\delta + 2M(t))$ and $\lambda_2(t, t_1) = \mu(t)(\delta + 1)/(\delta + M(t) + M(t_1))$. From this, one can derive the relation between the transition hazards as

$$\lambda_2(t, t_1) = \lambda_1(t)(1 + 1/\delta)\frac{2\exp(2\Lambda_1(t)/\delta)}{\exp(2\Lambda_1(t)/\delta) + \exp(2\Lambda_1(t_1)/\delta)},$$

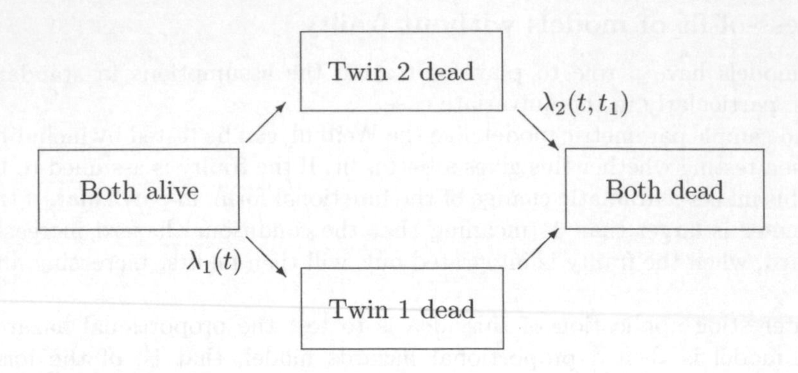

FIGURE 22.1
A multistate model for twin survival.

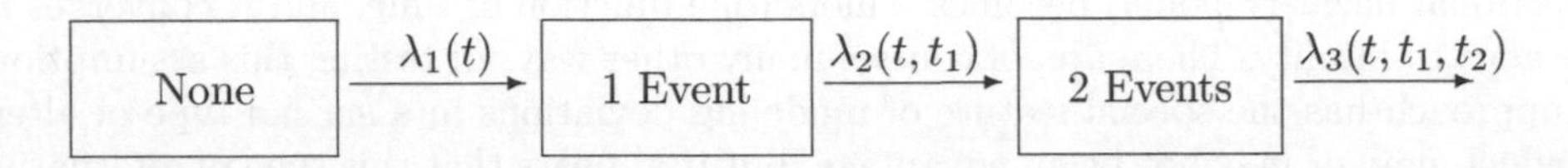

FIGURE 22.2
A multistate model for recurrent events.

where $\Lambda_1(t) = \int_0^t \lambda_1(u)du$ is the integrated hazard for the first event. By seeing the data as coming from the multistate model, it is possible to suggest an alternative formula for the transition hazards and check whether that model fits better than the expressions from the frailty model.

It is possible to do similar calculations in the recurrent events case, which is illustrated in Figure 22.2. In that case, the gamma frailty model leads to the transition hazards $\lambda_1(t) = \mu(t)\delta/(\delta + M(t))$ and $\lambda_2(t, t_1) = \mu(t)(\delta + 1)/(\delta + M(t))$, which turns out to be of the Markov form (that is, it does not depend on t_1). These formulas can be generalised to $\lambda_k(t) = \mu(t)(\delta + k - 1)/(\delta + M(t))$, from which one can derive $\lambda_k(t) = (\delta + k - 1)\lambda_1(t)/\delta$. This implies that the multistate transition hazards are proportional. Also in this case, alternative transition hazards can be used to test the goodness-of-fit of the frailty model. Other frailty distributions will give different expressions, but the multistate model is of the Markov type in any case.

22.11 Applications

As the shared frailty model has been available in standard software for many years, there are now plenty of applications of this model.

In terms of the various types of data, most applications cover the case of several individuals. They are related typically, either due to family (twins, sibships, parent-child) or due to center in clinical trials.

Some applications include several organs (components) on the same individuals. The most common example of this type of data is dental data, where the observation unit is either a tooth or a surface on a tooth. Lie et al. (2004) describe an application to hip replacement, where most patients receive only one, but some patients receive one on both sides.

Recurrent events are also presented in some applications. One example is Cui et al. (2008) that covers a number of parametric models for recurrent cardiovascular events in a clinical trial.

Finally, the cross-over type of experiment is not common. A classical example of this type is the exercise test times of Danahy et al. (1977). In these data, 21 patients try four different treatments in a cross-over fashion and in relation to that undergo 10 exercise tests. Some models for this dataset are presented in Hougaard (2000).

22.12 Key aspects of using frailty models

As this is a handbook, it should be quite explicit. Therefore, this is a list of aspects to consider in relation to the use of a frailty model. It includes items related to survival data as well as items related to random effects models and finally specific items for frailty models. In the univariate case, use of a frailty model is limited to an extension of the available models, whether it is going to be used as such or as a goodness-of-fit test for a simpler model.

So the real list of aspects applies to multivariate data:

- What kind of multivariate data is available? This could be multivariate parallel data, such as several individuals or several organs. It could be recurrent events, that is, events of the same kind for an individual? Is it a controlled experiment, where an individual goes through various pre-defined treatments?
- What is the purpose of the study? Is it to find the effect of covariates or to assess the dependence in the data?
- If the aim is to find the effect of covariates, what is then the reason of including multivariate data? Is it technically a nuisance, but necessary to include all available data? Is it a design feature to improve precision (such as a cross-over or matched pairs study)?
- Should the hazard function be modelled parametrically or non-parametrically?
- Is a shared frailty model appropriate or is some kind of extension relevant?
- What distribution model should be used for the frailty?
- Is censoring dependent on the development in the cluster?

22.13 Software

The software in the major packages only handles the shared frailty model. The univariate, multivariate and recurrent events data are handled in the same way. For the recurrent events, the so-called "start-stop coding" is needed. Hirsch & Wienke (2012) present a comparison of several programs.

The R-system (as well as S+) has for many years included the possibility of frailty terms in the proportional hazards model (the coxph procedure). This procedure allows for the frailty following a gamma distribution or a lognormal distribution, and the estimation procedure uses a penalized approach. The program does present standard errors, but these are not based on the full model and must therefore be expected to be lower than the real standard errors.

SAS (proc phreg) includes frailty terms from version 9.3. This procedure only allows frailties following a lognormal distribution and the estimation procedure uses a penalized approach.

22.14 Literature

Three books cover frailty models in details. Hougaard (2000) was the first and compares the frailty model with alternative models (particularly the multistate model), both from a theoretical, conceptual and applied point of view. Duchateau & Janssen (2008) is more computationally oriented. Finally, Wienke (2010) focuses more on the extensions of the model.

22.15 Summary

The frailty model as a hazard-based random effects model for survival data has matured over the last 30 years. Basic models in the shared frailty case have been developed, including some asymptotic results covering both parametric and non-parametric models and software is now available in several major packages. Extended models are certainly more complex. While there are several potential such models, the theory as well as the software is less well developed.

Bibliography

Clayton, D.G. (1978). A model for association in bivariate life tables and its application in epidemiological studies of familial tendency in chronic disease incidence. *Biometrika* **65**, 141–151.

Cui, J., Burwell, D.T., Aronow, W.S., and Prakash, R. (2008). Parametric conditional frailty models for recurrent cardiovascular events in the lipid study. *Clin. Trials* **5**, 565–574.

Danahy, D.T., Forbes, A., Kirby, A., Marschner, I., Simes, J., West, M., and Tonkin, A. (1977). Sustained hemodynamic and antianginal effect of high dose oral isosorbide dinitrate. *Circulation* **55**, 381–387.

Duchateau, L. and Janssen, P. (2008). *The frailty model.* Springer-Verlag, New York.

Gjessing, H.K., Aalen, O.O. and Hjort, N.L. (2003). Frailty models based on Levy processes. *Adv. Appl. Prob.* **35**, 532–550.

Glidden, D.V. (1999). Checking the adequacy of the gamma frailty model for multivariate failure times. *Biometrika* **86**, 381–393.

Gorfine, M. and Hsu, L. (2011). Frailty-based competing risks model for multivariate survival data. *Biometrics* **67**, 415–426.

Greenwood, M. and Yule, G.U. (1920). An inquiry into the nature of frequency distributions representative of multiple happenings with particular reference to the occurrence of multiple attacks of disease or of repeated accidents. *J. R. Statist. Soc.* **83**, 255–279.

Hirsch, K. and Wienke, A. (2012) Software for semiparametric shared gamma and log-normal frailty models: An overview. *Comp. Meth. Programs in Biomedicine* **107**, 582–597.

Hougaard, P. (1984). Life table methods for heterogeneous populations: Distributions describing the heterogeneity. *Biometrika* **71**, 75–84.

Hougaard, P. (1986a). Survival models for heterogeneous populations derived from stable distributions. *Biometrika* **73**, 387–396. (Correction **75**, 395.)

Hougaard, P. (1986b). A class of multivariate failure time distributions. *Biometrika* **73**, 671–678. (Correction, **75**, 395.)

Hougaard, P. (2000). *Analysis of multivariate survival data.* Springer-Verlag, New York.

Kosorok, M.R., Lee, B.L. and Fine, J.P. (2004). Robust inference for univariate proportional hazards frailty regression models. *Ann. Statist.* **32**, 1448–1491.

Lie, S.A., Engesaeter, L.B., Havelin, L.I., Gjessing, H.K., and Vollset, S.E. (2004). Dependency issues in survival analyses of 55782 primary hip replacements from 47355 patients. *Statist. Med.* **23**, 3227–3240.

Martinussen, T. and Pipper, C.B. (2005). Estimation in the positive stable shared frailty Cox proportional hazards model. *Lifetime Data Anal.* **11**, 99–115.

Murphy, S.A. (1995). Asymptotic theory for the frailty model. *Ann. Statist.* **23**, 182–198.

Oakes, D. and Manatunga, A.K. (1992). Fisher information for a bivariate extreme value distribution. *Biometrika* **79**, 827–832.

Parner, E. (1998). Asymptotic theory for the correlated gamma-frailty model. *Ann. Statist.* **26**, 183–214.

Shih, J.H. and Louis, T.A. (1995). Assessing gamma frailty models for clustered failure time data. *Lifetime Data Anal.* **1**, 205–220.

Vaupel, J.W., Manton, K.G. and Stallard, E. (1979). The impact of heterogeneity in individual frailty on the dynamics of mortality. *Demography* **16**, 439–454.

Wienke, A. (2010). *Frailty models in survival analysis.* Chapman & Hall, Boca Raton, FL.

Yamaguchi, T., Ohashi, Y. and Matsuyama, Y. (2002). Proportional hazards models with random effects to examine centre effects in multicentre cancer clinical trials. *Statist. Meth. Med. Res.* **11**, 221–236.

23

Bayesian Analysis of Frailty Models

Paul Gustafson
Department of Statistics, University of British Columbia

CONTENTS

23.1	Background	475
23.2	A basic frailty model	476
	23.2.1 Modeling the baseline hazard	476
	23.2.2 Modeling the frailties	477
	23.2.3 Example	479
	23.2.4 Model comparison	480
23.3	Recent developments	482
23.4	Final thoughts	484
	Appendix A	484
	Appendix B	485
	Bibliography	486

23.1 Background

In the application of statistical models across various scientific disciplines, data that are grouped or clustered in some way are commonly encountered. It is also extremely common to see random-effect models used to acknowledge such data structures. A cluster-specific random effect can reflect the reality that data observations arising in the same cluster are likely more similar to one another than data observations arising across different clusters. Moreover, by making these cluster-specific effects random rather than fixed, "borrowing of strength" is attained. Inference about what happens in a particular cluster is aided by information drawn from the other clusters as well.

The need for random-effect models to acknowledge clustered data certainly presents itself in the survival analysis context. As a motivating example, many clinical trials have a time-to-event as the primary outcome, and many of these are multi-center trials. Thus a random-effect model can acknowledge that such data are clustered by center. Other studies involve the measurement of two or more different event times per individual, in which case a random-effect model acknowledges that event times arising from the same individual may be strongly associated.

Outside of a survival analysis context, one commonly sees models in which random effects act additively on a function of the mean response variable. This specification is the backbone for the very large class of *generalized linear mixed models* (often just referred to by the acronym GLMM). With survival analysis, however, the vast majority of modeling activity does *not* take the mean to be the primary descriptor of the response variable. Rather, the *hazard rate* is taken as the fundamental characterization of an event time. Since hazard rates are by definition non-negative, an obvious strategy is to introduce random effects which are also non-negative, and have them act multiplicatively on hazard rates. This is a particularly obvious strategy given the ubiquitous use of the Cox proportional hazards model (Cox, 1972) to specify fixed effects as operating multiplicatively on the hazard. The general parlance has evolved such that a *frailty model* (or more specifically a "shared frailty" model) involves random effects acting multiplicatively on hazard rates. Equivalently, of course, this can be recast as random effects acting additively on log-hazard rates.

In this chapter, we describe the basic form of a Bayesian frailty model in Section 23.2, with particular emphasis on possible ways to model the baseline hazard function and possible models for the frailties themselves. This is followed up with a real-data example, and then some discussion of model selection in the Bayesian frailty model set-up. Section 23.3 turns to reviewing recent developments in Bayesian frailty modeling, focusing on work subsequent to the survey chapter of Ibrahim et al. (2001, Ch. 4). Some brief concluding thoughts are given in Section 23.4. This chapter also contains two appendices on technical issues arising in Section 23.2.

23.2 A basic frailty model

To fix some notation, say that random variable T is a time-to-event, while $\boldsymbol{X}$ is a vector containing the values of p covariates. We assume data are clustered such that m_i study units belong to the i-th of n groups or clusters. Our datapoints can then be written as $(t_{ij}, \delta_{ij}, \boldsymbol{x}_{ij})$ for $= i = 1, \ldots, n$, and $j = 1, \ldots m_i$, where δ_{ij} is the right-censoring indicator. That is, $\delta_{ij} = 1$ indicates that $T_{ij} = t_{ij}$ was observed for the (i, j)-th unit, while $\delta_{ij} = 0$ indicates that $T_{ij} > t_{ij}$ is all that is known. Generically, we shall write $\boldsymbol{d_i} = \{(t_{ij}, \delta_{ij}, \boldsymbol{x_{ij}}) : j = 1, \ldots, m_i\}$ as the observed data for the i-th cluster, and $\boldsymbol{d_{1:n}} = (\boldsymbol{d_1}, \ldots, \boldsymbol{d_n})$ as the entire observed dataset.

A basic frailty model would have the form

$$\begin{aligned} h_{ij}(t; w_i, \boldsymbol{x_{ij}}) &= \lim_{\Delta \downarrow 0} \Delta^{-1} Pr\left(T_{ij} < t + \Delta \mid T_{ij} > t, W_i = w_i, \boldsymbol{X_{ij}} = \boldsymbol{x_{ij}}\right) \\ &= w_i \exp\left(\boldsymbol{\beta}^T \boldsymbol{x}_{ij}\right) h_0(t; \boldsymbol{\phi}). \end{aligned} \tag{23.1}$$

The first equality here is simply a definition of the hazard rate for the (i, j)-th unit's response T_{ij}, conditioned on the cluster's (unobservable) frailty W_i, and the unit's (observable) covariate value $\boldsymbol{X_{ij}}$. The second equality is the model specification, which postulates that the non-negative frailty acts multiplicatively on the hazard, as do the covariates via the exponential of the linear predictor $\boldsymbol{\beta}^T \boldsymbol{X}$, i.e., the parameter vector $\boldsymbol{\beta} = (\beta_1, \ldots, \beta_p)$ describes the dependence of T on $\boldsymbol{X}$ given W, in the usual manner of proportional hazards regression. The term $h_0(\cdot; \boldsymbol{\phi})$ is the baseline hazard function. That is, $h_0(; \boldsymbol{\phi})$ is a family of legitimate hazard functions indexed by a vector of parameters $\boldsymbol{\phi}$. As is usual for survival modeling, we do *not* take the first element of $\boldsymbol{\beta}$ to serve as an intercept, since the "baseline" behavior of $(T|W = 1, \boldsymbol{X} = \boldsymbol{0})$ is completely described and modeled by $h_0(; \boldsymbol{\phi})$. For future reference we also define the baseline cumulative hazard, $H_0(t; \boldsymbol{\phi}) = \int_0^t h_0(s; \boldsymbol{\phi}) ds$.

23.2.1 Modeling the baseline hazard

The basic frailty model (26.2) is completed upon committing to a model for the unobservable frailty W, and a form for the baseline hazard model $h(\cdot; \boldsymbol{\phi})$. With regard to the latter issue, in non-Bayesian survival modeling with only fixed effects, it is extremely common to *not* make modeling assumptions about the baseline hazard. This is most commonly achieved by use of the partial likelihood (Cox, 1975), which is a function of regression coefficients only and can be maximized to estimate these coefficients. This approach can, in fact, be transplanted to the context of Bayesian frailty models. Using an idea from Kalbfleisch (1978), and extended by Sinha et al. (2003), if a diffuse gamma process prior is applied to the baseline cumulative hazard, then the baseline hazard parameters can be integrated away analytically. What remains is a joint posterior on $(\boldsymbol{\beta}, \boldsymbol{w}_{1:n}|\boldsymbol{d}_{1:n})$, in which the likelihood contribution of $(\boldsymbol{d}_{1:n}|\boldsymbol{\beta}, \boldsymbol{w}_{1:n})$ takes the form of the partial likelihood (without any distinction between which effects are fixed and which are random, so that $(\boldsymbol{\beta}, \boldsymbol{w}_{1:n})$ is simply treated as a vector of regression coefficients). Thus a posterior density can be formed via the product of the partial likelihood and the chosen joint prior density of $(\boldsymbol{\beta}, \boldsymbol{w}_{1:n})$. Such use of the partial likelihood inside Bayesian frailty models has been considered by a number of authors, including Gustafson (1997), Sargent (1998), and Zhang et al. (2008).

Another commonly seen approach in Bayesian survival models generally, and Bayesian frailty models specifically, is the use of a piecewise-constant baseline hazard. If we choose constants $0 < c_1 < \ldots < c_{r-1} < \infty$, and define $c_0 = 0$, $c_r = \infty$, then

$$h_0(t; \boldsymbol{\phi}) = \sum_{k=1}^{r} \phi_k I_{[c_{k-1}, c_k)}(t) \tag{23.2}$$

defines a piecewise-constant baseline hazard across r sub-intervals of the time axis. One reason this specification is popular is it makes (26.2) a *generative* model, i.e., for given $(w_i, \boldsymbol{x}_{ij}, \boldsymbol{\beta}, \boldsymbol{\phi})$, a probability distribution for T_{ij} is completely specified. Given that Bayesian analysis is nothing more than reasoning probabilistically by conditioning on the values of observed quantities, there is arguably a strong spirit within the Bayesian community that inference should indeed be based on generative models. This spirit is also embodied in software. Currently, versions of the BUGS software, such as WinBUGS (Lunn et al., 2000, 2009), are really the only general-purpose software options for Bayesian inference. That is, they are the only options for users who wish to be shielded from details and implementation choices concerning how the posterior calculations are carried out. With BUGS, the user provides a model and prior specification, and the software outputs a representation of the posterior distribution in the form of a Monte Carlo sample. As a result, BUGS is widely used by practitioners. The software demands fully specified probability models for data given parameters and proper prior distributions for parameters. Consequently, it is applicable to specifications such as (23.2) with explicit representation of baseline hazard parameters, but it does not support the partial likelihood approach.

23.2.2 Modeling the frailties

The basic frailty model (26.2) is completed by specifying a form of the distribution of the frailty W across clusters. Or, in more generic statistical language, a random-effect distribution must be specified. Generically, let κ be the unknown parameter or parameters in the frailty distribution. Most commonly, a single parameter suffices. A variety of distributional assumptions about frailties have been entertained in the literature, with attendant rationales concerned with scientific plausibility, parameter interpretability, and computational feasibility.

For a specific reason of interpretability, Hougaard (1986) suggests a positive stable distribution for the frailty. Along the lines of (26.2), say the association between T and $\boldsymbol{X}$ given W follows a proportional hazards relation. A positive stable distribution with parameter $\kappa \in (0, 1]$ is characterized by its Laplace transform, $E\{\exp(-sW)\} = \exp(-s^{\kappa})$. An immediate consequence of this form is that the unconditional $(T, \boldsymbol{X})$ association will also follow a proportional hazards structure. Specifically, if $(T|W, \boldsymbol{X})$ has hazard rate $W \exp(\boldsymbol{\beta}^T \boldsymbol{X}) h_0()$, then $(T|\boldsymbol{X})$ is easily seen to have hazard rate $\exp(\kappa\boldsymbol{\beta}^T \boldsymbol{X})\{h_0()\}^{\kappa}$. Thus one has the sleek interpretation that multiplicative hazard contrasts are constant across the time axis, both conditionally on frailty and unconditionally. Moreover, contrasts are attenuated by a factor κ when one moves from conditional (within-cluster) to unconditional hazard ratios.

Another nice property of the positive stable frailty distribution is a simple characterization of within-cluster dependence via Kendall's tau. For a bivariate distribution on $\boldsymbol{S} = (S_1, S_2)$, Kendall's tau is $2Pr\{(S_1^{(1)} - S_1^{(2)})(S_2^{(1)} - S_2^{(2)}) > 0\} - 1$, where the superscript indexes two independent and identically distributed realizations of $\boldsymbol{S}$. This can be viewed as describing the tendency for *both* components of one pair to outlive their counterparts in the other pair. Moreover, it can be shown that Kendall's tau simplifies to $1 - \kappa$ for any bivariate distribution induced by a positive stable (κ) frailty distribution. See Hougaard (2000) for an in-depth discussion of dependence measures for frailty models.

Unfortunately, the nice interpretations afforded by the positive stable frailty distribution are offset by computational challenges. Principally, the positive stable distribution lacks a closed-form expression for its density function. This is particularly challenging when designing MCMC algorithms that operate jointly on the distribution of parameters and latent frailties given observed data. Unsurprisingly, then, the positive stable distribution is not a supported distribution in the BUGS software. Nonetheless, algorithms for Bayesian inference in models based on positive stable frailties have been developed (Qiou et al., 1999; Chen et al., 2002), and even extended to mixtures of positive stable distributions (Ravishanker and Dey, 2000).

By far the most commonly used frailty distribution in Bayesian models is the gamma distribution. The specification $W \sim \text{Gamma}(\kappa^{-1}, \kappa^{-1})$ is commonly seen (where the second argument is the rate parameter rather than its reciprocal, the scale parameter). This gives $E(W|\kappa) = 1$ and $\text{Var}(W|\kappa) = \kappa$. The reduction to a single parameter and a known unit mean is appropriate, since the baseline hazard function can adapt arbitrarily to capture the absolute magnitude of the hazard for a typical unit, and the frailty distribution serves only in the relative sense of proportional variation across clusters. As with the positive stable case, the gamma frailty offers a simple characterization of within-cluster dependence in terms of Kendall's tau being $1/(1 + 2/\kappa)$.

In contrast to the positive stable model, the gamma frailty model does not preserve proportionality of hazards. Thus one is more wedded to interpreting parameters with reference to the conditional relationship of $(T|W, \boldsymbol{X})$, in which proportionality holds. The rationale for the gamma frailty specification involves computational ease, since *conditional conjugacy* arises. The likelihood arising from (23.4), viewed as a function of w_i with all other inputs fixed, is itself proportional to a gamma density function. When combined with the gamma density of $(w_i|\kappa)$ then, a further gamma density results. More formally, this identifies the *full conditional* distribution of each frailty. Letting $\boldsymbol{w_{-i}} = (w_1, \ldots, w_{i-1}, w_{i+1}, \ldots, w_n)$, we have

$$(w_i|\boldsymbol{w_{-i}}, \boldsymbol{\beta}, \boldsymbol{\phi}, \boldsymbol{d_{1:n}}) \sim \text{Gamma}\left\{\kappa^{-1} + \delta_i, \kappa^{-1} + g(\boldsymbol{\beta}, \boldsymbol{\phi}, \boldsymbol{d_i})\right\},$$

where $\delta_i = \sum_{j=1}^{m_i} \delta_{ij}$ is the number of observed failures in the i-th cluster, while

$$g(\boldsymbol{\beta}, \boldsymbol{\phi}, \boldsymbol{d_i}) = \sum_{j=1}^{m_i} \exp\left(\boldsymbol{\beta}^T \boldsymbol{x_{ij}}\right) H_0(t_{ij}; \boldsymbol{\phi}). \tag{23.3}$$

Hence MCMC algorithms which make updates to blocks of parameters and/or latent variables in turn can update $\boldsymbol{w_{1:n}}$ with all other quantities fixed, via n independent draws from the appropriate gamma distributions. Clayton (1991) was the first to take advantage of this particular conditionally conjugate structure, in what was in fact one of the earliest applications of the Gibbs sampling and MCMC revolution spawned by Gelfand and Smith (1990). Other early examples of Bayesian frailty modeling with gamma frailties include Sinha (1993), Sahu et al. (1997), and Aslanidou et al. (1998).

Outside of survival analysis, a huge majority of mixed modeling involves normally distributed random effects. This is certainly an option in the frailty model situation as well, via the log-normal specification $W \sim LN(0, \kappa^2)$, or equivalently $\log W \sim N(0, \kappa^2)$. This is slightly less computationally convenient than the gamma frailty model, though empirically, of course, it might provide a better (or worse) fit to a given dataset. There is also less in the way of simple interpretation available. For example, there isn't a closed-form expression for Kendall's tau to quantify the within-cluster dependence. However, an obvious advantage of the log-normal frailty model arises when we go beyond the basic frailty model, to consider multivariate frailties. For instance, consider the setting of a multi-center clinical trial, and say the first component of the covariate vector $\boldsymbol{X}$ is a binary treatment group indicator. Conceivably there could be center-to-center variation in the efficacy of treatment, as well as in the overall risk of failure. A natural generalization of the basic frailty model would then be to replace the multiplicative effect $\exp(\log w_i + \beta_1 x_{ij,1})$ acting upon the hazard with $\exp\{\log w_{i,1} + (\beta_1 + \log w_{i,2})x_{ij,1}\}$, in the spirit of having both a random intercept and random slope, as is commonly seen in GLMM applications. Thus $\boldsymbol{W} = (W_1, W_2)$ now constitutes a bivariate frailty. Whereas bivariate gamma distributions are not commonly encountered, an interpretable bivariate log-normal distribution is trivially obtained by applying the bivariate normal distribution as a model for $(\log W_1, \log W_2)$. Thus the log-normal frailty model extends quite readily to multivariate frailties. An example of a bivariate Bayesian frailty model along these lines is found in Gray (1994). Legrand et al. (2005) adopt a similar approach, though they make a strong assumption that the two frailties are independent of one another.

To some extent, the issue of selecting a frailty model for a given application may not be crucial. There is a general sentiment, supported by evidence, that somewhat misspecified random effect distributions still yield estimators with reasonable inferential properties. This is particularly thought to be the case, when, as is typical, fixed effect parameters are of inferential interest. Some evidence in this direction for the GLMM setting includes Neuhaus et al. (1992), Gustafson (1996), and Litière et al. (2008), while Pickles and Crouchley (1995) considers frailty models specifically (though not in a Bayesian context). Of course insensitivity of inference to the choice of frailty model does not yield a license to omit a frailty model. In a survival analysis context, the deleterious effects of not modeling across-cluster variation when it is present are examined by Henderson and Oman (1999).

23.2.3 Example

To give an example of fitting a basic Bayesian frailty model, we take data originally reported on by Byar and Green (1980), and used as an example by several authors, including Wei et al. (1989) and Kleinbaum and Klein (2011). In this study, 85 patients were followed after surgical excision of bladder cancer tumors, with sequential monitoring and removal of tumor recurrences. (The datafile includes an 86th patient, but since he/she is recorded as right-censored at time zero, he/she makes no contribution to the likelihood.) Each patient was randomized to either placebo or a chemotherapeutic agent, thiotepa. For the i-th patient, T_{i1} is the time, in months, from initial excision to first recurrence, T_{i2} is the gap time from first recurrence to second recurrence, and so on. Overall, the data comprise $\sum_{i=1}^{n} m_i = 190$ records and $\sum_{i=1}^{n} \sum_{j=1}^{m_i} \delta_{ij} = 112$ observed recurrences. The number of observations per subject is distributed as $\sum_{i=1}^{n} I\{m_i = j\} = 39, 19, 7, 8, 12$, for $j = 1, \ldots, 5$. For all but seven subjects, the cluster size of m_i arises from $m_i - 1$ observed recurrences along with a censoring time prior to an m_i-th recurrence. The seven exceptional subjects had m_i observed recurrences, but no further follow-up time after the last recurrence. There are $p = 3$ explanatory variables. The first is the treatment assignment ($X_1 = 0$ for placebo, $X_1 = 1$ for thiotepa), with 38 patients in the active treatment group. The number of tumors at study enrollment is given as X_2, which is treated as a numerical covariate (although the three subjects with $X_2 = 8$ are in actuality subjects with eight or more tumors initially). The final covariate X_3 is the initial tumor size (in centimeters).

We fit several versions of the basic frailty model (26.2) to these data. We fix the cutpoints for the piecewise-constant baseline hazard at $(c_1, \ldots, c_4) = (2.5, 4.5, 6.5, 12.5)$, which yields about equal numbers of observed recurrences in each the $r = 5$ subintervals. While not pursued here, a fuller analysis could involve an investigation of whether results are sensitive to the choice of r, or whether model selection criteria speak strongly to an appropriate choice. Diffuse normal priors (mean=0, SD=1000) are assigned to each of $\log \phi_j$, $j = 1, \ldots, r$, and β_j, $j = 1, \ldots, p$. To approach the gamma and log-normal frailty models in a comparable manner, we define $\tau = \tau(\kappa) = SD(W)$ to be the frailty standard deviation. Thus for the frailty model $W \sim \text{gamma}(\kappa^{-1}, \kappa^{-1})$, $\tau = \kappa^{1/2}$. Whereas for $W \sim \text{LN}(0, \kappa^2)$, $\tau = \{\exp(2\kappa^2) - \exp(\kappa^2)\}^{1/2}$. For either model, we induce a prior on κ by specifying a prior on τ, particularly $\tau \sim \text{Unif}(0, 1.5)$. Under either choice of frailty distribution, the upper bound of $\tau = 1.5$ corresponds to very large across-cluster variation in response time distribution, so this prior specification can be regarded as "weakly informative." See Gelman (2006) for some general discussion of priors for variance components.

The models are fit using MCMC as implemented in WinBUGS. Since this software does not have the piecewise-constant hazard distribution as one of its supported probability distributions, we resort to a latent variable trick to fit the model, as elaborated on in Appendix A. The reported results are based on two MCMC chains each with 20,000 iterations after 2,000 burn-in iterations, and every second iteration retained. The computed posterior quantities are stable across the two chains, lending credence to reported inferences based on pooling the chains. In addition to the gamma and log-normal frailty models, the no-frailty model is fit. This can be regarded as the model arising with all frailties fixed at one, so that responses are analyzed without regard to their clustering within patients. Posterior inferences for the treatment effect (β_1) and the frailty SD (τ) are given in Table 23.1. The point estimate of the treatment effect is insensitive to presence or choice of the frailty model, but the interval estimate is 15% wider when a frailty model is used. This makes sense, since an analysis which treats all observations as independent without regard to their clustering is over-confident, so that an artificially narrow interval estimate arises. The choice between a gamma frailty model and a log-normal frailty model has only a modest impact on the point estimate of the frailty SD, but has slightly more impact on the corresponding interval estimate.

23.2.4 Model comparison

Two popular techniques for comparing various Bayesian survival models are the deviance information criterion (DIC) (Spiegelhalter et al., 2002), and the conditional predictive ordinate (CPO) (Geisser, 1993).

Both are attractive on computational grounds relative to full Bayesian model comparison which requires computing the marginal density of the observed data under each model, quantities which are very challenging to obtain from MCMC output for the respective models. Before saying more about DIC and CPO, we note that in the present context both

TABLE 23.1
Parameter estimates in the bladder cancer data example. The no-frailty model, the gamma frailty model, and the log-normal frailty model are considered. Posterior means and 95% equal-tailed confidence intervals are reported for the treatment effect β_1 and the frailty SD τ.

Frailty model	treatment effect (β_1)		frailty SD (τ)	
None	-0.35	(-0.76, 0.04)		
Log-Normal	-0.37	(-0.84, 0.08)	0.41	(0.03, 1.14)
Gamma	-0.37	(-0.84, 0.08)	0.35	(0.02, 0.80)

require computation of the per-cluster likelihood with the frailty integrated out, i.e.,

$$\begin{aligned} L_i(\boldsymbol{\beta}, \boldsymbol{\phi}, \kappa) &= f(\boldsymbol{d_i}|\boldsymbol{\beta}, \boldsymbol{\phi}, \kappa) \\ &= E\{f(\boldsymbol{d_i}|W_i, \boldsymbol{\beta}, \boldsymbol{\phi}, \kappa)\} \\ &= c_i E_\kappa\{W^{\delta_i} \exp(-g_i W)\}, \end{aligned} \tag{23.4}$$

where

$$\log c_i = \sum_{j=1}^{m_i} \delta_{ij} \left\{\boldsymbol{\beta}^T x_{ij} + \log h_0(t_{ij}; \boldsymbol{\phi})\right\},$$

$g_i = g(\boldsymbol{\beta}, \boldsymbol{\phi}, \boldsymbol{d_i})$ is as per (23.3), and the expectation in (23.4) is understood to be with respect to the frailty distribution of $(W|\kappa)$.

We have already described the computational advantage of the gamma frailty model over the log-normal model in terms of MCMC algorithms to fit either model. A further advantage accrues in that (23.4) can be evaluated exactly in the gamma case, but some form of quadrature is required in the log-normal case. Further discussion on this point is given in Appendix B.

The per-cluster likelihood can be used to evaluate models according to the CPO criterion as follows. Let $\boldsymbol{d_{-i}} = (\boldsymbol{d_1}, \ldots, \boldsymbol{d_{i-1}}, \boldsymbol{d_{i+1}}, \ldots, \boldsymbol{d_n})$ be all the data except for that of the i-th cluster. Also, let $\boldsymbol{\theta}$ generically denote the parameters, i.e., $\boldsymbol{\theta} = (\boldsymbol{\beta}, \boldsymbol{\phi}, \kappa)$ at present. Then the i-th CPO is $f(\boldsymbol{d_i}|\boldsymbol{d_{-i}})$, the predictive density of the i-th cluster's data given all the other data, evaluated at the actual i-th cluster data values. Intuitively then, a higher CPO value corresponds to a model doing a better job predictively.

It is easy to check that the i-th CPO can be expressed as

$$\begin{aligned} f(\boldsymbol{d_i}|\boldsymbol{d_{-i}}) &= \int f(\boldsymbol{d_i}|\boldsymbol{\theta}) f(\boldsymbol{\theta}|\boldsymbol{d_{-i}}) d\boldsymbol{\theta} \\ &= \left(E\left[\{L_i(\boldsymbol{\theta})\}^{-1}|\boldsymbol{d_{1:n}}\right]\right)^{-1}. \end{aligned}$$

This gives a route to computing the i-th CPO directly from MCMC output based on fitting the model to all the data, i.e., for each i we need to compute a posterior expectation based on the posterior distribution of $(\boldsymbol{\theta}|\boldsymbol{d_{1:n}})$. It is customary to summarize the model's predictive performance via the log pseudomarginal likelihood (LPML), given by

$$LPML = \sum_{i=1}^{n} \log f(\boldsymbol{d_i}|\boldsymbol{d_{-i}}). \tag{23.5}$$

Note here the passing resemblance to the actual log marginal density of the data, given by

$$\log f(\boldsymbol{d}_{1:n}) \quad = \quad \log f(\boldsymbol{d}_1) + \sum_{i=2}^{n} \log f(\boldsymbol{d}_i|\boldsymbol{d}_{1:(i-1)}), \tag{23.6}$$

where the ordering of the n datapoints is actually arbitrary. Whereas (23.6) aggregates over predictions based on different amounts of data, (23.5) focuses only on predictions based on data from $n-1$ clusters. Otherwise, the two criteria operate in a very similar spirit. Note also that we can get some sense of the magnitude of differences in predictive performance, in that $\exp(n^{-1}LPML)$ can be interpreted as the typical height of the predictive density at the realized cluster data. Hence, for comparing two models, $\exp\{n^{-1}(LPML_2 - LPML_1)\}$ describes a density ratio sense in which one model predicts better than another.

In the present example, there is no appreciable difference in predictive performance between the no-frailty model (LPLM= -462.7) and the gamma frailty model (LPLM=-462.6). However, the log-normal frailty model is somewhat worse (LPLM=-465.6). In the density ratio sense described above, the gamma frailty model predicts 3.6% better than the log-normal frailty model.

In contrast, the DIC criterion has a similar spirit to the BIC criterion (Schwarz, 1978) that is often used as a heuristically Bayesian model selection scheme when maximum likelihood model fits are applied. Defining the deviance as $D(\boldsymbol{\theta}) = -2\sum_{i=1}^{n} \log L_i(\boldsymbol{\theta})$, the DIC criterion is based upon the deviance evaluated at the posterior mean parameter estimates, which we write as $D(\overline{\boldsymbol{\theta}})$, and the posterior mean of the deviance, which we write as $\overline{D(\boldsymbol{\theta})}$. In particular, $D(\overline{\boldsymbol{\theta}})$ is taken to summarize the fit of the model to the data (lower values are better), while the model complexity is reflected by the effective number of parameters, $p_D = 2\{\overline{D(\boldsymbol{\theta})} - D(\overline{\boldsymbol{\theta}})\}$. Overall, to trade off fit and complexity, theory suggests favoring the model with the lower value of $D(\overline{\boldsymbol{\theta}}) + 2p_D$.

While the DIC criterion is successful and widely used, our experience with it in the present context is less successful. One concern is that we obtain a somewhat lower value of $D(\overline{\boldsymbol{\theta}})$ from the no frailty model than from the gamma frailty model, even though the former is a sub-model of the latter. Though this is not precluded theoretically (as it is when maximum likelihood estimates are used), it makes it difficult to regard $D(\overline{\boldsymbol{\theta}})$ as a pure measure of fit. Another concern is that we obtain an implausibly large $p_D \approx 30$ from the no-frailty model, even though this model has no latent variables and only eight unknown parameters. Thus we do not have a great deal of confidence in the DIC results, and prefer to rely on the CPO results given above.

23.3 Recent developments

The state of the Bayesian frailty world in 2001 was surveyed by Ibrahim et al. (2001, Ch. 4). Thus for present purposes we define "recent" to be 2001 onwards, and mention some research themes in Bayesian frailty models over this period of time.

One interesting recent variation on the basic Bayesian frailty model arises when the clusters are based on geographic areas. For instance, Zhao et al. (2009) consider time to death due to breast cancer for a cohort of patients clustered by county, so that w_i reflects the extent to which subjects in the i-th county are at higher or lower risk than other counties, in a sense going beyond variation explained by county variation in the distribution of covariates. It is scientifically plausible that small-area frailties would vary smoothly across the large area being considered. Consequently, the specification that the n frailties are independent and identically distributed could be replaced with a specification permitting the correlations between frailties for pairs of close (say adjacent) small areas to be higher than those between distant pairs. Thus the rich literature on spatially structured prior distributions for random effects, much of which traces back to conditionally autoregressive models as popularized by Besag et al. (1991), can be brought to bear. This is demonstrated to good effect by Banerjee et al. (2003), Banerjee and Carlin (2003), and Zhao et al. (2009), amongst others. As a further extension, Diva et al. (2008) consider multiple event times per person in addition to clustering by small areas.

Another topic attracting considerable recent attention is the use of frailties in Bayesian modeling for recurrent event data. In Section 23.2.3 we saw a very simple approach to recurrent events, with an assumption of a unit's gap times being conditionally independent of one another given the unit's frailty which acts on the hazard of each such time. However, there are much more complex recurrent event settings, as surveyed, for instance, by Cook and Lawless (2007). The Bayesian frailty model approach can indeed be infused into more complicated settings. For instance, Sinha and Maiti (2004) consider situations with fixed inspection times, so that actual event times are not observed. Also, Sinha et al. (2008) explore the dependent termination issue, whereby association between the hazard rate of recurrent events and termination time is dealt with via frailties. The work of Manda et al. (2005) is motivated by dental applications, where *nested* frailties are useful. In the case of amalgam restoration, there is a recurrent event process for each tooth of each subject. Thus a tooth-level frailty accommodates dependence between gap times, while a subject-level frailty accommodates dependence of times across teeth but within subject.

Still in the recurrent event framework, a potentially big new idea is that of *time-dependent frailty.* This allows a unit's hazard rate or intensity rate for recurrent events to exhibit more change over time than is explained just by observable time-varying covariates. This notion is explored in a Bayesian manner by both Manda and Meyer (2005) and Pennell and Dunson (2006). For instance, in the former work, an autoregressive model is prescribed for the evolution of each cluster's frailty over time.

A nice feature of Bayesian frailty models is that they can be used as "modules" in tandem with other modeling features, to result in quite complex and flexible models. For instance, there has been quite a lot of recent work on relaxing the proportional hazard assumption in Bayesian survival modeling, via the use of a transformation involving an unknown parameter. Yin and Ibrahim (2005*b*) and Yin and Ibrahim (2005*a*) assume that frailty acts multiplicatively on the baseline hazard, and then use a Box-Cox transformation to express uncertainty about how the covariates act upon the cluster-specific hazard. Upon varying the transformation parameter one obtains a proportional hazards structure, or an additive hazards structure, or something in between. In a less parametric vein, Mallick and Walker (2003) include frailties in a model structure where a transformation is modeled nonparametrically, such that a proportional hazards model and a proportional odds models are but two points within a vast space of possibilities concerning how covariates act upon the hazard rate. Also in the spirit of considering alternatives to the proportional hazards model, Komárek et al. (2007) use cluster-specific random effects which act upon the location of the log failure time distribution, in a Bayesian treatment of an accelerated failure time model. And Hanson and Yang (2007) use cluster-specific random effects which act upon the survival odds, in a Bayesian treatment of a proportional odds model. One might debate whether these random effect models should be called frailty models or whether frailty should be reserved exclusively to describe a multiplicative action on a hazard rate. Terminology aside, however, the point is that there exists a rich array of possibilities for Bayesian treatment of clustered failure-time data.

Another modular use of Bayesian frailty models arises in the area of cure-rate models. Such models involve a survival function with a positive limit as time goes to infinity, to reflect the possibility that some subjects are cured by treatment and no longer susceptible to failing due to the cause being considered. Chen et al. (2002) use frailties combined with a cure rate model in a Bayesian context, with a further variant given by Yin (2005). Then Yin (2008) extends the model further to involve (i) frailties, (ii) cure rates, and (iii) transformation structure that ranges from a proportional hazards model to a proportional odds model. This proves an excellent example of how the Bayesian approach allows a number of sophisticated modeling features to be drawn together.

23.4 Final thoughts

A great part of the strong uptake of Bayesian methods seen over the last two decades is the application of hierarchically structured models across a wide array of application areas. And much of this hierarchical model milieu is effectively a random-effect model milieu, with a first-stage model that conditions on random effects, and a second stage that models the random effects. As outlined in this chapter, Bayesian frailty models comprise an important and well-studied class of Bayesian random-effect models.

One unclear issue is whether, going forward, work on Bayesian frailty models might become more integrated with research on other Bayesian random-effect models. On the one hand, in the spirit of generalized linear models, one might think that the distinction between failure-time outcomes and other kinds of outcomes is a minor technical distinction, and when we discuss random-effect models we should try to minimize the bandwidth consumed by distinctions between the type of response variable involved. On the other hand, in the survival analysis world the primacy of multiplicative effects acting upon hazard rates is strong, and does constitute something of a wedge between Bayesian frailty models and Bayesian GLMM instantiations. In fact, one might speculate that new developments such as time-varying frailties could even strengthen this wedge. In any event, regardless of the level of integration with other outcome types, it seems safe to forecast that new ideas for Bayesian analysis of clustered failure-times will continue to come forth.

Appendix A

One way to implement the frailty model (23.2) in WinBUGS is to associate latent variables $\boldsymbol{Z_{ij}} = (Z_{ij1}, \dots Z_{ijr})$ with the observable failure time T_{ij}. Conditioned on all the parameters and frailties, we presume that the Z_{ijk} are mutually independent of one another, with

$$Z_{ijk} \quad \sim \quad \text{Exponential}\left\{w_i \phi_k \exp\left(\boldsymbol{\beta}^T \boldsymbol{x_{ij}}\right)\right\}.$$

Then T_{ij} is determined from $\boldsymbol{Z_{ij}}$ as follows. Let $\Delta_k = c_k - c_{k-1}$ be the width of the k-th subinterval of the time axis, for $k = 1, \dots, r$, (with this definition implying that $\Delta_r = \infty$). We set $T_{ij} = c_{k^*-1} + Z_{ijk^*}$, where $k^* = \min\{k \in \{1, \dots, r\} : Z_k < \Delta_k\}$. That is, if Z_{ij1} does not exceed the first cutpoint c_1, we take $T_{ij} = Z_{ij1}$. Otherwise, we try to set $T_{ij} = c_1 + Z_{ij2}$ if this does not exceed the second cutpoint c_2, and so on. By this construction we see intuitively that T_{ij} will have the claimed piecewise-constant hazard function. Also, note that for $t \in (c_{a-1}, c_a)$, we have

$$\{T_{ij} = t\} \quad \leftrightarrow \quad \{Z_{ij1} > \Delta_1, \dots, Z_{ij,a-1} > \Delta_{a-1}, Z_{ija} = t - c_{a-1}\}.$$

Thus an observed failure at a value t in the a-th sub-interval of the time axis can be encoded as right censoring of Z_{ijk} at Δ_k for $k = 1, \ldots, a-1$, observed failure of Z_{ija} at time $t - c_{a-1}$, and right censoring of Z_{ijk} at time zero, for $k = a+1, \ldots, r$. Similarly, right censoring at a value t in the a-th sub-interval of the time axis can be encoded as right censoring of Z_{ijk} at Δ_k for $k = 1, \ldots, a-1$, right censoring of Z_{ija} at time $t - c_{a-1}$, and right censoring of Z_{ijk} at time zero, for $k = a+1, \ldots, r$. Bluntly, we can recast the frailty model with piecewise-constant baseline hazard applied to $n^* = \sum_{i=1}^{n} m_i$ datapoints into a frailty model with constant baseline hazard (i.e., an exponential baseline distribution) applied to rn^* datapoints. While this may not be the most computationally efficient scheme, it does yield an easy route to specifying the desired model in WinBUGS. Example code is given on the website.

As some notes about this implementation, with WinBUGS right censoring is treated stochastically, so that MCMC is applied to the joint posterior distribution of all parameters $(\boldsymbol{\beta}, \boldsymbol{\phi}, \boldsymbol{\kappa})$, the frailties $\boldsymbol{w}_{1:n}$, and all those Z_{ijk} to which right censoring applies. Thus initial values are required for all these, and for Z_{ijk} initial values exceeding the censoring times must be supplied. It should also be mentioned that in concept it is not actually necessary to include those Z_{ijk} which are right censored at zero, i.e., the number of effective datapoints could be taken to be smaller than rn^*. This, however, would require the use of a ragged array to represent the necessary Z_{ijk} components. While this is not a difficultly in WinBUGS, it presents challenges when using suitable R packages to call WinBUGS from R.

There are other routes to implementing a piecewise-constant hazard function in WinBUGS. One we are aware of is described in Yu (2010), using the "zeros trick" described in the WinBUGS manual (http://www.mrc-bsu.cam.ac.uk/bugs/winbugs/manual14.pdf). This is a general trick to implement a model which is not explicitly supported by the software. It is unclear whether this might be less or more computationally efficient than the approach described above, though the approach described above can be coded in a sleeker fashion.

Appendix B

To evaluate (23.4), when $W \sim \text{gamma}(\kappa^{-1}, \kappa^{-1})$ we immediately have

$$\begin{aligned} \log L_i &= \log c_i + \log \Gamma(\kappa^{-1} + \delta_i) - \log \Gamma(\kappa^{-1}) + \\ &\quad \kappa^{-1} \log(\kappa^{-1}) - (\kappa^{-1} + \delta_i) \log(\kappa^{-1} + g_i). \end{aligned}$$

In the case of $W \sim LN(0, \kappa^2)$, we can write

$$L_i = c_i E[\exp\{\delta_i \kappa Z - g_i \exp(\kappa Z)\}],$$

where $Z \sim N(0,1)$. If we write $t_i(z) = \exp\left[g_i \left\{\exp(\kappa z) - 1 - \kappa z - (1/2)\kappa^2 z^2\right\}\right]$, then

$$L_i = c_i e^{-g_i} E\left[t_i(Z) \exp\left\{-(1/2) g_i \kappa^2 Z^2 - \kappa(g_i - \delta_i) Z\right\}\right].$$

From here, we can "complete the square." If we set $\nu^2 = (1 + g_i \kappa^2)^{-1}$ and $\mu = \kappa^2(\delta_i - g_i)\kappa$, then

$$L_i = c_i \exp\{(1/2)\mu^2/\nu^2 - g_i\} E\{t_i(Z^*)\}, \tag{23.7}$$

where $Z^* \sim N(\mu, \nu^2)$. Thus (23.7) is in a form to which we can directly apply Gauss-Hermite quadrature.

Bibliography

Aslanidou, H., Dey, D. and Sinha, D. (1998), 'Bayesian analysis of multivariate survival data using Monte Carlo methods', *Canadian Journal of Statistics* **26**, 33–48.

Banerjee, S. and Carlin, B. (2003), 'Semiparametric spatio-temporal frailty modeling', *Environmetrics* **14**, 523–535.

Banerjee, S., Wall, M. and Carlin, B. (2003), 'Frailty modeling for spatially correlated survival data, with application to infant mortality in Minnesota', *Biostatistics* **4**, 123–142.

Besag, J., York, J. and Mollie, A. (1991), 'Bayesian image restoration with two applications in spatial statistics (with discussion)', *Annals of the Institute of Statistical Mathematics* **43**, 1–59.

Byar, D. and Green, S. (1980), 'The choice of treatment for cancer patients based on covariate information', *Bulletin du Cancer* **67**, 477–490.

Chen, M., Ibrahim, J. and Sinha, D. (2002), 'Bayesian inference for multivariate survival data with a cure fraction', *Journal of Multivariate Analysis* **80**, 101–126.

Clayton, D. (1991), 'A Monte Carlo method for Bayesian inference in frailty models', *Biometrics* **47**, 467–485.

Cook, R. and Lawless, J. (2007), *The Statistical Analysis of Recurrent Events*, Springer, New York.

Cox, D. (1972), 'Regression models and life-tables', *Journal of the Royal Statistical Society: Series B* **34**, 187–220.

Cox, D. (1975), 'Partial likelihood', *Biometrika* **62**, 269–276.

Diva, U., Dey, D. and Banerjee, S. (2008), 'Parametric models for spatially correlated survival data for individuals with multiple cancers', *Statistics in Medicine* **27**, 2127–2144.

Geisser, S. (1993), *Predictive Inference: An Introduction*, Chapman & Hall/CRC, Boca Raton, FL.

Gelfand, A. and Smith, A. (1990), 'Sampling-based approaches to calculating marginal densities', *Journal of the American Statistical Association* **85**, 398–409.

Gelman, A. (2006), 'Prior distributions for variance parameters in hierarchical models', *Bayesian Analysis* **1**, 515–533.

Gray, R. (1994), 'A Bayesian analysis of institutional effects in a multicenter cancer clinical trial', *Biometrics* **50**, 244–253.

Gustafson, P. (1996), 'The effect of mixing-distribution misspecification in conjugate mixture models', *Canadian Journal of Statistics* **24**, 307–318.

Gustafson, P. (1997), 'Large hierarchical Bayesian analysis of multivariate survival data', *Biometrics* **53**, 230–242.

Hanson, T. and Yang, M. (2007), 'Bayesian semiparametric proportional odds models', *Biometrics* **63**, 88–95.

Henderson, R. and Oman, P. (1999), 'Effect of frailty on marginal regression estimates in survival analysis', *Journal of the Royal Statistical Society: Series B (Statistical Methodology)* **61**, 367–379.

Hougaard, P. (1986), 'Survival models for heterogeneous populations derived from stable distributions', *Biometrika* **73**, 387–396.

Hougaard, P. (2000), *Analysis of Multivariate Survival Data*, Springer, New York.

Ibrahim, J., Chen, M. and Sinha, D. (2001), *Bayesian Survival Analysis*, Springer, New York.

Kalbfleisch, J. (1978), 'Non-parametric Bayesian analysis of survival time data', *Journal of the Royal Statistical Society: Series B* **40**, 214–221.

Kleinbaum, D. and Klein, M. (2011), *Survival Analysis: A Self-Learning Text, Third Edition*, Springer, New York.

Komárek, A., Lesaffre, E. and Legrand, C. (2007), 'Baseline and treatment effect heterogeneity for survival times between centers using a random effects accelerated failure time model with flexible error distribution', *Statistics in Medicine* **26**, 5457–5472.

Legrand, C., Ducrocq, V., Janssen, P., Sylvester, R. and Duchateau, L. (2005), 'A Bayesian approach to jointly estimate centre and treatment by centre heterogeneity in a proportional hazards model', *Statistics in Medicine* **24**, 3789–3804.

Litière, S., Alonso, A. and Molenberghs, G. (2008), 'The impact of a misspecified random-effects distribution on the estimation and the performance of inferential procedures in generalized linear mixed models', *Statistics in Medicine* **27**, 3125–3144.

Lunn, D., Spiegelhalter, D., Thomas, A. and Best, N. (2009), 'The BUGS project: Evolution, critique and future directions', *Statistics in Medicine* **28**, 3049–3067.

Lunn, D., Thomas, A., Best, N. and Spiegelhalter, D. (2000), 'WinBUGS-a Bayesian modeling framework: Concepts, structure, and extensibility', *Statistics and Computing* **10**, 325–337.

Mallick, B. and Walker, S. (2003), 'A Bayesian semiparametric transformation model incorporating frailties', *Journal of Statistical Planning and Inference* **112**, 159–174.

Manda, S., Gilthorpe, M., Tu, Y., Blance, A. and Mayhew, M. (2005), 'A Bayesian analysis of amalgam restorations in the Royal Air Force using the counting process approach with nested frailty effects', *Statistical Methods in Medical Research* **14**, 567–578.

Manda, S. and Meyer, R. (2005), 'Bayesian inference for recurrent events data using time-dependent frailty', *Statistics in Medicine* **24**, 1263–1274.

Neuhaus, J., Hauck, W. and Kalbfleisch, J. (1992), 'The effects of mixture distribution misspecification when fitting mixed-effects logistic models', *Biometrika* **79**, 755–762.

Pennell, M. and Dunson, D. (2006), 'Bayesian semiparametric dynamic frailty models for multiple event time data', *Biometrics* **62**, 1044–1052.

Pickles, A. and Crouchley, R. (1995), 'A comparison of frailty models for multivariate survival data', *Statistics in Medicine* **14**, 1447–1461.

Qiou, Z., Ravishanker, N. and Dey, D. (1999), 'Multivariate survival analysis with positive stable frailties', *Biometrics* **55**, 637–644.

Ravishanker, N. and Dey, D. (2000), 'Multivariate survival models with a mixture of positive stable frailties', *Methodology and Computing in Applied Probability* **2**, 293–308.

Sahu, S., Dey, D., Aslanidou, H. and Sinha, D. (1997), 'A Weibull regression model with gamma frailties for multivariate survival data', *Lifetime Data Analysis* **3**, 123–137.

Sargent, D. (1998), 'A general framework for random effects survival analysis in the Cox proportional hazards setting', *Biometrics* **54**, 1486–1497.

Schwarz, G. (1978), 'Estimating the dimension of a model', *The Annals of Statistics* **6**, 461–464.

Sinha, D. (1993), 'Semiparametric Bayesian analysis of multiple event time data', *Journal of the American Statistical Association* **88**, 979–983.

Sinha, D., Ibrahim, J. and Chen, M. (2003), 'A Bayesian justification of Cox's partial likelihood', *Biometrika* **90**, 629–641.

Sinha, D. and Maiti, T. (2004), 'A Bayesian approach for the analysis of panel-count data with dependent termination', *Biometrics* **60**, 34–40.

Sinha, D., Maiti, T., Ibrahim, J. and Ouyang, B. (2008), 'Current methods for recurrent events data with dependent termination', *Journal of the American Statistical Association* **103**, 866–878.

Spiegelhalter, D., Best, N., Carlin, B. and Van Der Linde, A. (2002), 'Bayesian measures of model complexity and fit', *Journal of the Royal Statistical Society: Series B (Statistical Methodology)* **64**, 583–639.

Wei, L., Lin, D. and Weissfeld, L. (1989), 'Regression analysis of multivariate incomplete failure time data by modeling marginal distributions', *Journal of the American Statistical Association* **84**, 1065–1073.

Yin, G. (2005), 'Bayesian cure rate frailty models with application to a root canal therapy study', *Biometrics* **61**, 552–558.

Yin, G. (2008), 'Bayesian transformation cure frailty models with multivariate failure time data', *Statistics in Medicine* **27**, 5929–5940.

Yin, G. and Ibrahim, J. (2005*a*), 'Bayesian frailty models based on Box-Cox transformed hazards', *Statistica Sinica* **15**, 781.

Yin, G. and Ibrahim, J. (2005*b*), 'A class of Bayesian shared gamma frailty models with multivariate failure time data', *Biometrics* **61**, 208–216.

Yu, B. (2010), 'A Bayesian MCMC approach to survival analysis with doubly-censored data', *Computational Statistics & Data Analysis* **54**, 1921–1929.

Zhang, W., Chaloner, K., Cowles, M., Zhang, Y. and Stapleton, J. (2008), 'A Bayesian analysis of doubly censored data using a hierarchical Cox model', *Statistics in Medicine* **27**, 529–542.

Zhao, L., Hanson, T. and Carlin, B. (2009), 'Mixtures of Polya trees for flexible spatial frailty survival modeling', *Biometrika* **96**, 263–276.

24

Copula Models

Joanna H. Shih

National Cancer Institute

CONTENTS

24.1	Introduction	489
24.2	Copula	491
	24.2.1 Definition	491
	24.2.2 Archimedean copula	491
	24.2.3 Bivariate association measures	491
	Global association measures	491
	Local association measures	492
	24.2.4 Examples	493
	Clayton	493
	Frank	494
	Gumbel-Hougaard	494
	Gaussian copula	495
24.3	Estimation	496
24.4	Model assessment	499
	Local association measures	499
	Concordance estimators	500
	Posterior mean frailty process	502
24.5	Example	503
	24.5.1 Data	503
	24.5.2 Analysis	504
	24.5.3 Goodness-of-fit	505
24.6	Summary	506
	Bibliography	508

24.1 Introduction

Multivariate failure time data arise when a sampling unit contains a cluster of multiple, possibly correlated failure times. The sampling unit may be an individual at risk of multiple failures or a cluster of multiple individuals such as twin and household. For example, in a diabetic study, treatment regimens were randomly assigned to right eye and left eye of an individual. In this case, the sampling unit is patient and times to blindness of the right and left eye from the same patient were matched and may be correlated because they share common patient characteristics. The purpose of the matched design was to improve efficiency of the treatment comparison (Ederer et al., 1984). In family studies such as twin studies with survival endpoint (Anderson et al., 1992; Hougaard et al., 1992), age at onset can be used to assess the familial aggregation of the disease of interest. The sampling unit here is family, and ages at onset of a certain disease of family members are correlated because they share some common genetic and environmental risk factors. In cancer studies patients may experience multiple tumor recurrences during the course of follow-up, whereas patients after receiving bone marrow transplantation may be at risk of a range of treatment-related complications such as acute graft-versus-host disease and cytomegalovirus. Recurrent event times in the former are of the repeated events of the same type and times to different complications in the latter belong to different failure types. Competing risks result in another type of correlated failure times. They occur when a subject is at risk of multiple distinct failure types and only the first failure type is observed. Since other failure types are censored by the first observed failure type, dependency of failure times of the multiple failure types is not directly observable, and will not be considered in this chapter.

Two approaches are commonly used in modeling multivariate data: conditional and marginal approaches. Conditional models induce the dependency structure by including random effects. One typically assumes multivariate failure times are independent conditional on a scalar non-negative random variable, the so-called frailty. Mixing over the distribution of frailty produces dependency. In contract, the marginal approach models the marginal distribution directly and then imposes a dependency structure to construct the multivariate distribution. For gaussian data, in presence of covariates, generally the interpretation of regression coefficients for the two approaches are the same, but for non-gaussian data, as typically the case for failure time data, conditional and marginal approaches usually lead to different distributional forms. For example, if one proposes a proportional hazards model for failure time data conditional on the frailty, then proportionality no longer holds for the marginal distribution unless the frailty term is degenerate or follows a stable law. On the other hand, if the marginal hazard is proportional, then the conditional model is different from the marginal model. Consequently, the choice of model depends on the research problem and inferential goals.

This chapter explores a marginal approach for modeling multivariate failure time data, where association between failure times is modeled by a copula function. A $k-$dimensional copula function is defined as the joint continuous cumulative distribution (survival) function on the unit cube $[0,1]^k$ with uniform marginals. Since an absolutely continuous distribution function can be transformed to a uniform distribution, a generalized copula function can have arbitrary margins. There is abundant literature contributed to the study of copula functions, including Sklar (1959); Nelsen (1999); Genest and MacKay (1986); Marshall and Olkin (1988), to name just a few. One attractive feature of a copula function is that the marginal distributions do not depend on the choice of dependency structure, and one can model and estimate the margins and dependency separately. This feature is analogous to the multivariate normal distribution, where the mean vectors are separable from the covariance matrix and jointly determine the multivariate distribution.

The chapter is organized as follows. In Section 24.2, the copula function is defined and a few examples of bivariate survival functions belonging to copula models are presented. Local as well as global association measured are introduced and used to compare these bivariate survival functions. In Section 24.3, a two-stage estimation procedure and its asymptotic properties are presented. Section 24.4 is devoted to diagnostic procedures for assessing the goodness-of-fit of a copula model. An analysis of veteran twins study data is provided in Section 24.5. The chapter is ended with a brief summary in Section 24.6.

24.2 Copula

24.2.1 Definition

Consider a vector of clustered failure times $(T_1, \cdots, T_k)$ with marginal survival functions $S_1, \cdots, S_k$ and joint survival function S. Based on Sklar's theorem (Sklar, 1959) that relates an arbitrary distribution function on $\mathcal{R}^k$ to a copula function via the univariate marginal distributions, there exists a copula C_α such that $S(t_1, \cdots, t_k) = C_\alpha(S_1(t_1), \cdots, S_k(t_k))$. Furthermore, if S is continuous, then the copula C_α is uniquely specified and is given by $C_\alpha(u_1, \cdots, u_k) = S(S^{-1}(u_1), \cdots, S^{-1}(u_k))$ for $(u_1, \cdots, u_k) \in (0,1)^k$ and $\alpha \in \mathcal{R}^1$, where $S^{-1}(u) = \inf\{x : S_i(x) \leq u\}, i = 1, \cdots, k$.

24.2.2 Archimedean copula

One special class of copula functions is Archimedean copula which has the representation

$$C_\alpha(u_1, \cdots, u_k) = \phi[\phi^{-1}(u_1) + \cdots + \phi^{-1}(u_k)], \tag{24.1}$$

where ϕ is an Archimedean generator. The representation in (24.1) is a $k-$dimensional copula if and only if ϕ is k-monotone on $[0, \infty]$(McNeil and Slehová, 2009). That is, the jth derivatives of ϕ satisfy $(-1)^j \phi^{(j)}(u) \geq 0$ for all $u \geq 0$ and $j = 0, 1, \cdots, k-2$, and $(-1)^{k-2}\phi^{k-2}(u)$ is non-increasing and convex. One such generator is the Laplace transform of a non-negative random variable. In that case, the Archimedean copula reduces to the proportional hazard frailty model (Marshall and Olkin, 1988; Oakes, 1989). The multivariate survival function generated by the proportional hazard frailty model takes the form

$$S(t_1, \cdots, t_k) = E[e^{-Z\sum_i H_i(t_i)}], \tag{24.2}$$

where the expectation is taken over frailty Z, and $H_i, i = 1, \cdots, k$ are the cumulative hazard functions given Z. Equating (24.1) and (24.2) results in $H_i = \phi^{-1}(S_i)$.

24.2.3 Bivariate association measures

Global association measures

The Pearson correlation coefficient is a commonly used statistic for measuring the strength of dependency of two random variables. It is a convenient and concise measure of association for Gaussian type. However, for survival data which is usually skewed and non-Gaussian, the correlation coefficient might not adequately represent the true association but is dominated by influential observations and outliers. To describe adequate association for correlated failure time data, a measure which does not depend on the absolute magnitude of the variates is desirable. Kendall's coefficient of concordance, τ, meets this requirement. It is defined by

$$\tau = p[(T_{ii} - T_{j1})(T_{i2} - T_{j2}) \geq 0] - p[(T_{ii} - T_{j1})(T_{i2} - T_{j2}) < 0],$$

where (T_{i1}, T_{i2}) and (T_{j1}, T_{j2}) are two random draws from a common bivariate distribution. Two pairs are concordant if $(T_{i1} - T_{j1})(T_{i2} - T_{j2}) \geq 0$ and discordant, otherwise. Kendall's τ is the difference between the probability of having concordant pairs and the probability of having discordant pairs. It is clear from the definition that τ depends only on ranks, not actual values of the variables. Kendall's τ is bounded between -1 and 1. It is positive when

T_1 and T_2 are positively associated, and negative when they are negatively associated, and equals 0 when they are independent.

There is a close relation between Archimedean copula and Kendall's τ. Genest and MacKay (1986) showed that τ is determined by a simple function in terms of the inverse of the Archimedean generator, denoted by $\psi = \phi^{-1}$,

$$\tau = 4\int_0^1 \kappa(s)\,ds + 1, \quad \kappa(s) = \psi(s)/\psi'(s). \tag{24.3}$$

There is a one-to-one correspondence between $\kappa(s)$ and the bivariate survival function generated by a Archimedean copula. By the definition of κ above, it is clear that the bivariate survival function S determines κ. By simple algebra, it can be shown that ψ is determined in terms of κ by

$$\psi_c(s) = \exp[\int_c^s 1/\kappa(u)\,du], \tag{24.4}$$

where c is inserted so that $1/\kappa(s)$ is bounded in $[c, 1)$. Since ϕ and equivalently ψ determines the bivariate survival function, by (24.4) κ determines the bivariate survival function. Consequently, κ contains all the information about the dependency of bivariate failure times. Genest and Rivest (1993) and Wang and Wells (2000) proposed a diagnostic procedure based on κ.

Another global association measure invariant to monotone transformation of variable is Spearman's correlation coefficient, defined by

$$\rho = 12E\{(S_1(T_1) - 1/2)(S_2(T_2) - 1/2)\} = 12\int_0^1\int_0^1 C(u,v)\,du\,dv - 3.$$

Local association measures

In some applications, interest centers on changes of the dependency over time. For example, in an identical twins study, researchers are interested in studying how the influence of the genetic and environmental factors shared by the twins changes as they age. For applications as such, local quantities describing change of strength of association over time are needed. Local association refers to the relationship of variables at local levels.

Clayton (1978) introduced the following cross-ratio function to measure the association in chronic disease incidence among family members,

$$\theta(t_1, t_2) = \frac{\lambda(t_1 \mid T_2 = t_2)}{\lambda(t_1 \mid T_2 \geq t_2)}, \tag{24.5}$$

where $\lambda(t_1 \mid T_2 = t_2)$ and $\lambda(t_1 \mid T_2 \geq t_2)$ are the conditional hazard function of T_1 given $T_2 = t_2$ and $T_2 \geq t_2$, respectively. Following the definition of the hazard function, $\theta(t_1, t_2)$ can be expressed by

$$\theta(t_1, t_2) = \frac{S(t_1, t_2)\frac{\partial^2 S(t_1,t_2)}{\partial t_1 \partial t_2}}{\frac{\partial S(t_1,t_2)}{\partial t_1}\frac{\partial S(t_1,t_2)}{\partial t_2}}.$$

The cross-ratio can be interpreted as the instantaneous odds ratio at t_1, t_2 (Anderson et al., 1992), and therefore is useful in measuring the local association between T_1 and T_2. The cross-ratio equals 1 if and only if T_1 and T_2 are independent. Oakes (1989) showed the unique relation between the bivariate survival function generated by Archimedean copula and cross-ratio. Suppose that Archimedean representation (24.1) holds for $k = 2$. Then

$\theta(t_1, t_2)$ can be denoted by $\theta(s), s = S(t_1, t_2)$, because it depends on t_1, t_2 only through s. Some simple algebra yields the following explicit formula

$$\theta(s) = -s\psi''(s)/\psi'(s).$$

Furthermore, Oakes (1989) related the cross-function to Kendall's coefficient of concordance by showing that

$$\theta(s) = \frac{p(\mathbf{T}_i, \mathbf{T}_j \text{concordant} \mid \tilde{\mathbf{T}}^{ij} = \mathbf{t})}{p(\mathbf{T}_i, \mathbf{T}_j \text{discordant} \mid \tilde{\mathbf{T}}^{ij} = \mathbf{t})}, \tag{24.6}$$

where $\mathbf{T}_i = (T_{i1}, T_{i2})$, $\mathbf{T}^{ij} = (T_{i1} \wedge, T_{j1}, T_{i2} \wedge T_{j2})$, and $\mathbf{t} = (t_1, t_2), s = S(t_1, t_2)$. To see this note that the numerator of Equation (24.6) equals $S(t_1, t_2)\frac{\partial^2 S(t_1,t_2)}{\partial t_1 \partial t_2}$ and the denominator equals $\frac{\partial S(t_1,t_2)}{\partial t_1}\frac{\partial S(t_1,t_2)}{\partial t_2}$. Consequently, the ratio $\frac{\theta(s)-1}{\theta(s)+1}$ is a conditional version of Kendall's coefficient of concordance. Due to the one-to-one correspondence between $\theta(.)$ and the bivariate survival function generated by Archimedean copula, Oakes (1989) suggested a diagnostic procedure based on θ for bivariate uncensored data.

24.2.4 Examples

Clayton

The multivariate survival function in Clayton's family has the representation

$$S(t_1, \cdots, t_k) = \{\sum_{i=1}^{k} S_i(t_i)^{-\alpha} - k + 1\}^{-1/\alpha}, \quad , \alpha > 0, t_i > 0, i = 1, \cdots, k. \tag{24.7}$$

The density function of (24.7) is

$$\prod_{i=1}^{k}\{\lambda_i(t_i)S_i(t_i)^{-\alpha}\}^{\delta_i} S(t_1, \cdots, t_k)^{1+\alpha\bar{\delta}}\alpha^{\bar{\delta}}\Gamma(1/\alpha + \bar{\delta})/\Gamma(1/\alpha),$$

where λ_i is the marginal hazard function, $\delta_i = 1$ if T_i is observed and 0 if T_i is censored, and $\bar{\delta} = \sum_i \delta_i$.

The Clayton family belongs to the Archimedean copula where the generator $\phi(u) = (1+u)^{-1/\alpha}, \alpha > 0$ is the Laplace transform of a gamma distribution. Hence, in the literature the Clayton model is also called the "gamma frailty model." The inverse of ϕ is $\psi(v) = v^{-\alpha} - 1$. The parameter α determines the dependency of the multivariate failure times with $\alpha > 0$ corresponding to positive dependency and $\alpha \to 0$ corresponding to independence. The cross-ratio is constant and equals $\alpha + 1$. According to (24.6), Kendall's $\tau = \alpha/(\alpha+2)$. Spearman's correlation coefficient can be evaluated by numerical integration or by the formula

$$\rho = 12\frac{(\alpha+1)}{(\alpha+2)^2}\, {}_3F_2((\alpha+1)/\alpha, 1, 1, 2(\alpha+1)/\alpha, 2(\alpha+1)/\alpha, 1), \tag{24.8}$$

where the hypergeometric function ${}_3F_2$ (Hougaard, 2000) is defined by

$${}_3F_2(a, b, c, d, e, x) = \sum_{m=0}^{\infty} \frac{\Gamma(a+m)\Gamma(b+m)\Gamma(c+m)\Gamma(d)\Gamma(e)x^m}{\Gamma(a)\Gamma(b)\Gamma(c)\Gamma(d+m)\Gamma(e+m)m!}.$$

Genest and MacKay (1986) extended the Clayton model to allow for negative dependency by the generator $\phi(u) = (1-u)^{-1/\alpha}, \alpha < 0$ with the inverse function $\psi(v) = 1 - v^{-\alpha}$.

Since ϕ is not a Laplace transform, it does not correspond to a frailty model. McNeil and Slehová (2009) showed that the above generator is k-monotone for a particular $k \geq 2$ if and only if $\alpha \geq -1/(k-1)$. Correspondingly, for a bivariate Clayton model with negative dependency, $\alpha \geq -1$. To avoid negative probability in some regions of sample space, the bivariate survival function is expressed as

$$S(t_1, t_2) = [max\{S_1(t_1)^{-\alpha} + S_2(t_2)^{-\alpha} - 1, 0\}]^{-1/\alpha}, \quad -1 \leq \alpha < 0.$$

Frank

The multivariate survival function in Frank's family has the representation

$$S(t_1, \cdots, t_k) = log_\alpha \left[1 + \frac{\prod_{i=1}^{k}(\alpha^{S_i(t_i)} - 1)}{(\alpha - 1)^{k-1}}\right], \quad 0 < \alpha < 1, t_i > 0, i = 1, \cdots, k. \tag{24.9}$$

No simple expression of the corresponding density exists for general k. For $k = 2$, the density function is

$$\left[\frac{(\alpha - 1)\log(\alpha)\alpha^{S_1(t_1)+S_2(t_2)}}{\{(\alpha - 1) + (\alpha^{S_1(t_1)} - 1)(\alpha_2^S(t_2) - 1)\}^2}\right]^{\delta_1\delta_2} \left[\alpha^{-S(t_1,t_2)}\alpha^{S_2(t_2)-1}\alpha^{S_1(t_1)}/(\alpha - 1)\right]^{\delta_1(1-\delta_2)}$$
$$\times \left[\alpha^{-S(t_1,t_2)}\alpha^{S_1(t_1)-1}\alpha^{S_2(t_2)}/(\alpha - 1)\right]^{\delta_2(1-\delta_1)} S(t_1, t_2)^{(1-\delta_1)(1-\delta_2)}.$$

The Frank family belongs to the Archimedean copula with $\phi(u) = \log_\alpha\{1 - (1-\alpha)e^{-u}\}$ which is a Laplace transform for $0 < \alpha < 1$. The inverse function of ϕ is $\psi(v) = -\log\{(1 - \alpha^v)/(1 - \alpha)\}$. When $\alpha > 1$, the generator is not d-monotone and hence cannot be used to generate multivariate survival function of any dimensions. When $k = 2$, the survival function in (24.9) is valid for $\alpha > 1$. The cross-ratio takes the form,

$$\theta(s) = -s\log\alpha\{1 + \frac{\alpha^s}{(1 - \alpha^s)}\},$$

where $\theta(s) > 1$ if $0 < \alpha < 1$, $\theta(s) < 1$ if $\alpha > 1$, and $\to 1$ when $\alpha \to 1$ corresponding to independence.

It can be shown that

$$\tau = 1 + 4[D_1(\gamma) - 1]/\gamma,$$

where $\gamma = -\log\alpha$ and for integer $k \geq 1$,

$$D_k(\gamma) = \frac{k}{\gamma^k}\int_0^\gamma \frac{t^k}{e^t - 1}\,dt.$$

Spearman's ρ is given by

$$\rho = 1 + 12\{D_2(\gamma) - D_1(\gamma)\}/\gamma.$$

Gumbel-Hougaard

The multivariate survival function in the Gumbel-Hougaard family has the representation

$$S(t_1, \cdots, t_k) = \exp[-\{\sum_{i=1}^{k}\Lambda_i(t_i)^{1/\alpha}\}^\alpha], \tag{24.10}$$

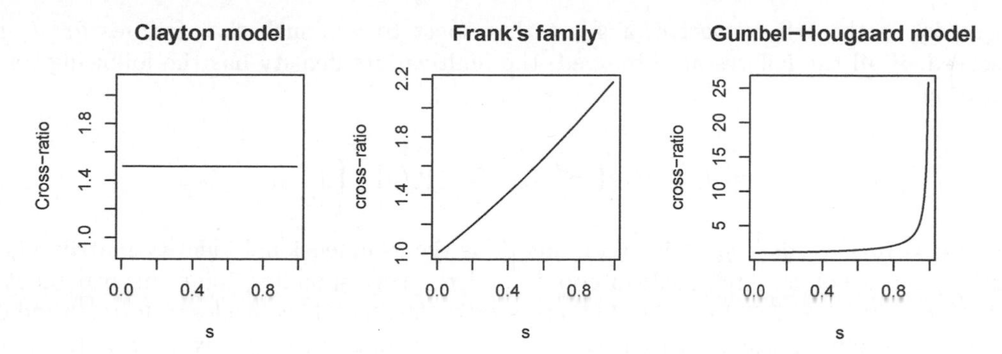

FIGURE 24.1
Cross-ratios of three Archimedean-copula models with $\tau = 0.2$

where $\Lambda_i = -\log S_i$. The multivariate density function has a complex form given by

$$\prod_i \{\lambda_i(t_i)\Lambda_i(t_i)^{1/\alpha-1}/\alpha\}^{\delta_i} QS(t_1, \cdots, t_k),$$

where $Q = \sum_{i=1}^{\bar{\delta}} c_{\bar{\delta},i}\alpha^i \{\sum_j \Lambda_j(t_j)^{1/\alpha}\}^{i\alpha-\bar{\delta}}$, and c_{ik} is polynomial in α given by the recursive formula

$$c_{p,1}(\alpha) = \Gamma(p-\alpha)/\Gamma(1-\alpha), c_{pp} = 1$$
$$p_{p,j(\alpha)} = c_{p-1,j-1}(\alpha) + c_{p-1,j}(\alpha)\{(p-1) - j\alpha\}.$$

The Gumbel-Hougaard family is also called the "positive stable frailty model" because it belongs to the Archimedean copula and the generator $\phi(s) = \exp(-s^\alpha), 0 < \alpha < 1$ is the Laplace transform of a positive stable distribution. The inverse of the generator is $\psi(v) = (-\log v)^{1/\alpha}$. Kendall's τ is simply $1-\alpha$. Spearman's ρ does not have a closed form.

One attractive feature of the multivariate survival function induced by positive stable frailty is that proportional hazard model holds for both the marginal and conditional approach ((Hougaard, 1986*a*,*b*)), although the proportionality constant is changed.

The cross-ratio is $\theta(s) = 1 + (1-\alpha)/(-\alpha \log s), 0 < s < 1$. It increases from 1 to ∞ as s increases from 0 to 1. Hence the Gumbel-Hougaard model implies high early dependency which diminishes as failure times increase. For comparison, the cross-ratio for the above three Archimedean copula models were plotted and displayed in Figure 24.1, where Kendall's τ is set at 0.2 in all the three models.

Gaussian copula

The Gaussian copula over the unit cube $[0,1]^k$ with correlation matrix Σ is given by

$$C_\Sigma(u_1, \cdots, u_k) = \Phi_\Sigma(\Phi^{-1}(u_1), \cdots, \Phi^{-1}(u_k)),$$

where Φ^{-1} is the quantile of the standard normal distribution and Φ_Σ is the multivariate normal distribution with mean 0 and correlation matrix Σ. By setting $S_i(T_i) = u_i$, the density function of $T_1, \cdots, T_k$ can be written as

$$\Phi_\Sigma^{\bar{\Delta}}(\Phi^{-1}(u_1), \cdots, \Phi^{-1}(u_k)) \prod_{i=1}^{k} \{f_i(t_i)/\phi(\Phi^{-1}(u_i))\}^{\delta_i},$$

where $\Phi_{\Sigma}^{\bar{\delta}}$ is the $\bar{\delta}$th derivative of Φ_{Σ} with respect to u_i's such that failures for T_i are observed. If all the failures are observed, the multivariate density has the following representation

$$|\Sigma|^{-1/2} \exp\{-q^t(I_k - \Sigma^{-1})q/2\} \prod_{i=1}^{k} f_i(t_i),$$

where $\mathbf{q} = (q_1, \cdots, q_k)^t$, $q_i = \Phi^{-1}(u_i)$, and I_k is the k-dimensional identity matrix. Gaussian copula is not an Archimedean copula and the corresponding joint survival function does not have a explicit formula. The cross-ratio does not have a closed form. Kendall's τ and Spearman's ρ equal $2 \arcsin(\alpha)/\pi$ and $6 \arcsin(\alpha/2)/\pi$, respectively, where α is the correlation coefficient of the bivaraite Gaussian copula model.

24.3 Estimation

Bivariate and multivariate survival functions in copula models are characterized by arbitrary continuous univariate survival functions and a dependency function C. This special structure suggests that one may estimate the marginals and association parameter α in C separately. Genest et al. (1995) proposed a semiparametric estimation procedure for dependency parameter where the marginal distributions were estimated by the empirical cumulative distribution function. Shih and Louis (1995*b*) proposed a similar two-stage estimation approach for possibly censored data without covariates. At the first stage, the marginal distribution was estimated assuming working independence. Any estimation procedure which produces consistent estimators for the marginal distributions may be used. At the second stage, the association parameter was estimated by fixing the margins at the estimates obtained from stage 1. One advantage of the two-stage estimation is that estimators of parameters in the marginal distributions are consistent upon correct specification of the marginal distributions and robust against mis-specification of the copula model. Glidden (2000*b*) extended the two-stage estimation approach to incorporate covariates in the marginal distributions, where covariates effects follow a marginal proportional hazards model and clustered failure times follow the Clayton model. For parametric marginal distributions, the two-stage parametric estimation is straightforward, and asymptotic properties of the estimators follow that of the GEE estimators, where scores and Fisher information can be derived analytically or approximated numerically. See, for example, Bjarnason and Hougaard (2000) and Wang et al. (1995) for the scores and Fisher information of the gamma and positive frailty model with marginal Weibull model, respectively. For the rest of the section, the focus is on semi-parametric estimation, where the marginal distribution is modeled as the proportional hazards model and the baseline hazard is unspecified. Direct full maximum likelihood estimation is complex and unstable because it involved possibly infinite number of parameters in the baseline hazard. In contrast, the two-stage semi-parametric estimation approach is computationally stable and efficient.

Suppose the cohort under study consists of n independent clusters, each of at most L subjects and K failures. Let T_{ijk} and $C_{ijk}, i = 1, \cdots, n, j = 1, \cdots, k_i$, denote failure and censoring times for observation j of failure type k in cluster i. Let $X_{ijk} = \min(T_{ijk}, C_{ijk})$ denote the observed failure time and δ_{ijk} the censoring indicator (1 for failure and 0 otherwise). The p-dimensional vector $\boldsymbol{Z}_{ijk}$ denotes a set of covariates. Throughout the estimation, it is assumed that failure times T_{ijk} are independent of censoring time C_{ijk} conditional on the observed covariates $\boldsymbol{Z}_{ijk}$. It is convenient to introduce additional counting process notation. Let $Y_{ijk}(t) = I(X_{ijk} \geq t)$ and $N_{ijk}(t) = \delta_{ijk} I(X_{ijk} \leq t)$. The maximum follow-up time is denoted by ν. Let $\mathbf{T}_i = (T_{ijk}, j = 1, \cdots, L, k = 1, \cdots, K), i = 1, \cdots, n$ denote n independent failure time vectors. Finally assume that $\mathbf{T}_i, i = 1, \cdots, n$ follow a copula model with marginal proportional hazards given by $\lambda_{ijk}(t \mid \boldsymbol{Z}_{ijk}) = \lambda_{0k}(t) e^{\boldsymbol{\beta}_0^T \boldsymbol{Z}_{ijk}}$, where $\lambda_{0k}(.)$ is an unspecified marginal baseline hazard function and $\boldsymbol{\beta}_0$ is a p-variate regression parameter with a population relative risk interpretation. If there is only one failure type, $K = 1$ and $\lambda_{01}(.) \equiv \lambda_0(.)$.

At stage 1, under working independence among failure times in each cluster, the estimator $\hat{\boldsymbol{\beta}}$ of $\boldsymbol{\beta}_0$ is the solution to the estimation equation $U(\boldsymbol{\beta}) = 0$, where

$$U(\boldsymbol{\beta}) = \sum_{i=1}^{n} \sum_{j=1}^{m_i} \sum_{k=1}^{K} \int_0^{\nu} \{\boldsymbol{Z}_{ijk} - \bar{\mathbf{e}}_k(\beta, u)\} \, dN_{ijk}(u),$$

$$\mathbf{e}_k(\boldsymbol{\beta}, t) = \frac{\mathbf{S}_k^{(1)}(\boldsymbol{\beta}, t)}{S_k^{(0)}(\boldsymbol{\beta}, t)}, \quad S_k^{(r)}(\boldsymbol{\beta}, t) = n^{-1} \sum_{i=1}^{n} \sum_{j=1}^{L} Y_{ijk}(t) e^{\boldsymbol{\beta}^T \boldsymbol{Z}_{ijk}} \boldsymbol{Z}_{ijk}^{\otimes r}, r = 0, 1, 2,$$

where for a column vector a, $a^{\otimes 0} = 1, a^{\otimes 1} = a$, and $a^{\otimes 2} = aa^T$. The Nelson-type estimator of the cumulative hazard, defined as $\Lambda_{0k}(t) = \int_0^t \lambda_{0k}(u) \, du, 0 < t \leq \nu$, has the form

$$\hat{\Lambda}_{0k}(t; \hat{\boldsymbol{\beta}}) = \int_0^t \frac{dN_{++k}(u)}{n S_k^{(0)}(\hat{\boldsymbol{\beta}}, u)},$$

where $N_{++k}(t) = \sum_{i=1}^{n} \sum_{j=1}^{L} N_{ijk}(t)$.

With given marginals, the contribution from each cluster to the likelihood of α is obtained by taking the derivative of the joint survival function with respect to uncensored failure times as shown in the previous section. The likelihood is given by

$$\prod_{i=1}^{n} \prod_{j=1}^{L} \prod_{k=1}^{K} \{\lambda_{ijk}(X_{ij}) S_{ijk}(X_{ijk})^{-\alpha}\}^{\delta_{ijk}} S_i(X_{ij1}, \cdots, X_{iLK})^{1+\alpha\delta_{i++}} \times \alpha^{\delta_{i++}} \Gamma(1/\alpha + \delta_{i++}) / \Gamma(1/\alpha), \tag{24.11}$$

where $\delta_{i++} = \sum_j \sum_k \delta_{ijk}$.

The logarithm of (24.11) can be simplified to the following form

$$\begin{aligned} l(\alpha) &= n^{-1} \sum_i l_i \\ &= n^{-1} \sum_{i=1}^{n} [\sum_{j=1}^{\delta_{i++}} \log\{1 + (j-1)\alpha\}] + [\sum_{j=1}^{L} \sum_{k=1}^{K} \alpha \delta_{ijk} \Lambda_{ijk}] - \\ &\quad (\alpha^{-1} + \delta_{i++}) \log\{R_i(\alpha)\}, \end{aligned} \tag{24.12}$$

where $\Lambda_{ijk} = e^{\boldsymbol{\beta}_0^T \boldsymbol{Z}_{ijk}} \Lambda_{0k}(X_{ijk})$, and $R_i(\alpha) = \sum_{j=1}^{L} \sum_{k=1}^{K} e^{\alpha \Lambda_{ijk}} - KL + 1$.

At stage 2, $\hat{\boldsymbol{\beta}}$ and $\hat{\Lambda}_{0k}(.; \hat{\boldsymbol{\beta}})$ calculated from stage 1 replace $\boldsymbol{\beta}_0$ and Λ_{0k} in (24.12), respectively. The estimation of α is obtained by maximizing the pseudo log-likelihood $\hat{l}(\alpha)$, where ˆ over l indicates that $\hat{\Lambda}_{0k}$ and $\hat{\boldsymbol{\beta}}_0$ are inserted in the likelihood.

The consistency and asymptotic normality of $(\hat{\boldsymbol{\beta}}, \hat{\Lambda}_{0k}(.), k = 1, \cdots, K)$ were established by Spiekerman and Lin (1998) for general dependency structures. Extending the results of Wei et al. (1989), Lee et al. (1992) and Spiekerman and Lin (1998) showed that the estimator $\hat{\boldsymbol{\beta}}$ converges in probability to $\boldsymbol{\beta}_0$, and $n^{-1/2}(\hat{\boldsymbol{\beta}} - \boldsymbol{\beta}_0)$ is asymptotically equivalent to the sum of n independent variables, $n^{-1/2}\mathbf{A}^{-1}\sum_{i=1}^n \mathbf{w}_{i++}$, where

$$\mathbf{w}_{i++} = \sum_{j=1}^{L}\sum_{k=1}^{K}\int_0^{\nu} \{\boldsymbol{Z}_{ijk} - \mathbf{e}_k(\boldsymbol{\beta}, u)\}\, dM_{ijk}(u),$$

and

$$\begin{aligned}\mathbf{A} &= \lim_{I\to\infty} \mathbf{A}_I \\ &= n^{-1}\sum_{i=1}^{n}\sum_{j=1}^{L}\sum_{k=1}^{K}\int_0^{\nu}\left\{\frac{\mathbf{S}_k^{(2)}(\boldsymbol{\beta}, u)}{S_k^{(0)}(\boldsymbol{\beta}, u)} - \mathbf{e}_k(\boldsymbol{\beta}, u)^{\otimes 2}\right\} dN_{ijk}(u).\end{aligned}$$

In addition, Spiekerman and Lin (1998) showed that the estimator $\hat{\Lambda}_{0k}(t;\hat{\beta})$ converges in probability to $\Lambda_{0k}(t)$ uniformly in $t \in [0, \tau]$ and $n^{1/2}(\hat{\Lambda}_{0k}(t) - \Lambda_{0k}(t))$ is asymptotically equivalent to $n^{-1}\sum_{i=1}^n \Psi_{ij}(t)$, where

$$\Psi_{ik}(t) = \int_0^t \frac{dM_{i+k}(u)}{s^{(0)}(\boldsymbol{\beta}, u)} + \boldsymbol{h}_k^T(t)\mathbf{A}^{-1}\mathbf{w}_{i+},$$

where

$$M_{ijk}(t) = N_{ijk}(t) - \int_0^t Y_{ijk}(u)e^{\boldsymbol{\beta}^T \boldsymbol{Z}_{ij}}\, d\Lambda_{0k}(u),$$

$$\boldsymbol{h}_k(t) = \int_0^t \mathbf{e}_k(\boldsymbol{\beta}, u)\, d\Lambda_{0k}(u), where$$

$s_k^{(r)}(\boldsymbol{\beta}, t) = \mathcal{E}\{S_k^{(r)}(\boldsymbol{\beta}, t)\}$ and $\mathbf{e}_k(\boldsymbol{\beta}, t) = \frac{\mathbf{s}_k^{(1)}(\boldsymbol{\beta},t)}{\mathbf{s}_k^{(0)}(\boldsymbol{\beta},t)}$. The asymptotic properties of $\hat{\alpha}$ with covariates were developed by Glidden (2000*b*). Under the regularity conditions detailed in Spiekerman and Lin (1998) and assuming the given copula model holds and α_0 belongs to the parameter space, the semi-parametric estimator $\hat{\alpha}$ of α_0 is consistent and $\sqrt{n}(\hat{\alpha} - \alpha_0)$ converges weakly to a zero-mean normal distribution. The asymptotic variance has the sandwich form $\sigma^2 = B^{-1}(\alpha_0)\sigma_\Phi^2 B^{-1}$, where B is the limit of minus second derivative of the mean log-likelihood of α given by

$$\begin{aligned}B(\alpha) &= \lim_{n\to\infty} -\frac{\partial^2 l_2(\alpha)}{\partial\alpha^2} \\ &= \lim_{n\to\infty} n^{-1}\sum_{i=1}^{I}\sum_{l=1}^{\delta_{i++}} \frac{(l-1)^2}{\{1+(l-1)\alpha\}^2} + 2\alpha^{-3}\log\{R_i(\alpha)\} - 2\alpha^{-2}U_i(\alpha)R_i^{-1}(\alpha) \\ &\quad + \{\alpha^{-1} + \delta_{i++}\}\{V_i(\alpha)R_{ij}^{-1}(\alpha) - U_i^2(\alpha)R_i^{-2}(\alpha)\},\end{aligned}$$

where $U_i(\alpha) = \sum_j \sum_{k=1} \Lambda_{ijk}e^{\alpha\Lambda_{ijk}}$ and $V_i(\alpha) = \sum_j \sum_{k=1} \Lambda_{ijk}^2 e^{\alpha\Lambda_{ijk}}$.

The term σ_Φ^2 is the asymptotic variance of the sum of n independent random variables Φ_i, where

$$\Phi_i = \phi_i + \sum_{k=1}^{K}\int_0^{\nu} \pi_k(s)\, d\Psi_{ik}(s) + \mathbf{F}^T\mathbf{A}^{-1}\mathbf{w}_{i++}, \tag{24.13}$$

where $\phi_i = \partial U_i/\partial\alpha$, $U_i = \partial l_i/\partial\alpha$, and

$$\pi_k(t) = \lim_{n\to\infty} \partial U(\alpha)/\partial d\Lambda_{0k}(t),$$

$$F = \lim_{n\to\infty} \partial U(\alpha)/\partial \boldsymbol{\beta},$$

where $U = n^{-1}\sum_{i=1}^{n} U_i$. The asymptotic variance can be consistently estimated by $\hat{\sigma}^2 = \hat{B}^{-1}(\hat{\alpha})\hat{\sigma}_{\Phi}^2 B^{-1}(\hat{\alpha})$, where $\hat{B}$ and and $\hat{\sigma}_{\Phi}^2$ are obtained by inserting $\hat{\boldsymbol{\beta}}$, $\hat{\Lambda_{0k}}$ for $\boldsymbol{\beta}_0$ and Λ_0, and $\hat{\alpha}$ for α_0. The terms ϕ_i and $\pi_k(t)$ and F have the following forms

$$\begin{aligned}\phi_i &= \{\sum_{l=1}^{\delta_{ij+}} \frac{l-1}{1+(l-1)\alpha}\} + \{\sum_{j=1}^{L}\sum_{k=1}^{K} \delta_{ijk}\Lambda_{ijk}\} + \alpha^{-2}\log\{R_i(\alpha)\} \\ &\quad -(\alpha^{-1}+\delta_{i++})R_{i+}^{-1}(\alpha)\sum_{j=1}^{L}\sum_{k=1}^{K}\Lambda_{ijk}e^{\alpha\Lambda_{ijk}},\end{aligned}$$

$$\begin{aligned}\pi_k(t) &= \lim_{n\to\infty} n^{-1}\sum_{i=1}^{n}\sum_{j=1}^{L} e^{\boldsymbol{\beta}^T \boldsymbol{Z}_{ijk}} Y_{ijk}(t)\Bigg[\alpha^{-1}R_i^{-1}(\alpha)e^{\theta\Lambda_{ijk}} - (\alpha^{-1}+\delta_{i++}) \\ &\quad \times(1+\alpha\Lambda_{ijk})R_i^{-1}(\alpha)e^{\alpha H_{ijk}} + (1+\alpha\delta_{i++})R_i^{-2}\sum_i\sum_j \Lambda_{ijk}e^{\alpha H_{ijk}} + \delta_{ijk}\Bigg],\end{aligned}$$

$$\begin{aligned}\mathbf{F} &= n^{-1}\sum_{i=1}^{n}\sum_{j=1}^{L}\sum_{k=1}^{K} H_{ijk}\boldsymbol{Z}_{ijk}\Bigg\{\alpha^{-1}R_i^{-1}(\alpha)e^{\alpha\Lambda_{ijk}} - (\alpha^{-1}+\delta_{i++})(1+\alpha\Lambda_{ijk}) \\ &\quad \times R_i^{-1}(\alpha)e^{\alpha\Lambda_{ijk}} + (1+\alpha\delta_{i+})\{\sum_j\sum_k \Lambda_{ijk}e^{\alpha\Lambda_{ijk}}\}R_i^{-2}(\alpha)e^{\alpha\Lambda_{ijk}} + \delta_{ijk}\Bigg\}.\end{aligned}$$

Fisher information with clustered censored data for the Gumbel-Hougaard model (Wang et al., 1995) can be modified to obtain the analytical expression of ϕ_i and $\pi_k(t)$ and $\mathbf{F}$. Using these terms together with terms in the marginal inference, the sandwich form of the asymptotic variance of the two-stage estimator $\hat{\alpha}$ for the positive stable frailty model can be obtained. Alternatively, numerical approximation to the first and second derivatives of the log-likelihood function may be used to calculate the estimates of ϕ_i, $\pi_k(.), k = 1, \cdots K$ and $\mathbf{F}$.

24.4 Model assessment

As the dependency structure of the copula model and the marginal distributions are modeled separately, goodness-of-fit assessment for the marginal distributions and dependency structure can be performed separately. Model checking procedures for the marginal Cox model for correlated failure times have been established (Spiekerman and Lin, 1996).

A number of goodness-of-fit testing and model selection procedures for the subclass Archimedean models have been developed and are described here.

Local association measures

As the cross-ratio under Archimedean model depends on bivariate failure times only through values of bivariate survival, Oakes (1989) proposed a diagnostic procedure for Archimedean models by plotting a discretized version of $\theta(s)$ vs. bivariate survival value s, where the bivariate survival function was estimated non-parametrically. Various non-parametric survival functions for bivariate censored data have been proposed including Campbell and Oldes (1982); Dabrowska (1988); Lin and Ying (1993); Prentice and Cai (1992), to name just a few. Such a plot can aid in assessing the appropriate relationship between θ and bivariate survival. For example, if there is no evidence of any smooth trend with increasing values of s, then the Clayton model might be appropriate. However, there are some challenges and disadvantages of this diagnostic approach. First data have to be grouped or discretized and different discretizing may yield different estimate of θ and that non-parametric estimate of S may be unstable for highly censored data.

Another Archimedean model selection procedure based on $\kappa(s) = \psi(s)/\psi'(s)$ in (24.3) was proposed by Genest and Rivest (1993) for bivariate uncensored data and by Wang and Wells (2000) for bivariate censored data. The main idea is to assess the goodness-of-fit based on the distance between model-based estimator of $\kappa(s)$ and the non-parametric counterpart, where the former can be obtained by the two-stage estimation procedure described in the previous section, and the latter is a function of the non-parametric bivariate survival function estimator. The distribution of the distance statistic, however, is quite complicated. As alternatives, some resampling based procedures have been proposed to approximate the null distribution of the distance statistic (Dobrič and Schmid, 2007; Nikoloulopoulos and Karlis, 2008).

Concordance estimators

A simple goodness-of-fit test to check the assumption of the Clayton model for bivariate right-censored data was developed by Shih (1998). The idea is to compare the unweighted and weighted concordance estimator, where the weighted estimator is the maximizer of a pseudo-likelihood (Clayton, 1978). If the Clayton model holds, both estimators are consistent, but the weighted estimator is more efficient and the difference of the two estimators converges to zero. The proposed test is consistent against alternatives under which the two concordance estimators converge to different values. An explicit formula for the asymptotic variance of the test statistic for uncensored data was developed. The test has been extended to general Archimedean models (Emura et al., 2010).

For a set of n paired failure times $\{(T_{i1}, T_{i2}); i = 1, \cdots, n\}$, let $\Delta_{ij} = I[(T_{i1} - T_{j1})(T_{i2} - T_{j2}) > 0], i \neq j$ denote the concordance indicator. Under an Archimedean model, according to (24.6),

$$E[\Delta_{ij} \mid \tilde{\mathbf{T}}^{ij} = \mathbf{t}] = \frac{\theta(s)}{\theta(s)+1}, \tag{24.14}$$

where $\mathbf{t} = (t_1, t_2)$ and $s = S(t_1, t_2)$. A class of estimation equations for the association parameter α in the Archimedean model based on (24.14) can be formulated by

$$U_k(\alpha) = \sum_{i<j} W_k(\tilde{\mathbf{T}}^{ij}, \alpha) \left[\Delta_{ij} - \frac{\theta(\hat{s}^{ij})}{\theta(\hat{s}^{ij})+1}\right], \tag{24.15}$$

where W_k is a weight function and $\hat{s}^{ij} = \hat{S}(\tilde{\mathbf{T}}^{ij})$ and $\hat{S}$ is an estimator of S. Similar to Shih (1998), Emura et al. (2010) compared the unweighted estimator, i.e., $W_1 = 1$ and a weighted estimator such that $U_2(\alpha)$ is the estimating equation of a pseudo-likelihood which generalizes the likelihood proposed by Clayton (1978). Define the set of grid points, $\omega = \{(t_1, t_2) \mid \sum_{i=1}^n I(T_{i1} = t_1, T_{i2} \geq t_2) = 1, \sum_{i=1}^n I(T_{i1} \geq t_1, T_{i2} = t_2) = 1\}$. Define $D(t_1, t_2) = \sum_{i=1}^n I(T_{i1} = t_1, T_{i2} = t_2), (t_1, t_2) \in \omega$, and $R(t_1, t_2) = \sum_{i=1}^n I(T_{i1} \geq t_1, T_{i2} \geq t_2)$ which is the size of the risk set at $(t_1, t_2) \in \omega$. In the absence of tie, conditional on

$R(t_1,t_2)$ and $(t_1,t_2) \in \omega$, $p(D(t_1,t_2) = 1 \mid R(t_1,t_2) = r, (t_1,t_2) \in \omega) = \frac{\theta(s)}{r-1+\theta(s)}$. Under the working independence assumption among the grids in ω, the pseudo-likelihood function of α can be written as

$$L(\alpha) = \prod_{t_1,t_2 \in \omega} \frac{\theta(S(t_1,t_2))^{D(t_1,t_2)}[R(t_1,t_2)-1]^{1-D(t_1,t_2)}}{R(t_1,t_2) - 1 + \theta(S(t_1,t_2))} \tag{24.16}$$

After some algebraic operation, the score function of the above pseudo-likelihood can be written as

$$U_2(\alpha) = \sum_{i<j} \frac{\dot{\theta}(\hat{S}(\tilde{\mathbf{T}}^{ij})(\theta(\hat{S}(\tilde{\mathbf{T}}^{ij}))+1)}{\theta(\hat{S}(\tilde{\mathbf{T}}^{ij}))[R_{ij} - 1 + \theta(\hat{S}(\tilde{\mathbf{T}}^{ij}))]} \left[\Delta_{ij} - \frac{\theta(\hat{S}^{ij})}{\theta(\hat{S}(\tilde{\mathbf{T}}^{ij}))+1}\right], \tag{24.17}$$

where $R_{ij} = R(\tilde{\mathbf{T}}^{ij}) = n\hat{S}(\tilde{\mathbf{T}}^{ij})$ and $\dot{\theta(s)} = \partial\theta(s)/\partial\alpha$. Hence, the weight function $W_2(\tilde{\mathbf{T}}^{ij},\alpha) = \frac{\dot{\theta}(\hat{s}^{ij})(\theta(\hat{s}^{ij})+1)}{\theta(\hat{s}^{ij})[R_{ij}-1+\theta(\hat{s}^{ij})]}$. For the Clayton model, W_2 is simplified to $\frac{(\theta+1)}{\theta[R_{ij}-1+\theta]}$, which assigns higher weight to later events. For the Gumbel model, $W_2 = \frac{2\log \hat{s}^{ij} - \alpha}{\{\log \hat{S}(\mathbf{T}^{ij})-\alpha\}\{\alpha - R_{ij}\hat{S}(\mathbf{T}^{ij})\}}$, which approaches ∞ at $\mathbf{T}^{ij} = 0$ and decreases over time. For the Frank family, $W_2 = \frac{\{1-\theta(\hat{s}^{ij})\alpha^{\hat{s}^{ij}}\}\{\theta(\hat{s}^{ij})+1\}}{R_{ij}-1+\theta(\hat{s}^{ij})}$, which approaches a finite value at $\mathbf{T}^{ij} = 0$ and also decreases over time. Let $\hat{\alpha}_k, k = 1,2$ denote the solution to $U_k = 0$. Thus, among the three copula models, the Clayton model emphasizes late dependency and the Gumbel-Hougaard model emphasizes early dependency, whereas the Frank family implies early dependency but to a less degree than the Gumbel-Hougaard model. If the association parameter α has positive value, taking $\gamma = \log(\alpha)$ can improve the normal approximation of its estimators. Emura et al. (2010) showed that under the correct model and the regularity conditions, $\log(\hat{\alpha}) = \hat{\gamma}_k, k = 1,2$ are consistent, and $n^{1/2}(\hat{\gamma}_k - \gamma_0)$ converges to a zero-mean normal distribution. Consequently, $n^{1/2}(\hat{\gamma}_1 - \hat{\gamma}_2)$ converges in distribution to a normal distribution with mean zero and variance $\sigma^2 = E[h(\mathbf{T}_1,\mathbf{T}_2)h(\mathbf{T}_1,\mathbf{T}_3)]$, where

$$h(\mathbf{T}_i,\mathbf{T}_j) = \frac{1}{\alpha}\left(\frac{\dot{\theta}(S(\tilde{\mathbf{T}}^{ij})(\theta(S(\tilde{\mathbf{T}}^{ij}))+1)}{A_L\theta(S(\tilde{\mathbf{T}}^{ij})S(\tilde{\mathbf{T}}^{ij})} - \frac{1}{A}\right)\left[\Delta_{ij} - \frac{\theta(S(\mathbf{T}^{ij}))}{\theta(S(\tilde{\mathbf{T}}^{ij}))+1}\right],$$

$$A \equiv E\left(\frac{\dot{\theta}(S(\tilde{\mathbf{T}}^{ij})}{[\theta(S(\tilde{\mathbf{T}}^{ij}))+1]^2}\right) \quad A_L \equiv E\left(\frac{\dot{\theta}(S(\tilde{\mathbf{T}}^{ij})^2}{\theta(S(\tilde{\mathbf{T}}^{ij}))[\theta(S(\tilde{\mathbf{T}}^{ij}))+1]}\right).$$

The asymptotic variance σ^2 can be estimated by averaging over all possible triples of empirical $h(\mathbf{T}_1,\mathbf{T}_2)h(\mathbf{T}_1,\mathbf{T}_3)$. Alternatively, σ^2 can be estimated by either jackknife or bootstrap resampling procedure.

In the presence of right censoring, let $\mathbf{C}_i = (C_{i1}, C_{i2})$ denote the censoring times and are independent of $\mathbf{T}_i$, $i = 1,\cdots,n$. Let $X_{ij} = \min(T_{ij}, C_{ij})$ Δ_{ij} is observed if and only if $\tilde{\mathbf{T}}_{ij} \le \tilde{\mathbf{C}}_{ij}$, where $\tilde{\mathbf{C}}_{ij} = (C_{i1} \wedge C_{j1}, C_{i2} \wedge C_{j2})$. Let Z_{ij} denote the indicator of this event. The modified estimation function $U_k, k = 1,2$ which includes only pairs of which Δ_{ij} is observed (i.e., $Z_{ij} = 1$) is given by

$$U_k(\alpha) = \sum_{i<j} Z_{ij} W_k(\tilde{\mathbf{T}}^{ij},\alpha)\left[\Delta_{ij} - \frac{\theta(\hat{s}^{ij})}{\theta(\hat{s}^{ij})+1}\right], \tag{24.18}$$

where $\hat{S}$ can be estimated non-parametrically or semi-parametrically by solving U_2 with the marginal survival functions estimated by Kaplan-Meier or Nelson estimators. The asymptotic normal distribution of the estimator and test statistic can be established using similar arguments as for the uncensored case. The asymptotic variance σ^2 of $n^{1/2}(\hat{\gamma}_2 - \hat{\gamma}_1)$ can be estimated by

$$\hat{\sigma}^2 = \frac{1}{n} \sum_{i \neq j \neq k} Z_{ij} Z_{ik} \hat{h}(\mathbf{T}_i, \mathbf{T}_j) \hat{h}(\mathbf{T}_i, \mathbf{T}_k),$$

where $\hat{h}(\mathbf{T}_i, \mathbf{T}_j)$ is estimated by replacing α by $\hat{\alpha}_2$ and S by $\hat{S}$.

Posterior mean frailty process

Shih and Louis (1995*a*) proposed a parametric graphical diagnostic procedure for assessing the gamma frailty model for clustered failure time data. The basic idea is to assess the gamma frailty assumption by its posterior mean process over time. If the gamma frailty model is correct, the average of posterior mean process converges in probability to the prior mean. Large discrepancies between the two quantities indicate lack of fit. Glidden (2000*a*) extended the approach to allow for non-parametric marginal distributions without covariates, and compared the observed posterior mean process to simulated realizations generated under the null hypothesis that the gamma frailty model is correct. Cui and Sun (2004) and Shih and Lu (2007) further extended the approach to incorporate covariates in the marginal proportional hazards model. Using the same notation as in Section (24.2.4), under the gamma frailty model with mean $1/\alpha$ and variance $1/\alpha$, it follows that the conditional distribution of the frailty given the data up to time $t, t \leq \tau$, is also gamma with posterior expectation proportional to $\bar{\gamma}_i(t; \boldsymbol{\beta}_0, \alpha_0) = \{1 + \alpha_0 N_{i++}(t)\}/R_i(t; \boldsymbol{\beta}_0, \alpha_0)$, where $R_i(t; \boldsymbol{\beta}_0, \alpha_0) = \sum_j \sum_k \exp[\alpha_0 \Lambda_{ijk}(t \wedge X_{ijk})] - KL + 1$ and $\Lambda_{ijk}(t \wedge X_{ijk}) = \exp\{\boldsymbol{\beta}_0^T \boldsymbol{Z}_{ijk}\} \Lambda_0(t \wedge X_{ijk})$. Under the gamma frailty model, the posterior mean process $\bar{\gamma}_i(t; \boldsymbol{\beta}_0, \alpha_0), i = 1, \cdots, n$ are independent with mean vector $\mathbf{1}$. Furthermore, the process

$$W(t) = n^{-1/2} \sum_{i=1}^{n} \{\bar{\gamma}_i(t; \boldsymbol{\beta}_0, \alpha_0) - 1\} \tag{24.19}$$

converges to a zero-mean gaussian process. Since $W(t)$ involves the unknown parameters $\boldsymbol{\beta}_0$ and α_0 and baseline cumulative hazard function $\Lambda_{0k}(.)$. Estimates obtained from the two-stage estimation approach described in the previous section were plugged in (24.19). Let $\hat{W}(t)$ denote the $W(t)$ process with the plug-in estimators. It follows that $\hat{W}(t)$ is asymptotically equivalent to $n^{-1/2} \sum_{i=1}^{n} \Upsilon_i(t)$ which converges to a zero-mean Gaussian process $\Omega(t)$ with covariance function $E\{\Omega(t)\Omega(s)\} = \lim_{n \to \infty} n^{-1} \sum_i E\{\Upsilon_i(t)\Upsilon_i(s)\}$, where $\Upsilon_i(t)$ is given by

$$\sum_{i=1}^{n} \epsilon_i(t) + B_2(\alpha_0)^{-1} \Phi_{2i} f(t) + \int_0^t g(u; t)\, d\Psi_i(u) + \mathbf{q}^T(t) \mathbf{A}^{-1} \mathbf{w}_{i++},$$

where $\epsilon_i(t) = \{\bar{\gamma}_i(t; \boldsymbol{\beta}_0, \theta_{20}) - 1\}$, $f(.)$ is the limit of $\tilde{f}(.)$ given by

$$f(t) = \lim_{n \to \infty} \tilde{f}(t) = n^{-1} \sum_i \sum_j R_i^{-1}(\alpha_0; t))\{N_{i++}(t) - \bar{\gamma}_i(t; \boldsymbol{\beta}_0, \alpha_{20}) U_i(\alpha_0; t)\},$$

where $U_i(\alpha_{20}; t) = \sum_j \sum_{k=1}^{K} \Lambda_{ijk}(t \wedge X_{ijk}) e^{\alpha_0 \Lambda_{ijk}(t \wedge X_{ijk})}$,

$g(.)$ is the limit of $\tilde{g}(.)$ given by

$$\begin{aligned} g(t) &= \lim_{n\to\infty} \tilde{g}(t) \\ &= \lim_{n\to\infty} n^{-1} \sum_i \alpha_0 R_i^{-1}(\alpha_0; t))\bar{\gamma}_i(t; \boldsymbol{\beta}_0, \alpha_0) \\ &\quad \times \sum_j \sum_k e^{\alpha_0 \Lambda_{ijk}(t\wedge X_{ijk}) + \boldsymbol{\beta}_0' \boldsymbol{Z}_{ijk}} Y_{ijk}(t), \end{aligned}$$

and $\mathbf{q}(.)$ is the limit of $\tilde{\mathbf{q}}(.)$ given by

$$\begin{aligned} \mathbf{q}(t) &= \lim_{n\to\infty} \tilde{q}(t) \\ &= \lim_{n\to\infty} n^{-1} \sum_i \alpha_0 R_i^{-1}(\alpha_{20}; t))\bar{\gamma}_i(t; \boldsymbol{\beta}_0, \alpha_0) \\ &\quad \times \sum_j \sum_k e^{\alpha_0 \Lambda_{ijk}(t\wedge X_{ijk})} \Lambda_{ijk}(t \wedge X_{ijk}) \boldsymbol{Z}_{ijk}. \end{aligned}$$

The covariance function may be consistently estimated by its empirical counterpart. The test statistic for assessing the gamma frailty model takes the form $V = \sup_t |\hat{W}(t)|$. Because the asymptotic distribution of V is not analytically tractable, the resampling method of Lin and Ying (1993) was used to simulate the distribution of the process $\hat{W}(.)$. Let $\tilde{W}(t) = I^{-1/2} \sum_i \hat{\Upsilon}_i(t) Z_i$, where $\hat{\Upsilon}_i(t)$ is the empirical counterpart of $\Upsilon_i(t)$, and $Z_i, i = 1, \cdots, n$ are independent standard normal random variables independent of the observed data. By the arguments of Spiekerman and Lin (1996), $\tilde{W}(.)$ and $\hat{W}(.)$ are asymptotically equivalent and both converge to $\Omega(.)$. Thus the null distribution of $\hat{W}(.)$ can be approximated by repeatedly generating random values of $(Z_1, \cdots, Z_I)$ from the standard normal distribution a large number of times with the observed data fixed. Once the approximated null distribution of $\hat{W}(.)$ is generated, the null distribution of a test statistic such as V is available, and the p-value of the V test is estimated by the proportion of simulated test statistics that exceed the observed value of V.

24.5 Example

24.5.1 Data

In 1955 the National Academy of Science-National Research Council (NAS-NRC) initiated the development of a white twin registry of U.S. Armed Forces veterans (Jablon et al., 1967). A panel of 16,000 pairs of twins, all of them are veterans, were obtained by matching Veterans Administration (VA) Master Index against the names given on record of multiple births in 42 vital statistics offices for the years 1917-1927. Of the 32,000 individuals in the Registry about 20,000, including both members of 7,000 twins pairs, responded to questionnaires mailed since 1965. The questionnaire was short, seeking follow-up medical information and the twins' opinions of their zygosity.

Covariates included in the analysis were history of hypertension and smoking status. Covariate hypertension is binary, taking value 1 if the individual was hypertensive at the time of survey and 0 otherwise. Covariate smoking status had three categories: never smoker(0), former smoker (1) and current smoker (2). The survival time was age of death which was censored if the individual was alive as of December 31, 1990. Deaths from this twin registry data were ascertained only partly through National Death Index (NDI) and mostly through military death benefits services. Only twin pairs who had complete data on the two covariates were included in the analysis, yielding 1875 MZ twin pairs and 1992 DZ twin pairs. Among these individuals, 17.2% of MZ twin members and 18.9% of DZ twin members died as of December 31, 1990, with 41 the minimal age of death. The finding of less mortality in MZ than in DZ pairs is consistent with that from the data with shorter follow-up (Hrubec and Neel, 1981). They and others attributed the reduction in mortality in MZ pairs relative to DZ pairs to selection effects at the induction screening. An MZ pair would more likely be excluded from military service than a DZ pair for factors that were more concordant in MZ twins in the population and which predicted subsequent morbidity and mortality, because both members of the pair had to pass the induction physical exam. Fifty-nine percent of the MZ twin members were current smokers at the time of responding to the questionnaire, compared to 61% of DZ twin members. Thirty-five percent of both MZ and DZ twin members were hypertensive.

24.5.2 Analysis

The MZ twins and DZ twins data were analyzed separately. The marginal distribution for each twin type was specified by the Cox proportional hazards model containing smoking status and hypertension with a common baseline hazard. Interaction terms between the two covariates were not significant for both MZ and DZ twins data. The estimates of the regression coefficients are listed in Table 24.1. For MZ twins, the difference in risk of death between former smokers and never smokers and the effect of hypertension were not significant. Correspondingly, never and former smokers were lumped into one single subgroup, and a final Cox model containing only one covariate, current smoker vs. not, was fitted. The estimate of the regression coefficient equaled 0.743 ($p < 0.0001$). For DZ twins, the three smoking groups had significantly different risks of death, and the effect of hypertension was also significant ($p < 0.01$).

The two-stage semi-parametric estimation procedure was applied to estimate the association of lifespans between MZ and DZ twin members. The estimates of the association parameters, Frank family, and Gumbel-Hougaard models are listed in Table 24.2. As expected, the association between the MZ pairs was stronger than between the DZ pairs.

TABLE 24.1
Estimates of marginal regression parameters for the veteran twins study.

Variable	Parameter estimates	Standard error	p-value
	MZ twins		
Former smoker	0.290	0.164	0.076
Current smoker	0.886	0.128	< 0.0001
Hypertension	0.097	0.085	0.255
	DZ twins		
Former smoker	0.394	0.152	0.010
Current smoker	1.092	0.122	< 0.0001
Hypertension	0.202	0.074	0.007

TABLE 24.2
Estimates of association parameters (α) for the veteran twins study.

Model	Parameter estimates	Standard error	τ
	MZ twins		
Clayton	1.29	0.234	0.392
Frank	0.090	0.033	0.253
Gumbel-Hougaard	0.878	0.022	0.122
	DZ twins		
Clayton	0.213	0.127	0.096
Frank	0.610	0.173	0.055
Gumbel-Hougaard	0.988	0.017	0.012

Kendall's tau equaled 0.39 for MZ twins and only 0.096 for DZ twins. It is interesting to note that the values of Kendall's tau were very different between the three estimated models. The smaller values of τ under Frank family and Gumbel-Hougaard model are due to the fact that these two models imply high early dependence whereas the Clayton model implies high late dependence and that not many twins in this cohort died at young ages.

24.5.3 Goodness-of-fit

The goodness-of-fit test of the marginal Cox model for multivariate failure times (Spiekerman and Lin, 1996) was applied. Since the covariates were dichotomous, only the assumption of proportional hazards was checked. For the MZ twins, the observed score process for current smoker (yes vs. no) vs. 20 realizations under the null distribution is displayed in Figure 24.2. The p-value for the proportional hazards test obtained based on 5,000 realizations of the null score process is 0.783. The plots for the DZ twins were similar. The p-values are 0.510, 0.809 and 0.241 with respect to former smoker, current smoker, and hypertension, respectively.

As the life times of DZ twin members were nearly independent, only the dependency structure of MZ twins was assessed. The observed posterior gamma frailty process vs. 20 realizations of the null process is displayed in Figure 24.3. The plot is indicative of some lack of fit of the dependency induced by gamma frailties, although the V-test based on the supremum of the absolute posterior mean frailty process was not significant (p=0.112). From the graphical posterior process, it is hard to tell how the dependency structure was misspecified. One plausible explanation for the negative (i.e., less-than-expected) posterior mean frailty process before age 60 is that not many deaths were observed at young ages. Since there was only one binary covariate in the marginal model, the test of constant cross-ratio for Clayton model was tested for each subgroup according to the smoking status of each twin (both members current smoker, one member current smoker, neither current smoker). The standard error of the test of concordance estimators for the Clayton model was based on Shih (1998). The p-value equaled 0.844, 0.131 and 0.029 for the three subgroups, respectively. Thus, the constant cross-ratio was rejected for the non-current smoking twins. The tests of concordance estimators for Frank's family and Gumbel-Hougaard were also

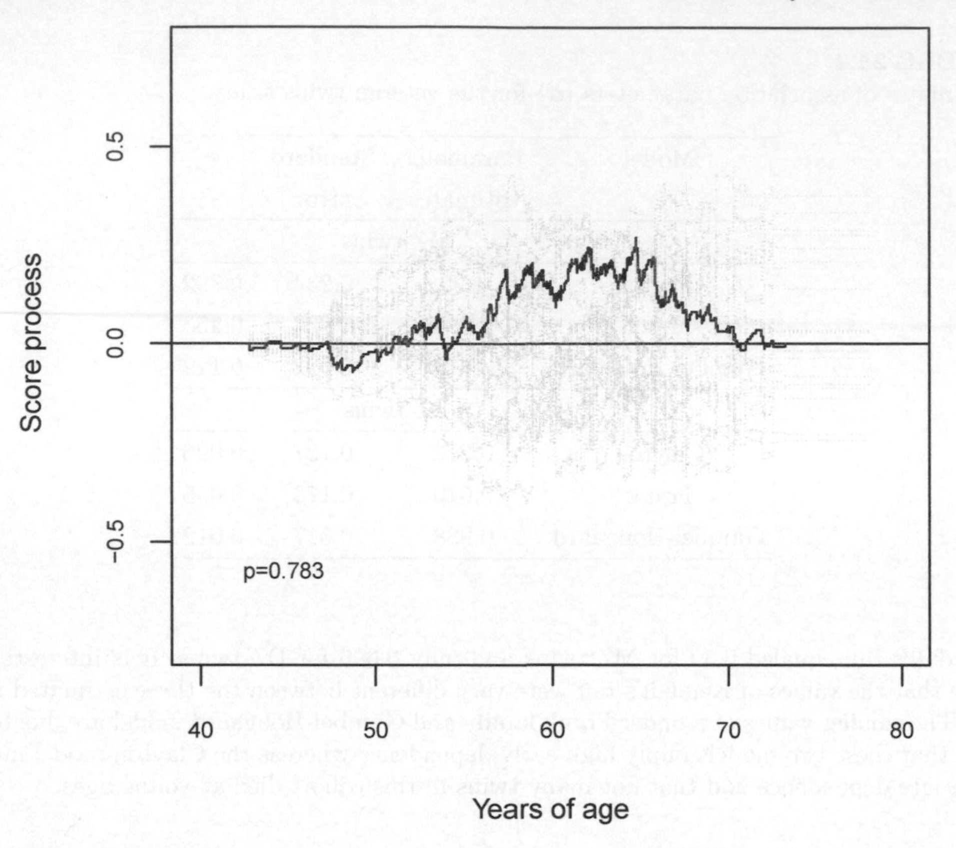

FIGURE 24.2
Score process of MZ twins.

applied. The standard errors of the test statistics for these two models were obtained by Jackknifing. The p-value equaled (0.704, 0.150, 0.075) and (0.121, 0.312, 0.032) for the three subgroups under Frank's family and Gumbel-Hougaard model, respectively. These p-values suggest that Frank's family fits the MZ twins data better than the other two models. Under Frank's family, the unweighted ($\hat{\alpha}_1$) and weighted estimate ($\hat{\alpha}_2$) of the association parameter equaled $(0.10, 0.095)$, $(0.037, 0.055)$ and $(0.019, 0.048)$ for the three subgroups of both current smokers, one current smoker, and neither current smokers, respectively. The associated Kendall's τ equaled $(0.243, 0.248)$, $(0.333, 0.298)$, and $(0.385, 0.310)$, respectively.

24.6 Summary

Copula models have become increasingly popular in modeling dependency of correlated continuous random variables. The wide spectrum of their applications encounters many fields, including finance (Cherubini et al., 2004; Embrechts et al., 2003), engineering (Genest and Favre, 2007; Yan, 2006), actuarial science (Frees and Valdez, 1998; Frees, 2005) and clinical studies (Shih and Louis, 1995*b*; Wang and Wells, 2000). Their popularity stems partly from the fact that multivariate normality may be in question in many applications and their flexibility in modeling dependency while allowing for separate specification of the

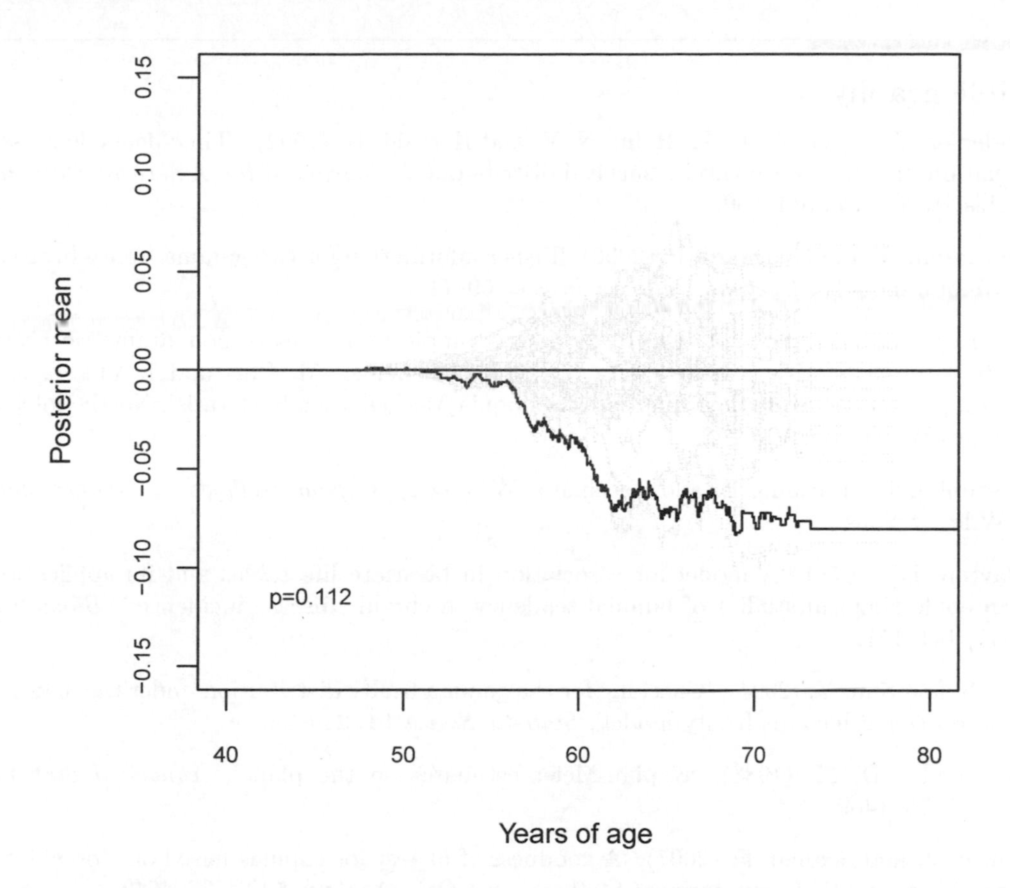

FIGURE 24.3
Posterior gamma frailty mean process of MZ twins.

possibly skewed marginal distributions. In this chapter, special attention was devoted to the subclass of Archimedean models for its appealing features for survival data and its link to proportional hazards frailty models. A number of local and global association measures were described and compared between three Archimedean copulas: Clayton, Frank and Gumbel-Hougaard models. As these Archimedean copulas exhibit different time-dependent dependency structures and result in different interpretations, model assessment is essential for copula model applications. A number of graphical diagnostic and non-parametric and semi-parametric testing procedures have been developed and presented in this chapter. While the graphical diagnostic procedures based on $\theta(.)$ and κ are complex for censored survival data, tests based on non-parametric concordance estimators and posterior mean frailty process appear promising and are relatively easy to implement.

The Akaike information criterion (AIC) has become a popular measure for model selection. While it provides relative goodness of fit of a statistical model, it does not test a specific null hypothesis. The original AIC is restricted to parametric models. Later Xu et al. (2009) developed AIC using profile likelihood for semi-parametric model selection with application to proportional hazards mixed model. However, statistical properties of AIC under pseudo-likelihood for semi-parmatric copula models described in this chapter are unknown and its applicability to these models is yet to be established.

All the analyses done in this chapter were implemented by the author's own GAUSS and R code, which are not for the moment publicly available.

Bibliography

Anderson, J. E., Louis, T. A., Holm, N. V. and Harvald, B. (1992), 'Time-dependent association measures for bivariate survival distributions', *Journal of the American Statistical Association* **87**, 641–650.

Bjarnason, H. and Hougaard, P. (2000), 'Fisher information for two gamma frailty bivariate Weibull models', *Lifetime Data Analysis* **6**, 59–71.

Campbell, G. and Oldes, A. F. (1982), Large-sample properties of nonparametric bivariate estimators with censored data, *in* B. V. Gnedenko, M. Puri and I. Vincze, eds., 'Nonparametric Statistical Inference, Colloquia Methematica-Societatis', North-Holland, János Bolyai, pp. 103–122.

Cherubini, U., Luciano, E. and Vecchiato, W. (2004), *Copula Methods in Finance*, John Wiley & Sons.

Clayton, D. (1978), 'A model for association in bivariate life tables and its applications in epidemilogical studies of familial tendency in chronic disease incidence.', *Biometrika* **65**, 141–151.

Cui, S. and Sun, Y. (2004), 'Checking for the gamma frailty distribution under the marginal proportional hazards frailty model', *Statistic Sinica* **14**, 249–267.

Dabrowska, D. M. (1988), 'Kaplan-Meier estimates on the plane', *Annals of Statistics* **16**, 1475–1489.

Dobriċ, J. and Schmid, F. (2007), 'A goodness of fit test for copulas based on Rosenblatt's transformation', *Computational Statistics and Data Analysis* **51**, 4633–4642.

Ederer, F., Podgor, M. J. and the Diatetic Retinopathy Study Group (1984), 'Assessing possible late treatment effects in stopping a clinical trial early: A case study. diabetic retinopathy study report no. 9', *Controlled Clinical Trials* **5**, 373–381.

Embrechts, P., Lindskog, F. and Mcneil, A. (2003), Modeling dependence with copulas and applications to risk management, *in* S. T. Rachev, ed., '*Handbook of Heavy Tailed Distributions in Finance*', Elsevier, Boston, pp. 329–384.

Emura, T., Lin, C.-W. and Wang, W. (2010), 'A goodness-of-fit test for Archimedean copula models in the presnece of right censoring', *Computational Statistics and Data Analysis* **54**, 3033–3043.

Frees, E. W. (2005), 'Credibility using copulas', *North American Actuarial Journal* **9**, 31–48.

Frees, E. W. and Valdez, E. A. (1998), 'Understanding relationships using copulas', *North American Actuarial Journal* **2**, 1–25.

Genest, C. and Favre, A.-C. (2007), 'Everything you always wanted to know about copula modeling but were afraid to ask', *Journal of Hydrologic Engineering* **12**, 347–368.

Genest, C., Ghoudi, K. and Rivest, L.-P. (1995), 'A semiparametric estimation procedure of dependence parameters in multivariate families of distributions', *Biometrika* **82**, 543–552.

Genest, C. and MacKay, J. (1986), 'The joy of copulas: bivariate distributions with uniform marginals', *The American Statistician* **40**, 280–283.

Genest, C. and Rivest, L.-P. (1993), 'Statistical inference procedures for bivariate Archimedean copulas', *Journal of the American Statistical Association* **88**, 1034–1043.

Glidden, D. (2000*a*), 'Checking the adequacy of the gamma frailty model for multivariate failure time', *Biometrika* **86**, 381–393.

Glidden, D. (2000*b*), 'A two-stage estimator for the dependence parameter for the Clayton-Oakes model', *Lifetime Data Analysis* **6**, 141–156.

Hougaard, P. (1986*a*), 'A class of multivariate failure time distributions', *Biometrika* **73**, 671–678.

Hougaard, P. (1986*b*), 'Survival models for heterogeneous populations derived from stable distributions', *Biometrika* **73**, 387–396.

Hougaard, P. (2000), *Analysis of Multivariate Survival Data*, Springer-Verlag, New York.

Hougaard, P., Harvald, B. and Holm, N. V. (1992), 'Measuring the similarities between the lifetimes of adult Danish twins born between 1881-1930', *Journal of the American Statistical Association* **87**, 17–24.

Hrubec, Z. and Neel, J. V. (1981), 'Familial factors in early deaths: twins followed 30 years to ages 51-61', *Human Genetics* **59**, 39–46.

Jablon, S., Neel, J. V., Gershowitz, H. and Atkinson, G. F. (1967), 'The NAS-NRC twin panel: methods and construction of the panel, zygosity diagnosis, and proposed use', *American Journal of Human Genetics* **19**, 133–161.

Lee, E., Wei, L. J. and Amato, D. A. (1992), Cox-type regression analysis for large numbers of small groups of correlated failure time observations, *in* J. P. Klein and P. Goel, eds., *Survival Analysis: State of the Arts*, Kluwer Academic, Dordrecht, pp. 237–247.

Lin, D. Y. and Ying, Z. (1993), 'A simple nonparametric estimator of the bivariate survival function under univariate censoring', *Biometrika* **80**, 573–581.

Marshall, A. W. and Olkin, I. (1988), 'Families of multivariate distribution', *Journal of the American Statistical Association* **83**, 834–841.

McNeil, A. J. and Slehová, J. N. (2009), 'Multivariate Archimedean copulas, d-monotone functions and l_1-norm symmetric distribuiton', *Annals of Statistics* **37**, 3059–3097.

Nelsen, R. B. (1999), *An Introduction to Copulas*, volume 139 of Lecture Notes in Statistics, Springer-Verlag, New York.

Nikoloulopoulos, A. K. and Karlis, D. (2008), 'Copula model evaluation based on parametric bootstrap', *Computational Statistics and Data Analysis* **52**, 3342–3353.

Oakes, D. (1989), 'Bivariate survival models induced by frailties', *Journal of the American Statistical Association* **84**, 487–493.

Prentice, R. and Cai, J. (1992), 'Covariance and survival function estimation using censored multivariate failure time data', *Biometrika* **79**, 495–512.

Shih, J. H. (1998), 'A goodness-of-fit test for association in a bivariate survival model', *Biometrika* **85**, 189–200.

Shih, J. H. and Louis, T. A. (1995*a*), 'Assessing gamma frailty models for clustered failure time data', *Lifetime Data Analysis* **1**, 205–220.

Shih, J. H. and Louis, T. A. (1995*b*), 'Inferences on the association parameter in copula models for bivariate survival data', *Biometrics* **51**, 1384–1399.

Shih, J. H. and Lu, S.-E. (2007), 'Analysis of failure time data with multi-level clustering, with application to the child vitamin an intervention trial in Nepal', *Biometrics* **63**, 673–680.

Sklar, A. (1959), 'Fonctions de répartition à n dimensions et leurs marges', *Publ. Inst. Stataist. Univ. Paris* **9**, 229–231.

Spiekerman, C. and Lin, D. Y. (1998), 'Marginal regression models for multivariate failure time data', *Journal of the American Statistical Association* **93**, 1164–1175.

Spiekerman, C. and Lin, D. Y. (1996), 'Checking the marginal Cox model for correlated failure time data', *Biometrika* **6**, 141–156.

Wang, S., Klein, J. and Moeschberger, M. L. (1995), 'Semi-parametric estimation of covariate effects using the positive stable frailty model', *Applied Stochastic Models and Data Analysis* **11**, 121–133.

Wang, W. and Wells, M. T. (2000), 'Model selection and semiparametric inference for bivariate failure-time data', *Journal of the American Statistical Association* **95**, 62–72.

Wei, L. J., Lin, D. Y. and Weissfeld, L. (1989), 'Regression analysis of multivariate incomplete failure time data by modeling marginal distributions', *Journal of the American Statistical Association* **84**, 1065–1073.

Xu, R., Vaida, F. and Harrington, D. P. (2009), 'Using profile likelihood for semiparametric model selection with application to proportional hazards mixed models', *Statistica Sinica* **19**, 819–842.

Yan, J. (2006), Multivariate modeling with copulas and engineering applications, *in* H. Pham, ed., 'Handbook in Engineering Statistics', Springer-Verlag, New York, pp. 973–990.

25

Clustered Competing Risks

Guoqing Diao

George Mason University

Donglin Zeng

University of North Carolina

CONTENTS

25.1	Introduction	511
25.2	Notation and definitions	512
25.3	Estimation of multivariate CSHs and CIFs	513
25.4	Association analysis	514
25.5	Regression analysis	516
	25.5.1 Fine and Gray model	516
	25.5.2 Conditional approach using Fine and Gray model	517
	25.5.3 Marginal approach using Fine and Gray model	518
	25.5.4 Mixture model with random effects	518
	25.5.5 Alternative approaches	519
25.6	Example	519
25.7	Discussion and future research	520
	Bibliography	520

25.1 Introduction

Competing risks data are commonly encountered in clinical trials and observational studies, when subjects are subject to failure from one of distinct causes (Kalbfleisch and Prentice, 1980). In several applications, competing risks data cannot be considered as independent because of a clustered design, for instance in multicentre clinical trials and family studies. Two challenges arise naturally in the analysis of clustered competing risks data. First, methodologies for standard survival data cannot be applied to competing risks data directly without modifications. Secondly, appropriate procedures are needed to account for the correlations of event times among subjects from the same cluster.

The aim of this article is to provide an overview of the statistical analysis of clustered competing risks data. We first describe the data structure and necessary definitions of important concepts. We then focus on three major issues arising in the analysis of clustered competing risks data, namely: (1) estimation of the multivariate cause-specific hazard function and cumulative incidence function; (2) estimation and inference of properly defined measure of association among the clustered competing risks data; and (3) regression analysis to assess the covariate effects on the cause-specific hazard function and cumulative incidence function. Extensions and future research directions will be discussed.

25.2 Notation and definitions

Suppose that there are n independent clusters, with n_i subjects in the ith cluster. We assume that there are K possible distinct competing risks for each subject. Let T_{ij} be the time to failure for the jth subject ($j = 1, ..., n_i$) in the ith cluster, ϵ_{ij} the cause of the failure, and $\mathbf{Z}_{ij}$ a $p \times 1$ vector of covariates. The failure time is subject to right censoring. We assume non-informative censoring and denote by C_{ij} the censoring time. Suppose that $\Delta_{ij} = 0$ indicates that the failure time is censored and that $\Delta_{ij} = 1$ otherwise. Therefore, the observed data consist of triplets $\{X_{ij} \equiv \min(T_{ij}, C_{ij}), \epsilon_{ij}\Delta_{ij}, \mathbf{Z}_{ij}\}$ for $i = 1, ..., n$ and $j = 1, ..., n_i$. Throughout this chapter, we assume conditional independent censoring, i.e., (T_{ij}, ϵ_{ij}) and C_{ij} are independent given $\mathbf{Z}_{ij}$ for each i and j.

Consider a single observation $\{X \equiv \min(T, C), \epsilon\Delta, \mathbf{Z}\}$. Several approaches have been used to describe the distribution of T and ϵ and assess their relationship with covariates $\mathbf{Z}$. One approach is through the cause-specific hazard (CSH) function defined as

$$\lambda_{k,C}(t|\mathbf{Z}) = \lim_{h \to 0} \Pr(T \in [t, t+h), \epsilon = k|\mathbf{Z}, T \geq t)/h, \quad k = 1, ..., K,$$

where $\lambda_{k,C}(t|\mathbf{Z})$ is the rate of failure at time t from cause k. One can in turn define the cause-specific cumulative hazard function

$$\Lambda_{k,C}(t|\mathbf{Z}) = \int_0^t \lambda_{k,C}(u|\mathbf{Z}) du.$$

Another approach is to use the so-called "subdistribution" or "cumulative incidence function" (CIF)

$$F_k(t|\mathbf{Z}) = \Pr(T \leq t, \epsilon = k|\mathbf{Z}), \quad k = 1, ..., K.$$

Note that $F_k(t|\mathbf{Z})$ is not a proper distribution in the presence of all causes of failure as the probability that the subject will fail from cause k, $F_k(\infty|\mathbf{Z}) < 1$. The subdistribution hazard described by Gray (1988) is defined as

$$\begin{aligned}
\lambda_k(t|\mathbf{Z}) &= \lim_{h \to 0} \Pr\{T \in [t, t+h], \epsilon = k|T \geq t \cup (T \leq t \cap \epsilon \neq k), \mathbf{Z}\}/h \\
&= \{dF_k(t|\mathbf{Z})/dt\}/\{1 - F_k(t|\mathbf{Z})\} \\
&= -d \log\{1 - F_k(t|\mathbf{Z})\}/dt.
\end{aligned}$$

Unlike in the case of standard survival data, there is no direct relationship between the cause-specific cumulative incidence function and the cause-specific hazard function. To connect $F_k(t|\mathbf{Z})$ and $\Lambda_{k,C}(t|\mathbf{Z})$, one needs to define the following cumulative hazard function from any cause

$$\Lambda_C(t|\mathbf{Z}) = \sum_{k=1}^{K} \Lambda_{k,C}(t|\mathbf{Z}),$$

and the survival function

$$S(t|\mathbf{Z}) = \Pr(T > t|\mathbf{Z}) = \exp\{-\Lambda_C(t|\mathbf{Z})\}.$$

Simple calculus yields

$$F_k(t|\mathbf{Z}) = \int_0^t \lambda_{k,C}(u|\mathbf{Z}) S(u|\mathbf{Z}) du.$$

The above equation implies that one needs the cause-specific hazard function from each cause in order to calculate the cumulative incidence function for a particular cause.

Now consider two observations in a cluster $\{X_1 \equiv \min(T_1, C_1), \epsilon_1\Delta_1, \mathbf{Z}_1\}$ and $\{X_2 \equiv \min(T_2, C_2), \epsilon_2\Delta_2, \mathbf{Z}_2\}$. Cheng et al. (2007) generalized the definition of univariate cause-specific hazard functions to the bivariate setup and defined the cause-specific bivariate hazard function

$$\lambda_{kl}(s,t) = \lim_{(h_1,h_2)\downarrow 0} \Pr(T_1 \in [s, s+h_1], \epsilon_1 = k, T_2 \in [t, t+h_2], \\ \epsilon_2 = l | T_1 \geq s, T_2 \geq t)/(h_1 h_2),$$

which can be interpreted as the instantaneous rate of a double failure of subject 1 from cause k event and subject 2 from cause l event at time (s, t) given that the subjects were at risk at times s and t. The bivariate cumulative cause-specific hazard function is then defined as

$$\Lambda_{kl}(s,t) = \int_0^s \int_0^t \lambda_{kl}(u,v)dudv.$$

The bivariate cumulative incidence function is defined as

$$\begin{aligned} F_{kl}(s,t) &= \Pr(T_1 \leq s, D_1 = k, T_2 \leq t, D_2 = l) \\ &= \int_0^s \int_0^t \lambda_{kl}(u,v) S(u-, v-)dudv, \end{aligned}$$

where $S(u, v) = \Pr(T_1 > u, T_2 > v)$ is the overall bivariate survival function. Note that $S(u, v)$ involves the bivariate cause-specific hazard functions for all combinations of causes (k, l). Hence, there is no direct relationship between $\Lambda_{kl}(s, t)$ and $F_{kl}(s, t)$.

25.3 Estimation of multivariate CSHs and CIFs

One important issue in clustered competing risks data analysis concerns the nonparametric estimation of multivariate cause-specific hazard and cumulative incidence functions. Most existing work focuses on the analysis of bivariate competing risks data. Particularly, Cheng et al. (2007) derived nonparametric estimators for Λ_{kl} and F_{kl} defined in Section 25.2 without making any assumptions about the dependence of the risks. We describe the nonparametric estimators below.

Consider a pair of observations $\{X_1 \equiv \min(T_1, C_1), \epsilon_1\Delta_1\}$ and $\{X_2 \equiv \min(T_2, C_2), \epsilon_2\Delta_2\}$. We first define counting processes $N_{jk}(s) = I(\epsilon_j\Delta_j = k)I(X_j \leq t)$ and at risk processes $H_j(t) = I(X_j \geq t)$, for $j = 1, 2$, and $k = 1, ..., K$. We then define $N(s, t) = N_{1k}(s)N_{2l}(t)$ and $H(s, t) = H_1(s)H_2(t)$. It can be shown that

$$\Lambda_{kl}(s,t) = \int_0^s \int_0^t \mathbb{E}N(du, dv)\{\mathbb{E}H(u,v)\}^{-1},$$

where $\mathbb{E}F$ denotes the expectation of a random function F. Naturally, the nonparametric estimator of $\Lambda_{kl}(s, t)$ can be obtained by replacing the theoretical quantities in the above equation by their empirical counterparts from the data, i.e.,

$$\widehat{\Lambda}_{kl}(s,t) = \int_0^s \int_0^t \mathbb{E}_n N(du, dv)\{\mathbb{E}_n H(u,v)\}^{-1},$$

where $\mathbb{E}_n F$ is the empirical process for F based n i.i.d. pairs of observations. The bivariate CIFs can in turn be estimated by

$$\widehat{F}_{kl}(s,t) = \int_0^s \int_0^t \widehat{S}(u-, v-)\widehat{\Lambda}(du, dv),$$

where $\widehat{S}(u-, v-)$ is an estimator of the overall bivariate survival function. The Breslow-type estimator of $S(u-, v-)$ can be given by $\exp\left\{-\sum_{k=1}^{K}\sum_{l=1}^{K}\widehat{\Lambda}_{kl}(u,v)\right\}$. Cheng et al. (2007) established the uniform consistency, weak convergence, and bootstrap validity for $\widehat{\Lambda}_{kl}$ and $\widehat{F}_{kl}$ through the use of empirical process theories and the functional Delta method.

25.4 Association analysis

The association analysis of multivariate survival data has been extensively studied in Hougaard (2000). However, there is relatively limited work on these analyses for multivariate competing risks data due to the complication that there may be within-subject dependence in the latent failure times leading to dependent censoring, in addition to the cross-subject dependence.

Almost all existing work on the association analysis with competing risks data concerns bivariate data with the exception of Cheng et al. (2010). Bandeen-Roche and Liang (2002) proposed to use Oakes (1989) cross-hazard ratio to assess cause-specific association among clustered times-to-disease-onset with bivariate competing risks data. The cause-specific hazard ratio is defined as

$$\theta_{CS}(s,t;k,l) = \frac{S(s,t)f(s,t;k,l)}{\{\int_t^\infty \sum_{h=1}^K f(s,v,k,h)dv\}\{\int_s^\infty \sum_{h=1}^K f(u,t;h,l)du\}},$$

where

$$f(s,t,k,l) = \lim_{(h_1,h_2)\downarrow 0} \Pr(T_1 \in [s,s+h_1], T_2 \in [t,t+h_2], , \epsilon_1 = k, \epsilon_2 = l)/(h_1 h_2)$$

is the cause-specific bivariate density and

$$S(x,y) = \int_x^\infty \int_y^\infty \sum_{k=1}^K \sum_{l=1}^K f(u,v,k,l)dudv$$

is the absolutely continuous joint survival function. Intuitively, when two events $(T_1, \epsilon_1 = k)$ and $(T_2, \epsilon_2 = l)$ are independent, $\theta_{CS}(s,t;k,l) = 1$ at all (s,t). In general, positive (negative) associations occur when $\theta_{CS}(s,t;k,l) > (<)1$. Bandeen-Roche and Liang (2002) developed a family of parametric models for θ_{CS}. A nonparametric method using ranks that localize Kendall's tau for the estimation of θ_{CS} was further developed by Bandeen-Roche and Ning (2008).

Cheng and Fine (2008) proposed an alternative representation of the cause-specific cross-hazard ratio for bivariate competing risks data by using bivariate hazard functions. Define

$$\lambda_{k0}(s,t) = \lim_{h_1 \downarrow 0} \Pr(T_1 \in [s,s+h_1], \epsilon_1 = k | T_1 \geq s, T_2 \geq t)/h_1,$$

$$\lambda_{0l}(s,t) = \lim_{h_2 \downarrow 0} \Pr(T_2 \in [t,t+h_2], \epsilon_2 = l | T_1 \geq s, T_2 \geq t)/h_2.$$

Cheng and Fine (2008) then defined a cause-specific association measure using these bivariate hazard functions

$$\xi(s,t;k,l) = \frac{\lambda_{kl}(s,t)}{\lambda_{k0}(s,t)\lambda_{0l}(s,t)}.$$

The above cause-specific association measure has appealing and intuitive interpretations.

The independence of two events $(T_1, \epsilon_1 = k)$ and $(T_2, \epsilon_2 = l)$will lead to $\xi(s,t;1,1) = 1$ for all s and t, whereas positive and negative associations occur when $\xi(s,t;1,1) > 1$ and $0 < \xi(s,t;1,1) < 1$, respectively. Furthermore, Cheng and Fine (2008) showed that the above association measure is identical to the cause-specific cross-hazard ratio θ_{CS} of Bandeen-Roche and Liang (2002).

The representation of the cause-specific association measure of Cheng and Fine (2008) leads to a simple nonparametric plug-in estimator. The basic idea to estimate $\xi(s,t;k,l)$ to replace the population parameters by their empirical counterparts. From the definition of $\xi(s,t;k,l)$, one can show that

$$\mathbb{E}N_{kl}(ds,dt)\mathbb{E}Y(s,t) = \xi(s,t;k,l)\mathbb{E}N_{k0}(ds,t)\mathbb{E}N_{0l}(s,dt),$$

where $N_{kl}(s,t) = I(X_1 \le s, \epsilon_1\Delta_1 = k, X_2 \le t, \epsilon_2\Delta_2 = l), N_{k0}(s,t) = I(X_1 \le s, \epsilon_1\Delta_1 = k, X_2 \ge t)$, $N_{0l}(s,t) = I(X_1 \ge s, X_2 \le t, \epsilon_2\Delta_2 = l)$, and $Y(s,t) = I(X_1 \ge s, X_2 \ge t)$. By assuming that $\xi(s,t;k,l)$ is a constant over a pre-defined region Ω, i.e., $\xi(s,t;k,l) \equiv \xi_\Omega$ for all $(s,t) \in \Omega$, the parameter ξ_Ω is a solution of

$$\int\int_\Omega w(s,t)\{\mathbb{E}N_{kl}(ds,dt)\mathbb{E}Y(s,t) - \xi_\Omega\mathbb{E}N_{k0}(ds,t)\mathbb{E}N_{0l}(s,dt)\} = 0,$$

where $w(s,t)$ is a known weight function satisfying

$$\int\int_\Omega w(s,t)dsdt = 1.$$

Plugging in the corresponding empirical processes into the above estimating equation, one can derive the following nonparametric estimator of ξ_Ω

$$\widehat{\xi_\Omega} = \frac{\int\int_\Omega \widehat{w}(s,t)\mathbb{E}_nN_{kl}(ds,dt)\mathbb{E}_nY(s,t)}{\int\int_\Omega \widehat{w}(s,t)\mathbb{E}_nN_{k0}(ds,t)\mathbb{E}_nN_{0l}(s,dt)},$$

where $\widehat{w}(s,t)$ is an estimator of $w(s,t)$ that satisfies

$$\int\int_\Omega \widehat{w}(s,t) = 1, ||\widehat{w} - w||_\infty \to 0, \sqrt{n}(\widehat{w} - w) = O_p(1).$$

Various choices of $w(s,t)$ can be considered in the estimation. The choice of Ω can be governed, in part, by scientific interest and in part by feasibility of estimation. A particular choice of $w(s,t) = \{\mathbb{E}Y(s,t)\}^{-2}$ leads to

$$\xi_\Omega = \left\{\int\int_\Omega \lambda_{kl}(u,v)dudv\right\}\left\{\int\int_\Omega \lambda_{k0}(u,v)\lambda_{0l}(u,v)dudv\right\}^{-1},$$

the merits of which without competing risks are discussed by Fan et al. (2000).

The consistency and asymptotic normality of $\widehat{\xi_\Omega}$ can be proved by applying the empirical process theories (van der Vaart and Wellner, 2000) and functional Delta method. When $\xi(s,t;k,l)$ is not constant in Ω, one can define a summary measure in Ω

$$\xi_\Omega = \frac{\int\int_\Omega w(s,t)\mathbb{E}N_{kl}(ds,dt)\mathbb{E}Y(s,t)}{\int\int_\Omega w(s,t)\mathbb{E}N_{k0}(ds,t)\mathbb{E}N_{0l}(s,dt)}.$$

The consistency and asymptotic normality of $\widehat{\xi_\Omega}$ as an estimator of ξ_Ω continue to hold.

Alternatively, Cheng et al. (2007) considered association measure derived from bivariate cause-specific hazard function Λ_{kl}

$$\phi(s,t;k,l) = \Lambda_{kl}(s,t)\{\Lambda_1^{(k)}(s)\Lambda_2^{(l)}(t)\}^{-1},$$

where $\Lambda_j^{(k)}$ is the univariate cause-specific cumulative hazard for cause k for subject j. Similarly, one can define the association measure based on the bivariate cumulative incidence function F_{kl}

$$\psi(s,t;k,l) = F_{kl}(s,t)\{F_1^{(k)}(s)F_2^{(l)}(t)\}^{-1},$$

where $F_j^{(k)}$ is the marginal cumulative incidence function for cause k for subject j. Both association measures ϕ and ψ can be estimated by plugging the nonparametric estimators of population parameters described in Section 25.3. When $(T_1, \epsilon_1 = k)$ and $(T_2, \epsilon_2 = l)$ are independent, both ϕ and ψ are equal to 1. However, it is worth to note that the independence for the cause-specific hazard function does not imply the independence for the corresponding cumulative incidence function and vice versa, because the cumulative incidence function involves cause-specific hazards for other causes.

Testing of association can be accomplished by Wald-type test statistics for testing whether the aforementioned measures of association are equal to 1. Functional Delta method and/or bootstrapping method can be used to estimate the variance of the test statistic.

All the aforementioned measures of association are defined for bivariate competing risks data. Cheng et al. (2010) extended the cause-specific association measure of Bandeen-Roche and Liang (2002) to accommodate more general competing risks family data with $n_i \geq 2$ and allow exchangeability among siblings. Cheng et al. (2010) derived the rank correlation estimator similar to that of Bandeen-Roche and Liang (2002) and established the large sample properties through the use of U-statistic theories.

25.5 Regression analysis

There are two main approaches in the regression analysis of clustered competing risks data, namely the conditional approach and the marginal approach. When the focus is on the covariate effects on the cause-specific hazards for a particular cause, techniques for standard clustered survival data can be applied by treating failures from other causes as censored observations. However, special treatments are needed for assessing the covariate effects on cumulative incidence functions.

25.5.1 Fine and Gray model

The most commonly used model for relating covariates and cumulative incidence function pertains to the so-called Fine and Gray model (Fine and Gray, 1999). The Fine and Gray model postulates the following proportional hazards expression for the subdistribution of a competing risk

$$\lambda_1(t|\mathbf{Z}) = \lambda_1(t)\exp(\boldsymbol{\beta}^T\mathbf{Z}),$$

where $\lambda_1(t|\mathbf{Z})$ is the conditional subdistribution hazard for cause 1 given covariates $\mathbf{Z}$ as defined in Section 25.3, $\lambda_1(t)$ is the completely unspecified baseline subdistribution hazard function and $\boldsymbol{\beta}$ is a $p\times 1$ vector of unknown regression parameters. The Fine and Gray model allows one to model the covariate effects on the cumulative incidence function directly.

Given n i.i.d. observations $\{X_i \equiv \min(T_i, C_i), \epsilon_i\Delta_i, \mathbf{Z}_i\}$, Fine and Gray (1999) proposed to estimate the regression parameters by maximizing a variation of the partial log-likelihood for standard survival data by adapting inverse probability of censoring weighting techniques (Robins and Rotnitzky, 1992)

$$\sum_{i=1}^{n} I(\epsilon_i \Delta_i = 1) \left\{ \boldsymbol{\beta}^T \mathbf{Z}_i - \log \left(\sum_{j=1}^{n} w_j(X_i) \exp(\boldsymbol{\beta}^T \mathbf{Z}_j) \right) \right\},$$

where $w_i(t) = I(X_i \geq t \cup \epsilon_i \Delta_i > 1)\widehat{G}(t)/\widehat{G}(X_i \wedge t)$ is the weight for subject i at time t, and $\widehat{G}$ is the Kaplan-Meier estimate of the survival function of the censoring time. Similar to the Breslow estimator for the baseline hazard function with standard survival data, the baseline cumulative subdistribution hazard can be estimated by

$$\widehat{\Lambda}_1(t) = \frac{1}{n} \sum_{i=1}^{n} \int_0^t \frac{1}{\widehat{S}^{(0)}(\widehat{\boldsymbol{\beta}}, u)} w_i(u) dN_i(u),$$

where $N_i(u) = I(T_i \leq u, \epsilon_i = 1)$, $\widehat{S}^{(0)}(\boldsymbol{\beta}, u) = \frac{1}{n} \sum_{i=1}^{n} w_i(u) Y_i(u) \exp(\boldsymbol{\beta}^T \mathbf{Z}_i)$, and $Y_i(u) = 1 - N_i(u-)$.

One key feature of the above partial log-likelihood is that subjects with failure from other causes remain in the risk at time t as long as $C_i > t$. If there is only a single cause of failure, the above partial likelihood reduces to the typical partial likelihood for the Cox model.

It is worth to note that although the Fine and Gray model is the standard model, it is somewhat difficult to interpret the regression coefficients within the context of the proportional hazard model for the subdistribution hazard function (Fine and Gray, 1999; Andersen and Keiding, 2012). In addition, there is no direct relationship between the effect of a covariate on the cause-specific hazard function and cumulative incidence function; see Fine and Gray (1999), Katsahian et al. (2006), and Beyersmann and Schumacher (2008).

25.5.2 Conditional approach using Fine and Gray model

To accommodate clustered competing risks data, Katsahian et al. (2006) extended Fine and Gray's model by including random effects or frailties in the subdistribution hazard. Specifically, the conditional subdistribution hazard of subject j in the ith cluster given the cluster-specific random effect u_i takes the form

$$\lambda_1(t|\mathbf{Z}_{ij}, u_i) = \lambda_1(t) \exp(\boldsymbol{\beta}^T \mathbf{Z}_{ij} + u_i).$$

The random effects u_i's are assumed to be Gaussian with mean 0 and variance θ. The mean of the frailties is fixed at 0 to ensure the identifiability of the model. A zero value of the variance parameter θ corresponds to the case that there is no within-cluster dependence. The conditional partial log-likelihood given the frailties is then expressed as

$$\sum_{i=1}^{n} \sum_{j=1}^{n_i} I(\epsilon_{ij} \Delta_{ij} = 1) \Bigg\{ \boldsymbol{\beta}^T \mathbf{Z}_{ij} + u_i - \log \left(\sum_{i'=1}^{n} \sum_{j'=1}^{n_i} w_{i'j'}(X_{ij}) \exp(\boldsymbol{\beta}^T \mathbf{Z}_{i'j'} + u_{i'}) \right) \Bigg\},$$

where $w_{ij}(t) = I(X_{ij} \geq t \cup \epsilon_{ij} \Delta_{ij} > 1)\widehat{G}(t)/\widehat{G}(Y_{ij} \wedge t)$. To obtain the estimators of the regression parameters, one can maximize the marginal partial likelihood by integrating out the frailties from the conditional partial likelihood. However, there is no closed-form solution for the marginal partial likelihood and numerical approximation is needed to calculate the marginal likelihood function. Katsahian et al. (2006) used the residual maximum likelihood approach to estimate the regression parameters. Alternatively, Katsahian and Boudreau (2011) proposed to use the penalized partial log-likelihood (PPLL) approach and used a Laplace approximation (Ripatti and Palmgren, 2000; Therneau et al., 2003) to the marginal partial likelihood. One advantage of the PPLL approach is that the maximization can be accomplished using existing statistical software for Gaussian frailty models. To test the presence of center effects, i.e., $H_0 : \theta = 0$ versus $H_1 : \theta > 0$, one can use the likelihood ratio test statistic, which asymptotically follows a 50:50 mixture of χ^2 with 1 d.f. and a degenerated random variable with mass at 0. Note that one can obtain the parameter estimates under H_0 by fitting the Fine and Gray model to the data directly.

The random effects model of Katsahian et al. (2006) allows subject-specific regression effects. However, as pointed by Scheike et al. (2010), the frailty parameter can only be identified from the marginal models and the marginal models depend on the distribution of the cluster-specific random effects.

25.5.3 Marginal approach using Fine and Gray model

The conditional approach of Katsahian et al. (2006) and Katsahian and Boudreau (2011) enables the assessment of both covariate effects and within-cluster associations. However, the conditional approach involves explicit distribution assumption of the frailties and computation can be intensive as in general there is no closed form for the marginal partial likelihood function and numerical approximation is required. Furthermore, the random effects model induces a restrictive positive correlation structure among failure times within a cluster, which may not always be true in practice.

Alternatively, Zhou et al. (2012) proposed a population average regression model to assess the marginal effects of covariates on the cumulative incidence function accounting for dependence across individuals within a cluster. In particular, they extended the Fine-Gray proportional hazards model for the subdistribution to the clustered competing risks setting. Under an independence working assumption, one can obtain the estimate of the cumulative incidence function and the covariate effects by following the Fine and Gray methodology described in Subsection 25.5.1 treating subjects within a cluster as independent observations. Zhou et al. (2012) then proposed to use Sandwich-type variance estimators to account for the correlation within clusters. In particular, the proposed variance estimators account for both correlations among failure times and correlations among censoring times within a cluster. The estimator of the cumulative incidence function can be obtained by plugging in the estimators of the regression parameters and the baseline cumulative subdistribution hazard in the expression of the cumulative incidence function. While one can derive the asymptotic properties of the estimator of the cumulative incidence function using functional Delta method, the variance is complicated and bootstrapping technique can be applied in practice (Zhou et al., 2012).

25.5.4 Mixture model with random effects

Naskar et al. (2005) proposed a semiparametric mixture model for analyzing clustered competing risks data. First, the cause of failure is assumed to have a multinomial distribution assuming a logistic model given by

$$\Pr(\epsilon_{ij} = l|\mathbf{Z}_{ij}) = \frac{\boldsymbol{\beta}_l^T \mathbf{Z}_{ij}}{\sum_{k=1}^K \exp(\boldsymbol{\beta}_k^T \mathbf{Z}_{ij})},$$

where $\boldsymbol{\beta}_l$ is the set of regression parameters corresponding to the l type of failure. Conditional on the failure from cause l and cluster-specific frailty u_{il} for the lth cause, the hazard function for the lth cause is modeled as

$$\lambda_{ijl}(t|\mathbf{Z}_{ij}) = \lambda_{0l}(t)u_{il}\exp(\boldsymbol{\gamma}_l^T\mathbf{Z}_{ij}),$$

where λ_{0l} is the unspecified baseline hazard function corresponding to the failure from cause l. Note that one may use different sets of covariates in the logistic model and the hazard model. To accommodate multimodality for the distribution of the frailties, Dirichlet process is used to model the unknown frailty distribution nonparametrically. Monte Carlo ECM algorithm is used to obtain the estimates of the parameters that assess the extent of the effects of the causal factors for failures of a certain type.

25.5.5 Alternative approaches

Scheike et al. (2008) and Scheike and Zhang (2008) studied a general semiparametric model for the cumulative incidence function, under which some covariates have time-varying effects and others have constant effects. This work was further extended by Scheike et al. (2010) to deal with clusters by providing robust standard errors through the standard GEE type approach. The approach of Scheike et al. (2010) also extended the work by Chen et al. (2008) that considered the k-sample case.

To overcome the drawback of the approach of Katsahian et al. (2006) that the marginal models depend on the distribution of the frailties, Scheike et al. (2010) considered an alternative random effects model, under which the marginal cumulative incidence functions follow a generalized semiparametric additive model that is independent of the frailty parameters. Furthermore, the frailities in this model reflect solely the amount of variation due to clusters as the frailties account only for correlations between competing risks failure times within clusters. Scheike et al. (2010) introduced the concept of cross-odds ratio to measure the degree of association between two cause-specific failure times within the same cluster. A two-stage estimation procedure was developed, where the marginal models are estimated in the first stage and the dependence parameters are estimated in the second stage. This approach has been implemented in the functions *random.cif* and *cor.cif* in the *mets* R package.

25.6 Example

Data are taken from the R package *crrSC* in which a random subsample of 400 patients were selected from the multicenter Bone Marrow transplantation data as described in Zhou et al. (2012). In this example, 400 patients were from 153 centers with center sizes ranging from 1 to 10. The event of interest was the time from transplantation to the occurrence of either acute or chronic Graft-versus-host disease (GvHD). Death and relapse free of GvHD are competing risks. Of the 400 patients, 194 patients experienced GvHD, 74 patients had relapse or died without GvHD, and 132 patients were censored. We considered two covariates, the source of stem cells and female donor to male recipient match was available. For the jth patient in the ith center, we define $Z_{ij1} = I$(source of stem cells is peripheral blood) and $Z_{ij2} = I$(female donor to male recipient). After excluding 17 patients with missing information on female donor to male recipient match, the final dataset contains 383 patients, of which 71 patients were female donor to male recipients and the source of stem cells for 206 patients were peripheral blood.

TABLE 25.1
Analysis of bone marrow transplantation data.

Covariate	$\widehat{\beta}$	$\exp(\widehat{\beta})$	$\mathrm{se}(\widehat{\beta})$	p-value
Zhou et al. (2012)				
Female donor to male recipients vs. others	0.289	1.346	0.148	0.050
Peripheral blood vs. BMT	-0.225	0.799	0.138	0.104
Scheike et al. (2010)				
Female donor to male recipients vs. others	0.361	1.435	0.170	0.034
Peripheral blood vs. BMT	-0.192	0.825	0.142	0.176

We analyzed the data using both the approach of Zhou et al. (2012) and the approach of Scheike et al. (2010). In particular, we assume both covariates have constant effects under the random effects model of Scheike et al. (2010). Table 25.1 presents the estimates for regression coefficients and the corresponding standard error estimates under both models. It appears that the gender match has a marginally significant effect on the cumulative incidence of GvHD with p-values of 0.05 and 0.034 under the approach of Zhou et al. (2012) and the approach of Scheike et al. (2010), respectively, whereas no significant effect of source of stem cell was detected. The estimated effects of female donor to male recipient match are 0.289 and 0.361 with the standard errors of 0.148 and 0.170 under the two approaches, respectively. The female donor to male recipients appeared to be at a higher risk of GvHD with the subdistribution hazard ratio of 1.34 and 95% confidence interval (1.00, 1.78) under the approach of Zhou et al. (2012) and the subdistribution hazard ratio of 1.435 and 95% confidence interval (1.02, 2.00) under the approach of Scheike et al. (2010). The frailty variance under the model of Scheike et al. (2010) was estimated at 0.407 with the standard error of 0.447 and the log-cross odds ratio was estimated at 0.334 with the standard error of 0.313, both suggesting non-significant within-cluster association between two cause-specific failure times.

25.7 Discussion and future research

In this section, we provide an overview of clustered competing risks data focusing on (1) estimation of multivariate CSHs and CIFs; (2) association analysis; and (3) regression analysis. Because of the page limit, it is not feasible to cover all important work in literature in this chapter, such as Shih and Albert (2010), Gorfine and Hsu (2011), Dixon et al. (2011), Logan et al. (2011), Dixon et al. (2012), Scheike and Sun (2012), Cheng and Fine (2012), among others. Particularly, Scheike and Sun (2012) studied the parametric regression modeling of the cross-odds ratio for multivariate competing risks data, which is a measure of association between the correlated cause-specific failure times within a cluster.

Future research directions may include (1) model diagnostics; (2) exploration of alternative survival models such as the proportional odds model, semiparametric transformation model, and cure rate models; (3) development of methods for more complicated family structures; and (4) development of user-friendly software.

Bibliography

Andersen, P. K. and Keiding, N. (2012), 'Interpretability and importance of functionals in competing risks and multistate models', *Statistics in Medicine* **31**, 1074–1088.

Bandeen-Roche, K. and Liang, K.-Y. (2002), 'Modeling multivariate failure time associations in the presence of a competing risk', *Biometrika* **89**, 299–314.

Bandeen-Roche, K. and Ning, J. (2008), 'Nonparametric estimation of bivariate failure time associations in the presence of a competing risk', *Biometrika* **95**, 221–232.

Beyersmann, J. and Schumacher, M. (2008), 'Time-dependent covariates in the proportional subdistribution hazards model for competing risks', *Biostatistics* **9**, 765–776.

Chen, B. E., Kramer, J. L., Greene, M. H. and Rosenberg, P. S. (2008), 'Competing risks analysis of correlated failure time data', *Biometrics* **64**, 172–179.

Cheng, Y. and Fine, J. (2012), 'Cumulative incidence association models for bivariate competing risks data', *Journal of the Royal Statistical Society, Series B* **74**, 183–202.

Cheng, Y. and Fine, J. P. (2008), 'Nonparametric estimation of cause-specific cross hazard ratio with bivariate competing risks data', *Biometrika* **95**, 233–240.

Cheng, Y., Fine, J. P. and Bandeen-Roche, K. (2010), 'Association analyses of clustered competing risks data via cross hazard ratio', *Biostatistics* **11**, 82–92.

Cheng, Y., Fine, J. P. and Kosorok, M. R. (2007), 'Nonparametric analysis of bivariate competing risks data', *Journal of the American Statistical Association* **102**, 1407–1416.

Dixon, S. N., Darlington, G. A. and Desmond, A. F. (2011), 'A competing risks model for correlated data based on the subdistribution hazard', *Lifetime Data Analysis* **17**, 473–495.

Dixon, S. N., Darlington, G. A. and Edge, V. (2012), 'Applying a marginalized frailty model to competing risks', *Journal of Applied Statistics* **39**, 435–443.

Fan, J., Prentice, R. and Hsu, L. (2000), 'A class of weighted dependence measures for bivariate failure time data', *Journal of the Royal Statistical Society, Series B* **62**, 181–190.

Fine, J. P. and Gray, R. J. (1999), 'A proportional hazards model for the subdistribution of a competing risk', *Journal of the American Statistical Association* **94**, 496–509.

Gorfine, M. and Hsu, L. (2011), 'Frailty-based competing risks model for multivariate survival data', *Biometrics* **67**, 415–426.

Gray, R. J. (1988), 'A class of k-sample tests for comparing the cumulative incidence of a competing risk', *The Annals of Statistics* **16**, 1141–1154.

Hougaard, P. (2000), *Analysis of Multivariate Survival Data*, Springer, New York.

Kalbfleisch, J. D. and Prentice, R. L. (1980), *The Statistical Analysis of Failure Time Data*, Wiley, New York.

Katsahian, S. and Boudreau, C. (2011), 'Estimating and testing for center effects in competing risks', *Statistics in Medicine* **30**, 1608–1617.

Katsahian, S., RescheRigon, M., Chevret, S. and Porcher, R. (2006), 'Analysing multicenter competing risks data with a mixed proportional hazards model for the subdistribution', *Statistics in Medicine* **25**, 4267–4278.

Logan, B. R., Zhang, M. J. and Klein, J. P. (2011), 'Marginal models for clustered time-to-event data with competing risks using pseudovalues', *Biometrics* **67**, 1–7.

Naskar, M., Das, K. and Ibrahim, J. G. (2005), 'Semiparametric mixture model for analyzing clustered competing risks data', *Biometrics* **61**, 729–737.

Oakes, D. (1989), 'Bivariate survival models induced by frailties', *Journal of the American Statistical Association* **84**, 487–493.

Ripatti, S. and Palmgren, J. (2000), 'Estimation of multivariate frailty models using penalized partial likelihood', *Biometrics* **56**, 1016–1022.

Robins, J. M. and Rotnitzky, A. (1992), Recovery of information and adjustment for dependent censoring using surrogate markers, *in* N. Jewell, K. Dietz and V. Farewell, eds., *AIDS Epidemiology-Methodological Issues*, Birkhauser, Boston, pp. 24–33.

Scheike, T. H., Sun, Y., Zhang, M. J. and Jensen, T. K. (2010), 'A semiparametric random effects model for multivariate competing risks data', *Biometrika* **97**, 133–145.

Scheike, T. H. and Zhang, M. J. (2008), 'Flexible competing risks regression modeling and goodness-of-fit', *Lifetime Data Analysis* **14**, 464–483.

Scheike, T. H., Zhang, M. J. and Gerds, T. (2008), 'Predicting cumulative incidence probability by direct binomial regression', *Biometrika* **95**, 205–220.

Scheike, T. and Sun, Y. (2012), 'On cross-odds ratio for multivariate competing risks data', *Biostatistics* **13**, 680–694.

Shih, J. H. and Albert, P. S. (2010), 'Modeling familial association of ages at onset of disease in the presence of competing risk', *Biometrics* **66**, 1012–1023.

Therneau, T. M., Grambsch, P. M. and Pankratz, V. S. (2003), 'Penalized survival models and frailty', *Journal of Computational and Graphical Statistics* **12**, 156–175.

van der Vaart, A. W. and Wellner, J. A. (2000), *Weak Convergence and Empirical Processes*, Springer, New York.

Zhou, B., Fine, J., Latouche, A. and Labopin, M. (2012), 'Competing risks regression for clustered data', *Biostatistics* **13**, 371–383.

26

Joint Models of Longitudinal and Survival Data

Wen Ye

Department of Biostatistics, University of Michigan

Menggang Yu

Department of Biostatistics and Medical Informatics, University of Wisconsin

CONTENTS

26.1 Introduction 523
26.2 The basic joint model 525
26.2.1 Survival submodel 526
26.2.2 Longitudinal submodel 526
26.2.3 Joint likelihood formulation and assumptions 527
26.2.4 Estimation 528
26.2.4.1 Maximum likelihood estimation 528
26.2.4.2 Bayesian methods 529
26.2.5 Asymptotic inference for MLEs 529
26.2.6 Example: an AIDS clinical trial 530
26.3 Joint model extension 532
26.3.1 Extension of survival submodel 532
26.3.1.1 Competing risks 532
26.3.1.2 Recurrent event data 533
26.3.1.3 Nonproportional hazards model 533
26.3.2 Extension of longitudinal submodel 534
26.3.2.1 Joint models with discrete longitudinal outcomes 534
26.3.2.2 Joint models with multiple longitudinal biomarkers 534
26.3.3 Variations of the link between survival and longitudinal submodels 535
26.3.4 Joint latent class models 536
26.4 Prediction in joint models 537
26.4.1 Prediction of future longitudinal outcome 537
26.4.2 Prediction of survival distribution 538
26.4.3 Performance of prediction accuracy 538
26.5 Joint model diagnostics 539
26.6 Joint model software 540
Bibliography 541

26.1 Introduction

In biomedical research, along with censored time-to-event data and baseline covariates, repeated measurements of biomarkers are also collected at a number of time points. A well-known example of this is HIV research (Wang and Taylor, 2001; Pawitan and Self, 1993), in which the biomarker CD4 lymphocyte count is measured at regularly scheduled intervals. In these studies patients are followed until an event, such as progression to AIDS or death. In addition to biomarker data, other covariates, such as treatment and demographic information, are recorded at baseline. In order to understand the natural history of the disease and to search for a "surrogate marker" for the time to AIDS or death, investigators are often interested in both modeling the progression of the CD4 count and estimating the relative risk of progressing to AIDS associated with different CD4 count levels. Data such as this are important for other studies including those of prostate cancer in which research interest lies in the association between level or rate of change of prostate specific antigen (PSA) and time to cancer recurrence (Ye et al., 2008*b*), as well as studies of cognitive aging (Proust et al., 2006) which investigate the relationship between cognitive functioning decline and time to dementia.

To illustrate the basic concept of joint modeling of longitudinal and time-to-event data, consider a sample of N subjects. For the ith subject, let T_i^* and C_i be the event and censoring times, respectively. Rather than observing T_i^*, we observe only $T_i = \min(T_i^*, C_i)$ along with an indicator δ_i that equals 1 if $T_i^* \leq C_i$ and 0 otherwise. Let $\bar{Y}_i^*(T_i) \equiv \{Y_i^*(u); 0 \leq u \leq T_i\}$ be the trajectory of a time-dependent covariate Y_i^*. Instead of observing the entire true history of the time-dependent covariate, only intermittent measurements with possible measurement error are observed on subject i, at n_i time points $t_{ij} (j = 1, ..., n_i)$:

$$Y_i(t_{ij}) = Y_i^*(t_{ij}) + e_i(t_{ij}) \tag{26.1}$$

Here $e_i(t_{ij})$ represents the measurement error.

As shown in the above examples, research interests often lie in both capturing change in patterns in the longitudinal data, including relationship between such change and relevant baseline factors $E\{Y_i^*(t)|Z_i\}$, and elucidating the association structure between the longitudinal data and event times through modeling the hazard function of T^* as

$$\lim_{dt \to 0} Pr\{t \leq T_i^* < t + dt | T_i \geq t, \bar{Y}_i^*(u), Z_i\}/dt. \tag{26.2}$$

where $\bar{Y}_i^*(u) = \{Y_i^*(s); 0 \leq s \leq u\}$ denotes the history of the true and unobserved longitudinal process up to time point u, and Z_i is a vector of fully observed time-independent covariates.

Both objectives in such studies pose difficulties. For instance, in HIV studies, subjects with sharper rates of CD4 decline may have more serious disease and higher risk of developing AIDS or death, and in turn are more likely to develop AIDS or die earlier and have fewer CD4 count measurements. This is known as "nonignorable" missing data (Little and Rubin, 2002). Attempts to make inferences for the longitudinal CD4 count process using ordinary longitudinal models can lead to biased estimates. To study the association between CD4 count and time to AIDS or death, if the true CD4 count trajectory $\bar{Y}_i^*(T_i)$ were known, the time to AIDS or death could be described by the proportional hazards model (Cox, 1972), with the CD4 count as a time-dependent covariate. However, there are two additional challenges. First, the CD4 count is typically subject to substantial measurement error due to both laboratory error and short-term biological variability (e.g., the coefficient of variation was approximately 50%, according to Tsiatis et al. (1995)). A commonly used naïve method substitutes the observed time-dependent covariate for the true covariate values in the proportional hazard model. Rabout (1991) argued that this method leads to biased estimates of the relative risk parameters, and the extent of the bias is proportional to the variance of the measurement error (Prentice, 1982). Second, the CD4 count is usually only measured intermittently, and thus may not be collected at the time when a failure event occurs for members in the corresponding risk set. To solve this problem, in early AIDS studies the

"Last Value Carried Forward (LVCF)" method was often used. It simply pulls forward the nearest preceding value of the marker and treats it as if it were the current value of the marker at the failure time. This naïve method not only ignores the measurement error, but also ignores the possible trend of the marker, yielding very poor imputes for the missing marker values and leading to poor estimates of relative risk parameters.

To eliminate the bias due to measurements error, two-stage analysis (Tsiatis et al., 1995; Bycott and Taylor, 1998; Dafni and Tsiatis, 1998; Ye et al., 2008*b*) using smoothing techniques is the earliest approach. The idea is to first use regression calibration to capture the underlying trend of the covariate over time and impute the missing covariate values for subjects in all risk sets, and then use these imputes in the disease risk model as if they were appropriate $Y_i^*(t)$ at the time of each failure. Although two-stage approaches can easily be implemented using standard software for mixed effects and proportional Cox models, under many circumstances, they produced biased results and underestimate the uncertainty of the risk parameters associated with the repeated measures, especially when the collected longitudinal data are relatively sparse.

To better model the longitudinal and time-to-event data, in the past two decades a class of statistical models known as joint models has been developed. In this chapter, we introduce the joint modeling framework. The idea of jointly modeling longitudinal measurements and time-to-event data can be traced back to Little's pattern-mixture models (Little, 1993) for multivariate incomplete data. Also relevant work is by Wu and Carroll (1988) developed a joint likelihood-ratio test on the basis of a latent variable model whereby they described longitudinal data with linear random effect and the right censoring with a probit model. Although their primary interest is to analyze the incomplete longitudinal data, this idea has been adopted and extended by other researchers to assess the dependence of failure time on a time-dependent covariate process.

Faucett and Thomas (1996) and Wulfsohn and Tsiatis (1997) laid the standard and basic framework for joint modeling of longitudinal and time-to-event data. Following their pioneer work, this area of research has received remarkable attention in the methodological literature over the past decade. By jointly maximizing a likelihood $[T_i, Y_i|Z_i]$ from both the covariate process and time-to-event data, information from both sources can be used to obtain parameter estimates for the two processes simultaneously. By doing so, the informative drop-out can be adjusted by borrowing information from the survival model, and, conversely, unbiased information from the longitudinal covariate can be incorporated to the survival model. This process allows the dependence of the failure time on the longitudinal marker to be correctly assessed. In addition to correcting biases, joint modeling can improve the efficiency of parameter estimates in either part of the model, because extra information is being used (Tsiatis and Davidian, 2004; Wu et al., 2012).

This chapter describes how joint modeling can be used to make appropriate inferences about the effect of a longitudinal process on survival. The models discussed in this chapter can also be used for modeling the longitudinal data with informative dropouts. For more information on this, interested readers can refer to Chapter IV in Fitzmaurice et al. (2008).

26.2 The basic joint model

Three broad classes of joint models have been discussed in the literature: selection models, pattern-mixture models and latent variable models (Xu and Zeger, 2001). In selection models, which is useful in predicting the survival time T_i^*, the likelihood is formulated as $[T_i|Y_i, Z_i][Y_i|Z_i]$ (Henderson et al., 2000). In pattern-mixture models (Little, 1993; Hogan and Laird, 1997*a*,*b*; Pawitan and Self, 1993), the decomposition $[Y_i|T_i, Z_i][T_i|Z_i]$ is used. Pattern-mixture models are employed to facilitate inference for the longitudinal process $Y^*(u)$.

Here we focus on latent variable models, also known as "shared parameter models," which are the mostly widely used and are good for inference for both marker and survival processes. The central feature of latent variable models is to introduce some unobserved latent variables η_i and assume that the marker process and failure times are conditionally independent given η_i and other baseline covariates Z_i. In latent variable models, likelihood is formulated in terms of

$$[T_i, Y_i|Z_i] = \int [T_i, Y_i|\eta_i, Z_i]d[\eta_i|Z_i] = \int [T_i|\eta_i, Z_i][Y_i|\eta_i, Z_i]d[\eta_i|Z_i] \quad (26.3)$$

To specify a latent variable joint model, we need to first specify a longitudinal submodel $[Y_i|\eta_i, Z_i]$, a survival submodel $[T_i|\eta_i, Z_i]$, and the latent variables η_i, which describe the association between the two submodels.

26.2.1 Survival submodel

The standard approach for the survival submodel is the proportional hazard model (Cox, 1972)

$$\lambda_i(t) = \lambda_0(t)exp\{\alpha^T q(t, \eta_i) + \gamma^T Z_i\} \quad (26.4)$$

where $\lambda_0(t)$ denotes the baseline risk function, and $q(\cdot, \cdot)$ is a vector function of time and the latent variables η_i. γ^T are a vector of regression coefficients associated with baseline covariates Z_i. The form of $q(\cdot, \cdot)$ used in the vast majority of joint models is $Y_i^*(t)$, which postulates that the risk of event at time t depends only on the current value of the longitudinal process.

Several options for specifying the baseline hazard function $\lambda_0(t)$ have been proposed in the literature, including leaving $\lambda_0(t)$ completely unspecified as in standard survival analysis (Chapter 1), using risk function corresponding to a known parametric distribution (e.g., the Weibull, the log-normal, the Gompertz, and the Gamma distributions), or using parametric but flexible specifications such as the piecewise-constant and regression spline approach (Rosenberg, 1995).

26.2.2 Longitudinal submodel

A standard approach to modeling repeated measures is to characterize the unobserved true marker process $Y_i^*(t)$ using a vector of subject specific random effects $\mathbf{b}_i$:

$$Y_i^*(t) = \mathbf{b}_i^T f_i(t); \qquad \mathbf{b}_i|Z_i \sim \mathcal{N}(\mu_{\mathbf{b}}, \Sigma) \quad (26.5)$$

where $f_i(t)$ is a vector function of time t with q elements. This equation and (26.1) specify a standard linear mixed effect model (Laird and Ware, 1982) where $\varepsilon(t_{ij}) \overset{\text{iid}}{\sim} \mathcal{N}(0, \sigma_\varepsilon^2)$ are independent of $\mathbf{b}_i$. In their pioneer work, Faucett and Thomas (1996) and Wulfsohn and Tsiatis (1997) considered a simple linear random effects model for $f_i(t)$ to capture true CD4 trajectories.

$$Y_i^*(t) = b_{0i} + b_{1i}t \quad (26.6)$$

where b_{0i} and b_{1i} are the subject-specific random intercept and slope, respectively. To capture nonlinear subject-specific longitudinal profiles, more flexible forms of $f_i(t)$, such as polynomials and splines, have also been considered (Rizopoulos et al., 2009; Brown et al.,

2005; Ding and Wang, 2008; Dimitris and Pulak, 2011). In this type of models, the random effects $\mathbf{b}_i$ can be viewed as the latent variables η_i in (26.3) that are time-invariant and determine the complete smoothing trajectory of the evolution of the subject-specific marker process.

An alternative approach to modeling repeated measures is to use stochastic models (Taylor et al., 1994; Lavalley and De Gruttola, 1996; Wang and Taylor, 2001; Henderson et al., 2000)

$$Y_i^*(t) = \mathbf{b_i}^T f_i(t) + U_i(t) \tag{26.7}$$

where $\mathbf{b}_i^T f_i(t)$ is defined similarly as in (26.5), and $U_i(t)$ is a stochastic process (e.g., integrated Ornstein-Uhlenbeck (IOU) in Wang and Taylor (2001) and Brownian motion in Henderson et al. (2000)) capturing additional serial correlation between measurements not captured by random effects. This model can also be used to capture highly nonlinear shapes of subject-specific trajectories. Tsiatis and Davidian (2004) pointed out that the choice between (26.7) and (26.5) is to some extent a philosophical issue and often relies on the modeler's belief about the underlying biological mechanism (For more complete understanding, see their insightful discussion in Tsiatis and Davidian (2004)).

26.2.3 Joint likelihood formulation and assumptions

The joint likelihood can be formulated using (26.3). For ease of exposition, we consider the joint model defined by (26.4) with $q(t, \eta_i) = Y_i^*(t)$, (26.5) and (26.1). Under this setup the complete data log-likelihood conditional on baseline covariates Z_i is

$$l_c = \sum_{i=1}^{N} \log \left[h(T_i, \delta_i|\mathbf{b}_i, \lambda_0, \alpha, \gamma, Z_i) \Big\{ \prod_{j=1}^{n_i} h(Y_{ij}|\mathbf{b}_i, \sigma^2) \Big\} h(\mathbf{b}|\mu_{\mathbf{b}}, \Sigma, Z_i) \right] \tag{26.8}$$

where $h(Y_{ij}|\mathbf{b}_i, \sigma^2) = (2\pi\sigma^2)^{-1/2} \exp\{-(Y_{ij} - \mathbf{b}_i^T f_i(t_{ij}))^2/2\sigma^2\}$ is the density of Y_{ij} at time t_{ij}, $h(\mathbf{b}_i|\mu_{\mathbf{b}}, \Sigma) = (2\pi|\Sigma|)^{-1/2q} \exp\{-(\mathbf{b}_i - \mu_{\mathbf{b}})^T \Sigma^{-1} (\mathbf{b}_i - \mu_{\mathbf{b}})/2\}$ is the density of the random effects $\mathbf{b}_i$, and $h(T_i, \delta_i|\mathbf{b}_i, \lambda_0, \alpha, \gamma, Z_i)$ is the density for the survival data and can be written as

$$\left[\lambda_0(T_i) \exp\{\alpha \mathbf{b}_i f_i(T_i) + \gamma^T Z_i\} \right]^{\delta_i} \exp\left[-\int_0^{T_i} \lambda_0(u) \exp\{\alpha \mathbf{b}_i f_i(u) + \gamma^T Z_i\} du \right]$$

Since random effects $\mathbf{b}_i$ are latent and not observed, the log-likelihood for the observed data is

$$\ell_o = \sum_{i=1}^{N} \log \left[\int h(T_i, \delta_i|\mathbf{b}_i, \lambda_0, \alpha, \gamma, Z_i) \Big\{ \prod_{j=1}^{n_i} h(Y_{ij}|\mathbf{b}_i, \sigma^2) \Big\} h(\mathbf{b}_i|\mathbf{b}, \Sigma, Z_i) d\mathbf{b}_i \right] \tag{26.9}$$

To justify the validity of inference based on (26.9) the following assumptions are typically used in the literature:

1. The observed marker measurement $Y_i(t)$ only depends on the observed history $\{Y_{i1}, ...Y_{ik}\}$(where $t_{ik} < t$) and latent random effects $\mathbf{b}_i$ but not additionally on the unobserved future event time T_i.

2. Given the observed marker history $\{Y_{i1}, ...Y_{ik}\}$(where $t_{ik} < t$) and covariates Z_i, the risk of being censored and chance of observing the error-prone marker process $Y_i(t)$ at time t do not depend on the latent marker process $Y_i^*(t)$(which can be summarized by $\mathbf{b}_i$) and the unobserved future event time T_i^*.

Practically speaking, the first assumption implies that given the latent subject characteristics associated with prognosis, measurement errors do not provide additional information on the risk of event. The second assumption implies that decisions on whether a subject withdraws from the study or appears at the clinic for a longitudinal measurement depend on the observed past history (longitudinal measurements and baseline covariates), but there is no additional dependence on underlying, latent subject characteristics associated with prognosis or the risk of event. For details on formal elucidation of these assumptions refer to Tsiatis and Davidian (2004).

26.2.4 Estimation

The two most commonly used methods for estimation in the joint analysis of longitudinal and time-to-event data are maximum likelihood estimation (MLE) and Bayesian-Markov chain Monte Carlo (MCMC).

26.2.4.1 Maximum likelihood estimation

Maximization of the observed log-likelihood function (26.9) with respect to model parameters can be achieved using standard algorithms such as the Expectation-Maximization (EM) algorithm (Dempster et al., 1977) or the Newton-Raphson algorithm (Press et al., 1992).

Viewing the unobserved latent variables $\mathbf{b}_i$ as missing data, it is natural to consider using the EM algorithm to maximize function (26.9). For simplicity, we combine all parameters into one vector $\boldsymbol{\theta}$. The intuition underlying the EM algorithm is that the log-likelihood corresponding to the complete data (26.8), often in closed form, is typically much easier to maximize. To take advantage of this feature the EM algorithm iterates between two steps, the Expectation (E) step and the Maximization (M) step. The E step yields the conditional expectation of each function of $\mathbf{b}_i$ appearing in the complete data log-likelihood function l_c (26.8), given the observed data $(Y_i, Z_i, T_i, \delta_i)$, and using the current estimates of $\boldsymbol{\theta}$, namely $E(g(\mathbf{b}_i)|Y_i, Z_i, T_i, \delta_i, \boldsymbol{\theta}^{m-1})$ in the mth iteration. In the M step, each function of $\mathbf{b}_i$ appearing in Equation (26.8) is replaced by its conditional expectation, and maximized to update $\boldsymbol{\theta}$.

The M step is straightforward, even when a nonparametric baseline function is used in survival submodel (26.4). In the E step, the main computation involves finding the conditional expectation of each function of $\mathbf{b}_i$. In the literature these expectations are routinely evaluated using the numerical integration technique of Gaussian quadrature (Press et al., 1992; Wulfsohn and Tsiatis, 1997; Henderson et al., 2000; Ratcliffe et al., 2004) and Monte Carlo method (Law et al., 2002).

However, when the dimensionality of the random effects $\mathbf{b}_i$ increases (e.g., when spline functions are used in the longitudinal submodel (26.5) to capture nonlinear subject specific trajectories) the computation of the E-step can become highly demanding and thus becomes a computational bottleneck for fitting joint models. For this reason, inspired by numerical techniques proposed for approximating the integrated log-likelihood in generalized nonlinear mixed models (Pinheiro et al., 1995), some authors turned to the Newton-Raphson algorithm using Laplace approximations (Ye et al., 2008*a*; Rizopoulos et al., 2009; Guo and Carlin, 2004; Vonesh et al., 2006). Ye et al. (2008*a*) and Rizopoulos et al. (2009) showed that their methods are computationally more efficient compared to the EM algorithm using Gaussian quadrature or Monte Carlo. To further decrease the computational burden, Rizopoulos (2012*a*) proposed the use of pseudo-adaptive quadrature where posterior distribution of the random effects obtained from separate fit of the longitudinal model is used to appropriately re-scale the subject-specific integrand. However, note that the order of standard Laplace approximation error is $O(n_i^{-1})$, which cannot be made arbitrarily accurate, which means that the approximation might not work well when n_i is small.

26.2.4.2 Bayesian methods

The Bayes approach has also been proposed for parameter estimation for joint models (Faucett and Thomas, 1996; Wang and Taylor, 2001; Brown and Ibrahim, 2003*a*,*b*; Ibrahim et al., 2004; Guo and Carlin, 2004; Yu et al., 2004; Brown et al., 2005; Chi and Ibrahim, 2006).

Denote $h(\boldsymbol{\theta}|\boldsymbol{\theta}_0)$ the prior distribution of $\boldsymbol{\theta}$, given some known hyperparameters $\boldsymbol{\theta}_0$. The idea is to estimate the joint posterior distribution of all unknown parameters and latent random effects

$$h(\boldsymbol{\theta}, \mathbf{b}_i|Y_i, Z_i, T_i, \delta_i; i = 1, \ldots, N) \tag{26.10}$$
$$\propto \prod_{i=1}^{N} \left[h(T_i, \delta_i|\mathbf{b}_i, \lambda_0, \alpha, \gamma, Z_i) \left\{ \prod_{j=1}^{n_i} h(Y_{ij}|\mathbf{b}_i, \sigma^2) \right\} h(\mathbf{b}_i|\mathbf{b}, \Sigma, Z_i) \right] h(\boldsymbol{\theta}|\boldsymbol{\theta}_0)$$

using a Markov chain Monte Carlo algorithm, often the Gibbs sampler (Gelfand and Smith, 1990). The Gibbs sampler involves iteratively sampling from the full conditional distribution of each parameter given the current assignment of all other parameters and data. When the process converges, the results can be described in terms of means, medians and variance of the Gibbs samples, and graphs of the empirical distributions.

A main challenge in the maximum likelihood approach for joint models is the complex numerical integration and maximization of the likelihood over a large number of parameters, especially when the dimension of the random effects is not small or when the submodels are expanded to nonlinear or non-normal forms (Section 26.3). In some situations, such as when the convergence of the EM is an issue and using MLE method is impossible, Bayesian methods are effective for fitting the joint model. Two other advantages also motivate investigation of a Bayesian alternative. First, Bayesian methods can borrow additional information from similar studies or from experts and incorporate this information in the current analysis, in the forms of prior distributions for the current model parameters. Second, Bayesian methods also permit full and exact posterior inference for any parameter or predictive quantity of interest.

26.2.5 Asymptotic inference for MLEs

When parameteric functions are used to model baseline hazard $\lambda_0(t)$ in the survival submodel, it is straightforward that parameter estimates obtained by maximizing the observed joint log-likelihood (26.9) have asymptotic properties of MLEs (consistency, asymptotic normality, and efficiency), and the standard likelihood inference tests, i.e., the Wald, score, and likelihood ratio tests are directly available. The covariance matrix of parameter estimates can be calculated as the negative of the inverse of the Hessian matrix which is calculated as the second derivative of ℓ_o (26.9) with respect to $\boldsymbol{\theta}$. In practice, numerical derivative routines such as the forward or the central difference approximations (Press et al., 1992) are often used.

When an unspecified baseline risk function is defined for $\lambda_0(t)$ in the survival submodel, calculation of the likelihood is based on nonparameteric maximum likelihood arguments in which the unspecified cumulative incidence function $\Lambda_0(t) = \int_0^t (\lambda_0(s)ds$ is replaced by a step function with jumps at the unique event times (van der Vaart, 1998). Zeng and Cai (2005) rigorously proved the consistency and efficiency of the nonparameteric maximum likelihood estimators (NPMLEs) proposed in the literature (Wulfsohn and Tsiatis, 1997; Henderson et al., 2000) and derived their asymptotic distribution.

An argument similar to that of Parner (1998) can be used to show that the inverse of the observed information matrix is a consistent estimator of the covariance of the NPMLEs $\hat{\boldsymbol{\theta}}$. To obtain the observed information matrix, a few methods have been proposed in the literature. Law et al. (2002) applied the formula from Louis (1982) and calculated the observed data information matrix by extracting the information for the missing data (unobserved latent random effects under joint modeling framework) from the information for the complete data. Alternatively, Lin et al. (2002*b*) considered the formula in McLachlan and Krishnan (1997) and approximated the observed information matrix by the observed empirical information matrix $I(\hat{\boldsymbol{\theta}}) = \sum_{i=1}^{N} s(Y_i, Z_i, T_i, \delta_i; \hat{\boldsymbol{\theta}}) s^T(Y_i, Z_i, T_i, \delta_i; \hat{\boldsymbol{\theta}})$, where $s(Y_i, Z_i, T_i, \delta_i; \hat{\boldsymbol{\theta}})$ is the observed score which equals the conditional expectation of the complete-data score $s(Y_i, Z_i, T_i, \delta_i; \hat{\boldsymbol{\theta}}) = E[\frac{\partial}{\partial \boldsymbol{\theta}^T}\{\ell_c(\boldsymbol{\theta})\}|Y_i, Z_i, T_i, \delta_i]|_{\boldsymbol{\theta}=\hat{\boldsymbol{\theta}}}$. In practice the observed information matrix is typically of very high dimension due to the large number of sub-parameters in the unspecified $\lambda_0(t)$. As such, the calculation of the inverse of the matrix may be intimidating and often unfeasible.

To overcome the above limitation of using the observed information matrix of NPLMEs, variance estimation methods using profile likelihood have been proposed. In particular, since inferences for the baseline hazard function $\lambda_0(t)$ are of less interest in most applications, obtaining standard errors for the remaining parameter estimates $\hat{\boldsymbol{\theta}}_{-\lambda} = (\alpha, \gamma, \sigma^2, \Sigma)$ are often the main focus. One profile likelihood approach is proposed by Wulfsohn and Tsiatis (1997) and adopted by others (Song et al., 2002; Henderson et al., 2002; Ratcliffe et al., 2004). This method obtains the variance-covariance matrix of $\hat{\boldsymbol{\theta}}_{-\lambda}$ by inverting the information matrix of a profile log-likelihood $p\ell_c(\hat{\boldsymbol{\theta}}_{-\lambda}, \hat{\lambda}_0(\hat{\boldsymbol{\theta}}_{-\lambda}))$ based on the complete data log-likelihood l_c (26.8) in the M step of EM algorithm, in which each function of $\mathbf{b}_i$ is replaced by its conditional expectations to simplify computation. However, Hsieh et al. (2006) showed both theoretically and empirically that such replacement may lead to underestimation of the standard error of the parameter estimate for risk coefficients. Alternatively, Zeng and Cai (2005) showed that by treating $\lambda_0(t)$ as nuisance parameters, a profile likelihood function $p\ell_o(\hat{\boldsymbol{\theta}}_{-\lambda}, \hat{\lambda}_0(\hat{\boldsymbol{\theta}}_{-\lambda}))$ based on the observed log-likelihood ℓ_o (26.9) can be used to give a consistent estimator for the asymptotic variance of the regression coefficients. The profile likelihood approach based on the complete likelihood is computationally more convenient because the related information matrix can be readily derived from the complete data score function in the M-step. In contrast, the profile likelihood approach using the observed likelihood requires calculating the negative second-order difference of $p\ell_o(\hat{\boldsymbol{\theta}}_{-\lambda})$ at $\theta = \hat{\theta}$ numerically, and thus is computationally more intensive.

26.2.6 Example: an AIDS clinical trial

We use an AIDS study (Abrams et al., 1994) to illustrate the idea of joint modeling and demonstrate its properties. The data used for analysis in this chapter is available in the R joint model package *JM* (Rizopoulos (2012*b*); for more detail see Section 26.6). Interested readers can play with this data to get hands on experience with joint models.

This study was a multi-center, open-label, randomized clinical trial designed to compare two antiviral treatments, didanosine (ddI) and zalcitabine (ddC), in 467 HIV patients in whom zidovudine treatment had failed or intolerance of the drug had developed. After randomization, these patients were followed until the time of death or the end of the study, and their absolute CD4 cell counts were scheduled to be measured at baseline, 2, 6, 12, and 18 months. The primary goal of this study was to determine whether ddI or ddC was a better treatment in these patients using time-to-death as the primary endpoint. After

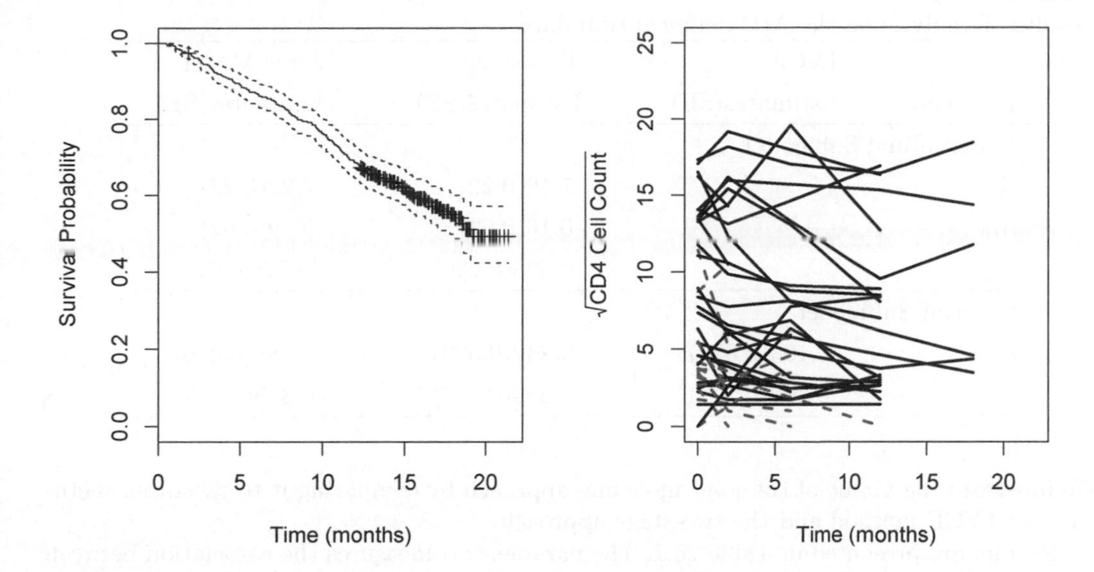

FIGURE 26.1
Left panel: Kaplan-Meier estimate of the survival function for time-to-death in the AIDS trial. The dashed lines correspond to 95% pointwise confidence intervals. Right panel: Longitudinal trajectories for square root CD4 counts for 50 randomly selected patients in the AIDS study. Dashed lines represent patients who died during the study and solid lines indicate censored patients. (Reprinted with permission from Rizopoulos (2012*b*).)

a median of follow-up of 16 months, 188 (40.3%) patients had died. For demonstration purposes, our focus here examines how CD4 count was associated with the risk of death in these advanced HIV patients.

Figure 26.1 shows the Kaplan-Meier estimate of the survival function for time to death and the longitudinal trajectories of the square root of the CD4 cell count for a randomly selected sample of 50 patients. Among the longitudinal trajectories, the red lines represent patients who had died during the study while black lines indicate patients whose death time was censored, suggesting subjects with lower CD4 counts tended to have higher risk of death.

For the AIDS trial data, we assume the following submodels. For the longitudinal submodel, we assume a linear mixed effects model

$$
\begin{aligned}
Y_i^*(t) &= b_{0i} + b_{1i}t \\
b_{0i} &= \beta_{00} + r_{0i} \\
b_{1i} &= \beta_{10} + \beta_{11} ddI_i + r_{1i} \\
(r_{0i}, r_{1i}) &\sim \mathcal{N}(\mathbf{0}, \Sigma)
\end{aligned}
$$

where ddI_i is an indicator which equals 1 for a subject who received ddI treatment, and equals 0 otherwise. For the time-to-death submodel, we assume the hazard of death at time t is a function of treatment and the true CD4 value at time t.

$$\lambda_i(t) = \lambda_0(t) exp\{\alpha Y^*(t) + \gamma^T ddI_i\}$$

TABLE 26.1
Results of analysis on the AIDS clinical trial data.

Parameter	LVCF Estimates(SE)	Two-stage Estimates(SE)	Joint Model Estimates(SE)
Longitudinal Submodel			
β_{00}		7.19(0.22)	7.22(0.22)
β_{10}		-0.16(0.02)	-0.19(0.02)
β_{11}		0.028(0.030)	0.012(0.030)
Survival Submodel			
α	-0.193(0.024)	-0.242(0.029)	-0.288(0.036)
γ	0.309(0.147)	0.344(0.147)	0.335(0.157)

We illustrate the virtue of the joint modeling approach by comparing it to two other methods, the LVCF method and the two-stage approach.

Results are presented in Table 26.1. The parameter α measures the association between the CD4 counts and risk of death. Estimates obtained from fitting the joint model shows a $exp(-\alpha) = 1.33$ (95%CI: 1.24: 1.43) fold increase in risk of death associated with one unit of decrease in square root of CD4 level. In contrast, estimates obtained from the LVCF and two-stage methods show a $exp(-\alpha) = 1.21$ (95%CI: 1.16: 1.27) fold and a $exp(-\alpha) = 1.27$ (95%CI: 1.20: 1.35) fold increase in risk of death associated with one unit of decrease in square root of CD4 level, respectively. This is consistent with results of simulation studies reported in the joint modeling literature such as Tsiatis and Davidian (2004) and Wu et al. (2012). Although the results from LVCF and two-stage methods are often attenuated and biased, joint model approach typically gives unbiased results and larger estimate of the risk coefficient. In addition, simulation studies in the above two articles also showed that joint models are more efficient than the LVCF and two-stage method for estimating the risk coefficients related to the longitudinal biomarkers.

26.3 Joint model extension

26.3.1 Extension of survival submodel

26.3.1.1 Competing risks

Standard methods for joint modeling of longitudinal and survival data allow for one event with a single mode of failure and an assumption of independent censoring. When several reasons can explain the occurrence of an event or some informative censoring occurs, this is known as "competing risks." For example, Williamson et al. (2008) investigated the effect of titration on the relative effects of two anti-epileptic drugs on treatment failure. In this study, during the course of treatment, patients may have experienced one of the two types of treatment failures, switching to an alternative drug because of inadequate seizure control or withdrawal from a treatment because of an unacceptable adverse effect.

Consider a general scenario under which each subject may experience one of K distinct failure types or could be right censored during follow-up. For such situations two types of general approaches have been proposed for the survival submodel (Elashoff et al., 2007; Williamson et al., 2008; Li et al., 2009; Yu and Ghosh, 2010; Li et al., 2010), namely cause-specific hazards model and mixture model as discussed in Chapter 6. The estimation of the joint model is based on the same principle introduced in Section 26.2.3 and Section 26.2.4, except that the construction of the survival model part of the likelihood function is different.

When the research interest focuses on the covariate effects on the marginal probability of the competing risk, the mixture model approach is appropriate. When the research interest is in the stochastic behavior of the competing risks process, it is natural to consider the cause-specific hazards models. Joint models with competing risk can also be used to account for informative censoring by treating it as one of the K types of failures.

26.3.1.2 Recurrent event data

Recurrent event data arise when study subjects experience multiple events during follow-up. For example, HIV patients may experience multiple episodes of infections and patients with chronic disease may be hospitalized multiple times. The observation of the events can be censored administratively or by a terminal (different type of) event such as death. Given data of this type, we observe $T_{ik}, k = 1, 2, \ldots, n_i$ events from subject i, together with a sequence of longitudinal data. Assume that the observation process is stopped by C_i. This censoring time C_i can be a planned study stop time or lost to follow up time. When C_i is the dropout or death time, it may create dependent censoring, especially there are reasons to suspect that unmeasured confounding variables other than the baseline covariates and longitudinal data collected from the study may influence dropout or death. In this case, it is important to account for the dependent censoring. One such attempt has been made by Liu and Huang (2009) and Liu et al. (2008) where death, recurrent events, and longitudinal data are jointly modeled using frailty terms. This work has been extended by Kim et al. (2012) to more general survival models. Han et al. (2007) used an intensity-based formulation of the recurrent event and introduced a parametric latent class joint model.

26.3.1.3 Nonproportional hazards model

Tseng et al. (2005) considered the accelerated failure time (AFT) model for the survival data. The AFT model takes the form

$$\lambda(t) = \lambda_0 \left[\int_0^t \exp\{\beta Y^*(s)\} \mathrm{d}s \right] \exp\{\beta Y^*(s)\},$$

where $\lambda_0(\cdot)$ is a baseline hazard function. The functional form of $\lambda_0(\cdot)$ is unknown, similar to the Cox model. Thus if we let U be a random variable with $\lambda_0(\cdot)$ as its hazard function, then we can write the above AFT model as $U = \int_0^T \exp\{\beta Y^*(s)\} \mathrm{d}s$, which is the AFT model with time-dependent covariate $Y^*(s)$.

Inference in the above model is based on an EM algorithm. In the Cox model, the time points for the nonparametric estimates of the baseline hazard depends on unknown parameters. Handling the AFT structure in the joint modeling setting is more difficult than for the Cox model because the baseline hazard is nonseparable with other parameters. To deal with this issue, Tseng et al. (2005) used a step function to model the baseline hazard.

Transformation models include the Cox model as a special case, but also other nonproportional hazards models such as the proportional odds model (Cheng et al., 1995; Murphy et al., 1997). Zeng and Lin (2007) considered transformation models in joint modeling setting and use the EM algorithm for inference.

Yet another popular extension for the survival data is the cure model as described in

Chapter 29, which assumes that a proportion of subjects may be cured and thus are not subject to risks of failure. Two types of cure models are commonly used. The mixture cure model (Kuk and Chen, 1992) assumes that there is a latent variable D for cure status. Only when $D = 1$, is the subject susceptible to disease and his/her time to event follows a survival model. The cure status D is usually assumed to follow a logistic regression model that can depend on baseline covariates (Yu et al., 2004, 2008). Another formulation of the cure model involves improper distribution functions (Chen et al., 2004; Brown and Ibrahim, 2003*a*). Due to the complexity of these cure models, Bayesian approaches were adopted.

26.3.2 Extension of longitudinal submodel

26.3.2.1 Joint models with discrete longitudinal outcomes

Standard joint models of longitudinal and time-to-event data focus on continuous longitudinal outcomes. However, discrete outcomes are also often encountered in medical studies. For example, in a randomized clinical trial on pain relief medication after wisdom teeth extraction (Pulkstenis et al., 1998), the response is a binary variable indicating a lack of reduction in pain relative to baseline. The extension of the basic joint model to handle discrete longitudinal response is straightforward by substituting the longitudinal submodel described in Section 26.2.2 with generalized linear mixed models (GLMM) (Breslow and Clayton, 1993). Assuming the n_i repeated outcome measures Y_i for subject i are independent with densities from the exponential family of distributions given a subject-specific random effect $\mathbf{b}_i$, the longitudinal submodel is formulated as

$$\begin{aligned} p(Y_i|\mathbf{b}_i) &= \exp\left(\sum_{j=1}^{n_i}\left[Y_{ij}\psi_{ij}(\mathbf{b}_i) - c\{\psi_{ij}(\mathbf{b}_i)\}\right]/a(\varphi) - d(Y_{ij},\varphi)\right) \\ E(Y_{ij}|\mathbf{b}_i) &= \frac{\partial c\{\psi_{ij}(\mathbf{b}_i)\}}{\partial \psi_{ij}(\mathbf{b}_i)} = g^{-1}\{\mathbf{b}_i^T f_i(t)\} \\ \mathbf{b}_i|Z_i &\sim \mathcal{N}(\mu_{\mathbf{b}}, \Sigma) \end{aligned}$$

where $\psi_{ij}(\mathbf{b}_i)$ and φ denote the natural and dispersion parameters in the exponential family, respectively, and $a(\cdot)$, $c(\cdot)$, and $d(\cdot)$ are known functions specifying the member of the exponential family. Several authors have studied such extensions (Faucett et al., 1998; Pulkstenis et al., 1998; Ten Have et al., 2000, 2002; Yao, 2008; Li et al., 2010). The survival submodel can be formulated similarly as Function (26.4) or extended appropriately as described in Section 26.3.1.

26.3.2.2 Joint models with multiple longitudinal biomarkers

Often in clinical longitudinal studies, several characteristics of a set of study participants are measured repeatedly over time. For example, in AIDS studies, both CD4 counts (immunologic measure) and viral loads (virologic measure) are considered to be important biomarkers and are measured simultaneously throughout the follow-up period.

To extend the longitudinal submodel to the multivariate case, let p be the number of longitudinal outcomes. For ease of exposition, consider the case where all longitudinal response outcomes are continuous. Let $Y_{ik}(t_{ij})$ be an assessment of the kth outcome for the ith subject at time t_{ij} and $Y_{ik}^*(t_{ij})$ be the corresponding trajectory function representing its true value, where $k = 1, ..., p, j = 1, ..., n_i$. The longitudinal multivariate model can be specified as

$$\begin{aligned} Y_{ik}(t_{ij}) &= Y_{ik}^*(t_{ij}) + e_{ijk} \\ Y_{ik}^*(t) &= \mathbf{b}_{ik}^T f_{ik}(t), \end{aligned}$$

where $\mathbf{b}_{ik}^T$ is a $1 \times r_k$ vector of random effects associated with the kth outcome, and f_{ik} is a vector of functions of time t with r_k elements. To account for correlations within and between the p longitudinal outcomes, one approach is to assume $e_{ijk} \overset{\text{iid}}{\sim} \mathcal{N}(0, \sigma_k^2)$ and $\mathbf{b}_i = (b_{i1}^T, b_{i2}^T, ..., b_{ip}^T)^T \sim \mathcal{N}_r(\mu_{\mathbf{b}}, G)$ with dimension $r = r_1 + r_2 + ... + r_p$ (Xu and Zeger, 2001; Lin et al., 2002*a*). The same general form of the linear predictor of the relative risk model for the survival process defined in (26.4) can be used, in which $\mathbf{b}_i$ is η_i.

The main difficulties encountered in practice when fitting joint longitudinal models with multiple longitudinal outcomes are the estimation of the large number of parameters for the unstructured covariance matrix G and the requirement for numerical integration respective to a large number of random effects. To ease the computational burden, Chi and Ibrahim (2006) and Brown et al. (2005) assumed $e_{ij\cdot} \overset{\text{iid}}{\sim} \mathcal{N}_p(0, V)$ and $\mathbf{b}_{ik} \overset{\text{ind}}{\sim} \mathcal{N}_{r_k}(\mu_{\mathbf{b}_k}, G_k)$ for $k = 1, ..., K$, but this approach only partially eases the computational burden. Fieuws et al. (2008) further proposed a pairwise modeling strategy, where all possible pairs of bivariate mixed models are fitted and used to obtain parameter estimates for the multivariate longitudinal submodel.

When the research interest is to study the association between these multiple longitudinal biomarkers and the risk of a certain event, a joint analysis of all relevant longitudinal biomarkers simultaneously with survival data is attractive in several respects. As shown by Fieuws et al. (2008), joint analysis of many markers substantially improved predictions compared to separate analysis of each marker. In addition, joint models accommodating multiple markers can be used to evaluate whether multiple biomarkers are a better substitute for the clinical endpoint than a single marker is Xu and Zeger (2001). In the literature, attempts have also been made to accommodate multivariate discrete longitudinal outcomes (Ten Have et al., 2002) and different types of longitudinal outcomes (Tsonaka et al., 2006; Fieuws et al., 2008) in a joint modeling framework.

Several authors have also considered nonlinear mixed effects models (NLMM) for modeling longitudinal data in joint models (Wu, 2002; Yu et al., 2004; Fieuws et al., 2008; Guedj et al., 2011). The linear mixed effects models defined in Section 26.2.2 and the GLMM models defined in Section 26.3.2.1 are empirical models. Unlike these models, NLMM models are often mechanistic models based on biological knowledge of longitudinal biomarkers. This type of model often has the advantage of meaningful interpretation for model parameters, but presents computational challenges.

26.3.3 Variations of the link between survival and longitudinal submodels

In Section 26.2.1 we stated that $Y_i^*(t)$ is the most commonly used form of function $q(t, \eta_i)$, which governs the form of association between the longitudinal biomarker and the event of interest. This means that the risk of event at time t depends only on the current true level of the marker. Although this parametrization is easy to interpret, it might be insufficient to capture the true association. To address this issue three different types of extension/variation for $q(t, \eta_i)$ have been explored in literature.

The first type of function represents features of the longitudinal trajectory at only a single time point. Besides current level of biomarker, the rate of change of biomarker $\frac{\partial Y_i^*(t)}{\partial t}$ has also been considered in many cases. One example is a study on association between prostate specific antigen (PSA) and the risk of prostate cancer recurrence (Ye et al., 2008*b*), in which the current rate of change in PSA level has shown to be a significant risk predictor independent of the current value of PSA. Yet in some other cases the true level of biomarker at a certain time ahead, $Y(t - c)$ (c is a positive constant) is preferred to the current level of biomarker (Cavender et al., 1992).

The second type of association function involves only random effects and no time function is involved. This type of association function is frequently used in the missing data framework (Follmann and Wu, 1995; Pulkstenis et al., 1998; Ten Have et al., 1998; Vonesh et al., 2006). In such models, the hazard function can directly depend on random effects in the longitudinal submodel $\lambda_i(t) = \lambda_0(t)\exp(\alpha^T \mathbf{b}_i + \gamma^T Z_i)$ (Pulkstenis et al., 1998; Pauler and Finkelstein, 2002). A slightly different but related approach is to allow a frailty term W_i in the hazard function $\lambda_i(t) = \lambda_0(t)\exp(W_i + \gamma^T Z_i)$, and model the association between the longitudinal submodel and the survival model through a stochastic relationship between W_i and $\mathbf{b}_i$ (Henderson et al., 2000). Because this type of association is time-independent and thus often leads to closed-form solutions in the integral of the survival function (26.9), the computation is often simplified. The simpler structure also enables some manageable theoretical development (Tsiatis and Davidian, 2001; Rizopoulos et al., 2008; Huang et al., 2009). In addition, when parametric models such as piecewise exponential models are used for the survival submodel, the likelihood can be marginalized with respect to the random effects using Gaussian quadrature (Liu and Huang, 2009) or Laplace approximation (Vonesh et al., 2006). However, this type of models can impose quite restrictive correlation structure among the outcomes (Verbeke and Davidian, 2008). A further limitation is that interpreting the dependence of the risk of event on longitudinal marker is difficult when there are more random effects in the longitudinal submodel, participially when splines are used to capture the nonlinear trajectory of the longitudinal biomarker.

The third type of association was considered by Gao et al. (2011) to assess the impact of biomarker variability on the risk of developing a clinical outcome. Their model was motivated by a study on ocular hypertension treatment, in which intraocular pressure fluctuation was posited to impact the risk of primary open-angle glaucoma. They modeled the trajectory of intraocular pressure with a linear mixed effects model that incorporates subject-specific variance (Lyles et al., 1999), and allowed the hazard to be a function of subject-specific intercept, subject-specific slope, and subject-specific variance.

26.3.4 Joint latent class models

The extended joint models given in Sections 26.3.1, 26.3.2, and 26.3.3 all belong to the large category of shared random-effect models, which assume a homogeneous population with a single average trajectory of longitudinal submodel, and a continuous relationship between the biomarker and the risk of an event. An alternative approach for joint modeling of longitudinal and time-to-event data is to use joint latent class models (Lin et al., 2002*b*; Proust-Lima et al., 2007, 2009, 2012). This method considers the population of subjects as heterogeneous and assumes that the original population of N subjects can be divided into G homogeneous subpopulations that share the same marker trajectory and the baseline hazard function.

Let c_i be the variable indicating class membership for subject i, which equals g if subject i belongs to latent class g ($g = 1, ..., G$). Assuming that given c_i the longitudinal and survival processes are independent, a general joint latent class model can be specified as

$$\lambda_i(t|c_i = g) = \lambda_{g0}(t) exp\{\gamma_g^T Z_{gi}^{(s)}\} \tag{26.11}$$

$$Y_i(t_{ij}) = Y_i^*(t_{ij}) + e_{ij}, \quad \varepsilon(t_{ij}) \overset{\text{iid}}{\sim} \mathcal{N}(0, \sigma_\varepsilon^2) \tag{26.12}$$

$$Y_i^*(t|c_i = g) = \mathbf{b}_{ig}^T f_{ig}(t), \quad \mathbf{b}_{ig}|Z_g^{(l)} i \sim \mathcal{N}(\mu_{\mathbf{b}_g}, \Sigma_g) \tag{26.13}$$

$$Pr(c_i = g) = exp(\lambda_g^T Z_{ci}) / \sum_{l=1}^{G} exp(\alpha_l^T Z_i^{(c)}) \tag{26.14}$$

where γ_g is the risk coefficients associated with a vector of covariates $Z_{gi}^{(s)}$ for the gth subpopulation. $\mathbf{b}_{ig}$ are the random effects in the longitudinal submodel for the gth subpopulation and follow a multivariate normal distribution with mean related to a vector of covariates $Z_{gi}^{(l)}$. The latent class membership c_i is specified by a multinomial submodel (26.14) with $\alpha_1 = 0$ for identifiability.

In joint latent class models, the survival submodel (26.12) does not involve any time-dependent component and random effects $\mathbf{b}_i$ are involved only in the longitudinal submodel (26.13 and 26.14). Therefore, the log-likelihood under these models often has closed form and is more tractable. However, like other latent class models, the log-likelihood function of joint latent class models may have multiple local maxima, which requires multiple fittings of the model with different sets of initial values to find the global maxima.

Since joint latent class models do not model the association between the survival and longitudinal submodel directly through biomarker level, they are not suitable for evaluating specific assumptions regarding the link between longitudinal markers and the risk of event. Despite this disadvantage, they offer a flexible framework to model the joint distribution of the longitudinal marker and the time-to-event, and may be particularly useful for prediction problems (Proust-Lima et al., 2012).

26.4 Prediction in joint models

One important application of joint model is for prediction. Because the longitudinal profile is in some sense unique to its corresponding subject, prediction of the time-to-event outcome for censored or future patients utilizing observed longitudinal data provides an individualized prognosis. We consider two types of predictions here. One is for future longitudinal outcomes and the other is for the probability distribution of a future event.

26.4.1 Prediction of future longitudinal outcome

For a subject who is followed until time t_i with no event ($\delta_i = 0$) and with observed longitudinal data, $y_{i1}, \ldots, y_{ik}$, his/her longitudinal outcome at time $u > t_i$ has a predictive distribution $h\{Y^*(u)|T > t_i, Y(t_{i1}) = y_{i1}, \ldots, Y(t_{ik}) = y_{ik}, Z_i\}$. This distribution is obtained by decomposing

$$
\begin{aligned}
&h\{Y^*(u)|T > t_i, Y(t_{i1}) = y_{i1}, \ldots, Y(t_{ik}) = y_{ik}, Z_i\} \\
&= \int h\{Y^*(u)|\mathbf{b}_i, \boldsymbol{\theta}, Z_i\} h\{\mathbf{b}_i, \boldsymbol{\theta}|T > t_i, Y(t_{i1}) = y_{i1}, \ldots, Y(t_{ik}) = y_{ik}, Z_i\} \mathrm{d}\mathbf{b}_i \mathrm{d}\boldsymbol{\theta}
\end{aligned}
$$

where $\boldsymbol{\theta}$ represents all fixed-effects parameters, including those from the survival submodel.

When inferences are based on the Bayesian approach, draws of the posterior distribution of $\mathbf{b}_i$ and $\boldsymbol{\theta}$ in the above integrand are obtained, and predictive distribution of $Y^*(u)$ is easily evaluated by averaging over $h\{Y^*(u)|\mathbf{b}_i^{(m)}, \boldsymbol{\theta}^{(m)}, Z_i\}$ for the draws $m = 1, \ldots, M$. The posterior predictive mode, mean, and credible intervals can be obtained accordingly. For example, if the mean of $Y^*(u)$ is the quantity of interest, then it can be estimated as $\sum_{m=1}^{M} \mathbf{b}_i^{(m)} f_i(u)$ from Model (26.5).

When inferences are based on a non-Bayesian approach such as the MLE, the mean of $Y^*(u)$ may be estimated by $E\{Y^*(u) \,|\, \hat{\mathbf{b}}_i, \hat{\boldsymbol{\theta}}, Z_i\} = \hat{\mathbf{b}}_i f_i(u) + \hat{\boldsymbol{\theta}} f_i(u)$ where $\hat{\mathbf{b}}_i$ predicts $\mathbf{b}_i$ and $\hat{\boldsymbol{\theta}}$ estimates $\boldsymbol{\theta}$. In this approach the variations in both $\hat{\mathbf{b}}_i$ and $\hat{\boldsymbol{\theta}}$ need to be accounted when calculating a confidence interval of the prediction. Usually these variations are hard to estimate, leading to the introduction of approximations. In particular, Rizopoulos (2012*b*) considered a resampling-based approach which has a strong Bayesian flavor.

26.4.2 Prediction of survival distribution

For a subject who is followed until time t_i with no event ($\delta_i = 0$) and with observed longitudinal data, $y_{i1}, \ldots, y_{ik}$, his/her chance of surviving $u > t_i$ can be expressed as

$$\begin{aligned} &P\{T > u|T > t_i, y_{i1}, \ldots, y_{ik}, Z_i\} \\ &= \int P\{T > u|T > t_i \mathbf{b}_i, \boldsymbol{\theta}, Z_i\} h\{\mathbf{b}_i, \boldsymbol{\theta}|T > t_i, y_{i1}, \ldots, y_{ik}, Z_i\} \mathrm{d}\mathbf{b}_i \mathrm{d}\boldsymbol{\theta} \\ &= \int \frac{P\{T > u|\mathbf{b}_i, \boldsymbol{\theta}, Z_i\}}{P\{T > t_i|\mathbf{b}_i, \boldsymbol{\theta}, Z_i\}} h\{\mathbf{b}_i, \boldsymbol{\theta}|T > t_i, y_{i1}, \ldots, y_{ik}, Z_i\} \mathrm{d}\mathbf{b}_i \mathrm{d}\boldsymbol{\theta} \end{aligned}$$

Similar to the prediction of longitudinal outcomes, the Bayesian approach leads to straightforward calculation of the posterior predictive mode, mean, and credible intervals. Non-Bayesian approaches usually need to rely on resampling techniques due to the complexity of the above formula.

26.4.3 Performance of prediction accuracy

Performance of prediction accuracy is typically evaluated in two ways: discrimination and calibration. Discrimination refers to the ability of a model to correctly distinguish outcomes. With binary outcomes, outcomes are clearly defined as cases and controls. For time-to-event outcomes, due to censoring, controls are not well defined. Also it may be necessary to distinguish early failures from late failures. In the literature, several definitions of cases and controls have been given. The cumulative sensitivity and dynamic specificity approach defines cases at each time point t as those with $T^* \le t$ and controls as those with $T^* > t$ (Heagerty and Zheng, 2005), which means that late failures may serve as controls for early failures. The incident sensitivity approach considers only failures at the given time t as cases (Heagerty and Zheng, 2005). Static specificity considers only those with $T^* > \tilde{t}$ for some fixed and large $\tilde{t}$ as controls. In general, assume that the disease status $D(t)$ such that $D(t) = 1, 0$ indicates cases and controls, respectively. Then a monotonic rule based on observed longitudinal marker Y can be constructed. Denote this rule as $R(t, \bar{Y}(t))$ for any given time t. The prediction for positive diagnosis occurs when $R(t, \bar{Y}(t)) \ge c$ for a given c. Accordingly the sensitivity is $P(R(t, \bar{Y}(t)) \ge c|D(t) = 1)$ and specificity is $P(R(t, \bar{Y}(t)) < c|D(t) = 0)$; see Heagerty et al. (2000). Time-dependent receiver operating characteristic (ROC) curves can be constructed with a varying c (Zheng and Heagerty, 2007; Heagerty and Zheng, 2005). A subject who is censored is never a case, but may have missing status due to censoring. Nevertheless, they still contribute to estimation of both sensitivity and specificity. Estimation under the joint modeling framework is considered in Rizopoulos (2011). A model with good discrimination ability should produce higher predicted probabilities to subjects who had events than subjects who did not have events. The area under the ROC curve is a popular measure for evaluating model discrimination (Hanley and McNeil, 1982).

Calibration describes how closely the predicted probabilities agree numerically with the actual outcomes. A model is well calibrated when predicted and observed values agree for any reasonable grouping of the observation, ordered by increasing predicted values. The calibration of a model can be compared with sample frequencies, especially for validation data. For example, categories may be assigned according to predicted event-free probabilities and then compared with the number of observed events for agreement (Yu et al., 2008).

Although a model with a good calibration will tend to have good discrimination and vice versa, a given model may be good on one measure but weak on another. Harrell et al. (1996) recommended that good discrimination is always to be preferred to good calibration since a model with a good discrimination can always be recalibrated, but the rank orderings of the probabilities cannot be changed to improve discrimination.

26.5 Joint model diagnostics

The research on joint model diagnostics is relatively scant. Recently Jacqmin-Gadda et al. (2010) considered a score test in a joint latent class model setting. The null hypothesis is independence between the marker and the outcome given the latent classes, and the alternative hypothesis is that the risk of event depends on one or several random effects from the mixed effets model in addition to the latent classes.

Recently, reports of residual based analysis for joint modeling have appeared in the literature based on longitudinal data. Two types of residuals can be defined (Nobre and Singer, 2007). Marginal residuals use $Y_i(t) - \hat{\mu}_{\mathbf{b}} f_i(t)$ based on the longitudinal Submodel (26.5) to predict the marginal errors where only fixed effects estimates $\hat{\mu}_{\mathbf{b}}$ are used. Conditional residuals use $Y_i(t) - \hat{\mathbf{b}}_i f_i(t)$ to predict the conditional errors where the random effects estimates $\hat{\mathbf{b}}_i$ are used, which involve implicitly the fixed effects estimates $\hat{\mu}_{\mathbf{b}}$. The random effects estimates can be based on procedures such as best linear unbiased prediction (BLUP) or draws from Bayesian approaches (Nobre and Singer, 2007). Dobson and Henderson (2003) considered joint modeling for longitudinal data and dropout time. They developed graphical tools based on conditional distributions of longitudinal marginal residuals on the dropout pattern when the time points for longitudinal data collected are fixed. The focus was on accounting for the effect of informative dropout on longitudinal residuals. In particular they showed that after fitting a joint model residuals between observed and expected responses can be markedly affected by knowledge of the dropout time and type. A multiple imputation-based approach has also been implemented to examine longitudinal marginal and subject-specific residuals in joint modeling setting (Rizopoulos et al., 2010). The approach imputes missed longitudinal data due to censoring from the survival event. Imputation is based on both fixed time and random time schemes for longitudinal data. For the fixed-time scheme, the imputation is based on posterior distribution of missed longitudinal data conditioning on both observed longitudinal data and survival data. The evaluation of the posterior distribution involves both fixed and random effects parameters. For the random time scheme, it is necessary to model the visiting times to facilitate the imputation. The visiting process is modeled with gamma frailty and Weibull baseline hazard. The visiting times are then built into the imputation procedure for missed longitudinal data.

It has been observed that the shared parameter joint models enjoy a certain robustness property. Hsieh et al. (2006) demonstrated robustness of the MLE to random effect model specification when there is enough information from the longitudinal data. Rizopoulos et al. (2008) observed a similar phenomenon and showed that for survival models with finite dimensional parameter space that the score vector under the misspecified model is close to the correct score vector when the number of repeated measurements per subject is large enough. Huang et al. (2009) provided an alternative explanation and showed that when the longitudinal data information is rich enough, the random effects can be well estimated by their ordinary least square estimates so that it is as if the random effects were observed like fixed effects instead of being latent quantities. The near sufficiency of the least square estimates for the random effects as the longitudinal information increases then ensures robustness of the survival parameters too.

In order to take advantage of this robustness feature of shared parameter joint models, it is important to know when the available longitudinal information in a particular dataset is rich enough to yield such robustness. To this end, Huang et al. (2009) proposed a diagnostic tool based on the well-known Simulation Extrapolation (SIMEX). In particular, the procedure artificially introduced perturbations on the longitudinal data indexed by a perturbation parameter ζ such that larger ζ indicates more perturbation whereas $\zeta = 0$ corresponds to the no perturbation. The resulting SIMEX plot is then used to assess the survival parameter estimates sensitivity to the perturbations. In particular, a curve relatively flat at the origin, $\zeta = 0$, indicates robustness. In a Bayesian framework, Zhu et al. (2012) developed a variety of influence measures for carrying out sensitivity analysis to joint models. In particular, they considered perturbation models for individual and global sensitivity for all components of the joint models, including the data points, the prior distribution, and the sampling distribution. Local influence measures were proposed to quantify the degree of these perturbations. The proposed methods also allow the detection of outliers or influential observations and the assessment of the sensitivity of inferences to various unverifiable assumptions on the Bayesian analysis of joint models.

26.6 Joint model software

Although joint modeling research has received considerable attention in the methodological literature in the past two decades, its applications in clinical studies are scarce. The largest hurdle of using joint modeling in practice has been lack of software. Recently, however, some progress has been made to solve this critical issue. Two general packages are now available in R and Stata. In addition, codes for a few special models in SAS, Winbug, and Fortune have also been published.

An R package called *JM* developed by Rizopoulos (2012*b*) allows fitting joint models with a single continuous longitudinal outcome and time-to-event data using the maximum likelihood approach. This package has some excellent features that make it by far the most comprehensive joint modeling software. First, various options for modeling the baseline hazard function are available in *JM* (e.g., unspecified nonparametric function, piecewise exponential, hazard function corresponding to commonly used parametric survival models). Second, modelers can choose among a list of forms of association between longitudinal and survival submodels (e.g., effect of the current biomarker value, effect of the rate of change of the biomarker value, etc.). In addition, *JM* can be used to fit joint models with several types of extended survival submodels, including AFT models and competing risk models. Lastly, *JM* provides diagnostic and predictions tools. To learn more about *JM*, see Rizopoulos (2012*b*). Yet another R package developed by Proust-Lima and associates called *lcmm* provides the *Jointlcmm* function that fits joint latent class mixed models (JLCM) using a maximum likelihood method.

In Stata, Crowther et al. (2012) developed a user written command, *stjm*, for fitting the basic joint models (one continuous longitudinal response and one time-to-event outcome). This development allows four choices for the survival submodel and several forms of association between the survival and longitudinal submodels as well as provides some basic diagnostic and prediction tools. A special feature of *stjm* is its flexibility in modeling the longitudinal trajectory through the use of fixed and/or random fractional polynomials of time. One drawback, however, is that the current version of *stjm* cannot handle any extension in either survival or longitudinal submodel.

Both Guo and Carlin (2004) and Vonesh et al. (2006) provided SAS code for fitting joint models using SAS PROC NLMIXED. Guo and Carlin (2004) also showed how to fit joint models in a Bayesian framework using R and Winbug code.

Bibliography

Abrams, D. I., Goldman, A. I., Launer, C., Korvick, J. A., Neaton, J. D., Crane, L. R., Grodesky, M., Wakefield, S., Muth, K., Kornegay, S., Cohn, D. L., Harris, A., Luskin-Hawk, R., Markowitz, N., Sampson, J. H., Thompson, M. and Deyton, L. (1994), 'A comparative trial of didanosine or zalcitabine after treatment with zidovudine in patients with human immunodeficiency virus infection', *New England Journal of Medicine* **330**(10), 657–662. **URL:** *www.nejm.org/doi/full/10.1056/NEJM199403103301001*

Breslow, N. E. and Clayton, D. G. (1993), 'Approximate inference in generalized linear mixed models', *Journal of the American Statistical Association* **88**, 9–25.

Brown, E. R. and Ibrahim, J. G. (2003*a*), 'Bayesian approaches to joint cure-rate and longitudinal models with applications to cancer vaccine trials', *Biometrics* **59**(3), 686–693.

Brown, E. R. and Ibrahim, J. G. (2003*b*), 'A Bayesian semiparametric joint hierarchical model for longitudinal and survival data', *Biometrics* **59**(2), 221–228.

Brown, E. R., Ibrahim, J. G. and DeGruttola, V. (2005), 'A flexible B-spline model for multiple longitudinal biomarkers and survival', *Biometrics* **61**(1), 64–73.

Bycott, P. and Taylor, J. (1998), 'A comparison of smoothing techniques for CD4 data measured with error in a time-dependent Cox proportional hazards model', *Statistics in Medicine* **17**, 2061–2077.

Cavender, J. B., Rogers, W. J., Fisher, L. D., Gersh, B. J., Coggin, C. J. and Myers, W. O. (1992), 'Effect of smoking on survival and morbidity in patients randomized to medical or surgical therapy in the coronary artery surgery study (cass): 10-year follow-up', *Journal of the American College of Cardiology* **20**(2), 287–294.
URL: *dx.doi.org/10.1016/0735-1097(92)90092-2*

Chen, M.-H., Ibrahim, J. G. and Sinha, D. (2004), 'A new joint model for longitudinal and survival data with a cure fraction', *Journal of Multivariate Analysis* **91**(1), 18–34.

Cheng, S., Wei, L. and Ying, Z. (1995), 'Analysis of transformation models with censored data', *Biometrika* **82**, 835–45.

Chi, Y.-Y. and Ibrahim, J. G. (2006), 'Joint models for multivariate longitudinal and multivariate survival data', *Biometrics* **62**(2), 432–445.

Cox, D. R. (1972), 'Regression models and life-tables (with discussion)', *Journal of the Royal Statistical Society, Series B: Methodological* **34**, 187–220.

Crowther, M., Abrams, K. and Lambert, P. (2012), 'Joint modeling of longitudinal and survival data', *The State Journal* **13**(1), 165–184 .

Dafni, U. G. and Tsiatis, A. A. (1998), 'Evaluating surrogate markers of clinical outcome when measured with error', *Biometrics* **54**, 1445–1462.

Dempster, A. P., Laird, N. M. and Rubin, D. B. (1977), 'Maximum likelihood from incomplete data via the EM algorithm (C/R: P22-37)', *Journal of the Royal Statistical Society, Series B: Methodological* **39**, 1–22.

Dimitris, R. and Pulak, G. (2011), 'A bayesian semiparametric multivariate joint model for multiple longitudinal outcomes and a time-to-event', *Statistics in Medicine* **30**(12), 1366–1380.

Ding, J. and Wang, J.-L. (2008), 'Modeling Longitudinal Data with Nonparametric Multiplicative Random Effects Jointly with Survival Data', *Biometrics* **64**(2), 546–556.

Dobson, A. and Henderson, R. (2003), 'Diagnostics for joint longitudinal and dropout time modeling', *Biometrics* **59**, 741–751.

Elashoff, R. M., Li, G. and Li, N. (2007), 'An approach to joint analysis of longitudinal measurements and competing risks failure time data', *Statistics in Medicine* **26**(14), 2813–2835.

Faucett, C. L., Schenker, N. and Elashoff, R. M. (1998), 'Analysis of censored survival data with intermittently observed time-dependent binary covariates', *Journal of the American Statistical Association* **93**, 427–437.

Faucett, C. L. and Thomas, D. C. (1996), 'Simultaneously modeling censored survival data and repeatedly measured covariates: A Gibbs sampling approach', *Statistics in Medicine* **15**, 1663–1685.

Fieuws, S., Verbeke, G., Maes, B. and Y., V. (2008), 'Predicting renal graft failure using multivariate longitudinal profiles', *Biostatistics* **9**, 419–431.

Fitzmaurice, G., Davidian, M., Verbeke, G. and Molenberghs, G. (2008), *Longitudinal Data Analysis*, Chapman & Hall/CRC, Boca Raton, FL.

Follmann, D. and Wu, M. (1995), 'An approximate generalized linear model with random effects for informative missing data (Corr: 97V53 p384)', *Biometrics* **51**, 151–168.

Gao, F., Miller, J. P., Xiong, C., Beiser, J. A., Gordon, M. and Group, T. O. H. T. S. O. (2011), 'A joint-modeling approach to assess the impact of biomarker variability on the risk of developing clinical outcome', *Statistical Methods and Applications* **20**(1), 83–100.

Gelfand, A. E. and Smith, A. F. M. (1990), 'Sampling-based approaches to calculating marginal densities', *Journal of the American Statistical Association* **85**, 398–409.

Guedj, J., Thiébaut, R. and Commenges, D. (2011), 'Joint modeling of the clinical progression and of the biomarkers' dynamics using a mechanistic model', *Biometrics* **67**(1), 59–66.

Guo, X. and Carlin, B. P. (2004), 'Separate and joint modeling of longitudinal and event time data using standard computer packages', *The American Statistician* **58**(1), 16–24.

Han, J., Slate, E. and Pena, E. (2007), 'Parametric latent class joint model for a longitudinal biomarker and recurrent events', *Stat Med* **26**, 5285–5302.

Hanley, J. and McNeil, B. (1982), 'The meaning and use of the area under a receiver operating characteristic (ROC) curve', *Radiology* **143**, 29–36.

Harrell, F., Lee, K. and Mark, D. (1996), 'Multivariate prognostic models: issues in developing models, evaluating assumptions and adequacy, and measuring and reducing errors', *Stat Med* **15**, 361–387.

Heagerty, P., Lumley, T. and Pepe, M. (2000), 'Time-dependent roc curves for censored survival data and a diagnostic marker', *Biometrics* **56**, 337–344.

Heagerty, P. and Zheng, Y. (2005), 'Survival model predictive accuracy and roc curves', *Biometrics* **61**, 92–105.

Henderson, R., Diggle, P. and Dobson, A. (2000), 'Joint modeling of longitudinal measurements and event time data', *Biostatistics (Oxford)* **1**(4), 465–480.

Henderson, R., Diggle, P. and Dobson, A. (2002), 'Identification and efficacy of longitudinal markers for survival', *Biostatistics (Oxford)* **3**(1), 33–50.

Hogan, J. W. and Laird, N. M. (1997*a*), 'Mixture models for the joint distribution of repeated measures and event times', *Statistics in Medicine* **16**, 239–257.

Hogan, J. W. and Laird, N. M. (1997*b*), 'Model-based approaches to analysing incomplete longitudinal and failure time data', *Statistics in Medicine* **16**, 259–272.

Hsieh, F., Tseng, Y.-K. and Wang, J.-L. (2006), 'Joint modeling of survival and longitudinal data: Likelihood approach revisited', *Biometrics* **62**(4), 1037–1043.

Huang, X., Stefanski, L. A. and Davidian, M. (2009), 'Latent-model robustness in structural measurement error models', *Biometrika* **93**, 53–64.

Ibrahim, J. G., Chen, M.-H. and Sinha, D. (2004), 'Bayesian methods for joint modeling of longitudinal and survival data with applications to cancer vaccine trials', *Statistica Sinica* **14**(3), 863–883.

Jacqmin-Gadda, H., Proust-Lima, C., Taylor, J. and Commenges, D. (2010), 'Score test for conditional independence between longitudinal outcome and time to event given the classes in the joint latent class model', *Biometrics* **66**, 11–19.

Kim, S., Zeng, D., Chambless, L. and Li, Y. (2012), 'Joint models of longitudinal data and recurrent events with informative terminal event', *Statistics in Biosciences* **4**, 262–281.

Kuk, A. and Chen, C. (1992), 'A mixture model combining logistic regression with proportional hazards regression', *Biometrika* **79**, 531–541.

Laird, N. M. and Ware, J. H. (1982), 'Random-effects models for longitudinal data', *Biometrics* **38**, 963–974.

Lavalley, M. P. and De Gruttola, V. (1996), 'Models for empirical Bayes estimators of longitudinal CD4 counts (Disc: P2337-2340)', *Statistics in Medicine* **15**, 2289–2305.

Law, N. J., Taylor, J. M. G. and Sandler, H. (2002), 'The joint modeling of a longitudinal disease progression marker and the failure time process in the presence of cure', *Biostatistics (Oxford)* **3**(4), 547–563.

Li, N., Elashoff, R. M. and Li, G. (2009), 'Robust joint modeling of longitudinal measurements and competing risks failure time data', *Biometrical Journal* **51**(1), 19–30.

Li, N., Elashoff, R. M., Li, G. and Saver, J. (2010), 'Joint modeling of longitudinal ordinal data and competing risks survival times and analysis of the NINDS rt-PA stroke trial', *Statistics in Medicine* **29**(5), 546–557.

Lin, H., McCulloch, C. E. and Mayne, S. T. (2002*a*), 'Maximum likelihood estimation in the joint analysis of time-to-event and multiple longitudinal variables', *Statistics in Medicine* **21**(16), 2369–2382.

Lin, H., Turnbull, B. W., McCulloch, C. E. and Slate, E. H. (2002*b*), 'Latent class models for joint analysis of longitudinal biomarker and event process data: Application to longitudinal prostate-specific antigen readings and prostate cancer', *Journal of the American Statistical Association* **97**(457), 53–65.

Little, R. J. A. (1993), 'Pattern-mixture models for multivariate incomplete data', *Journal of the American Statistical Association* **88**, 125–134.

Little, R. and Rubin, D. (2002), *Statistical Analysis with Missing Data*, second edition, Wiley, New York.

Liu, L. and Huang, X. (2009), 'Joint analysis of correlated repeated measures and recurrent events processes in the presence of death, with application to a study on acquired immune deficiency syndrome', *Journal of the Royal Statistical Society, Series C* **58**, 65–81.

Liu, L., Huang, X. and O'Quigley, J. (2008), 'Analysis of longitudinal data in the presence of informative observational times and a dependent terminal event, with application to medical cost data', *Biometrics* **64**, 950–958.

Louis, T. A. (1982), 'Finding the observed information matrix when using the EM algorithm', *Journal of the Royal Statistical Society, Series B: Methodological* **44**, 226–233.

Lyles, R. H., Munoz, A., Muñoz, A., Xu, J., Taylor, J. M. G. and Chmiel, J. S. (1999), 'Adjusting for measurement error to assess health effects of variability in biomarkers', *Statistics in Medicine* **18**, 1069–1086.

McLachlan, G. J. and Krishnan, T. (1997), *The EM Algorithm and Extensions*, John Wiley & Sons, New York.

Murphy, S., Rossini, A. and Van der Vaart, A. (1997), 'Maximum likelihood estimation in the proportional odds model', *J. Am. Statist. Assoc.* **92**, 968–76.

Nobre, J. and Singer, J. (2007), 'Residuals analysis for linear mixed models', *Biometrical Journal* **6**, 863–875.

Parner, E. (1998), 'Asymptotic theory for the correlated gamma-frailty model', *The Annals of Statistics* **26**, 183–214.

Pauler, D. K. and Finkelstein, D. M. (2002), 'Predicting time to prostate cancer recurrence based on joint models for non-linear longitudinal biomarkers and event time outcomes', *Statistics in Medicine* **21**(24), 3897–3911.

Pawitan, Y. and Self, S. (1993), 'Modeling disease marker processes in AIDS', *Journal of the American Statistical Association* **88**, 719–726.

Pinheiro, J. C., Pinheiro, J. C. and Bates, D. M. (1995), 'Approximations to the log-likelihood function in the nonlinear mixed-effects model', *Journal of Computational and Graphical Statistics* **4**, 12–35.

Prentice, R. L. (1982), 'Covariate measurement errors and parameter estimation in a failure time regression model (Corr: V71 p219)', *Biometrika* **69**, 331–342.

Press, W., Flannery, B. and Teukolsky SA, V. W. (1992), *Numerical Recipes in C: The Art of Scientific Computing, Second Edition*, Cambridge University Press, Cambridge.

Proust, C., Jacqmin-Gadda, H., Taylor, J. M. G., Ganiayre, J. and Commenges, D. (2006), 'A nonlinear model with latent process for cognitive evolution using multivariate longitudinal data', *Biometrics* **62**(4), 1014–1024.

Proust-Lima, C., Joly, P., Dartigues, J.-F. and Jacqmin-Gadda, H. (2009), 'Joint modeling of multivariate longitudinal outcomes and a time-to-event: A nonlinear latent class approach', *Computational Statistics & Data Analysis* **53**(4), 1142–1154.

Proust-Lima, C., Letenneur, L. and Jacqmin-Gadda, H. (2007), 'A nonlinear latent class model for joint analysis of multivariate longitudinal data and a binary outcome', *Statistics in Medicine* **26**(10), 2229–2245.

Proust-Lima, C., Séne, M., Taylor, J. M. and Jacqmin-Gadda, H. (2012), 'Joint latent class models for longitudinal and time-to-event data: A review', *Statistical Methods in Medical Research* doi:10.1177/0962280212445839.

Pulkstenis, E. P., Ten Have, T. R. and Landis, J. R. (1998), 'Model for the analysis of binary longitudinal pain data subject to informative dropout through remedication', *Journal of the American Statistical Association* **93**, 438–450.

Rabout, J. (1991), 'Errors in measurement in survival analysis', *Doctoral Dissertation.*

Ratcliffe, S. J., Guo, W. and Ten Have, T. R. (2004), 'Joint modeling of longitudinal and survival data via a common frailty', *Biometrics* **60**(4), 892–899.

Rizopoulos, D. (2011), 'Dynamic predictions and prospective accuracy in joint models for longitudinal and time-to-event data', *Biometrics* **67**, 819–829.

Rizopoulos, D. (2012*a*), 'Fasting fitting of joint models for longitudinal and event time data using a pseudo-adaptive gaussian quadrature rule', *Computational Statistics & and Analysis* **56**, 491–501.

Rizopoulos, D. (2012*b*), *Joint Models for Longitudinal and Time-to-Event Data: With Applications in R*, Chapman & Hall/CRC, Boca Raton, FL.

Rizopoulos, D., Verbeke, G. and Lesaffre, E. (2009), 'Fully exponential Laplace approximations for the joint modeling of survival and longitudinal data', *Journal of the Royal Statistical Society, Series B: Statistical Methodology* **71**(3), 637–654.

Rizopoulos, D., Verbeke, G. and Molenberghs, G. (2008), 'Shared parameter models under random effects misspecification', *Biometrika* **95**, 1–12.

Rizopoulos, D., Verbeke, G. and Molenberghs, G. (2010), 'Multiple imputation-based residuals and diagnostic plots for joint models of longitudinal and survival outcomes', *Biometrics* **66**, 20–29.

Rosenberg, P. S. (1995), 'Hazard function estimation using *B*-splines', *Biometrics* **51**, 874–887.

Song, X., Davidian, M. and Tsiatis, A. A. (2002), 'A semiparametric likelihood approach to joint modeling of longitudinal and time-to-event data', *Biometrics* **58**(4), 742–753.

Taylor, J. M. G., Cumberland, W. G. and Sy, J. P. (1994), 'A stochastic model for analysis of longitudinal AIDS data', *Journal of the American Statistical Association* **89**, 727–736.

Ten Have, T. R., Kunselman, A. R., Pulkstenis, E. P. and Landis, J. R. (1998), 'Mixed effects logistic regression models for longitudinal binary response data with informative drop-out', *Biometrics* **54**, 367–383.

Ten Have, T. R., Miller, M. E., Reboussin, B. A. and James, M. K. (2000), 'Mixed effects logistic regression models for longitudinal ordinal functional response data with multiple-cause drop-out from the longitudinal study of aging', *Biometrics* **56**(1), 279–287.

Ten Have, T. R., Reboussin, B. A., Miller, M. E. and Kunselman, A. (2002), 'Mixed effects logistic regression models for multiple longitudinal binary functional limitation responses with informative drop-out and confounding by baseline outcomes', *Biometrics* **58**(1), 137–144.

Tseng, Y., Hsieh, F. and Wang, J. (2005), 'Joint modeling of accelerated failure time and longitudinal data', *Biometrika* **92**, 587–603.

Tsiatis, A. A. and Davidian, M. (2001), 'A semiparametric estimator for the proportional hazards model with longitudinal covariates measured with error', *Biometrika* **88**(2), 447–458.

Tsiatis, A. A. and Davidian, M. (2004), 'Joint modeling of longitudinal and time-to-event data: An overview', *Statistica Sinica* **14**(3), 809–834.

Tsiatis, A. A., DeGruttola, V. and Wulfsohn, M. S. (1995), 'Modeling the relationship of survival to longitudinal data measured with error. Applications to survival and CD4 counts in patients with AIDS', *Journal of the American Statistical Association* **90**, 27–37.

Tsonaka, R., Rizopoulos, D. and Lesaffre, E. (2006), 'Power and sample size calculations for discrete bounded outcome scores', *Statistics in Medicine* **25**(24), 4241–4252.

van der Vaart, A. W. (1998), *Asymptotic Statistics*, Cambridge University Press, Cambridge.

Verbeke, G. and Davidian, M. (2008), Joint models for longitudinal data: introduction and overview, *in* '*Longitudinal Data Analysis*', G. Fitzmaurice, M. Davidian and G. Molenberghs, eds., Chapman & Hall/CRC, Boca Raton, FL.

Vonesh, E. F., Greene, T. and Schluchter, M. D. (2006), 'Shared parameter models for the joint analysis of longitudinal data and event times', *Statistics in Medicine* **25**(1), 143–163.

Wang, Y. and Taylor, J. M. G. (2001), 'Jointly modeling longitudinal and event time data with application to acquired immunodeficiency syndrome', *Journal of the American Statistical Association* **96**(455), 895–905.

Williamson, P. R., Kolamunnage-Dona, R., Philipson, P. and Marson, A. G. (2008), 'Joint modeling of longitudinal and competing risks data', *Statistics in Medicine* **27**(30), 6426–6438.

Wu, L. (2002), 'A joint model for nonlinear mixed-effects models with censoring and covariates measured with error, with application to AIDS Studies', *Journal of the American Statistical Association* **97**(460), 955–964.

Wu, L., Liu, W., Yi, G. Y. and Huang, Y. (2012), 'Analysis of longitudinal and survival data: Joint modeling, inference methods, and issues', *Journal of Probability and Statistics* **2012 Article ID 640153**.

Wu, M. C. and Carroll, R. J. (1988), 'Estimation and comparison of changes in the presence of informative right censoring by modeling the censoring process (Corr: V45 p1347; V47 p357)', *Biometrics* **44**, 175–188.

Wulfsohn, M. S. and Tsiatis, A. A. (1997), 'A joint model for survival and longitudinal data measured with error', *Biometrics* **53**, 330–339.

Xu, J. and Zeger, S. L. (2001), 'The evaluation of multiple surrogate endpoints', *Biometrics* **57**(1), 81–87.

Yao, F. (2008), 'Functional approach of flexibly modeling generalized longitudinal data and survival time', *Journal of Statistical Planning and Inference* **138**(4), 995–1009.

Ye, W., Lin, X. and Taylor, J. M. G. (2008*a*), 'A penalized likelihood approach to joint modeling of longitudinal measurements and time-to-event data', *Statistics and Its Interface* **1**, 34–45.

Ye, W., Lin, X. and Taylor, J. M. G. (2008*b*), 'Semiparametric modeling of longitudinal measurements and time-to-event data – A two-stage regression calibration approach', *Biometrics* **64**(4), 1238–1246.

Yu, B. and Ghosh, P. (2010), 'Joint modeling for cognitive trajectory and risk of dementia in the presence of death', *Biometrics* **66**(1), 294–300.

Yu, M., Law, N. J., Taylor, J. M. G. and Sandler, H. M. (2004), 'Joint longitudinal-survival-cure models and their application to prostate cancer', *Statistica Sinica* **14**(3), 835–862.

Yu, M., Taylor, J. and Sandler, H. (2008), 'Individual prediction in prostate cancer studies using a joint longitudinal-survival-cure model', *Journal of the American Statistical Association* **103**, 178–187.

Zeng, D. and Cai, J. (2005), 'Asymptotic results for maximum likelihood estimators in joint analysis of repeated measurements and survival time', *The Annals of Statistics* **33**(5), 2132–2163.

Zeng, D. and Lin, D. (2007), 'Maximum likelihood estimation in semiparametric models with censored data (with discussion)', *Journal of the Royal Statistical Society B* **69**, 507–564.

Zheng, Y. and Heagerty, P. (2007), 'Prospective accuracy for longitudinal markers', *Biometrics* **63**, 332–341.

Zhu, H., Ibrahim, J., Chi, Y. and Tang, N. (2012), 'Bayesian influence measures for joint models for longitudinal and survival data', *Biometrics* **68**, 954–964.

27

Familial Studies

Karen Bandeen-Roche

Department of Biostatistics, Johns Hopkins Bloomberg School of Public Health

CONTENTS

27.1 Overview 549
27.2 Notation 550
27.3 Analyses aimed exclusively at determining relationships of individuals' failure times to predictor variables 551
27.4 Characterizing familial associations 552
27.4.1 Summary measures of dependence 552
27.4.2 Association through frailty modeling 553
27.4.3 Association through copula modeling and relation to frailty modeling 553
27.4.4 Association modeling specific to familial data: Simple random family sampling 555
27.4.5 Association modeling specific to familial data: Case-control designs . 556
27.5 Age- and time-dependence of failure time associations 558
27.5.1 Checking the fit of parametric copula models for association 558
27.5.2 Nonparametric estimation of the conditional hazard ratio as a function of time 559
27.6 Competing risks 559
27.6.1 Approaches generalizing the conditional hazard ratio function 560
27.6.2 Alternative approaches to describing and estimating failure time associations subject to competing risks 562
Bibliography 563

27.1 Overview

In many ways, familial studies simply are special cases of sampling designs in which multivariate failure times arise; however, they also have special emphases and features which are the topic of this chapter. To assist in exemplifying these, we shall carry the following example throughout this chapter: the study of dementia in the Cache County Study on Memory in Aging. Initiated in 1995, this investigation was designed to examine the prevalence of various dementias (Breitner et al., 1999). Its study design targeted the entire 65-year and older population of Cache County, Utah, U.S.A., thus provides a unique opportunity for the investigation of familial factors underlying dementia. In brief, a population-based sample of study participants was ascertained and assessed for specific dementias, as well as many other characteristics and outcomes. Then, for each participant, information was collected from all first-degree relatives, and dementia diagnoses also made on these. In all, the study provides information on dementia occurrence, age at onset among those experiencing the disease, and either current age or age at death for more than 5,000 family clusters.

Among the various applications of multivariate failure-time analysis, familial studies are distinguished as follows: Associations among family members' failure times often are of at least equal interest as the relationships of individuals' failure times to risk factors. Such associations may reflect disease heritability, congenital frailty or robustness, influences on health of a shared home environment, and other such constructs that are difficult to ascertain otherwise than by a family study. This is not to say that familial studies do not find usefulness in methods to account for clustering in making inferences on relationships of individuals' failure times to risk factors, but only that the study of association frequently is of interest and not merely a nuisance to be handled.

Once one addresses these primary features of failure-time analyses in family studies-handling, characterizing, and interpreting within-family associations, a second feature frequently comes into play: that the failure event of primary interest may be subject to competition from the occurrence of one or more other events. This is universally the case in familial studies of disease onset, where death free of disease precludes the possibility of future disease onset: a truly significant issue in studies of older adults and persons with serious comorbid disease. Competing risks pose all the issues for multivariate failure time analysis as they do for the analysis of univariate failure times, and then additionally challenge not only the analysis of within-family associations but the definition of these.

Following a brief review of analysis primarily aiming to characterize relationships of individuals' failure times to potential determinants in familial settings, this article proceeds to review methodology focused on the characterization of associations and analysis in the presence of competing risks. Three well-studied frameworks for estimating failure time associations are considered along with methods for estimating these: simple empirical measures, frailty models, and copulas. Models specifically directed to familial settings are considered. Wherever models are applied, methods to evaluate their fitness for describing analytic data are needed. A good number of these have been developed for models of failure time association in familial data. The article concludes with a survey of methods to accommodate competing risks in the study of failure-time association.

27.2 Notation

Let us consider a sample of families $i = 1, \ldots, n$ with the ith family represented in the sample by m_i members. Per family the idealized outcomes to be studied are times to an event of interest, say ages at dementia onset, $T_{i1}, \ldots, T_{im_i}$ with associated survival function $S_i(t_{i1}, \ldots, t_{im_i})$, cumulative distribution function $F_i(t_{i1}, \ldots, t_{im_i})$ and, for absolutely continuous F, density function $f_i(t_{i1}, \ldots, t_{im})$ and hazard function $\lambda_i(t_{i1}, \ldots, t_{im})$. Corresponding marginals for each jth family member are $S_{ij}(t_{ij})$, $F_{ij}(t_{ij})$, and so on. Frequently we suppress the i subscript in the succeeding text. As is usually the case for failure time analysis, outcomes are observed as the occurrence of the event itself or of a censoring event such that data collection ceases before the event of interest occurs. Then, the observable data are times $(Y_{i1}, \ldots, Y_{im})$ to occurrence of either the target or censoring event and indicators $\delta_{ij} = 1$ if one observes the event of interest for member j of family i and is 0 otherwise, for $j = 1, \ldots, m_i$; $i = 1, \ldots, n$.

27.3 Analyses aimed exclusively at determining relationships of individuals' failure times to predictor variables

If the primary goal is to explicate the relationship of individuals' event timings to risk factors or other determinants, and within-family associations are considered as nuisance features that must be accommodated for the sake of making correct inferences but are not otherwise of interest, then marginal models provide a reasonable way to proceed. These describe the population distribution of individual failure times T_{ij}, conditional possibly on covariates $X_{ij} = (X_{ij1}, \ldots, X_{ijp})$ but not on unobservable factors through which family members' failure times may otherwise be inter-related $j = 1, \ldots, m_i$; $i = 1, \ldots, n$.

Arguably the most widely implemented method for accomplishing such analysis is the marginal Cox modeling approach as introduced by Lee, Wei & Amato (1992) and elaborated by Lin (1994) and Spiekerman & Lin (1998). Of interest are the person-wise marginal hazard functions: for each jth member of each ith family,

$$\lambda_j(t; X_{ij}) = \lambda_{oj}(t)e^{\beta' X_{ij}(t)}, \tag{27.1}$$

where $\lambda_{oj}(t)$ are baseline hazard functions and the notation $X_{ij}(t)$ highlights that the covariates may be time-varying. Frequently in familial studies it makes sense to assume $\lambda_{oj}(t) = \lambda_o(t)$ for all j, and indeed (27.1) makes clearest sense in cases of equal cluster sizes with fixed family relationships such as a sample of mother-father-eldest-child trios. Analysis then proceeds by standard Cox modeling with a robust estimator variance correction akin to generalized estimating equations methodology (Liang & Zeger, 1986). Specifically, one obtains regression coefficient estimators via the standard partial likelihood approach, imposing an independence working covariance model. Then, the variance estimator is obtained as the sandwich product $I^{-1}VI^{-1}$, where I^{-1} is the inverse of the observed partial likelihood information matrix (divided by n)-the usual variance estimate, and V is the empirical variance of the partial score function; see for example, Spiekerman and Lin (1998), Section 2.5. Increasingly many software packages fit the needed robust variance correction as a matter of course; for example, the book by Therneau and Grambsch (2000) elucidates code for Splus and SAS.

Two streams of work have augmented the set of tools available for marginal modeling of failure times in familial studies. The first has addressed distributional specifications other than the Cox proportional hazards model. Alternatives addressed include a class of linear transformation generalizations to the proportional hazards model (Cai, Wei & Wilcox, 2000), the accelerated failure time model (Lee, Wei & Ying, 1993), discrete-time proportional hazards models (Guo & Lin, 1994; Ross & Moore, 1999) and quantile regression models (Yin & Cai, 2005). The second stream is motivated in that the independence working model approach is convenient but may lose efficiency relative to an analysis that leverages the association structure underlying the data. Cai & Prentice (1995; 1997) and Clegg, Cai & Sen (1999) addressed this issue by incorporating weights into partial likelihood score equations otherwise like those utilized by the methodologies of the previous paragraph. With appropriately selected weight matrices, considerable efficiency gains are possible, particularly when within-family failure time associations are strong and censoring is light. The work of Clegg, Cai & Sen (1999) extended the applicability of the marginal approach to compromises between family member-specific baseline hazards as in (27.1) and a single baseline hazard applicable to all family members, in which some members share baseline hazards and others do not. Gray & Li (2002) extended the work of Cai and Prentice to provide optimal weights; only quite small efficiency gains were obtained except under extremely strong dependence. Alternative approaches addressing the marginal hazards (27.1) but requiring full specification of the association structure have also been developed (e.g., Glidden & Self, 1999; Pipper & Martinussen, 2003).

27.4 Characterizing familial associations

Liang et al. (1995) provided an early review of multivariate failure-time analysis; the review herein adapts considerably from this article to the familial setting. For convenience of notation, development is primarily in the bivariate setting, but in most cases the methodology is applicable to larger and to variably sized clusters.

27.4.1 Summary measures of dependence

Within-family failure time correlations and Kendall's tau coefficients are two common measures of dependence that have been applied to study dependence among failure times clustered within families. Estimation of each is challenged when censoring is present. MacLean et al. (1990) outlined a method for correcting the Pearson correlation for censoring through inverse-weighting by the estimated joint survival function to account for disproportionate loss of later onset cases. Prentice & Cai (1992) alternatively proposed evaluation of correlations between cumulative hazard-transformed failure times as less sensitive to influencing by disparate shapes of the marginal failure time distributions and developed methods to rigorously estimate this quantity. However, as we outline below, Kendall's tau has particularly nice connections to models and measures of association that have come to predominate in multivariate failure time analysis, and so its appearance in the multivariate failure time literature has outstripped that of correlation as a summary measure of association. Recall that Kendall's tau estimates the difference in probabilities that two randomly selected, say, failure time pairs are "concordant" (both individuals of the second pair experience later dementia onset, or both earlier onset, than the members of the first pair) versus "discordant" (one member of the second pair experiences an earlier onset than, and the other later than, the individuals of the first pair). Then its appeal for failure time analysis is clear: between-pair concordance and discordance may be adjudicated in some cases in which some failure times are censored.

Still, censoring may render complete adjudication of concordance and discordance impossible, and so methods to accommodate it in estimating tau are needed in failure time studies. For years the most common accommodation for censoring was the method introduced by Brown et al. (1974), in which pairings whose concordance adjudication is inconclusive are discarded in the same way that the Kendall "tau-b" coefficient discards ties. However, the resulting estimator is not consistent when tau is non-null (Wang & Wells, 2000*a*); there are various alternative measures which achieve improved accuracy, for example, one proposed by Wang & Wells (2000*b*) exploiting that tau $= 4\int_0^\infty \int_0^\infty F(x,y)dF(x,y) - 1$ and then plugging in empirical estimators of F. Additionally one summary measure of association tailored specifically to the failure time context has gained particularly widespread attention: the so-called "cross-ratio" or "conditional hazard ratio" function $\theta(t_1, t_2)$. Introduced in time-invariant form by Clayton (1978) and elucidated more generally by Oakes (1989), this measure is defined by

$$\theta(t) = \frac{S(t_1,t_2)/f(t_1,t_2)}{\frac{\partial}{\partial s_1}S(s_1,t_2)|_{s_1=t_1}\frac{\partial}{\partial s_2}S(t_1,s_2)|_{s_2=t_2}} = \frac{\lambda_j(t_j|T_{j'}=t_{j'})}{\lambda_j(t_j|T_{j'}>t_{j'})} \tag{27.2}$$

invariantly for $j = 1,2$ and $j' = 2,1$. In the dementia example, $\theta(t_1, t_2)$ is the factor by which one's risk of dementia onset at age t_1 is increased if his familial relative is known to have been diagnosed with dementia versus remain dementia free as of age t_2 and vice versa.

The widespread usage of the conditional hazard ratio owes at least in part to its direct, natural parameterization, shortly to be described, within the two model classes most frequently employed for the analysis of clustered failure time associations, frailty models and copula models. It may also owe to the fact that $\{\theta(t)-1\}/\{\theta(t)+1\}$ equals a localized Kendall's tau computed from all time pairings with pairwise minimum $= t$. Hence Kendall's tau estimates can be used to estimate θ, as outlined by Oakes (1989) and described in more detail by Viswanathan & Manatunga (2001) and Chen & Bandeen-Roche (2005).

27.4.2 Association through frailty modeling

The frailty model is a random effects formulation for within-cluster association among times-to-events. In their univariate form, frailties are cluster-specific factors that multiply the hazard of each cluster member to be higher or lower than for typical persons otherwise like them. Though first proposed for characterizing subject-specific heterogeneity in demographic models (Vaupel et al., 1979), these provide a natural conceptualization in the familial context where they may conveniently be thought of as summarizing the contributions of shared genetic and environmental factors to longevity, age-at-disease-onset, and the like. In an early paper elucidating application of frailty models to family data, Thomas et al. (1990) likened these to liabilities in genetics and susceptibility in epidemiology.

According to the univariate frailty model, association is generated through non-negative random effects, denoted by A_i, carried by families $i = 1, \ldots, n$ and distributed identically and independently with distribution G typically having mean $= 1$. We now suppress the i notation: conditional on A, a family's, say, ages at dementia onset are assumed to be independent with survival functions $\{S_j^*(t)\}^A$ corresponding to hazard functions $A\lambda_j^*(t)$, for some continuous survival functions S_j^* and family members $j = 1, \ldots, m$. Thus, a family having strong genetic risk for dementia might share onset hazard $\lambda(t|A) = A\lambda^*(t)$ with $A >>1$. Then the marginal survival functions $S_j(t) = \int\{S_j^*(t)\}^a dG(a)$. Many choices of distribution G have been studied; among the most commonly used are Gamma (Clayton, 1978), positive stable (Hougaard, 1986), lognormal (McGilchrist & Aisbett, 1991), and Gumbel (Johnson & Kotz, 1981). In particular taking G as gamma with mean$= 1$ and variance$= \theta - 1$ results in familial failure times with conditional hazard ratio$= \theta$, such that a family member's event risk at t_1 is increased θ-fold if his relative is known to have experienced the event versus not yet experienced it by t_2 invariantly over (t_1, t_2).

27.4.3 Association through copula modeling and relation to frailty modeling

As natural as it is to conceptualize familial failure time associations as arising through a shared frailty induced by genetics and/or shared environment, there have been at least two drawbacks to data analysis using frailty models. The first arises because studies frequently aim to study relationships of individuals' failure times to predictor variables and not only familial failure time associations. Then, frailty models are specified in terms of individual hazard functions conditional on the familial frailty, λ_{ij}^*, whereas interest may be in the marginal hazards λ_{ij}. For only one frailty distribution, proportional hazards form can be maintained for both of these: the positive stable (Hougaard, 1986). The second drawback arises if, additionally, one seeks a semiparametric specification of the individual survival functions, such as a Cox proportional hazards model. Maximum and profile likelihood procedures to fit such models appeared relatively early in the history of frailty modeling (Klein,

1992; Nielsen et al, 1992), but fitting proved slow (Therneau & Grambsch, 2000, p. 232) and sampling distribution development proved complex (e.g., Murphy, 1995). In fact the shared frailty model of the last section can be fit for common frailty distributions such as normal and gamma quite conveniently (e.g., Therneau & Grambsch, 2000, pp. 232-238; Fine et al., 2003), but in the meantime the alternative approach of copula modeling proliferated.

Copulas (Schweizer & Sklar, 1983) are multivariate distributions with marginals that are uniform on the unit interval. These offer a clear device for constructing multivariate survival distributions with easily specifiable marginal components: if $C(u_1, \ldots, u_J)$ is the copula, then under mild regularity conditions $C(S_{i1}, \ldots, S_{iJ})$ is a legitimate multivariate survival function with jth margin $= S_{ij}$.

The Archimedean copula family (Genest & MacKay, 1986) has attracted particularly active attention for the modeling of failure time associations because certain ones parameterize easily in terms of Kendall's tau and the conditional hazard ratio and can be derived as frailty models (Oakes, 1989). Consider a frailty model

$$S(t_1, \ldots, t_J) = \int \prod_{j=1}^{J} \{S_j^*(t_j)\}^a dG(a) \tag{27.3}$$

with $p(x) = E[\exp(-xA)]$ equal to the Laplace transform of the frailty distribution, G. The familial survival function is then an Archimedean copula as follows:

$$\begin{aligned} S(t_1, \ldots, t_J) &= \int a \sum_{j=1}^{J} \log\{S_j^*(t_j)\} dG(a) \\ &= p[-\sum_{j=1}^{J} \log\{S_j^*(t_j)\}] = p[\sum_{j=1}^{J} q\{S_j(t_j)\}] \end{aligned} \tag{27.4}$$

where q is the inverse function of p and the final equation reflects that $S_j(t) = p[-\log\{S_j^*(t)\}]$. Equation (27.4) so elegantly nests the marginal models within the model for association as to evoke a two-stage strategy to analyzing both relationships of individuals failure times to predictor variables and familial failure time associations:

1. Analyze the individual failure distributions using approaches as in Section 27.3
2. Analyze the familial associations using pseudo-maximum likelihood (Gong & Samaniego, 1981) plugging in survival function estimates from step 1 into (27.4) as if they were known and then correct the likelihood-based estimates of variability to account for the uncertainty of estimation from the first step.

Shih & Louis (1995*b*) and Genest, Ghoudi & Rivest (1995) independently elucidated this approach for i.i.d. survival functions (within component), in which the plug-ins from Step 1 to Step 2 are the component-wise Kaplan-Meier estimators. Glidden (2000) extended it to accommodate marginal Cox modeling in the first step.

As a nice by-product, the conditional hazard ratio measure of familial association follows directly as a functional of models of form (27.4) as the symmetric and uniquely defined quantity

$$\theta(t_1, t_2) = \{vq''(v)q'(v)\}|_{v=S(t_1,t_2)}. \tag{27.5}$$

As a result the conditional hazard ratio function may be directly or nearly directly parameterized in (27.4) for a number of the commonly used defining distributions; for example, for the Clayton (1978) model characterized by constant hazard ratio $\theta(t_1, t_2) = \theta$

corresponds to gamma frailty with $p(u) = (1+u)^{1/(1-\theta)}$ resulting in $S(t_1,\dots,t_j) = [\sum_{j=1}^{J}\{S_j(t_j)\}^{1-\theta} - J + 1]^{\frac{1}{1-\theta}}$.

By way of an example, Bandeen-Roche and Liang (2002) applied a gamma copula model and the two-stage estimating approach just described (componentwise i.i.d. case) to a subset of the Cache County data comprising eldest children and their mothers. They obtained a conditional hazard ratio estimator $\hat{\theta}$=2.44 with bootstrap confidence interval 1.78 to 3.33. The point estimate corresponds to a nearly two-and-a-half-fold increase in the hazard of dementia onset at each age for eldest children of mothers diagnosed with dementia as compared to those with mothers remaining dementia free as of a given age, invariantly over the choice of maternal index age (and vice versa).

27.4.4 Association modeling specific to familial data: Simple random family sampling

Failure time analysis explicitly tailored to accommodating associations in families goes back at least to a 1974 paper addressing survival analyses in twin studies (Holt and Prentice, 1974). That paper elucidated stratified Cox modeling with marginal likelihood inference, with families as strata. Restricted models they considered essentially were frailty models with frailties as fixed effects. Most modeling for family data has occurred in the context of either frailty or copula models, and so we restrict our attention to these cases.

Both univariate frailty and Archimedean copula models have natural generalizations that are particularly well suited to familial data. Beginning with frailty models, consider partitioning the overall cluster of the family into K subclusters $I_1,\dots,I_K$ such as all siblings versus each parent ($K=3$). Then, a multivariate survival function for the family is readily defined through a multivariate frailty $A=(A_1,\dots,A_k)$ as

$$S(t_1,\dots,t_m) = \int \prod_{k-1}^{K} \prod_{j \epsilon I_k} \{S_j^*(t_j)\}^{a_k} dG(a_1,\dots,a_k), \tag{27.6}$$

Equation (27.6) permits individual family members to carry different, possibly dependent random effects. Far too many authors to enumerate have proposed models akin to (27.6) to characterize failure times in families. As one example in a substantial body of work by the primary author, Yashin and colleagues used correlated frailties to estimate heritability by distinguishing monozygotic and dizygotic twins (Yashin et al, 1995). Xue & Brookmeyer (1996) proposed a bivariate lognormal version of (27.6). Petersen (1998) parameterized a variety of family models (e.g., genetic versus adoption) by building individuals' frailties from overlapping sums over independent subfrailties. Two properties of the formulation (27.6) have motivated the development of alternative, copula-based models for multi-tiered familial data. First, whereas the survival function for members of any single subcluster (e.g., siblings) has the univariate frailty form of Equation (27.3), the survival function for members $j, j\prime$ of distinct subclusters $I_k, I_{k'}$ (e.g., a child and mother) takes a more general, complex form. Therefore, for convenient multivariate frailty distribution choices, between- and within-subcluster associations may differ substantially both in time dependence and complexity of form. Secondly, the continuing development of multivariate frailty distributions notwithstanding, the diversity and computational tractability of existing univariate frailty distributions remains appealing.

The basic idea of copula-based family models, introduced by Joe (1993) and generalized by Bandeen-Roche and Liang (1996), is to build the model up recursively from the closest familial relationships to more distant ones. Subclusters $I_1,\dots,I_K$ provide the base level of this process; then, at each "level" of familial closeness, subclusters are aggregated from the previous level. Increasing levels then correspond to increasingly dissociated subclusters. For example, a family cluster of two sets of siblings who all share a grandfather would have siblings as the first level and first cousins as the second. The idea can then be elucidated by considering an example with five family members split as $I_1 = \{1, 2\}$ and $I_2 = \{3, 4, 5\}$ two sets of siblings who between them are cousins. Then, the proposed model is

$$S(t) = p_2 \left[q_2 \left\{ p_1 \left(\sum_{j=1}^{2} q_1 S_j(t_j) \right) \right\} + q_2 \left\{ p_1 \left(\sum_{j=3}^{5} q_1 S_j(t_j) \right) \right\} \right], \tag{27.7}$$

where p_1 is the Laplace transform of the sibling association-defining distribution and p_2 is the Laplace transform of the cousin association-defining distribution. Each bivariate marginal of (27.7) follows the Archimedean copula structure and has associated conditional hazard ratio functions as defined by (27.5) above, so that the form of all the bivariate associations can be defined by the researcher. Equation (27.7) and its generalizations incorporate a meaningful constraint: roughly, that the higher-level (e.g., cousin) associations be no stronger than the lower-level (e.g., sibling) associations. In some cases this constraint is reasonable, while in others it may be considered overly restrictive. Partly with this consideration in mind, Andersen (2004) proposed an alternative two-stage method for estimating these models than pseudo-maximum likelihood, which one might think of as pseudo-composite-likelihood. In a first step, the marginal survival functions are estimated just as in the pseudo-maximum likelihood approach. In a second step, each copula represented in Equation (27.7) is estimated through a composite log-likelihood function constructed as sums of bivariate log likelihood contributions for all pairs of individuals relevant to the copula being estimated (e.g., pairs of siblings versus pairs of cousins in separate sums) specifically:

$$\log L^*(\theta, \beta) = \sum_{i=1}^{n} \sum_{(j,h) \epsilon M_i} w_{jh} \log L_{jh}(\theta, \beta)$$

where (j, h) are two family members, w_{jh} are positive weights, θ and β respectively represent the association and marginal parameters, the second sum is taken over the various types of relationships of interest (e.g., siblings and cousins), and L_{jh} is the likelihood corresponding to the bivariate copula hypothesized for the relationship type targeted by the inner term. One chooses weights to account for differences in family size, solves score functions resulting after fixing β at estimates from the first step, and makes inferences accounting for the variability in both steps as described by Andersen (2004). Whereas the copula approach to modeling of multi-tiered failure time associations in families has had occasional application (e.g., Li & Huang, 1998), a literature scan suggests that the frailty approach has been considerably more commonly applied. This having been said, it is arguable that neither approach has had the extent of application one might expect. Matthews et al. (2007) cite as barriers that estimation remains nontrivial and requires specialized software, the models do not easily accommodate complex ascertainment such as proband-based sampling, the conditional hazard ratio measure is non-intuitive for health investigators, covariate modeling of the strength of association is not easily accommodated, and tests for association are formulated in terms of a null parameter that lies on the boundary of the parameter space. They propose discretizing time and then modeling marginal relationships and intra-familial associations via second-order generalized estimating equations akin to those developed by Heagerty & Zeger (1996). If the failure times are fully observed, the models can be fit by standard second-order GEE software. Otherwise, because association is described by explicitly conditioning on other family members outcomes, an additional step to marginalize over the potential outcomes of family members who become censored is required.

27.4.5 Association modeling specific to familial data: Case-control designs

Frequently in familial studies, sampling does not aim to achieve representativeness of families in a target population as achieved, for example, in the Cache County study. Rather, sampling may be indexed on cases and controls or a proband through whom other family members are identified, or is otherwise complex. We exemplify the issues through the simplest version of the case-control proband design.

In this design, cases and controls are ascertained as probands, and one relative is identified for each proband. For example, such a design could have arisen in the Cache County study if sampling had identified a set of prevalent dementia cases and a second set of individuals free of dementia, and then ascertained dementia for one first-degree relative of each sampled individual. In this special scenario, the conditional hazard ratio equals the risk ratio for dementia onset at age t comparing relatives with case probands to those with control probands, conditional on X_0: = the proband's age at dementia onset or age at sampling into the study for case and control probands, respectively. As elucidated by Hsu et al. (1999), this can be seen as

$$\theta(t, X_0) = \frac{\lambda(t|X_0, \delta_0 = 1)}{\lambda(t|X_0, \delta_0 = 0)},$$

where $\delta_0 = 1$ denotes that the proband is a case (and his/her age at dementia onset equals X_0) and $\delta_0 = 0$ denotes that the proband is a control (and he/she is dementia-free as of X_0). If one is willing to parameterize this quantity as $\exp\{\alpha(t, X_0)\}$ for some parametric function $\alpha(t, X_0)$, then the conditional hazard ratio becomes estimable via Cox proportional hazards regression of relatives' failure times on probands' times and an indicator of whether one's proband is a case or control. Appealingly, additional covariates Z may be included as predictors of risk, and the α-function can be parameterized to model dependence of the familial association on time.

A drawback of the strategy just described is reliance on a parametric function to capture the potentially complicated variation of $\lambda(t|X_0, \delta)$ with the ascertainment time. Accordingly Hsu et al. (1999) considered a matched case-control design in which baseline hazards could be stratified by matches $h = 1, \ldots, H$, resulting in $\lambda_{hj}(t|X_{ho}, \delta_{jho}, Z_{hj}) = \lambda_{oh}(t) \exp\{Z'_{hj}\beta + \delta_{hjo}\alpha(t|X_{ho}, \theta)\}$. They proposed estimation by a pseudo-partial likelihood equation constructed as if individuals within families and matches were independent but equipped with a sandwich-type covariance estimator to assure valid inferences in the same way as is accomplished by GEE. In cases of no covariates Z, and bivariate family data, it was noted that a simple ratio of concordance and discordance counts within bins could be used; a localized Kendall's tau like that is proposed by Oakes (1989) and detailed in Section 27.5.1. below.

Methodology has also been developed for analyses including not only the relative data as outcomes but the proband data in the design just described. To this end, Li, Yang & Schwartz (1998) proposed a likelihood combining conditional likelihood contributions for the probands, as one would have in a standard matched case-control failure time analysis (Prentice & Breslow, 1978), with gamma Archimedean copula contributions for the case relatives and the control relatives. Estimation by maximum likelihood was proposed. This approach requires that the baseline hazard function $\lambda_{oh}(t)$ be invariant over matches and follow a parametric or step-function form. Subsequently it was generalized to allow for non-parametric specification of $\lambda_{oh}(t)$ (Shih & Chatterjee, 2002). This was accomplished through sequential two-stage fitting iterating between nonparametric estimation of the baseline hazard function and pseudo-maximum likelihood estimation of Li et al. (1998) estimating function with baseline hazard estimates fixed from the prior step.

27.5 Age- and time-dependence of failure time associations

Much of the literature on the modeling of familial failure time association has relied on the gamma shared frailty/copula model, which imposes a conditional hazard ratio for pairwise association that is constant in all time dimensions. In fact, considerably many investigations of familial association have in mind explicitly a time varying association function. For example, an important question for the field of cognitive aging is whether early-onset dementia is more strongly heritable than late-onset dementia (Silverman et al., 2005). Shared frailty/copula distributions other than gamma do indeed generate time-varying conditional hazard ratio functions, as per Equation (27.5). For example, positive stable frailty imposes association that is strongest early on and declines progressively as time goes along, with a conditional hazard ratio function asymptoting to infinity as the bivariate survival function approaches one and to unity (no association) as the bivariate survival function approaches 0. There is a considerable literature on methods to diagnose and test the fit of various copula choices to clustered failure times, which we now summarize. We then proceed to outline nonparametric approaches to directly describing conditional hazard ratio function dependence on time. All of these are easiest to describe for bivariate failure times (e.g., sib pairs).

27.5.1 Checking the fit of parametric copula models for association

A hallmark has been to identify a function that has a signature shape for a given choice of Archimedean copula (AC) family, estimate it empirically, and then assess whether the given AC choice is consistent with the data. Genest & Rivest (1993) implemented this idea through the bivariate probability integral transformation distribution $K(v) = pr[C\{F_1(T_1), F_2(T_2)\} \leq v]$. $K(v)$ has a signature shape as a function of v for each given AC family, and for fully observed data, its empirical counterpart $K_n(v)$ is easily estimable as the empirical distribution function of the number of pairs (T_{h1}, T_{h2}) with which an index pair (T_{i1}, T_{i2}) is concordant, taken over index pairs $i = 1, \ldots, n$. Genest and Rivest provided experiment-wise confidence bands for $v - K_n(v)$ which could then be plotted together with the signature functions $v - K(v)$, versus v, over a variety of AC choices. Later Wang & Wells (2000*b*) extended this idea to allow for censored failure times albeit without confidence bands; instead they proposed comparison of empirical and hypothesized versions of $K(v)$ by L^2-norm distance.

In contrast, Viswanathan & Manatunga (2001) and Chen & Bandeen-Roche (2005) both developed diagnostic displays exploiting that $\{\theta(t) - 1\}/\{\theta(t) + 1\}$ equals a localized Kendall's tau, so that $\theta(t)$ can be estimated as $\hat{\theta}(t)$ equal to the ratio of the number of concordances to the number of discordances among all inter-familial comparisons with pairwise minimum failure times $= t$. This idea has the disadvantage of having to deal with sparseness: for an absolutely continuous survival function, we expect at most one comparison with pairwise minimum failure time $= t$. The situation is helped a little when $S(t_1, t_2)$ has AC form. In this case, $\theta(t)$ relates to t only through $S(t_1, t_2)$ hence varies only univariately rather than bivariately, but in any case some form of smoothing is needed. Viswanathan & Manatunga (2001) proposed to smooth the plot of $\hat{\theta}(t)$ versus the Dabrowska (1988) nonparametric survival function estimator. They then evaluated the slope in a plot of the resulting smooth for each survival function estimate v versus $-1/\log(v)$ so as to explicitly compare positive stable (positive slope) versus gamma (flat) copulas. Chen & Bandeen-Roche (2005) proposed to bin the bivariate survival distribution domain and then compute the ratio of the number of concordances to the number of discordances among all inter-familial comparisons with pairwise minimum failure times in each bin. They then compared the step function resulting by plotting these ratios versus the Dabrowska (1988) estimator of the bivariate survival function domain to the signature curve of $\theta(t)$ versus $S(t)$ determined by various AC choices visually and in L^2-norm. Their idea has the advantage of visualizing the conditional hazard ratio, hence both the strength and time dependence of association.

In addition, a number of methodologies have been developed to assess the fit of bivariate survival data to specific AC models, primarily, gamma. Shih & Louis (1995*b*) proposed a graphical method for checking the adequacy of a gamma AC assumption, which was subsequently augmented by Glidden (1999) to include numerical techniques. Shih (1998) proposed an alternative test for gamma AC form, which employed the concordance counting strategy we outlined above, and she derived the asymptotic distribution of the test statistic. Manatunga & Oakes (1996) suggested a diagnostic procedure for the goodness-of-fit of uncensored data to positive stable frailty choice.

27.5.2 Nonparametric estimation of the conditional hazard ratio as a function of time

Rather than to fit a parametric model for the conditional hazard ratio and assess its adequacy in describing the data, one might prefer to describe the conditional hazard ratio as a function of time directly and nonparametrically. There have been essentially two approaches to this task. The first is the binning approach employed by Chen & Bandeen-Roche (2005) or its more general application in two dimensions (Bandeen-Roche & Ning, 2008). The latter approach was primarily developed in the context of competing risks data; we defer its discussion to the respective section of this article.

A second approach first plugs into the general formulation of the conditional hazard ratio function (27.2), or a function thereof, nonparametric estimators of the multivariate survival function or cumulative hazard function (Fan, Hsu & Prentice, 2000; Fan, Prentice & Hsu, 2000). Taken literally this approach encounters the same sparseness difficulties as highlighted for the local Kendall's tau in the previous section; to address this, Fan and colleagues proposed to integrate with respect to the bivariate failure time density over finite failure time regions $[0, t_1]$x$[0, t_2]$. Time dependence can then be explored by varying (t_1, t_2). The work spanned a variety of association measures including the conditional hazard ratio, its inverse, and Kendall's tau.

Consideration of time-variation in the conditional hazard ratio may be exemplified by application of the binning approach of Chen & Bandeen-Roche (2005) to the Cache County data on mothers and their eldest children, as reported by Bandeen-Roche & Liang (2002). Conditional hazard ratios were estimated by quintiles of bivariate survival (dementia-free) probability. These indicated stronger familial association in dementia onset by far in the quintile of earliest failure ($\hat{\theta} = 8.86$ for t: $S(t) > 0.8$) than later ($\hat{\theta} = 2.58$, 2.20, 2.92, 2.39 for $S(t)$ decreasing through the subsequent quintiles).

27.6 Competing risks

Competing risks are frequent in familial studies of age-to-disease onset, where after all any disease may be censored by the semi-competing risk of death prior to disease onset. Yet up to 2000 or so, there had been virtually no consideration of competing risks in the multivariate failure time literature. To ignore censoring as competing may be reasonable if the main goal is to describe the individual failure hazards as in Section 27.3; then the cause-specific quantities that are characterized by such analyses (Prentice et al., 1978) may be scientifically interpretable and useful. In contrast, incomplete observation due to a competing risk poses serious conceptual challenges for determining associations.

Bandeen-Roche & Liang (2002) elucidated the challenges as follows. The description is taken in large part from that article, with only minor changes. Consider the simplest setting of onset of a disease such as dementia with death as a semi-competing risk. Then, the failure quantities to be analyzed for a given person are (a) the failure time, T, to the first event among dementia onset or death, and (b) a code, K, for cause of failure, say, 1 if onset of dementia is observed and 2 if death is observed prior to dementia onset. The primary approaches to describing failure time associations—copula and shared frailty models—employ a joint survival function defined in terms of survival and density functions that marginalize over causes other than the target cause. For a failure type that is not inherently subject to competition, such as death, such marginalized functions are well defined and unambiguous. Specifically, it is sensible to define a death time T_D that equals the targeted failure time T if the failure type is death $K = 2$, and is distinct otherwise. Then, $pr\{T_D > t\} = pr\{T_D > t|K = 2\}pr\{K = 2\} + pr\{T_D > t|K = 1\}pr\{K = 1\}$, where $pr\{T_D > t|K = 1\}$ describes the death time distribution given disease onset prior to death, and $pr\{T_D > t|K = 2\}$ equals the cause-conditional survival function $pr\{T_D > t|K = 2\}$. In contrast, suppose that T_O defines a dementia onset time that equals T if $K = 1$. Because onset of dementia cannot occur after death, it could be argued that:

A. $pr\{T_O > t|K = 2\} = 1$ for all t; e.g., dementia onset never occurs if death occurs first.

B. $pr\{T_O > t|K = 2\} = 0$ for all t; e.g., the probability of dementia onset is 0 if death occurs first. This equates the marginal and cause-specific dementia onset time distributions.

C. The marginal and cause-conditional dementia onset distributions are equivalent; e.g., dementia onset time is only well defined on a support conditioned on $K = 1$.

D. $pr\{T_O > t|K = 2\}$ follows some form other than those already specified; e.g., T_O is latent everywhere other than the support conditioned on $K = 1$ (e.g., Zheng & Klein, 1995).

Whereas none of the above is uniformly correct, the approach one selects must be targeted clearly to the scientific question and goals. In the years since 2000, the study of multivariate failure time associations in the face of competing risks has become among the most active areas of research in multivariate failure time analysis.

27.6.1 Approaches generalizing the conditional hazard ratio function

In Section 12.6 of his volume on the analysis of multivariate survival data, Hougaard (2000) suggested that the univariate cause-specific hazard function could be adapted to extend the conditional hazard ratio to competing risks data. Bandeen-Roche & Liang (2002) proposed such a measure, terming it the conditional cause-specific hazard ratio (CCSHR):

$$\theta_{CS}(t; k_1, k_2) = \frac{\lambda_{1,k_1}(t_1|T_2 = t_2, K_2 = k_2)}{\lambda_{1,k}(t_1|T_2 < t_2)} \quad (27.8)$$
$$= \frac{S(t_1, t_2)/f(t_1, t_2, k_1, k_2)}{\left\{\int_{t_2}^{\infty}\sum_{k=1}^{M} f(t_1, t, k_1, k)dt\right\}\left\{\int_{t_1}^{\infty}\sum_{k=1}^{M} f(t, t_2, k, k_2)dt\right\}}$$

where f is the bivariate cause-specific incidence function defined by $f(t, k)$

$$= \lim_{(\Delta t_1, \Delta t_2)\downarrow 0} pr(t \leq T \leq t + \Delta t, K = k)/\left(\prod_{m=1}^{2}\Delta t_m\right), \quad (27.9)$$

where T and K denote vector quantities and M denotes the number of competing causes. The CCSHR is the factor by which an individual's risk of failure at t_1 due to cause k_1 is increased if a specific cluster partner is known to have failed at t_2 due to cause k_2 versus not yet failed at all by t_2. Bandeen-Roche & Liang preferred it (27.8) to an obvious competitor, a cause-conditional CHR $= \lambda_1(t_1, k_1|T_2 = t_2, K_2 = k_2)/\lambda_1(t_1, k_1|T_2 > t_2, K_2 = k_2)$, for being defined in terms of quantities that are unambiguous and empirically identifiable. The Cache County study "conditional hazard ratios" reported in Section 27.5.2 in fact were CCSHRs—factors by which an eldest child's hazard of dementia onset at age t_1 is increased if his mother was diagnosed with, versus living without, dementia at age t_2.

Bandeen-Roche & Liang (2002) proceeded to propose an Archimedean copula-like model built up from two frailties: a scalar frailty to modify the (overall) risk of first failure among the competing causes, and an M-variate shape process to induce familial heterogeneity in the tendency to fail from one as opposed to the other of the M causes. Their formulation inherited many of the desirable features of AC models, for instance, a direct connection to the CCSHR very much like Equation (27.5) for the conditional hazard ratio in the noncompeting setting. However straightforward implementations of the modeling framework proved highly sensitive to mis-specification. Shih & Albert (2010) also proposed partitioning of the CCSHR into factors representing association of times to first failure and associations among failure types but did so through direct modeling of the conditional hazard ratio for times to first failure followed by estimation of the odds ratio for association of failure due to one versus the other cause with the failure cause of one's relative, conditional on the pairs' first failure times. Their proposed estimation strategy was nonparametric; we outline it two paragraphs below. Gorfine & Hsu (2011) proposed a formulation in which each cause-specific hazard was multiplied by its own frailty process and the M processes per family may be inter-correlated. It includes the model of Bandeen-Roche & Liang (2002) as a special case but also a more general class appealingly including the time-invariant M-variate normal. Cox-like regression modeling of the individual hazards was also incorporated. Both parametric and non-parametric maximum likelihood implementations exhibited solid performance and robustness to mis-specification.

Much of the work subsequent to that of Bandeen-Roche & Liang (2002) has relied on non- or semiparametric estimation of the association function. A particularly straightforward nonparametric method of CCSHR estimation was proposed by Bandeen-Roche & Ning (2008). These authors exploited the close relationship of the conditional hazard ratio to Kendall's tau as suggested by Oakes (1989) and developed in detail by Viswanathan & Manatunga (2001) and Chen & Bandeen-Roche (2005), whereby $\theta(t)$ can be estimated as the ratio of the number of concordances to the number of discordances among all inter-familial comparisons with pairwise minimum failure times $= t$. They noticed, and proved using U-statistic theory, that if one further filters the inter-familial comparison to not only have pairwise minimum failure times equal to t but also to have desired causes k_1, k_2 associated with those times, one has a quantity that consistently estimates the CCSHR under certain conditions. To accomplish the smoothing whose need was outlined in Section 27.5.1, they proposed to bin much as in the paper by Chen & Bandeen-Roche (2005), but in two dimensions.

Shih & Albert (2010) also employed binning for the modeling and estimation of their CCSHR factorization described in the previous paragraph. However, they proposed a quite different estimation strategy in which all failure types are pooled to estimate a conditional hazard ratio for first failure time, and odds ratios for association of failure types are computed empirically among pairs with both failures observed within a given bin. Conditional hazard ratio modeling was proposed by a sequential two-stage strategy in which an estimator of the bivariate survival function, hence the conditional hazard ratio, can be built recursively across bins as per Nan et al. (2006).

Cheng & Fine (2008) alternatively proposed an empirical process plug-in estimator of the CCSHR. This can be seen as feasible upon considering an equivalent formulation for the CCSHR derived by these authors: a ratio of the bivariate cause-specific hazard to the product of the respective marginal quantities, $\theta_{CS}(t_1, t_2; k_1, k_2) = \lambda(t_1, t_2, k_1, k_2)/\{\lambda_1(t_1, k_1)\lambda_2(t_2, k_2)\}$. The method exhibits similar accuracy and precision as that proposed by Bandeen-Roche & Ning (2008) but enjoys a considerable advantage in computing time to derive inferences due to its provision of a valid plug-in standard error.

The work described in the preceding paragraphs made no meaningful attempt to extend applicability beyond bivariate clustering to larger and variably sized clusters, a crucial capability if the methodology is to have broad applicability to family data. The extension for the estimator of Bandeen-Roche & Ning (2008) was provided by Cheng, Fine & Bandeen-Roche (2010), together with an easily implemented test for constancy of association over different time regions.

Many of the estimators just described have been exemplified using the Cache County data. Both the Bandeen-Roche & Ning ("BRN"; 2008) and Cheng & Fine ("CF"; 2008) estimators were applied to obtain time-invariant CCSHRs for association in ages of dementia onset between eldest children and their mothers. The two methods agreed closely, with BRN estimate (95% confidence interval) of 2.98 (1.98, 4.30) (Bandeen-Roche & Liang, 2002) and CF estimate (95% confidence interval) of 2.90 (1.72, 4.08) (Cheng & Fine, 2008). Bandeen-Roche & Ning (2008) proceeded to obtain CCSHRs age-varying by quadrants defined by eldest child 75 years of age or less versus older than 75 and mother 80 years of age or less versus older than 80. To their surprise, they found these to be actually less for shared early onset ($\hat{\theta} = 3.81$) than for shared later onset ($\hat{\theta} = 5.89$), although confidence intervals overlapped. Cheng, Fine & Bandeen-Roche (2010) addressed this inconsistency with the bulk of existing literature through the estimation of sibling associations ("CFBR-A") and sibling-mother associations ("CFBR-B") in analyses including all siblings, and not only the eldest. Here the estimates of CCSHR for shared early onset ($\hat{\theta} = 3.45$ and 3.55 for CFBR-A and CFBR-B) did exceed those for shared later onset ($\hat{\theta} = 2.16$ and 2.90 for CFBR-A and CFBR-B). The comparison argues the utility of the broader inclusion of data the more fully multivariate methodology of Cheng, Fine & Bandeen-Roche (2010) allows.

27.6.2 Alternative approaches to describing and estimating failure time associations subject to competing risks

Cheng, Fine & Kosorok (2007) proposed two alternative measures: one comparing bivariate cumulative cause-specific hazards to the product of the marginal counterparts, and the second doing likewise for the cumulative incidence function (i.e., the bivariate integral through (t_1, t_2) of Equation 27.9). Specifically the measures are the cumulative hazards ratio $\Lambda(t_1, t_2, k_1, k_2)/\{\Lambda_1(t_1, k_1)\Lambda_2(t_2, k_2)\}$ and the cumulative incidence function ratio $F(t_1, t_2, k_1, k_2)/\{F_1(t_1, k_1)F_2(t_2, k_2)\}$, each a ratio of a bivariate quantity to the product of its respective marginals. These measures have the appealing property of being easily estimable, and their inferential properties then characterized, by replacing the population distributional quantities with nonparametrically estimated plug-ins. Cheng and colleagues delineated this approach.

The methodology was subsequently extended to accommodate clusters of more than two exchangeable individuals (Cheng, Fine & Kosorok, 2010). Ultimately, Cheng & Fine (2012) developed an Archimedean copula framework for modeling the bivariate cumulative incidence function in terms of its marginals and equipped it with two-stage fitting strategies, inference procedures, and goodness-of-fit testing. Shih & Albert (2010) also applied their two-component modeling strategy outlined in the previous section to bivariate cumulative incidence function estimation. Scheike and colleagues (2010) proposed association analysis for cause-specific failure times through a cross-odds ratio-type measure, stated here for members j and h of family i with respect to cause "1": $\pi_{j|h}(t) =$

$$\frac{pr(T_{ij} \leq t_1, K_{ij} = 1 | T_{ih} \leq t, K_{ih} = 1)/pr\{T_{ij} \leq t, K_{ij} = 1)^c | T_{ih} \leq t, K_{ih} = 1\}}{pr(T_{ij} \leq t, K_{ij} = 1)/pr\{(T_{ij} \leq t, K_{ij} = 1)^c\}}$$

where A^c denotes the complement to event "A". They originally specified this quantity through a model multiplying cumulative incidence functions by frailties thusly as to yield a semiparametric generalized additive model for the marginal cumulative incidence functions, $-log\{1 - F_k(t|x_{ij}Z_{ij})\} = \eta(t)^T x_{ij} + (Y^T z_{ij})t$. Equivalently, their formulation is an Archimedean copula model for the joint cause-specific survival function, $pr\{(T_{ij} \leq t, K_{ij} = 1, T_{ih} \leq s, K_{ih} = 1)^C | x_{ij}, z_{ij}, x_{ih}, z_{ih}\}$. Estimation was proposed by a two-stage approach. In the first, marginal cumulative incidence regression is carried out via an estimating equations approach as proposed by Scheike et al. (2008). The association parameters are estimated in a subsequent stage by a plug-in estimating equation. Because the joint cumulative incidence function can be expressed as a function of the marginal cumulative incidence functions and the cross-odds ratio, estimation for the cross-odds ratio then follows. In a subsequent paper, Scheike & Sun (2012) developed an estimating equations approach to estimate the cross-ratio directly, possibly as a function of covariates. The work was motivated by familial data from the Danish Twin Registry. A model similar to the frailty-based formulation of Scheike et al. (2010) was proposed by Katsahian et al. (2006), except with cumulative incidence regression given conditionally on the frailty rather than marginally and Cox-like proportional hazards form for the subdistribution hazard instead of semiparametric additive for the cumulative incidence function itself. This work generalizes the competing risk subdistribution regression methodology of Fine & Gray (1999) from the univariate to the multivariate setting. Dixon et al. (2011) later extended this work to show, in the case of gamma frailty, that the conditional regression of the subdistribution hazard at issue can be stated simply in terms of the parameters of a Fine & Gray (1999) marginal subdistribution hazard regression. Following on this insight, the authors proposed estimation by a three-step procedure generalizing an idea of Pipper & Martinussen (2003) to the competing risks setting. The procedure iterates amongst prediction of the family-wise frailties, nonparametric estimation of the cumulative subdistribution hazard function, and a plug-in estimator for the parametric part of the model, the regression and gamma frailty variance terms.

Bibliography

Andersen, E. W. (2004). Composite likelihood for family studies. *Biostatistics*, 5:15-30.

Bandeen-Roche, K. J. and Liang, K. Y. (1996). Modeling failure-time associations in data with multiple levels of clustering. *Biometrika*, 83:29-39.

Bandeen-Roche, K. and Liang, K. Y. (2002). Modeling multivariate failure time associations in the presence of a competing risk. *Biometrika*, 89:299-314.

Bandeen-Roche, K. and Ning, J. (2008). Non-parametric estimation of bivariate failure time associations in the presence of a competing risk. *Biometrika*, 95:221-232.

Breitner, J. C. S., Wyse, B. W., Anthony, J. C., Welsh-Bohmer, K. A., Steffens, D. C., Norton, M. C., Tschanz, J. T., Plassman, B. L., Meyer, M. R., Skoog, I., and Khachaturian,

A. (1999). APOE-e4 count predicts age when prevalence of AD increases, then declines: The Cache County Study. *Neurology*, 53:321-31.

Brown, W. B. Jr., Hollander, M., and Korwar, R. M. (1974). Nonparametric tests of independence for censored data with applications to heart transplant studies. *Reliability and Biometry: Statistical Analysis of Lifelength*, 327-54.

Cai, J. and Prentice, R. L. (1995). Estimating equations for hazard ratio parameters based on correlated failure time data. *Biometrika*, 82:151-184.

Cai, J. and Prentice, R. L. (1997). Regression estimation using multivariate failure time data and a common baseline hazard function model. *Lifetime Data Analysis*, 3:197-213.

Cai, T., Wei, L. J., and Wilcox, M. (2000). Semiparametric regression analysis for clustered failure time data. *Biometrika*, 87:867-878.

Chen, M. C. and Bandeen-Roche, K. (2005). A diagnostic for association in bivariate survival models. *Lifetime Data Analysis*, 11:245-64.

Cheng, Y. and Fine, J. P. (2008). Nonparametric estimation of cause-specific cross hazard ratio with bivariate competing risks data. *Biometrika* 95:233-240.

Cheng, Y. and Fine, J. P. (2012). Cumulative incidence association models for bivariate competing risks data. *Journal of the Royal Statistical Society, Series B*, 74:183-202.

Cheng, Y., Fine, J. P., and Bandeen-Roche, K. (2010). Association analyses of clustered competing risks data via cross hazard ratio. *Biostatistics*, 11:82-92.

Cheng, Y., Fine, J. P., and Kosorok, M. R. (2007). Nonparametric association analysis of bivariate competing-risks data. *Journal of the American Statistical Association*, 102:1407-1415.

Cheng, Y., Fine, J. P., and Kosorok, M. R. (2010). Nonparametric association analysis of exchangeable clustered competing risks data. *Biometrics*, 65:385-393.

Clayton, D. G. (1978). A model for association in bivariate life tables and its application in epidemiological studies of familial tendency in chronic disease incidence. *Biometrika*, 65:141-151.

Clegg, L. X., Cai, J., and Sen, P. K. (1999). A marginal mixed baseline hazards model for multivariate failure time data. *Biometrics*, 55:805-812.

Dabrowska, D. M. (1988). Kaplan-Meier estimate on the plane. *Annals of Statistics*, 16:1475-1489.

Dixon, S. N., Darlington, G. A., and Desmond, A. I. (2011). A competing risks model for correlated data based on the subdistribution hazard. *Lifetime Data Appl.*, 17:473-495.

Fan, J. J., Hsu, L., and Prentice, R. L. (2000). Dependence estimation over a finite bivariate failure time region. *Lifetime Data Analysis*, 6:343-355.

Fan, J. J., Prentice, R. L., and Hsu, L. (2000). A class of weighted dependence measures for bivariate failure time data. *Journal of the Royal Statistical Society, Series B*, 62:181-190.

Fine, J. P., Glidden, D. V., and Lee, K. E. (2003). A simple estimator for a shared frailty regression model. *Journal of the Royal Statistical Society, Series B*, 65:317-329.

Fine, J. P. and Gray, R. J. (1999). A proportional hazards model for the subdistribution of a competing risk. *Journal of the American Statistical Association*, 94:496-509.

Genest, C., Ghoudi, K., and Rivest, L. P. (1995). A semiparametric estimation procedure of dependence parameters in multivariate families of distributions. *Biometrika*, 82:543-552.

Genest, C. and MacKay, J. (1986). The joy of copulas: bivariate distributions with uniform marginals. *The American Statistician*, 40:280-283.

Genest, C. and Rivest, L. P. (1993). Statistical inference procedures for bivariate Archimedean copulas. *Journal of the American Statistical Association*, 88:1034-1043.

Glidden, D. V. (1999). Checking the adequacy of the gamma frailty model for multivariate failure times. *Biometrika*, 86: 381-393.

Glidden, D. V. (2000). A two-stage estimator of the dependence parameter for the Clayton-Oakes model. *Lifetime Data Analysis*, 6:141-156.

Glidden, D. V. and Self, S. G. (1999). Semiparametric likelihood estimation in the Clayton Oakes failure time model. *Scandinavian Journal of Statistics*, 26:363-372.

Gong, G. and Samaniego, F. (1981). Pseudo maximum likelihood estimation: Theory and applications. *Ann. Statist.*, 9(14): 861-869.

Gorfine, M. and Hsu, L. (2011). Frailty-based competing risks model for multivariate survival data. *Biometrics*, 67:415-426.

Gray, R. J. and Li, Y. (2002). Optimal weight functions for marginal proportional hazards analysis of clustered failure time data. *Lifetime Data Analysis*, 8:5-19.

Guo, S. W. and Lin, D. Y. (1994). Regression analysis of multivariate grouped survival data. *Biometrics*, 50:632-639.

Heagerty, P. J. and Zeger, S. L. (1996). Marginal regression models for clustered ordinal measurements. *Journal of the American Statistical Association*, 91:1023-1036.

Holt, J. D. and Prentice, R. L. (1974). Survival analyses in twin studies and matched pair experiments. *Biometrika*, 61:17-30.

Hougaard, P. (1986). Survival models for heterogeneous populations derived from stable distributions. *Biometrika*, 73: 671-678.

Hougaard, P. (2000). *Analysis of Multivariate Survival Data*. Springer, New York.

Hsu, L., Prentice, R. L., Zhao, L. P., and Fan, J. J. (1999). On dependence estimation using correlated failure time data from case-control family studies. *Biometrika*, 86:743-753.

Joe, H. (1993). Parametric families of multivariate distributions with given margins. *Journal of Multivariate Analysis*, 46:262-282.

Johnson, N. L. and Kotz, S. (1981). Dependent relevations: Time-to-failure under dependence. *American Journal of Mathematical and Management Sciences*, 1:155-165.

Katsahian, S., Resche-Rigon, M., Chevret, S., and Porcher R. (2006). Analysing multicentre competing risks data with a mixed proportional hazards model for the subdistribution. *Statistics in Medicine*, 25:4267-4278.

Klein, J. P. (1992). Semiparametric estimation of random effects using the Cox model based on the EM algorithm. *Biometrics,* 48:795-806.

Lee, E. W., Wei, L. J., and Amato, D. A. (1992). Cox-type regression analysis for large numbers of small groups of correlated failure time observations. In J. P. Klein and P. K. Goel (eds.), *Survival Analysis: State of the Art*, pp. 237-247, Kluwer Academic Publishers, Dordrecht, The Netherlands.

Lee, E. W., Wei L. J., and Ying, Z. (1993). Linear regression analysis for highly stratified failure time data. *Journal of the American Statistical Association*, 88:422.

Li, H. and Huang, J. (1998). Semiparametric linkage analysis using pseudolikelihoods on neighbouring sets. *Annals of Human Genetics*, 62:323-336.

Li, H., Yang, P., and Schwartz, A. G. (1998). Analysis of age of onset data from case-control family studies. *Biometrics*, 54:1030-1039.

Liang, K. Y., Self, S. G., Bandeen-Roche, K. and Zeger, S. L. (1995). Some recent developments for regression analysis of multivariate failure time data. *Lifetime Data Analysis*, 1:403-415.

Liang, K. Y. and Zeger, S.L. (1986). Longitudinal data analysis using generalized linear models. *Biometrika*, 73:13-22.

Lin, D. Y. (1994). Cox regression analysis of multivariate failure time data: The marginal approach. *Statistics in Medicine*, 13:2233-2247.

MacLean, C. J., Neale, M. C., Meyer, J. M., and Kendler, K. S. (1990). Estimating familial effects on age at onset and liability to schizophrenia. II. Adjustment for censored data. *Genetic Epidemiology*, 7:419-426.

Manatunga, A. K. and Oakes, D. (1996). A measure of association for bivariate frailty distributions. *Journal of Multivariate Analysis*, 56:60-74.

Matthews, A. G., Finkelstein, D. M., and Betensky, R. A. (2007). Multivariate logistic regression for familial aggregation in age at disease onset. *Lifetime Data Analysis*, 13:191-209.

McGilchrist, C. A. and Aisbett, C. W. (1991). Regression with frailty in survival analysis. *Biometrics*, 47:461-466.

Murphy, S.A. (1995). Asymptotic theory for the frailty model. *The Annals of Statistics*, 23:182-198.

Nan, B., Lin, X., Lisabeth, L. D., and Harlow, S. D. (2006). Piecewise constant cross-ratio estimation for association of age at a marker event and age at menopause. *Journal of the American Statistical Association*, 101:65-77.

Nielsen, G. G., Gill, R. D., Andersen, P. K., and Sorensen, T. I. A. (1992). A counting process approach to maximum likelihood estimation in frailty models. *Scandinavian Journal of Statistics*, 19:25-44.

Oakes, D. (1989). Bivariate survival models induced by frailties. *Journal of the American Statistical Association*, 84:406.

Petersen, J. H. (1998). An additive frailty model for correlated life times. *Biometrics*, 54:646-661.

Pipper, C. B. and Martinussen, T. (2003). A likelihood based estimating equation for the Clayton-Oakes model with marginal proportional hazards. *Scandinavian Journal of Statistics*, 30:509-521.

Prentice, R. L. and Breslow, N. E. (1978). Retrospective studies and failure time models. *Biometrika*, 65:153-158.

Prentice, R. L. and Cai, J. (1992). Covariance and survivor function estimation using censored multivariate failure time data. *Biometrika*, 79:495-512

Prentice, R. L., Kalbfleisch, J. D., Peterson, A. V., Flournoy, N., Farewell, V. T., and Breslow, N. E. (1978). The analysis of failure times in the presence of competing risks. *Biometrics*, 34:541-554.

Ross, E. A. and Moore, D. (1999). Modeling clustered, discrete, or grouped time survival data with covariates. *Biometrics*, 55:813-819.

Scheike, T. H. and Sun, Y. (2012). On cross-ratio for multivariate competing risks data. *Biostatistics*, 13:680-694.

Scheike, T. H., Sun, Y., Zhang, M. J., and Jensen, T. K. (2010). A semiparametric random effects model for multivariate competing risks data. *Biometrika*, 97:133-145.

Scheike, T. H., Zhang, M. J., and Gerds, T. (2008). Predicting cumulative incidence probability by direct binomial regression. *Biometrika*, 95:205-220.

Schweizer, B. and Sklar, A. (1983). *Probabilistic Metric Spaces*. North-Holland, New York.

Shih, J. H. (1998). Modeling multivariate discrete failure time data. *Biometrics*. 54:1115-1128.

Shih, J. H. and Albert, P. S. (2010). Modeling familial association of ages at onset of disease in the presence of competing risk. *Biometrics*, 66:1012-1023.

Shih, J. H. and Chatterjee, N. (2002). Analysis of survival data from case-control family studies. *Biometrics*, 58:502-509.

Shih, J. H. and Louis, T. A. (1995*a*). Inferences on the association parameter in copula models for bivariate survival data. *Biometrics*, 51:1584-1599.

Shih, J. H. and Louis, T. A. (1995*b*). Assessing gamma frailty models for clustered failure time data. *Lifetime Data Analysis*, 1:205-220.

Silverman, J. M., Ciresi, G., Smith, C. J., Marin, D. B., Schnaider-Beeri, M. (2005). Variability of familial risk of Alzheimer disease across the late life span. *Arch. Gener. Psychol.*, 7: 143-155.

Spiekerman, C. F. and Lin, D. Y. (1998). Marginal Regression Models for Multivariate Failure Time Data. *Journal of the American Statistical Association*, 93:1164-1175.

Therneau, T. M. and Grambsch, P. M. (2000). *Modeling Survival Data: Extending the Cox Model*. Springer-Verlag, New York.

Thomas, D. C., Langholz, B., Mack, W., and Floderus, B. (1990). Bivariate survival models for analysis of genetic and environmental effects in twins. *Genetic Epidemiology*, 7:121-135.

Vaupel, J. W., Manton, K. G., and Stallard, E. (1979). The impact of heterogeneity in individual frailty on the dynamics of mortality. *Demography*, 16:439-454.

Viswanathan, B. and Manatunga, A. K. (2001). Diagnostic plots for assessing the frailty distribution in multivariate survival data. *Lifetime Data Analysis*, 7:143-155.

Wang, W. and Wells, M. T. (2000*a*). Estimation of Kendall's Tau under censoring. *Statistica Sinica*, 10:1199-1215.

Wang, W. and Wells, M. T. (2000*b*). Model selection and semiparametric inference for bivariate failure time data. *Journal of the American Statistical Association*, 95:62-72.

Xue, X. and Brookmeyer, R. (1996). Bivariate frailty model for the analysis of multivariate survival time. *Lifetime Data Analysis*, 2:277-289.

Yashin, A. I. and Iachine, I. A. (1995). Genetic analyses of durations: Correlated frailty model applied to survival of Danish twins. *Genetic Epidemiology*, 12:529-538.

Yin, G. and Cai, J. (2005). Quantile regression models with multivariate failure time data. *Biometrics*, 61:151-161.

Zheng, M. and Klein, J. P. (1995). Estimates of marginal survival for dependent competing risks based on an assumed copula. *Biometrika*, 82:127-138.

Part VI

Clinical Trials

Part VI looks at three topics in survival analysis useful in the design and analysis of clinical trials where the outcome is the time to some event. Clinical trials where the time to event is the primary endpoint are difficult to design since one needs to wait for the primary event to occur all the while entering new patients who may not be needed to draw a conclusion. Sequential clinical trials are especially difficult since decisions on enrolment of new patients must be made in many cases before the outcome of the previous batch of patients can be ascertained.

Typically in a survival time clinical trial an accrual time for patients is set. In this accrual period, patients become available for study and randomly assigned to treatment. The accrual period is followed by an observation period in which patients are followed for the endpoint of interest and no new patients are enrolled. The sample size depends on the length of the accrual and observational windows, the rate of entry into the study, and the censoring fraction as well as the usual power and detectable difference parameters.

In Chapter 28 Ohneberg and Schumacher examine sample size calculations for survival clinical trials. They look at a number of sampling situations including the no censoring case, censored survival comparing survival at a fixed point and nonparametric and parametric tests comparing the entire survival curve. Extensions of these basic sample size calculations to multi-arm tests, non-inferiority tests, non-randomized tests and adjustments for prognostic factors, left-truncated data and cluster randomized trials are also surveyed.

In Chapter 29 Jennison and Turnbull survey results on group sequential clinical trials. Here patients are analyzed at a series calendar times during the course of the study. A decision to stop the study concluding that either the null or alternative hypothesis is true can be made at any of these times if there is sufficient evidence in favor of one of the hypotheses. At the interim analysis time some patients may be censored if they are still alive or if they were lost to follow-up prior to this time. Patients are allowed to enter the study between interim analysis points. Given α and the number of interim analysis, k, one needs to find (a_k, b_k) where null hypothesis is accepted if a test statistic is in $\{-\infty, a_k\}$ or rejected if $\{b_k, \infty\}$ and one continues testing otherwise. This is considered for the one-sided test for superiority. The authors also examine the sequential log rank test for the testing equality of two survival curves using a grouped sequential design.

The final chapter, Chapter 30, by Le-Rademacher and Brazauskas looks at paired survival data. Here we have survival time on two matched pairs of survival times. Matching can be on two highly correlated individuals such as brothers, two organs within the person such as eyes or on two individuals retrospectively matched on a set of risk factors. The two responses are survival times with possibly right-censored data. The censoring time can be the same for both units or there can be different censoring times for each individual. When there is no censoring, paired data is handled by paired t-tests, sign or sign rank tests. With censored data, some modification is needed such as tests using a generalized rank in place of the raw data, tests using differences in survival times, tests based on a weighted difference in Kaplan-Meier estimates with an appropriately adjusted variance estimator, marginal and/or stratified Cox models to adjust for covariates and frailty models to account for correlations between observations. Pseudo-value regression models can be adapted to account for association between survival times when testing equality at a fixed time point.

28

Sample Size Calculations for Clinical Trials

Kristin Ohneberg and Martin Schumacher
Institute of Medical Biometry and Medical Informatics, University Medical Center Freiburg

CONTENTS

28.1 Clinical trials and time-to-event data 571
 28.1.1 Binomial sample size formula 572
 28.1.2 Noncensored time-to-event endpoints 573
 28.1.3 Exponential model 573
28.2 Basic formulas 574
 28.2.1 Schoenfeld's formula 575
 28.2.2 Alternative formula by Freedman 576
28.3 Sample size 577
 28.3.1 Parametric estimation 577
 28.3.2 Nonparametric approximation 578
 28.3.3 Competing risks 579
28.4 Data example 581
 28.4.1 4D trial 581
  28.4.1.1 Two-state model 581
  28.4.1.2 Competing risks analysis 583
28.5 Extensions 586
 28.5.1 Multi-arm survival trials 586
 28.5.2 Test for non-inferiority/superiority and equivalence 587
 28.5.3 Prognostic factors and/or non-randomized comparisons 588
 28.5.4 Left truncation 589
 28.5.5 Proportional subdistribution modeling 589
 28.5.6 Cluster-randomized trials 590
 28.5.7 Cox regression with a time-varying covariate 591
28.6 Summary 591
 Acknowledgments 592
 Bibliography 592

28.1 Clinical trials and time-to-event data

In clinical research, investigators are often interested in the occurrence of certain events such as disease progression, relapse or death. One objective may be to evaluate the effect of a new treatment or a new drug on the prevention or reduction of such undesired events. The time to the occurrence of an event is referred to as "time-to-event." When the event is death, the time-to-event is the patient's survival time, hence analysis of time-to-event data is referred to as "survival analysis." Statistical methods for analysis of time-to-event data are different from those used for continuous outcomes, e.g., comparing means or proportions, as time-to-event is usually subject to censoring. In the presence of censoring for some patients the exact event time is not known, but only that the event time is larger than an observed censoring time. Furthermore normality assumptions of standard statistical methods such as the t-test usually do not hold for time-to-event data.

In clinical trials, patients are often followed for a specified length of time in order to record whether or not the event of interest occurred. A first insight in the data is provided by the analysis of the proportion of patients on whom the event was observed, that is analyzing dichotomous responses, which was essentially the only analysis prior to the 1950s (Peace, 2009). Essential drawback of this approach is that the time at which the event was observed as well as information on patients censored prior to the end of the study is not incorporated in the analysis. These criticisms have been overcome by survival analysis, prompted decisively by Kaplan and Meier (1958) and Cox (1972) in particular. Instead of analyzing dichotomous responses, simply indicating whether a subject has experienced an event or not, survival analysis requires careful definition of the event time, censored observations, the accrual and follow-up period and whether one deals with independent or noninformative censoring.

Sample size calculation plays an important role in planning clinical trials, as it ensures the validity of studies and guarantees that the intended trial will have a desired power for correctly detecting a clinically meaningful difference if such a difference between treatments or drugs truly exists. Purpose of sample size calculation is to estimate the required number of patients who have to be enrolled into the study in order to achieve a prespecified power at a given level of significance.

This chapter aims at deriving sample size formulas for analyzing time-to-event data. In a first step the binomial sample size formula is used in order to gain an idea of the approximate number of patients required by treating outcome data not as survival or censoring times, but as dichotomous responses, simply indicating whether a subject has experienced an event or not. In the case of noncensored time-to-event endpoints a sample size formula for log-transformed event times is briefly mentioned, using the t-test for two independent samples. In the presence of censoring sample size calculation actually consists of two steps: first, one determines the total number of events required for the trial; this quantity is often called the "effective sample size." After that the total number of subjects that have to be enrolled to get the necessary number of observed events in a given time frame is determined. Schoenfeld's formula, named after its first publication by Schoenfeld (1981), for sample size computation within the log rank test as well as under a proportional hazards model will be given. An alternative formula was proposed by Freedman (1982) and is the basis of a comprehensive analysis of sample sizes for clinical studies published by Machin et al. (2008). The framework is then extended to the situation of competing risks. A concrete clinical study is used to explain all necessary considerations in a step-by-step manner. Finally we summarize various extensions with reference to the corresponding publications.

28.1.1 Binomial sample size formula

Regarding initial trial planning, Peace (2009) proposed to gain an idea of the approximate number of patients required by treating the outcomes data not as survival or censoring times, but as dichotomous responses, simply indicating whether a subject has experienced an event or not. Assume two groups: an experimental, indicated by $i = 1$, and a control group indicated by $i = 0$. Let $Y_i, i \in \{0, 1\}$ be independent binomial random variables, $Y_i = 1$ identifying an event in group i. Denote $P(Y_i = 1) = \pi_i$, that is $\pi_0 = P(Y_0 = 1)$ possibility of an event in the control group 0 and $\pi_1 = P(Y_1 = 1)$ the possibility of an event in the treatment group 1. Assume equal sample sizes in both groups. Then the hypothesis testing situation is as follows:

$$H_0 : \pi_0 = \pi_1 =: \pi \quad \text{vs.} \quad H_1 : \pi_0 \neq \pi_1, \text{ say } \pi_0 - \pi_1 =: \delta$$

Finally the binomial sample size formula to determine the total number of patients required for a two-tailed test is given by

$$n = \frac{\left(u_{1-\alpha/2}\sqrt{2\pi(1-\pi)} + u_{1-\beta}\sqrt{\pi_0(1-\pi_0)+\pi_1(1-\pi_1)}\right)^2}{(\pi_0 - \pi_1)^2},$$

where u_γ denotes the γ-quantile of the standard normal distribution. This formula can provide a first insight into the sample size required for the trial, but it relies on identical event probabilities in both groups. This is only fulfilled if the follow-up periods for all patients is the same; thus it is inappropriate to dwell on the binomial formula in general if the intent were indeed to use survival methods.

28.1.2 Noncensored time-to-event endpoints

Peace (2009) proposed sample size calculation for noncensored time-to-event data using the t-test. Aim is to test the null hypothesis of equal means of the time-to-event for the control and experimental group, respectively. The t-test based on the normality assumption uses a logarithmic transformation of the event times so that the transformed data will be closer to the normal distribution. A sample size formula can be derived in order to achieve a power of $1 - \beta$ at significance level α in testing the difference between means based on log-transformed data. This formula resembles Schoenfeld's formula (28.3), given later in this chapter, with the anticipated treatment effect expressed in terms of difference between means of the log-transformed time-to-event endpoints.

28.1.3 Exponential model

In certain circumstances it is possible to postulate a distributional form for the survival distribution. As the simplest case we assume constant hazard rates for both groups, that is the hazard rate within group $i \in \{0, 1\}$ does not change over time, $\lambda_i(t) \equiv \lambda_i$. The survival function then follows an exponential distribution:

$$S_i(t) = \exp(-\lambda_i t), \quad t \geq 0$$

The exponential model constitutes a simple parametric model for time-to-event data. The hypothesis to be tested is

$$H_0 : \lambda_0 = \lambda_1 \quad \text{vs.} \quad H_1 : \lambda_0 \neq \lambda_1 \quad \text{say} \quad \frac{\lambda_1}{\lambda_0} =: \theta \neq 1$$

The sample estimate of the log hazard rate $\log(\hat{\lambda})$ asymptotically follows a normal distribution. Lachin (2000) uses the test statistic $T = \log(\hat{\lambda}_1/\hat{\lambda}_0)$ and derives with its expectation and variance under H_0 and H_1, respectively, the equation for the sample size depending on the prespecified power. Assume patients are randomized to receive one of two treatments, then the total number of events required to give a test with significance α and power $1 - \beta$ is approximately

$$d = \frac{(u_{1-\alpha/2} + u_{1-\beta})^2}{p\,(1-p)\,(\log\theta)^2} \tag{28.1}$$

with p the proportion allocated to the control group and $\theta = \frac{\lambda_1}{\lambda_0}$ the anticipated value of the hazard ratio which is of interest to detect (Lachin, 2000; Machin and Campbell, 1997).

Therefore for equal allocation, that is patients are randomized to the treatment groups in the ratio $1:1$ resulting in $p = \frac{1}{2}$, the total number of events is approximated by

$$d = \frac{4(u_{1-\alpha/2} + u_{1-\beta})^2}{(\log\theta)^2} \tag{28.2}$$

Equation (28.2) is derived by George and Desu (1974) based on the fact that the logarithm of a sum of independent, identically distributed exponential random variables has an approximately normal distribution.

The other way around, in case one is interested in the power $1-\beta$ of a trial for a given number of observed events d, Equation (28.1) is inverted and results in

$$u_{1-\beta} = \sqrt{p\,(1-p)\,d}\,\log\theta - u_{1-\alpha/2}$$

$$\text{which yields} \quad 1-\beta = \Phi\left(\sqrt{p\,(1-p)\,d}\,\log\theta - u_{1-\alpha/2}\right)$$

for the power with Φ the cumulative distribution function of the standard normal distribution.

28.2 Basic formulas

Schoenfeld (1981) studied the asymptotic properties of nonparametric tests for comparing survival distributions and derived a sample size formula for the log rank test. Later on, Schoenfeld (1983) showed that this sample size formula is valid also for the proportional hazards regression model. The log rank test is a nonparametric test of the hypothesis that two groups have the same survival distributions. This is equivalent to the fact that the corresponding hazard functions are identical.

In the following we will give a brief introduction to the basic quantities of survival analysis. A classical attempt within survival analysis is to investigate the time from randomization to a specific treatment until death occurs. Let $(X_t)_{t\geq 0}$ denote whether the individual is alive or dead over the course of time, namely for $t \geq 0$

$$X_t = \begin{cases} 0 & \text{if alive at time } t, \\ 1 & \text{if dead at time } t. \end{cases}$$

The failure time or event time T then is specified as $T := \inf\{t : X_t \neq 0\}$ and is therefore a nonnegative random variable with distribution which can be characterized by the hazard function, the instantaneous failure rate among those at risk:

$$\lambda(t) = \lim_{dt \searrow 0} \frac{P(T \in [t, t+dt) | T \geq t)}{dt}.$$

Another characterization of the distribution of T is given by the survival function, which represents the probability of surviving beyond time t:

$$S(t) := P(T > t) = \exp\left(-\int_0^t \lambda(s)ds\right).$$

Suppose the aim is to test the equality of two survival distributions in a randomized clinical trial with a heterogeneous patient population. The characteristics of the patients are taken into account by adjusting for covariates X such as gender or age. The two groups can then be compared using Cox proportional hazards model, the standard regression model for time-to-event data. Let $\lambda(t|Z)$ denote the hazard rate at time t given covariate value Z. For comparing two different treatments assume Z to be binary with $Z = 0$ indicating the control versus $Z = 1$ the new treatment. Then Cox proportional hazards model is given by

$$\lambda(t|Z, X) = \lambda(t|0) \exp(\beta Z + \gamma X)$$

The proportional hazards model specifies that the ratio of the hazard function for a patient given the new treatment to the same patient given the control treatment will be a constant, namely

$$\frac{\lambda(t|Z = 1, X)}{\lambda(t|Z = 0, X)} = \exp(\beta) =: \theta$$

i.e., the regression coefficient β is in line with the log-hazard ratio. In addition, randomization ensures that covariates X are orthogonal to the treatment indicator Z.

28.2.1 Schoenfeld's formula

Determining the sample size when the outcome is time-to-event in the presence of censoring is really a two-step process. One has to calculate

- firstly the *total number of events* required
- and in a second step the *total number of subjects* who must be followed to obtain the required number of events.

Assume constant hazard rates λ_0, λ_1 within the control and the treatment group, respectively. Then Schoenfeld's formula, named after its first publication by Schoenfeld (1981), for the number of events needed in a two-group trial for a two-sided log rank test to detect a hazard ratio of $\frac{\lambda_1}{\lambda_0} =: \theta$ at the α level of significance and power $1 - \beta$ is given by

$$d(\theta, \alpha, \beta, p) := \frac{(u_{1-\alpha/2} + u_{1-\beta})^2}{p\,(1-p)\,(\log\,\theta)^2} \tag{28.3}$$

with p the proportion of subjects allocated to the control group. Note that the formula is the same as under the assumption of an exponential model; see Section 28.1.3. The total number of observed events d is called "effective sample size" and turns out to be minimal under equal allocation, i.e., $p = 0.5$, and the formula is simplified to

$$d(\theta, \alpha, \beta, 0.5) = \frac{4\,(u_{1-\alpha/2} + u_{1-\beta})^2}{(\log\,\theta)^2}.$$

In a second step, we have to take censoring into account. The effective sample size d has to be adjusted by an estimate of the overall probability of an uncensored observation by the end of the study, denoted by Ψ, to get the *total number of subjects* who have to be enrolled into the trial:

$$n = \frac{d}{\Psi} = \frac{(u_{1-\alpha/2} + u_{1-\beta})^2}{p\,(1-p)\,(\log\,\theta)^2\,\Psi}. \tag{28.4}$$

The determination of Ψ will be described in Section 28.3. Obviously in the case of complete time-to-event data, that is neither administrative censoring (subjects without an event by

TABLE 28.1
Some values of the factor $(u_{1-\alpha/2} + u_{1-\beta})^2$ depending on α and β.

	$\beta = 20\%$	$\beta = 10\%$	$\beta = 5\%$
$\alpha = 5\%$	7.849	10.507	12.995
$\alpha = 1\%$	11.679	14.879	17.814

the end of the study) nor loss-to-follow-up (only the minimum time a patient has been without an event is known, afterwards the contact is disrupted, for whatever reason), the overall probability of an uncensored observation by the end of the study is $\Psi = 1$ and the effective sample size equally represents the total sample size.

Schoenfeld (1983) points out that the sample size formula assuming a proportional hazards model and using a test based on the partial likelihood is the same as when the log rank test is used to compare treatments with proportional hazards without covariates.

28.2.2 Alternative formula by Freedman

Freedman (1982) provided tables of the number of patients required in clinical trials using the log rank test to compare survival times of two treatment groups. Assume patients are randomized to one of two treatments, then the total number of events needed to be observed in order to detect a hazard ratio of θ is

$$d = \frac{(u_{1-\alpha/2} + u_{1-\beta})^2}{p(1-p)} \left(\frac{p + (1-p)\theta}{1-\theta} \right)^2 \tag{28.5}$$

with p the proportion allocated to the control group. For a 1:1 randomization, that is equal allocation with $p = \frac{1}{2}$, the formula is simplified to

$$d = (u_{1-\alpha/2} + u_{1-\beta})^2 \left(\frac{1+\theta}{1-\theta} \right)^2 . \tag{28.6}$$

Machin et al. (2008) use Freedman's formula in order to derive sample sizes for comparing survival curves in clinical studies using the log rank test. Their work offers different examples and describes the use of the given sample size tables.

Peace (2009) points out that Schoenfeld's formula (28.3) tends to slightly underestimate the required number of events. When a 1:1 randomization is used, both methods yield approximately the same number of events required with Freedman's approximation being slightly greater than Schoenfeld's.

Hsieh (1992) compares different sample size formulas for the log rank test, including the formula provided by Schoenfeld and Freedman, and describes their differences on optimal allocation of sample sizes. In order to get more experience with an experimental treatment, some investigators prefer an unbalanced design with more patients allocated to the new treatment. Since the inequality between both groups will reduce the power of the test, it is advisable to take a ratio of 2:1 or 3:2 for experimental versus control treatment. To achieve the power of the test with unequal allocation, the sample size has to be increased accordingly. The optimal design leading to a minimal number of events required with respect to Schoenfeld's formula is given by $p = \frac{1}{2}$, whereas Freedman's formula implies $p = \frac{\theta}{1+\theta}$. Hsieh uses Monte Carlo simulations to compare the powers of equal sample size designs with equal event designs.

28.3 Sample size

In many clinical trials it is not possible to wait until every patient has experienced an event, especially in cancer trials when the event of interest is relapse, some patients will stay without relapse, hence for them no event time will be observed. These patients are subject to so-called "administrative censoring," i.e., no event occurred by the end of the trial. As a result one has to estimate the number of patients n to be recruited corresponding to the required number of events d.

The number of patients needed to enroll into the study in order to get the required number of events depends on:

- length of the accrual period t_1
- length of the additional follow-up period t_2
- distribution of entry times
- distribution of survival times

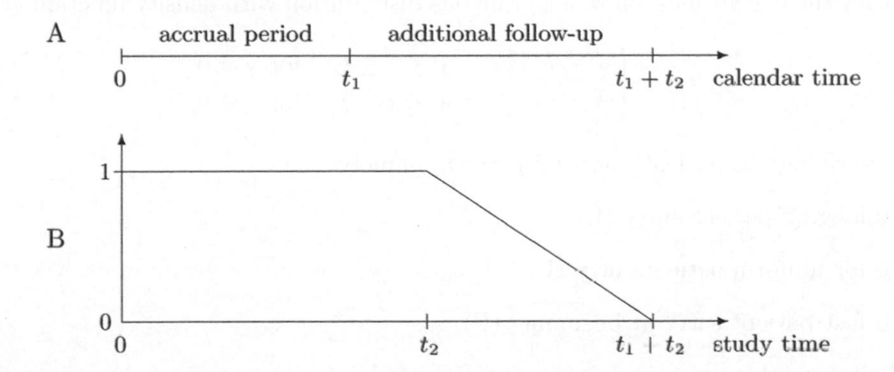

FIGURE 28.1
Accrual and additional follow-up period in calendar time (A) and resulting probability of being uncensored in study time (B) assuming no loss-to-follow-up.

28.3.1 Parametric estimation

We consider the simple case where patients enter the study uniformly during an accrual period $[0, t_1]$; this results in administrative censoring as shown in Figure 28.1, when there is no loss-to-follow-up. Assume survival times are exponentially distributed with hazard rates λ_0 and λ_1 for the control and the experimental group, respectively. Then the probability of observing an event until the end of the follow-up is given by

$$\Psi(\lambda) = 1 - \frac{\exp(-\lambda t_2) - \exp(-\lambda(t_1 + t_2))}{\lambda t_1}. \tag{28.7}$$

This probability has to be computed for control and the experimental group separately and then combined to get the overall probability of observing an event

$$\Psi = p\Psi(\lambda_0) + (1 - p)\Psi(\lambda_1)$$

with p the proportion allocated to the control group. For equal allocation, that is $p = 0.5$, the formula yields

$$\Psi = \frac{1}{2}(\Psi(\lambda_0) + \Psi(\lambda_1)).$$

Finally the number of patients n that have to be enrolled into the trial in order to get the required number of events d to detect a pre-specified treatment effect θ at the α level of significance with a power of $1 - \beta$ is given by

$$n = \frac{d}{\Psi},$$

by ignoring further variability due to the randomness of d given n and Ψ. Abel et al. (2012) point out that the power of a trial with given sample size need not be fully correct as its calculation depends on two random elements, namely the random length of accrual as well as the number of observed events. Additional censoring due to loss-to-follow-up could be accounted for by assuming yet another independent exponential distribution.

Instead of a uniform distribution of the entry times, Chow et al. (2008) assumed that each patient enters the trial independently with entry time following a more flexible distribution. As shown in Figure 28.1 suppose that the accrual period is of length t_1 and the additional follow-up period of length t_2 so the total length of the trial is $t_1 + t_2 =: \tilde{t}$. Chow et al. assume that the entry times follow a continuous distribution with density function given by

$$g(t) := \begin{cases} \frac{\gamma \exp(-\gamma t)}{1-\exp(-\gamma t_1)}, & 0 \le t \le t_1 \quad \text{for } \gamma \ne 0, \\ \frac{1}{t_1}, & 0 \le t \le t_1 \quad \text{for } \gamma = 0, \end{cases}$$

with γ describing the patient's accrual pattern, namely

- $\gamma < 0$ lagging patient entry (L)
- $\gamma = 0$ for uniform patient entry (U)
- $\gamma > 0$ fast patient entry at beginning (F).

Explicit formulas assuming an exponential model for the survival time are given by Chow et al. (2008).

28.3.2 Nonparametric approximation

Schoenfeld (1983) proposed to use Simpson's rule in order to estimate the overall probability of observing an event Ψ. Use an estimate of the survival function for the control group $\hat{S}_0(t)$, e.g., the Kaplan–Meier estimator, to approximate the survival function under a proportional hazards model for the experimental group as

$$\hat{S}_1(t) = \hat{S}_0(t)^{\theta}.$$

Subjects are assumed to be recruited uniformly during the accrual period t_1 and then are followed for an additional time period t_2. Simpsons's rule to estimate the overall probability of observing an event by the end of the study, i.e., after a time period of $t_1 + t_2$, then is given by

$$\Psi_0 = 1 - \frac{1}{6}[\hat{S}_0(t_2) + 4\hat{S}_0(0.5t_1 + t_2) + \hat{S}_0(t_1 + t_2)].$$

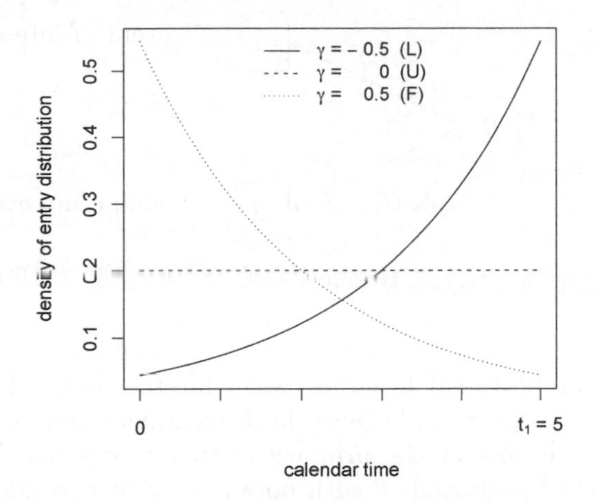

FIGURE 28.2
Patient's accrual pattern (Chow et al. (2008)).

for the control group. For the experimental group Ψ_1 is calculated separately by using $\hat{S}_1(t) = \hat{S}_0(t)^\theta$ resulting in

$$\Psi_1 = 1 - \frac{1}{6}[\hat{S}_1(t_2) + 4\hat{S}_1(0.5t_1 + t_2) + \hat{S}_1(t_1 + t_2)].$$

Freedman (1982) simplifies the calculation of Ψ by taking into account only one time point, say $0.5t_1 + t_2$, which leads to

$$\Psi_0 = 1 - \hat{S}_0(0.5t_1 + t_2) \quad \text{and} \quad \Psi_1 = 1 - \hat{S}_1(0.5t_1 + t_2)$$

for the control and experimental group, respectively. Finally the overall probability of an uncensored observation is given by

$$\Psi = p\Psi_0 + (1 - p)\Psi_1.$$

28.3.3 Competing risks

In clinical trials it is often the case that observation of the event of interest is precluded by the incidence of another event, the so-called "competing event." For example, if interest is in death of a cardiovascular cause, death of any other cause constitutes a competing event. A competing risks process is displayed in Figure 28.3 with cause-specific hazard rate $\lambda_I(t)$ for the event of interest I and $\lambda_C(t)$ for the competing event C.

Denote $\lambda_{0I}(t)$ the hazard for the event of interest I in the control group 0 and $\lambda_{1I}(t)$ the hazard for the event of interest I in the experimental group 1. Correspondingly $\lambda_{0C}(t), \lambda_{1C}(t)$ the hazard for the competing event c in the control 0 and in the experimental group 1.

The assumed treatment effect on the event of interest then is expressed in terms of cause-specific hazard rates as

$$\theta = \frac{\lambda_{1I}(t)}{\lambda_{0I}(t)}$$

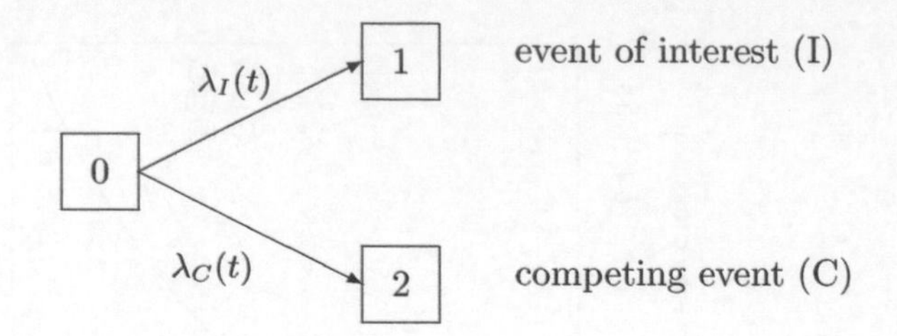

FIGURE 28.3
Competing risks process.

Note that as before a proportional hazards model for the event of interest is assumed, equivalent to a constant hazard ratio between both treatment groups.

The total number of events of the primary endpoint required to detect a clinically relevant treatment effect of magnitude θ with power $1-\beta$ at two-sided level α is given by Schoenfeld's formula (28.3) as

$$d = \frac{(u_{1-\alpha/2} + u_{1-\beta})^2}{p(1-p)(\log\ \theta)^2}$$

equivalent as for the classical survival model.

Actually the presence of competing risks affects the second step of sample size calculation, namely determination of Ψ, the overall probability of observing a primary event of interest.

As in the simple survival model Ψ has to be computed separately for the control and the treatment group. Assume constant cause-specific hazard rates $\lambda_{0I}, \lambda_{0C}$ for the control and $\lambda_{1I}, \lambda_{1C}$ for the experimental group, respectively. An equivalent of Formula (28.7) in the presence of competing risks then is given for the control group as

$$\Psi(\lambda_{0I}) = \frac{\lambda_{0I}}{\lambda_0}\left[1 - \frac{\exp(-\lambda_0 t_2) - \exp(-\lambda_0(t_1+t_2))}{\lambda_0 t_1}\right]; \quad \lambda_0 = \lambda_{0I} + \lambda_{0C}$$

and analogous for the experimental group as

$$\Psi(\lambda_{1I}) = \frac{\lambda_{1I}}{\lambda_1}\left[1 - \frac{\exp(-\lambda_1 t_2) - \exp(-\lambda_1(t_1+t_2))}{\lambda_1 t_1}\right]; \quad \lambda_1 = \lambda_{1I} + \lambda_{1C}.$$

These calculations are valid under the assumption that the treatment has no effect on the hazard of the competing risk; see Schulgen et al. (2005).

Finally $\Psi(\lambda_I)$ is given by

$$\Psi(\lambda_I) = p\Psi(\lambda_{0I}) + (1-p)\Psi(\lambda_{1I})$$

with p the proportion allocated to the control group 0.

28.4 Data example

28.4.1 4D trial

The 4D trial (Die Deutsche Diabetes Dialysis Studies) enrolled patients with type 2 diabetes on hemodialysis to evaluate the efficacy of antihyperlipidemic treatment with atorvastatin, a HMG-CoA reductase inhibitor, in reducing death of cardiovascular cause and the frequency of non-fatal myocardial infarction in the presence of competing other causes of death (Schulgen et al., 2005; Wanner et al., 2005).

The aim of this section is executing sample size determination, in a first step simply assuming a two-state model with death of any cause as outcome variable, in a second step considering a multistate model explicitly taking into account competing events.

28.4.1.1 Two-state model

Initially consider the classic survival model, where interest is in time to event, for the 4D trial time until death of any cause as illustrated in Figure 28.4.

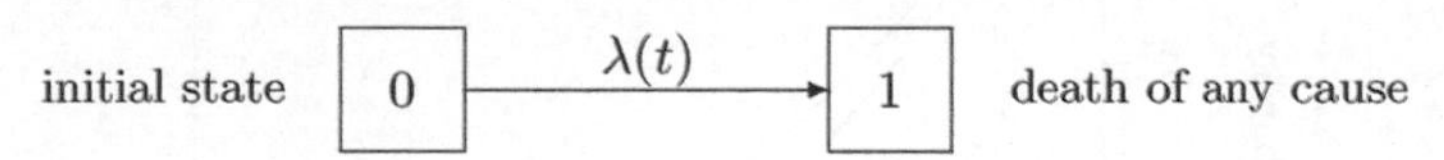

FIGURE 28.4
Two-state model for the 4D trial.

Assume patients enter the study uniformly during an accrual period of length t_1 which is followed by a follow-up period of length t_2, so the whole trial is of length $t_1 + t_2 = \tilde{t}$.

The 4D trial aimed at testing whether the experimental treatment, i.e., antihyperlipidemic treatment with atorvastatin, is superior compared to a control treatment, in this case placebo.

The hypothesis for the two-sided test is formulated as:

$$\begin{aligned} H_0 &: \lambda_0(t) = \lambda_1(t) \quad \text{for all } 0 < t < \tilde{t} \quad \text{vs.} \\ H_1 &: \lambda_0(t) \neq \lambda_1(t) \quad \text{for at least some } 0 < t < \tilde{t} \end{aligned}$$

where $\lambda_0(t), \lambda_1(t)$ denote the hazard functions for the control and the experimental group, respectively.

For sample size determination one has to specify a clinically relevant treatment effect one wishes to detect within the study. Therefore, if available results from previous studies are taken into account. Schulgen et al. (2005) relied on a historical cohort study of 412 diabetic patients treated in 35 dialysis centers in Germany in the time period from 1985 to 1994 to collect information about the survival pattern of the trial population associated with the reference treatment.

Table 28.2 of hazard rates under control and experimental treatment is adopted from the work of Schulgen et al. (2005) assuming a time-homogeneous Markov process, that is constant cause-specific hazard rates.

For analysis of time to death of any cause, the hazard rates of lines 2 and 3 have to be combined, resulting in $\lambda_0 = 0.21+0.14 = 0.35$ for the control and $\lambda_1 = 0.15+0.14 = 0.29$ for the experimental group. Patients experiencing a non-fatal myocardial infarction are treated as censored observations. The expected hazard ratio between both groups then is given by

$$\theta = \frac{\lambda_1}{\lambda_0} = \frac{0.29}{0.35} = 0.83.$$

TABLE 28.2
Hazard rates under control and experimental treatment.

	control group	experimental group
non-fatal myocardial infarction	0.05	0.04
death of cardiovascular cause	0.21	0.15
death of other causes	0.14	0.14

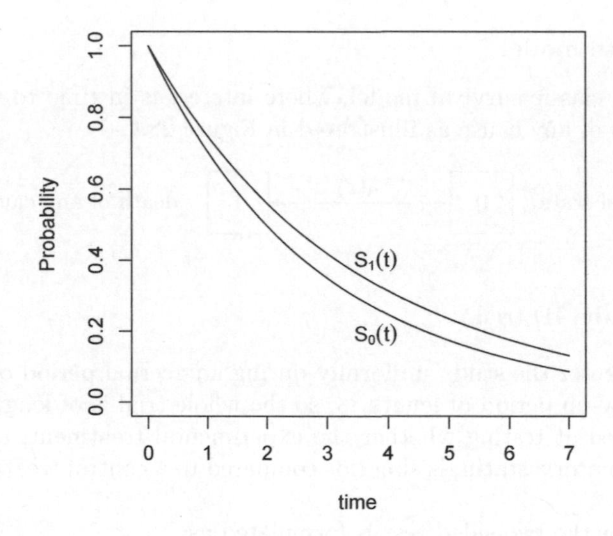

FIGURE 28.5
Theoretical survival function for the control group $S_0(t)$ with $\lambda_0 = 0.35$ and the experimental group $S_1(t)$ with $\lambda_1 = 0.29$.

Schoenfeld's formula (28.3) is used to derive the effective sample size to achieve a power of 90% at a two-sided level of significance 5% to detect a clinically relevant treatment effect of $\theta = 0.83$:

$$d = \frac{4(u_{0.975} + u_{0.9})^2}{(\log\ 0.83)^2} = 1211$$

Note that equal allocation between both treatment groups is assumed.

Taken the formula provided by Freedman (28.6) the effective sample size is given by

$$d = \frac{(u_{0.975} + u_{0.9})^2(1 + 0.83)^2}{(1 - 0.83)^2} = 1218,$$

that is 1,218 events of the primary endpoint have to be observed to detect a hazard ratio of $\theta = 0.83$ at the $\alpha = 5\%$ level of significance with power $1 - \beta = 90\%$.

To derive the total number of patients needed for the trial, one has to estimate the probability of observing an event Ψ. Because constant hazard rates are assumed, Ψ can be calculated assuming exponentially distributed survival times resulting in

$$\Psi(\lambda) = 1 - \frac{\exp(-\lambda t_2) - \exp(-\lambda(t_1 + t_2))}{\lambda t_1}.$$

For the 4D trial an accrual period of $t_1 = 1.5$ years and a follow-up of $t_2 = 2.5$ years are assumed. Ψ has to be determined separately for the control and the experimental group using the respective hazard rates for each group:

$$\Psi(\lambda_0) = 1 - \frac{\exp(-0.35 \cdot 2.5) - \exp(-0.35(1.5 + 2.5))}{0.35 \cdot 1.5} = 0.676$$
$$\Psi(\lambda_1) = 1 - \frac{\exp(-0.29 \cdot 2.5) - \exp(-0.29(1.5 + 2.5))}{0.29 \cdot 1.5} = 0.607$$

and finally for equal allocation, that is $p = \frac{1}{2}$,

$$\Psi = \frac{1}{2}(\Psi(\lambda_0) + \Psi(\lambda_1)) = 0.641$$

The total number of patients needed for the trial analyzed by a two-state model then is given by

$$n = \frac{d}{\Psi} = \frac{1211}{0.641} = 1890.$$

Note that under the assumption of a stronger treatment effect, in our example an even smaller hazard ratio than the one taken so far, the effective sample size d and hence the sample size n decreases considerably. For example, instead of $\theta = 0.83$ assume $\theta = 0.73$, then for the same α and β the effective sample size decreases from $d = 1211$ to $d = 424$. In general one gets the maximum sample size when the hazard ratio is close to one. On the contrary the bigger difference between the hazard for the experimental and the control group is, the smaller sample size is required.

28.4.1.2 Competing risks analysis

The 4D trial with time-to-event endpoints in the presence of competing risks is best described in the framework of multi-state models, as displayed in Figure 28.6. Consider as event of interest non-fatal myocardial infarction or cardiovascular death, whereas death of other causes constitutes the competing event. Let $\lambda_C(t)$ denote the hazard for the competing event and $\lambda_I(t)$ the hazard for the event of interest. The all-cause-hazard $\lambda(t) = \lambda_C(t) + \lambda_I(t)$ denotes the hazard function for the intensity of moving to any one of the outcome states at time t.

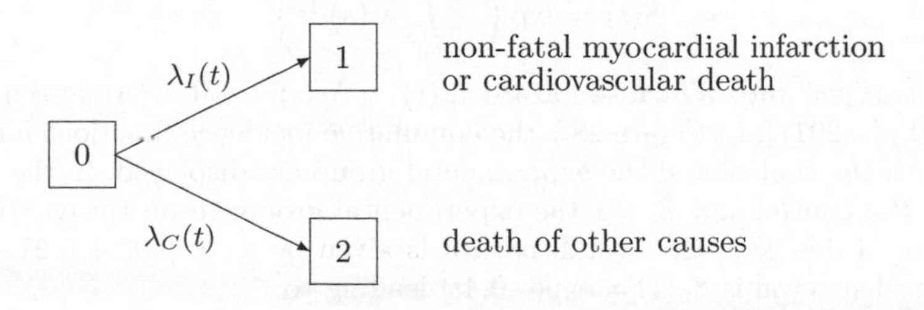

FIGURE 28.6
Competing risks model for the 4D trial.

The hypothesis in the given competing risks setting is formulated as

$$H_0 : \lambda_{0I}(t) = \lambda_{1I}(t) \quad \text{for all } 0 < t < \tilde{t} \quad \text{vs.}$$
$$H_1 : \lambda_{0I}(t) \neq \lambda_{1I}(t) \quad \text{for at least some } 0 < t < \tilde{t}$$

with $\lambda_{0I}(t)$ the hazard for the event of interest in the control and $\lambda_{1I}(t)$ the hazard for the event of interest in the experimental group.

Following Table 28.2 the hazard rate for the event of interest is given by

$$\lambda_{0I} = 0.05 + 0.21 = 0.26 \quad \text{for the control group and}$$
$$\lambda_{1I} = 0.04 + 0.15 = 0.19 \quad \text{for the experimental group.}$$

The assumed treatment effect in terms of the hazard ratio for the event of interest then is given by

$$\theta = \frac{\lambda_{1I}(t)}{\lambda_{0I}(t)} = \frac{0.19}{0.26} = 0.73.$$

We now briefly introduce some notation with respect to the competing risk framework. Figure 28.6 represents a competing risk setting with two possible transitions $0 \to 1$ and $0 \to 2$. Hence X_t takes value 0 for $t < T$ and at failure time T either moves to state 1 or to state 2:

$$X_t = \begin{cases} 0 & \text{at risk at time } t, \\ 1 & \text{event of interest at time } t, \\ 2 & \text{competing event at time } t. \end{cases}$$

The failure time is given by $T = \inf\{t : X_t \neq 0\}$, but in a competing risk setting additionally the type of the event $X_T = j, j \in \{1, 2\}$ has to be recorded. Cause-specific hazard rates are defined as

$$\lambda_j(t) := \lim_{dt \searrow 0} \frac{P(T \in [t, t + dt), X_T = j | T \geq t)}{dt}, \quad j \in \{1, 2\}.$$

which represent the hazard from failing from a given cause in the presence of the competing risks.

The probability of experiencing an event of type j over the course of time is represented by the cumulative incidence function (CIF)

$$F_j(t) := P(T \leq t, X_T = j) = \int_0^t S(u)\lambda_j(u)du$$

where

$$S(t) = \exp\left(-\int_0^t \lambda.(s)ds\right)$$

is the overall survival with all-cause hazard $\lambda.(t) = \lambda_1(t) + \lambda_2(t)$ (Andersen et al., 2002; Beyersmann et al., 2011). In Figure 28.7 the cumulative incidence functions for the event of interest for both the control and the experimental group are displayed. In the following, let $Z = 0$ denote the control and $Z = 1$ the experimental group, respectively. For the control group, following Table 28.2, the overall hazard is given by $\lambda_0 = 0.05 + 0.21 + 0.14 = 0.4$, hence the overall survival is $S_0(t) = \exp(-0.4t)$ leading to

$$F_0(t) = P(T \leq t, X_T = 1, Z = 0) = \int_0^t S_0(u)\lambda_{0I}du = 0.65(1 - \exp(-0.4t)).$$

For the experimental group, the overall hazard is $\lambda_1 = 0.04 + 0.15 + 0.14 = 0.33$, the

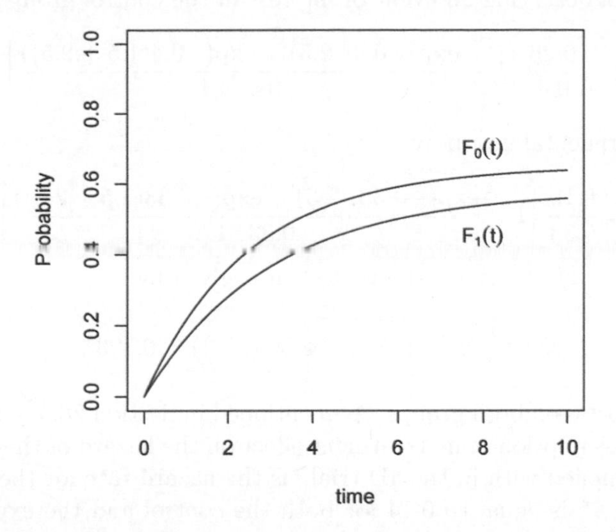

FIGURE 28.7
Theoretical cumulative incidence function for the event of interest for the control $F_0(t)$ and the experimental group $F_1(t)$, respectively.

overall survival $S_1(t) = \exp(-0.33t)$, hence the cumulative incidence function for the event of interest

$$F_1(t) = P(T \leq t, X_T = 1, Z = 1) = \int_0^t S_1(u)\lambda_{1I} du = 0.58(1 - \exp(-0.33t)).$$

Back to sample size determination, for equal allocation between the experimental and the control group, i.e., $p = 0.5$, the required number of events of the primary endpoint to detect a hazard ratio of $\theta = 0.73$ at the $\alpha = 5\%$ level of significance with a power of $1 - \beta = 90\%$ is given by Schoenfeld's formula (28.3) as

$$d = \frac{4(u_{0.975} + u_{0.9})^2}{(\log\ 0.73)^2} = 424.$$

For comparison, taking the formula provided by Freedman (28.6) the effective sample size is given by

$$d = \frac{(u_{1-\alpha/2} + u_{1-\beta})^2(1+\theta)^2}{(1-\theta)^2} = \frac{(u_{0.975} + u_{0.9})^2(1+0.73)^2}{(1-0.73)^2} = 431,$$

that is 431 events of the primary endpoint have to be observed to detect a hazard ratio of $\theta = 0.73$ at the $\alpha = 5\%$ level of significance with a power of $1 - \beta = 90\%$.

The next step is to determine the number of patients n that have to be enrolled into the study in order to get the anticipated number of events of the primary outcome. The probability of observing an event of interest Ψ is given by

$$\Psi(\lambda_I, \lambda) = \frac{\lambda_I}{\lambda}\left[1 - \frac{\exp(-\lambda t_2) - \exp(-\lambda(t_1 + t_2))}{\lambda t_1}\right]$$

with all-cause hazard $\lambda = \lambda_I + \lambda_C$ and an accrual period of $t_1 = 1.5$ years and a follow-up of $t_2 = 2.5$ years.

The probability of observing an event of interest in the control group then is given by

$$\Psi(\lambda_{0I}, \lambda_0) = \frac{0.26}{0.4}\left[1 - \frac{\exp(-0.4 \cdot 2.5) - \exp(-0.4(1.5 + 2.5))}{0.4 \cdot 1.5}\right] = 0.47$$

and within the experimental group by

$$\Psi(\lambda_{1I}, \lambda_1) = \frac{0.19}{0.33}\left[1 - \frac{\exp(-0.33 \cdot 2.5) - \exp(-0.33(1.5 + 2.5))}{0.33 \cdot 1.5}\right] = 0.376$$

The overall probability of observing an event then is given by

$$\Psi = \frac{1}{2}(\Psi(\lambda_{0I}, \lambda_0) + \Psi(\lambda_{1I}, \lambda_1)) = 0.423$$

for equal allocation between both groups. As mentioned in Section 28.3.3, these calculations are valid under the assumption of no treatment effect on the hazard of the competing event. This condition is complied with in the 4D trial, as the hazard rate for the competing event "death of other causes" is equal to 0.14 for both the control and the experimental group; see Table 28.2. Finally the total number of patients required for the competing risks setting is given by

$$n = \frac{d}{\Psi} = \frac{424}{0.423} = 1002.$$

Indeed the trial protocol of the 4D trial assumed to recruit about 1,200 patients, that is about 20% patients more than the calculated minimum number required. This was to take into account loss-to-follow-up and to protect against potential incorrect assumptions in sample size calculation (Schulgen et al., 2005).

28.5 Extensions

28.5.1 Multi-arm survival trials

In some trials more than one experimental treatment should be evaluated against a standard therapy. Barthel et al. (2006) give an overview on extensions of Schoenfeld's formula for multi-arm trials, based on the work of Ahnn and Anderson (1998). Assume n patients randomized to one of K treatment groups and compare the K treatments globally in terms of time to failure using a log-rank test. Denote $\lambda_k(t)$ the hazard rate in treatment group $k \in \{1, 2, \ldots, K\}$, then the null hypothesis of equality of the K survival distributions is expressed as

$$\begin{aligned} H_0 &: \lambda_1(t) = \lambda_2(t) = \cdots = \lambda_K(t) \quad \text{vs.} \\ H_1 &: \lambda_k(t) \neq \lambda_l(t) \quad \text{for at least one } k \neq l, \end{aligned}$$

that is the global alternative hypothesis, that at least one survival distribution is different from the others.

Consider the simple case of a log-rank test under proportional hazards, that is the hazard ratio in group k relative to group 1 is constant over time:

$$\frac{\lambda_k(t)}{\lambda_1(t)} =: \theta_k.$$

Assume equal allocation within the different treatment groups, then the sample size is given by

$$n = \frac{K\tau}{\Psi\left[\frac{K-1}{K}\sum_{k=2}^{K}(\log\theta_k)^2 - \frac{2}{K}\sum_{k=2}^{K}\sum_{k\leq q}^{K}\log\theta_k\log\theta_q\right]} \tag{28.8}$$

with τ the non-centrality parameter obtained for a given power $1-\beta$ and significance level α from tables of the cumulative non-central chi-squared distribution (Barthel et al., 2006). Ψ denotes the probability of an uncensored observation by the end of the trial; see Section 28.3. For only two treatment groups, that is $K = 2$, $\tau = (u_{1-\alpha/2} + u_{1-\beta})^2$ and (28.8) is identical with Schoenfeld's Formula (28.4) for equal allocation.

28.5.2 Test for non-inferiority/superiority and equivalence

Assume interest is in comparing two treatments, namely one new, in the following referred to as experimental treatment $Z = 1$, compared to a control treatment $Z = 0$. Cox proportional hazards model is given by

$$\lambda(t|Z) = \lambda(t|Z = 0)\exp(\beta Z)$$

with $\lambda(t|Z)$ the hazard rate at time t given covariate value $Z \in \{0, 1\}$, β the regression coefficient. The treatment effect then can be expressed in terms of the hazard ratio between both groups:

$$\frac{\lambda(t|Z = 1)}{\lambda(t|Z = 0)} = \exp(\beta) =: \theta.$$

Schoenfeld's Formula (28.4) given in Section 28.2.1 to determine the total number of patients required for a clinical trial to detect a treatment effect θ at the α level of significance with power $1-\beta$ is given by

$$n = \frac{(u_{1-\alpha/2} + u_{1-\beta})^2}{p\,(1-p)\,(\log\,\theta)^2\,\Psi}$$

with p the proportion allocated to the control group and Ψ the overall probability of observing an event; see Section 28.3.

Actually this formula holds in order to test for equality of two survival curves, given by the following hypothesis testing situation:

$$H_0 : \lambda(t|Z = 1) = \lambda(t|Z = 0) \quad \text{vs.} \quad H_1 : \frac{\lambda(t|Z = 1)}{\lambda(t|Z = 0)} =: \theta \neq 1$$

Instead of testing for equality, one could be interested in testing for non-inferiority of a new treatment compared to the standard, expressed by

$$H_0 : \log(\theta) \leq \delta \quad \text{vs.} \quad H_1 : \log(\theta) > \delta \tag{28.9}$$

with $\delta < 0$. Rejection of the null hypothesis then indicates non-inferiority of the experimental over the control treatment.

The same hypothesis testing situation (28.9) with $\delta > 0$ results in testing superiority, namely rejection of the null hypothesis indicating superiority of the experimental treatment over the control treatment.

The two-sided sample size formula with significance level α and power $1-\beta$ assuming a test for non-inferiority or superiority then is given by

$$n = \frac{(u_{1-\alpha/2} + u_{1-\beta})^2}{p\,(1-p)\,(\log\,\theta - \delta)^2\,\Psi} \tag{28.10}$$

see Chow et al. (2008).

In order to test for equivalence of two treatments, consider the following hypothesis situation:

$$H_0 : |\log(\theta)| \geq \delta \quad \text{vs.} \quad H_1 : |\log(\theta)| < \delta$$

The sample size needed for the one-sided test for equivalence in order to achieve a power of $1-\beta$ at the α level of significance then is given by

$$n = \frac{(u_{1-\alpha} + u_{1-\beta/2})^2}{p\,(1-p)\,(|\log\,\theta| - \delta)^2\,\Psi}. \tag{28.11}$$

28.5.3 Prognostic factors and/or non-randomized comparisons

Studies on prognostic factors attempt to determine a prediction of the course of the disease for groups of patients defined by the values of prognostic factors, and to rank the relative importance of various factors (Schmoor et al., 2000; Crowley and Hoering, 2012).

Consider the situation that the prognostic relevance of a certain factor Z_1 should be studied in the presence of second factor Z_2. Let the analysis of the main effects of Z_1 and Z_2 be performed with a Cox proportional hazards model given by

$$\lambda(t|Z_1, Z_2) = \lambda_0(t)\exp(\beta_1 Z_1 + \beta_2 Z_2)$$

where $\lambda_0(t)$ denotes an unspecified baseline hazard function and β_1, β_2 are the unknown regression coefficients representing the effects of Z_1 and Z_2.

Consider binary covariates with $p := P(Z_1 = 1)$ and $q := P(Z_2 = 1)$. Assume that the effect of Z_1 is to be tested by an appropriate two-sided test based on the partial likelihood derived from the Cox model given above with significance level α and power $1-\beta$ to detect an effect which is given by a relative risk of $\theta_1 = \exp(\beta_1)$.

For independent Z_1 and Z_2 Schoenfeld's formula for the total number of patients required (28.4) holds:

$$n = \frac{(u_{1-\alpha/2} + u_{1-\beta})^2}{p\,(1-p)\,(\log\,\theta_1)^2\Psi}$$

with $p = P(Z_1 = 1)$ the prevalence of $Z_1 = 1$. That is two groups are defined by the value of $Z_1 \in \{0, 1\}$ and p is the proportion allocated to group $Z_1 = 1$. As before Ψ denotes the overall probability of observing an event; see Section 28.3.

In case that the second factor Z_2 is correlated with the interesting prognostic factor Z_1, Schoenfeld's Formula given above is not valid. Schmoor et al. (2000) give an extension of Schoenfeld's Formula (28.4) for the situation when Z_1 and Z_2 are correlated with correlation coefficient ρ given by

$$\rho = \frac{\mathrm{Cov}(Z_1, Z_2)}{\sqrt{p(1-p)q(1-q)}} = (p_1 - p_0)\sqrt{\frac{q(1-q)}{p(1-p)}}$$

with $p_0 := P(Z_1 = 1|Z_2 = 0)$ and $p_1 := P(Z_1 = 1|Z_2 = 1)$. The total number of patients required then is given by

$$n = \frac{(u_{1-\alpha/2} + u_{1-\beta})^2}{\Psi\,p\,(1-p)\,(\log\,\theta_1)^2}\left(\frac{1}{1-\rho^2}\right) \tag{28.12}$$

with $\frac{1}{1-\rho^2}$ the so-called "variance inflation factor."

This formula can also be used for a treatment comparison within a non-randomized study with all the caveats of such an enterprise. The treatment indicator is then Z_1 and the covariates correlated with treatment allocation are summarized by Z_2. The formula above was derived for two binary covariates by Schmoor et al. (2000) and simultaneously by Bernardo et al. (2000). An algorithm to calculate power and sample size for correlated covariates with a multivariate normal distribution was presented by Schoenfeld and Borenstein (2005).

In a second step Schmoor et al. (2000) provide an approximate formula for sample size and power to detect an interaction between the interesting prognostic factor and a second correlated factor. An interaction between Z_1 and Z_2 may be analyzed by a Cox model

$$\lambda(t|Z_1, Z_2) = \lambda_0(t)\exp(\beta_1 Z_1 + \beta_2 Z_2 + \beta_{12} Z_1 Z_2).$$

An approximation of the number of patients needed to detect an interaction of size $\tau = \exp(\beta_{12})$ by a two-sided level α test with power $1-\beta$ then is given by

$$n = \frac{(u_{1-\alpha/2} + u_{1-\beta})^2}{\Psi\,(\log\,\tau)^2}\left(\frac{1}{p_{00}} + \frac{1}{p_{01}} + \frac{1}{p_{10}} + \frac{1}{p_{11}}\right) \tag{28.13}$$

with $p_{ij} = P(Z_1 = i, Z_2 = j)$ (Schmoor et al., 2000; Peterson and George, 1993). Schmoor et al. (2000) show that Formula (28.13) is just an extension of Formula (28.12) for independent effects.

28.5.4 Left truncation

Xu and Chambers (2011) show that Schoenfeld's Formula (28.3), although initially derived for right-censored data, also holds for left-truncated data. They study the effects of drugs or vaccines on pregnancy outcomes and therefore take spontaneous abortion as endpoint. In this case often prospective exposure cohort studies are conducted, where women are recruited after recognizing their pregnancy. Since the exact date of conception is unknown, such data typically is subject to left truncation. Assume a proportional hazards model between the experimental and the control group and suppose interest is in detecting a treatment effect θ. The number of observed events required to detect a treatment effect θ using a log-rank test which takes into account left truncation is given by Schoenfeld's Formula (28.3). This formula initially was derived for right-censored data, yet Xu and Chambers (2011) use a sequential conditioning argument to show that this formula also holds for left-truncated data. As they point out, a key point is that the number of events is asymptotically proportional to the Fisher information while the size of the risk set which is affected by truncation or censoring, does not enter into the calculation.

28.5.5 Proportional subdistribution modeling

Latouche et al. (2004) present a sample size formula for the proportional hazards modeling of competing risks subdistribution. In a competing risks setting, the instantaneous risk of a specific failure type sometimes is of less interest than the overall probability of this specific failure, called "cumulative incidence function." The cumulative incidence function is a combined quantity of all cause-specific hazards, hence separate Cox models of all cause-specific hazards are required. In order to facilitate the interpretation, Fine and Gray (1999) proposed the subdistribution hazard, a hazard "attached" to the cumulative incidence function as it re-establishes the one-to-one relation between hazard and distribution function known from conventional survival analysis.

Latouche et al. (2004) provide a sample size formula for the subdistribution hazards model to design a two-arm randomized clinical trial with a right-censored competing endpoint. Let $F_1(t|Z)$ denote the cumulative incidence function for the event of interest assuming a binary covariate $Z \in \{0,1\}$. The subdistribution hazard ratio between the experimental ($Z = 1$) and the control group ($Z = 0$) then is given by

$$\theta = \frac{\log(1 - F_1(t|Z = 0))}{\log(1 - F_1(t|Z = 1))}$$

Latouche et al. (2004) derive a sample size formula similar to that proposed by Schoenfeld (1983)

$$n = \frac{(u_{1-\alpha/2} + u_{1-\beta})^2}{p\,(1-p)\,(\log\,\theta)^2\,\Psi} \tag{28.14}$$

where θ denotes the subdistribution hazard ratio instead of the cause-specific hazard ratio and Ψ the proportion of failures of the event of interest by the end of the study $\tilde{t}$. In case of complete data, that is in the absence of censoring, Ψ is given by the cumulative distribution function for the event of interest by the end of the study $F_1(\tilde{t})$.

For completely understanding a competing risks process, all cause-specific hazards have to be analyzed. Assuming each cause-specific hazard to follow a proportional hazards model, precludes the assumption of proportional subdistribution hazards. Yet Grambauer et al. (2010) recommend employing both types of analysis, as even a misspecified proportional subdistribution hazards analysis offers a summary analysis in terms of a time-averaged effect on the cumulative event probabilities.

Latouche et al. (2004) also consider correlated covariates when presenting sample size formulas for the proportional hazards modeling of competing risks subdistribution. The required sample size to detect a relevant effect in a prognostic study is similar to Formula (28.12) in Section 28.5.3, with θ the subdistribution hazard ratio instead of the cause-specific hazard ratio.

28.5.6 Cluster-randomized trials

Jahn-Eimermacher et al. (2013) give an extension of Schoenfeld's Formula (28.4), accounting for a clustered design of a clinical trial. In cluster-randomized trials, treatments are randomized not to individuals themselves, but to groups of individuals, the so-called "clusters." These trials are indicated when it is logistically too difficult to administer treatments to individuals separately. Observations within the same cluster, like patients of the same hospital, tend to be more similar than observation in different clusters, causing correlations. Correlation reduces the statistical information in the data and thus the effective sample size. Hence the clustered design has to be considered when planning the sample size of a trial to ensure an adequate power to detect intervention effects.

To address the case where observations within the same cluster are correlated, Jahn-Eimermacher et al. (2013) give a correction of Schoenfeld's formula. They derive a sample size formula for clustered time-to-event data with constant marginal baseline hazards and correlation within clusters induced by a shared frailty term.

Assume a balanced trial with N clusters per group, each of size K. Then Schoenfeld's formula, designed for uncorrelated data, yields

$$N = \frac{2(u_{1-\alpha/2} + u_{1-\beta})^2}{K(\log\,\theta)^2\,\Psi}$$

clusters required per group to detect a treatment effect θ at the α level of significance with a power of $1-\beta$. Ψ denotes the overall probability of observing an event, see Section 28.3. For two groups, that is a control and an experimental group, this yields a total of

$$n = 2KN = \frac{4(u_{1-\alpha/2} + u_{1-\beta})^2}{(\log\,\theta)^2\;\Psi}$$

patients that have to be enrolled into the trial, in accordance with (28.4).

Jahn-Eimermacher et al. (2013) point out that sample size calculation had to account for the correlation within the data. Hence they use an additional summand to account for the clustered design resulting in a total of

$$N = \frac{2(u_{1-\alpha/2} + u_{1-\beta})^2}{K(\log\,\theta)^2\;\Psi} + (u_{1-\alpha/2} + u_{1-\beta})^2\nu^2\frac{1+\theta^2}{(1-\theta)^2}$$

clusters required per group to detect a hazard ratio of θ between both treatment groups. This correction summand depends on the anticipated hazard ratio and on the coefficient of variation ν of survival times between clusters within each intervention group. The coefficient of variation can be calculated using a distributional assumption for the frailty variable, such as the gamma or log-normal distribution; see Jahn-Eimermacher et al. (2013).

Duchateau and Janssen (2008) give several alternatives for the frailty model that take into account the clustering of observations, for example a stratified model. They suggest to use different and unspecified baseline hazards for each of the clusters, that is for each stratum, and a common regression coefficient.

28.5.7 Cox regression with a time-varying covariate

Austin (2012) describes data-generating processes for the Cox proportional hazards model in the presence of time-varying covariates. Simulations are used in order to examine the performance of statistical procedures. Austin points out that in complex settings in which analytic calculations are either very difficult or not feasible, data-generating processes allow to select an appropriate sample size to analyze survival data in the presence of both time-invariant and time-varying covariates.

28.6 Summary

The aim of this chapter was to give an insight and brief overview of sample size calculation in clinical trials with time-to-event data.

Schoenfeld (1981) presented a sample size formula for comparing survival curves using a log-rank test and subsequently pointed out that the same formula is also valid assuming a proportional hazards model (Schoenfeld, 1983). An alternative formula is given by Freedman (1982). Hsieh (1992) provides an overview of different sample size formulas especially accounting for unbalanced allocation using the log-rank test.

We pointed out that determining the sample size when the outcome is time-to-event in the presence of right censoring is actually a two-step process. First one has to calculate the total number of events required and in a second step the total number of patients who have to be enrolled into the study to obtain the required number of observed events. The total number of observed events required for the trial is called the "effective sample size." In order to get the total sample size, the probability of an uncensored observation has to be determined. Therefore the anticipated length of the accrual and the additional follow-up time has to be specified, as well as the distribution of entry and survival times. We consider the simple case where patients enter the study uniformly and survival times are exponentially distributed in Section 28.3.1. Chow et al. (2008) assume more flexible distributions of a patient's entry time. A nonparametric approximation of the probability of an uncensored observation is provided by Schoenfeld using Simpson's rule.

As we pointed out the presence of competing risks does not affect the effective sample size, but has to be taken into account within the second step, when the total sample size is determined. We assume constant cause-specific hazards and present formulas assuming that the treatment has no effect on the hazard of the competing event. An alternative approach is to fit a proportional subdistribution hazards model as proposed by Fine and Gray (1999). Latouche et al. (2004) demonstrate that the sample size formula in this case is given by Schoenfeld's formula including the anticipated subdistribution hazard ratio instead of the cause-specific hazard ratio.

Extensions of the simple two-group randomized trial are given, for example if more than two groups are compared, for sample size calculations in cluster-randomized trials, in case that prognostic factors are of interest or time-varying covariates are incorporated into the analysis. It turns out that the corresponding formulas are all more or less modifications of the formula derived by Schoenfeld (1981). That emphasizes the outstanding role of Schoenfeld's formula.

Peace (2009) offers a comprehensive book on design and analysis of clinical trials with time-to-event endpoints. On sample size calculations in clinical research we also refer to the book of Chow et al. (2008) as well as the book written by Machin et al. (2008). Lachin (2000) provides a chapter on the evaluation of sample size and power assuming exponential survival as well as a Cox proportional hazards model.

For sample size calculations in more complex situations, in which analytic calculation is either very difficult or not feasible, the use of simulations offers a beneficial alternative. An example is provided by Austin (2012), who simulates event times assuming a Cox proportional hazards model to estimate the statistical power to detect a statistically significant effect of different types of binary time-varying covariates. Allignol et al. (2011) propose a simulation point of view for understanding competing risks. They point out that an empirical simulation approach provides a flexible tool for study planning in the presence of competing risks.

Acknowledgments

Kristin Ohneberg was supported by Grant BE 4500/1-1 of the German Research Foundation.

Bibliography

Abel, U., Jensen, K. and Karapanagiotou-Schenkel, I. (2012), 'Sample sizes for time-to-event endpoints: Should you insure against chance variations in accrual?', *Contemporary Clinical Trials* **33**(2), 456–458.

Ahnn, S. and Anderson, S. (1998), 'Sample size determination in complex clinical trials comparing more than two groups for survival endpoints', *Statistics in Medicine* **17**(21), 2525–2534.

Allignol, A., Schumacher, M., Wanner, C., Drechsler, C. and Beyersmann, J. (2011), 'Understanding competing risks: a simulation point of view', *BMC Medical Research Methodology* **11**, 86.

Andersen, P., Abildstrom, S. and Rosthøj, S. (2002), 'Competing risks as a multi-state model', *Statistical Methods in Medical Research* **11**(2), 203–215.

Austin, P. (2012), 'Generating survival times to simulate cox proportional hazards models with time-varying covariates', *Statistics in Medicine* **31**, 3946–3958.

Barthel, F., Babiker, A., Royston, P. and Parmar, M. (2006), 'Evaluation of sample size and power for multi-arm survival trials allowing for non-uniform accrual, non-proportional hazards, loss to follow-up and cross-over', *Statistics in Medicine* **25**(15), 2521–2542.

Bernardo, M., Lipsitz, S., Harrington, D. and Catalano, P. (2000), 'Sample size calculations for failure time random variables in non-randomized studies', *Journal of the Royal Statistical Society: Series D (The Statistician)* **49**(1), 31–40.

Beyersmann, J., Allignol, A. and Schumacher, M. (2011), *Competing risks and multistate models with R*, Springer, New York.

Chow, S., Shao, J. and Wang, H. (2008), *Sample size calculations in clinical research*, Vol. 20, Chapman & Hall, Boca Raton, FL.

Cox, D. (1972), 'Regression models and life-tables', *Journal of the Royal Statistical Society. Series B (Methodological)* **34**(2), 187–220.

Crowley, J. and Hoering, A. (2012), *Handbook of statistics in clinical oncology*, Chapman & Hall, Boca Raton, FL.

Duchateau, L. and Janssen, P. (2008), *The frailty model*, Springer, New York.

Fine, J. and Gray, R. (1999), 'A proportional hazards model for the subdistribution of a competing risk', *Journal of the American Statistical Association* **94**(446), 496–509.

Freedman, L. (1982), 'Tables of the number of patients required in clinical trials using the logrank test', *Statistics in Medicine* **1**(2), 121–129.

George, S. and Desu, M. (1974), 'Planning the size and duration of a clinical trial studying the time to some critical event', *Journal of Chronic Diseases* **27**(1), 15–24.

Grambauer, N., Schumacher, M. and Beyersmann, J. (2010), 'Proportional subdistribution hazards modeling offers a summary analysis, even if misspecified', *Statistics in Medicine* **29**(7-8), 875–884.

Hsieh, F. (1992), 'Comparing sample size formulae for trials with unbalanced allocation using the logrank test', *Statistics in Medicine* **11**(8), 1091–1098.

Jahn-Eimermacher, A., Ingel, K. and Schneider, A. (2013), 'Sample size in cluster-randomized trials with time to event as the primary endpoint', *Statistics in Medicine* **32**(5), 739–751.

Kaplan, E. and Meier, P. (1958), 'Nonparametric estimation from incomplete observations', *Journal of the American Statistical Association* **53**(282), 457–481.

Lachin, J. (2000), *Biostatistical Methods: The Assessment of Relative Risks*, Wiley (New York).

Latouche, A., Porcher, R. and Chevret, S. (2004), 'Sample size formula for proportional hazards modeling of competing risks', *Statistics in Medicine* **23**(21), 3263–3274.

Machin, D. and Campbell, M. (1997), *Sample Size Tables for Clinical Studies*, Wiley-Blackwell, Hoboken, NJ.

Machin, D., Campbell, M., Tan, S. and Tan, S. (2008), *Sample Size Tables for Clinical Studies, 3rd Edition*, Wiley-Blackwell, Hoboken, NJ.

Peace, K. (2009), *Design and Analysis of Clinical Trials with Time-to-Event Endpoints*, Chapman and Hall/CRC, Boca Raton, FL.

Peterson, B. and George, S. (1993), 'Sample size requirements and length of study for testing interaction in a 1 x k factorial design when time-to-failure is the outcome', *Controlled Clinical Trials* **14**(6), 511–522.

Schmoor, C., Sauerbrei, W. and Schumacher, M. (2000), 'Sample size considerations for the evaluation of prognostic factors in survival analysis', *Statistics in Medicine* **19**(4), 441–452.

Schoenfeld, D. (1981), 'The asymptotic properties of nonparametric tests for comparing survival distributions', *Biometrika* **68**(1), 316–319.

Schoenfeld, D. (1983), 'Sample-size formula for the proportional-hazards regression model', *Biometrics* **39**(2), 499–503.

Schoenfeld, D. and Borenstein, M. (2005), 'Calculating the power or sample size for the logistic and proportional hazards models', *Journal of Statistical Computation and Simulation* **75**(10), 771–785.

Schulgen, G., Olschewski, M., Krane, V., Wanner, C., Ruf, G. and Schumacher, M. (2005), 'Sample sizes for clinical trials with time-to-event endpoints and competing risks', *Contemporary Clinical Trials* **26**(3), 386–396.

Wanner, C., Krane, V., März, W., Olschewski, M., Mann, J., Ruf, G. and Ritz, E. (2005), 'Atorvastatin in patients with type 2 diabetes mellitus undergoing hemodialysis', *New England Journal of Medicine* **353**(3), 238–248.

Xu, R. and Chambers, C. (2011), 'A sample size calculation for spontaneous abortion in observational studies', *Reproductive Toxicology* **32**(4), 490–493.

29

Group Sequential Designs for Survival Data

Chris Jennison
University of Bath

Bruce Turnbull
Cornell University

CONTENTS

29.1	Introduction	595
29.2	Canonical joint distribution of test statistics based on accumulating data	596
29.3	Group sequential boundaries and error spending	599
29.4	The group sequential log-rank test	604
29.5	Example: A clinical trial for carcinoma of the oropharynx	605
29.6	Monitoring a hazard ratio with adjustment for strata and covariates	608
29.7	Further work	609
29.8	Concluding remarks	611
	Bibliography	612

29.1 Introduction

Consider an experiment or study where entry of subjects is staggered over time. We are interested in a survival or "time to event" response, measured from entry into the trial. The subjects are followed for a certain duration until their event time is observed or censored. The situation is depicted in Figure 29.1 with the horizontal lines in the diagram representing

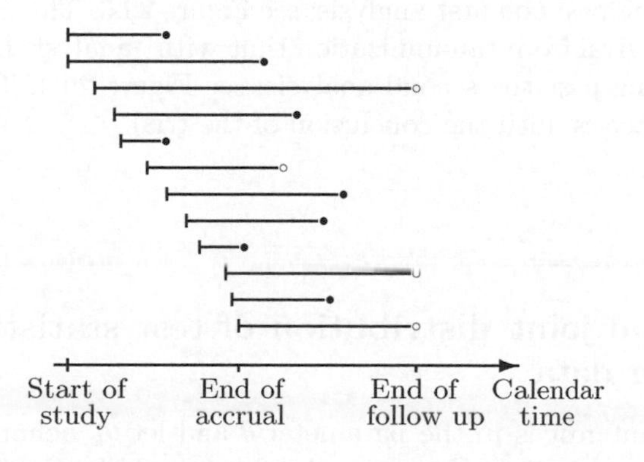

FIGURE 29.1
Accrual and follow-up in a survival study.

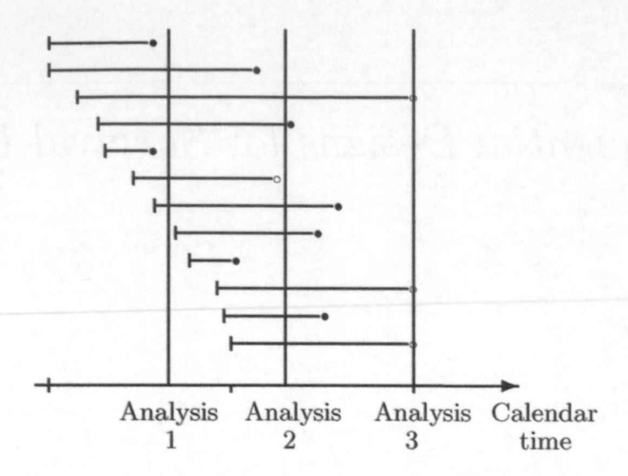

FIGURE 29.2
Interim analyses.

survival times of twelve subjects. A solid circle at the right-hand end designates an exact observation (subjects 1, 2, 4, 5, 7, 8, 9 and 11), whereas a hollow circle indicates that the survival time is censored. Note that censoring can occur because of end-of-study (subjects 3, 10 and 12) or for some other reason such as competing risk or loss to follow-up (subject 6). This situation is common in the conduct of clinical trials. Of course, the situation where all subjects start together at the beginning is a special case and this is more common in engineering or product life-testing experiments.

Consider the problem of testing between two hypotheses H_0 and H_1 concerning some parameter θ. The data are analysed not just at the planned end of the study, but also at interim times at calendar time points during the course of the study, with a maximum of $K > 1$ analyses. At the interim analyses, the decision can be made to stop the study concluding either H_0 or H_1, or to continue on to the next analysis. Figure 29.2 illustrates the case of three analyses. At an interim analysis, subjects are censored if they are still known to be alive at this point. Information on such subjects will continue to accrue at later analyses.

At the first interim analysis, we analyze data on elapsed survival times from randomization. These times have a common starting point of zero and "analysis time" censoring occurs for subjects surviving past the first analysis; see Figure 29.3. Then, at interim analysis 2, we analyze data on survival from randomization time with "analysis time" censoring occurring for subjects surviving past the second analysis; see Figure 29.4. This process continues on through further analyses until the conclusion of the trial.

29.2 Canonical joint distribution of test statistics based on accumulating data

Suppose our main interest is in the parameter θ and let $\widehat{\theta}_k$ denote an estimate of θ based on data available at analysis k. For survival data, θ could be the hazard ratio between two survival distributions, assumed constant over time, or the coefficient for a treatment effect in a Cox (1972) regression model or other type of failure time model.

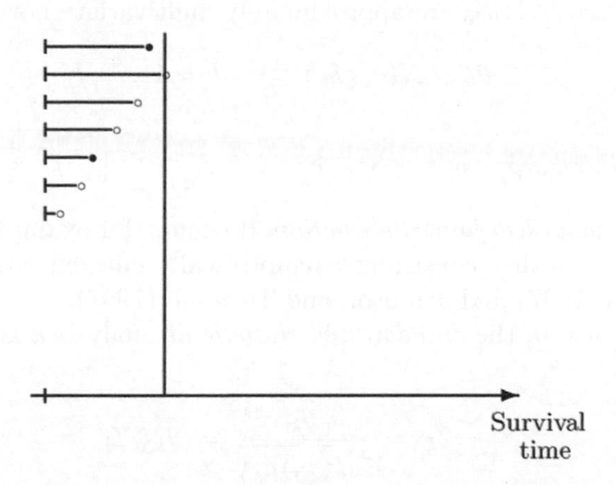

FIGURE 29.3
Interim analysis 1.

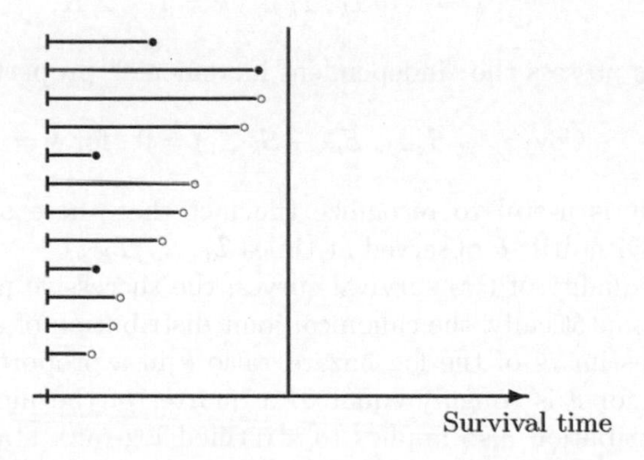

FIGURE 29.4
Interim analysis 2.

The information for θ at analysis k is

$$\mathcal{I}_k = \{\text{Var}(\widehat{\theta}_k)\}^{-1}, \quad k = 1, \ldots, K.$$

In many situations, $\widehat{\theta}_1, \ldots, \widehat{\theta}_K$ are approximately multivariate normal,

$$\widehat{\theta}_k \sim N(\theta, \{\mathcal{I}_k\}^{-1}), \quad k = 1, \ldots, K,$$

and

$$\text{Cov}(\widehat{\theta}_{k_1}, \widehat{\theta}_{k_2}) = \text{Var}(\widehat{\theta}_{k_2}) = \{\mathcal{I}_{k_2}\}^{-1} \quad \text{for } k_1 < k_2.$$

This is termed the *canonical joint distribution*. It occurs, for example, when $\widehat{\theta}$ is a maximum likelihood estimate or other consistent asymptotically efficient estimator; see Scharfstein, Tsiatis and Robins (1997) and Jennison and Turnbull (1997).

For testing H_0: $\theta = 0$, the *standardized statistic* at analysis k is

$$Z_k = \frac{\widehat{\theta}_k}{\sqrt{\text{Var}(\widehat{\theta}_k)}} = \widehat{\theta}_k \sqrt{\mathcal{I}_k}.$$

For this statistic, the canonical joint distribution of $(\widehat{\theta}_1, \ldots, \widehat{\theta}_K)$ implies that

$(Z_1, \ldots, Z_K)$ is multivariate normal,

$Z_k \sim N(\theta\sqrt{\mathcal{I}_k}, 1), \quad k = 1, \ldots, K,$

$\text{Cov}(Z_{k_1}, Z_{k_2}) = \sqrt{\mathcal{I}_{k_1}/\mathcal{I}_{k_2}}$ for $k_1 < k_2$.

The *score statistics*, $S_k = Z_k\sqrt{\mathcal{I}_k}$, are also approximately multivariate normal with

$$S_k \sim N(\theta\,\mathcal{I}_k, \mathcal{I}_k), \quad k = 1, \ldots, K.$$

The score statistics possess the "independent increments" property,

$$\text{Cov}(S_k - S_{k-1}, S_{k'} - S_{k'-1}) = 0 \quad \text{for } k \neq k'.$$

For computation it is useful to recognize the fact that these score statistics behave as Brownian motion with drift θ observed at times $\mathcal{I}_1, \ldots, \mathcal{I}_K$.

In testing the equality of two survival curves, the successive non-standardized log-rank statistics have, asymptotically, the canonical joint distribution of a sequence of score statistics. Here $\widehat{\theta}$ is an estimate of the log hazard ratio θ in a proportional hazards model and the information $\mathcal{I}$ for θ is roughly equal to a quarter of the number of observed events. The canonical distribution also applies to stratified log-rank statistics; see Jennison and Turnbull (2000, Sec. 13.6.2).

If a Cox (1972) proportional hazards regression model is fitted by maximum partial likelihood, the canonical joint distribution holds approximately for successive estimates of a regression coefficient. Kaplan-Meier (1958) estimates of survival probabilities at a fixed time point or of a specified quantile (e.g., the median) also follow the canonical joint distribution; see Section 29.7.

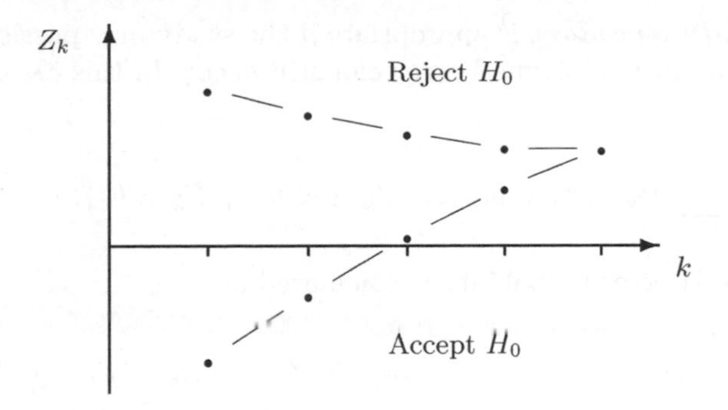

FIGURE 29.5
A group sequential boundary.

29.3 Group sequential boundaries and error spending

Suppose we are interested in testing the null hypothesis H_0: $\theta = 0$ versus a one-sided or two-sided alternative hypothesis H_1. At each interim analysis or "stage," we must decide whether to continue the study or to terminate, concluding either H_0 or H_1. At each stage k, $k = 1, \ldots, K$, this decision is based on a statistic Z_k according to the rule

If $Z_k \in \mathcal{C}_k$, continue on to stage $k+1$,
if $Z_k \in \mathcal{A}_k$, stop and conclude H_0,
if $Z_k \in \mathcal{B}_k$, stop and conclude H_1,

where $\mathcal{A}_k$, $\mathcal{B}_k$ and $\mathcal{C}_k$ are disjoint and exhaustive subsets of the real line, so $\mathcal{A}_k \cup \mathcal{B}_k \cup \mathcal{C}_k = (-\infty, \infty)$, and we set $\mathcal{C}_K = \emptyset$ in order that the procedure terminates at stage K.

Here, we shall consider the case of one-sided tests for superiority. Results for tests of non-inferiority, two-sided tests and equivalence tests can be developed analogously; see Jennison and Turnbull (2000). In a one-sided test where positive θ values are desirable the hypotheses are H_0: $\theta \leq 0$ and H_1: $\theta > 0$. The type 1 error probability constraint is

$$P_{\theta=0}\{\text{Reject } H_0\} = \alpha \tag{29.1}$$

and the type 2 error probability is specified through the power requirement at effect size δ,

$$P_{\theta=\delta}\{\text{Reject } H_0\} = 1 - \beta. \tag{29.2}$$

In this case, the continuation and stopping regions are $\mathcal{A}_k = (-\infty, a_k)$, $\mathcal{B}_k = (b_k, \infty)$ and $\mathcal{C}_k = (a_k, b_k)$, where $a_k \leq b_k$ for $k = 1, \ldots, K-1$, and $a_K = b_K$. A typical boundary with critical values $\{(a_k, b_k)\}$ is depicted in Figure 29.5.

The upper boundary, $\{b_k\}$, is often termed the *efficacy* boundary and the lower boundary, $\{a_k\}$, the *futility* boundary. The role of the futility boundary and whether it will be used for guidance or as a binding rule affects the construction of the boundaries. With a ***binding futility boundary***, it is assumed that crossing the lower boundary will definitely lead to stopping and acceptance of H_0, and the type I error probability is calculated as

$$\sum_{k=1}^{K} P_{\theta=0}\{a_1 < Z_1 < b_1, \ldots, a_{k-1} < Z_{k-1} < b_{k-1}, Z_k > b_k\}.$$

A ***non-binding futility boundary*** is appropriate if the study may possibly continue after crossing the lower boundary, so a type I error can still occur. In this case, the type I error probability is calculated as

$$\sum_{k=1}^{K} P_{\theta=0}\{Z_1 < b_1, \ldots, Z_{k-1} < b_{k-1}, Z_k > b_k\}.$$

In either case the type II error probability is calculated as

$$\sum_{k=1}^{K} P_{\theta=\delta}\{a_1 < Z_1 < b_1, \ldots, a_{k-1} < Z_{k-1} < b_{k-1}, Z_k < a_k\}.$$

The Pampallona and Tsiatis (1994) family provides a selection of one-sided group sequential tests. The test with index Δ has critical values of the form

$$\begin{aligned} b_k &= \tilde{C}_1 \left(\mathcal{I}_k/\mathcal{I}_K\right)^{\Delta-0.5}, \\ a_k &= \delta\sqrt{\mathcal{I}_k} - \tilde{C}_2 \left(\mathcal{I}_k/\mathcal{I}_K\right)^{\Delta-0.5}, \quad k = 1, \ldots, K. \end{aligned}$$

Given a specified pattern of information levels, for example, equally spaced values $\mathcal{I}_k = (k/K)\,\mathcal{I}_K$, $k = 1, \ldots, K$, and a choice of binding or non-binding futility boundary, constants $\mathcal{I}_K, \tilde{C}_1$ and $\tilde{C}_2$ can be found such that $a_K = b_K$ and the error probability constraints (29.1) and (29.2) are satisfied.

However, for survival data statistics such as those mentioned above, it is impractical to schedule the interim analyses at equal or pre-specified increments of information. Indeed, the increments in information will be both unequal and unpredictable. For example, the information for the log-rank statistic (approximately one quarter of the number of observed events) will only become known at the time of an analysis. Information for a treatment effect in a Cox (1972) regression model or a survival probability or quantile is similarly unpredictable. Thus we shall need to use the error spending approach of Lan and DeMets (1983) in which types I and II error probabilities are "spent" as functions of the observed information.

For a one-sided test of H_0: $\theta \le 0$ against H_1: $\theta > 0$, we need two functions to spend

Type I error probability α under $\theta = 0$,

Type II error probability β under $\theta = \delta$.

A *maximum information design* works towards a target information level $\mathcal{I}_{\max}$. The type I error probability α spending function $f(\mathcal{I})$ rises from zero to α as $\mathcal{I}$ increases from zero to $\mathcal{I}_{\max}$. Similarly, the type II error spending function $g(\mathcal{I})$ rises from zero at $\mathcal{I} = 0$ to β at $\mathcal{I} = \mathcal{I}_{\max}$.

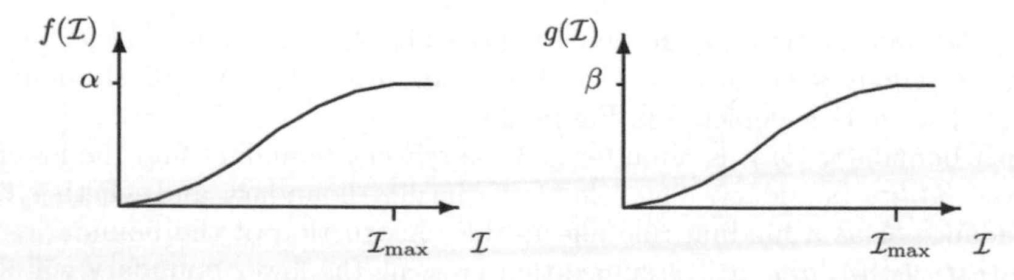

In implementing this error spending design, boundaries at each interim analysis, k, are constructed so that the cumulative type I error probability thus far is $f(\mathcal{I}_k)$ and the cumulative type II error probability is $g(\mathcal{I}_k)$. This calculation can be carried out treating the futility as binding or non-binding, as required.

At analysis 1:

The observed information is $\mathcal{I}_1$.

We reject H_0 if $Z_1 > b_1$, where

$$P_{\theta=0}\{Z_1 > b_1\} = f(\mathcal{I}_1)$$

and we accept H_0 if $Z_1 < a_1$, where

$$P_{\theta=\delta}\{Z_1 < a_1\} = g(\mathcal{I}_1).$$

Solving these equations determines the critical values a_1 and b_1.

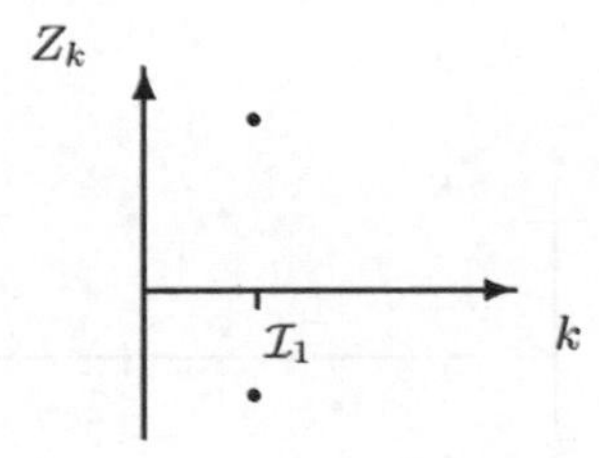

At analysis 2:

The observed information is $\mathcal{I}_2$.

We reject H_0 if $Z_2 > b_2$ where, for a binding futility boundary,

$$P_{\theta=0}\{a_1 < Z_1 < b_1, Z_2 > b_2\} = f(\mathcal{I}_2) - f(\mathcal{I}_1).$$

or, for a non-binding futility boundary,

$$P_{\theta=0}\{Z_1 < b_1, Z_2 > b_2\} = f(\mathcal{I}_2) - f(\mathcal{I}_1).$$

We accept H_0 if $Z_2 < a_2$, where

$$P_{\theta=\delta}\{a_1 < Z_1 < b_1, Z_2 < a_2\} = g(\mathcal{I}_2) - g(\mathcal{I}_1).$$

In either case, since a_1 and b_1 have been fixed at the previous analysis, we can solve these equations for a_2 and b_2.

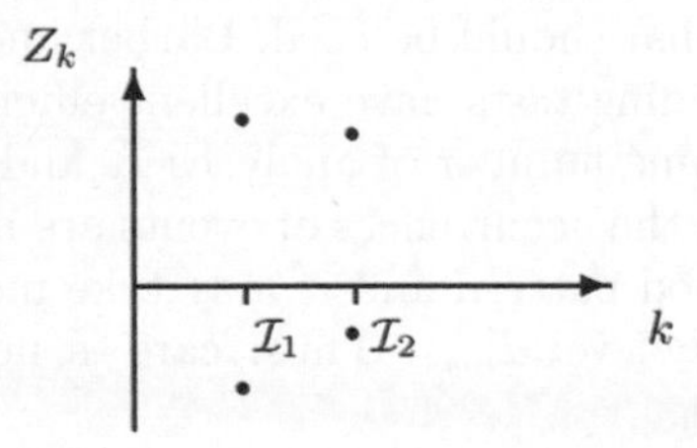

At a general analysis k:

The observed information is $\mathcal{I}_k$.

We reject H_0 if $Z_k > b_k$ where, for a binding futility boundary,

$$P_{\theta=0}\{a_1 < Z_1 < b_1, \ldots, a_{k-1} < Z_{k-1} < b_{k-1}, Z_k > b_k\} = f(\mathcal{I}_k) - f(\mathcal{I}_{k-1}),$$

or, for a non-binding futility boundary,

$$P_{\theta=0}\{Z_1 < b_1, \ldots, Z_{k-1} < b_{k-1}, Z_k > b_k\} = f(\mathcal{I}_k) - f(\mathcal{I}_{k-1}).$$

We accept H_0 if $Z_k < a_k$, where

$$P_{\theta=\delta}\{a_1 < Z_1 < b_1, \ldots, a_{k-1} < Z_{k-1} < b_{k-1}, Z_k < a_k\} = g(\mathcal{I}_k) - g(\mathcal{I}_{k-1}).$$

Since $a_1, \ldots, a_{k-1}$ and $b_1, \ldots, b_{k-1}$ were determined at analysis $k-1$, these equations can be solved for a_k and b_k.

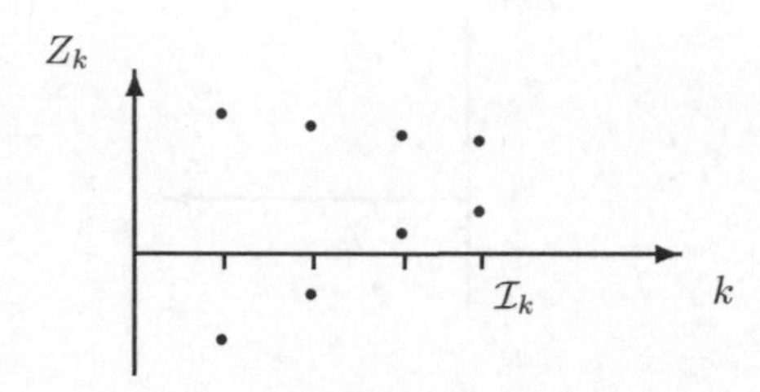

We remark that in the above description, the computation of a_k and b_k does ***not*** depend on future information levels, $\mathcal{I}_{k+1}$, $\mathcal{I}_{k+2}$, ... The error spending design is fully determined once the maximum information, $\mathcal{I}_{\max}$, and the spending functions $f(\mathcal{I})$ and $g(\mathcal{I})$ have been specified, although the critical values will depend on the information levels actually observed. One would like the upper and lower boundaries to meet at a single point at the concluding analysis where $f(\mathcal{I}) = \alpha$ and $g(\mathcal{I}) = \beta$. The maximum information $\mathcal{I}_{\max}$ and functions $f(\mathcal{I})$ and $g(\mathcal{I})$ can be chosen so that this will happen when observed information levels follow a particular pattern, but it is important to be able to handle other observed sequences $\mathcal{I}_1$, $\mathcal{I}_2$, ...

A convenient choice of error spending functions is provided by the so-called ρ-family, for which

$$f(\mathcal{I}) = \alpha \min\{1, (\mathcal{I}/\mathcal{I}_{\max})^\rho\} \quad \text{and} \quad g(\mathcal{I}) = \beta \min\{1, (\mathcal{I}/\mathcal{I}_{\max})^\rho\}.$$

Values $\rho > 0$ can be used and common choices are $\rho = 1$, 2 or 3. Lower values of ρ correspond to plans with more aggressive early stopping. The value of $\mathcal{I}_{\max}$ should be chosen so that boundaries converge with $a_K = b_K$ at the final analysis under a typical sequence of information levels. So, for design purposes we might plan for a maximum of K analyses at equally spaced information levels, $\mathcal{I}_k = (k/K)\,\mathcal{I}_{\max}$, $k = 1, \ldots, K$. Then, for each value of ρ there is an associated $\mathcal{I}_{\max}$ that should be used. Barber and Jennison (2002) show that the resulting ρ-family error spending tests have excellent efficiency properties when compared with other designs for the same number of analyses K and maximum information $\mathcal{I}_{\max}$.

Once the trial is running, the occurrences of events are unpredictable. Information levels may not follow the anticipated pattern and it may take more or fewer than K analyses to reach the target information level $\mathcal{I}_{\max}$. Thus, care is needed at the final analysis of a one-sided error spending test.

Over-running: If an analysis is reached with $\mathcal{I}_K > \mathcal{I}_{\max}$, solving the equations for a_K and b_K is liable to give $a_K > b_K$.

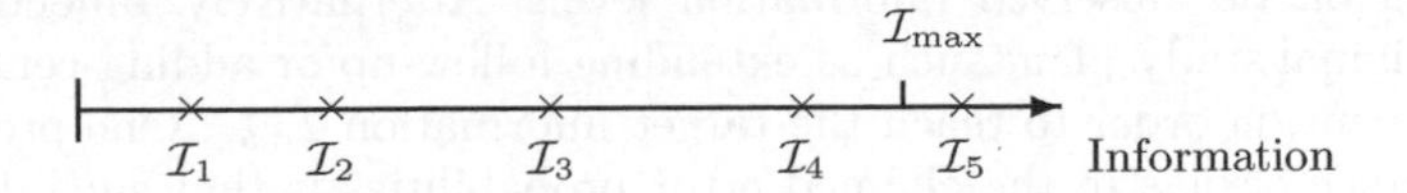

The value calculated for b_K will guarantee that the type I error probability is equal to α. So, in this case, we can reduce a_K to b_K and the power attained under $\theta = \delta$ will be greater than $1 - \beta$.

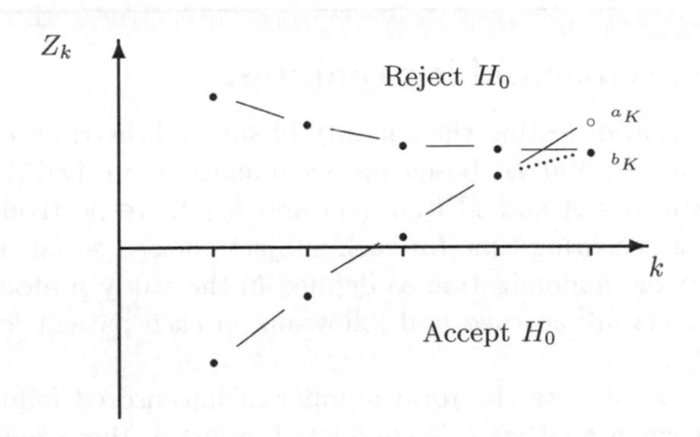

Even when $\mathcal{I}_K = \mathcal{I}_{\max}$, over-running may occur if information deviates from the pattern of, say, equally spaced values used in choosing $\mathcal{I}_{\max}$.

Under-running: A final information level $\mathcal{I}_K < \mathcal{I}_{\max}$ may be imposed as part of the trial design when a final planned analysis is reached, for example, after a maximum length of follow-up of subjects' survival.

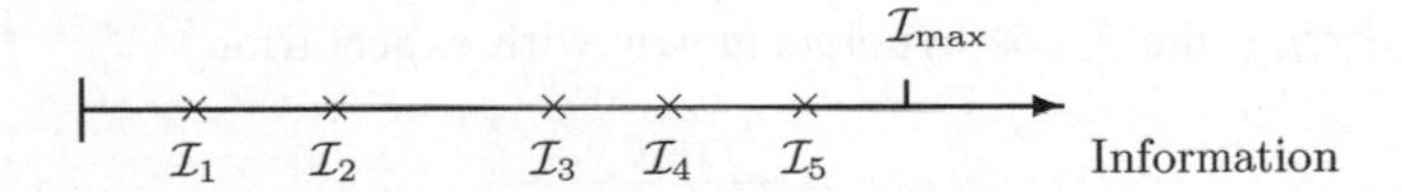

In this situation, values $f(\mathcal{I}_K) = \alpha$ and $g(\mathcal{I}_K) = \beta$ are used in the equations for a_K and b_K. Since the information level at this point is lower than $\mathcal{I}_{\max}$, the solutions of these equations are liable to have $a_K < b_K$.

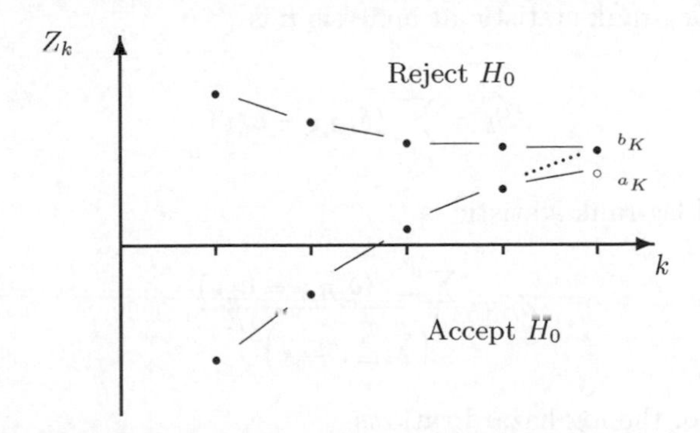

Again, with b_K as calculated, the type I error probability is exactly α. Here, we increase a_K to b_K in order to protect the type I error rate and the attained power at $\theta = \delta$ will be below the planned $1 - \beta$.

There is considerable freedom in implementing error spending group sequential designs. A series of analyses can be stipulated at fixed calendar times and the attained power will vary, depending on the observed information levels. Alternatively, amendments may be made to the original study plan, such as extending follow-up or adding centres to increase patient recruitment, in order to reach the target information $\mathcal{I}_{\max}$. One proviso to protect against any chance of bias in the claimed error probabilities is that such decisions should be made in response to observed information levels and not estimated treatment effects.

29.4 The group sequential log-rank test

We return to the problem of testing the equality of survival distributions $S_A(t)$ and $S_B(t)$ for two treatment arms, A and B, based on accumulating survival data. We denote the hazard rates on treatments A and B by $h_A(t)$ and $h_B(t)$, respectively. At each analysis we observe a failure or censoring time for each subject entered so far, measured from that subject's date of entry or randomization as defined in the study protocol. The way the set of data grows as patients are accrued and follow-up on each patient lengthens was shown in Figures 29.1 to 29.4.

Let d_k, $k = 1, \ldots, K$, denote the total number of uncensored failures observed across both treatment arms when analysis k is conducted. Some of these times may be tied and we suppose that d'_k of the d_k failure times are distinct, where $1 \le d'_k \le d_k$. We denote these distinct failure times by $\tau_{1,k} < \tau_{2,k} < \ldots < \tau_{d'_k,k}$ and let $r_{iA,k}$ and $r_{iB,k}$ be the numbers at risk on treatment arms A and B, respectively, just before time $\tau_{i,k}$. Finally, we denote by $\delta_{iA,k}$ and $\delta_{iB,k}$ the numbers on treatment arms A and B that fail at time $\tau_{i,k}$ and define $\delta_{i,k} = \delta_{iA,k} + \delta_{iB,k}$ for $i = 1, \ldots, d'_k$. If there are no ties, then $\delta_{i,k} = 1$ and either $\delta_{iA,k} = 1$ and $\delta_{iB,k} = 0$ or $\delta_{iA,k} = 0$ and $\delta_{iB,k} = 1$ for each pair i and k.

If the survival distributions $S_A(t)$ and $S_B(t)$ are equal, the conditional distribution of $\delta_{iB,k}$ given $r_{iA,k}$, $r_{iB,k}$ and $\delta_{i,k}$ is hypergeometric with expectation

$$e_{i,k} = \frac{r_{iB,k}\,\delta_{i,k}}{r_{iA,k} + r_{iB,k}}$$

and variance

$$v_{i,k} = \frac{r_{iA,k}\, r_{iB,k}\, \delta_{i,k}\, (r_{iA,k} + r_{iB,k} - \delta_{i,k})}{(r_{iA,k} + r_{iB,k} - 1)\,(r_{iA,k} + r_{iB,k})^2}. \tag{29.3}$$

The unstandardized log-rank statistic at analysis k is

$$S_k = \sum_{i=1}^{d'_k} (\delta_{iB,k} - e_{i,k})$$

and the standardized log-rank statistic is

$$Z_k = \frac{\sum_{i=1}^{d'_k} (\delta_{iB,k} - e_{i,k})}{\left(\sum_{i=1}^{d'_k} v_{i,k}\right)^{1/2}}. \tag{29.4}$$

The information $\mathcal{I}_k$ for the log hazard ratio is

$$\mathcal{I}_k = \sum_{i=1}^{d'_k} v_{i,k}. \tag{29.5}$$

The log-rank test has optimal power properties to detect alternatives when hazard rates in the two treatment arms are proportional, so $h_A(t) = \lambda h_B(t)$. The sequence of log-rank statistics defined by (29.4) then has, approximately, the canonical joint distribution for a sequence of Z-statistics, given $\mathcal{I}_1, \ldots, \mathcal{I}_K$, with $\theta = \log(\lambda)$, the log hazard ratio.

Since the canonical joint distribution holds, the methods described in Section 29.3 can be used to construct group sequential error spending tests from the sequence of statistics Z_k and information levels $\mathcal{I}_k$. In designing a maximum information trial to meet a given power requirement, it is necessary to predict the information levels that will arise, especially that at the final possible analysis. Here, it is helpful to note from (29.3) that each $v_{i,k}$ is approximately $\delta_{i,k}/4$ if $r_{iA,k} \approx r_{iB,k}$ and either $\delta_{i,k} = 1$ or $\delta_{i,k}$ is small relative to $r_{iA,k} + r_{iB,k}$. Hence, $\mathcal{I}_k$ will be approximately equal to $d_k/4$ and the final information level will be close to one quarter of the total number of observed failures. The illustrative example in the next section will show the usefulness of this approximation in planning the sample size and length of follow-up that may be necessary in a survival study.

29.5 Example: A clinical trial for carcinoma of the oropharynx

We illustrate the methods we have described by applying them to a clinical trial conducted by the Radiation Therapy Oncology Group in the U.S. to investigate treatments of carcinoma of the oropharynx. We use the data from six of the larger institutions participating in this trial as recorded by Kalbfleisch and Prentice (2002, Appendix II). Subjects were recruited to the study between 1968 and 1972 and randomized to a standard radiotherapy treatment or an experimental treatment in which the radiotherapy was supplemented by chemotherapy. The major endpoint was patient survival and patients were followed until around the end of 1973. Several baseline covariates, thought to have strong prognostic value, were also recorded.

TABLE 29.1
Summary data for oropharynx cancer clinical trial.

Analysis		Number of subjects entered		Number of deaths	
k	Date	Treatment A	Treatment B	Treatment A	Treatment B
1	12/69	38	45	13	14
2	12/70	56	70	30	28
3	12/71	81	93	44	47
4	12/72	95	100	63	66
5	12/73	95	100	69	73

The conduct of the study did not follow a group sequential plan but, for purposes of illustration, we have reconstructed patients' survival times and their status, dead or censored, at times 720, 1080, 1440, 1800 and 2160 days from the beginning of 1968. This "reconstructed" dataset was used by Jennison and Turnbull (2000, Ch. 13). A summary of the reconstructed data is given in Table 29.1: we used the precise death or censoring times in the reconstructed data to compute the statistics and information values in applying retrospectively a group sequential error spending design. As the central survival records would not have been updated continuously, our constructed datasets most likely resemble the information that would have been available at interim analyses conducted a month or two after these times, and so they are an approximation to the data that could have been studied by a monitoring committee meeting at dates a little after 2, 3, 4, 5 and 6 years from the start of the study. The longer waiting period to the first interim analysis is intended to compensate for the slow initial accrual of survival information while only a few patients had been entered to the trial.

Since the experimental treatment involved chemotherapy as well as radiotherapy, the researchers would have been looking for a substantive improvement in survival on this treatment in return for the additional discomfort and short-term health risks. A one-sided testing formulation is, therefore, appropriate and we shall conduct our retrospective interim analyses as a group sequential test of the null hypothesis of no treatment difference against the one-sided alternative that the new combination therapy is superior to the standard treatment of radiotherapy alone. For the sake of illustration, we suppose the experiment was designed to achieve a type I error probability of $\alpha = 0.025$ and power $1 - \beta = 0.8$ when the log hazard ratio for the experimental treatment versus the standard is equal to 0.5.

We have supposed that at the design stage a maximum of $K = 5$ interim analyses were planned with equally spaced information levels. We use ρ-family error spending functions with index $\rho = 2$ to create efficacy and futility boundaries. Thus, type I error probability α and type II error probability β are "spent" in proportion to $(\mathcal{I}/\mathcal{I}_{\max})^2$. With these specifications we compute the target maximum information, $\mathcal{I}_{\max}$, which gives $a_K = b_K$ when the $\{a_k, b_k\}$ are calculated as described in Section 29.3. The computations depend on whether a binding or a non-binding futility (lower) boundary is to be employed. Table 29.2 displays the design parameters for both situations under the planning assumptions of equally spaced information levels, culminating in $\mathcal{I}_{\max}$. Note the familiar "curved triangular" shape of the boundaries as seen in Figure 29.5.

The maximum information $\mathcal{I}_{\max}$ needed in the group sequential trial design is 34.48 if a binding futility boundary is used or 35.58 if the futility boundary is non-binding. Under the approximation $\mathcal{I} \approx d/4$, the maximum numbers of failures that may need to be observed, $d_f = 4\mathcal{I}_f$, are 138 and 143, respectively.

A fixed sample study with no interim monitoring but the same type I error rate $\alpha = 0.025$ and power $1 - \beta = 0.8$ at $\theta = 0.5$ requires information

$$\mathcal{I}_f = \frac{\{\Phi^{-1}(0.975) + \Phi^{-1}(0.8)\}^2}{0.5^2} = 31.40.$$

Under the approximation $\mathcal{I} \approx d/4$, the total number of failures to be observed is $d_f = 4\mathcal{I}_f \approx 126$. Clearly this is smaller than the maximum event numbers of 138 or 143 for the five-stage design. However, the group sequential procedure benefits from the opportunity to stop before the last stage.

Figure 29.6 shows the expected number of events for the group sequential design with a binding futility boundary under different values of the hazard ratio. Plotted values for hazard ratios away from one (and log hazard ratios away from zero) are less accurate since the approximation $\mathcal{I} \approx d/4$ is less reliable in these cases, particularly in later stages of the trial when numbers at risk on the two treatment arms become unequal. Figure 29.6 shows that the group sequential design with a binding futility boundary has an expected number of events under H_0 of 72.9 and under H_1 this becomes 94.5; the maximum expected number of events, which occurs for log hazard ratio $\theta = 0.35$, is 100.9, still considerably less than the 126 events for a fixed sample design. With a non-binding boundary the corresponding

TABLE 29.2
Design parameters for a group sequential procedure assuming equally spaced information levels, $\mathcal{I}_k = (k/5)\,\mathcal{I}_{\max}$, $k = 1, \ldots, 5$.

Design parameter	Binding futility boundary		Non-binding futility boundary	
$\mathcal{I}_{\max}$	34.48		35.58	
Maximum number of deaths	138		143	
a_1, b_1	−1.096	3.090	−1.075	3.090
a_2, b_2	−0.053	2.714	−0.023	2.714
a_3, b_3	0.722	2.473	0.758	2.473
a_4, b_4	1.387	2.276	1.429	2.280
a_5, b_5	2.055	2.055	2.114	2.114

numbers are 74.3, 96.4 and 103.3, assuming for purposes of this calculation that the futility boundary is in fact obeyed.

We now turn to the task of applying the monitoring boundaries to the reconstructed dataset summarized in Table 29.1. The boundary values $(a_1, b_1), \ldots, (a_5, b_5)$ are calculated using the observed information levels $\mathcal{I}_1, \ldots, \mathcal{I}_5$ rather than the equally spaced ones of the initial design. In doing this, we apply the formulae of Section 29.3 at each interim analysis in succession. From here on, we shall apply designs with binding futility boundaries, noting that the exposition would be very similar if we were to use non-binding futility boundaries instead.

The sequence of standardized log-rank statistics, $Z_1, \ldots, Z_5$, and the corresponding critical values $(a_1, b_1), \ldots, (a_5, b_5)$ are displayed in Table 29.3.

We can see from this table that, had this design been used, the trial would have stopped for futility at analysis 2, about three years earlier than the original trial, reaching the same conclusion with only 126 subjects accrued instead of 195. Of course, those last three years may have produced further valuable information about other aspects of the treatments such as toxicity or quality of life. With this in mind, investigators might have opted to continue the trial despite unpromising interim results. It is in anticipation of such eventualities that a non-binding futility boundary could be chosen since it allows a subsequent positive result for efficacy to be reported without concern that the type 1 error rate is inflated above the specified α.

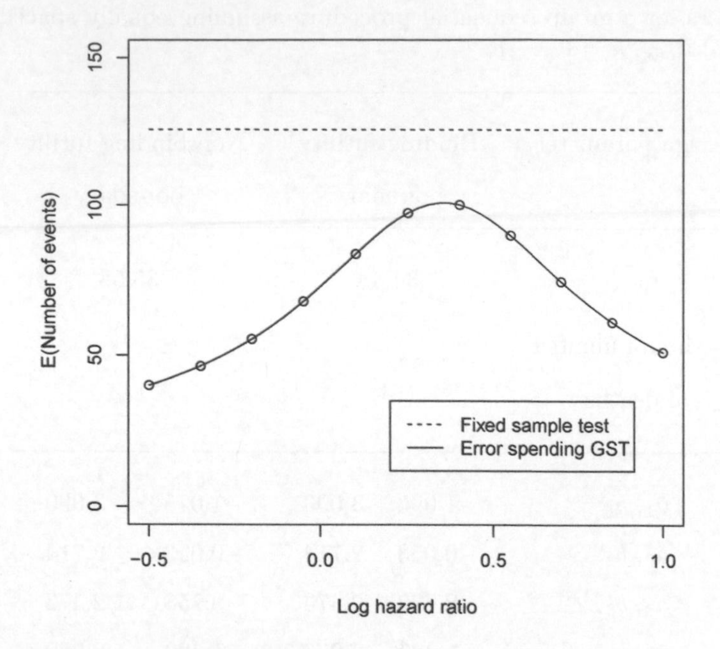

FIGURE 29.6
Expected number of events on termination of the group sequential log-rank test with a binding futility boundary and equally spaced information levels.

29.6 Monitoring a hazard ratio with adjustment for strata and covariates

The Oropharynx Cancer dataset contained information on a number of baseline covariates for each subject. These included gender, initial condition, T-staging, N-staging and two indicator variables describing the tumor site. Each patient was treated at one of six participating institutions and we shall treat institution as a stratifying variable. We model the data by means of a stratified proportional hazards regression model (Cox, 1972) in which the hazard rate for patient i is modeled as

$$h_{il}(t) \;=\; h_{0l}(t)\,\exp\{\beta_1 I(\text{Patient } i \text{ on Treatment B}) + \Sigma_{j=2}^{7}\, x_{ij}\beta_j\}.$$

The parameter β_1 represents the log hazard ratio between treatments after adjustment for the other covariates and stratification. We take the objective to be to test H_0: $\beta_1 \le 0$ against the one-sided alternative $\beta_1 > 0$.

Standard software for Cox regression will provide the maximum partial likelihood estimate of the parameter vector, β, and its estimated variance matrix. We are interested in the treatment effect represented by the first component, β_1. At analysis k we have

$$\widehat{\beta}_1^{(k)}, \quad v_k = \widehat{\text{Var}}\,(\widehat{\beta}_1^{(k)}), \quad \mathcal{I}_k = v_k^{-1} \quad \text{and} \quad Z_k = \widehat{\beta}_1^{(k)}/\sqrt{v_k}.$$

The standardized statistics $Z_1, \ldots, Z_5$ have, approximately, the canonical joint distribution

TABLE 29.3
Summary data for the oropharynx trial and critical values for the error spending design with binding futility boundary.

Analysis k	Number entered	Number of deaths	$\mathcal{I}_k$	a_k	b_k	Z_k
1	83	27	5.43	−1.41	3.23	−1.04
2	126	58	12.58	−0.21	2.76	−1.00
3	174	91	21.11	0.78	2.44	−1.21
4	195	129	30.55	1.68	2.16	−0.73
5	195	142	33.28	2.14	2.14	−0.87

of Section 29.2. Thus we may apply the group sequential designs and error spending method of Section 29.3 to monitor the adjusted log hazard ratio at successive interim analyses. In fact we can take exactly the same method that we described in Section 29.5 and simply use the above statistics Z_k and information values $\mathcal{I}_k$, $k = 1, \ldots, 5$, in place of those for the log-rank statistic.

Calculation gives the values (a_1, b_1), ... , (a_5, b_5) shown in Table 29.4 for the error spending group sequential design, again with a binding futility boundary, to be applied to $Z_1, \ldots, Z_5$. Under this model and stopping rule, the study would — just — have stopped for futility at the second analysis.

29.7 Further work

In this chapter, we have concentrated on the use of an error spending group sequential design for monitoring a log-rank statistic or a regression coefficient in a Cox regression model. The methods we have presented form a good introduction to other group sequential methods for survival data. The ideas have been extended in two directions:

A. To other features of group sequential designs;

B. To other features of survival analysis.

A. *Further group sequential methods that can be applied to the collection and analysis of survival data.* We have considered the "curved triangular" testing boundaries that arise in one-sided hypothesis tests. These are commonly used in superiority trials where it is hoped to show that a new treatment improves on the current standard; the same forms of boundary also arise in non-inferiority trials where hypotheses H_0: $\theta \leq 0$ and H_1: $\theta > 0$ are replaced by H_0: $\theta \leq -\delta$ and H_1: $\theta > \delta$, where δ represents an acceptable "margin of inferiority." Other boundary shapes are applicable for testing a null hypothesis against a two-sided alternative or in tests of equivalence, where it is hoped to demonstrate that the effect of a new treatment is within a specified tolerance of that of an existing treatment.

TABLE 29.4
Covariate-adjusted group sequential analysis of the oropharynx data.

Analysis k	$\mathcal{I}_k$	a_k	b_k	$\widehat{\beta}_1^{(k)}$	Z_k
1	4.11	−1.75	3.39	−0.79	−1.60
2	10.89	−0.44	2.85	−0.14	−0.45
3	19.23	0.59	2.50	−0.08	−0.33
4	28.10	1.45	2.24	0.04	0.20
5	30.96	2.23	2.23	0.01	0.04

In addition to the positive or negative outcome of a hypothesis test, it is usually required to give point or interval estimates of the treatment effect at the termination of a trial or to provide a P-value summarizing the strength of evidence against a null hypothesis. Special methods are needed to construct such quantities, taking into account the sequential nature of the design; see Jennison and Turnbull (2000, Ch. 8).

Repeated confidence intervals permit an interval estimate of a treatment effect to be stated at any stage of the trial (not just the last), with the property that the coverage probability of all the intervals is simultaneously controlled at a given confidence level, $1-\gamma$ say. Such confidence intervals are wider than naïve, fixed sample size intervals computed at each stage, but they are free from the "multiple looks" bias of sequential testing. This obviates the problem of "over-interpretation of interim results"; see Jennison and Turnbull (1989).

B. *Further techniques for survival data to which group sequential methods can be applied.* First consider a one-sample problem, where we are interested in the time to an event such as death or the disease recurrence in a homogeneous population. Sometimes a binary outcome is defined to indicate whether failure has occurred after an elapsed time, τ say. If not all subjects are followed for time τ, the simple proportion of those surviving to time τ will be a biased estimate of the survival rate, while omitting subjects with potential censoring times less than τ is inefficient. These difficulties are overcome by use of the Kaplan-Meier estimate (Kaplan and Meier, 1958) of the survival function $S(t)$. Let $\widehat{S}_k(t)$ denote the Kaplan-Meier estimate of the survival probability $S(t)$ at time t based on data available at analysis k. For a given value of τ, suppose $0 < S(\tau) < 1$ and there is a positive probability for each observation to be uncensored and greater than τ, then Jennison and Turnbull (1985) show that the sequence

$$Z_k = \frac{\{\widehat{S}_k(\tau) - S(\tau)\}}{\sqrt{\mathrm{Var}\{\widehat{S}_k(\tau)\}}}, \quad k = 1, \ldots, K, \tag{29.6}$$

has, asymptotically, the canonical joint distribution of Section 29.2 with $\theta = S(\tau)$ and information levels $\mathcal{I}_k = [\mathrm{Var}\{\widehat{S}_k(\tau)\}]^{-1}$. A consistent estimate of the variance of $\widehat{S}_k(\tau)$ is provided by Greenwood's formula — see, for example, Jennison and Turnbull (1985). Hence, a group sequential test of the hypothesis H_0: $S(\tau) = p_0$, where τ and p_0 are specified, can be based on the standardized statistics

$$Z_k = \frac{\{\widehat{S}_k(\tau) - p_0\}}{\sqrt{\{\widehat{V}_k(\tau)\}}}, \quad k = 1, \ldots, K,$$

and associated information levels $\mathcal{I}_k = \{\widehat{V}_k(\tau)\}^{-1}$, where $\widehat{V}_k(\tau)$ denotes a consistent estimate of $\text{Var}\{\widehat{S}_k(\tau)\}$. Since information depends on the number and times of observed failures, the error spending approach of Section 29.3 is needed for the construction of such tests. The Greenwood estimate is straightforward to calculate and is typically available in the output of standard statistical computer software for estimating survival curves. Alternatively, the "constrained" variance estimator introduced by Thomas and Grunkemeier (1975, Sec. 4) can be used in place of the Greenwood formula: simulations reported by Thomas and Grunkemeier and by Barber and Jennison (1999) show this should lead to more accurate attainment of error rates and coverage probabilities for repeated confidence intervals. Barber and Jennison (1999) go on to propose further methods to achieve error rates and coverage probabilities more accurately in smaller sample sizes.

Sometimes, interest is in a certain *quantile* of the survival distribution. For $0 < p < 1$, we define the pth quantile of the survival distribution $S(t)$ to be $t_p = \inf\{t: S(t) \geq p\}$. Assuming $S(t)$ to be strictly decreasing in t, a group sequential test of H_0: $t_p = t^*$ for specified t^* and p is equivalent to a test of H_0: $S(t^*) = p$ and the same Kaplan-Meier test statistics can be used with $\tau = t^*$ and $p_0 = p$. Jennison and Turnbull (1985) have investigated repeated confidence intervals for the median survival time.

Analogous methods can also be used in a two-sample comparison. If $S_A(t)$ and $S_B(t)$ denote survival functions on treatments A and B in a randomized trial, a test of H_0: $S_A(\tau) = S_B(\tau)$, for a given choice of τ, can be based on successive statistics

$$Z_k = \frac{\{\widehat{S}_{Ak}(\tau) - \widehat{S}_{Bk}(\tau)\}}{\sqrt{\{\widetilde{V}_{Ak}(\tau) + \widetilde{V}_{Bk}(\tau)\}}}, \quad k = 1, \ldots, K,$$

where $\widehat{S}_{Ak}(\tau)$ and $\widehat{S}_{Bk}(\tau)$ are Kaplan-Meier estimates of $S_A(\tau)$ and $S_B(\tau)$, respectively, at analysis k and $\widetilde{V}_{Ak}(\tau)$ and $\widetilde{V}_{Bk}(\tau)$ are their estimated variances. The problem of comparing the pth quantiles of two survival distributions has been addressed by Keaney and Wei (1994).

29.8 Concluding remarks

A variety of software packages is now available to implement the methods we have described. One choice that can compute the error spending boundaries described in Section 29.3 and that has a dedicated module for planning and analyzing survival trials is East (Cytel, 2012). Another choice is the **gsDesign** package in R.

It should be noted that not all sequences of standardized statistics follow the canonical joint distribution of Section 29.2. As an example, Slud and Wei (1982) have shown that this property does *not* hold for some weighted log-rank test statistics when there is staggered entry. These statistics include those arising in Gehan's (1965) procedure for modifying the Wilcoxon test to allow censored data.

This chapter has provided a basic overview of the use of group sequential methods for survival data. There is a large literature on the subject which we have not attempted to summarize here; some more references can be found in Jennison and Turnbull (2000, Ch. 13). In particular, there is an emerging literature on the adaptive clinical trial designs for survival data. The availability at interim analyses of partial information about patients' continuing survival causes particular problems in adaptive designs; for one example with correlated survival endpoints, and a solution to the adaptive design problem, see Jenkins, Stone and Jennison (2011).

Bibliography

Barber, S. and Jennison, C. (1999), 'Symmetric tests and confidence intervals for survival probabilities and quantiles of censored survival data', *Biometrics*, **55**, 430–436.

Barber, S. and Jennison,C. (2002), 'Optimal asymmetric one-sided group sequential tests' *Biometrika*, **89**, 49–60.

Cox, D. R. (1972), 'Regression models and life-tables', *Journal of the Royal Statistical Society - Series B* **34**, 187–220.

Cytel Software Corporation, (2012), *EaSt v.6: A software package for the design and interim monitoring of group-sequential clinical trials*, Cytel Software Corporation, Cambridge, Massachusetts.

Gehan, E. A. (1965), 'A generalized Wilcoxon test for comparing arbitrarily singly censored samples', *Biometrika*, **52**, 203–223.

Jenkins, M., Stone, A. and Jennison, C. (2011), 'An adaptive seamless phase II/III design for oncology trials with subpopulation selection using correlated survival endpoints', *Pharmaceutical Statistics*, **10**, 347–356.

Jennison, C. and Turnbull, B.W. (1985), 'Repeated confidence intervals for the median survival time', *Biometrika*, **72**, 619–625.

Jennison, C. and Turnbull, B. W. (1989), 'Interim analyses: the repeated confidence interval approach (with discussion)', *Journal of the Royal Statistical Society, Series B*, **51**, 305–361.

Jennison, C. and Turnbull, B. W. (1997), 'Group sequential analysis incorporating covariate information', *Journal of the American Statistical Association*, **92**, 1330–1341.

Jennison, C. and Turnbull, B. W. (2000), *Group Sequential Methods with Applications to Clinical Trials*, Chapman & Hall/CRC, Boca Raton.

Kalbfleisch, J. D. and Prentice, R. L. (2002), *The Statistical Analysis of Failure Time Data, Second Edition*, Wiley, New York.

Kaplan, E. and Meier, P. (1958), 'Nonparametric estimation from incomplete observations', *Journal of the American Statistical Association* **43**, 457–481.

Keaney, K.M. and Wei, L-J. (1994), 'Interim analyses based on median survival times', *Biometrika*, **81**, 279–286.

Lan, K. K. G. and DeMets, D. L. (1983), 'Discrete sequential boundaries for clinical trials', *Biometrika*, **70**, 659–663.

Pampallona, S. and Tsiatis, A. A. (1994), 'Group sequential designs for one-sided and two-sided hypothesis testing with provision for early stopping in favor of the null hypothesis', *Journal of Statistical Planning and Inference*, **42**, 19–35.

Scharfstein, D. O., Tsiatis, A. A. and Robins, J. M. (1997), 'Semiparametric efficiency and its implication on the design and analysis of group sequential studies', *Journal of the American Statistical Association*, **92**, 1342–1350.

Slud, E. V. and Wei, L-J. (1982), 'Two-sample repeated significance tests based on the modified Wilcoxon statistic', *Journal of the American Statistical Association*, **77**, 862–868.

Thomas, D.R. and Grunkemeier, G.L. (1975), 'Confidence interval estimation of survival probabilities for censored data', *Journal of the American Statistical Association*, **70**, 865–871.

30

Inference for Paired Survival Data

Jennifer Le-Rademacher and Ruta Brazauskas
Division of Biostatistics, Medical College of Wisconsin

CONTENTS

30.1 Introduction 615
30.2 Example 616
30.3 Notation 617
30.4 Tests for paired data 618
  30.4.1 Rank-based tests 618
  30.4.2 Within-pair comparison 620
  30.4.3 Weighted Kaplan-Meier comparison 622
30.5 Regression models for paired data 624
  30.5.1 Stratified Cox models 624
  30.5.2 Marginal Cox models 625
  30.5.3 Shared frailty models 626
30.6 Comparing survival probabilities at a fixed time point 627
30.7 Discussion 630
Acknowledgments 631
Bibliography 631

30.1 Introduction

A matched pairs study design is used in medical research to minimize variability caused by extraneous variables. Many such studies result from individuals being artificially matched on a set of known covariates. When individuals in a pair are randomized to receive two different treatments, then the comparison is done between two subjects that are alike. Hence, the difference in outcome can be directly attributed to treatment effect. Paired data can also be obtained from observations with a biological link such as pairs of organs (eyes, kidneys, knees, etc.) from one person or pairs of twins or siblings. Another type of paired data arise from pairs of observations with different baseline survival or hazards for each pair, for example, pairs of patients treated at the different centers Regardless of the pairing mechanism, outcomes between individuals might be correlated. Comparison of outcomes between treatments in such a study must account for this correlation. For positively correlated outcomes, methods for unpaired data ignoring correlation between individuals within pairs may underestimate treatment effect.

Besides allowing comparison of like-to-like, a matched study design can provide logistical advantage in retrospective case-control studies using data from large databases. If the number of cases are few in these studies, a prespecified number of matched controls can be selected for each case. When additional information is needed, it is only necessary to collect the additional data on the cases and their matched controls instead of on all control subjects in the database. On the other hand, one drawback of a matched study design is that it might not be possible to find matched controls for cases with uncommon characteristics. Cases without matched controls will be excluded from the study. For studies with time-to-event outcomes, another drawback of a matched study design is that the effective sample size might be reduced if the smaller of the paired times is censored as will be discussed in Section 30.4.

Methods to analyze paired continuous or categorical data are well established. However, analysis of paired time-to-event data is complicated by the fact that the event times for some patients are not observed due to loss to follow-up or subjects not having experienced the event by the end of the study. Individuals with an unobserved event time are censored at the last follow-up. The focus of this exposition is to present current inference methods to analyze paired survival data subject to right-censoring.

Several tests have been proposed specifically for this type of paired data. Many of these tests assume that individuals within a pair have a common censoring time. Such an assumption is reasonable for pairs with biological link, for example, in a study designed to compare the effect of two treatments for diabetic retinopathy where a patient's right and left eyes are randomly assigned to different treatments (Diabetic Retinopathy Study Research Group, 1976). In this case, loss to follow-up in a patient prevents observation of failure in both eyes. Here members of the pair have a common censoring time.

This common censoring time assumption might not be reasonable for pairs artificially matched on a set of covariates. For example, in an observational study evaluating the effect of chemotherapy versus radiation therapy for breast cancer, comparing patients receiving chemotherapy to patients receiving radiotherapy matched on age, disease stage, and other factors minimizes the potential bias caused by imbalance in baseline characteristics. In this case, loss to follow-up for one member does not automatically imply loss to follow-up for the other member of the pair. A common censoring time is a strong assumption in this situation. One argument for common censoring times in artificially matched pair design is to prevent bias resulting from differential follow-up time between treatment groups. This assumption can cause significant reduction in effective sample size used for inference as illustrated in Section 30.4 where we describe tests developed specifically for paired survival data. Alternatively, paired survival data can be analyzed using regression methods developed for clustered survival data as described in Section 30.5. In Section 30.6, we will describe methods to compare survival probabilities at a fixed time point. A real data example will be used to illustrate methods discussed in the paper.

30.2 Example

Throughout this chapter we will utilize data from a hematopoietic cell transplantation (HCT) study conducted at the Center for International Blood and Marrow Transplant Research (CIBMTR). The CIBMTR collects data on essentially all allogeneic HCTs done in the United States. An estimated 80% of autologous transplants done in the U.S. are reported to the center. CIBMTR also receives data on allogeneic and autologous stem cell transplants from participating international centers worldwide. Extensive data on patient risk factors and outcomes are collected at the time of transplantation, six months post-transplant, and subsequently every year during patients' follow-up visits. The data is reported to a Statistical Center housed at the Medical College of Wisconsin in Milwaukee, WI, USA.

The data used in this chapter is a subset of a larger study examining the effect of stem cell source on the outcomes of patients receiving HCT to treat severe aplastic anemia (SAA) first reported by Eapen et al. (2011). Aplastic anemia is a disorder where the bone marrow stops making enough red blood cells, white blood cells, and platelets for the body. People with severe aplastic anemia are at risk for life-threatening infections or bleeding. For many patients with severe aplastic anemia, a bone marrow transplant is the preferred standard treatment. Outcomes after unrelated donor bone marrow (BM) transplantation for SAA patients have improved, with a five-year survival rate now being approximately 65%. However, in recent years the use of peripheral blood stem cells (PBSC) instead of bone marrow as a graft source has increased and the aforementioned study by Eapen et al. (2011) attempted to compare the outcomes of patients receiving either BM or PBSC transplants. Using a full dataset of 296 patients (225 BM recipients and 71 PBSC recipients), Eapen et al. (2011) shows that transplant using PBSC for severe aplastic anemia results in higher risk of post-transplant complications and higher risk of mortality after adjusting for age compared to BM transplant. Therefore, BM is the preferred graft type in unrelated donor HCT for severe aplastic anemia.

To illustrate the methods presented in this chapter, we select a subset of patients from this study by finding a BM recipient for each PBSC recipient within 18 months of age to create matched pairs. The paired dataset consists of 108 individuals receiving HCT (54 BM and 54 PBSC) from unrelated donors matched at human leukocyte antigen -A, -B, -C, -DRB1 for severe aplastic anemia. We use this dataset to determine the effect of graft source on overall survival. The analyses presented here are only for illustration of the statistical methodology. The results from our analyses should not be taken as clinical conclusion.

30.3 Notation

Let (X_{1i}, X_{2i}) be the failure times for pair i where $(X_{1i}, X_{2i}), i = 1, \ldots, n$, are independent and identically distributed. In right-censored data, X_{ki} is potentially unobservable for some member $k = 1, 2$ of pair $i = 1, \ldots, n$. Let (C_{1i}, C_{2i}) be the censoring times for pair i and assume that (C_{1i}, C_{2i}) are independent of (X_{1i}, X_{2i}). For each subject, one observes (T_{ki}, γ_{ki}) where $T_{ki} = \min(X_{ki}, C_{ki})$ is the event time or time to the last follow-up and $\gamma_{ki} = I(X_{ki} < C_{ki})$ is the indicator whether the event was observed. If a common censoring time is required, define $C_i = \min(C_{1i}, C_{2i})$ as the censoring time for pair i. Then $T_{ki} = \min(X_{ki}, C_i)$ and $\gamma_{ki} = I(X_{ki} < C_i)$.

Methods for paired survival data analysis often rely on the differences between two survival functions describing survival experience of individuals in each of the two populations. Let $S_1(t)$ and $S_2(t)$ be the survival function for the individuals in group 1 and 2, respectively, where $S_k(t) = P(X_k > t), k = 1, 2$. The Kaplan-Meier estimator can be used to estimate $S_1(t)$ and $S_2(t)$. Let $t_1, \ldots, t_D$ be the distinct ordered failure times in the pooled sample, d_{kj} be the number of events at time t_j, and y_{kj} be the number of subjects at risk in the treatment group k at time t_j. The Kaplan-Meier estimators are then given by

$$\hat{S}_k(t) = \prod_{t_j \leq t} \left(1 - \frac{d_{kj}}{y_{kj}}\right), \quad k = 1, 2.$$

The estimated variances of the Kaplan-Meier estimators are calculated by Greenwood's formula:

$$\text{Var}[\hat{S}_k(t)] = \hat{S}_k^2(t) \sum_{t_j \le t} \frac{d_{kj}}{y_{kj}(y_{kj} - d_{kj})}, \quad k = 1, 2.$$

Survival experience can also be compared using the hazard functions $h_1(t)$ and $h_2(t)$ of group 1 and 2, respectively, where

$$h_k(t) = lim_{\Delta t \to 0} \frac{P(t \le X_k < t + \Delta t | X_k \ge t)}{\Delta t}, \quad k = 1, 2.$$

30.4 Tests for paired data

Tests developled for paired survival data fall into three broad categories. It should be noted that while all of them compare the survival experience between the two groups, exact formulation of the hypothesis of interest varies. First we will consider tests based on ranks in the pooled sample using scoring schemes designed to accommodate censored data and then the test statistics are computed using the differences in ranks in the two samples. Another class of tests relies on within-pair differences in survival times. Lastly, we will consider a class of tests comparing the weighted Kaplan-Meier estimates between the treatment groups. In this section, several tests from each of the three categories are discussed in greater detail.

30.4.1 Rank-based tests

In this section, we describe two rank-based tests proposed by O'Brien and Fleming (1987) and Akritas (1992). Both of these approaches compare the survival times between the two treatment groups. The O'Brien-Fleming tests assume a common censoring time for both members of a pair, whereas the Akritas test only requires that the censoring time distributions are equal between the two treatment groups.

O'Brien and Fleming (1987) developed a class of rank-based tests using the generalized rank statistics. The first step involves finding appropriate score η for each individual in the pooled sample. Two score functions considered are the paired Prentice-Wilcoxon score generalized from Prentice (1978) and the Gehan-Wilcoxon score adapted from Gehan (1965).

The Prentice-Wilcoxon score is defined as follows. Let $t_j, j = 1, \ldots, D$, be the ordered distinct failure times from the pooled sample and y_j be the number of subjects still at risk in the pooled sample at time t_j and define $s_j = \prod_{l=1}^{j} y_l/(y_l + 1), j = 1, \ldots, D$. When there is exactly one death at observed failure time t_j, the Prentice-Wilcoxon score is defined as $\eta = (1 - 2s_j)$ for an individual having an observed death at time t_j and $\eta = (1 - s_j)$ for individuals censored between times t_j and t_{j+1}. When there are multiple deaths, say m_j deaths, at time t_j, arbitrarily order these individuals by assigning distinct times infinitesimally to the left of t_j. The scores are then computed as in the case without ties. Individuals failing at time t_j are then assigned the average of m_j scores.

Alternatively, the scores can be computed using the Gehan-Wilcoxon score function. For each individual, the Gehan-Wilcoxon score is the proportion of $2n$ observations known to have survival time smaller than that particular individual's minus the proportion of observations known to have survival times larger than that individual.

Let (η_{1i}, η_{2i}) be the scores for the ith pair computed from the pooled sample (using either the Prentice-Wilcoxon or the Gehan-Wilcoxon score function). The test is based on the difference in scores within pair members. Define $\Delta_i = \eta_{1i} - \eta_{2i}$. Given the magnitude of the score diffences $|\Delta_i|, i = 1, \ldots, n$, and considering only the pairs such that $|\Delta_i| > 0$, the test statistic is given by

$$Z_n = \frac{\sum_{i=1}^{n} \Delta_i}{\sqrt{\sum_{i=1}^{n} \Delta_i^2}}. \tag{30.1}$$

The p-value for the proposed test is obtained by comparing Z_n in (30.1) to tables of the standard normal distribution. A test based on Z_n computed from the Prentice-Wilcoxon score is called a "paired Prentice-Wilcoxon" (PPW) test. The test derived from the Gehan-Wilcoxon score is called a "paired Gehan-Wilcoxon" test.

O'Brien and Fleming (1987) recommend PPW over paired Gehan-Wilcoxon test since PPW is less sensitive to heavy censoring. Moreover, the Prentice-Wilcoxon score is relatively insensitive to outlier pairs. In their simulation study, they showed that PPW is almost as powerful as the two-sample Prentice-Wilcoxon test when the data are uncorrelated and more powerful when correlation increased. They also demonstrated that PPW is more powerful in most cases when compared to the generalized signed rank test of Woolson and Lachenbruch (1980) which will be described in Section 30.4.2.

We now illustrate the paired Prentice-Wilcoxon test using the HCT dataset from CIBMTR. Survival curves by graft type are shown in Figure 30.1 suggesting that survival is superior after bone marrow transplantation as compared to peripheral blood transplant with survival probabilities of 0.766 (Standard Error, SE = 0.059) at one year and 0.719 (SE = 0.064) at three years compared to 0.689 (SE = 0.065) and 0.587 (SE = 0.073), respectively. A two-sample logrank test ignoring potential correlation between pair members gives a chi-square statistic of 1.707 with a p-value of 0.191. Table 30.1 summarizes the test results contrasting survival times between bone marrow transplant versus peripheral blood transplant in severe aplastic anemia patients using the paired Prentice-Wilcoxon test and the Akritas test. The observed Z_n statistic from the PPW is 1.139. Under the null hypothesis, Z_n is asymptotically normal resulting in a two-sided p-value of 0.255.

TABLE 30.1
Results of the O'Brien-Fleming and the Akritas tests.

Test	n_e	**Test Statistic**		**P-value**
		Distribution	Observed value	
Paired Prentice-Wilcoxon	26	$N(0, 1)$	1.139	0.255
Akritas test	54	t_{53}	1.179	0.244

Note that all observations were used to compute the scores from the pooled sample. However, due to the assumption of a common censoring time, pairs with the smaller time being censored will become doubly censored pairs and they do not contribute to the test statistic Z_n. Applying the censoring restriction, 28 of the 54 pairs in this dataset were doubly censored. The observed test statistic came from the score differences of the 26 remaining pairs shown under the column heading n_e (number of effective pairs) in Table 30.1. Thus the effective sample size in this example is reduced by more than 50%.

Another rank-based test which is comparable to the paired Prentice-Wilcoxon test was proposed by Akritas (1992). Akritas test is a paired t-test on the rank transformation of the survival times. In this test, separate Kaplan-Meier estimates are computed for each treatment group. Let $\hat{S}_1(t)$ be the Kaplan-Meier estimate for group 1 and $\hat{S}_2(t)$ the estimate for group 2 at time t and $\bar{S}(t) = [\hat{S}_1(t) + \hat{S}_2(t)]/2$. Each of the $2n$ observations is then replaced by its rank defined by $R(t_i) = 2n[1 - \bar{S}(t_i)]$ for uncensored observation and censored

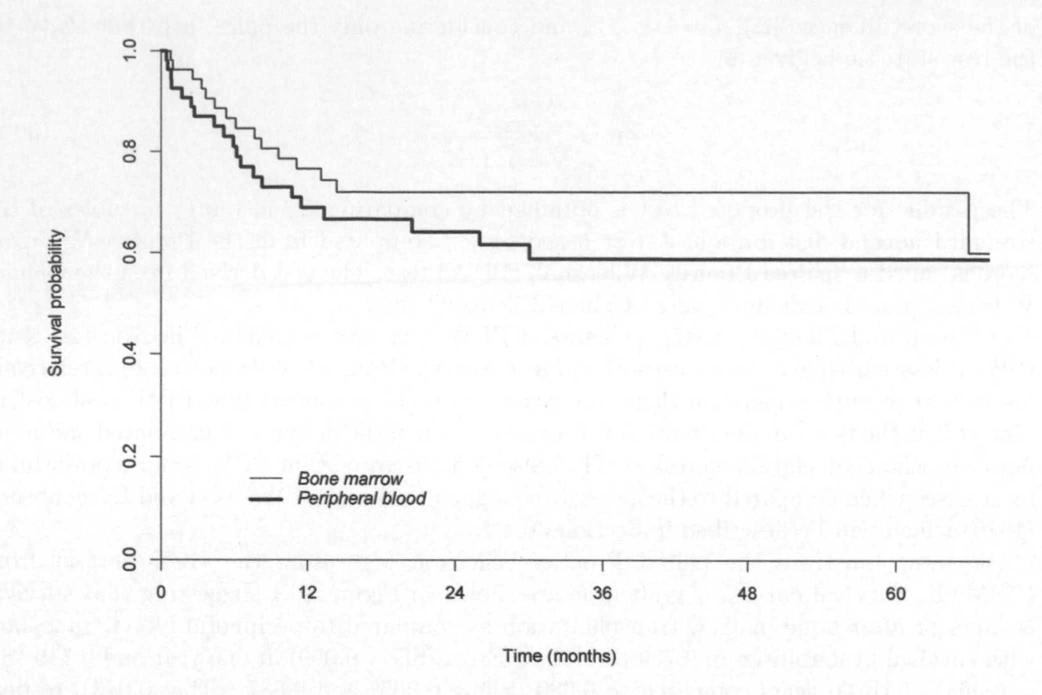

FIGURE 30.1
Survival curves post-transplant by graft type.

observations are replaced by $R(t_i) = 2n[1 - 0.5\bar{S}(t_i)]$. A paired t-test is then performed on the transformed ranks. As shown in Table 30.1, paired t-test on rank transformation of the survival times after bone marrow transplant versus peripheral blood transplant gives an observed t statistic of 1.179 and a two-sided p-value of 0.244.

Theoretical justification for the paired t-test requires that censoring times in both groups come from a common distribution. This assumption is less stringent than the assumption of a common censoring time for members of each pair in the O'Brien-Fleming tests. In the Akritas test, all pairs including doubly censored pairs contribute to the test statistic. In this example, inference was based on information from 54 pairs compared to the 26 pairs used in the O'Brien-Fleming tests.

Woolson and O'Gorman (1992) compared the performance of the O'Brien-Fleming paired Prentice-Wilcoxon, the paired Gehan-Wilcoxon, and the Akritas tests along with the generalized sign rank test on the survival time or the log survival time of Woolson and Lachenbruch (1980). Note that the Woolson and Lachenbruch tests are based on within-pair differences which are more closely related to Dabrowska's test and will be described in Section 30.4.2 below. Their simulation study shows that, among all tests considered, the PPW and the Akritas tests perform consistently well under various distributions. Furthermore, the powers of these two tests are approximately equal in all simulation settings. Woolson and O'Gorman (1992) recommend the PPW and the Akritas tests for general application.

30.4.2 Within-pair comparison

Unlike rank-based tests described in Section 30.4.1 which use differences in ranks computed from the pooled sample, Woolson and Lachenbruch (1980) and Dabrowska (1990) proposed a class of tests based on within-pair differences. These tests compare the survival times by testing the hypothesis of bivariate symmetry via inference about the distribution of the paired differences in survival times. That is, the random variable of interest here is $D = X_1 - X_2$. Assuming D has the same distribution as $W + \theta$ where W is an absolutely continuous random variable symmetrically distributed around 0, the test hypothesis is H_0 : $\theta = 0$ vs. $H_a : \theta > 0$. Both of these tests require a common censoring time for pair members similar to the tests of O'Brien and Fleming (1987),

One of the earliest tests extending general linear rank order statistics to right-censored data was proposed by Woolson and Lachenbruch (1980). In Woolson-Lachenbruch's generalized signed rank (GSR) test, the ranking of the observations is done in the following way. First, for those pairs where $\gamma_{1i} = \gamma_{2i} = 1$, that is, both observations are uncensored, compute $D_j = T_{1j} - T_{2j}$. Let r denote the number of these uncensored pairs. Then denote the ordered sequence of the absolute values of $D_1, \ldots, D_r$ by $Z_{(1)}, \ldots, Z_{(r)}$ and define $Z_{(0)} = 0$ and $Z_{(r+1)} = \infty$. For $j = 1, \ldots, r$, define $d_{(j)} = 1$ if $Z_{(j)} > 0$ and $d_{(j)} = 0$ if $Z_{(j)} < 0$. Next, compute the differences D_i for pairs where one of the members of the pair is censored. Then, for $j = 0, \ldots, r$, let

n_j be the number of censored absolute differences in $[Z_{(j)}, Z_{(j+1)})$ and
p_j be the number of positive censored differences in $[Z_{(j)}, Z_{(j+1)})$.

Compute $m_j = \sum_{i=j}^{r}(n_i + 1)$ and $P_j = \prod_{i=1}^{j} \frac{m_i}{m_i+1}$, $j = 1, \ldots, r$. The test statistic is defined as follows:

$$Z_{WL} = \frac{\sum_{j=1}^{r}(2d_{(j)} - 1)(1 - P_j) + 0.5(2p_0 - n_0) + \sum_{j=1}^{r}(2p_j - n_j)(1 - 0.5P_j)}{[\sum_{j=1}^{r}(1 - P_j)^2 + \sum_{j=1}^{r} n_j(1 - 0.5P_j)^2 + 0.25n_0]^{1/2}}. \quad (30.2)$$

Under the null hypothesis, Z_{WL} is asymptotically standard normal. Alternatively, the test statistic Z_{WL} in (30.2) can be computed based on the within-pair difference of the logarithms of the observed survival times. The first two rows of Table 30.2 show the results from the Woolson-Lachenbruch generalized signed rank test applied to the HCT dataset. The observed test statistic comparing the survival times is -1.433 with a two-sided p-value of 0.152. The observed test statistic comparing the log of the survival times is -1.410 with a two-sided p-value of 0.158. Note that the pairs with both observations censored do not contribute anything to the test statistic reducing the inference sample size to 26.

A simulation study by Woolson and O'Gorman (1992) shows that the GSR test is more powerful than the paired Gehan-Wilcoxon, the paired Prentice-Wilcoxon, and the Akritas tests when comparing the survival times between two groups where the survival times for each group follow an exponential distribution with a different scale parameter. However, when outliers are added to the exponential survival times or in comparisons where the survival times for one group is a location shift from the survival times for the other group with the same scale parameter, the PPW test and the Akritas perform consistently better than the GSR test.

Dabrowska (1990) proposes another class of tests based on the distribution of $D = X_1 - X_2$ but applying different scores to observed and censored differences. Let $D_i = T_{1i} - T_{2i}, i = 1, \ldots, n$, and let $\epsilon_i = 1$ if $D_i > 0$, $\epsilon_i = -1$ if $D_i < 0$, and $\epsilon_i = 0$ if $D_i = 0$. Inferences for this class of tests are based on the distribution of D_i. Based on the assumption of common censoring time for members within a pair, $\gamma_{1i} = \gamma_{2i} = 0$ means $\epsilon_i = 0$ and $D_i = 0$. That is, doubly censored pairs do not contribute to the test. Pairs with failure time for at least one member observed can be divided into four groups as defined below:

$B_1 = \{i : \epsilon_i = 1, \gamma_{1i}\gamma_{2i} = 1\}$ - both failures observed with positive difference,
$B_2 = \{i : \epsilon_i = -1, \gamma_{1i}\gamma_{2i} = 1\}$ - both failures observed with negative difference,

TABLE 30.2
Results of the Woolson-Lachenbruch and the Dabrowska tests.

Test	Test Statistic	P-value
Woolson-Lachenbruch test on survival time	−1.433	0.152
Woolson-Lachenbruch test on log of survival time	−1.410	0.158
Dabrowska sign test	−0.775	0.439
Dabrowska signed Wilcoxon test	−1.069	0.285
Dabrowska signed normal test	−1.069	0.285

$B_3 = \{i : \gamma_{1i} = 1, \gamma_{2i} = 0\}$ - one failure observed with positive difference, and
$B_4 = \{i : \gamma_{1i} = 0, \gamma_{2i} = 1\}$ - one failure observed with negative difference.

Now let $N_j(t) = \sum_{i=1}^{n} N_{ij}(t), j = 1, \ldots, 4$, such with $|D_i| \leq t$ where $N_{ij}(t) = I[|D_i| \leq t, i \in B_j]$. That is, $N_j(t)$ is the number of pairs whose magnitude of within-pair difference is less than or equal to t. The test statistic is given by

$$T = \int K_u d(N_1 - N_2) + K_c d(N_3 - N_4)$$

where K_u and K_c are scoring processes for fully observed and singly censored pairs, respectively.

Three score functions considered in this chapter are

1. $K_u = K_c = 1$ which corresponds to the sign test,
2. $K_u = 1 - \hat{F}$ and $K_c = 1 - \hat{F}/2$ which corresponds to the signed Wilcoxon test, and
3. $K_u = \Phi^{-1}(1 - \hat{F}/2)$ and $K_c = 2\hat{F}^{-1}\phi\{\Phi^{-1}(1 - \hat{F}/2)\}$ which corresponds to the signed normal test

where $\hat{F}(t) = \prod_{s \leq t}\{1 - \Delta\hat{\Lambda}(s)\}$ with $\hat{\Lambda}(s) = \int_0^t U^{-1}I(U > 0)d(N_1 + N_2)$ where $U(t) = \sum I[|D_i| \geq t, \epsilon_i \neq 0]$ and ϕ and Φ are the density and the cumulative distribution function, respectively, of the standard normal distribution. Without censoring, $\hat{\Lambda}(s)$ is the Nelson-Aalen estimator of the within-pair difference $|X_1 - X_2|$ and $\hat{F}(s)$ is the empirical survival function.

Under the null hypothesis of bivariate symmetry, $n^{-1/2}T$ converges weakly to a mean-zero normal distribution with a variance consistently estimated by

$$\hat{\sigma}_T^2 = \sum_{j=1}^{2} \int K_u^2 dN_j + \sum_{j=3}^{4} \int K_c^2 dN_j.$$

Results of Dabrowska tests comparing post-transplant survival between bone marrow and peripheral blood using the three score functions are given in Table 30.2. In this example, the signed Wilcoxon test and the signed normal test give the same observed statistics whereas the signed test is more conservative.

30.4.3 Weighted Kaplan-Meier comparison

As an alternative to rank-based tests, Murray (2001) presents a test based on weighted Kaplan-Meier statistics for paired survival data. This test is an extension of the two-sample

Pepe-Fleming test (Pepe and Fleming, 1989) using the integrated difference between the Kaplan-Meier curves which gives an interpretation to the test statistics. This test allows the inclusion of singletons which increases its power compared to tests that only use paired observations such as the O'Brien-Fleming tests, the Akritas test, the Woolson-Lachenbruch tests, and the Dabrowska tests. Furthermore, because it compares the integrated survival curves, this test is more robust against crossing hazards than rank-based tests. Inference for the weighted Kaplan-Meier approach is for the null hypothesis $H_0 : S_1(t) = S_2(t)\ \forall t$.

Define n_1, n_2 the number of observations in group 1 and group 2, respectively. The weighted difference between the two groups is defined by

$$\tau = \left(\frac{n_1 n_2}{n_1 + n_2}\right)^{1/2} \int_0^\infty \hat{w}(u)[\hat{S}_1(u) - \hat{S}_2(u)]du \tag{30.3}$$

where $\hat{w}(t)$ is a predictable weighting process such that

$$\sup_{u \in [0,t)} |\hat{w}(u) - w(u)| \xrightarrow{P} 0.$$

As $n \to \infty$, τ has asymptotic normal distribution with variance

$$\begin{aligned}\sigma^2 &= \sum_{k=1}^{2} \frac{\pi_1 \pi_2}{\pi_k} \left[\int_0^\infty \frac{\{A_k(u)\}^2 h_k(u)}{P(T_{ki_k} \geq u)} du\right] \\ &\quad - \theta \int_0^\infty \int_0^\infty A_1(u) A_2(v) G_{12}(u,v) dv du \end{aligned} \tag{30.4}$$

where π_k is the probability of belonging to group k, $A_k(t) = \int_t^\infty w(u)S_k(u)du$ and $G_{12}(t_1,t_2) = P(T_{1i} \geq t_1, T_{2i} \geq t_2)\{h_{12}(t_1,t_2) - h_{1|2}(t_1|t_2)h_2(t_2) - h_{2|1}(t_2|t_1)h_1(t_1) + h_1(t_1)h_2(t_2)\}/\{P(T_{1i} \geq t_1)P(T_{2i} \geq t_2)\}$ where $h_k(t)$ is the marginal hazard function of group $k = 1,2$ at time t, $h_{12}(t_1,t_2)$ is the joint hazard function of group 1 and 2, and $h_{1|2}(t_1|t_2)$ is the hazard function of group 1 at time t_1 given that other member of the pair is still alive at time t_2. The first term of σ^2 in Equation (30.4) is the variance of the Pepe-Fleming test statistic for independent samples. The second term in (30.4) accounts for the dependence within pairs. When the outcomes between groups are uncorrelated, this term reduces to zero.

The variance (30.4) can be estimated using the pooled sample or unpooled sample. Asymptotic closed forms for the variance estimators are given in Murray (2001). It is recommended that the pooled sample data be used to estimate the variance σ^2 for hypothesis testing and for confidence interval construction, unpooled sample estimates should be employed. Note that all observations including singletons in the sample contribute to the variance estimation but only paired observations are used to estimate the joint and conditional hazards in $G_{12}(\cdot,\cdot)$.

Let $J(t) = 1$ if $\sum_{i_1=1}^{n_1} I(T_{1i_1} \geq t) \sum_{i_2=1}^{n_2} I(T_{2i_2} \geq t) > 0$ and $J(t) = 0$ otherwise. Possible weight functions include $\hat{w}_1(t) = J(t)$ and $\hat{w}_2(t) = J(t)P(T_{1i_1} \geq t)P(T_{2i_2} \geq t)/\{\hat{\pi}_1 P(T_{1i_1} \geq t) + \hat{\pi}_2 P(T_{2i_2} \geq t)\}$. The statistic τ in (30.3) with weight function w_1 has an interpretation of years of life saved on study and τ with weight function w_2 has an interpretation as the difference in mean survival times.

Applying Murray's test to our dataset using the weight $\hat{w}_1(\cdot)$ suggests that bone marrow transplant for severe aplastic anemia saves 8 months of life on study compared to transplant with peripheral blood stem cells. The test statistic using the pooled variance estimate is 0.813 resulting in a two-sided p-value of 0.416. The observed test statistic using weight $\hat{w}_2(\cdot)$ is 0.944 with a two-sided p-value of 0.345. Note that, the weighted Kaplan-Meier tests do not assume equal censoring. Therefore, all 54 pairs contribute information to the test statistics. Moreover, using this method, observations from the original dataset without a match could be included in the analysis which would significantly increase the power of the tests.

30.5 Regression models for paired data

In addition to the tests discussed in Section 30.4, regression models provide an alternative approach to analyze paired survival data. In this chapter we discuss three regression models developed for correlated data: the stratified Cox model, the marginal Cox model, and the shared frailty model. Regression models are more flexible than the inference procedures described in Section 30.4 as they allow comparison of treatment effect while adjusting for other covariates that might affect the outcome. They also provide an interpretation for the treatment effect in terms of the hazard ratio. Since these methods were developed for correlated (clustered) data in general, they do not require one-to-one pairing nor do they require that members of the pair receive different treatments.

All three regression models described in this section are extensions of the Cox (1972) proportional hazards model. Let X be the survival time and C be the censoring time where X and C are independent. In regression problems, observable data consist of the triplets $(T, \gamma, \boldsymbol{Z})$ where $T = min(X, C)$, $\gamma = I(X < C)$, and $\boldsymbol{Z}$ is the vector of covariates. Define the conditional hazard function by

$$h(t|\boldsymbol{Z}) = lim_{\Delta t \to 0} \frac{P(t \leq X < t + \Delta t | X \geq t, \boldsymbol{Z})}{\Delta t}.$$

The proportional hazards regression model is expressed as

$$h(t|\boldsymbol{Z}) = h_0(t) \exp(\boldsymbol{\beta}' \boldsymbol{Z}) \tag{30.5}$$

where $h_0(t) = h(t|\boldsymbol{Z} = \boldsymbol{0})$ is the baseline hazard. When the only covariate of interest is the main treatment effect where $Z = 0$ for treatment group 1 and $Z = 1$ for treatment group 2, model (30.5) reduces to $h(t|Z) = h_0(t) \exp(\beta Z)$ where $h_0(t)$ is the hazard rate at time t for treatment group 1 and $\exp(\beta)$ is the hazard ratio of treatment 2 compared to treatment 1. The null hypothesis of interest in this model is $H_0 : \beta = 0$. Inference for $\boldsymbol{\beta}$ and regression diagnostics for the Cox regression model are discussed in Chapter 1 of this handbook. Note that when applied to paired data, model (30.5) does not account for potential correlation between paired outcomes.

Table 30.3 summarizes the estimated effect of graft type on mortality after transplant for severe aplastic anemia using the Cox model, the stratified, the marginal, and the shared frailty regression models which will be discussed later. For the HCT data, ignoring the within-pair association, the risk of mortality after receiving peripheral blood transplant is estimated to be 1.555 times (95% confidence interval 0.797 – 3.304) the risk of mortality after bone marrow transplant with a p-value of 0.195. In the remainder of this section, we will describe three approaches that extend the Cox model to account for the correlation in different ways.

30.5.1 Stratified Cox models

The first extension considered in this section is the stratified Cox model discussed in Chapter 1. In the context of paired data, the stratified Cox model accounts for the association

TABLE 30.3
Estimated treatment effect from extended Cox regression models.

Model	n_e	HR (95% CI)	Chi-square	P-value
Cox model	54	1.555 (0.797 – 3.034)	1.677	0.195
Marginal	54	1.555 (0.807 – 2.998)	1.739	0.187
Stratified	26	1.167 (0.540 – 2.522)	0.154	0.695
Gamma frailty	54	1.556 (0.797 – 3.035)	1.680	0.195
Positive stable frailty	54	1.555 (0.797 – 3.034)	1.677	0.195

between paired outcomes by estimating the within-pair hazard ratio. It assumes a separate baseline hazard function $h_{0i}(t)$ for each pair i, $i = 1, \dots, n$. That is,

$$h_i(t|\boldsymbol{Z}) = h_{0i}(t) \exp(\boldsymbol{\beta}'\boldsymbol{Z}). \tag{30.6}$$

Note that inference for $\boldsymbol{\beta}$ in model (30.6), given by Holt and Prentice (1974), use only pairs where the smaller of the two times is not censored. The effective sample size for inference in this approach is the same as in the O'Brien-Fleming, the Woolson-Lachenbruch, and the Dabrowska tests. Furthermore, the stratified tests have good power when the effect of treatment is in the same direction for all pairs and the large-sample properties of the estimator for $\boldsymbol{\beta}$ are only valid when the number of pairs is large.

Results from the stratified Cox model (Table 30.3) suggest the risk of mortality for a severe aplastic anemia patient receiving peripheral blood transplant is 1.167 times higher than the risk of mortality after bone marrow transplant with 95% confidence interval for the hazard ratio of (0.540 – 2.522). The Wald chi-square statistic of the treatment effect is 0.154 with a p-value 0.695. Note that the hazard ratio given in the stratified model estimates the within-pair effect and the within-pair correlation is accounted for by the common baseline hazard h_{0i} for members of pair i.

30.5.2 Marginal Cox models

Another extension of the Cox model is the marginal approach proposed by Lee, Wei, and Amato (1992). In this approach, the treatment effect is estimated using the Cox model with the marginal hazard function

$$h(t|\boldsymbol{Z}_{ki}) = h_0(t) \exp(\boldsymbol{\beta}'\boldsymbol{Z}_{ki}) \tag{30.7}$$

for $k = 1, 2; i = 1, \dots, n$. Model (30.7) is fitted using the generalized estimating equation (GEE) approach of Liang and Zeger (1986) under the independent working model.

Fitting the marginal Cox model to our data indicates the risk of mortality after peripheral blood transplant is 1.555 times higher than the risk of mortality after bone marrow transplant with a 95% confidence interval of (0.807 – 2.998). The Wald chi-square statistic for the effect of graft type is 1.739 with a p-value of 0.187. Unlike the stratified Cox model which estimates the within-pair treatment effect, the marginal model estimates the average treatment effect across all pairs and the within-pair correlation is accounted for by the robust covariance estimated by the GEE approach. Furthermore, in the marginal model, all pairs contribute information to the inference of treatment effect. That is 54 pairs contribute to the inference in the marginal model compared to 26 pairs in the stratified model. Marginal models for artificially matched pairs may include the matching covariates in the regression models.

30.5.3 Shared frailty models

Another regression approach commonly used to analyze paired (clustered) survival data is the shared frailty model. Shared frailty models assume members of a pair share a common risk which is constant over time and has multiplicative effect on the baseline hazard function. The shared frailty induces dependence between members within a pair. Conditional on the frailty, members within a pair are independent. The shared frailty model is a random effects model with the conditional hazard function for pair i expressed as

$$h_{ki}(t|\boldsymbol{Z}_{ki}, W = w_i) = w_i h_0(t) \exp(\boldsymbol{\beta}' \boldsymbol{Z}_{ki}) \tag{30.8}$$

where W is the frailty random variable. In the shared frailty model, the baseline hazard for pair i is $w_i h_0(t)$. When the only covariate of interest is the treatment effect where $Z = 0$ for treatment group 1 and $Z = 1$ for treatment group 2, model (30.8) is simplified to $h_{ki}(t|Z_{ki}, W = w_i) = w_i h_0(t) \exp(\beta' Z_{ki})$ and β is the conditional treatment effect. The model in (30.8) further assumes proportional conditional hazards. See Hougaard (2000) for a comprehensive treatment of the shared frailty models.

Similar to the stratified model, the shared frailty model estimates the within-pair hazard ratio and it assumes different baseline hazard functions for each pair i. However, unlike the stratified model where the baseline hazards are arbitrary and are considered fixed, the shared frailty model assumes that the frailty is a random variable and it has a multiplicative effect on the hazard functions. Two distributions often assumed for W are the $Gamma(\theta, \theta)$ (or simply $Gamma(\theta)$) and the positive stable $PS(\alpha)$.

For $W \sim Gamma(\theta)$, the marginal hazard function is given by

$$h_0(t) \exp(\boldsymbol{\beta}' \boldsymbol{Z}_{ki}) / \{1 + H_0(t) \exp(\boldsymbol{\beta}' \boldsymbol{Z}_{ki} / \theta)\} \tag{30.9}$$

where $H_0(t) = \int_0^t h_0(u) du$ is the cumulative baseline hazard function at time t. The marginal hazard function in (30.9) indicates that the marginal hazards are nonproportional in the gamma frailty model.

Assuming $W \sim PS(\alpha)$, the marginal hazard function is given by

$$h_0(t) \exp(\alpha \boldsymbol{\beta}' \boldsymbol{Z}_{ki}) \alpha H_0(t)^{\alpha - 1}. \tag{30.10}$$

As indicated by (30.10), the marginal hazard function in a shared positive stable frailty preserves proportionality.

Fitting the shared gamma frailty model to the HCT data suggests that the risk of mortality after peripheral blood transplant is 1.556 times higher than after bone marrow transplant with a 95% confidence interval of $(0.797 - 3.035)$ for the treatment effect β. The chi-square statistic for the test of treatment effect is 1.680 with a p-value of 0.195. The gamma frailty paramter θ is estimated to be 5×10^{-5} with a corresponding Kendall's τ of 2.5×10^{-5} suggesting that survival times are indepedent between pair members in this study.

Similar results were found after fitting the shared positive frailty model with the estimated conditional hazard ratio, β, of 1.555, a 95% confidence interval of $(0.797 - 3.034)$, a chi-square statistic of 1.677, and a p-value 0.195. The frailty parameter α is estimated to be 1 with a corresponding Kendall's τ of 0, again, reiterating that the paired survival times are independent.

30.6 Comparing survival probabilities at a fixed time point

The tests presented in Section 30.4 and regression models of Section 30.5 compare the entire survival curves. There are instances where the comparison of survival probabilities at a fixed time point is of interest. In these instances, the emphasis is placed on the effect of treatment on outcome at that specific time point regardless of the process leading to that time. Comparison of outcome at a fixed time point is especially important when a treatment is designed to improve long-term outcome although early mortality might be higher. In this case the survival curves may cross in the early period.

One example of this situation is the comparison between allogeneic versus autologous hematopoietic cell transplant. In autologous transplant, a patient's own hematopoietic cells are collected prior to chemotherapy. After chemotherapy treatment, these cells are reinfused into the patient. In allogeneic transplant, hematopoietic cells from a related or unrelated donor are given to the patients after chemotherapy. Since the hematopoietic cells in allogeneic transplant come from another person, these cells launch an attack on the host body causing complications known as "graft-versus-host disease." These complications can be fatal and result in higher mortality in the early post-transplant period. In contrast, autologous transplant recipients are not affected by graft-versus-host disease and therefore expected to have better early survival. However, the relapse rate after autologous transplant is higher than after allogeneic transplant which leads to higher long-term mortality. Klein et al. (2007) provides an example comparing survival between autologous transplant versus allogeneic transplant from a matched sibling donor for leukemia. In their example, the survival curves appear to cross between 12 and 18 months and level off at two years after transplant. Tests comparing the entire curves have low power to detect differences in the treatment effect in situations where early and late differences are in opposite directions. Comparison at a fixed time point after the curves have flattened out gives a better estimate of long-term effect on survival.

Su et al. (2011) propose techniques to compare survival probabilities at a fixed point in time for paired and clustered data. Their approaches extend methods studied by Klein et al. (2007) to account for potential dependence in paired or clustered outcomes. Both Klein et al. (2007) and Su et al. (2011) consider tests based on various transformations of the Kaplan-Meier estimator and a regression approach using the pseudo-values of Andersen et al. (2003) and Klein and Andersen (2005). Logan et al. (2011) also propose a marginal model approach for clustered competing risks data using pseudo-values that could be applied to paired survival data. The null hypothesis under consideration here is $H_0 : S_1(t_0) = S_2(t_0)$ for a fixed time t_0.

In general terms, the test statistic is defined as

$$\frac{\varphi[\hat{S}_2(t_0)] - \varphi[\hat{S}_1(t_0)]}{\sqrt{V[(\varphi(\hat{S}_2(t_0)) - \varphi(\hat{S}_1(t_0))]}} \tag{30.11}$$

where $\hat{S}_k(t_0)$ is the Kaplan-Meier estimator for group $k = 1, 2$ and $\varphi(\cdot)$ is a transformation of s. The square of the denominator in (30.11) is the variance of $\varphi[\hat{S}_2(t_0)] - \varphi[\hat{S}_1(t_0)]$ which equals

$$V[\varphi(\hat{S}_1(t_0))] + V[\varphi(\hat{S}_2(t_0))] - 2\text{Cov}[\varphi(\hat{S}_1(t_0)), \varphi(\hat{S}_2(t_0))]. \tag{30.12}$$

The covariance term in (30.12) accounts for the dependence in paired data. When all observations come from independent samples, this term becomes zero and the variance in (30.12) reduces to $V[\varphi(\hat{S}_2(t_0))] + V[\varphi(\hat{S}_1(t_0))]$.

By employing the delta method, one can show that $\varphi(\hat{S}_k(t_0))$ is asymptotically normal with mean $\varphi(S_k(t_0))$ and variance estimated by $V(\hat{S}_k(t_0))[\varphi'(\hat{S}_k(t_0))]^2$ where $\varphi'(t) = d\varphi(t)/dt$. When the observations are paired, using the results of Murray (2001), $\text{Cov}(\varphi(\hat{S}_1(t_0)), \varphi(\hat{S}_2(t_0)))$ can be estimated by

$$\varphi'(\hat{S}_1(t_0))\varphi'(\hat{S}_2(t_0))\frac{1}{n}\sum_{u\leq t_0}\sum_{v\leq t_0}\hat{G}_{12}(u,v) \tag{30.13}$$

where

$$\hat{G}_{12}(u,v) = \frac{\pi_{uv}/n}{\pi_u\pi_v/n^2}\left(\frac{q_{uv}}{\pi_{uv}} - \frac{q_{u|v}q_v}{\pi_{uv}\pi_v} - \frac{q_{v|u}q_u}{\pi_{uv}\pi_u} + \frac{q_u q_v}{\pi_u\pi_v}\right)$$

and π_{uv} is the number of pairs whose member in group 1 is still at risk at time u and member in group 2 is still at risk at time v; π_u is the number of individuals in the pooled sample who are still at risk at time u; q_{uv} is the number of pairs whose member in group 1 fails at time u and member in group 2 fails at time v; and $q_{v|u}$ is the number of pairs whose member in group 2 fails at time v while the member in group 1 is still at risk at time u; q_u is the number of individuals in the pooled sample failing at time u. Under the null hypothesis, the test statistic in (30.11) is asymptotically standard normal.

The five transformations considered by Klein et al. (2007) and Su et al. (2011) include:

1. identity: $\varphi(x) = x$,
2. log transformation: $\varphi(x) = \log(x)$,
3. complementary log-log transformation: $\varphi(x) = \log[-\log(x)]$,
4. arcsine-square-root transformation: $\varphi(x) = \arcsin(\sqrt{x})$, and
5. logit transformation: $\varphi(x) = \log[x/(1-x)]$.

Another technique to compare survival at a fixed point in time is a regression analysis using pseudo-values (Andersen et al., 2003; Klein and Andersen, 2005) . The pseudo-value at time t for each of the $2n$ observations is computed as

$$\hat{\theta}_{ki}(t) = 2n\hat{S}(t) - (2n-1)\hat{S}^{-ki}(t) \tag{30.14}$$

where $\hat{S}(t)$ is the Kaplan-Meier estimate at time t using all observations in the pooled sample and $\hat{S}^{-ki}(t)$ is the Kaplan-Meier estimate with observation k in group i removed. Without censoring, $\hat{\theta}_{ki}(t)$ is the indicator whether patient k in group i is alive at time t.

The pseudo-values can be used in generalized linear models to evaluate the treatment effect as well as to adjust for other covariates. Common link functions used to model survival data include the logit link $g(\theta) = \log[(\theta/(1-\theta)]$ and the complementary log-log link $g(\theta) = \log(-\log(\theta))$. Let $\boldsymbol{Z}_{ki} = [Z^1_{ki}, Z^2_{ki}, \ldots, Z^p_{ki}]$ be a vector of covariates for patient k in group i where Z^1 is the treatment indicator, i.e., $Z^1 = 0$ for treatment group 1 and $Z^1 = 1$ for treatment group 2.

The pseudo-values obtained from (30.14) can be fit using a marginal model

$$g(\theta_{ki}) = \beta_0 + \sum_{j=1}^{p}\beta_j Z^j_{ki} = \boldsymbol{\beta}'\boldsymbol{Z}_{ki}. \tag{30.15}$$

The parameters $\boldsymbol{\beta}$ in (30.15) can be estimated using the marginal GEE approach of Liang and Zeger (1986) as described by Logan et al. (2011). Let $\boldsymbol{\mu}_i = [\mu_{1i}, \mu_{2i}]$ where $\mu_{ki} = g^{-1}(\boldsymbol{\beta}'\boldsymbol{Z}_{ki})$. Let $\boldsymbol{V}_i$ be a 2×2 working correlation matrix of $\hat{\boldsymbol{\theta}}_i = [\hat{\theta}_{1i}, \hat{\theta}_{2i}]$. The estimating equation is given by

$$\boldsymbol{U}(\boldsymbol{\beta}) = \sum_{i=1}^{n} \boldsymbol{U}_i = \sum_{i=1}^{n} \left(\frac{\delta \boldsymbol{\mu}_i}{\delta \boldsymbol{\beta}}\right)' \boldsymbol{V}_i^{-1}(\hat{\boldsymbol{\theta}}_i - \boldsymbol{\mu}_i).$$

The maximum likelihood estimator $\hat{\boldsymbol{\beta}}$ of $\boldsymbol{\beta}$ can be found by solving $\boldsymbol{U}(\boldsymbol{\beta}) = \mathbf{0}$. Following Liang and Zeger (1986), $\sqrt{n}(\hat{\boldsymbol{\beta}} - \boldsymbol{\beta})$ is asymptotically normal with mean zero and variance consistently estimated by

$$\widehat{\boldsymbol{\Sigma}} = \boldsymbol{I}^{-1}(\hat{\boldsymbol{\beta}})\{\hat{\boldsymbol{V}}[\boldsymbol{U}(\boldsymbol{\beta})]\}\boldsymbol{I}^{-1}(\hat{\boldsymbol{\beta}})$$

where

$$\boldsymbol{I}(\hat{\boldsymbol{\beta}}) = \sum_{i=1}^{n} \left(\frac{\delta \boldsymbol{\mu}_i}{\delta \boldsymbol{\beta}}\right)' \boldsymbol{V}_i^{-1} \left(\frac{\delta \boldsymbol{\mu}_i}{\delta \boldsymbol{\beta}}\right)$$

and

$$\hat{\boldsymbol{V}}[\boldsymbol{U}(\boldsymbol{\beta})] = \sum_{i=1}^{n} \boldsymbol{U}_i \boldsymbol{U}_i'.$$

Note that the GEE approach fits a marginal model and accounts for the within-pair correlation by a robust covariance estimator. Model (30.15) specifies a common intercept term β_0 for all pairs and β_1 is the treatment effect across all pairs.

The pseudo-values from (30.14) can also be fit using the following stratified model

$$g(\theta_{ki}) = \beta_{0i} + \sum_{j=1}^{p} \beta_j Z_{ki}^{j} \tag{30.16}$$

where each pair i has a different intercept term β_{0i} while the members of a pair have a common intercept. Unlike the marginal model (30.15), here the within-pair correlation is accounted for by the common intercept term β_{0i} and β_1 is the within-pair treatment effect. The β's in model (30.16) can be estimated using the GEE approach as described for unpaired survival data by Klein et al. (2007). However, with a large number of nuisance parameters β_{0i}, $i = 1, \ldots, n$, the estimates for $\beta's$ in (30.16) might not be consistent. More research is needed to explore this approach further.

We applied the methods discussed in this section to compare survival probability at three years post transplant for patients receiving bone marrow graft and those receiving peripheral blood graft. The results are summarized in Table 30.4. As seen in Figure 30.1, patients receiving BM graft have a three-year survival probability of 0.719 (SE = 0.064), while this probability is 0.587 (SE = 0.073) for peripheral blood transplant recipients. Comparisons of three-year survival between the two treatments using the log, complementary log-log, and the arcsine-square-root transformation of the Kaplan-Meier estimator lead to similar conclusions. Pseudo-value regression using marginal model approach with the complementary log-log link shows that the risk of mortality at three years after peripheral blood transplant is 1.612 times higher than the mortality risk after bone marrow transplant for severe aplastic anemia with a 95% confidence interval for the relative risk of (0.803 − 3.235). However, there is no significant evidence to show that bone marrow transplant results in superior survival at three years post-transplant.

Klein et al. (2007) note that, since $\hat{S}(t)$ is a consistent estimator for the marginal survival distribution, the pseudo-value approach works better when the censoring distribution is the same in both groups.

TABLE 30.4
Comparison of three-year survival.

Method	Test statistic	Two-sided p-value
Transformation		
Log	1.164	0.244
Complemenatry log-log	-1.171	0.242
Arcsine-square-root	1.180	0.238
Pseudo-value regression	1.344	0.179

30.7 Discussion

In this chapter, we reviewed a number of approaches to compare paired survival data. Survival curves can be compared using hypothesis tests developed specifically for paired data using various rank functions to handle right censoring. Most of the methods presented in this exposition give comparable results in our example which is consistent with various simulation studies. It is important to note that inferences for different classes of tests are based on a slightly different hypothesis. Specifically, the tests of O'Brien-Fleming (1987) and Akritas (1992) evaluate the within-pair differences in pooled ranks. The tests of Woolson and Lachenbruch (1980) and Dabrowska (1990) make inference based on the distribution of the within-pair differences in survival times. These are conditional approaches, whereas the test proposed by Murray (2001) which compares the integrated Kaplan-Meier curves and allows us to estimate the number of years life saved is a marginal approach. This test accounts for the association between paired survival times by adding a covariance term in the variance estimator. The choice of test for a specific study depends on the question of interest. Another factor that should be considered when choosing a test is the amount of censoring in the data and whether it is appropriate to assume a common censoring time for members of a pair. The O'Brien-Fleming, the Woolson-Lachenbruch, and the Dabrowska tests require a common censoring time for pair members. As illustrated in our example, this assumption can significantly reduce the effective sample size used for inference especially in the case of moderate to heavy censoring. The censoring requirement for the Akritas and the Murray tests is less stringent, hence may increase the power of the hypothesis testing procedure. Murray's weighted Kaplan-Meier test also allows the inclusion of singletons which can further increase the study power.

When it is necessary to adjust for additional covariates beyond the treatment effect, extensions of the Cox regression models are available. These models test different hypotheses and their parameter estimates represent slightly different quantities which may influence the choice of model to use in a study. The stratified Cox model and the shared frailty model provide estimates for the within-pair hazard ratios. These are conditional approaches while the model proposed by Lee et al. (1992) is a marginal approach. The marginal model estimates the average treatment effect across all pairs. Of these regression models, the stratified model may be the least efficient when there is moderate to heavy censoring since pairs with the smaller of the two times being censored do not contribute to the test statistic.

Besides approaches to compare the entire survival curves, we also presented tests available to compare the survival probabilities at a fixed time point using various transformations of the Kaplan-Meier estimator. The tests using transformations perform better than the untransformed test. All transformations produce comparable results. Generalized linear models using the pseudo-values is another approach to analyze the effect of treatment on survival at a fixed time. An advantage of the pseudo-value approach is that it allows comparison of the survival probabilities at a fixed time point while adjusting for other covariates. Depending on the link function used in the regression model, the estimates from the pseudo-value approach yield the results in terms of the odds ratio or the hazard ratio to quantify the difference between the treatments at that time point.

Acknowledgments

This research was supported by supplement (3 UL1 RR031973-02S1) to the Medical College of Wisconsin's Clinical and Translational Science Award (CTSA) grant.

Bibliography

Akritas, M. (1992). Rank transform statistics with censored data. *Statistics and Probability Letters* **13**, 209-221.

Andersen, P. K., Klein, J. P., and Rosthj, S. (2003). Generalized linear models for correlated pseudo-observations with applications to multi-state models. *Biometrika* **90**, 15-27.

Cox, D. R. (1972). Regression Models and Life Tables (with Discussion). *Journal of the Royal Statistical Society B* **34**, 187-220.

Dabrowska, D. M. (1990). Signed-rank tests for censored matched pairs. *Journal of the American Statistical Association* **85**, 476-485.

Diabetic Retinopathy Study Research Group (1976). Preliminary report on effects of pho tocoagulation therapy. *Am J Ophthalmology* **81**, 383-396.

Eapen M., Le-Rademacher J., Antin J., Champlin R. E., Carreras J., Fay J., Passweg J. R., Tolar J., Horowitz M. M., Marsh J. C., and Deeg H. J. (2011). Effect of stem cell source on outcomes after adult unrelated donor transplantation in severe aplastic anemia. *Blood* **118**, 2618-2621.

Gehan, E. A. (1965). A generalized Wilcoxon test for comparing arbitrarily singly-censored samples. *Biometrika* **52**, 203-223.

Holt, J.D. and Prentice, R.L. (1974). Survival analyses in twin studies and matched pair experiments. *Biometrika* **61**, 17-30.

Hougaard, P. (2000). *Analysis of Multivariate Survival Data*. Springer, New York.

Klein, J. P. and Andersen, P. K. (2005). Regression modeling of competing risks data based on pseudo-values of the cumulative incidence function. *Biometrics* **61**, 223-229.

Klein, J. P., Logan, B., Harhoff, M., and Andersen, P. K. (2007). Analyzing survival curves at a fixed point in time. *Statistics in Medicine* **26**, 4505-4519.

Lee, E. W., Wei, L. J., and Amato, D. A. (1992). Cox-type regression analysis for large numbers of small groups of correlated failure time observations. In: *Survival Analysis: State of the Art* (eds. Klein and Goel), Kluwer Academic Publishers, 237-247.

Liang K.Y. and Zeger S.L. (1986) Longitudinal data analysis using generalized linear models. *Biometrika*, **73**: 13-22.

Logan, B., Zhang, M. J., and Klein, J. P. (2011). Marginal models for clustered time-to-event data with competing risks using pseudovalues. *Biometrics* **67**, 1-7.

Murray, S. (2001). Using weighted Kaplan-Meier statistics in nonparametric comparisons of paired censored survival outcomes. *Biometrics* **57**, 361-368.

O'Brien, P.C. and Fleming, T. R. (1987). A paired Prentice-Wilcoxon test for censored paired data. *Biometrics* **43**, 169-180.

Pepe, M. S. and Fleming, T. R. (1989). Weighted Kaplan-Meier statistics: A class of distance tests for censored survival data. *Biometrics* **45**, 497-507.

Prentice, R. L. (1978). Linear rank tests with right censored data. *Biometrika* **65**, 167-179.

Su, P. F., Chi, Y., Li, C. I., Shyr, Y., and Liao, Y. D. (2011). Analyzing survival curves at a fixed point in time for paired and clustered right-censored data. *Computational Statistics and Data Analysis* **55**, 1617-1628.

Woolson, R. F. and Lachenbruch, P. A. (1980). Rank tests for censored matched pairs. *Biometrika* **67**, 597-606.

Woolson, R. F. and O'Gorman, T. W. (1992). A comparison of several tests for censored paired data. *Statistics in Medicine* **11**, 193-208.

Index

G^ρ family, 331
L measure, 290
U-statistic, 516

Aalen-Johansen estimator, 206, 430
  competing risks, 161
Absolute risk, 222, 359
Accelerated failure time (AFT) model, 58, 377, 399, 533
  Buckley-James estimator, 62
  Gehan rank estimator, 60
  high-dimensional data, 103
Adaptive design, 612
Adaptive lasso, 305
Adaptive shrinkage, 295
Additive hazard regression, 49, 210, 374, 400, 401
  Aalen additive gamma frailty model, 55
  competing risks, 166
  direct effect, 54
  goodness-of-fit procedures, 53
  high-dimensional regressors, 56
  Lin and Ying model, 50
  semiparametric model, 50
  structural properties, 54
Additive models, 256
Additive risk model
  high-dimensional data, 103
Adjusted cumulative incidence curves, 258
Adjusted p-value, 314, 316
Adjusted survival curves, 248
AIC, 100, 287, 507
AIDS, 192
Algorithm
  EM-ICM algorithm, 375
  iterative convex minorant algorithm, 375
  pool-adjacent-violator algorithm, 375
  self-consistency algorithm, 375
Avalanche data, 392
Average regression effect, 324

Bagging, 275
Baseline hazard
  piecewise-constant, 477, 480
Bayes estimator of $S(t)$, 38
Bayesian inference, 476
Bayesian methods, 529
  Bayes factor, 287
  Bayesian model averaging, 294
  Bayesian nonparametrics, 295
  CPO, Conditional Predictive Ordinate, 290, 480
  LPML, Log Pseudo-Marginal Likelihood, 289
  posterior model probabilities, 286
  pseudo-Bayes factor, 290
Beta process, 36
Beta-min-condition, 304
BIC, 100, 288
Binomial regression, 222
Bivariate failure times, 180
Bonferroni, 312
Bootstrap, 229, 514, 516, 518
Box-Cox transformation, 483
Breslow estimator, 10
  case-cohort studies, 360
  nested case-control studies, 359

C-statistic, 269
Calibration distribution, 291
Canonical joint distribution, 596–598
Case-cohort studies, 344, 348–353
  absolute risk estimation, 360
  alternative models, 353, 360
  biomarkers, 354
  Breslow type estimator, 360
  calibration, 353
  comparison with nested case-control studies, 353–354
  covariance estimation, 350, 351
  doubly weighted estimators, 353
  IPW estimator, 350, 351
  maximum likelihood estimation, 361–362
  misspecified models, 364

multiple endpoints, 354
post-stratification, 352
Prentice estimator, 349
pseudo-likelihood, 349
relative efficiency, 350, 363
Self-Prentice estimator, 349, 357
software, 352, 353
stratified, 351, 354
subcohort, 348, 351
weighted likelihood, 350
Case-control sampling, 406
Case-control studies, 557–558
case-cohort, *see* case-cohort studies
classical case-control design, 363
nested case-control, *see* nested case-control studies
Causal inference, 136
Confounding, 138, 145
Consistency, 139, 146
Feedback, 142
Positivity, 139, 146
Cause-specific hazard, 158, 159, 161, 181, 183, 224, 511–513, 516, 517, 533
regression models, 163–166
Censoring, 572
administrative censoring, 5, 448, 575
by a competing risk, 161, 164, 165
conditionally independent, 327
covariate dependent, 327
current status data, 370
function, 6
independent, 325
left-censored data, 370
loss-to-follow-up, 576
Right-censored data, 223, 369, 512
Clayton model, 493, 553, 554
Clinical trial, 333, 572
Closed testing procedure, 313
Cluster-randomized trials, 590
Clustered competing risks, 511, 513, 517, 518, 520
Clustered data, 106, 200, 209, 218
Collapsed Gibbs sampler, 190
Comparison of survival functions, 376
Competing risks, 157–173, 180, 196, 200, 202, 204, 225, 244, 252, 369, 380, 404–406, 468, 511, 513–516, 518–520, 532–533, 559–563, 579, 583, 589
cause removal, 160
conditional probability function, 173
latent failure time model, 159–160
missing cause of failure, 173
mixture model, 172
multistate model for, 158–160
vertical model, 172
Composite time-to-event endpoint, 157, 165
Conditional baseline hazard function, 185
Conditional cause-specific hazards ratio, 560–562
Conditional distribution of covariate, 325
Conditional hazards ratio, 552–554, 558–559
nonparametric estimation, 559
Conditional independence, 459
Conditional model, 490, 626
Conditional residual, 539
Consistency, 514, 515
Copula, 490–506, 553–559
Archimedean copula, 491, 554
Archimedean generator, 491
Frank family, 494
Gaussian copula, 495
Goodness-of-fit test, 500
Sklar's theorem, 491
Weighted concordance estimator, 500
Correlated failure times, 489
Correlation, 552
coefficient, 588
correlated prior, 32
survival times, 466
Counter-matching, *see* nested
Counterfactual variables, 136
Counting processes, 513
Cox regression model, 9, *see* Proportional hazards, 28, 33, 226, 244, 253, 420, 517, 574, 596, 624
Cox-Aalen model, 56
Cross ratio, 492, *see* Conditional hazards ratio
Cross-hazard ratio, 514, 515
Cross-odds ratio, 549, 563
Cross-over, 471
Cross-validation, 99, 274
Cumulative incidence, 200, 202–204, 208, 210–212, 214, 217, 222
Cumulative incidence function, 159, 161, 165, 166, 187, 202–204, 209, 214, 218, 244, 252, 431, 512, 513, 516, 518, 562–563, 584, 589
estimate, 203
regression models, 168–170

inverse probability of censoring weighting, 169
link function, 168
Cure model, 113–129, 374, 408, 520, 533
additive cure model, 125
bounded cum. hazard model, 123
Box-Cox transformation, 125
joint model with cured fraction, 127
longitudinal measurement, 127
mixture cure model, *see* Mixture cure (MC) model
non-mixture cure model, 129
promotion time, 123
prop. hazards cure (PHC) model, 122–125
prop. odds cure (POC) model, 125
relative survival cure model, 128
semiparametric transformation cure model, 126
unified cure model, 126
Cure rate model, *see* Cure model
Current status data, 373, 391–408
Current study, 41

Delayed entry, 16–17, 450
Dependence
negative, 467
survival times, 458
Derived variables, 100
Deviance
deviance function, 41
DIC, 40–42, 191, 480
DIC, 289
Direct binomial modeling, 256
Directed acyclic graphs, 137
Dirichlet process, 38, 519
Discretized beta process prior, 36
Doubly censored data, 382
Dynamic prediction, 445

Effective sample size, 575
Elastic net, 97, 307
Empirical processes, 515
Error spending, 600–604
Estimating equation, 199, 207–209, 211, 228, 324–337
weighted, 326
Excess mortality rate, 128
Expectation-Maximization (EM) algorithm, 466, 528
Exponential family, 461
Exponential model, 30, 573

Failure function, 6
Failure time, 574
False discovery
false discovery proportion, 311
False Discovery Rate (FDR), 311
Familial studies, 549–568
Family-wise Error Rate, 311
Feature selection, 95
Fine-Gray model, 168–171, 226, 253, 257, 433, 517
stratified, 258
subdistribution hazard, 169–170
Fisher information, 11, 467, 499
Frailty model, 40, 458–471, 476, 490, 553–555, 626
asymptotics, 466–468
extension, 468
gamma, 331, 458, 463–469, 472, 478
goodness-of-fit, 468–470
inverse Gaussian, 465, 466
log-normal, 465–468, 472, 479
positive stable, 464–467, 469, 477
PVF, 465–467, 469
shared, 458, 459, 462
software, 471
spatial, 483
stochastic process, 468
updating, 461, 464
Freedman's formula, 576
Functional Delta-method, 514–516
Futility boundary
binding, 599
non-binding, 600

Gamma process, 33
Gene-sets, 98
Generalized linear mixed models, 534
Generalized linear model (GLM), 199–201, 207, 211, 212, 214, 217
Genetic effect, 459
Gibbs sampler, 190, 529
Graft-versus-host-disease, 244
Graphical diagnostic procedure, 502
Gray's test, 168
Greenwood's variance estimator, 7
Group sequential boundary, 599–604
Group sequential log-rank test, 604–605
Grouped data, 371
Grouped/discretized survival data, 32

Gumbel-Hougaard family, 494

Hazard function, 7, 458, 574
  all-cause, 583
  cause-specific, 579, 584
  constant, 573
  subdistribution, 589
Hazard ratio, 13
Hematopoietic stem cell transplantation, 243
High-dimensional data, 93
Historical
  data, 39, 42
  study, 41
HIV infection, 192
Hochberg method, 314
Holm method, 312
Hommel method, 314
HPD interval, 42

Illness-death model, 407
Immortal time bias, 442
Improper cumulative hazard function, 189
Incidence rate, 162
Inference, 265, 270
  generalizability, 277
Influence function, 15, 228
Informative drop-out, 525
Informatively interval-censored data, 381
Interaction, 589
Interim analysis, 596
Interval-censored data, 223, 391, 393, 408, 431
  bivariate, 379
  clustered data, 379
  doubly censored data, 370
  informative censoring, 371
  missing covariates, 371, 382
  multi-state models, 371, 382
  multivariate, 369, 378
  noninformative censoring, 371, 372
  univariate, 369
Inverse probability of censoring, 147
Inverse probability of treatment estimation, 141, 145
Inverse probability weighting, 328
  IPCW, 223, 227, 232
Irrepresentable condition, 304
ISIS, 306

Jacod's likelihood formula, 421
Joint latent class model, 536–537
Joint likelihood, 527–528
Joint modeling, 454

Kaplan-Meier estimator, 6, 38, 200, 201, 206, 208, 209, 466, 517, 598
  jump of, 326
Kendall's tau, 466, 478, 491, 514, 552–554, 557–559

Landmark, 222, 442
  landmark super model, 449
Laplace approximation, 518
Laplace transform, 459–466
Lasso, 96, 303
  adaptive, 97
  Bayesian, 295
  group, 98
  hierarchical group, 99
  relaxed, 305
Last Value Carried Forward (LVCF), 525, 532
Latent class joint model, 533
Latent variable model, 525
Left truncation, 589
Likelihood ratio confidence intervals, 397
Linear random effects models, 526
Local association, 492
Log-binomial regression model, 225
Log-logistic model, *see* Proportional odds model
Log-rank test, 574, 604
  adjusting for ties, 604
Logistic regression model, 225
LPML, 191

Marginal model, 551–552, 556, 625
Marginal rate function, 423
Marginal residual, 539
Marginal structural model, 142, 145
Markov assumption, 418–419
  Test for, 420
Markov chain Monte Carlo, 529
Markov model, 470
Markov prior process, 37
Matched pairs, 458, 471
Matched pairs study design, 615
Maximum information design, 600
Maximum likelihood estimation, 528–529
Measurement error, 524
Melanoma data, 232

Micro-simulation, 447
Misclassification, 406
Mixture cure (MC) model, 114–122
  acc. fail. time MC (AFTMC), 115
  accel. hazard MC (AHMC), 115
  clustered survival data, 120
  identifiability, 118
  incidence model, 114
  latency model, 114
  marginal model, 122
  prop. hazards MC (PHMC), 115
  prop. odds MC (POMC), 115
  with random effects, 122
Mixture model, 181, 184, 533
Mixture of Dirichlet Process (MDP), 38
Model
  averaging, 275
  building, 265, 266
  checking, 558–559
  diagnosis, 209–211, 217
  diagnostics, 539–540
  prediction capability, 269
  selection, 265, 266
  selection/building proposal, 275, 280
Model checking procedures , 499
Model complexity, 288
Model space, 286
Monitoring grid, 29
Monte Carlo ECM algorithm, 519
Multi-arm trials, 586
Multi-state model, 199, 204–206, 215–217, 225, 407, 447, 460, 470
  illness-death model, 447
Multi-state process, 417
Multiple longitudinal biomarkers, 534–535
Multiple testing, 302
Multiplicity adjustment, 295
Multivariate data, 181, 458, 459, 462, 464, 465, 467, 471, 489

Negative binomial distribution, 458, 463
Neighbourhood stability condition, 304
Nelson-Aalen estimator, 7, 206, 421, 466
  competing risks, 101
Nested case-control studies, 344–348
  absolute risk estimation, 359
  biomarkers, 354
  Breslow type estimator, 359
  comparison with case-cohort studies, 353–354
  counter-matched, 354, 358–359
  counting process formulation, 356
  IPW estimator, 354
  matching, 348
  maximum likelihood estimation, 361–362
  misspecified models, 364
  multiple endpoints, 354
  partial likelihood, 346, 356, 359
  relative efficiency, 347
  reuse of controls, 354–355
  stratified, *see* counter-matched
  weighted likelihood, 354
Newton-Raphson algorithm, 528
Non-linear mixed effects model (NLMM), 535
Non-proportional hazards, 325, 464
Nonparametric hazard model
  high-dimensional data, 103
Nonparametric maximum likelihood estimator (NPMLE), 336, 373, 394, 404, 530

Observational study, 333
One-dimensional submodel, 336
Optimal cut point, 247
Oracle property, 302
Over-running, 602
Overdispersion, 458, 460, 463
Overfitting, 99, 268, 273

Paired rank test, 618–620
  Akritas, 619
  Gehan-Wilcoxon, 618
  generalized sign rank test, 620
  Prentice-Wilcoxon, 618
Paired survival data, 615–631
  fixed time-point, 627
    Kaplan-Meier estimator, 627
    pseudo-values, 628
  marginal Cox model, 625
  shared frailty model, 626
    gamma, 626
    positive stable, 626
  stratified Cox model, 624
  weighted Kaplan-Meier statistics, 622
  within-pair comparison, 620–622
Parsimony, 95
Partial least squares, 102
Partial likelihood, 10, 34, 186, 345, 422, 477, 517, 518

nested case-control studies, 346, 356, 359
Pathways, 98
Pattern-mixture model, 525
PBC data, 266
Pedigrees, 468
Penalized likelihood, 303
partial likelihood, 336
Piecewise constant hazard model, 32, 41
Poisson model, 422, 458, 463
Poly-Weibull model, 32
Pool-adjacent-violators algorithm, 394
Positive stable distribution, 333
Posterior mean frailty process, 502
Power prior, 39, 41
Prediction, 537–539
Predictive bias, 290
Predictive criteria, 289
Predictive variance, 290
Principal components, 101
Product partition models, 295
Profile likelihood, 188, 530
Prognostic factor studies, 588
Prognostic index, 9
Proportional hazards, 8–9, 201, 205, 210–213, 217, 246, 266, 344, 374, 399–404, 448, 460, 464, 469, 490, 526, 574
baseline hazard, 9
checking, 333
competing risks, 165
frailty model, 491
mixed-effects (PHMM), 335
not nested, 333
reliable prediction model, 266, 270
stratified, 19–20, 247, 348, 608, 624
structure, 184, 187
time-varying covariate effects, 448
transformation of covariate, 324
Proportional odds model, 374, 399, 464, 520, 533
Proportional subdistribution hazards model, 168, *see* Fine-Gray model, 518, 589
Pseudo-likelihood, 500
case-cohort studies, 349
Pseudo-values, 199–218, 244, 253, 257, 628
PVF distribution
Laplace transform, 465

Quality adjusted survival, 200, 206–207
Quantile regression for survival analysis, 62
global linear model, 62
linear model, 62
local linear model, 62
martingale-based estimation, 66
model checking, 67
Powell's estimator, 63
quantile calculus, 66
second-stage inference, 67
self-consistent estimation, 64
variance estimation, 67

Random effects model, 458, 468, 475, 490
Randomized controlled trials, 136
Receiver operating characteristic (ROC) curves, 538
Recurrent events, 463–471, 483, 533
Related individuals, 458
Relative survival function, 128
Replicate experiment, 290
PDRE, Predictive Density, 290
Restricted eigenvalue condition, 305
Restricted mean survival, 200–202, 211
Reversible jump MCMC, 293
Ridge regression, 96, 307
Risk prediction
high-dimensional data, 94
Risk set, 6, 17
sampled, 345, 358
Robustness, 323

Sample size, 575
censored data, 572, 575
cluster-randomized trials, 591
competing risks, 580
left-truncated data, 589
multi-arm trials, 587
prop. subdistribution model, 590
studies on prognostic factors, 588
test for equality, 587
test for equivalence, 588
test for non-inferiority/superiority, 587
time-to-event data, 575
time-varying covariates, 591
Sandwich variance estimator, 207, 209, 449
SCAD, 97, 305
Schoenfeld residuals, 20
Schoenfeld's formula, 572, 573, 575, 576
Score
equation, 325–336
function, 20

Selection model, 525
Semi-competing risks, 180, 196
Semi-parametric estimation, 496
Shared parameter model, 526
Shrinkage, 96
Simes inequality, 313
Simulation, 591, 592
Simulation Extrapolation (SIMEX), 540
Simultaneous confidence sets, 317
Software, 540–541
  East, 611
  R, 17, 21, 211, 232, 383, 453
    Epi, 383
    Icens, 383
    MLEcens, 383
    coxme, 336
    dynpred, 453
    dynsurv, 383
    geepack, 211, 231
    gfcure, 129
    glrt, 383
    gsDesign, 611
    intcox, 383
    interval, 383
    nltm, 129
    phmm, 336
    pseu.r, 211
    pseucheck.r, 212
    pseudo.r, 211
    riskRegression, 232
    semicure, 129
    smcure, 129
    survBayes, 383
    survival, 10, 17, 347, 352, 353
    timereg, 232
    Causal model, 149
    frailty models, 471
  SAS, 17, 21, 232, 383
    frailty models, 472
    GENMOD, 211, 232
    IML, 42
    LIFEREG, 31, 383
    MCMC, 28
    PHREG, 15, 28
    PSPMCM, 129
  SPLUS, 383
    kaplanMeier, 383
  SPSS, 21
  Stata, 17, 21
    CUREREH, 129
    STRSMIX, 129
    STRSNMIX, 129
  OpenBUGS, 28, 31
  WinBUGS, 477, 480, 484
Sparsity assumption, 95, 304
Spearman's rho, 466, 492
State occupation probability, 418
  Kaplan-Meier differences, 432
Stepwise selection
  high-dimensional data, 96
Stochastic model, 527
Sub-survival function, 182–184
Subdistribution, *see* Fine-Gray model
  hazard, 169, 185, 203, 204, 209, 214, 253, 512, 589
  hazard model, 181, 204, 210, 211, 213
  hazard ratio, 204
  model, 185, 188
    fully specified, 181, 187, 188
Sure Independence Screening, 98, 306
Survival function, 6, 574
Survival regression models, 222

t-test, 573
Test
  for equality, 587
  for equivalence, 588
  for non-inferiority/superiority, 587
Ties
  Breslow method, 10
Time-dependent covariate, 17–18, 163–164, 246, 247, 253, 257, 324–326, 442, 524
Time-varying effect, 20–21, 247, 253, 255, 257, 259, 324–334
Tonsil cancer, 117
Training data, 99
Transformation model, 77–88, 520
  asymptotics, 82
  Box-Cox transformation, 78
  counting processes, 78
  high-dimensional data, 104
  joint models, 80
  linear, 374
  NPMLE, 81
  random effects, 80
  Software, 88
Transition intensity, 418
Transition probability, 418
  Meira-Machado estimator, 432
  Regression models, 432–433
Treatment-related mortality, 244

Tumor kinetics, 123
Tuning parameters, 99
Twin survival, 458, 459, 468–470
Two-stage analysis, 525, 532
Two-stage estimation, 496
Two-stage modeling, 554–557, 561, 563

Ultra-high dimensional data, 93
Under-running, 603
Univariate selection
    high-dimensional data, 95

Validation data, 99
Variable screening property, 304
Variable selection, 302
    high-dimensional data, 95
    stochastic search, 292
Variance component, 337

Wald-type test, 516
Weak convergence, 514
Weibull model, 31, 394, 399, 401–404, 464, 467, 469
Working independence, 497

T - #0015 - 011124 - C0 - 254/178/37 - PB - 9780367330965 - Gloss Lamination

JAVA Programming Fundamentals

Problem Solving Through Object Oriented Analysis and Design